Lecture Notes in Mechanical Engineering

Lecture Notes in Mechanical Engineering (LNME) publishes the latest developments in Mechanical Engineering—quickly, informally and with high quality. Original research reported in proceedings and post-proceedings represents the core of LNME. Volumes published in LNME embrace all aspects, subfields and new challenges of mechanical engineering. Topics in the series include:

- Engineering Design
- Machinery and Machine Elements
- Mechanical Structures and Stress Analysis
- Automotive Engineering
- Engine Technology
- Aerospace Technology and Astronautics
- Nanotechnology and Microengineering
- Control, Robotics, Mechatronics
- MEMS
- Theoretical and Applied Mechanics
- Dynamical Systems, Control
- Fluid Mechanics
- Engineering Thermodynamics, Heat and Mass Transfer
- Manufacturing
- Precision Engineering, Instrumentation, Measurement
- Materials Engineering
- Tribology and Surface Technology

To submit a proposal or request further information, please contact the Springer Editor of your location:

China: Dr. Mengchu Huang at mengchu.huang@springer.com
India: Priya Vyas at priya.vyas@springer.com
Rest of Asia, Australia, New Zealand: Swati Meherishi at swati.meherishi@springer.com
All other countries: Dr. Leontina Di Cecco at Leontina.dicecco@springer.com

To submit a proposal for a monograph, please check our Springer Tracts in Mechanical Engineering at http://www.springer.com/series/11693 or contact Leontina.dicecco@springer.com

Indexed by SCOPUS. All books published in the series are submitted for consideration in Web of Science.

More information about this series at http://www.springer.com/series/11236

V. H. Saran · Rakesh Kumar Misra
Editors

Advances in Systems Engineering

Select Proceedings of NSC 2019

Springer

Editors
V. H. Saran
Department of Mechanical and Industrial Engineering
Indian Institute of Technology Roorkee
Roorkee, India

Rakesh Kumar Misra
Department of Electrical Engineering
Indian Institute of Technology (BHU) Varanasi
Varanasi, India

ISSN 2195-4356 ISSN 2195-4364 (electronic)
Lecture Notes in Mechanical Engineering
ISBN 978-981-15-8024-6 ISBN 978-981-15-8025-3 (eBook)
https://doi.org/10.1007/978-981-15-8025-3

This Springer imprint is published by the registered company Springer Nature Singapore Pte Ltd.
The registered company address is: 152 Beach Road, #21-01/04 Gateway East, Singapore 189721, Singapore

Dedicated with utmost humility
and gratefulness
in the Lotus Feet of Most Revered
Rai Sahib Gurcharn Das Mehta Sahab,
The perfect spiritual Master,
A systems Expert par excellence

and

An illustrious Thomsonian

Foreword

Forty-third National Systems Conference (NSC 2019) was organized by Department of Mechanical and Industrial Engineering, Indian Institute of Technology Roorkee, during 6–8 December 2019. Indian Institute of Technology Roorkee is the oldest technical institution in the country and shall be completing 175 years in 2022. The institute ranks amongst the best technological institutions in the world and has contributed to all sectors of technological development. It has been a trendsetter in the area of education and research in the field of science, technology and engineering.

The theme of this National Conference "Innovative and Emerging Trends in Engineering Systems" explores cutting-edge ideas for handling complex systems through holistic systems thinking. The papers of the conference touch state-of-the-art application of the most recent scientific knowledge of a range of systems in the field of human language technologies, information retrieval, health and welfare, engineering and transportation, environment and sustainable development, business and economics and energy resources and policy. Besides application in different types of systems, few papers have used AI/machine learning concepts in system analysis.

The papers of the conference explain the relevance of systems engineering ideas in interpreting different complex problems relevant to the contemporary science and technology development. This proceedings is a systematic documentation of research undertaken by different researchers on the theme of the conference. An overview of the papers indicates strong interdisciplinary character of systems engineering research. Future efforts of interdisciplinarity will help in improved understanding of complex engineering systems.

I sincerely compliment all authors and participants for their contributions.

Manoranjan Parida
Deputy Director
Indian Institute of Technology Roorkee
Roorkee, India

Preface

In the words of renowned systems thinker Dr. Russell L. Ackoff, "A problem never exists in isolation; it is surrounded by other problems in space and time. The more of the context of a problem that a scientist can comprehend; the greater are his chances of finding a truly adequate solution.". This underlying principle of systems thinking makes it a versatile and unique tool for solving a problem of any degree of complexity and therefore has become indispensable in today's fast-changing world of emergent complexity.

This book is a collection of selected cutting-edge research articles from the forty-third National Systems Conference on "Innovative and Emerging Trends in Engineering Systems" (NSC 2019). The Department of Mechanical and Industrial Engineering, IIT Roorkee, and the Systems Society of India jointly organized the conference from 6 December to 8 December 2019 at IIT Roorkee, India. This book gives the participants a vehicle for disseminating their ideas and is envisioned to serve as reference guide for researchers and practitioners in holistic systems thinking.

The NSC 2019 received 153 full-length papers against the announcement, which were peer-reviewed, each by two experts from IITs and other eminent institutions. Out of these, 86 papers have been finally included in this book. It comprises the state of the art in systems engineering covering a spectrum of research areas such as robotic systems, dynamical systems, materials and systems, manufacturing systems, space systems, thermal systems, green energy systems, data analytics systems, mathematical systems, solar energy systems, literary and consciousness systems and miscellaneous systems.

We remain ever grateful to the Founder of Systems Society of India, Most Revered Prof. Prem Saran Satsangi, who has been a source of inspiration for all systems thinkers, for his unbounded Grace and Guidance, without which this conference would not have been successfully organized. We further wish to acknowledge all the authors and reviewers who have contributed to this book. We also deeply express our gratitude to the generous support provided by IIT Roorkee, the Council of Scientific and Industrial Research, the Defence Research and Development Organisation, M/s Indian Sugar and General Engineering Corporation Heavy Engineering Ltd., Yamunanagar, and M/s Steel Exchange India Ltd., Hyderabad.

Roorkee, India
Varanasi, India

Prof. V. H. Saran
Prof. Rakesh Kumar Misra

Contents

About the Editors

Dr. V. H. Saran is Professor in the Department of Mechanical and Industrial Engineering at Indian Institute of Technology, Roorkee with a teaching and research experience of more than 30 years. He completed his B.Sc. Engineering (Mechanical Engineering) from Dayalbagh Educational Institute, Agra in 1984, M. E. (Machine Design) from IIT Roorkee in 1987 and Ph.D. degree from IIT Roorkee in 2004. He has published over 70 research papers in reputed refereed international journals and conferences. He has organised several international and national conferences, including the NSC 2019 and has also delivered many invited talks in India and abroad. His research interests include vehicle dynamics, ergonomics, human response to vibrations and fault diagnosis and prognosis of roller bearings. He is member of The American Society of Mechanical Engineers, life member of The Institution of Engineers (India), Indian Society for Mechanical Engineers and Systems Society of India.

Dr. Rakesh Kumar Misra is Professor in the Department of Electrical Engineering, Indian Institute of Technology (BHU) Varanasi. He is also heading the Department of Humanistic Studies, Indian Institute of Technology (BHU) Varanasi since 2015. He was awarded Ph.D. in March 2005 from Indian Institute of Technology (BHU) Varanasi, India. He did his M.Tech. in Engineering Systems and B.Sc. (Engineering) in Electrical Engineering in 1997 and 1995, respectively from Dayalbagh Educational Institute Dayalbagh, Agra, India. His fields of interest include systems engineering, power systems, and applications of computational intelligence in power systems. He has presented papers in reputed conferences and has published papers in reputed journals such as IET Generation, Transmission & Distribution, Journal of Intelligent & Fuzzy Systems, International Transactions on Electrical Energy Systems, International Journal of Electrical Power & Energy Systems, IEEE Transactions on Power Systems, and Electric Power Components and Systems. An optimization algorithm, EBO with CMAR (EBO_with_CMAR), proposed by his team in a paper entitled “Improving the local search capability of

Effective Butterfly Optimizer using Covariance Matrix Adapted Retreat phase" was ranked first in a competition on real-parameter bound-constrained single objective optimization at IEEE Congress in Evolutionary Computation (CEC-2017), Spain. He is Member of IEEE, Institution of Engineers (India), and Life Member of ISTE and Systems Society of India.

Literary and Consciousness Systems

Gaussian Local Ternary Co-occurrence Pattern for Image Retrieval

Amit Singhal and Megha Agarwal

1 Introduction

Content-based image retrieval (CBIR) systems are meant for fetching multiple images similar to an input chosen by the user. These systems are gaining popularity across a variety of applications ranging from health care to entertainment and many more. To find similar images, feature set obtained from the images is compared to each other. Pattern-based features capture information using intensity variations in the local neighborhood of a pixel and produce good results in various retrieval scenarios.

Local binary pattern (LBP) [13] is the first pattern-based feature proposed in the literature. Local neighborhood difference pattern (LNDP) [15] augments these binary patterns by accounting for the relation among neighborhood pixels also. Center symmetric local binary pattern (CSLBP) [7] is computed using pixels in the region of interest. On the contrary, local ternary pattern (LTP) is computed by considering ternary edges instead of binary edges. Co-occurrence of ternary edges is obtained for local ternary co-occurrence pattern (LTCoP) [12]. Authors in [4] utilize a group of Gaussian filtered images to construct 3D-LTCoP.

Second-order intensity differences in the local neighborhood of a pixel are processed to obtain local tetra pattern (LTrP) [11]. Maximum edges across four directions are procured for computing directional local extrema pattern (DLEP) [10]. Local vector pattern (LVP) [6] captures intensity variations between adjacent pixels from different directions. Authors in [2] pass the input image through Haar-like filters and then apply LTP to obtain Haar-like local ternary pattern (HLTP).

A. Singhal
Bennett University, Greater Noida, India
e-mail: singhalamit.iitd@gmail.com

M. Agarwal (✉)
Jaypee Institute of Information Technology, Noida, India
e-mail: meghagarwal29@gmail.com

V. H. Saran and R. K. Misra (eds.), *Advances in Systems Engineering*,
Lecture Notes in Mechanical Engineering,
https://doi.org/10.1007/978-981-15-8025-3_1

The features discussed so far focus on the local texture information but ignore the color content of the image. Histograms computed from the color channels [14] help in capturing this significant information. In multichannel local ternary co-occurrence pattern (MCLTCoP) [1], a color texture feature is proposed by accessing RGB and HSV color space simultaneously. Authors in [5] employ wavelet coefficients to construct local extrema peak valley pattern (LEPVP) and use RGB histograms for the color information. Color and local edge pattern for image retrieval (LEPINV) are investigated in [17]. Cross-channel color texture information is captured by multichannel local ternary pattern [3].

In this paper, we use three Gaussian filtered images, interleave them with the original image and then compute LTCoP values to form Gaussian local ternary co-occurrence pattern (GLTCoP). The proposed method extracts coarse as well as fine details from the image to capture local texture information using GLTCoP, in addition to the global information incorporated using the color histograms, making it an effective method for image retrieval applications. The paper is organized as follows: Sect. 2 discusses the proposed feature, and simulation results are presented in Sect. 3. Section 4 concludes the paper.

2 Proposed Method

Take an image and convert it into grayscale image $I(x, y)$. Obtain three smoother versions with $\sigma = 0.45, 0.5, 0.55$ using Gaussian filtering (σ denotes the standard deviation), i.e.,

$$G(x, y) = \frac{1}{2\pi\sigma^2}\exp\left(\frac{-(x^2 + y^2)}{2\sigma^2}\right) * I(x, y). \tag{1}$$

Extract 3×3 neighborhood ($R = 1$) for each pixel of $I(x, y)$, where R is the radius of the neighborhood. Get 5×5 neighborhood ($R = 2$) for each pixel of $G(x, y)$. Interleave 3×3 neighborhood of $I(x, y)$ into $G(x, y)$ at the corresponding positions. In the 5×5 interleaved image, compute ternary patterns by calculating the difference of $R = 1$ neighbors with the center pixel and $R = 2$ neighbors with $R = 1$ neighbors. Replace each pixel value of the neighborhood with the signs of difference at the corresponding positions and separate them as upper for $+ve$ and lower for $-ve$ patterns. $R = 1$ upper/lower and $R = 2$ upper/lower patterns are compared, respectively, to get auto and cross co-occurrence patterns as shown in Fig. 1. Finally, GLTCoP feature is constructed by computing histograms of the obtained GLTCoP values. In order to include global color outlook of the image, histograms of H and S channels of HSV color space are concatenated with GLTCoP. The dimension of the proposed feature is 14×256.

We use the d_1 distance metric [2, 3] to compute similarity between the images. Retrieved images are sorted in ascending order of their distances from the input image.

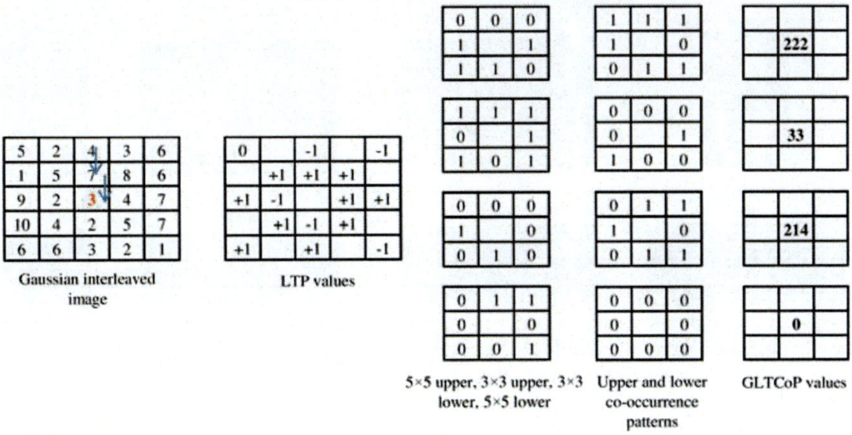

Fig. 1 Computation of GLTCoP values

3 Results

The retrieval results are computed for two popular databases: Corel 1000 [16] and Salzburg texture (STex) [9] database and are compared with the existing methods by calculating values for average precision and average recall as a function of number of images retrieved. These values are obtained as follows:

$$\text{Average Precision} = \frac{1}{|DB|}\sum_{k=1}^{|DB|} P_k, \tag{2}$$

$$\text{Average Recall} = \frac{1}{|DB|}\sum_{k=1}^{|DB|} R_k, \tag{3}$$

where P_k and R_k denote the precision and recall for kth database image, respectively, and $|DB|$ represents the cardinality of the database. By choosing kth database image as the query image, we obtain

$$P_k = \frac{\text{Number of retrieved images belonging to relevant class}}{\text{Total number of retrieved images}}, \tag{4}$$

$$R_k = \frac{\text{Number of retrieved images belonging to relevant class}}{\text{Total number of images in that class}}. \tag{5}$$

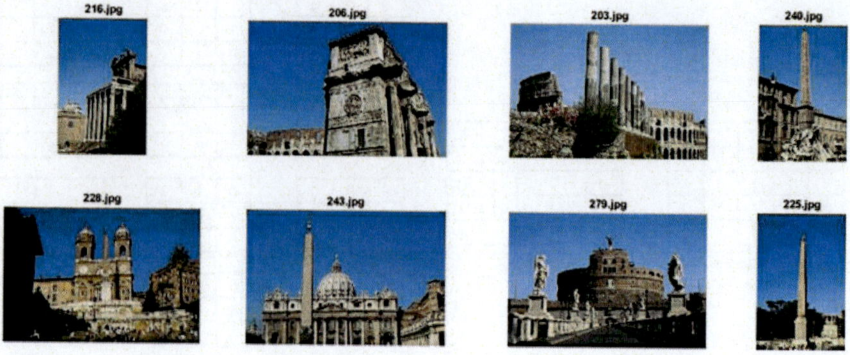

Fig. 2 Top eight retrieved images for a sample query from Corel 1000 database. The first image is same as the query image

Fig. 3 Average precision for Corel 1000 database

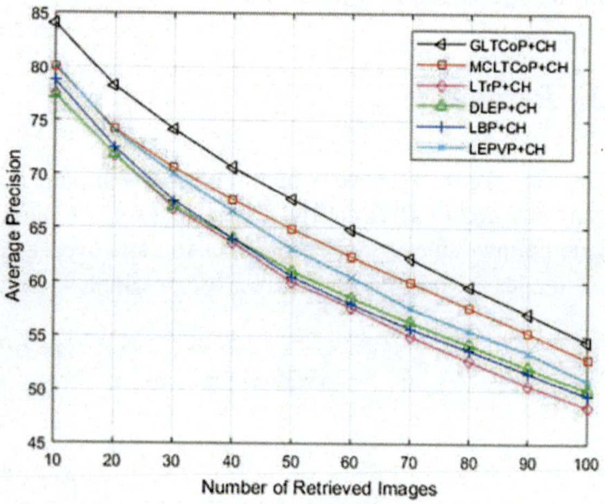

3.1 *Corel 1000 Database*

We first consider the Corel 1000 database [16] of natural images. It consists of 100 images in each of the ten classes. Some of the images retrieved for a sample query are shown in Fig. 2. Figure 3 depicts average precision values for the proposed feature in comparison with various techniques discussed in the literature. The number of retrieved images is increased from 10 to 100, in steps of 10. While retrieving ten images, we obtain an average precision of 84.05% and an average recall of 8.41%, and clearly outperform existing techniques such as MCLTCoP + CH [1], LTrP + CH [11], DLEP + CH [10], LBP + CH [13] and LEPVP + CH [5].

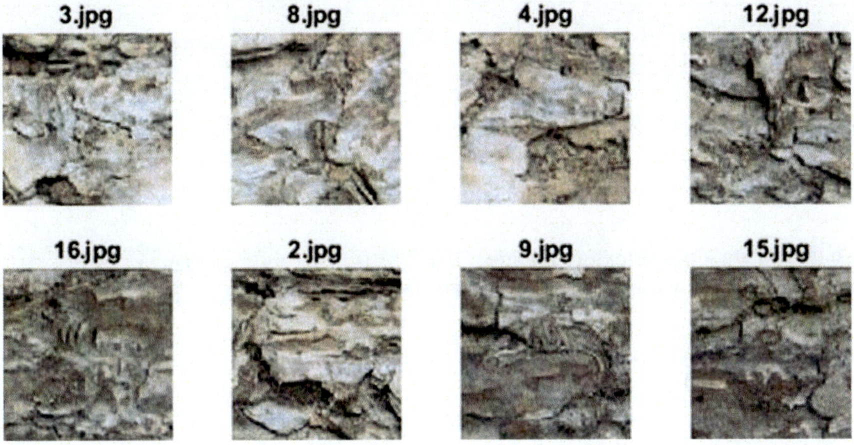

Fig. 4 Top eight retrieved images for a sample query from STex database. The first image is same as the query image

Fig. 5 Average precision for STex database

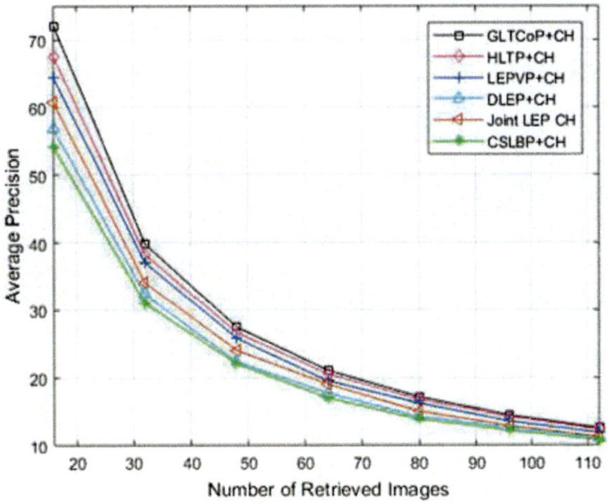

3.2 *STex Database*

Secondly, retrieval performance is tested on a large database of texture images comprising 7616 images divided into 476 categories, i.e., Salzburg texture (STex) database [9]. Figure 4 shows an example of retrieved images corresponding to a sample query image. Average precision values obtained by various methods for this database are depicted in Fig. 5. The number of retrieved images is varied from 16 to 112, in steps of 16. The proposed scheme produces superior results in contrast to the existing schemes including HLTP + CH [2], LEPVP + CH [5], DLEP + CH

[10], LEPINV + CH [17] and CSLBP + CH [8], with an average precision/recall of 72.15% for 16 retrieved images.

4 Conclusion

In this paper, a novel texture feature, GLTCoP, is proposed for image retrieval. GLTCoP utilizes the simple Gaussian filter, in order to explore fine and also the coarse information of an image. The similarity and differences between the original image and its smoother version are explored by interleaving them together. Furthermore, both smaller and larger neighborhoods are considered to compute the co-occurrence among pixel intensities. Global information content of the image is incorporated by combining the feature with color histograms. Thus, overall information gets captured making it a robust feature for image retrieval. The proposed method produces better precision as compared to the state-of-the-art methods.

References

1. Agarwal M, Maheshwari RP (2019) Multichannel local ternary co-occurrence pattern for content-based image retrieval. Iranian J Sci Technol Trans Elect Eng https://doi.org/10.1007/s40998-019-00219-1
2. Agarwal M, Singhal A (2018) Haar-like local ternary pattern for image retrieval. In: Proceedings of international conference on idustrial and information system, pp 1–5
3. Agarwal M, Singhal A (2019) Multi-channel local ternary pattern for content-based image retrieval. Pattern Anal Appl. https://doi.org/10.1007/s10044-019-00787-2
4. Agarwal M, Singhal A, Lall B (2018) 3D local ternary co-occurrence patterns for natural, texture, face and bio medical image retrieval. Neurocomputing 313:333–345
5. Dey M, Raman B, Verma M (2016) A novel colour- and texture-based image retrieval technique using multi-resolution local extrema peak valley pattern and RGB colour histogram. Pattern Anal Appl 19(4):1159–1179
6. Fan KC, Hung TY (2014) A novel local pattern descriptor: local vector pattern in high-order derivative space for face recognition. IEEE Trans Image Proce 23(5):2877–2891
7. Guo Z, Zhang L, Zhang D (2010) A completed modeling of local binary pattern operator for texture classification. IEEE Trans Image Proc 19(6):1657–1663
8. Heikkilä M, Pietikäinen M, Schmid C (2006) Description of interest regions with center-symmetric local binary patterns. Springer Berlin Heidelberg, pp 58–69
9. Kwitt R, Meerwald P (2012) Salzburg texture image database. http://www.wavelab.at/sources/STex//. Online; Accessed 20 Dec-2017
10. Murala S, Maheshwari R, Balasubramanian R (2012) Directional local extrema patterns: a new descriptor for content based image retrieval. Int J Multimedia Inf Retrieval 1(3):191–203
11. Murala S, Maheshwari R, Balasubramanian R (2012) Local tetra patterns: a new feature descriptor for content based image retrieval. IEEE Trans Image Proc 21(5):2874–2886
12. Murala S, Wu QMJ (2013) Local ternary co-occurrence patterns: a new feature descriptor for MRI and CT image retrieval. Neurocomputing 119:399–412
13. Ojala T, Pietikainen M, Harwood D (1996) A comparative study of texture measures with classification based on feature distributions. Pattern Recog 29:51–59

14. Swain MJ, Ballard DH (1990) Indexing via color histograms. In: Proceedings third international conference on computer vision, pp 390–393
15. Verma M, Raman B (2017) Local neighborhood difference pattern: a new feature descriptor for natural and texture image retrieval. Multimedia Tools Appl 1–24
16. Wang JZ, Li J, Wiederhold G (2001) SIMPLIcity: semantics-sensitive integrated matching for picture libraries. IEEE Trans Pattern Anal Mach Intell 23(9):947–963
17. Yao CH, Chen SY (2003) Retrieval of translated, rotated and scaled color textures. Pattern Recogn 36(4):913–929

Identifying Neural Correlates of Insightful Problem Solving with EEG Evidence

Raman Kumar and C. M. Markan

1 Introduction

Experience of 'Eureka!' and 'Aha!' moments while solving unknown problem experienced in past for example when Archimedes shouted 'Eureka!' ('I have found it') that he suddenly discovered that water displacement could be used to calculate density. Since then, 'Eureka' or 'Aha' moments often have been associated with the insight feeling during unknown problem solving. Insight can be defined as a sudden understanding of how to solve problem. Definition of insight varies from researcher to researcher, or we can say person to person; till now, there is no specific definition of insight. Basically, it is a feeling or an experience that could be same, but level of it varies from person to person. This results into various definitions of insight. But the common thing in all definitions is that it is a sudden and clear understanding of how to solve a problem when you leaset expect it. Most of the problems in our day-to-day life are solved using analytical method (like applying different methods or formulas to solve a problem) which one had learnt from their childhood. Insight is experienced when we are unable to solve problem then we push it on the backburner which makes our brain to think from a different point of view or restructuring takes place and then deeper understanding of the restructured occurs, which leads to solution with insight or sudden understanding of problem leading to solution [1]. Insight moment mostly occurs when people are not aware that they are not directly thinking about problem while resting or doing other work.

R. Kumar · C. M. Markan (✉)
Department of Physics and Computer Science, Dayalbagh Educational Institute, Agra, Uttar Pradesh, India
e-mail: cm.markan@dei.ac.in

R. Kumar
e-mail: ramandhakrey63@gmail.com

V. H. Saran and R. K. Misra (eds.), *Advances in Systems Engineering*,
Lecture Notes in Mechanical Engineering,
https://doi.org/10.1007/978-981-15-8025-3_2

2 Previous Work on Insight

Earlier studies with compound remote associate task paradigm, Jung-Beeman et al. [1] showed that high activation (approximately 300 ms prior to response) at right superior temporal in gamma band and at right parietal occipital (approximately 700 ms prior to response) in alpha band. Subramaniam et al. [2] showed that the people solve more problems insightfully when they are in positive mood as compare to normal mood. The Anterior Cingulate Cortex (ACC) showed sensitivity to both positive mood and insight (using fMRI). Kounios et al. [3] showed that activation associated with insight at mid-frontal cortex and bilateral temporal cortices using EEG and fMRI supported the EEG study by isolating ACC as the mid-frontal region and also showed increased activity within the bilateral temporal cortical areas. Recently, a rigorous study on insight by Santarnecchi et al. [4] shows that using transcranial Alternating Current Stimulation (tACS) it is possible to externally amplify α (10 Hz) and γ (40 Hz) activity in the right parietal and the temporal lobes, respectively. This has the potential of enhancing the probability of generating *Eureka*! moments in human being by means of frequency-specific noninvasive brain stimulation.

3 Current Study

This study is about identifying the neural correlates of insightful problem solving using Hindi version of remote associates test and Rebus puzzle. Previous studies showed the involvement of right superior temporal region and right parieto occipital region for insightful problem using compound remote associate test. Here, we are checking whether it is language effect or a true insight by using Hindi version of remote associate. Designing RAT in different languages is a big challenge as Hindi RAT must match to standard English RAT both in terms experientially (with 'aha' feeling) and experimentally (characteristic Neural Signature). Here, we analyzed the data in time, frequency and spatial domain using the following conditions (i) mental impasse of the brain that is when our brain is unable to solve the problem, (ii) rethinking about the problem that is restructuring the given problem, (iii) a deeper understanding of the restructured problem and then a sudden understanding of the problem with insightful solution [5].

4 Methods

4.1 Participants and Procedure

Twenty-five subjects were taken of same age group for EEG study from our own institution. They all were right handed with correct eye vision, and their age varied from 19 to 22 years (average age = 19.8 years, SD = 1.7). Firstly, we perform the Hindi remote associates test and Rebus puzzle on 25 subjects without using 32-EEG to check whether subjects are able to report insight verbally; for this, we have recorded the response of subjects in the form of rating scale which consists of rating from 1 to 3 in numerical form, 1 for when subject did not know the answer, 2 for when subject solved the problem through analytical method and 3 for when the problem is solved through insight. Then, we performed both experiments on 15 subjects to identify the neural correlates of insightful problem solving. Firstly, the instruction slide is shown to subjects; then, a fixation cross for 0.5s is followed by a problem slide for 6 s; this is the time within which response button also needs to be pressed. Again a fixation cross for 0.5 s is followed by answer slide for 2 s; this is the time within which response button also needs to be pressed; then feedback slide is displayed. For problem slide, the subject has to make one response out of two options, '1' when answer strikes in their mind and '2' for proceeding toward next one. Similarly, for answer slide, subject has to make one response out of two options, first when answer is matched with the striked answer in their mind at the time of problem slide shown otherwise second when answer is not matched with striked answer in their mind at the time of problem slide shown. For feedback slide, subjects have to report whether the problem is solved through insight or analytical means. Feedback slide consists of two symbols I (insightfully) and A (analytically). The screen is placed at 2 feet away from the subject. Answer is also recorded using microphone after every response.

4.2 Experiments

Here, we have designed two tests, one is *remote associates test* and another is rebus puzzle using E-prime 3.0 to identify neural correlates of insightful problem solving. Both the experiments consist of 20 problems with easy, medium and hard levels of questions in it. The Hindi version of remote associates test is similar to standard remote associates test which was developed by Mednick and Mednick [6]. Here, it also consists of three seemingly unrelated word to each other and subject is tested or (asked) to think of fourth word which somehow relates to each of the three words, for example **मोटर –बैल- माल** the 4th word is **गाड़ी** which associates with the above three words (**मोटरगाड़ी,, बैलगाड़ी, मालगाड़ी**). For Hindi remote association test, the words are taken from daily life problem. Due to limited English vocabulary, students find difficulty to solve RAT in English language. So, we designed RAT from commonly

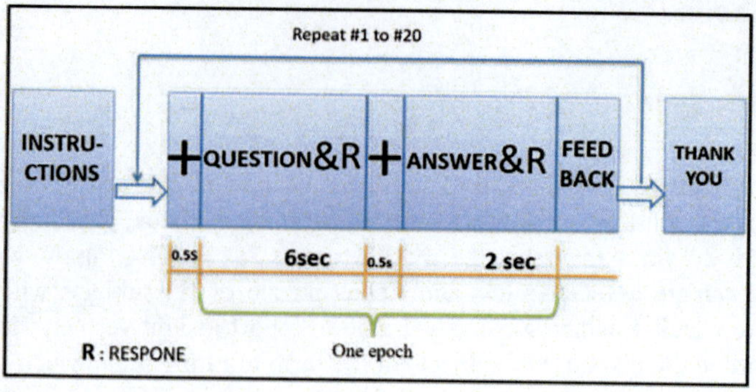

Fig. 1 Basic experiment design, first instruction slide is shown to subject then a fixation cross for 0.5 s is displayed, this is followed with problem slide for 6 s, this is the time within which response button also need to be pressed. Again a fixation cross for 0.5 s is followed by answer slide for 2 s; this is the time within which response button also needs to be pressed; then, feedback slide is displayed

used words in native language (Hindi). Hindi remote associate problems that we have developed have several advantages, (i) one can solve rapidly, (ii) one can solve the problem without any specific knowledge, (iii) one can record more data in single session and (iv) one can solve in a short time.

Rebus puzzle was designed by MacGregor and Cunningham [7]. Here, Rebus puzzle consists of words or pictures or both in some sequence or pattern and subject is being tested to think of common word or sentence or phrase which relates with the given words or pictures like MO ONCE ON (ONCE IN A BLUE MOON) (Fig. 1).

4.3 Data Acquisition

Brain signals were recorded using 32-channel EEG (due to high temporal resolution as compared to brain imaging techniques [7, 8]) and using Brain Vision recorder software. Artifacts related to eye ball movement and eye blink were recorded using channels FP1 and FP2. Initially, EEG data was band passed from 0 to 70 Hz for recording purpose for which sampling rate was set at 1000 Hz. Responses for the stimuli were recorded using Chronos containing five keys.

4.4 EEG Data Analysis

Data is analyzed in Brainstorm software which is an open-source software dedicated for the analysis of MEG/EEG data [9]. Firstly, the artifacts related to eye ball and eye blink data were removed using single-space projection method. Artifacts related to muscles movement and cardiac are removed manually. Data is then band passed through 0–49 Hz. Further, the data is divided into two bands (i) alpha (8–13 Hz) and (ii) gamma bands (30–45 Hz). The whole data was divided into epochs from data marker of problem slide to the data marker of answer slide. Based on response, epochs were divided into two categories 'Aha' and 'Non-Aha' feelings. Here, we have analyzed into time, frequency and spatial domain.

5 Behavioral Results

Statistical test (ANOVA) is applied over insight and non-insight over five subject out of 15 having experimental verification of insight for $p < 0.05$ (hypothesis passed with 95% of confidence level). Participants report insight for Hindi remote associates test. Participants solved correctly 60% of the total problems and solved 29% of as insight (mean response time = 5.2 s) and solved 31% as analytically (mean response time = 6.1 s) which is slower than the insight response. The behavioral analysis of data is conducted to test whether the subjects reported insight for Hindi remote associates test. Behavioral results of Hindi remote associates test showed same results which we got for rebus puzzle which means Hindi remote associates test can be conducted using EEG.

6 Results

In our study, we found the following neural signatures for insight. The first signature for insight is that we have found gamma power at right temporal regions (T8 and FC6 electrode) which is 350 ms prior to response and alpha power at right parietal occipital regions (P4 and P8 electrode) which is 750 ms prior to response (as shown in Fig. 2). As activation of alpha prior to response signifies shutting down of all the activities in the brain related to thinking (or in mental impasse condition) during insightful problem solving, there after there is a sharp activation at right superior temporal in gamma band which clearly signifies that the solution is insightful solution. We have also observed that the neural correlates for non-insightful solution (analytical solution) are same as insight for alpha band, but there is no activation associated with right temporal in gamma band for non-insightful solution which means that the problem can be solved by any other means which can be analytical. Here, we found alpha insight effect weaker than gamma insight effect for insight problems. Getting alpha insight effect weaker than gamma insight effect means brain activities did not shut down properly; this may be due any reason (some are effect of mood or selective attention) during experiment. Previous studies on insight showed that alpha power is stronger than gamma power [1].

Second signature for insight that we have found is (shown Fig. 2) that there is significant difference between gamma power for insightful solution as compared

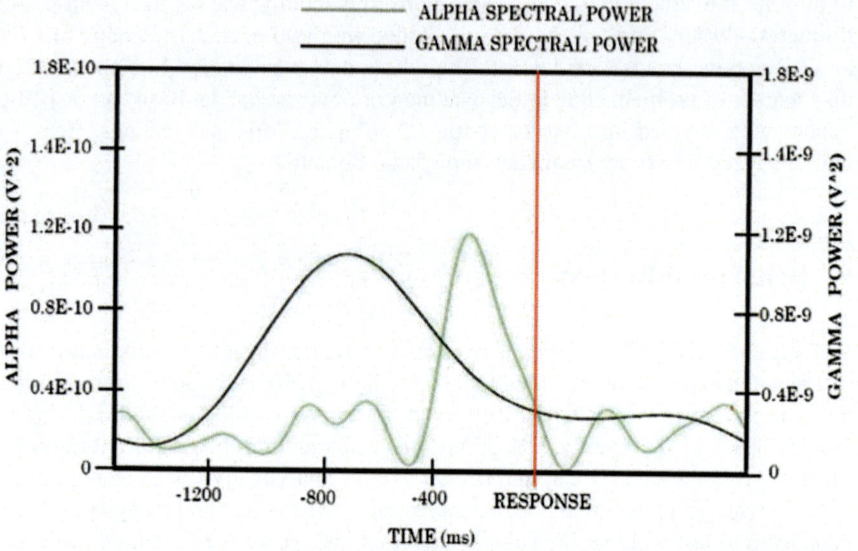

Fig. 2 Black line indicates average of alpha power (9–13 Hz at right parietal occipital) which is 0.75 s prior to response, and green line indicates average of gamma power (39–42 Hz at right temporal) which is 0.35 s prior to response; *X*-axis represents the time in milliseconds, and *Y*-axis represents spectral power

to non-insightful solution. The gray marker shows the point at which participant feels aha moment which started approximately 0.35 s prior to response taking into consideration that button press requires 0.3 s to initiate and execute [10]. Results we have observed for EEG power in gamma band for insight and non-insight with time (Fig. 2) are matched with the earlier studies [1, 11]. But, there is no comparison of EEG power in alpha band for insightful solution and non-insightful solution because we observed that activation at right parietal occipital region in alpha band is activated mostly for both insightful and for non-insightful solution.

Third signature for insight that we found is that there is gamma burst at right superior temporal region (0.35 s prior to button press) just after the activation of alpha peak at right parietal occipital region (0.75 s prior to button press) which is shown in Fig. 3. The red marker represents the response at which participants felt 'Aha' feeling. Another thing we have found is that sometimes participants experienced aha feeling but of lower amplitude as compared to others; it may be because getting alpha insight effect weaker than gamma insight effect which leads to lowering the magnitude of gamma burst/activity.

The interesting thing which we found from above neural signature for insight for Hindi version of remote associates test clearly signifies that this is not language effect and this somehow appears us true insight (as previous studies showed similar results for English remote associate test). Similarly, we got same neural signature for Rebus puzzle which confirmed us that this is not language effect and this somehow

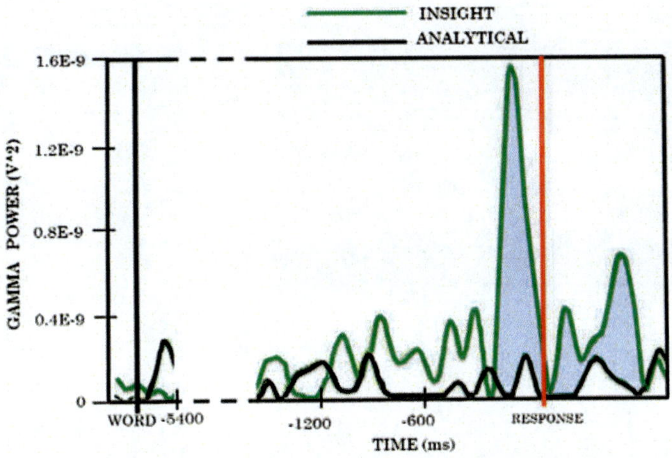

Fig. 3 Comparison of gamma insight effect for insight and non-insight in gamma band EEG power recorded at right temporal electrode; the shaded region indicates the difference between gamma power of insight and non-insight. Black line represents the non-insight gamma effect, and green line represents insight gamma effect

appears us true insight for Hindi version remote association test. Here, we got is a positive correlation of activation in right parietal occipital area for aha and non-aha feelings; there is a negative correlation of activation of temporal area in the right hemisphere, clearly indicating that T8 and FC6 are activated when the problem is solved insightfully. Spectral analysis of EEG indicates there is a sudden increase in gamma activity/gamma burst in right superior temporal region. The results which we obtained from the EEG experiments were performed on 15 healthy right-handed subjects with positive mood and in same environmental condition for all (Fig. 4).

7 Discussion

In our study, we test our Hindi remote associates test both as experiential that is associated with 'Aha' feeling similar to standard English remote associates test but also test it experimentally that is on the basis of neural signature. Hindi remote associates test matched fully with English remote associates test. One more interesting thing we found due to native language is that subjects reported more insight for Hindi remote associates test as compared to English remote associates test.

Previous studies have made efforts to prove that in solving a problem, semantic integration is necessary as it acts as the connection of various problems and helps in connecting the problem with solution. They coded that semantic integration occurred at Superior Temporal Gyrus (STG), especially while solving insightful problems; especially for verbal problems, they also showed that right temporal is responsible

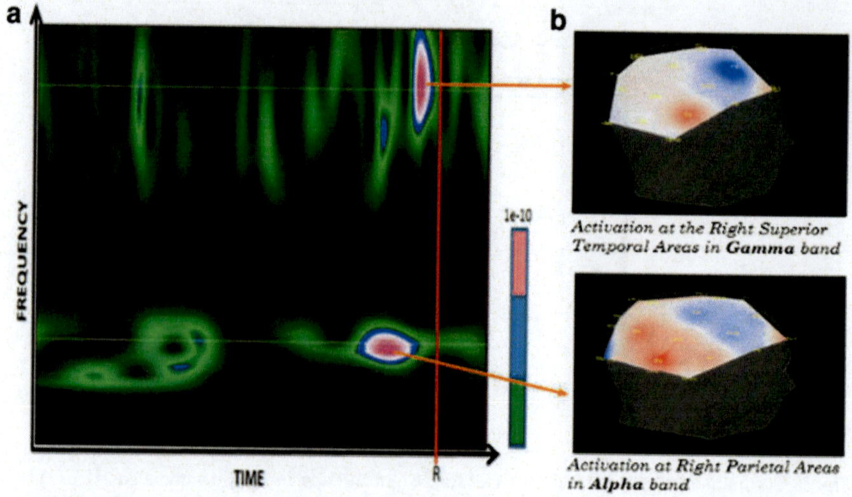

Fig. 4 **a** Left side figure shows time–frequency plot for averaged data of insight showing gamma burst at right superior temporal (0.35 s prior to response) and just before the alpha activation at right parietal occipital (0.75 s prior to response). **b** Right side figure shows activation of right hemisphere at right parietal occipital for alpha oscillation and at right superior temporal for gamma oscillation in 3D sensor cap view

for insightful problem [1]. Further studies made efforts to identify whether mental preparation exhibits different brain states before they see a verbal problem being solved through insight or non-insight and revealed the fact that it so happened that during preparatory stage neural activity of Anterior Cingulate Cortex (ACC) occurs while solving insightful problem and the semantic activation was at Superior Temporal Gyrus (STG) similar to previous studies, and during non-insightful preparatory stage, the neural activation was found at occipital region resulting in increased externally directed visual activity [3]. Furthermore, attempts were made by changing the variable for example what effects the preparatory stage i.e. does the mood of the subject effects the individual in problem solving. They revealed the fact that positive mood effects positively in problem solving, and dorsal Anterior Cingulate Gyrus (dACC) is activated for a person in positive mood at preparatory stage matched with previous studies. They further concluded that mood biases the insightful problem solving [2].

The present study has made efforts to prove that by changing the language variable, i.e., Hindi Remote Associates Test (RAT) in place of English RAT, we got same neural signature for insight during insightful problem solving as previous studies found by Jung-Beeman et al. [1]. The neural signature for insight we found is that there is an activation in alpha band at right parieto occipital region (0.75 s prior to button press) followed by an activation in gamma band at right superior temporal region (0.35 s prior to button press). A sudden gamma burst is observed for insightful response about 350 ms before button press which can be understood as the generation of sudden spark of solution to the robust problem; this result was strongly supported by the previous study which observed a gamma burst only for Aha moment [1, 12]. There is no such neural signature for insight as compared to non-insightful solution in gamma band; this result was supported by a previous study on verbal problem solving with insight using fMRI and EEG [1] which showed activation in T8 channel records signal from right temporal lobe for insightful problem solving.

As per Gary Klein, reduction in error and increase in insight substitutes the improvements in performance. During the process of meditation, accumulated stress is removed, energy is increased and health is positively affected overall [13], which actually increases the efficiency, and this leads to increase in insight. The sattva states involve delightfulness, happiness, selflessness, gratitude, fearlessness, and in the book 'the joy of yoga', sattvic guna is described as a 'quality of insight.' If we take the case of laziness, then it is a feeling when we are fully satisfied and we do not care to bother about anything else. As per our literature review, it has been observed that the feelings that we get from insight quite resembles to the feelings generated during sattva. Broadly, we can explain this as that insight is actually an 'aha moment' combined with happiness, surprise and sudden understanding in problem solving situations which gets in touch with sattva.

The reaction time of students with sattva guna is more as compared to student with tamasa while those with rajas guna have better selective attention [14]. This signifies that the people with sattva guna has more insightful ability. In the book 'the joy of yoga', sattvic guna is described as a 'quality of insight.' It has also been found out in our study as well as the previous studies conducted that response time for insightful problems is less as compared to non-insightful problems.

The 'Aha' moment/eureka moment are the solutions to problem presented with unexpected clarity; results in the burst of gamma. There were brain oscillations which were found in α and Υ bands in right parietal and right temporal lobes. By these results, Santarnecchi and other concluded that "the possibility of enhancing the probability of generating *Eureka*! moments in humans is by means of frequency-specific noninvasive brain stimulation" [4].

Previous study by Ren et al. [15] showed that group who practice meditation solved more insightful problem as compared to control group. It would be interesting to see in future studies whether there is any correlation between neural correlates of insightful problem solving with that of advanced meditationist for Hindi remote association test.

Acknowledgements I would like to thank the participants in this study, as well as useful feedback at the end of experiment. Finally, I would like to thank my colleagues D. Geeta Prem Chandoo and Sakshi Jain for their valuable suggestions, useful comments and assistance for the improvement of experimental design and methods for analysis.

References

1. Jung-Beeman M et al (2004) Neural activity when people solve verbal problems with insight. PLoS Biol 2(4):e97
2. Subramaniam K et al (2009) A brain mechanism for facilitation of insight by positive affect. J Cogn Neurosci 21(3):415–432
3. Kounios J et al (2006) The prepared mind: neural activity prior to problem presentation predicts subsequent solution by sudden insight. Psychol Sci 17(10):882–890
4. Santarnecchi E et al (2019) Gamma tACS over the temporal lobe increases the occurrence of Eureka! Moments. Sci Rep 9(1):5778
5. Sandkühler S, Bhattacharya J (2008) Deconstructing insight: EEG correlates of insightful problem solving. PLoS ONE 3(1):e1459
6. Mednick SA, Mednick M (1971) Remote associates test: examiner's manual. Houghton Mifflin. Q
7. MacGregor JN, Cunningham JB (2008) Rebus puzzles as insight problems. Behav Res Methods 40(1):263–268
8. Burle B et al (2015) Spatial and temporal resolutions of EEG: is it really black and white? A scalp current density view. Int J Psychophysiol Off J Int Organ Psychophysiol 97(3):210–220
9. Tadel F, Baillet S, Mosher JC et al (2011) Brainstorm: a user friendly application for MEG/EEG analysis
10. Smith RW, Kounios J (1996) Sudden insight: all-or-none processing revealed by speed-accuracy decomposition. J Exp Psychol Learn Mem Cogn 22:1443–1462
11. Kounios J, Beeman M (2014) The cognitive neuroscience of insight. Annu Rev Psychol 65(1):71–93
12. Bowden EM et al (2005) New approaches to demystifying insight. Trends Cogn Sci 9(7):322–328
13. Sharma H (2015) Meditation: process and effects. Ayu 36(3):233–237. https://doi.org/10.4103/0974-8520.182756
14. Kumar S, Satsangi A (2019) Impact of gunas on cognitive abilities, Int J Scientific Research and Review 07(03):624–633.https://ijsrr.co.in/images/full_pdf/1552704643_303.pdf
15. Ren J, Huang Z, Luo J et al (2011) Sci China Life Sci 54:961.https://doi.org/10.1007/s11427-011-4233-3

Identifying Neural Correlates in Stroop Task

Sakshi Jain, C. M. Markan, Sona Ahuja, and Manjari Tripathi

1 Introduction

Stroop color word task [23] can be confidently considered as the most reliable for the study of cognitive conflict resolution [2] and selective attention [16].

Stroop task consists of written word of color names, and participants were instructed to press right button when they saw a congruent stimulus, i.e., same color word with same ink, e.g., red written in red ink, and to press left button when they saw an incongruent stimulus, i.e., a color word with a different ink, e.g., red written in blue ink implied slow response for incongruent comparative to congruent [14] verified Stroop interference effect [8].

This study identifies neural correlates of Stroop task over a category of control subjects and meditators who had reached a different level of meditation following transcendental meditation [15, 21, 25]. Identification of the exact source of resolving conflict is still a subject of research. Previous studies have identified the brain regions responsible for dealing with conflict. Functional magnetic resonance imaging (fMRI) studies suggested that most of the activations are through ACC and through left dorsolateral prefrontal cortex (DLPFC) [5].

S. Jain (✉) · C. M. Markan
Department of Physics and Computer Science, Dayalbagh Educational Institute, Agra, Uttar Pradesh, India
e-mail: sakshi1612jain@gmail.com

C. M. Markan
e-mail: cm.markan@dei.ac.in

S. Ahuja
Faculty of Education, Dayalbagh Educational Institute, Agra, Uttar Pradesh, India
e-mail: sonaahujadei@gmail.com

M. Tripathi
Department of Neurology, AIIMS, New Delhi, India

V. H. Saran and R. K. Misra (eds.), *Advances in Systems Engineering*, Lecture Notes in Mechanical Engineering,
https://doi.org/10.1007/978-981-15-8025-3_3

Previous PET and fMRI meta-analysis on Stroop effect conclude left lateralized activations in the ACC as well as in pre-SMA [17]. Apart from PET, fMRI, EEG techniques, magnetoencephalography (MEG) offers better source localization than EEG. It is a completely noninvasive technique which records magnetic field generated through electrical neuronal activity within millisecond timescale [9].

This study investigates the effect of meditation on conflict resolution ability, it was hypothesized in this study that meditators performed Stroop task with better efficiency than control subjects, and a number of recent studies have suggested that meditation training may change brain morphology and function, particularly in areas related to attention and response selection [10, 11].

2 Method

2.1 *Participants*

Participants for this experiment were recruited from Dayalbagh Educational Institute, Agra through a notice, interested students then filled a questionnaire containing questions of required information. A category of subjects, i.e., ultra-transcendental meditators, were recruited from the faculty members of DEI (because it is hard to find practiced meditator in young age group).

This study includes 8 female, 5 male control subjects (21–42 years), 4 female, 3 male TMs (23–34 years), 7 female, 4 male aspirants (15–24 years) and 5 female, 7 male UTMs (37–59 years), native Hindi speaking, right-handed without any history of neuropsychiatric neurological disease, clear or corrected normal vision and normal color vision. Fifteen participants were excluded only for the analysis of source localization because of lack of data. The final analysis includes 7 subjects in each category, i.e., 7 in control (mean age 23.4 years, S.D 5.2 years), 7 in aspirants (mean age 20 years, S.D 2.8 years), 7 in TM (mean age 29 years, S.D 6.5 years) and 7 in UTM (mean age 48.4 years, S.D 6.2 years). Subjects from each category were randomly assigned to perform experiment.

2.2 *Experimental Paradigm*

Magnetic signals generated through electrical current occurring in brain were recorded using a whole scalp MEG using arrays of superconducting quantum interference sensors [9]. Participants performed a computerized Stroop task, designed on E-Prime software, with red, green, yellow and purple color. Task contains only congruent and incongruent stimuli as shown in Fig. 1. The stimuli were projected on a back-projection screen placed at comfortable distance from the subject, fitting in an area of 3 $\times$ 3 of the central visual field. To avoid negative priming effect stimuli

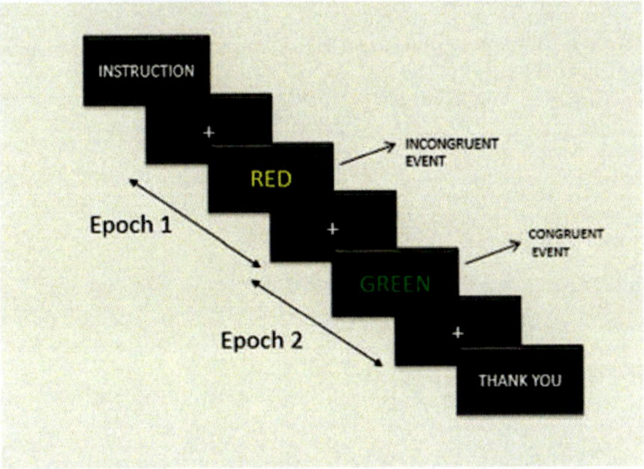

Fig. 1 Layout of experimental paradigm of stroop task, 1000 ms stimuli and 500 ms fixation cross

was presented in a pseudo-random order. No relevant word for congruent trial (i.e., red) is followed by an irrelevant word for incongruent trial (i.e., red written in any color).

Each epoch consists of fixation cross '+' for 500 ms; this is followed by stimulus colored word for 1000 ms in this time response button needed to be pressed. Stroop color word task is designed on E-Prime software (a packaged paid software). Responses were bimanual, i.e., right-hand button to be pressed for congruent stimuli and left-hand button to be pressed for incongruent stimulus. All the participants were instructed to press button as fast as possible. They perform a session of Stroop task having 15 congruent stimuli and 15 incongruent stimuli after 10 min of rest done by control subjects and 10 min of transcendental meditation by meditators (Fig. 1).

2.3 Data Acquisition

MEG signals were recorded using advanced brain imaging machine Elekta Neuromag TRIUX MEG. Head positioning was monitored by using HPI coils inside MEG helmet. The location of the coils was determined using an electromagnetic tracker before the acquisition.

Artifacts related to heartbeats, eye movements and eye blinks were recorded by ECG and EOG, respectively. Band pass of MEG data was through 0.1–500 Hz, and sampling rate was set as 1000 Hz. MEG offered high temporal resolution [29] of the order of millisecond, beneficial in direct measurement of neural activity compared to hemodynamic change.

Table 1 Comparative difference of congruent RT, incongruent RT and incorrect recognition over control subject, aspirants, TM and UTM

	Control ($N = 7$)	Aspirants ($N = 7$)	TM ($N = 7$)	UTM ($N = 7$)
Congruent RT	582 ± 57	534 ± 57	583 ± 67	653 ± 89
Incongruent RT	790 ± 58	709 ± 68	757 ± 111	821 ± 95
Incorrect recognition (for all participants)	6.5 ± 3.7	4.54 ± 3.04	4.71 ± 2.71	4.36 ± 3.07

2.4 *Behavioral Data Analysis*

Response time (RT) was analyzed only for the correct recognition of congruent and incongruent trials over the categorized subjects. Trial with incorrect and no response was not considered in this discussion. ANOVA test done over the categorized subject as independent variable, RT for congruent and incongruent as dependent variable disclose significant effect of category on congruent RT ($p < 0.05$) and on incongruent RT ($p < 0.05$), analyzed on 7 subject in each category. There exists a significant difference between the RT of incongruent and congruent trial maximum for control and minimum for UTM (incongruent > congruent). UTM performed more accurately than control subject, and data can be found in Table 1.

2.5 *MEG Data Analysis*

Data recorded using MEG 306 channel (204 planar gradiometers and 102 magnetometer) preprocessed offline using signal space separation method (SSS) beneficial in compensation of external interference and sensor artifacts [26]. Further offline preprocessing was performed on brainstorm software (an open-source software) [24].

The MEG signals were first band passed through 0.1–30 Hz. The stimuli were marked with symbol E, left response was marked with symbol L, and right response was marked with symbol R. All multiple responses were detected, and those were removed within the range of 300 ms by considering first response to be the obvious response. Artifacts generated through eye blink and heartbeat were detected and removed having amplitude above 150 μV through signal space projection (SSP) method. Poorly captured signals were marked as bad channels. The whole data was classified into epochs of 1500 ms subsequently extracted from 100 to 1000 ms relative to stimulus onset, 100 ms pre-stimulus used as baseline. Congruent and incongruent stimulus was separately named. Source localization was done using linearly constrained minimum variance (LCMV) beamformer method [29]. "The LCMV beamformer method generalizes the existing volumetric magnetic resonance inverse imaging (InI) technique, achieving higher detection sensitivity while maintaining whole- brain spatial coverage and 100 ms temporal resolution, LCMV offers better localization accuracy and superior detection sensitivity" [13]. The activations were observed from 300 ms, considering about 300 ms required for information processing in brain [1].

3 Results

3.1 Behavioral Results

Behavioral analysis showed significant difference in RT for congruent and incongruent trials. Significant difference is observed in difference in RT over the category of subjects, 208 ± 147 ms for control subjects, 174 ± 123 ms for aspirants, 174 ± 123 ms for TM and 167 ± 118 ms for UTM, and this clearly verified that UTM performed equivalently for congruent and incongruent stimulus. RT for congruent and incongruent trial was more for UTM, and they performed more accurately; this result verified that meditators take more considered decision.

3.2 MEG Data Results

Spatial analysis of the MEG data provides some evidence for uniquely activated part of brain in conflicting situation. Sequential activation was observed through source localization process. Observed from 300 ms post-stimulus, right occipital lobe activated from 300 to 450 ms and then activation shifts to parietal lobe from 400 to 500 ms; after that, frontal lobe got activated from 300 to 500 ms, observing further temporal lobes showed activation from 400 to 700 ms and finally the activation ends up at motor cortex from 400 to 600 ms for congruent trials. For incongruent trial, activation followed similar pattern; i.e., right occipital got activated from 300 to 500 ms post-stimulus and then parietal lobe from 300 to 600 ms; then, the activation shifts to right frontal lobe from 300 to 600 ms, then posterior cingulate gyrus and anterior cingulate cortex got activated from 400 to 700 ms, then temporal from 400 to 700 ms and finally the activations end at motor cortex from 400 to 700 ms. Activation is observed through all sensors, i.e., 204 planar gradiometer and 102 magnetometer.

These results showed a positive correlation of the regions occipital lobe, parietal lobe, frontal lobe, temporal lobe and motor cortex for congruent and incongruent trial and showed a negative correlation for the region of posterior cingulate gyrus and anterior cingulate cortex represented in Fig. 2. These observations verified PCG and ACC responsible for dealing with conflicting situation.

This analysis also revealed about more spread of activation in incongruent stimuli as compared to congruent (incongruent > congruent) stimuli for control in comparison with meditators (control > meditators). Activations were also observed in subcortical regions, i.e., insula and corpus callosum.

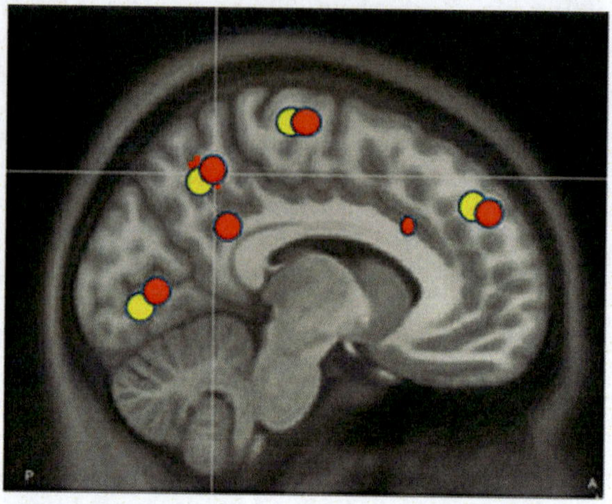

Fig. 2 Region of activation for congruent and incongruent stimuli, observed over 28 subjects, 7 in control group, 7 in pre-initiates, 7 in first initiates and 7 in second initiates after 300 ms post-stimulus. Congruent. Incongruent

4 Discussion

In this study, spatiotemporal analysis associated with Stroop color word task verified Stroop interference effect at behavioral level, by a positive difference in response time between incongruent trial and congruent trial.

This MEG study finds a unique activated region for incongruent than congruent trial for a sustained period of time, one possible explanation for activation of PCG and ACC to deal with conflicting resolution. Previous studies of source localization of Stroop task using MEG support this result; "ACC involvement would reflect error detection, whereas pre-supplementary motor area (SMA) activation would reflect response conflict in selecting the response" [6]. Sequential activation observed for congruent as well as for incongruent trials revealed the pattern, i.e., activation of occipital lobe responsible for visual action [7], parietal lobe activation for language processing [3], frontal lobe activation for thinking [4] and then activation of PCG only in case of incongruent trial for error recognition; after that, activation shifts to temporal lobe responsible for memory processing [27] and at last the activation shifts to motor cortex responsible for the execution of movement for button press [28]. These results were supported by previous study; "Stroop task involves circuit subserving attention, working memory, response selection and inhibition, motor planning and motor response among others" [20].

Behavioral result of this study shows that meditators performed with less error in comparison with control; this justifies the hypothesis that meditation improves decision making, and "studies suggest that meditation training may change brain morphology and function, particularly in areas related to attention and response selection" [10, 11]. Difference between RT of incongruent and RT of congruent trial,

208 ± 147 ms for control subject, 174 ± 123 ms for aspirants, 174 ± 123 ms for TM and 167 ± 118 ms for UTM identified that control subject was slower to resolve conflict as compared to meditator; a study reveals that regular meditators tend to have higher activation levels, compared to non-meditators, for congruent stimuli, and lower activation levels for incongruent stimuli. "It seems that regular meditators, unlike non-meditators, tend to process incongruent and congruent stimuli quite similarly" [12, 19]. In this study Spatial analysis revealed more spread of activation for control subject comparative to meditators (control > meditators) verified a previous study, "Regular meditators activated fewer brain regions than non-meditators in order to achieve the same performance during an attentional task. This is an evidence that meditation training can increase brain efficiency in attention and impulse control" [12] matched over a variety of meditators such as Zazen, mantra mediation and mindfulness of breathing. Behavioral data shows that the control subjects responds faster as compared to UTM can be justified by age effect, a study showed impact of age on cognition verified "Multiple cross-sectional studies have shown that there is an improvement in crystallized abilities until approximately age 60 followed by a plateau until age 80, and there is steady decline in fluid abilities from age 20 to age 80" [18]. Fluid abilities can be understood as processing speed.

5 Conclusion

Stroop interference effect is verified through the positive time difference between the incongruent and congruent trials. This study concludes positive correlates of meditation in cognitive task. Meditators take more considered decision followed by slow (perhaps due to age effect) but more accurate response. More brain areas were activated for control subjects rather than meditators. Difference in RT between incongruent and congruent trials was significantly less for meditators as compared to control group.

6 Future Work

Shortcoming in this study was the vast age gap between the group of UTM and other category; to overcome this shortcoming, we will select participant of similar age group and IQ and of similar educational background for further studies in this domain. Further analysis would be done for incorrect recognition.

Acknowledgements My sincere thank goes to my research colleague Ms. D. Geeta Prem Chandoo and Mr. Raman Kumar for their selfless support and motivation.

References

1. Amor TA et al (2016) Persistence in eye movement during visual search. Sci Rep 6:20815
2. Cohen JD, Dunbar K, McClelland JL (1990) On the control of automatic processes: a parallel distributed processing account of the stroop effect. Psychol Rev 97(3):332–361
3. Coslett HB, Schwartz MF (2018) The parietal lobe and language. Handb Clin Neurol 151:365–375
4. de Souza LC et al (2014) Frontal lobe neurology and the creative mind. Front Psychol 5(761)
5. Fan J, Flombaum JI, McCandliss BD, Thomas KM, Posner MI (2003) Cognitive and brain consequences of conflict. Neuroimage 18(1):42–57
6. Galer S et al (2014) Investigating the neural correlates of the stroop effect with magnetoencephalography. Brain Topogr 28
7. Galetta, S.L.. (2017). Occipital Lobe. 10.1016/B978-0-12-809324-5.03810-4.
8. Geukes S et al (2015) Stroop effects from newly learned color words: effects of memory consolidation and episodic context. Front Psychol 6
9. Hämäläinen M et al (1993) Magnetoencephalography: theory, instrumentation, and applications to noninvasive studies of the working human brain. Rev Mod Phys 65:413
10. Holzel B et al (2011) How does mindfulness meditation work? Proposing mechanisms of action from a conceptual and neural perspective. Perspect Psychol Sci 6:537–559
11. Jang JH et al (2011) Increased default mode network connectivity associated with meditation. Neurosci Lett 487(3):358–362
12. Kozasa EH, Sato JR, Lacerda SS, Barreiros MA, Radvany J, Russell TA, Sanches LG, Mello LE, Amaro E. (2012) Meditation training increases brain efficiency in an attention task. Neuroimage 59(1):745–749. ISSN 1053–8119, doi: https://doi.org/10.1016/j.neuroimage.2011.06.088.
13. Lin F-H et al (2008) Linear constraint minimum variance beamformer functional magnetic resonance inverse imaging. Neuroimage 43(2):297–311
14. Macleod C (1991) Half a century of research on the stroop effect—an integrative review. Psychol Bull 109:163–203
15. Maharishi (1969) Maharishi Mahesh Yogi on the Bhagavad Gita. Penguin, New York
16. Melara RD, Mounts JRW (1993) Selective attention to stroop dimensions: effects of baseline discriminability, response mode, and practice. Mem Cogn 21(5):627–645
17. Minzenberg MJ et al (2009) Meta-analysis of 41 functional neuroimaging studies of executive function in schizophrenia. Arch Gen Psychiatry 66(8):811–822
18. Murman DL (2015) The impact of age on cognition. Semin Hear 36(3):111–121
19. Norris C et al (2018) Brief mindfulness meditation improves attention in novices: evidence from ERPs and moderation by neuroticism. Front Hum Neurosci 12:315
20. Peterson BS et al (1999) An fMRI study of stroop word-color interference: evidence for cingulate subregions subserving multiple distributed attentional systems. Biol Psychiatry 45(10):1237–1258
21. Roth R (1994) TM—transcendental meditation: a new introduction to Maharishi's easy, effective and scientifically proven technique for promoting better health, unfolding your creative potential, and creating peace in the world. Plume Books, New York
22. Satsangi PS (2013) Towards integrating arts and science of consciousness. In: Vision talk, toward a science of consciousness (TSC 2013), DEI, Agra, Mar 2013
23. Stroop JR (1935) Studies of interference in serial verbal reactions. J Exp Psychol 18:643–662
24. Tadel F et al (2011) Brainstorm: a user-friendly application for MEG/EEG analysis. Comput Intell Neurosci 879716
25. Travis FT, Tecce J, Arenander A, Wallace RK (2002) Patterns of EEG coherence, power, and contingent negative variation characterize the integration of transcendental and waking states. Biol Psychol 61:293–319
26. Wen X, Ru Y (2011) Research of blind signal sepration algorithm based on ICA method. In: Jin D, Lin S (eds) Advances in multimedia, software engineering and computing, vol 1. Advances in intelligent and soft computing, vol 128
27. Yee LT, Warren DE, Voss JL, Duff MC, Tranel D, Cohen NJ (2014) The hippocampus uses information just encountered to guide efficient ongoing behavior. Hippocampus 24(2) 154–164

28. Zang Y-F et al (2003) Functional organization of the primary motor cortex characterized by event- related fMRI during movement preparation and execution. Neurosci Lett 337:69–72
29. Zhang J et al (2014) Source localization with MEG data: a beamforming approach based on covariance thresholding. Biometrics 70(1):121–131

Processing Multilingual Word Embeddings of Indian Languages with Semi-supervised Learning

Harsh Saxena

1 Introduction

Multilingual word embeddings have recently got attention with the rise of text-speech synthesis. Processing a direct application in cross-lingual tasks like machine translation and cross-lingual entity linking parsing and document classification. Nearly, all methods to learn these multilingual word embeddings use a parallel corpora. But with the scarcity of corpora of several languages, the conventional methods cease to exist.

For preventing this issue, we can make use of document-aligned corpora or label-aligned comparable corpora.

2 Embedding Mappings

Methods to workout the bilingual mappings work by independently learning the embeddings for every language using monolingual corpora, learning a transformation from one embedding plethora to the other based on a dictionary.

[1,2] proposed the first such method, learning the linear transformation that minimizes the sum of squared Euclidean distances for the dictionary entries. Incorporation of length normalization in the training of word embeddings and maximize the cosine similarity instead can be done.

H. Saxena (✉)
Department of Electrical Engineering, Faculty of Engineering, Dayalbagh Educational Institute, Agra, Uttar Pradesh, India
e-mail: harsh98saxena@gmail.com

V. H. Saran and R. K. Misra (eds.), *Advances in Systems Engineering*,
Lecture Notes in Mechanical Engineering,
https://doi.org/10.1007/978-981-15-8025-3_4

A strong arena of bilingual embedding mappings is in the field of machine translation and bilingual lexicon extraction, which is one of the main ways of evaluation. This mapping is used to formulate the translation of source language words, missing in the original dictionary, by taking their nearest neighbour word in the target language as per cosine similarity [3].

3 Unsupervised Bilingual Embeddings

As per the abstract, the method works with as less as 30-word pairs, whereas the methods adopted till date use more than a hundred pairs. An exception to this is the work by Zhang who used 10-word pairs with exceptional results on transfer learning for part-of-speech tagging. Here, the experiments will show that the traditional method captures coarse-grained relations; it fails on finer-grained tasks like bilingual lexicon induction and machine translation.

Algorithm 1: Traditional Framework

Input: S (source embeddings)
Input: T (target embeddings)
Input: D (seed dictionary)
1: W ← LEARN MAPPING (S, T, D)
2: D ← LEARN DICTIONARY (S T, W)

4 Self-learning Framework

We use the output dictionary in Algorithm 1 as the input of the continuing system in a self-learning approach which, assuming the output dictionary is indeed better than the original one.

The process can then be repeated iteratively to obtain a hopefully better mapping and dictionary each time until some convergence criterion is met. Algorithm 1 summarizes this alternative framework that we propose.

Our method can be combined with any embedding mapping and dictionary induction technique. For this orthogonality constraint, we minimize the squared Euclidean distance that becomes equal to maximizing the dot product.

5 Error Analysis

Understanding the behaviour of our system, an error analysis was performed of the output in English–Hindi bilingual lexicon induction when starting with the 2000 entry, the 30 entry because the dictionaries in comparison with this method are the 2000 entry dictionary.

I took 50 random examples from the test set in the [1–2k] frequency bin, another 50 from the [2–5k] frequency bin and 30 from the frequency bin and observe each of the errors made by all the three different fields [1].

6 Conclusion and Future Work

In this work, the proposal of a simple self-learning framework to learn bilingual word embedding mappings of Indian languages was presented. The experiments on bilingual lexicon induction and crosslingual word similarity present that the method is fully throttled to learn multilingual embeddings with as little multilingual 30-word dictionary or an automatically generated list of words, compiling results that are competitive with systems using much richer bilingual resources like larger dictionaries and corpora. Moreover, a more detailed analysis shows that the method is obtaining an objective function that is independent from any bilingual data which with better bilingual word embeddings in a completely semi-supervised manner.

References

1. Mikolov(2013): Distributed Representations of Words and Phrases and their Compositionality
2. Artetxe M, Labaka G, Agirre E (2018) A robust self-learning method for fully unsupervised cross-lingual mappings of word embeddings. In: Proceedings of the 56th annual meeting of the Association for Computational Linguistics, vol 1: long papers
3. Artetxe M, Labaka G, Agirre E (2016) Learning principled bilingual mappings of word embeddings while preserving monolingual invariance. In: Proceedings of the 2016 conference on empirical methods in natural language processing, pp 2289–2294

Developing Trends in Consciousness Among Adolescents

Preet Kumari and Jincy Cherian

1 Introduction

An adolescent is a young individual growing into an adult. The transitional period of psychological and somatic development which occurs during the phase from puberty to adulthood is known as adolescence. The World Health Organization defines an adolescent as a person between ages 10 and 19 years [38] and says distinctive changes may ensue during this phase and usually various adult patterns are also recognized. The America Academy of Child and Adolescent's Facts for Families have classified adolescents into three major stages: early adolescence (10–13 years), middle adolescence (14–17 years) and late adolescence (18–21 years) [2]. During early adolescence period, they have limited ability of abstract thinking, but develop profound moral thinking. In middle adolescence, they develop their capacity of abstract thinking, start setting long-term goals for themselves and also become more interested in finding the meaning of life and living it with morals. And in late adolescence, they enter adulthood with which they gain rationality, increased emotional stability and individuality. Hall [21] mentioned that adolescents go through a phase of 'storm and stress' during their evolution from childhood to maturity. This is a sensitive phase of life. Here, self-control level is decreased (storm) and sensitivity levels are increased (stress). Conflicts with parents, mood disruptions and risky behaviour can be seen among adolescents at this stage of life. Self-consciousness peaks up and this never-ending series of both physical and psychological adjustments have significant influence on their interactive and behavioural facets of life. Thus, to survive, they need to be encouraged to openly speak about their issues, stop criticizing themselves and point

P. Kumari · J. Cherian (✉)
Dayalbagh Educational Institute, Agra, Uttar Pradesh, India
e-mail: jincycherian.01@gmail.com

P. Kumari
e-mail: preetkumari1972@gmail.com

V. H. Saran and R. K. Misra (eds.), *Advances in Systems Engineering*,
Lecture Notes in Mechanical Engineering,
https://doi.org/10.1007/978-981-15-8025-3_5

towards positives of life and nurture a balanced perspective. Being conscious about their positive traits inside and their environment may help them nurture themselves and overcome the obstacles faced by them during this turmoil filled phase of life. An assortment of mindfulness, spiritual and emotional awareness is important because it improves the quality of life of an individual [17].

Consciousness can be defined as an individual's awareness about his/her unique thoughts, memoirs, emotions and surroundings. It can be called a state/quality to feel, comprehend or experience subjectively about the internal and external subsistence. Brazdau and Mihai [8] mention consciousness as the most mysterious individual experience ever undergone by human beings. Philosophers like Locke and Descartes had struggled in comprehending the makeup of consciousness and how it corresponds in the world. After psychology was recognized as a separate discipline from biology and philosophy, early psychologists became interested in studying conscious experience. Introspection was used by structuralism theorists to analyze and account conscious experiences, thoughts, etc. James [23] an American psychologist associated consciousness with a stream; despite of constant shifts, it remains unbroken and continuous. Recently, it has become an important topic for interdisciplinary exploration in cognitive psychology which includes linguistics, etc. In the present, consciousness is frequently observed as a person's awareness of his own interior states as well as the incidents going around him. Modern theorists focus on the neuroscience behind these conscious experiences. Scientists have used brain-scanning technologies to look for explicit neurons that may be associated with various conscious events. Integrated information theory [3] view consciousness through learning more about physical activities that trigger the conscious experience. It attempts to construct a measure of integrated information that shapes consciousness. The essential focus remains on understanding about what it means for knowledge to exist in consciousness. The global workspace theory [4] suggests existence of a memory bank, and it is from this bank that the brain obtains information which in turn forms the understanding of the conscious awareness.

To be conscious, one needs to be aware and mindful about various aspects in life. An aspect of consciousness that has much been conversed is mindfulness and awareness is said to be the background of consciousness, constantly scrutinizing the interior and exterior environment. Mindfulness is integrally a status of consciousness. Awareness to the present events and happenings are given characteristics of human beings. Mindfulness is a concept that has emerged from Buddhism and other envisaged traditions where awareness and consciousness are cultivated attentively Brown and Ryan [6]. Mindfulness is commonly defined as a state of being aware of the happenings in the present. Thera [33] described mindfulness as a clear awareness of what actually happens to us and in us at consecutive instants of perception. Hanh [22] specified mindfulness as the act of holding one's consciousness active to the reality of the present. Awareness and mindfulness is the base to experience consciousness.

Consciousness is the quality of the mind regarded to encompass virtues such as cognizance, awareness of the self, subjectivity and the aptitude to comprehend the relationship between self and the environment [34]. Major psychological, spiritual and philosophical backgrounds accentuate the significance of the eminence of consciousness for maintenance and enrichment of well-being [37]. Developing consciousness is found to have positive affect on major aspects of professionalism like emotional intelligence of tutor trainees [31] and also job satisfaction levels among bank managers [36]. Emotional consciousness is the level to which one is aware and sensitive and the attitude they hold towards their sentiments and emotions. Rearing emotional consciousness level will be able to advance and expand one's personal as well as social life. Being aware and conscious about the emotional conditions of self and others are found valuable in leadership practices [10]. Moral consciousness is the understanding, mindfulness towards the moral guidance system one has. Enhancing moral consciousness helps in life-building, character building in a person. It develops various values like love, empathy and nonviolence and social order based on the truth. Cognitive consciousness refers to the awareness one has about his ideas, opinions and other cognitive activities. Enhancing the cognitive consciousness may have optimistic effects on managing thoughts. Training the attention through mindfulness can help in identifying less with the flow of thinking and enhance the abilities to better manage the thoughts. Social consciousness includes the responsiveness towards a person's surrounding and the important social issues around him. It refers to the awareness of the connections and relations with other people and the communities one is a part of. Improving social consciousness helps in forming strong relationships and friendships. It also stimulates benevolence and empathy while interrelating with others. At last, spiritual consciousness signifies the cognizance of a person with his spiritual nature, a level beyond the body. It can be called the ability of a person to view his own body, thoughts and emotions and awareness as a neutral observer, from outside.

Consciousness appears to be diverse from all the other scientific conceptions and has been extremely difficult to be treated as a variable [5]. In psychology, assessing consciousness experience with valuation tools and statistical techniques has been an uncommon topic, maybe because psychologists find it difficult to produce a method that might reach a suitable level of magnitude of the results [25]. Yet, there are some assessment instruments that have been in use giving good results in evaluating some aspects of the conscious experience. In the process of developing a new consciousness scale for the targeted population, the researcher reviewed and analyzed some psychological instruments: Self-Consciousness Scale (SCS) developed by Fenigstein et al. [16], Consciousness Quotient Inventory by Brazdau [9], Mindfulness Attention and Awareness Scale-MAAS by Brown and Ryan [6], Spiritual Belief Scale by Deshmukh and Deshmukh [14], The Emotion Awareness Questionnaire (EAQ30) developed by Rieffe et al. [27] and The Moral Foundations Questionnaire (MFQ-30) by Graham et al. [18]. A necessity of developing a more concise, succinct and precise tool for consciousness for the adolescents was observed; thus, the researchers self-constructed a consciousness scale for the targeted population.

These results with 600 more studies performed in the dynamic field of consciousness motivated the researcher to explore the effects of various demographic aspects on consciousness in Indian settings. Also, very less research has been done on consciousness as a variable with adolescents as population sample, especially in Indian settings.

2 Problem

To study the effect of gender and age on consciousness among adolescent students.

3 Objectives of the Study

1. To study the effect of age on consciousness among adolescents.
2. To study the effect of gender on consciousness among adolescents.
3. To study the interaction effect of age and gender on consciousness among adolescents.

4 Hypotheses

1. There will be no significant effect of age on consciousness among adolescents.
2. There will be no significant effect of gender on consciousness among adolescents.
3. There will be no significant interaction effect of age and gender on consciousness among adolescents.

5 Method

5.1 Design

A 3 × 2 factorial design was used to study the effect of the independent variables (age and gender) as well as their interaction effect on the dependent variable (consciousness).

5.2 Variables

Independent Variables:

- Age (10–13 years, 14–17 years, 18–20 years)
- Gender (Male and Female)

Dependent variable:

- Consciousness

5.3 Sample

The sample included 300 adolescent students selected by convenient sampling method from various schools and colleges of Agra city. The students' age ranged from 10 to 20 years. 100 students (50 males and 50 females) from age range 10–13 years, 100 students (50 males and 50 females) from age range 14–17 years, and 100 students (50 males and 50 females) from age range 18–20 years were included in the sample (Table 1).

5.4 Tools Used

A self-constructed consciousness scale was used. The scale consists of 25 items divided into five dimensions—emotional consciousness (5 items), moral consciousness (5 items), cognitive consciousness (5 items), social consciousness (5 items) and spiritual consciousness (5 items). The scale was prepared on three-point Likert scale—agree, sometimes and disagree. The positive questions were scored 3, 2 and 1, respectively, and opposite scoring were done for the negative items of the scale. The reliability of the scale was estimated by Cronbach's Alpha to be 0.63. The internal consistency was determined by significant correlation of all the dimensions with the total score of the scale (0.45, 0.60, 0.33, 0.48, 0.65, respectively).

Table 1 Description of the sample

Gender/Age	Early adolescence (10–13 years)	Middle adolescence (14–17 years)	Late adolescence (18–20 years)	Total
Male	50	50	50	150
Female	50	50	50	150
Total	100	100	100	300

6 Result and Discussion

Table 2 shows the ANOVA table of the present study. It reveals the F value for age to be 19.192 indicating the effect of age on consciousness is significant at 0.01 level. This implies that there is a significant effect of age on consciousness among adolescents. The hypothesis made stating no significant effect of age on consciousness is rejected. Thus, it can be recognized that the age of a person has substantial effect on his/her consciousness level. Alispahic and Anic [1] confirm the findings through their study on 441 participants, who revealed that older participants had higher scores as compared to younger participants for aspects of mindfulness. As it is evident that mindfulness is associated and is an important path towards consciousness [15], it can be concluded that as mindfulness enhances the consciousness of a person will also develop. Gross et al. [20] provided research evidence that older adults demonstrated better levels in emotional control, which is an important dimension along with subjectivity in understanding consciousness [11].

The F value for gender is found to be 19.544 which is significant at 0.01 level. This indicates that gender has a significant effect on consciousness among adolescents. The hypothesis stating no significant effect of gender on consciousness was rejected. The results reveal that the gender of a person will have a considerable effect on his/her consciousness. Females were found more conscious as compared to males participants among the sample. This is found consistent with prior studies of Bryant [7], Underwood [35], and Peres [26], where they found gender difference in regard to various components of consciousness in their studies. A study by Rojiani et al. [28] supports the findings that women may have better conscious experiences and also benefit more in response to a college-based meditation training than men. Statistically significant gender difference was found between men and women participants on the subscales of mindfulness among Bosnian general population [1].

The F value for the interaction effect of age and gender on consciousness is found to be 3.455 which is significant at 0.05 level. This helps generalizing that there is a significant interaction effect of age and gender together on consciousness among adolescents. Thus, the third hypothesis of no significant interaction effect of age and gender on consciousness was also rejected in the study. Thus, it can be pointed

Table 2 Summary of two-way analysis of variance

ANOVA				
Source	Sum of squares	df	Mean square	F
Age	853.167	2	426.583	19.192**
Gender	434.403	1	434.403	19.544**
Age * Gender	153.607	2	76.803	3.455*
Error	6534.620	294	22.227	
Total		299		

*$p < 0.05$ **$p < 0.01$

Table 3 Two-way table of means

Gender/Age	Levels		
	10–13 years (Early Adolescence)	14–17 years (Middle Adolescence)	18–20 years (Late Adolescence)
Female	53.98	58.40	59.38
Male	53.8	55.92	55.28

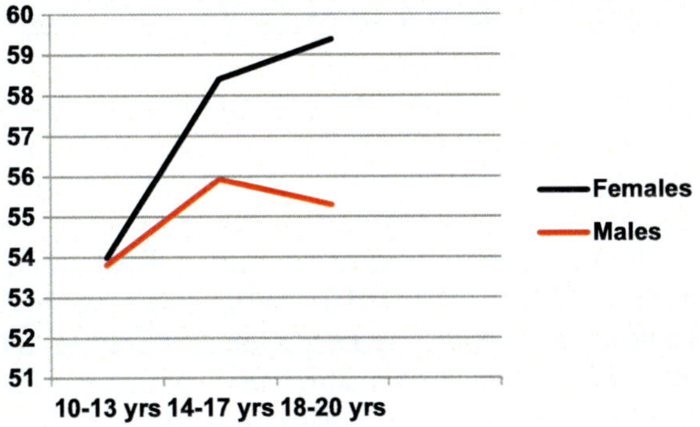

Fig. 1 Graph showing significant interaction effect of age and gender on consciousness

out that the interaction of age and gender of a person can have expressive effect on his/her consciousness level. Mankus et al. [24] confirmed gender and age to be significantly associated with emotional aspects of consciousness. As older participants were recorded scoring higher as compared to the younger participants. Sharma et al. [30] confirm the verity that there is a positive correlation between educational level and consciousness among university students. This means consciousness levels rise with the increase in educational level of individuals.

Table 3 shows the mean scores of consciousness among adolescents. The results reveal that females have high consciousness at the age range of 18–20 years and males have high consciousness at the age range of 14–17 years. It can also be inferred that females are found more conscious as compared to male adolescents throughout the sample. The table also discovers that as age increases, there is a rise in the consciousness level of adolescents. Thus, it can be concluded that consciousness rises as a person grows older. Graph 1 shows significant interaction effect of age and gender on consciousness among adolescents.

7 Conclusion

Consciousness is a very extensive word. It is not just awareness, mindfulness or alertness, but it also includes the attitude held by the person towards his day-to-day sentience. It cannot be confined to one single dimension but can be divided into many more domains and measured overall. It is a very vital part of human being and needs to be explored and enhanced at right levels. The present study reveals age and gender aspects of life as essential influences towards consciousness among adolescents. Grant and Jones [19] confirmed the boosting capacity of consciousness experiences. It helps developing cognitive abilities like memory, creativity, vigilance, attentiveness and intelligence. It also expands various affective qualities like well-being and self-concept.

Age was found to be an important factor affecting the consciousness levels among the sample. The possible explanation for the results demonstrating age difference can be maturational change or developmental clarification which suggests that as human beings age, they develop adaptive ways increasingly to manage their emotions; thus, they become less judging towards themselves and also others. As they are not interjected by the intensity of their emotions and society, they are able to focus on the 'here and now' [13]. Another psychological construct may be savouring which may be used to elucidate mindfulness [32], as older adults have a tendency of savouring the moment, have control over emotions and be focused on the present life; they manage to be more mindful. Effect of gender was also found significant in the present study. Women are found more conscious as compared to men among the sample, significantly. This might be because women may give more attention to their emotions to steer their traditional and stereotypical roles in the society. They are expected to be more cooperative and compassionate than men [29]. Such expectations may motivate women to show up their emotions so as to manage their daily and enduring tasks [12].

This study presents important verification about effects of age and gender on consciousness and it should be considered in future research. The research reveals that among Indian adolescents females and grownup adolescents were more conscious as in comparison with males and younger adolescents. These results offer basis for further massive research about consciousness not only on general but also on clinical populations. Helping adolescents understand their consciousness levels will help them in utilizing their limited time more efficiently towards a brighter future. It can help them comprehend the areas they lack knowledge or need improvement so as to excel. It will be helpful in identifying the problems faced in relation to the outside world. Recognizing the consciousness levels of the students will help the teachers and other elders to have a better empathetic approach towards the augmentation and enrichment of the young generation. This would also help the authority panels to help bring up various methods to boost such elements of the youth's personality and develop a better future for the students and young adults of the country.

References

1. Alispahic S, Anic EH (2017) Mindfulness: age and gender differences on a Bosnian sample. Psychol Thought 10(1):155–166
2. Allen B (2019) Stages of adolescence. Healthychildren.org. American Academy of Paediatrics. Retrieved 15 Nov 2019, from https://www.healthychildren.org/English/ages-stages/teen/Pages/Stages-of-Adolscence.aspx
3. Anderson NH (1971) Integration theory and attitude change. Psychol Rev 78:171–206
4. Baars BJ (1988) A cognitive theory of consciousness. Cambridge University Press, Cambridge
5. Baars BJ (1997) In the theater of consciousness: the workspace of the mind. J Conscious Stud 4:292–309
6. Brown KW, Ryan RM (2003) The benefits of being present: mindfulness and its role in psychological well-being. J Pers Soc Psychol 84:822–848
7. Bryant AN (2007) Gender differences in spiritual development during college years. Springer Science. https://doi.org/10.1007/s11199-007-9240-2
8. Brazdau O, Mihai C (2011) The consciousness quotient: a new predictor of students' academic performance. Procedia Soc Behav Sci 11:245–250
9. Brazdau O (2014) The consciousness quotient: construct development and content validity research. Procedia Soc Behav Sci 187:244–249
10. Brazdau O, Ahuja S (2016) The consciousness quotient: introducing the conscious experience as a research variable in psychological assessment. In: Higher-order thoughts, neural realization, and the metaphysics of consciousness, pp 187–216
11. Ciompi L (2003) Reflections on the role of emotions in consciousness and subjectivity, from the perspective of affect-logic. Conscious Emot 4:181–196
12. Chaplin TM, Cole PM, Zahn-Waxler C (2005) Parental socialization of emotion expression: sex differences and relations to child adjustment. Emotion 5:80–88
13. Charles ST, Reynolds CA, Gats M (2001) Age-related differences and change in positive and negative affect over 23 years. J Pers Soc Psychol 80(1):136–151
14. Deshmukh, Deshmukh (2012) Spiritual belief scale. National Psychological Corporation, Agra
15. Fabbro A, Crescentini C, Matiz A, Clarici A, Fabbro F (2017) Effects of mindfulness meditation on conscious and non-conscious components of mind. MDPI J Appl Sci 7(4):1–12. https://doi.org/10.3390/app7040349
16. Feningstein A, Scheier MF, Buss AH (1975) Public and private self-consciousness: asssesment and theory. J Consult Clin Psychol 43(4):522–527
17. Goleman D (1995) Emotional intelligence. Bantam, New York
18. Graham J, Nosek BA, Haidt J, Iyer R, Koleva S, Ditto PH (2011) Mapping the moral domain. J Pers Soc Psychol 101(2):366–385
19. Grant DJ, Jones CH (2008) The significance of pure consciousness for education. Retrieved 15 Nov 2019, from http://www.mum.edu/pdf_msvs/v07/education.pdf
20. Gross JJ, Carstensen LL, Pasupathi M, Tsai J, Skorpen C, Hsu AYC (1997) Emotion and aging: experience, expression, and control. Psychol Aging 12(4):590–599
21. Hall GS (1904) Adolescence: its psychology and its relations to physiology, anthropology, sociology, sex, crime, religion, and education. D. Appleton & Co, New York
22. Hanh TN (1976) Miracle of mindfulness. Beacon, Boston
23. James W (1890) The principals of psychology. Henry Holt, New York
24. Mankus AM, Boden MT, Thompson RJ (2016) Sources of variation in emotional awareness: age, gender and socioeconomic status. Personality Individ Differ 89:28–33
25. Natsoulas T (1990) Is consciousness what psychologists actually examine? Am J Psychol 105:363–384
26. Peres JA (2012) Gender difference in psychological well-being among Filipino college student sample. Int J Humanit Soc Sci 2(13):84–93
27. Rieffe C, Oosterveld P, Miers AC, Meerum Terwogt M, Ly V (2008) Emotion awareness and internalising symptoms in children and adolescents; the emotion awareness questionnaire revised. Personality Individ Differ 45:756–761

28. Rojiani R, Santoyo JF, Rarig H, Roth HD, Britton WB (2017) Women benefit more than men in response to college-based meditation training. Front Psychol 8:1–11. https://doi.org/10.3389/fpsyg.2017.00551
29. Rose AJ, Rodolph KD (2006) A review of sex differences in peer relationship processes: potential trade-offs for emotional and behavioural development of boys and girls. Psychol Bull 132:98–131
30. Sharma S, Ahuja S, Satsangee D (2016) A study of consciousness of university students in relation to the academic achievement and level of education. MIER J Educ Stud Trends Practices 6(1):91–97
31. Shivhare N (2007) Effect of science of consciousness on emotional intelligence of teacher trainees. Unpublished master's dissertation. Dayalbagh Educational Institute, Agra, India
32. Sturgess MA (2012) Psychometric validation and demographic differences in two recent developed trait mindfulness measures. Unpublished master's thesis. Victoria University of Wellington, Wellington, New Zealand
33. Thera N (1972) The power of mindfulness. Unity Press, San Fransisco, CA
34. Trehub C (2006) The science of consciousness: where it is and where it should be. Retrieved 10 Oct 2019, from people.umass.edu/trehub/Revonsuo.pdf
35. Underwood L (2011) The daily spiritual scale: overview and results. Religions 2:29–50
36. Usman A (2010) Spiritual consciousness in banking managers and its impact on job satisfaction. Int Bus Res 3(2):65–72
37. Wilber K (2000) Integral psychology: consciousness, spirit, psychology, therapy. Shambhala, Boston
38. World Health Organization (2007) Adolescent health at a glance in South-East Asia region. Retrieved 14 Aug 2019, from https://www.who.int/southeastasia/health-topics/adolescent-health

Neuromorphic Computing in Speech Recognition Using Nano-devices

Vishva Bhatt, Hetvi Shah, Kathan Shah, Jigarkumar Shah, and Manan Shah

1 Introduction

With the advancements made in technology, artificial intelligence has emerged to be an eye-catcher in the past few years. The analytic algorithm power that is used in artificial intelligence helps in making increasingly intelligent machines that can perform advanced human tasks. A major part of this wide branch is neuromorphic computing that is inspired by deep neural networks (DNN). Neuromorphic computing has managed to revolutionize the current Von Neumann architecture with its high efficiency and potentially low power consumption capabilities [1]. It uses AI to develop systems that attempt to mimic human intelligence and make neural networks that are biologically inspired and based on the analogy of the working of the human brain. With the help of non-volatile memory devices like memristor [2], neuromorphic computing has managed to shift the current programming paradigm to brain-inspired computing. These memristors offer both high connectivity and high density required for efficient computing [3, 4]. Neuromorphic computing itself has wide real-world applications. A few major applications are speech and pattern recognition, intelligence drug design, anti-aging therapies, block chain-enabled hospitals in healthcare sectors [5], financial industry, agriculture and many more [6].

Speech is the primary form of communication among humans; neuromorphic computing in speech recognition models has been one of the most demanding areas of research over the past few years [6]. Speech recognition technology provides numerous benefits like enhanced interactive voice response (IVR) analytics and conversational technologies in gaming, improved call center performance with AI and much more. Inspired by the neurons, it aims to develop nano-neurons using large-scale nonlinear oscillators to process information. These nonlinear oscillators [7] play

V. Bhatt (✉) · H. Shah · K. Shah · J. Shah · M. Shah
School of Technology, Pandit Deendayal Petroleum University, Gandhinagar, Gujarat, India
e-mail: vishva_pb@yahoo.com

V. H. Saran and R. K. Misra (eds.), *Advances in Systems Engineering*,
Lecture Notes in Mechanical Engineering,
https://doi.org/10.1007/978-981-15-8025-3_6

a crucial role in speech recognition machines. The paper talks about the working, proposed model, future trends and applications of neuromorphic computing in speech recognition using spiking neural networks [8]. The goal of speech recognition is to provide a means for human verbal communication to machine communication, by closely simulating the way biological neurons function. Besides, the neuromorphic chips used for speech recognition can detect complex input using relatively less energy and are good at adjusting its behavior as it interacts with the world using rich visual or auditory data. It changes the "weights" of the connections to learn the relationships between inputs, thereby improving performance.

The oscillators used to provide an output are similar to the current magnetic memory cells and have high endurance, operate at room temperature and can be fabricated in large numbers and hence can be easily used on a chip [9, 10]. Moreover, one of the biggest advantages of using these nano-scale oscillators is its stability in the presence of noise and other perturbations. Spintronic oscillators could pave the way to building specialized chips for large-scale neural networks. It provides functionality in a single unit and saves both money and time. The future trends that will come out by the application of the ability of speech recognition will be very helpful.

2 History of Approaches in Speech Recognition and Benefits of AI-Based Approach

Existing approaches to speech recognition.

1. **Template-based approach**

The template-based approach used for speech-to-text conversion relies mainly on the observed training sequence based on certain invariant properties. Each word is stored as a vector, and the speech recognition module provides a sequence of feature vector as its output. A collection of such reference templates is made each having its own feature vector. An unknown speech is then compared with these references, in order to find the best match.

2. **Pattern recognition approach**

The pattern recognition approach has two main steps—pattern training and recognition. It follows the analogy that if something close to the pattern is seen by the system, it recognizes it. A direct comparison is made between the unknown speech with every possible pattern stored in the system, during its training period, in order to recognize speech.

3. **Knowledge-based approach**

The knowledge-based approach (or the artificial intelligence-based approach) uses very large-scale integration (VLSI) technology for speech recognition. Such technology helps in lowering the cost and increases the capability of devices. Such a model requires high expertise knowledge making it difficult to use such models.

Approaches	Features	Advantages	Disadvantages	References
Template-based approach	Each word is stored as a vector	The pre-recorded templates make the model more accurate The errors caused due to smaller units such as phonemes can be avoided	The templates are pre-defined; hence, the variation in speech,pronunciation, makes the model inefficient	[6, 11]
Pattern recognition approach	Speech is defined as a pattern	It is relatively insensitive to the choice of words, syntaxes or semantics, making it capable of recognizing entire phrases	– The speech delivered to the system may get affected by the background noise or transmission characteristics of the medium where the speech is spoken, making the system inefficacious – The number of patterns stored in the system is limited, making it difficult for the system to recognize large number of phrases	[6, 12]
Knowledge-based approach	Data-driven approach to phoneme classification	The problems in the above models caused due to speech variations of the speaker and speech rate can be solved by the use of large speech database stored in the system – can generate decision trees using combination of parametric or acoustic features	– Requires high memory storage – Consumes a lot of space – High level of expertise is required to put such AI models into use	[6, 11, 13, 14]

Hence, as seen in the above table, AI proves to be a better approach to speech recognition than the other two existing approaches. Computers recognize speech by digitization, acoustic signal analysis and then linguistic representation. Humans do it by articulation which produces sound wave which the ear conveys to the brain for processing. A combination of both of these works the best, and hence, neuromorphic computing comes into picture. The main idea behind the neural network paradigm is to simulate the behavior of the brain using interconnected abstractions of the real neurons, which gives us complex systems that exhibit intelligent behavior by VLSI. In an artificial neuron, numerical values are used as inputs to the "dendrites." All the inputs are multiplied by a value called "weight," mimicking behavior of a real "dendrite," which are then added and thresholded. This can be expressed in mathematical terms as $y = f(\sum_{i=1}^{n} w_i x_i)$, where x_i is the input received by a dendrite, w_i the weight associated to the ith "dendrite," $f(\sum_{i=1}^{n} w_i x_i)$ is a threshold function, and y the output of the neuron [11].

Also, AI using neurocomputing works better in recognizing speech in noisy environment. Neural networks are robust to noise and faults and turn out to be ideal candidates for speech recognizing systems, as neuromorphic implementations have the ability to overcome these issues because of systems like spiking neural networks as they can operate asynchronously and can deal with noise and unreliability because of their inherent temporal processing component. Neuromorphic systems have also been applied to noise filtering applications as well to translate noisy speech or other signals into clean versions of the signal because of active noise canceling using analog neurochip with on-chip learning capability [15]. A basic layout of how neural networks recognizes speech is shown in Fig. 1.

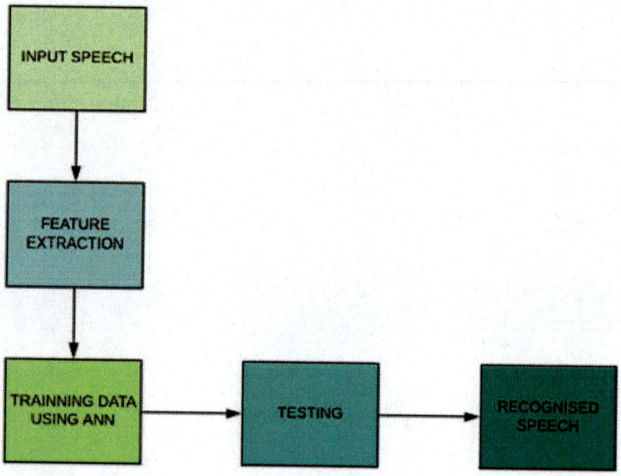

Fig. 1 Basic working of neural networks for speech recognition

3 Proposed Model—Basic Working of Spintronics in Speech Recognition and Proposed Idea

The model proposed shows how the idea of neural networks can practically be applied to the real world. This is done by using nano-scale oscillators. The dynamic features of spin-torque nano-oscillators have the potential of giving a promising future for the development of neurocomputing with coupled oscillators. These spintronic oscillators are composed of a normal non-magnetic disk placed between two ferromagnetic layers with a width of about 10–500 nm. Just as the frequency of a neuron is modified by the spikes received from other neurons, the frequencies of spin-torque nano-oscillators are highly sensitive to the magnetization dynamics of neighboring oscillators to which they are coupled [10].

The basic working of spin-torque nano-oscillator is shown in Fig. 2. An input DC current along with an input voltage through an arbitrary waveform generator is passed to the oscillator. The microwave voltage then produced by the oscillator in response to the excitation is measured. The amplitude dynamics of an oscillator like nonlinearity can be used to achieve waveform classification for input signal encoding, by using reservoir computing idea. It works by decreasing the frequency noise, amplitude and phase of the oscillator. Classification of sine and square waveforms increases the accuracy of encoding neural networks above 99% when decoding the output from the oscillator [16]. This is observed by control trials where input audio signals, after filtering to frequency channels like the cochlear model, are made to pass through pre-processed input without a spintronic oscillator. The output is directly reconstructed by a computer. Hence, a single oscillator can be used to replicate a full neural network characterizing them to work best when it comes to realizing day-to-day neuromorphic tasks like speech recognition.

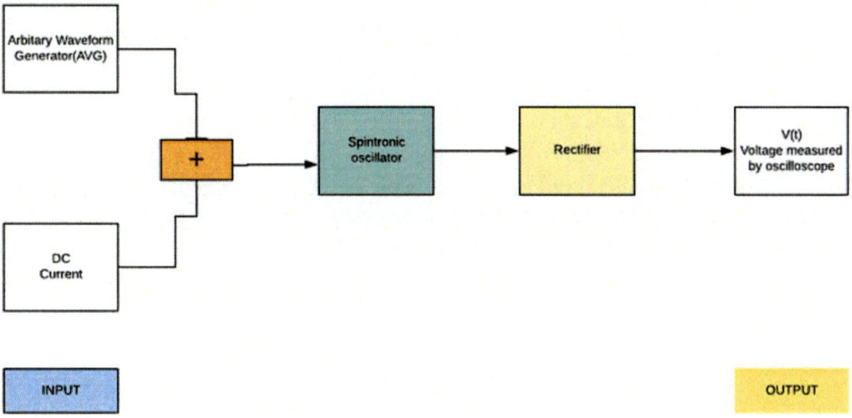

Fig. 2 Basic working of spin-torque oscillator in speech recognition

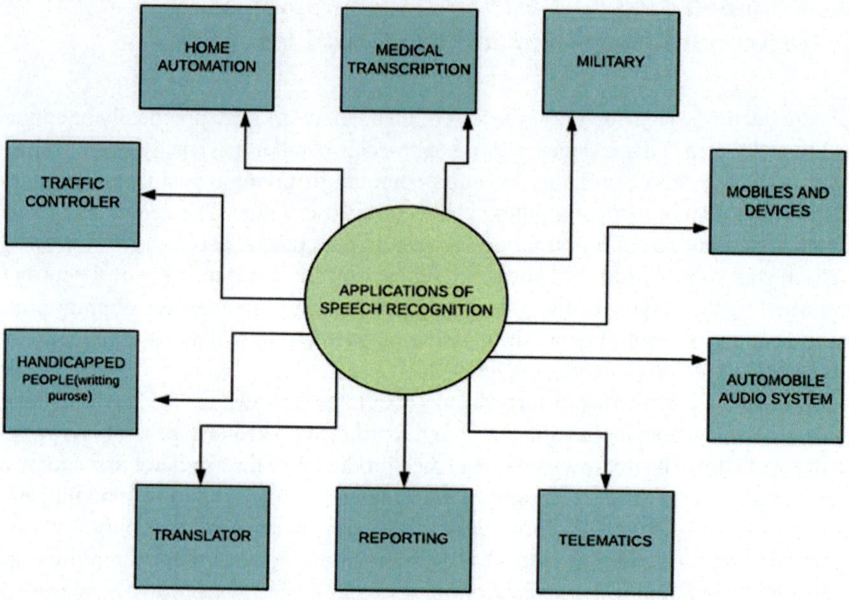

Fig. 3 Various applications of neurocomputing

As shown in Fig. 3, there are many fields in which the proposed model can be applied. The authors propose to extend this idea in future to make a speech recognizing device to be used in the fashion industry. The objective of this paper is to prove that the implementation and use of this device would enable the device to "talk" to the user. As most of the shopping is done online in recent times, it will store relevant information regarding what clothing was bought before and accordingly suggest clothes to wear for a particular event as predesigned. Voice interfaces offer some unique values related to voice search: being available 24/7, a conversational style, personalized recommendations, linking digital and physical reality, brand identity on platform and off-platform, personalized experiences. It would also make it easier to purchase products and make customers buy exactly what they want. This would also have a huge impact on the retail industry and benefit it financially too. The validation technology will be developed further to refine and design the algorithm. After the basic methodology, prototype and hypothesis will be developed; a change in the management, building the required and available data, testing and implementation and finally aftercare.

4 Future Scope and Challenges

Neuromorphic computing is still in its nurturing stage of development and there is a lot more to explore. The major factor driving the growth of neuromorphic computing is the use of supercomputers in everyday life, ranging from health monitoring to the making of brain-based robots. Computers nowadays are developing complex algorithms enabling it to perform cognitive tasks. One of the major areas of development lies in the making of brain-inspired intelligent robots. These neuromorphic computers, after reaching an operational stage, have the potential to change the way humans interact with intelligent machines [17]. Inspired by the data-driven cognitive computing [18], scientists are trying to create robots that not only help humans fulfill their basic needs but also can recognize human speech and follow commands. Moreover, smartphones can use chips that mimic about a million neurons which communicate with each other and enable the cognitive power of neural networks to respond within split seconds and help them promptly decide on actions such as lights, security system and many more.

Nonetheless, whatever comes with merits has demerits too. Neuromorphic computing is very complex and hence difficult to execute. Moreover, it is a challenging task for a computing system to cope up with human flexibility and task decisions involving human emotions. One of the major drawbacks of the speech recognizing systems is the word error rate (WER). It is a common metric used to compare the accuracy of the transcripts produced by speech recognition APIs. The WER is majorly affected by different accents, lack of training data, room acoustics and disturbing noises in the environment.

One of the major problems affecting the word error rate [19] and its remedies as discussed in the paper is noisy speech recognition using DNN. It aims at estimating a clean signal from the noisy input signals and minimizes the distortion between the estimated and original signal. Other problems include spontaneous [20] and multimodal speech recognition. Speech summarization method is used to interpret spontaneous speech. It extracts important information and removes any incorrect or unnecessary information. For multimodal recognition, studies have shown that having both visual and audio information increases the rate of successful transfer of information and shows that this gives better recognition performances than using only the audio or only the visual information, particularly, in noisy environment. The solutions to these problems can be seen as a future scope and extension.

A crucial challenge also relates to engineering interface properties. Moreover, there is a need for advanced tools and facilities to collect specific information associated with spintronic devices [21].

5 Conclusion

In this paper, a device for the recognition of speech is introduced. With the help of neuromorphic computing, this technique proves to be very useful in the ever-changing environment. Compared with Von Neumann bottleneck architecture, spintronic oscillators which are based on neurocomputing and are nonlinear are found to be more popular and easy to be fabricated. The proposed model which uses nano-devices such as nano-neurons improves the speech recognition process and also works efficiently on low power. The proposed idea shows how this methodology could be used in fashion technology. The paper also widely discusses how neuromorphic computing impacts speech recognition and other areas of cognitive tasks. It also discusses the future trends that will follow in robotics and other day-to-day activities. Furthermore, we wish to implement this hypothesis to our future works to achieve an optimal result when we have access to more resources and model it into a user-friendly interface. Overall, the proposed model can synchronize activities the like brain does. It aims to prove that the device will be a boon if implemented in real life.

References

1. Burr GW, Shelby RM, Sebastian A, Kim S, Kim S, Sidler S, Virwani K, Ishii M, Narayanan P, Fumarola A, Sanches LL, Boybat I, Gallo ML, Moon K, Woo J, Hwang H, Leblebici Y (2016) Neuromorphic computing using non-volatile memory. Adv Phys X 2(1):89–124. https://doi.org/10.1080/23746149.2016.1259585
2. Nafea SF, Dessouki AS, Rabaie SE (2015) Memristor overview upto 2015. Menoufia J Electron Eng Res Article 5, 24(1):79–106. https://doi.org/10.21608/MJEER.2015.64132
3. Jo SH, Chang T, Ebong I, Bhadviya BB, Mazumder P, Lu W (2010) Nanoscale memristor device as synapse in neuromorphic systems. Nano Lett 10(4):1297–1301
4. Truong SN, Ham S, Min K (2014) Neuromorphic crossbar circuit with nanoscale filamentary-switching binary memristors for speech recognition. Nanoscale Res Lett 9(629):1–9
5. Pastur-Romay LA, Cedrón F, Pazos A, Porto- AB (2016) Deep artificial neural networks and neuromorphic chips for big data analysis: pharmaceutical and bioinformatics applications. Int J Mol Sci 17:1–26
6. Anusuya MA, Katti SK (2009) Speech recognition by machine: a review. Int J Comput Sci Inf Secur 6(3):181–205
7. Stewart I (1992) Introduction to nonlinear oscillators. In: Taylor JG, Mannion CLT (eds) Coupled oscillating neurons. Perspectives in neural computing. Springer, London, pp 1–20
8. Dong M, Huang X, Xu B (2018) Unsupervised speech recognition through spiketiming-dependent plasticity in a convolutional spiking neural network. PLoS ONE 13(11):e0204596. https://doi.org/10.1371/journal.pone.0204596
9. Dixit D, Konishi K, TomyCV SY, Tulapukar AA (2012) Spintronic oscillator based on magnetic field feedback. Appl Phys Lett 101:122410
10. Torrejon J, Riou M, Araujo FA, Tsunegi S, Khalsa G, Querlioz D, Bortolotti P, Cros V, Fukushima A, Kubota H, Yuasa S, Stiles MD, Grollier J (2017) Neuromorphic computing with nanoscale spintronic oscillators. Nature 547:428–431
11. Tebelskis J (1995) Speech recognition using neural networks, CMU-CS-95-142
12. Atal B, Rabiner L (1976) A pattern recognition approach to voiced—unvoiced—silence classification with applications to speech recognition. IEEE Trans Acoust Speech Sig Process ASSP-24(3)

13. Samouelian A (1994) Knowledge based approach to speech recognition.Computer Science. Corpus ID: 16104404
14. Haton JP (1988) Knowledge-based approaches in acoustic-phonetic decoding of speech. NATO ASI series, vol F46. Springer, Berlin
15. Schuman CD, Potok TE, Patton RM, Douglas Birdwell J, Dean ME, Rose GS, Plank JS (2017) A survey of neuromorphic computing and neural networks in hardware. arXiv:1705.06963v1 [cs.NE]
16. Markovic D, Leroux N, Riou M, Abreu Araujo F, Torrejon J, Querlioz D, Fukushima A, Yuasa S, Trastoy J, Bortolotti P, Grollier J (2019) Reservoir computing with the frequency, phase, and amplitude of spin-torque nano-oscillators. Appl Phys Lett 114(1). https://doi.org/10.1063/1.5079305
17. Munch C, Bishnoi R, Tahoori MB (2019) Reliable in-memory neuromorphic computing using spintronics. In: Proceedings of the 24th Asia and South Pacific design automation conference, pp 230–236. https://doi.org/10.1145/3287624.3288745
18. Wang G (2017) DGCC: data-driven granular cognitive computing. Granular Comput 2(4):343–355. https://doi.org/10.1007/s41066-017-0048-3
19. Xiaodong H, Deng L, Acero A (2011) Why word error rate is not a good metric for speech recognizer training for speech translation task? IEEE 978-1-4577-0539-7/11
20. Furui S (2005) Spontaneous speech recognition and summarization. Article on https://www.researchgate.net/publication/228348581_Spontaneous_speech_recognition_and_summarization
21. Dieny B, Prejbeanu LL, Garello K, Gambardella P, Freitas P, Lehndorff R, Raberg W, Ebels U, Demokritov SO, Akerman J, Deac A, Pirro P, Adelmann C, Anane A, Chumak AV, Hiroata A, Mangin S, Cengiz M, Onbaşlı, Aquino MD, Prenat G, Finocchio G, Diaz LL, Chantrell R, Chubykalo-Fesenko O, Bortolotti P (2019) Opportunities and challenges for spintronics in the microelectronics industry.Nature Electronics (IF 27.500) Pub Date: 2020-08-18. https://doi.org/10.1038/s41928-020-0461-5

Materials and Systems

Investigation on Design of New Techniques in Mechanical Rebar Coupler as an Alternative to Lap Splices

Sanjivkumar Harinkhede and Valsson Varghese

1 Introduction

The RCC is broadly used in building construction industry worldwide. The extensively use of reinforced concrete in construction results in development of newest technologies. It helps to expand the superiority of structure and minimize the time consumption of building construction works. Many structural engineers, architects have discovered, some reinforcing bars must be spliced. In either case, rebar installers finish up with two or more pieces of steel that has got to be spliced together. We do not avoid lapping for the reason that the bars are available in market at a standard length of 18–12 m. On the other hand, lap splices cannot be used for bars having diameter greater than 36 mm; therefore, the bars having diameter greater than 36 mm could also be welded or coupled by using reinforcement couplers [1]. Overlapping of bars has been accepted as an effective, economical splicing method. Whereas welding requires trained workmanship and uninterrupted power supply. Therefore, to beat the above disadvantages and complexity of welded and spliced connection, the rebar couplers are preferred for connection.

Steel rebar of large diameter used in concrete members requires about 15% more steel than that used in a single bar. The lapping of big diameter bars has been avoided taking into account congestion of rebar and economy, by providing "Mechanical Couplers." Indian building construction industry has feel instantaneous requirement and is cheering builders to use mechanical couplers for use in many major infrastructure and multistoried construction projects. The application of mechanical couplers for connecting rebar is a promising technology; it is continuing to develop in terms of types of couplers available and their performance [2]. The designing and distribution

S. Harinkhede (✉) · V. Varghese
Department of Civil Engineering, K.D.K. College of Engineering, Nagpur, India
e-mail: sanju7471@gmail.com

V. H. Saran and R. K. Misra (eds.), *Advances in Systems Engineering*,
Lecture Notes in Mechanical Engineering,
https://doi.org/10.1007/978-981-15-8025-3_7

of rebar couplers is becoming a global business, and because of variety in construction practices design codes and specifications, standardization testing of coupler performance has been sluggish. In practice of couplers systems, the joint is made either in fabricators work or on construction site. As a result, there is a prerequisite for control of both the coupler development operation, and also the production of splice itself, which will normally necessitate several end preparation of the bar [3].

2 Research Methodology

The research was separated into different parts are as follows,

1. Structural analysis from IS 456: 2000 which content the bond stress and development length. According to IS code, Tensile strength of mechanical coupler not less than 690 N/mm^2 and percentage elongation should be minimum 3% before the failure of test piece.
2. The desire specification of different composite materials along with their properties are used for preparing the couplers. In general, the mechanical rebar coupler is fabricated from mild steel. For the development of rebar couplers, cutting of materials, boring and threading and at the end finishing are required. Couplers are manufactured on CNC or at conventional metal lathe machine (Fig. 1).
3. Estimation and comparison made between mechanical rebar coupler and conventional lap splicing much needed. The lapping of bars requires extra time and unwanted steel. Hence, this process of lapping consume more reinforcing bar which leads to the wastage of money.

3 Test Results

The tensile test carried on 16 mm diameter bar with conventional lap slice, welded slices and available mechanical threaded coupler as per recommended in IS code 1786–2008. The average tensile stress obtained was 198.94 N/mm^2 for conventional lap splice in 16 mm diameter bar (Fig. 2). Similarly, the average tensile stress obtained from weld splice is 505.65 N/mm^2 (Fig. 3) and the average tensile stress obtained for mechanical coupler is 690.46 N/mm^2 (Fig. 4).

4 Conclusion

Mechanical rebar couplers which content high carbon ratio have more strength and also the couplers having greater thickness are more efficient.

The comparison and test results are given below,

1. Tensile strength—Mechanical rebar coupler has more tensile strength as compared to conventional lap splice. Mechanical coupler has normally performs 125–150% more than conventional lapping strength.
2. Ultimate load—The ultimate load carrying capacity of lapped bar was less than mechanically spliced bars (Fig. 5).

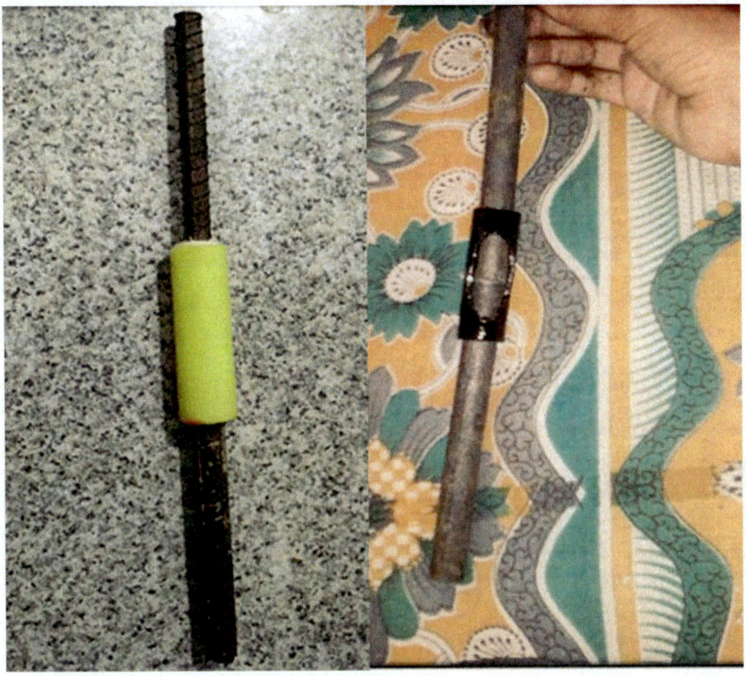

Fig. 1 A new design of mechanical rebar coupler model

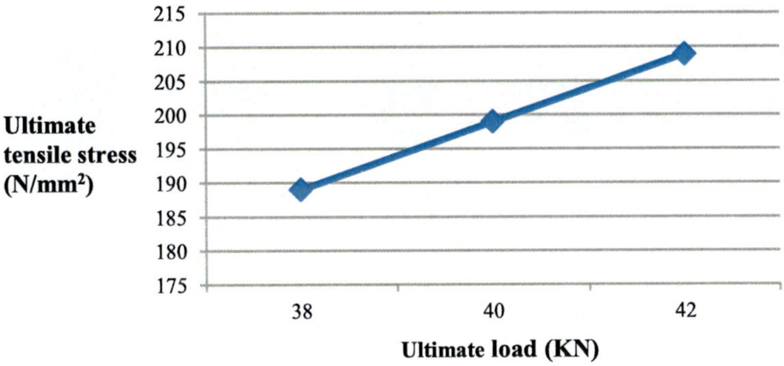

Fig. 2 Ultimate tensile stress on conventional lapping of 16 mm dia. Bar

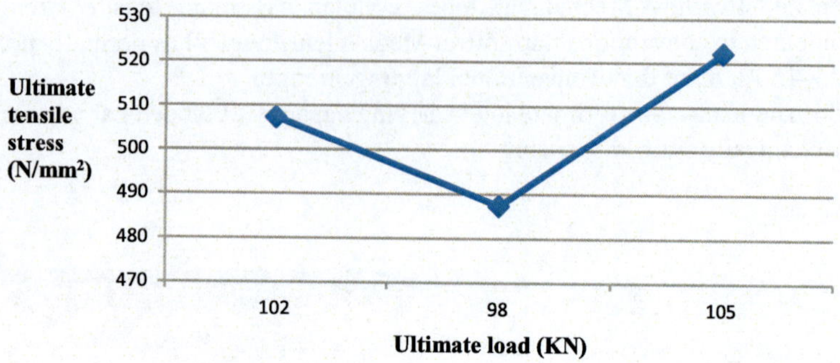

Fig. 3 Ultimate tensile stress on welding of 16 mm dia. Bar

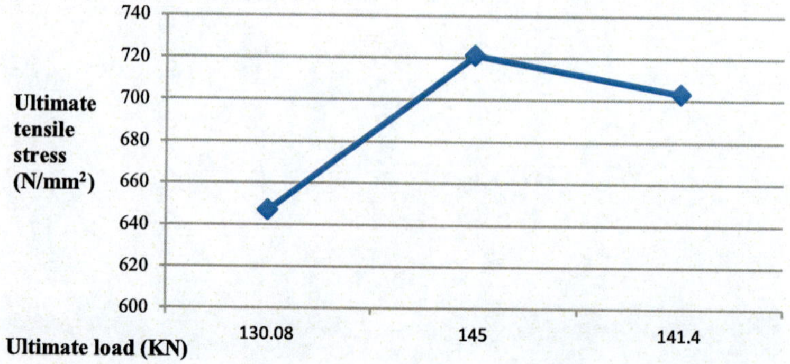

Fig. 4 Ultimate tensile stress on mechanical coupler of 16 mm dia. Bar

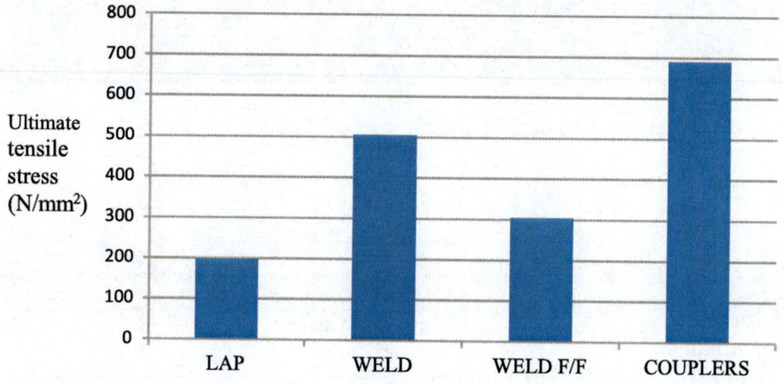

Fig. 5 Ultimate tensile stress for 16 mm dia. Bar

References

1. Harinkhede SN et al (2016) Investigation of new techniques in mechanical rebar coupler as an alternative to lap splices. Imperial J Interdisc Res (IJIR) 2(6):1039–1041. ISSN: 2454-1362. https://doi.org/10.15680/IJIRSET.2017.0604096
2. Harinkhede SN, Varghese V (2019) Performance of reinforcement connections in concrete structures as an alternative to lap splices. SSRN: https://ssrn.com/abstract=3368893 or https://doi.org/10.2139/ssrn.3368893
3. Sreenath HG (2013) Mechanical couplers for splicing steel reinforcements in concrete structures. ICI J 14(3)

Blending Methanol in Automobiles Fuels Towards Pollution Prevention in Indian Metros: Potentialities and Prospects

A. G. Matani, S. K. Doifode, and Ashish Mali

Objectives:

- to improve vehicle engine performance by blending methanol.
- to advance technologies to convert IC engines to methanol blending-based fuel engines.
- to develop modifications in engine setups to reduce air pollution.

1 Introduction

Including methanol into gas or diesel permits the fuel to have a total burning with the present of oxygen which expands its ignition effectiveness and lessens ozone depleting substance such as greenhouse gas outflow. The enthusiasm on methanol as an elective fuel in the automobile fuel showcase is believed to be developed quickly in the following decades. As far as application, Brazil has effectively and generally utilized methanol as a fuel for S.I. engine operation. In the USA, 10% methanol mixed with gasoline known as gasohol is accessible at a great number of petroleum stations as a vehicle fuel. Methanol has emerged up as a sustainable fuel for IC

A. G. Matani (✉)
Mechanical Engineering, Government College of Engineering, Amravati, Maharashtra 444604, India
e-mail: dragmatani@gmail.com; ashokgm333@rediffmail.com

S. K. Doifode
Chemistry Department, Government College of Engineering, Amravati, Maharashtra 444604, India
e-mail: shamal_doifode123@gmail.com

A. Mali
Thermal Engineering, Government College of Engineering, Amravati, Maharashtra 444604, India
e-mail: ashishmali04@gmail.com

V. H. Saran and R. K. Misra (eds.), *Advances in Systems Engineering*,
Lecture Notes in Mechanical Engineering,
https://doi.org/10.1007/978-981-15-8025-3_8

Table 1 Chemical contents of methanol and conventional fuels

Kind of fuel	Mass content			
	Oxygen	Sulphur	Carbon	Hydrogen
Methanol	50	–	37.5	12.5
Diesel oil	–	0.5	87	12.5
Gasoline	–	–	85	15

engines in past few decades due to its characteristics of soot free burning and higher efficiencies at less cost. In India, production and utilization of methanol has become a necessary factor and should be implemented [1].

The percentage substance of methanol, gasoline and diesel oil is given in the accompanying Table 1. Because of the reason that 50% (mass) of methanol content is oxygen, the stoichiometric measure of air essential for methanol burning is lesser and its calorific value is lower in correlation with petroleum fuels. Methanol shows a prosperous market at its minimal price and simple acquisition. Coal is relatively more extravagant than oil and petroleum gas in India. It plays 70% role of annual energy balance. Since it is simple and affordable to be liquefied to methanol, more considerations are paid to the application in SI engines [2].

2 Modifications Required in Gasoline Engine for Using Methanol Blends

The engine which is designed for the use of gasoline fuel cannot be used with methanol fuel. Methanol has higher octane rating than petrol. Hence, compression ratio to be achieved while using methanol is more than that while using petrol. Also the calorific value of methanol is less than petrol; hence, more fuel is needed to be burnt to obtain certain amount of energy, as compared to petrol. Methanol is a cleansing agent. When used in engine, it cleans the dirt and filth is formed in the engine. This may damage some parts of the engine and make them useless. Taking into account these properties of methanol, certain modifications are required in engine. The modifications vary depending on the maker of engine, technologies used percentage of fuel to be used, etc. [3].

Major areas where modification is necessary are as follows:

1. **Main Jet Modification**

Alcohol fuel requires a more richer air fuel proportion. It is important to bore out the main jet opening when utilizing methanol fuel. For working of the engine effectively on alcoholic fuel, it is important to increase this opening from 20 to 40%. There are a few carburetors that do not utilize fixed size jets alone, yet in addition use gadget known as a metering bar. This is normally a slender tapered or ventured bar that is suspended inside the stream opening, which might be removable. The fuel right now

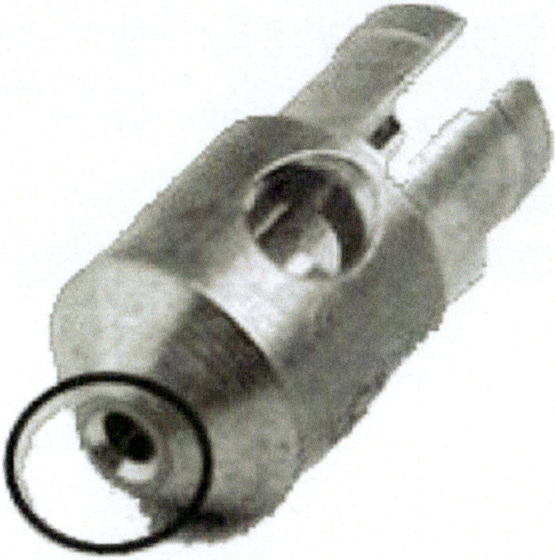

Fig. 1 Modified main jet

drawn through the space between the bar and its casing. Contingent upon how far the throttle is opened, the metering bar is lifted out of the gap. Since the bar is thick at its base and logically more slender at its tip—the more distant it is drawn out of the gap, the more fuel is permitted to stream between the central bar and opening. The transformation on this kind of metering framework is essentially equivalent to the fixed jet conversion. To increase the Orifice, either evacuate the metering bar and cautiously drill the jet or turn the brass rod slightly. The diameter, if increased in size, should be increased anywhere between 15 and 35% (Fig. 1).

2. **Idle Orifice Changes**

Most carburetors will require extra idle circuit growth all together for the engine to run at slowest or idle velocities. This is on the grounds that the circuit which is taken care of by the main jet works completely just when the throttle plate inside the throat of the carburetor is opened past the idle position. At the point when the plate is in the idle position, the air/fuel blend is permitted to enter the manifold just through the idle orifice. On the contrary if this hole is not sufficiently huge, it would not give the required measure of air/fuel mixture to keep the engine running. In certain engines, it is conceivable that the seat itself, into which the slendered screw advances, must be developed so as to achieve something very similar to above discussed. In most of the instances, if the seat must be moved out, it should be increased in size up to 50%. This will permit a full scope of modification with the idle blend screw regardless of whether one needs to return to gasoline fuel [4].

As a safeguard against the idle screw's vibrating free from its strung opening, the idle blend screw spring can be combined with a few little lock washers. This will

Fig. 2 Idle orifice

keep the screw from turning regardless of whether it is drawn out more remote from the seat than it typically would be [5] (Fig. 2).

3. **Power Valve Changes**

Most present-day autocarburetors have a valve, known as power valve that permits additional fuel to mix with the air/fuel blend when the accelerator is depressed, so as to enhance the blend under load conditions. This is a vacuum-controlled valve, which is spring stacked, and stops when it is not required all together to preserve fuel. The power valve utilized in few of the carburetors is some degree hard to change and additionally is adequate for alcohol use in its typical arrangement on the off chance that it is working appropriately. In some cases, there are a few carburetors that have effectively replaceable power valves which are accessible in different sizes. On the off chance that a power valve with a 25. Enlarged measurement or so is utilized, the air/alcohol blend will be adequately improved to give engine needs more force when required [6] (Fig. 3).

4. **Accelerator Pump Changes**

In addition to power valve, practically all car carburetors use an accelerator pump. This is a mechanically activated plunger or diaphragm that injects a stream of raw fuel directly down the throat of the carburetor when the accelerator is suddenly depressed. The explanation the accelerator pump is joined into present-day carburetors is that as the accelerator pump is depressed and more air/fuel blend is brought into the chambers, a portion of the fluid particles in the mixture will in general adhere to the walls of the inlet manifold, successfully inclining out the blend when it arrives at the combustion chambers. The additional squirt of fuel that is included by the accelerator pump compensates for this underlying lean condition. So as to adjust the accelerator pump to utilize alcohol successfully, the size of the injection hole should

Fig. 3 Changes in power valve

Fig. 4 Changes in accelerator pump

be expanded (somewhere in the range of 10–25% is fine). As an option in contrast to expansion of orifice, basic alteration of the stroke length of the pump arm so as to take care of more fuel is supplied [7] (Fig. 4).

5. Compression Ratio Changes

The compression ratio (CR) is calculated by the following formula:

$$\mathrm{CR} = \frac{\frac{\pi}{4}b^2 + V_c}{V_c}$$

where,

b = cylinder bore (diameter)
s = Piston stroke length
V_c = Clearance volume.

Compression ratio (CR) is the ratio cylinder volume when piston is at bottom dead centre (BDC) to the cylinder volume when piston is at top dead centre (TDC). This change will do a lot to improve engine performance and economy. The compression ratio climb will expose the potential of methanol to be used as a fuel. Ideally, the proportion can be expanded to I4:1 or I5:l. In any case, even an ostensible increment to I2:1 will bring about a huge improvement over the standard 8:1 or 8.5:1 that most manufacturers consolidate into their engines today.

The most reasonable approach to expand the compression ratio is to introduce a group of high compression pistons. The fabricated units are intended to pack the air/fuel charge firmly into the burning chamber for hiked power and have exceptional help scores incorporated with their heads for valve clearance. Yet few engines may not endure a 15:1 compression ratio with standard connecting rods and bearings. These segments also may be supplanted with high-quality competitive grade parts.

Another method for expanding compression ratio marginally is by "milling" the surfaces of the chamber head or block. A third—and maybe the most flexible—method for adequately expanding the compression ratio is by introducing a turbocharger. These units, in spite of the fact that go high in cost, give a drive push in the burning chamber corresponding to the engine, RPM. Henceforth, compression would not be over the limit during engine turn over as it would be with previous techniques. No issue is experienced with an extreme compression ratio increment, except if exchanging back to gasoline fuel, to switch back to gasoline, introducing a water infusion cooling system that would permit working the vehicle even on normal fuel unafraid of explosion would be useful [8].

6. Cold Weather Starting

Since methanol does not disintegrate as effectively as does gasoline, cold weather starting can be an issue, particularly if the engine itself is cold. To reduce this unwanted circumstance, a combined cold start/double-fuel system can be utilized, which can work with any engine. The various other remedies to overcome this problem are as follows:

- *Fuel Preheating*

In extremely cold climates, it might be important to preheat the alcoholic fuel before it enters the carburetor. This can be practised effectively by introducing a fuel preheater.

This can be accomplished by supplying the fuel through a pipe, which passes over the engine radiator. The hot engine parts will warm the fuel adequately. Yet, this is certainly not a compelling technique, as while turning over the cool engine, passing the fuel over the radiator makes no impact. The engine should be started by some alternative ways.

- *Air Preheating*

Most trucks and automobiles have air filter casings which are intended to permit warmed air from around the exhaust manifold to channel through a pipe and enter the carburetor when the engine first turns over from a cool state. As the engine heats up, a fold inside the air cleaner "snorkel" stops this stock of warm air and permits atmospheric air from the engine compartment to enter. This fold is typically either thermostatically or vacuum controlled.

External resistance-type air pre-heaters may also be used for this purpose.

3 Conclusions

Methanol has come up as a low-emission low polluting fuel and considered as profoundly effective through their smooth working capacity. India by adopting methanol can have its own indigenous fuel at the expense of roughly 19 Rs. per litre at any rate 30% less expensive than any other fuel. Methanol fuel can bring about incredible natural advantages because of its properties and the best fuel to follow every environmental condition and pollution standards. At any rate, 20% diesel utilization can be decreased in next 5–7 years and will bring up reserve funds of 26,000 Crores every year. Rs. 6000 Crores can be every year spared from diminished bill in LPG in the following three years itself. The methanol mixing program with gasoline will additionally lessen our fuel bill by 5000 crores every year in next the years. Methanol-mixed car fuel (petroleum and diesel) is as of now being used in a couple of business sectors in different blending rates directly from 3 to 85%.

References

1. Elfasakhany A, Tao T, Espenas B, Larfeldt J, Bai XS (2013) Pulverized wood combustion in a vertical furnace: experimental and computational analyses. Appl Energy 112:454–464
2. Masum BM et al (2013) Effect of ethanol–gasoline blend on NO_x emission in SI engine. Renew Sustain Energy Rev 24:209–222
3. Zhang F et al (2009) Influence of methanol gasoline blend fuel on engine and catalyst performance. SAE 2009-0101182. Tsinghua University, China
4. Bang-Quan H, Liu M, Zhao H (2015) Comparison of combustion characteristics of nbutanol/ethanol–gasoline blends in a HCCI engine. Energy Convers Manage 95:101–109
5. Sharudin H, Abdullah NR, Najafi G, Mamat R, Masjuki HH (2017) Investigation of the effects of iso-butanol additives on spark ignition engine fuelled with methanol-gasoline blends. Appl Therm Eng 114:593–600

6. Storey JM, Barone T, Norman K, Lewis S (2010) Ethanol blend effects on direct injection spark-ignition gasoline vehicle particulate matter emissions. SAE Int J Fuels Lubr 3:650–659
7. Shenghua L, Clemente ERC, Tiegang H, Yanjv W (2007) Study of spark ignition engine fuelled with methanol/gasoline. Appl Therm Eng 27:1904–1910
8. Sileghem L, Ickes A, Wallner T, Verhelst S (2015) Experimental investigation of a disi production engine fuelled with methanol, ethanol, butanol and isostoichiometric alcohol blends. SAE technical paper, No. 2015-01-0768
9. Elfasakhany A (2013) Investigation on performance and emissions characteristics of an internal combustion engine fuelled with petroleum gasoline and a hybrid methanol–gasoline fuel. Int J Eng Technol 13:24–43
10. Elfasakhany A (2014) The effects of ethanol–gasoline blends on performance and exhaust emission characteristics of spark ignition engines. Int J Automot Eng 4:608–620
11. Laing B et al (2013) Comparison of PM emissions from a gasoline direct injected (GDI) vehicle and a port fuel injected (PFI) vehicle measured by electrical low pressure impactor (ELPI) with two fuels: gasoline and M15 methanol gasoline. J Aerosol Sci 57:22–31
12. Sezer I, Bilgin A (2008) Exergy analysis of SI engines. Int J Exergy 5:204–217
13. Bromberg L, Cheng WK (2010) Methanol as an alternative transportation fuel in the U.S.: options for sustainable and/or energy secure transportation, 28 Nov 2010, Sloan Automotive Laboratory, Massachusetts Institute of Technology
14. Çelik MB (2008) Experimental determination of suitable ethanol–gasoline blend rate at high compression ratio for gasoline engine. Appl Therm Eng 28:396–404
15. Report of Press Information Bureau Government of India Ministry of Road Transport & Highways dtd, 28 Dec 2017
16. Jafarmadar S (2014) Exergy analysis of hydrogen/diesel combustion in a dual fuel engine using three-dimensional model. Int J Hydrogen Energy 39:9505–9514
17. US Department of Health and Human Services (2011) Report on carcinogens. Department of Health and Human Services, National Toxicology Program
18. Report: Methanol blending in China—trip report. Gal Luft, International Analysis for the Global Security, May 2012

Selection of Optimum Working Fluids for Low-Power Output Organic Rankine Cycle

Rustum Dhalait, Suhas Jagtap, and Bajirao Gawali

Nomenclature

h	Enthalpies of refrigerant, kJ/kg
$\dot{m}$	Mass flow rate, kg/s
P	Pressure, Pa
Q	Heat transfer rate, kW
V	Specific volume, m^3/kg
T	Temperature, °C
W	Work output, kW

Subscripts

in	Input
out	Output
is	Isentropic
p	Pump
t	Turbine

R. Dhalait (✉) · S. Jagtap · B. Gawali
Department of Mechanical Engineering, Walchand College of Engineering, Sangli, Maharashtra 416415, India
e-mail: rustamdhalait1694@gmail.com

V. H. Saran and R. K. Misra (eds.), *Advances in Systems Engineering*,
Lecture Notes in Mechanical Engineering,
https://doi.org/10.1007/978-981-15-8025-3_9

Abbreviation

ORC Organic Rankine cycle
EES Engineering equation solver

1 Introduction

Energy is a key factor in the development of human society and the economy. Nowadays, the energy demand is accelerated, so people thinking of the choice of low-grade energy sources. An important variety of recent solutions is planned to get various heat supply like low-temperature or low-power heat sources. Among the proposed solution, the Organic Rankine Cycle is mostly used. Nowadays, micro-scale ORC systems are attracting subject when electrical energy is converted from the small capacity heat source in waste heat recovery [1]. Only a few research projects can be found on the low-power output cycle. ORC primarily resembles the steam cycle in keeping with operating principles, replacing water with low boiling temperature fluids which has high molecular mass. Characteristics of these fluid make ORC favorable application of heat recovery (normally less than 400 °C) [2].

Yamanato et al. [3] have developed a small test rig of the Organic Rankine Cycle. He has done the numerical and experimental study by using the water and HCFC-123 as a working fluid. He suggested that an increase in turbine inlet temperature will increase the system power output. Agromayor and Nord [4] have screened the organic fluids for the power cycle. They found out the optimum cycle configuration for the ORC cycle. 80 number of fluids are studied in this work. From [1–6], R245fa is the most common working fluid for the Organic Rankine Cycle.

Muhammad et al. [5] have developed the 1 kW Organic Rankine Cycle by using R245fa as a working fluid. The system generates an electrical output of 1.016 kW with 5.64 and 4.66% of thermal efficiency and net efficiency, respectively. Glover et al. [6] have carried out the simulation work of a multiple heat source for vehicle waste heat recovery. 18 different types of fluids are used for performance comparison by using code based on the NIST properties. In this paper, the thermodynamic model of the Organic Rankine Cycle is developed by using the Engineering Equation Solver (EES) tool. The different types of fluids are studied under this model and the working fluid is selected by observing the simulation results based on energy input, power output and system thermal efficiency.

2 Screening of Working Fluid

Selection of the working fluid is a very complicated process in Organic Rankine Cycle. It requires the trade-off depending on the objectives of system designer [7] Fluids, i.e., R123, R245fa, R600, R600a and R245ca, give the best results in for thermal and exergy efficiency, in case of maximum heat recovery from heat source R41, R218 and R125 are good options [7]. by using R143a, R134a, R152a and R600 reduce the size of the heat exchanger, whereas R600a, R600, R152a, R143a, R134a R123 and R141b all behaves well in terms of overall levelized energy cost for the system under the conditions analyzed.

2.1 *Dry and Wet Fluids*

The wetness or dryness is a common criterion to differentiate the working fluid types. Fluids having $ds/dT \approx 0$ have an almost vertical saturation vapor curve and are classified as isentropic. Whereas fluids having a positive saturation vapor curve ($ds/dT > 0$) and are classified as dry fluids. Also, fluids having a negative saturation vapor curve ($ds/dT < 0$) and are classified as wet fluids. Traditional Rankine power plants usually use water as a working fluid. But due to drawbacks of the water and wet fluid, we cannot use it in ORC. Dry and isentropic fluids are the best options for capturing the heat from the low-grade energy source (Fig. 1).

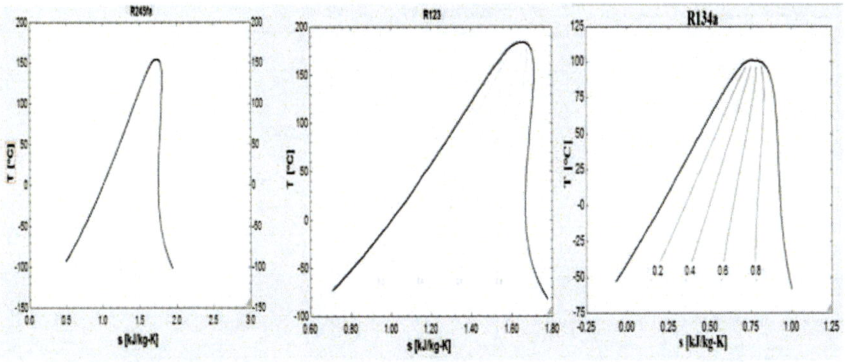

Fig. 1 T-s diagrams of various fluids from left to right R245fa (isentropic fluid), R123 (dry fluid), R134a (wet fluid)

2.2 *Thermo-Physical Properties*

The thermo-physical properties for good working fluid include high density, smaller volumetric expansion ratio, high latent heat of vaporization and high critical temperature. Though working fluid has to satisfy the above properties, there is no single fluid which meets all the requirements, so we have to think about the fluid which gives optimum results are selected.

3 Mathematical Model of the System

A mathematical model is developed in Engineering Equation Solver (EES) to analyze a small-scale Organic Rankine Cycle. This model developed on principles that appear in literature frequently. The ORC consists of the component as pump, boiler, expander and condenser. The basic ORC system is shown in Fig. 2.

Before carrying out the performance analysis of the ORC system, we have to consider certain assumptions for our model; these are as following.

- Expander efficiency is assumed as 80%
- Pump overall efficiency is assumed as 75%
- Heat loss through the piping and other types of equipment is assumed to be negligible.

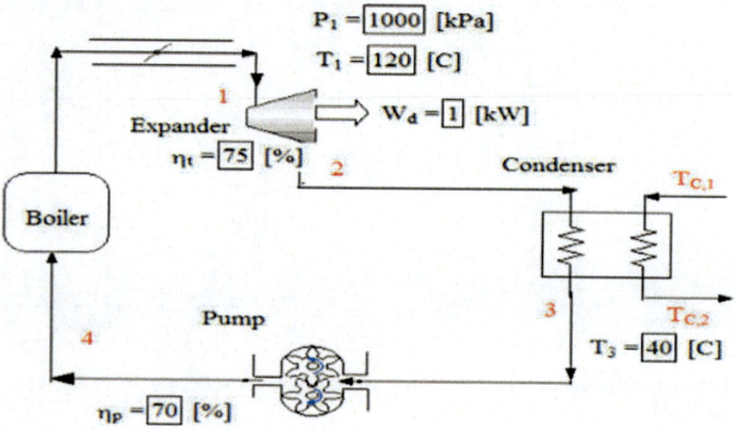

Fig. 2 The basic structure of organic Rankine cycle

Following is the energy balance equation at steady state for any component of system

$$\Sigma m_{\text{in}} = \Sigma m_{\text{out}} \tag{1}$$

$$Q^{\cdot} - W^{\cdot} + \Sigma m_{\text{in}} h_{\text{in}} - \Sigma m_{\text{out}} h_{\text{out}} = 0 \tag{2}$$

3.1 Pump Stage

$$W_{\text{p,is}} = v_3 \times (P_4 - P_3) \tag{3}$$

$$W_{\text{p}} = \frac{W_{\text{p,is}}}{\eta_{\text{p}}} \tag{4}$$

To calculate (h_4) specific enthalpy at pump outlet

$$h_4 = h_3 + W_{\text{p}} \tag{5}$$

$$W_{\text{p}} = \frac{W_{\text{p,is}}}{\eta_{\text{p}}} \tag{6}$$

3.2 Boiler Stage

The heat addition at constant pressure is taking place in the boiler or evaporator. The working fluid is heated from the pump outlet to the turbine inlet state condition. The heat added in the boiler is:

$$Q_{\text{in}} = h_1 - h_4 \tag{7}$$

3.3 Expander Stage

The working fluid enters in the turbine at a superheated or saturated vapor condition. The expansion of the vapor takes place in the turbine and fluid came out from turbine at low pressure P_{out} and at low temperature T_{out}. Then, the turbine work output is given by (Fig. 3)

$$W_t = (h_1 - h_2) \times \eta_t \tag{8}$$

$$W_{net} = W_t - W_p \tag{9}$$

4 Result and Discussion

Simulation is carried out to study the performance of the system by using various working fluid in the model. The working fluids selected are R123, R245fa,R236ea, R141b,R245fa, R600, R601a and R1233zd (e). The results of simulation are shown in Table 1.

The constraint for the simulation is inlet pressure of turbine ($P_1 = 1000$ kPa), inlet temperature of turbine ($T_1 = 125$ °C). With these input parameters, system simulation is done for the mentioned fluids.

The detailed comparison of the simulation is as shown in Figs. 4, 5, 6 and 7. The fluids from different categories are studied for the parameter like heat input in the boiler, network output, required pump work and cycle efficiency. From Figs. 4, 5, 6 and 7, it is interpreted that R600 and R600a give high network output but at the same time, the heat input required for this substance is also high, which results in the low-thermal efficiency. R141b gives moderate work output with comparatively low heat input. For the same power output, the R1233zd(e) requires a high mass flow rate which results in the more pump work. R141b gives good results in network output, efficiency but the pump work required is more. R123, R245fa, R236ea fluids show good thermal efficiency for the same input conditions. Now, further work is carried out on R123, R245fa, R141b and R236ea with different parameters.

As shown in Fig. 8, by increasing the inlet pressure of turbine (P_1), the efficiency of the cycle is also increasing. R141b possesses higher efficiency than the other three fluids but it is wet fluid as shown in Table 1. So it is avoided to use in the low-power cycle. From Fig. 9, it is understood that by increasing the condenser pressure, the efficiency of the cycle decreases; though by decreasing the condenser pressure, efficiency of the system increases; at the same time, saturation temperature for the condenser is 40 °C, so there is a limitation to go down below the saturation pressure of the fluid. R123 shows the higher efficiency so further investigation is carried out on the R123 (Fig. 10).

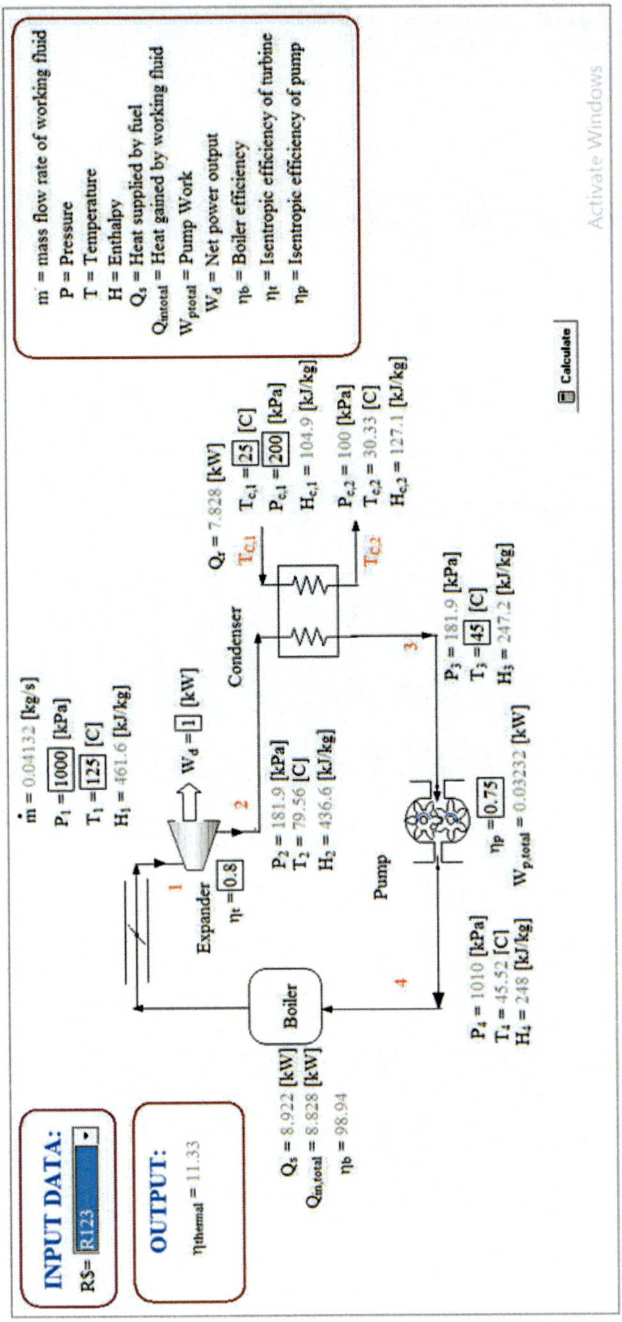

Fig. 3 EES results of ORC

Table 1 Comparison of the studied working fluids

Substance name	Condenser pressure (kPa)	Mass flow rate (kg/s)	Heat input (kJ/kg)	Net output (kJ/kg)	Pump work (kJ/kg)	η (%)
R123	130.7	0.035	219.5	28.3	0.8061	12.89
R141b	112.2	0.025	277	38.71	0.97	13.97
R600	329	0.023	502.6	42.09	1.596	8.37
R600a	463.3	0.034	486.6	29.05	1.33	6.19
R245fa	211	0.039	257.8	25.46	0.8025	9.87
R236ea	288.8	0.055	223.3	18.07	0.6804	8.08
R1233zd(e)	183	0.036	249	27.78	0.8798	11.16

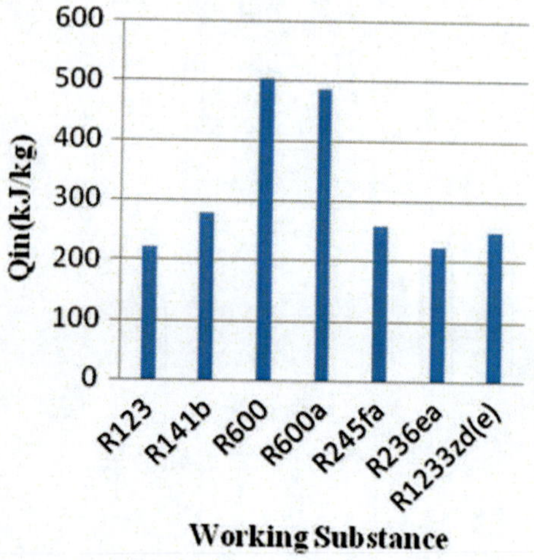

Fig. 4 Comparison of working fluids for heat input

5 Conclusion

This paper has analyzed the different types of working fluid used in ORC. Seven different fluids were studied and simulated following are outcomes:

- There is a lot of heat wasted in an industrial process which can be utilized effectively by using low-power output ORC.
- The selection of the correct working fluid is important, as it significantly affects the low-power output ORC.
- Among the selected fluids, R123 and R245fa are the good choices for the low-power output ORC.

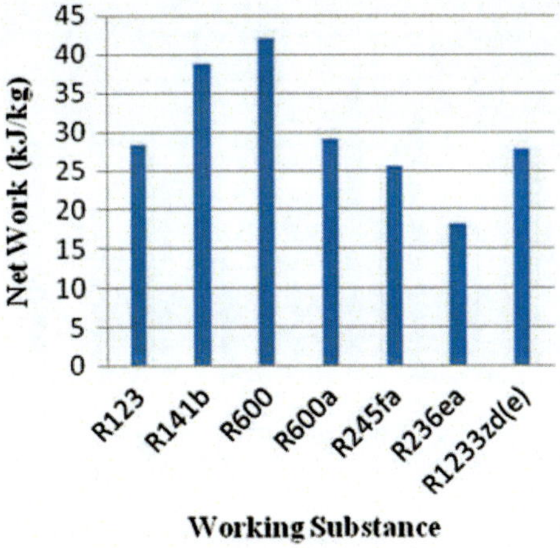

Fig. 5 Comparison of working fluids for network output

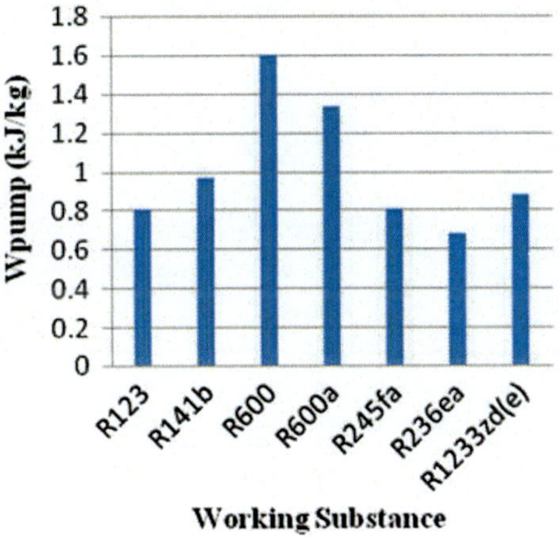

Fig. 6 Comparison of working fluids for pump work

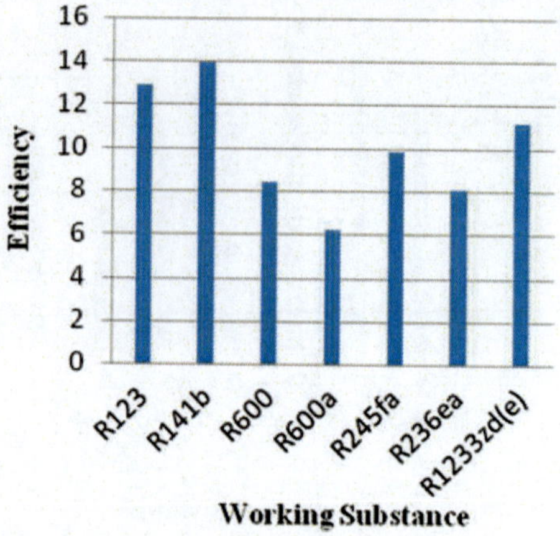

Fig. 7 Comparison of working fluids for efficiency

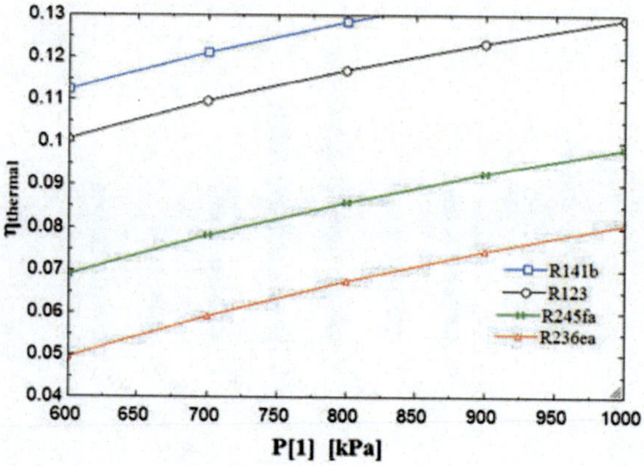

Fig. 8 Effect of turbine inlet pressure on efficiency

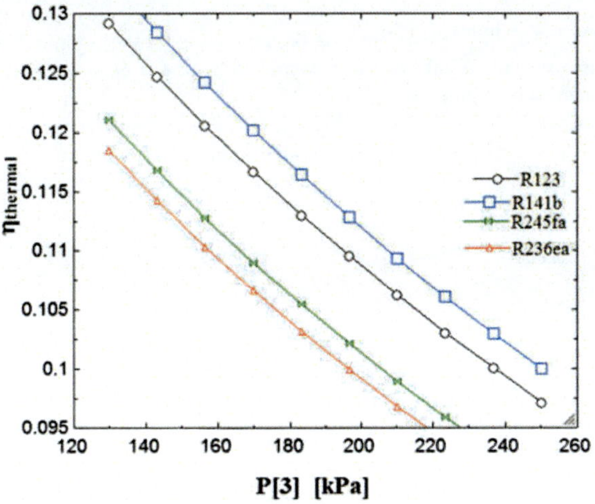

Fig. 9 Effect of condenser pressure on efficiency

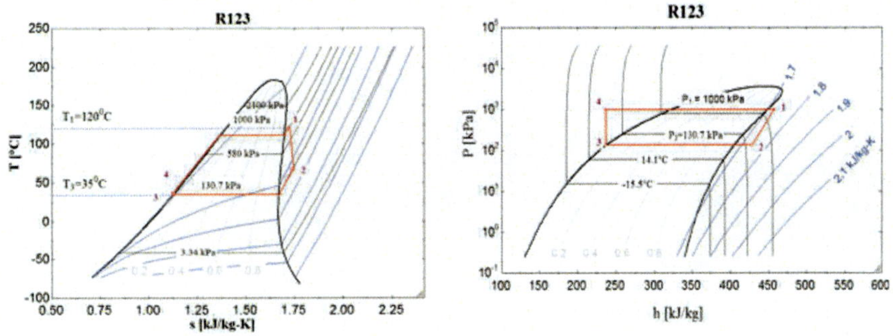

Fig. 10 T-s and P-h diagram of R123

References

1. Astolfi TTM (2017) A world overview of the organic Rankine cycle. Energy Procedia 129:2–9
2. Bonk C, Laux C, Rodder M, Neef M (2017) Design of 1 kW organic Rankine cycle for teaching and research issues. Energy Procedia 129:931–938
3. Yamamoto T, Furuhata T, Arai N, Mori K (2001) Design and testing of the organic Rankine cycle. Energy 26:239–2511
4. Agromayor R, Nord L (2017) Fluid selection and thermodynamic optimization of organic Rankine cycles for waste heat recovery applications. Energy Procedia 129:527–534
5. Muhammad U, Imran M, Lee D, Park B (2015) Design and experimental investigation of a 1 kW organic Rankine cycle system using R245fa as working fluid for low-grade waste heat recovery from steam. Energy Convers Manage 103:1089–1100
6. Glover S, Douglas R, Rosa M, Zhang X, Glover L (2015) Simulation of the multiple heat source supercritical ORC (organic Rankine cycle) for vehicle waste heat recovery. Energy 93:1568–1580

7. Douvartzides S, Karmalis I (2016) Working fluid selection for the organic Rankine cycle (ORC) exhaust heat recovery of an internal combustion engine power plant. In: IOP conference series: materials science and engineering
8. Li P, Li J, Pei G, Ji J (2016) A cascade organic Rankine cycle power generation system using hybrid solar energy and liquefied natural gas. Sol Energy 127:136–146

Hydrogen Gas Production from Paper–Pulp Industry Wastewater by Electrodeposited Cathodes in MECs

Amit Kumar Chaurasia and Prasenjit Mondal

1 Introduction

Microbial electrolysis cells (MECs) are extended versions of microbial fuel cells technologies that recover the energy from organics present in waste material using electrochemically active bacteria. MFCs use the action of exoelectrogenic microbes to oxidize organic compounds present in the actual wastewater and generate electricity over the transfer of electrons to the anode, while in MECs the electrons produced by exoelectrogens are diffused at cathode for the production of H_2 from H+ [1]. MECs are getting much attention in last decade due to it is a promising approach for the generation of renewable energy and valuable products from almost all types of waste materials. MECs combined the removal of organic material with higher hydrogen production (>60% energy recovery) at ambient atmospheric conditions but recovery of hydrogen/energy needs to be improved [2]. In MECs, cathode catalysts for proton reduction to hydrogen at cathode have a significant role in decreasing energy input and increasing hydrogen production or energy recovery [3]. Most of the literature reported that Pt. as standard cathode material utilized in MECs due to low hydrogen-over potential but noble metal platinum cannot be used for extensive practical application of MECs due to poisoning by sulphides and high cost [4]. Economical application of MECs required development of alternative cathodes materials, which have low hydrogen-over potential, low cost, more active surface area, good electrocatalytic activity and reuse capability. Many research group reported that first-row transition metals, stainless steel, nickel alloys, Fe, Zr, Mo, TiO_2 nanocrystals, titanium–nickel

A. K. Chaurasia · P. Mondal (✉)
Department of Chemical Engineering, Indian Institute of Technology Roorkee, Roorkee 247667, India
e-mail: pmondfch@iitr.ac.in

A. K. Chaurasia
e-mail: akcnitj@gmail.com

V. H. Saran and R. K. Misra (eds.), *Advances in Systems Engineering*, Lecture Notes in Mechanical Engineering,
https://doi.org/10.1007/978-981-15-8025-3_10

and electrodeposits of graphite felt nickel, Ni–Ti, Ni–Fe–Co, Ni–Fe–Co–P, Ni–Mo, Ni–Fe and Ni–W are promising cathodes catalyst due to their low overpotential and improved electrocatalytic activity [5]. Recent literature reported that the electrodeplating/addition or grouping of two or more transitions metal improved the cathode catalyzed reaction in MECs; it reduces the overpotential losses of electrode, mechanical stability with high catalytic activity by combined influence of their catalytic properties.

Most of the literature using these cathodes are investigated in synthetic wastewater using a single substrate. There are very few literature reported that the effectiveness of these alternative cathode catalyst using actual or real industrial wastewater. Among these studies, some tests were done using industrial and domestic wastewaters such as winery, potato, dairy wastewater. The most probable reason for it because these fabricated cathodes has not been employed in actual wastewater is due to the complex nature of actual wastewater, and also due to inorganic constituents of actual wastewater. The inorganic constituents of actual wastewater can greatly affects the cathodes electrocatalytic activity but also leads to the catalyst poisoning due to irreversible absorbance of cofactor/co-enzyme or metals ions at catalyst active site.

In this study, electrodeposits of Ni, Ni–Co and Ni–Co–P on SS and Cu were selected to evaluate hydrogen production and energy recovery of the paper–pulp industry actual wastewater in MECs. The aim of current study was to evaluate the developed cathodes in actual wastewater and viability of practical MECs operation. The paper–pulp industry wastewater was chosen because it generates large quantity of wastewaters during processing of agricultural raw materials and its contained high COD (100 g/L), BOD (35–40 g/L) and dyes (colour) with biodegrability index <0.4. The fabricated cathodes have been explored as a cathode in MEC with anode (graphite rod) under controlled temperature of 30 ± 2 °C at input voltage of 0.6 V.

2 Materials and Methods

Paper and pulp industry wastewater samples were collected from Star paper Mills Saharanpur, India and wastewater characteristics are shown in Table 1. Samples

Table 1 Physicochemical characteristics of paper and pulp industry actual wastewater

Parameters	Values (actual wastewater)
COD (g/L)	5.88
Soluble COD (g/L)	5.18
BOD5 (g/L)	1.8
pH	6.72
TS (g/L)	1.44
TDS	15.84
TSS	0.08

were collected at on-site from the primary clarifier of the industry and placed in ice after that stored at 4 °C overnight at processing laboratory. Bacterial source used in this study as mixed culture was taken from municipal sewerage treatment plant Haridwar, India. The eight different electrodeposited cathodes used in this study are the same cathodes, which are described, in our previous published study as Chaurasia et al. [5]. Fabricated cathodes represented in this manuscript as SS2 (Ni deposit), SS3 (Ni–Co deposit) and SS4 (Ni–Co–P deposit) and similarly Cu2, Cu3 and Cu4. The hydrogen gas produced in MECs was measured on a daily basis using water displacement techniques and collected in Tedlar gas sampling bags via silicon tube. It qualitatively analyzed by gas chromatography (GC NEWCHROME 6800) and more than 96% hydrogen gas fraction was detected in GC. Each MECs batch cycles consists of seven days and all the calculation with analysis were done as described in our previous study [5].

3 Reactor Configuration and Operation

Eight double chamber MECs (10 cm * 10 cm * 10 cm) with 800 ml working volume were made by using Plexiglas material consisting of graphite rod as an anode and electrodeposited electrode as the cathode at optimum electrode spacing (4 cm). All the experimentation in this study was carried out as in similar pattern with same working procedure of our previous study [5]. All the MECs experimental setup fed with 700 ml actual wastewater with 100 ml concentrated activated sludge as in column in controlled temperature 30 ± 2 °C at the applied voltage of 0.6 V. For the maximization of hydrogen production rate need to be inhibition of methonogens and it can be done by adding 10 ml of 100 μM of BES (2-Bromo ethane sulfonate) in each MECs reactor. Two times, in a day, sample was analyzed for COD and further chemical analysis and volume of produced gas were quantified by water displacement techniques on daily basis.

4 Results and Discussion

4.1 Current Generation

The current generation from different electrodeposited cathodes across the MEC reactors is shown in Fig. 1. Current density behaviour indicates that the electrodeposited cathodes give higher current rates than without electrodeposited cathodes such as SS1 and Cu1 and it is primarily due to the effects of electrodeposits and activity of exoelectrogens. The SS4 cathode achieves highest current density of 2.45 A/m^2 and Cu4 has 2.39 A/m^2 in comparison with bare cathodes using actual wastewater. The rate of hydrogen production predominantly depends on the availability of

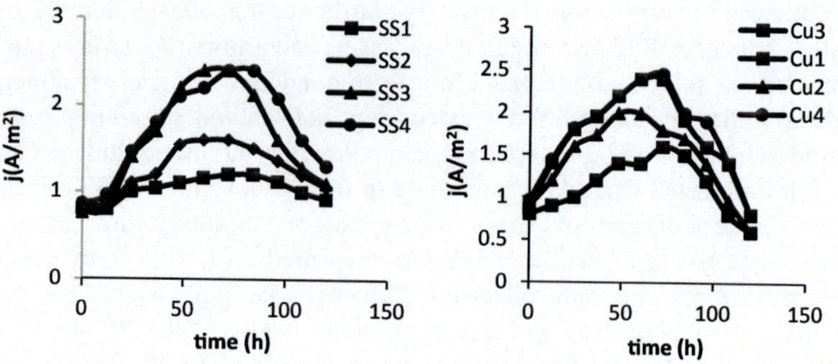

Fig. 1 Current density profile of developed cathode of paper and pulp industry actual wastewater in MECs. **a** Electrodeposits on SS and **b** electrodeposits on copper

electrons in the cathode (current density) and the current flows in the circuit triggered by microbial activity. As the growth of microorganism increases, simultaneously increases the current density. The current density profile on actual wastewater has slight variation to our previous study [5] on synthetic wastewater (single substrate acetate). The development of the current density profile also suggests that fast oxidation of simple organics then complex organic compound present in actual wastewater. This clearly revels that MECs performance has less impacts of substrate/nutrient medium. No hydrogen gas production was detected when the current tends to zero, which demonstrates that the current and gas produced in the MEC reactor are due to the microbial activity and their metabolism.

4.2 COD Removal

The percentage of COD removals for different cathodes is ranging from 54–65% as shown in Fig. 2b with respect to initial COD of wastewater (5.88 g/L) and it indicates COD removal performances depend on cathode catalyst as well as nature of the industrial wastewater. The SS3 cathodes containing MECs show 64 ± 2% COD removal; on the other hand, SS4 has 60 ± 2% and (61 ± 2%) in case of Cu2. Lowest COD removal detected in the case of Cu4 (54 ± 2%) cathode containing MEC reactor. The high COD removals achieved with SS3 and Cu2 cathodes indicated the paper and pulp industry real wastewater is highly biodegradable.

4.3 Energy Analysis

The columbic efficiency is used to express the energy yield compared with the energy input of the MEC reactor. Columbic efficiency of the electrodeposited cathode

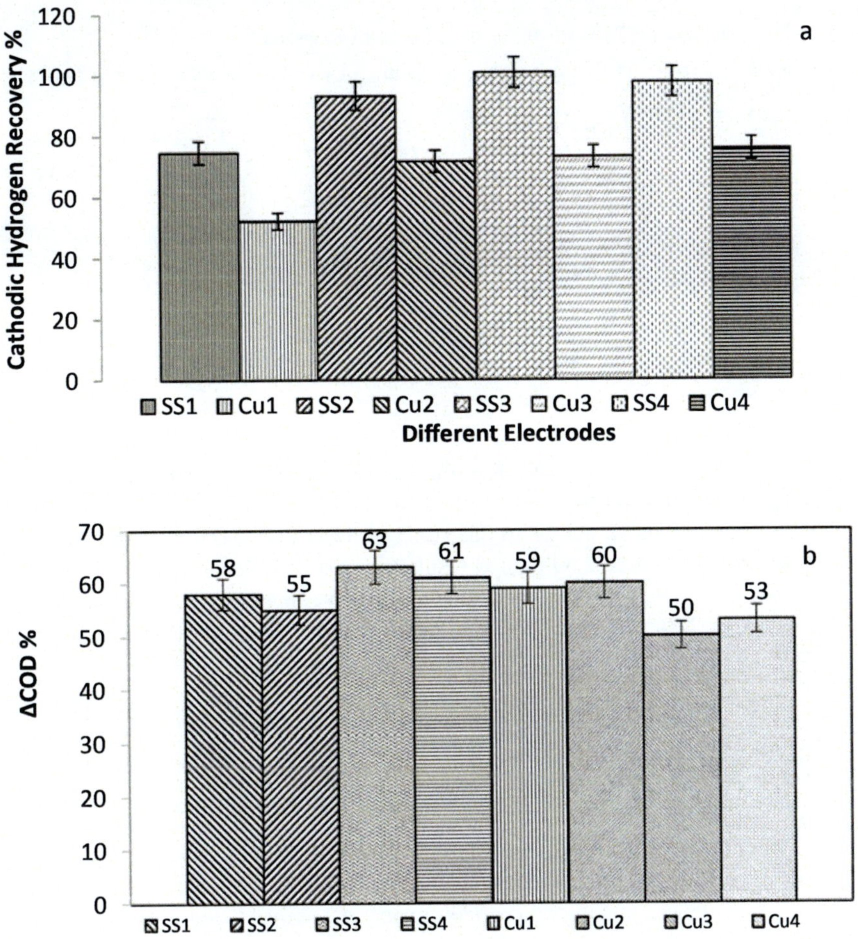

Fig. 2 Cathodic hydrogen recovery of different cathodes on actual wastewater (**a**), percentage of COD removal of actual wastewater with different electrodeposited cathodes (**b**)

reflects the trends observed in hydrogen production as shown in Fig. 2a, and estimated columbic efficiency is given in Table 2. The electrodeposited on SS cathodes improved the cathodic efficiency but it is insignificant in the case of Cu electrodeposited cathode. It is occured due to the fixed value of the applied voltage of 0.6 V for all the cathodes.

Obtained results show that COD removal and energy recoveries highly depend on cathodes catalyst as well as the nature of constituents present in actual wastewater; it also indicates that the electrodeposition has positive effects on COD removal and energy recoveries. The energy efficiency is the ratio of hydrogen produced to electrical energy input. The energy efficiency (ηE) of the MEC with different fabricated cathodes is summarized in Table 2. Estimated results suggest that a significant

Table 2 Comparisons of MECs performance on actual wastewater

Developed electrode/cathode	Q (m^3 H_2 m^{-3} d^{-1})	V_{H2} (ml)	C_E (%)	η_E (%)	Y_{H2} (n_{H2}/ns)
SS1	1.4 ± 0.1	96 ± 2	77.5 ± 2	77.3 ± 4	1.12 ± 0.1
SS2	2.1 ± 0.4	126 ± 2	88.3 ± 2	92.5 ± 4	1.42 ± 0.1
SS3	2.8 ± 0.5	164 ± 2	93.1 ± 2	97.6 ± 4	1.82 ± 2
SS4	3.8 ± 1	212 ± 2	94.9 ± 2	81.21 ± 4	2.88 ± 2
Cu1	1.1 ± 0.1	84 ± 2	53.2 ± 2	52.36 ± 4	1.06 ± 0.1
Cu2	1.98 ± 0.4	121 ± 2	69.6 ± 2	61.45 ± 4	1.32 ± 0.1
Cu3	2.6 ± 0.4	153 ± 2	72.2 ± 2	71.25 ± 4	1.47 ± 0.1
Cu4	3.87 ± 1	257 ± 2	76.7 ± 2	74.52 ± 4	2.43 ± 0.2

amount of energy was harvested from the substrate, not from electricity in the case of electrodeposited cathodes.

All the calculations are done in this study as described in our previous study [5] and represented in Table 2. Obtained results suggest that electrodeposited cathodes show stable and enhanced performance on actual wastewater. Among all the developed cathodes, SS4 cathodes show the maximum hydrogen production and energy recovery as compared to bare cathodes (SS1 and Cu4).

5 Conclusion

MECs parameter determined from the experimental results is suggested that the produced electrodeposited cathodes have increased the hydrogen production performances from two to fourfold than palladium and Pt. catalyst using actual paper–pulp industry wastewater. It is also observed that the electrodeposited cathodes enhanced the cathode performance towards HER on actual wastewater until more than five batch cycle (a batch cycle consist of seven days). It reveals that a combination of two or three non-noble metal cathode catalysts has potential to replace the Pt. for the commercial application of MEC as well as in real wastewater.

References

1. Tenca A, Cusick RD, Schievano A, Oberti R, Logan BE (2013) Evaluation of low cost cathode materials for treatment of industrial and food processing wastewater using microbial electrolysis cells. Int J Hydrogen Energy 38:1859–1865
2. Kadier A, Abdeshahian P, Simayi Y, Ismail M, Hamid AA, Kalil MS (2015) Grey relational analysis for comparative assessment of different cathode materials in microbial electrolysis cells. Energy 90:1556–1562. https://doi.org/10.1016/j.energy.2015.06.108

3. Mondal P, Kumari P, Singh J, Verma S, Chaurasia AK, Singh RP (2017) Oil from algae. In: Sustainable utilization of natural resources
4. Rani M, Shanker U, Chaurasia AK (2017) Catalytic potential of laccase immobilized on transition metal oxides nanomaterials: degradation of alizarin red S dye. J Environ Chem Eng 5:2730–2739
5. Chaurasia AK, Goyal H, Mondal P (2019) Hydrogen gas production with Ni, Ni– Co and Ni–Co–P electrodeposits as potential cathode catalyst by microbial electrolysis cells. Int J Hydrogen Energy. https://doi.org/10.1016/j.ijhydene.2019.07.175

Improvement in the Mechanical Properties of Paraffin-Based Propellant

Jhumki Nandy and P. C. Joshi

1 Introduction

In the category of jet propulsion, rocket propulsion is also a well-known class that stored burning matter by the ejection of this stored matter producing thrust. This kept inert charge within the rocket chamber is called propellant. The conventional propellants are divided into categories depending on the variety of the source of energy. Conventional chemically charged propulsion systems were distinguished well-known groups by their physical states. Nuclear and solar rocket are the type of different energy sources. Chemical rockets are further classified depending on the physical state of the propellant either solid, liquid or the combination of both solid and liquid. The current research era attenuated to the well-renounced proposed fuel groups only. Many of the disadvantages and demerits are responsible for the application of solid and liquid propellants. The solid propellant using the high energetic propellant ingredients for the better thrust. For the solid propellant also the start-stop and restart operation is not possible. Hybrid rockets became the best answer to all the problems. Hybrid rocket propulsion systems are such a together combined system where the decided fuel and oxidizer are kept within the rocket chamber separately without making change in their state. The newly proposed propulsion system where oxidizer and fuel are basically kept in liquid or gaseous and liquid state for ease of application. But sometime hybrid propulsion system liquid fuel and solid oxidizer. Examining based on performance, hybrid propulsion system yields the density specific impulse of 514 gm-s/cm^3 as compared to best known solid propulsion system which produces density specific impulse around 473 gm s/cm^3 and

J. Nandy (✉)
Aeronautical Engineering, PCE, Nagpur, Maharashtra 440019, India
e-mail: aero123fly@gmail.com

P. C. Joshi
Department of Space Engineering & Rocketry, BIT, Mesra, Ranchi, Jharkhand 835215, India

V. H. Saran and R. K. Misra (eds.), *Advances in Systems Engineering*,
Lecture Notes in Mechanical Engineering,
https://doi.org/10.1007/978-981-15-8025-3_11

best-known liquid propulsion system which yields the density specific impulse 192 gm s/cm^3. The comparison made above clearly shows that the hybrid propulsion system has an edge over their solid and liquid counterparts.

Bartel and Rannie [1] redirected to a combustion model for hybrid combustion at the beginning. The model was contemplated the fuel by 1D tube like model, airflow with turbulence come into with bulk amount of air molecules, density ρ_o and T_{to}. The main phenomenon was found that the spreading of oxygen molecules through a gigantic spacious manner, though the surface of the gas film was depending on the rate of the diffusion controlling factor and no assumptions were made by the researcher about the flame behaviour. Penner [2] observed in his work about the hollow cylindrical burning model with the flow of oxygen through it to study the internal burning. Diffusion controlled burning rate problem was proposed by Penner's about to know the state of the concentrated oxygen radial and axial functional relationship as governed by the linear equation.

$$\frac{\partial K_o}{\partial x} = \left(\frac{D}{u}\right)\left[\frac{\partial^2 K_o}{\partial r^2} + \left(\frac{1}{r}\right)\frac{\partial K_o}{\partial r}\right] \quad (1)$$

Leon Green [3] approached to the case of slow burning of the hybrid system with the help of the definition of Spalding's mass transfer number and with Chilton–Colburn modified model for Reynold's analogy between mass and momentum transfer. Houser and Peck [4] carefully planned the combustion model with the shape of hollow cylindrical fuel within the flow stream of the gaseous oxygen and forwarded the following equation for instantaneous radius versus time. Including the conventionally proposed oxygen droplet formation mechanism, Karabeyuglu [5] et al. put forward spreading of the molecules in bulk amount inclined to the phenomena to draw droplet fuel from the upper hot liquefied layer. Karabeyuglu showed in a very significant way that the formation of the liquid droplet because of the greater value of surface tension on the liquid layer; this instabilities present within this melted liquid layer was created because of the high speed gaseous flow presence in the fuel port. Because the proposed nature of the problem is so complex, the modelling was accomplished with three stages of burning phenomena. At first of the requirement of the stages is the transformation of the upper surface of the solid grain to the hot liquid layer occurrence over existing solid stored material which was considered for this study. In the next stage of the model, the formed thin low viscous melt layer under high concentrated gaseous flow of oxygen with strong action of shear force was considered because of maintaining linear stability (Fig. 1).

Marxman [6] has carried out a complete regression rate analysis using a flat plate combustion model that resembles with a diffusive type of flame with strong turbulence, where within the boundary layer the flame was formed. As soon as the molecules of the fuel going towards the boundary layer since the existing solid charge absorbs heat from the surrounded wall so quickly, and on the other hand oxidizer is supplied from existing main flow into the boundary layer. The process of burning of the charge started in a region under boundary layer where flame has occurred, with an achieved suitable mixture ratio. The formed boundary layer has considered

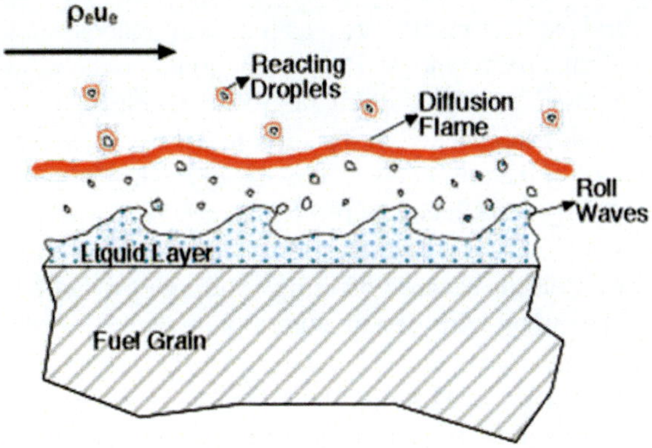

Fig. 1 Liquid layer instability

comprising of the upper zone augmented above the region of the propagating flame, where both of the gradients of the profile of raised temperature and increases velocity profile co-existing with opposite behaviour. In the lower surface area existing under the propagating flame, where both of the gradients of the temperature profile and the increased velocity profile co-existing with the same behavioural nature of increment. Ramohalli, Stickler10 introduced the hybrid charge burning model which is delineated to the pressure sensitivity. For this proposed model of combustion, this is especially accepted that the solid grain will be breaking into small pieces by depolymerization except for the presence of the monomers. The idea of the small fragmented size material was raised. He suggested that surface degradation is the cause of the regression rate. The depolymerized thermal reaction surface is supplemented by surface depolymerization by active species. Estey et al. [7] investigated that radiant heat transfer term empirically improved the result by including in the thermal radiation equation. On the other hand, the convective heat transfer term is much more important for pure hydrocarbon fuels.

Sarigul-Klijn et al [8] studied the advantages of a rocket motor stored with hybrid propellant bunt with cold nitrous oxide. Since the oxidizer is cold a turbo-pump was used to bring the oxidizer into the inner casing of the motor. As cold nitrous oxide (at temperature about −100 °F) is highly dense oxidizer; therefore, the oxidizer tank has less volume and less weight. Under this condition, the oxidizer burnt at a higher ratio with the fuel. This type of arrangement of the oxygen system will contribute the minimal size to the motor case which will reduce the weight of the motor. With the higher ratio of the fuel to oxidizer would also be contributed to the less unburnt fuel sliver which will reduce the residual propellant weight. Using of the turbo-pump increases the pressure inside the chamber and increases overall specific impulse which will prominently decrease the pressure of the oxidizer tank and reduce the weight of the tank. Oxidizer enhanced hybrid rocket engine has been investigated by Jacob [9]. The problem in the quick entrainment of the fuel in the gaseous flow of the oxygen and proper mixing of the fuel-oxidizer droplets results in low consumption of the propellants compromising with the condition of poor performance for the mostly used hybrid propellant-based rocket motor. This quick mixing mechanism of the solid oxidizer results into a combustion zone resembles with the combustion zone of the solid rocket motor. The proper mixing of the solid oxidizer with the fuel only can be possible under the stoichiometric condition. Otherwise, a complete combustion cannot take place with the solid fuel. Less rapid combustion reaction reducing the scope of the unsafe characteristics of the hybrid propellant-based rocket motor. The investigation which is purely based on the experiment has also shown that the safety of the hybrid rocket and the performance parameters of the hybrid rocket motor come into concern because of using solid oxidizer with the fuel grain. The practicality in the study with design study optimization was conducted by Ihnseok Rhee and Changjin Lee et al. [10], in the HTPB/LOX hybrid rocket motor application for air-launch system with first stage. So, the investigation showed that the popularity of using the solid rocket motor can be substituted with hybrid propellant-based system if the improvement can be possible in the value of the average specific impulse approximately about 350 s. And the design optimization for the proposed design variables like port design, initial oxidizer flux, the inside chamber pressure and area ratio of the nozzle were considered. Also, it was proposed from the design that the optimized result provides nearly same design of the length of the engine.

To give an improvement to the existing regression rates for the existing conventional hybrid rocket fuel several techniques were adopted by karabeyoglu et al. [11], by placing the net-like metal screens and the insertion of the different shape of mechanically operated device to control the turbulence level up to a desirable limit which can give desirable improvement in the rate of heat transfer from the wall of the grain to the combustion zone. By using swirl injector in-front of the flow of the oxidizer mass flux can give a favourable increment in the effective mass flux. Improved mass flux will provide a better heat transfer rate. By incorporating the different enhancing agents of self-oxidation and self-decomposed solid dispersed particle size materials in the hybrid fuel, which will help in creating rough burning surface which will create burning progression at different rate from all the exposed surface better than the conventional matrix material. Wilson and Keller [12] et al. examined through investigation that the combustion of the hybrid solid fuels and burning rate of the solid fuels of hybrid rocket and other discharged combusted hot smokes by small, portable ultraviolet and visible spectrometers. Throughout this type of investigation, the two spectrometers of the range in ultraviolet level and the visible range were taken into consideration for the visibility of hydroxyl radical spectrum, and under different oxidizer to fuel ratios combinations broadband visible emission spectrum was used for the visibility of the spectrum for the different hybrid rocket motors. Regression rates were measured both in the time and spatial coordinates.

Moutet and Barrere [13] took the adoption of the concluded results of the previous experimental and theoretical analysis from different combustion model of the liquid melted fuel in the hybrid-based propellant problem. The adopted concept regarding thermodynamic parameters introduced by Emmons from previous combustion analysis. The proposed model of Emmons was developed by Spalding. As an alternative, a flat plate model was considered, and a laminar zone of finite thickness was analysed. A flat plate model further analysed with the assumption of the finite concentration of fuel and oxidizer in it. Karabeyoglu and Cantwell [14] used boundary conditions from

Benjamin to extend the theory of Craik. From the boundary conditions, it was showed that the liquid–gas interface is linearly unstable under the flow conditions found in hybrid rockets. To deal with the nonlinear problem, an empirical mass transfer correlation was constructed by using the data of Gater and L'Ecuyer. The model manifested the fact about the paraffin that it consisted of low viscosity of the liquid layer leads to entrainment of droplets from the liquid–gas interface from Cantwell, Karabeyoglu and Altman. The entrainment mechanism of the droplet substantially increases the rate of fuel mass transfer leading to a much higher regression rate than the regression rate of the polymeric fuels by a factor of three to five. Green [15] evaluated the reported analysis of the burning model of the hybrid propellant. This model was compared the four different thermodynamic parameters and fluid-mechanical assumptions involved with it. The early result of Bartel and Rannie gives an indication based upon the analogical aspect between mass and momentum transfer, which can be generalized with Spalding's mass transfer number to predict the burning rate of slow burning. The oxygen system made of Plexiglass shows satisfactory with the published experimental data without any adjustment of empirical constants. With the extension of the existing theory involving high mass, addition rates would require in ducted flows would require more experimental research with a mass transfer at the wall and chemical reaction.

Fineman [16] was taken into account both laminar flow behaviour and turbulent flow behaviour in a duct. Mass transfer expression proposed by Spalding's in the stagnant film, a significant combustion theory has proposed. For the case of the laminar film, the thickness was arbitrarily established in-front of the duct by taken into account the thickness of the boundary layer. George et al. [17] was carried out ordered experimental investigations for regression rate enhancement for HTPB/GOX system. The effects of the addition of Ammonium Perchlorate (AP) or metal particle-like aluminium in the solid fuel, the variation of oxidizer fuel ratio and the effective variation of the characteristic dimensions of the fuel grain had been reported. With the significant addition of AP/A*l* and reduction of the grain port diameter quickly enhances the regression rate. The exponents of oxidizer mass flux and port diameter were found experimentally to be significantly different from those of the conventional turbulent diffusion-limited theory. Experimental investigations on combustion characteristics of a cylindrical multi-port grain of a 96 mm diameter multi-port grain hybrid rocket motor were conducted by Kim et al. [18]. This experimental investigation brought out the effect of the port number and the distance between ports on a regression rate of solid fuel. The significant results also indicate that the regression rates become fast as the port number increases to a typical number of up to four ports. Many experiments were conducted to investigate the enhancement of the regression rate with helically shaped grain and with the use of the metal wire embedded within the solid fuel.

Chiaverini and Kuo [19] introduced a discussion with aspects of all the starting theories of the hybrid rocket application. The combustion process regarding evaporated fuels from inert solid surface takes place through diffusive mixing between vaporized oxidizer flowing through the port. The thickness of the flame zone and location in the boundary layer are shown roughly to scale. It was observed that the flame zone is relatively deep in the boundary layer and according to Marxman and Gilbert and Marxman, Wooldridge and Muzzy. The observed flame zone tends to be fuel rich based on the observed flame position and comparatively low-flame temperatures measured in the boundary layer. Changjin6, et al. [20] did a series of experimental investigations to optimize the conditions of oxidizer swirl flow and different grain configuration for enhancement of the regression rate of solid fuel. PMMA solid fuel was considered with gaseous oxygen for investigation. The experimental test results also revealed that with the helical grain configuration of the solid fuel the regression rate enhanced up to 50%. The main concluding part was that the regression rate of paraffin-based fuel enhanced two to three-time folds compared to polyethylene of ultra-high molecular weight.

The combustion flow process achieved for a tested rocket motor with hybrid-based combination depending on many factors which include: combustion chamber geometry, nozzle geometry, air–fuel flow rates, and the properties of the respective fuel and oxidizers chosen. These methods, developed in conjunction with the Zeldovich-Novozhilov [21] solid-phase energy conservation method, provide a calculation strategy for transient solid-propellant burning. The diffusion of the solid fuel molecules is directly proportional to the rate-limiting factor. This rate-limiting factor is the function of the heat transfer rate occurring at interacting zone between the solid and gaseous components. In dealing with hybrid combustion, methods have been devised which increase the heat release rate adding the metal particles which are energetic in nature. These metal particles contributing promptly to increase release rate of the heat from the surface. This phenomenon occurs when the aluminium particles are oxidized to aluminium oxide which stabilizes the velocity boundary layer and provides a more energetic interaction surface.

Experimental studies on combustion characteristics of a cylindrical multi-port grain of a 96 96-mm mm-diameter hybrid rocket motor were conducted by Kim, et al. [22]. This work brought out the effect of the port number and the distance between ports on a regression rate of solid fuel. The results indicate that the regression rates become fast as the port number increases to a typical number of up to 4 ports. Chiaverini and Kuo [23] provide a comprehensive discussion of the fundamentals of hybrid rocket operation. Combustion takes place through diffusive mixing between vaporized oxidizer flowing through the port and fuel evaporating from the solid surface. Changjin6, et al. [24]. Conducted conducted a series of experimental investigations to optimize the conditions of oxidizer swirl flow and grain configuration for enhancement of the regression rate of solid fuel. The test results also revealed that with helical grain configuration the regression rate enhanced up to 50 percent. The ignition of the paraffin/polyethylene as solid fuel, regression rates, and burning performance with gaseous oxygen were studied by Santos et al. [25]. The main conclusion of their work was that the regression rate of paraffin-based fuel enhanced twofold to three folds compared to the ultra-high molecular weight polyethylene. The combustion flow process achieved in a hybrid rocket motor is a function of many factors which include: combustion chamber geometry, nozzle geometry, air fuel flow rates, and the properties of the respective fuel and oxidizers chosen. These methods, developed in conjunction with the Zeldovich-Novozhilov [26] solid phase energy conservation method, provide a calculation strategy for transient solid-propellant burning.

The experiments done for preceding the status of the investigation on hybrid rocket motors were performed at I. G. Farben in Germany by Andrussow et al. [27]. The experiments were done with a 10 kN hybrid motor constituting of coal as the solid fuel and nitrous oxide as the gaseous oxidizer. However, the hot-fire motor test was unsuccessful due to the significant heat required to vaporize the solid coal. Due to coal sublimation resulted in an undesirable low burning rate. General Electric Company in New York carried out investigations for the prospective use of hypergolic propellants. The work, spearheaded by Moore and Berman [28], intricate the burning of 90% hydrogen peroxide and polyethylene as oxidizer and fuel, respectively. A silver screen catalyst bed was embedded in the pre-combustion chamber to decompose the hydrogen peroxide, which in turn spontaneously ignited the polyethylene fuel. Making use of a unique rod and tube grain design configuration, over 300 motor static tests were performed to characterize the combustion reaction. Moore and Berman described uniform surface burning, grain cracks caused the combustion insensitivity, stable combustion and high-combustion efficiency. The most appreciated out coming observed by the authors was that unsatisfactorily low regression rate and insignificant entrainment of the fuel droplet to the oxidizer flow rate.

NASA Wallops, Stanford University and the Space Propulsion Group (SPG) in an utmost attempt to flight test a liquefying fuel hybrid sounding rocket to an altitude of 100 km. Dyer et al. [29] started initiating the propulsion system comprises of liquid nitrous oxide and paraffin-based fuel, which delivers a 5 kg payload to its predetermined altitude. Purdue University was aimed to develop a series of hybrid sounding rockets capable of carrying the payload like microgravity experimental instruments. The intermediate low-altitude rocket was powered by hypergolic propellants with 90% hydrogen peroxide oxidizer and low-density polyethylene (LDPE) four-port fuel grain for the launch. Larson et al. [30] incorporated the various types of agents to paraffin wax for use in a hybrid rocket motor in their study. In this investigation, for some paraffin-based fuels were doped with various percentages of LiAlH4 (up to 10%). By adding of LiAlH4 at 10% was found to give a favourable increment in the regression rates between 7 and 10% over baseline paraffin through tests in a gaseous oxygen hybrid rocket motor. Mass consumption at the time of burning for the grains of paraffin wax with 10% LiAlH4 was also higher than those of the virgin paraffin.

2 Experimental Methodology

One of the major shortcoming with paraffin wax fuel grain is its poor mechanical strength. The present work is directed towards an attempt to improve the mechanical behaviours of the virgin paraffin wax by incorporating certain additives like stiffening agent stearic acid, low-density polyethylene, ethylene-vinyl acetate copolymer in a small percentage and then study. Paraffin wax is a solid soft plastic-like substance at any temperature. Due to nature like plastic, the paraffin wax gets deformation under the action of pressure, not with the application of heat; at the temperature range from 45 to 95 °C, the normal paraffin wax is combustible and insoluble in water. Stearic acid supplied by CDH was used in the present work. Its molecular formula is $C_{18}H_{36}O$, and the molecular weight is 284.5 g. It is a hard white or faintly yellowish crystalline solid. Repeated epoxy groups are presented in the epoxy resin. Epoxy resin is got hardened with low grade usable plastic. Resin epoxy is constituted as a molecule with more than one epoxy group. This epoxy group is acted like the glycidyl group. This grey, the thixotropic adhesive has been developed to achieve a handling strength after 1 h at room temperature (253 °C). The high functionality of the resins increases cross-link density and improves thermal stability and chemical resistance. Cures with negligible shrinkage and without releasing volatiles.

Epichlorohydrin: (hardener) Epichlorohydrin used for this study is yellowish, volatile at ambient temperature and flammable with emitted toxic fumes when heated for decomposition. It is miscible with alcoholic-based solutions. The hardener is slightly soluble in water but insoluble in petroleum hydrocarbons. Epichlorohydrin starts the hydrolysis process slowly at room temperature but more rapidly in the presence of heat or traces of acid. Whenever it is heated for decomposition, epichlorohydrin emits toxic fumes of hydrochloric acid and other chlorinated compounds. The commercial epichlorohydrin is the product with 98% purity with a maximum of 0.2% water.

Eva-copolymer: EVA-copolymer is a high VA content ethylene-vinyl acetate copolymer with high flowability. EVA has possessed with several excellent properties like flexibility in physical character, elastic properties, stability in thermal behaviour, excellent low-temperature resistivity and nicely compatible with plastic-like materials and non-toxic. EVA recognizing with good thermally stable behaviour and it is nicely compatible with virgin paraffin wax. The hot melt semi-liquid solution made by EVA adheres with petroleum wax but as good as with some stiffening agent also.

LDPE: $(C_2H_4)_n$ H_2: Low graded polyethylene is a polymer with thermoplastic behaviour consisting of long hydrocarbon chains. Due to crystal structure the exact molecular weight, the exact melting point and glass transition temperature always not been observable. The point of temperature where glass transition property occurs varies strongly depending on the chain of the polyethylene group. For LDPE, which has average polyethylene chains the described melting point and use for commercial use is typical of the range 105–115 °C. Several fuel compositions were prepared using the various ingredients in different proportions and subjected to stress–strain analysis. Table 1 presents the proportionate quantity of ingredients used in the fuel composition. The experimental data have been obtained by analyzing the stress–strain curve obtained using Universal Test Machine (UTM made by INSTRON)

Table 1 Composition of the test sample

Sample No.	Paraffin wax in (%)	Stearic acid in (%)	Araldite in (%)	Hardener in (%)	Carbon Black in (%)	LDPE in (%)	EVA in (%)
1.	100	0	0	0	0	0	0
2.	90	10	0	0	0	0	0
3.	90	9	0	0	0.9	0	0
4.	85.47	8.54	4	0.85	0.85	0	0
5.	93.45	0	4.67	0.93	0.93	0	0
6.	81.87	8	4	0.81	1	4	0
7.	81.87	8	4	0.81	1	0	4
8.	78.36	7.83	3.91	0.78	1	4	4

3 Result and Discussion

Test Procedure: Instron Universal testing machine was used to obtain stress–strain curves. The sample size of $8 \times 2.5 \times 0.5$ cm of each composition was used in the present investigation.

These fuel samples as prepared by the process described in the previous section and were subjected to stress–strain analysis using Universal Tensile Test Machine (UTM). A total of eight samples were processed using ingredients such as stiffening agent stearic acid, Araldite, hardener, low-density polyethylene (LDPE), ethylene-vinyl acetate (EVA) and carbon black. Figures 4.1 to 4.8 present the stress–strain curve of Paraffin wax, Paraffin wax- Stearic Acid, PW- SA- Carbon Black, Paraffin wax- Stearic Acid- Araldite- Hardener- Carbon Black, Paraffin wax- Araldite- Hardener- Carbon Black, Paraffin wax- Stearic Acid- Araldite- Hardener- LDPE- Carbon Black, Paraffin wax- Stearic Acid- Araldite- Hardener- EVA- Carbon Black, Paraffin wax- Stearic Acid- Araldite- Hardener-LDPE- EVA- Carbon Black , respectively.

Table 2 Comparative results from tensile test experiment for all the compositions

Sample No.	P_{max}	σmax	E	σ_{UTS}	σ_e	E_U	E_t	Necking strain
1.	20	2.847	109.52	0.64	1.00	2.45	4.06	1.608
2.	9	4.48	157.72	0.58	1.17	1.55	3.09	1.174
3.	19	2.74	136.51	0.65	0.66	0.94	2.65	1.704
4.	15	2.82	75.52	0.79	0.44	1.10	2.57	1.471
5.	15	3.17	56.41	0.48	0.53	1.16	1.93	0.761
6.	46	3.07	50.990	0.80	0.58	1.89	2.49	0.6
7.	43	3.29	22.62	0.90	0.67	2.11	9.70	0.670
8.	41	5.86	225.62	1.386	1.42	1.65	1.98	0.329

Table 2 presents the numerical values of each parameter obtained for individual samples. As seen from Table 2, the calculated values for ultimate tensile strength and total elongation of pure paraffin wax are 0.42 MPa, 4.06, respectively. After the addition of the first composition of additives (8–9% stearic acid + 4% Araldite + Hardener + 1% of carbon black by weight), the ultimate tensile strength and total elongation come out to be 0.42 MPa and 2.57, respectively. The addition of second composition of additives (8–9% stearic acid + 4% Araldite + Hardener + 1% of carbon black + 4% LDPE + 4% EVA by weight) results in the ultimate tensile strength and total elongation of 0.25 MPa and 9.70, respectively. It is observed that the value of total elongation obtained after the addition of polymers LDPE and EVA is twice than that of pure paraffin wax. Young's modulus for pure paraffin wax is 109.52 MPa whereas the Young's modulus obtained after the addition of the second composition of additives comes out to be 225.62 MPa, which is twice the value of Young's modulus of virgin wax.

In this experimental study, an initiative has taken to the mechanical behaviour of paraffin wax by the addition of certain additives like stiffening agents like stearic acid, Araldite, hardener, different types of polyethylene. The propellant compositions thus prepared were subjected to stress–strain analysis. On comparing the different parameters concerning to mechanical properties of the fuel formulations resulted in the outcome that the total elongation, Young's modulus and maximum tensile strength of pure paraffin wax has been changed to a great extent with the addition of LDPE–EVA, stiffening agent stearic acid and Araldite- hardener in 5:1 proportion. It was observed in Table 2 that the flexural strength is of the range 3–6 MPa. It also observed that the lowest value of flexural strength for wax and the highest value for wax with different additives. Tensile strength is of range 0.6–1.5 MPa. The lowest value has obtained for wax, and higher value ranges have observed for wax with different additives. Young's modulus is of range 109–225.62 MPa. This highly favourable range was observed only for wax with the addition of different additives. Uniform elongation, total elongation is of ranges 0.9–2.45 and 1.9–9.70.

From Tables 3, 4, 5, 6, 7, 8, 9 and 10, the mechanical properties of the different samples were presented. Tensile strength, elongation in Table 3 for wax for the present work is deviated from mean value by 1.7, 11.8%, and Young's modulus is of 109.52 MPa. By adding the different additives with wax, Young's modulus are of values from Tables 4, 5, 6, 7, 8, 9 and 10, 157.72 MPa, 136.51 MPa, 75.52 MPa, 56.41 MPa, 50.990 MPa, 22.62 MPa and 225.62 MPa, respectively. From Figs. 2, 3, 4, 5, 6, 7, 8, 9, 10, 11, 12, 13, 14, 15, 16 and 17, the stress – % strain characteristics and mean stress-mean % strain characteristics have been presented. Wax is a brittle type of material. So, the stress – %strain characteristics will be in a zigzag pattern, which will pronounce for the breaking than the elongation. Wax with the different additives has reduced remarkably this type of breaking, bending and necking. Overall the different additives in different proportions will improve the mechanical parameters of the wax.

Table 3 Comparison of the values for the mechanical parameters of the wax used in this study with the wax used for the different literature

Comparison of the data for different wax	Properties		
	Tensile stress	(%) Elongation	*E* in (MPa)
Type of wax used for this study	0.4407 0.43 (0.426, 1.7%) 0.44	2.753 4.387 (4.064, 11.83) 5.053	109.52
DeSain et al. [15]	0.86 0.88(0.91, 8.8%) 0.99	0.51 0.55 (0.48, 19%) 0.39	207(219.7, 18%) 260
Maruyama et al. [16, 17]	1.0 0.7 (0.8, 25%) 0.7	1.5 (1.4, 28%) 1.7	– – –
Rajiv Kumar* [a, b]	1.42 1.47(1.48, 5.6%) 1.57	4.3 4.6(4.5, 4.4%) 4.5	66.0 65.1(64.8, 2.8%) 63

Table 4 Comparison of the values for the mechanical parameters of the wax with 10% S.A

Comparison of the data for different wax	Properties		
	Tensile stress	(%) Elongation	*E* in (MPa)
Type of wax used for this study	0.4407 0.43(0.426, 1.7%) 0.44	2.753 4.387(4.064, 11.83) 5.053	109.52
Wax + 10%S.A	0.570 0.634(0.588, 4%) 0.560	4.837 2.737(3.098, 15.89%) 1.720	157.72

Table 5 Comparison of the values for the mechanical parameters of the wax with 09% S.A + 0.9%C.B

Comparison of the data for different wax	Properties		
	Tensile stress	(%) Elongation	*E* in (MPa)
Type of wax used for this study	0.4407 0.43(0.426, 1.7%) 0.44	2.753 4.387(4.064, 11.83) 5.053	109.52
Wax + 10%S.A + 0.9%C.B	0.419 0.253(0.342, 8.4%) 0.354	3.087 2.470(2.653, 37.7%) 2.403	136.51

4 Conclusion

In this present investigation, eight samples were explored and were also validated with the available literature the following are the important findings of the present work,

Table 6 Comparison of the values for the mechanical parameters of the wax with 8.54% S.A + 0.85%C.B + 4% Araldite + 0.85 epichlorohydrin + 0.85 C.B

Comparison of the data for different wax	Properties		
	Tensile stress	(%) Elongation	*E* in (MPa)
Type of wax used for this study	0.4407 0.43(0.426, 1.7%) 0.44	2.753 4.387(4.064, 11.83) 5.053	109.52
Wax + 8.54%S.A + 0.85%C.B + 4%Araldite + 0.85 epichlorohydrin + 0.85 C.B	0.384 0.338(0.398, 9.6%) 0.509	3.470 2.037(2.576, 78%) 2.220	75.52

Table 7 Comparison of the values for the mechanical parameters of the wax with + 0.93%C.B + 4.67% Araldite + 0.93% epichlorohydrin

Comparison of the data for different wax	Properties		
	Tensile stress	(%) Elongation	*E* in (MPa)
Type of wax used for this study	0.4407 0.43(0.426, 1.7%) 0.44	2.753 4.387(4.064, 11.83) 5.053	109.52
Wax + 0.93%C.B + 4.67%Araldite + 0.93% epichlorohydrin	0.384 0.338(0.398, 9.6%) 0.509	3.470 2.037(2.576, 78%) 2.220	56.41

Table 8 Comparison of the values for the mechanical parameters of the wax with + 8% S.A + 4% Araldite + 0.81% epichlorohydrin + 1%C.B + 4% LDPE

Comparison of the data for different wax	Properties		
	Tensile stress	(%) Elongation	*E* in (MPa)
Type of wax used for this study	0.4407 0.43(0.426, 1.7%) 0.44	2.753 4.387(4.064, 11.83) 5.053	109.52
Wax + 8%S.A + 4%Araldite + 0.81%epichlorohydrin + 1%C.B + 4% LDPE	0.214 0.324(0.282, 5.9%) 0.307	2.753 2.303(2.498, 23.1%) 2.437	50.990

Table 9 Comparison of the values for the mechanical parameters of the wax with + 8% S.A + 4% Araldite + 0.81% epichlorohydrin + 1%C.B + 4% EVA

Comparison of the data for different wax	Properties		
	Tensile stress	(%) Elongation	*E* in (MPa)
Type of wax used for this study	0.4407 0.43(0.426, 1.7%) 0.44	2.753 4.387(4.064, 11.83) 5.053	109.52
Wax + 8%S.A + 4%Araldite + 0.81%epichlorohydrin + 1%C.B + 4% EVA	0 0.634(0.295, 31.9%) 0.251	17.672 7.087(9.704, 70.34%) 4.353	22.62

Table 10 Comparison of the values for the mechanical parameters of the wax with + 7.83% S.A + 3.91% Araldite + 0.78% epichlorohydrin + 1%C.B + 4% EVA + 4% LDPE

Comparison of the data for different wax	Properties		
	Tensile stress	(%) Elongation	*E* in (MPa)
Type of wax used for this study	0.4407 0.43(0.426, 1.7%) 0.44	2.753 4.387(4.064, 11.83) 5.053	09.52
Wax + 7.83%S.A + 3.91% Araldite + 0.78% epichlorohydrin + 1%C.B + 4% EVA + 4% LDPE	0.632 0.256(0.418, 19.3%) 0.366	2.737 1.737(1.981, 66.8%) 1.470	225.62

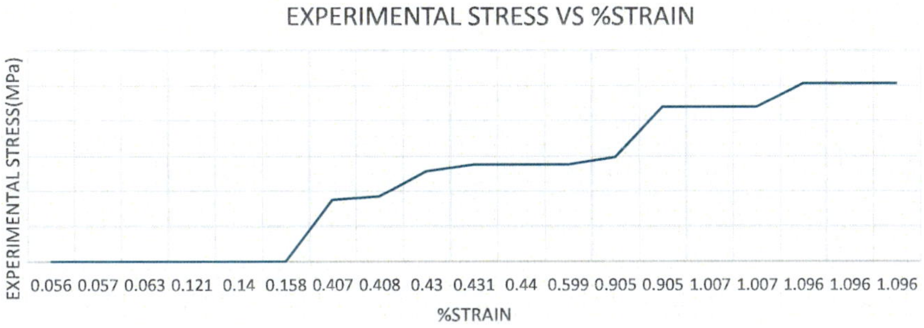

Fig. 2 Experimental stress versus %strain graph for wax for this study

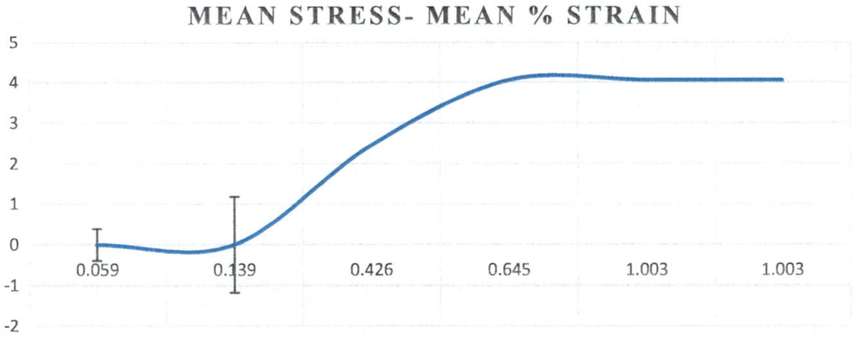

Fig. 3 Mean stress versus mean %strain graph for wax for this study

1. Further, from the analysis of mechanical properties, it is seen that the tensile strength and Young's modulus for these samples are in order of sample.3 > sample.1 > sample.2.
2. Highest Young's modulus has observed for sample-8 for sample dimension 8 × 2.5 × 0.5 cm.
3. Flexural strength of the wax with different additives was improved.

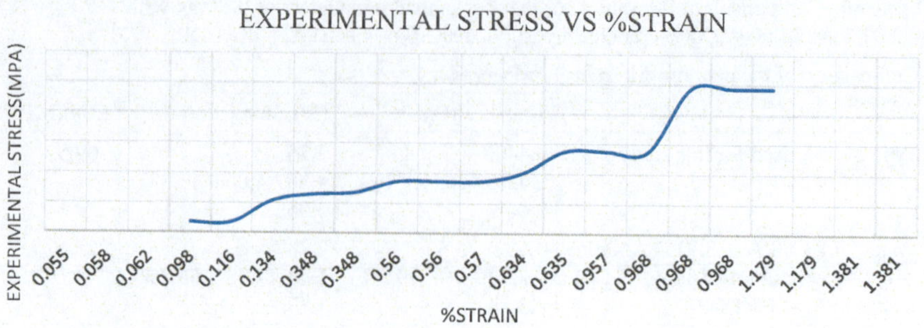

Fig. 4 Stress versus %strain graph for wax + 10%S.A

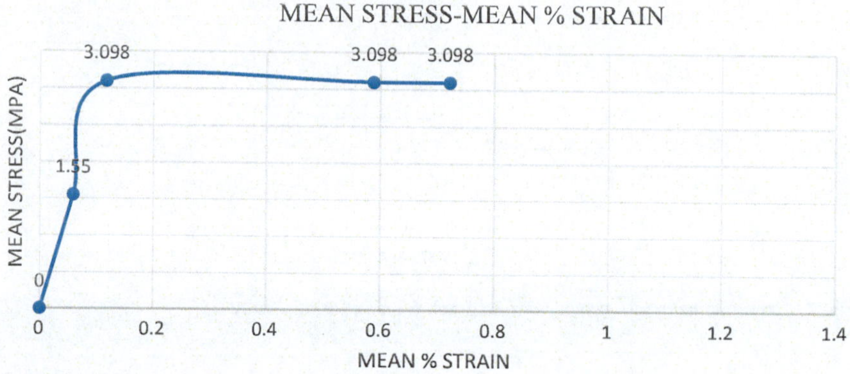

Fig. 5 Mean stress versus mean %strain graph for wax + 10%S.A

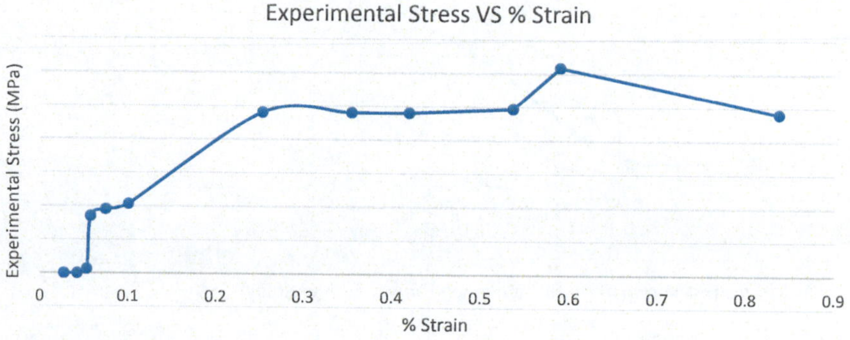

Fig. 6 Experimental stress versus %strain graph for wax + 10%S.A + 0.9%C.B

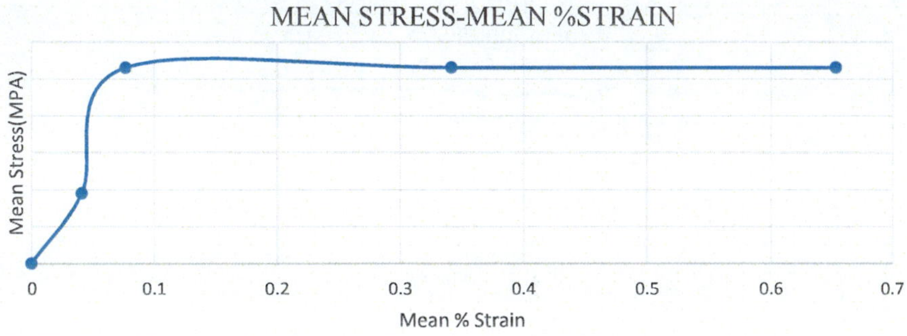

Fig. 7 Mean stress versus mean %strain graph for wax + 10%S.A + 0.9%C.B

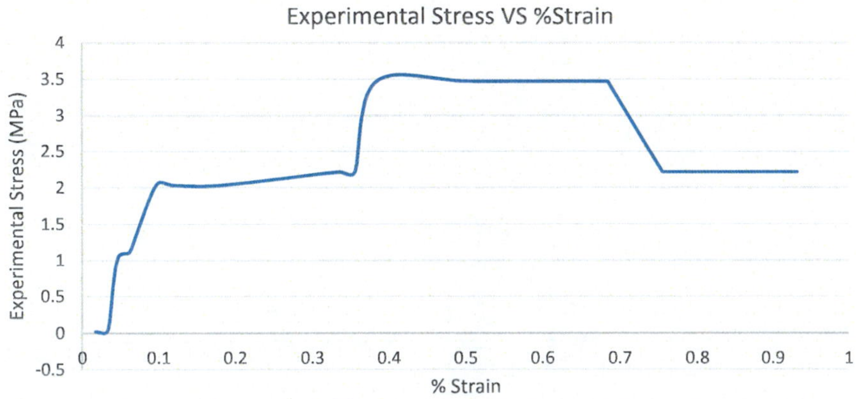

Fig. 8 Experimental stress versus %strain graph for wax + 8.54% S.A + 0.85%C.B + 4% Araldite + 0.85 epichlorohydrin + 0.85 C.B

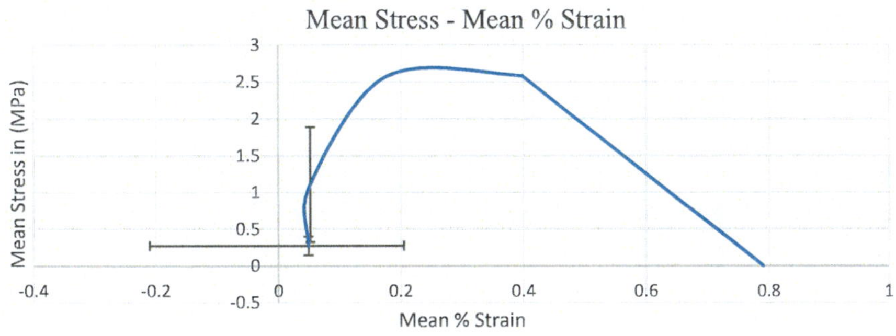

Fig. 9 Stress versus %strain graph for wax + 8.54% S.A + 0.85%C.B + 4%Araldite + 0.85 epichlorohydrin + 0.85 C.B

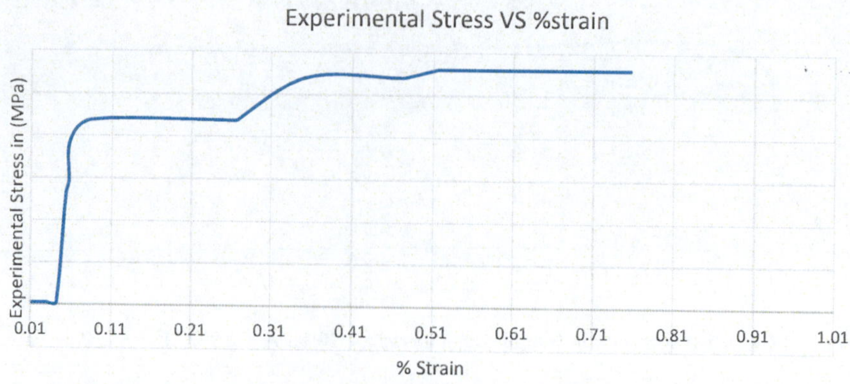

Fig. 10 Experimental stress versus %strain graph for wax + 0.93%C.B + 4.67% Araldite + 0.93% epichlorohydrin

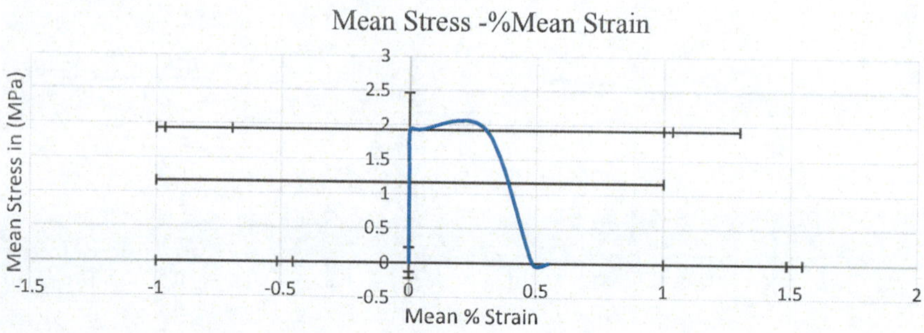

Fig. 11 Mean stress versus Mean %strain graph for wax + 0.93%C.B + 4.67% Araldite + 0.93%epichlorohydrin

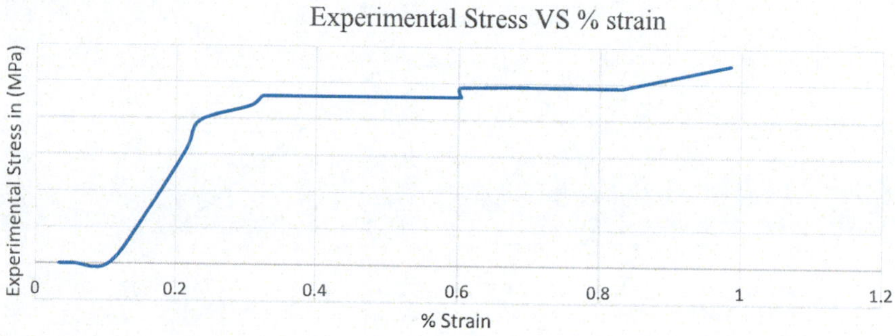

Fig. 12 Stress versus %strain graph for wax + 8% S.A + 4% Araldite + 0.81% epichlorohydrin + 1%C.B + 4% LDPE

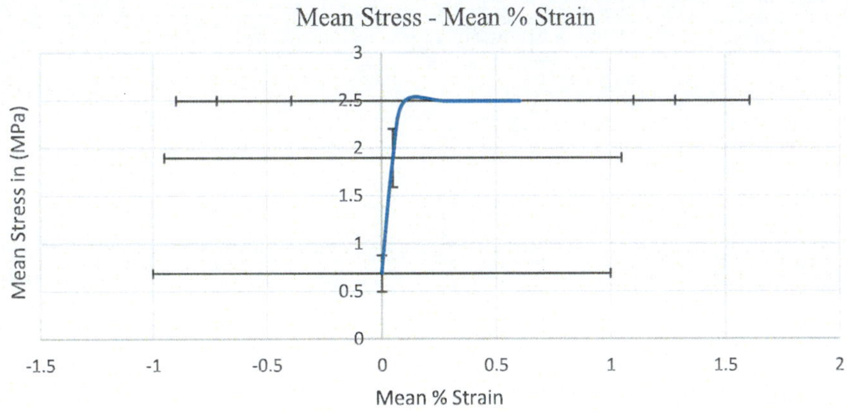

Fig. 13 Stress versus %strain graph for wax + 8% S.A + 4% Araldite + 0.81% epichlorohydrin + 1%C.B + 4% LDPE

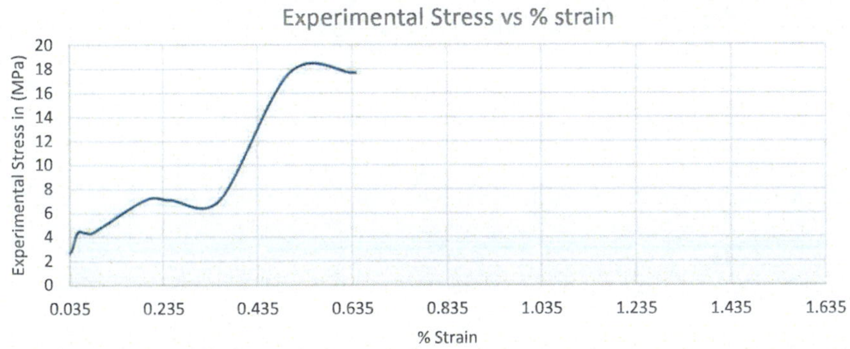

Fig. 14 Stress versus %strain graph for wax + 8% S.A + 4% Araldite + 0.81%epichlorohydrin + 1%C.B + 4% EVA

4. The tensile strength of the wax with additives was double in value than the wax used for this study.
5. Young's modulus of the wax with additives was increased by 116.62 MPa in value.
6. By adding LDPE, EVA the total elongation and uniform elongation have improved by five times and three times, respectively. This will indicate that the brittle nature of wax was decreased by adding additives. It will reduce the tendency of bending, necking of the wax significantly.

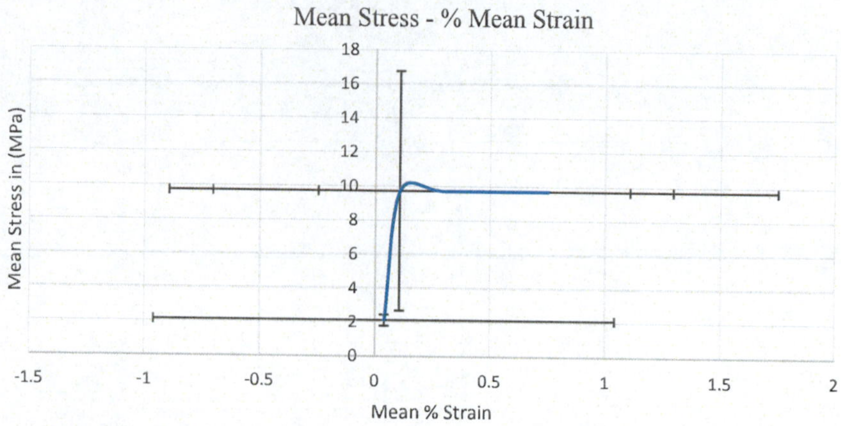

Fig. 15 Stress versus %strain graph for wax + 8% S.A + 4% Araldite + 0.81%epichlorohydrin + 1%C.B + 4% EVA

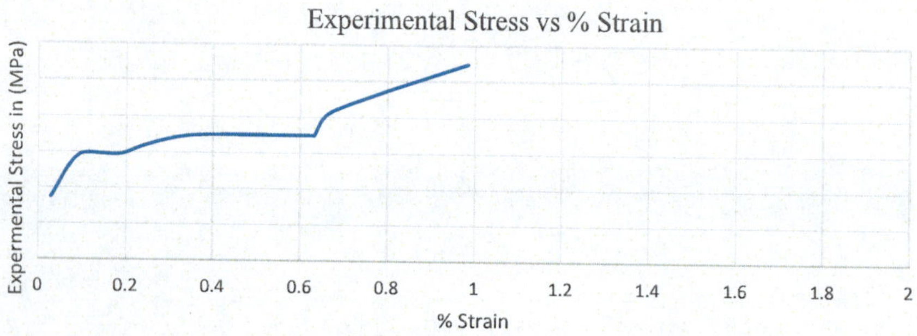

Fig. 16 Stress versus %strain graph for wax + 7.83% S.A + 3.91% Araldite + 0.78% epichlorohydrin + 1%C.B + 4% EVA + 4% LDPE

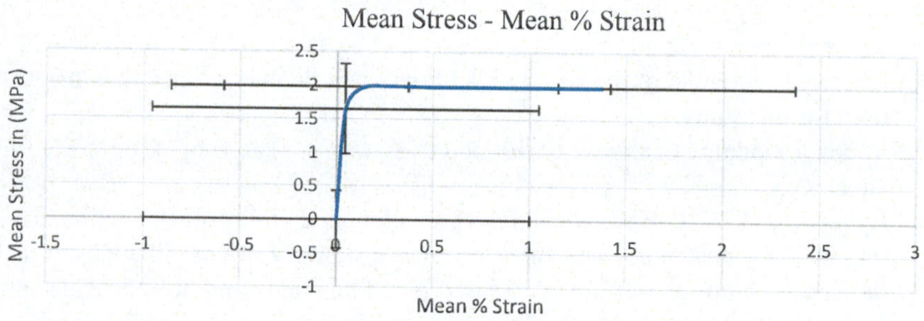

Fig. 17 Mean stress versus mean %strain graph for wax + 7.83% S.A + 3.91% Araldite + 0.78% epichlorohydrin + 1%C.B + 4% EVA + 4% LDPE

References

1. Bartel HR, Rannie WD (1964) Solid Fuel Combustion as applied to Ramjets, Progress report. Jet propulsion Laboratory, California Institute of Technology, pp 3–12
2. Penner SS (1962) Chemical rocket propulsion and combustion research. Gordon and Breach, New York, pp 145–153
3. Green L Jr (1964) Introductory Consideration on Hybrid Rocket Combustion, vol. 15. Progress in Aeronautics and Astronautics, AIAA, New York, pp 451–484
4. Houser TJ, Peck MV (1964) Research in hybrid combustion, vol 15. Progress in Aeronautics and Astronautics, AIAA, New York, pp 559–581
5. Karabeyoglu MA, Cantwell BJ, Altman D (2001) Development and testing of paraffin-hybrid rocket fuels. AIAA, pp 2001-4503 paper
6. Marxman G, Gilbert M (1963) Turbulent boundary layer combustion in hybrid rocket. In: Ninth symposium (international) on combustion, Academic Press New York, pp 371–372
7. Turbulent Boundary Layer and the Role of Chemical Kinetics, Hydrogen Energy Progress IV, Pergammon Press, 1982, pp 961–979
8. Estey PN, Altman D, McFarlane JS (1991) An evaluation of scaling effects for hybrid rocket motors. In: Proceedings of the 27th joint propulsion conference (AIAA/SAE/ASME'91), Sacramento, Calif, USA, June 1991, AIAA Papcr 91-2517
9. Sarigul-Klijn N et al (2004) Air Launching Earth-to-orbit vehicles: delta V gains from launch conditions and vehicle aerodynamics. AIAA Paper 2004-872, Jan 2004
10. Jacob EJ (2009) Nonlinear combustion instability generalized framework. Ph.D. Dissertation, Department of Aerospace Engineering, The University of Tennessee Knoxville, TN
11. Lee C, Lee JW, Byun DY (2003) Transient analysis of hybrid rocket combustion by the Zeldovich-Novozhilov method. KSME Int J 17(10):1572–1582
12. Karabeyoglu MA (2007) Thermal transients in hybrid rocket fuel grains - nonlinear effects. In: 43rd AIAA/ASME/SAE/ASEE joint propulsion conference and exhibit, 8–11 July 2007, Cincinnati, OH
13. Wilson EW, Keller BD, Haknes KM, Smeal CS, Easterly MS (2006) Ultraviolet-visible spectrometry characterization of combustion in hybrid rocket motor. In: 42nd AIEE/ASME/SAE/ASEE joint propulsion conference & exhibit, July 2006, pp 9–12
14. Moutet A, Barrere M (1960) Les fuses a lithergol on fuses hybrids. Rech Aeron 75:23–25
15. Karabeyoglu MA, Cantwell BJ, Altman D (2001) Development and testing of paraffin-based hybrid rocketfuels. In: Proceedings of the 37th AIAA/ASME/SAE/ASEE joint propulsion conference and exhibit, July 2001, AIAA Paper 2001–4503
16. Green L Jr (1964) Introductory consideration on hybrid rocket combustion, Progress in Astronautics and Aeronautics, vol 15, Heterogeneous Combustion, p 45 1, Academic Press
17. Fineman S (1962) Some analytical consideration of the hybrid combustion problem. MS. Thesis, Princeton University
18. George P, Krishnan S, Varkey PM, Ravindran M, Lalitha R (2001) Fuel regression rate in hydroxyl-terminated polybutadiene/gaseous-oxygen hybrid rocket motors. J Propul Power 17(1):35–42
19. DeSain JD, Brady BB, Metzler KM, Curtiss TJ. Albright TV Tensile tests of paraffin wax for hybrid rocket fuel grains, AIAA Paper 2009–5115
20. Maruyama S, Ishiguro T, Shinohara K, Nakagawa I Study on mechanical characteristic of paraffin-based fuel, AIAA Paper 2011–5678
21. Nandy J (2012) Slide Share Published 2012 too. https://www.slideshare.net/nandy_chutki/development-of-paraffin-based-fuel-for-hybrid-rocket-motor
22. Emmons HW (1953) Film combustion of a liquid fuel, Combustion Aerodynamics Lab., Interim Technical Rep. 6, Harvard University, Cambridge, MA
23. Karabeyoglu MA, Zillac G, Cantwell BJ, De Zilwa SRN, Castelucci P (2004) Scale-up tests of high regression rate paraffin-based hybrid rocket fuels. J Propuls Power, 20(6):1037–1045
24. Smoot LD, Price CF (1965) Regression rate mechanism of non-metalized hybrid fuel system. AIAA J 3(8):1408–1413

25. Ramhalli K, Stickler DB Combustion of hydrogen-nitrogen mixtures in
26. Maruyama S, Ishiguro T, Shinohara K, Nakagawa I (2011) Study on the mechanical characteristics of paraffin-based fuel. In: 8th international conference on flow dynamics, Sendai, Japan, 9–11 Nov 2011
27. Novozhilov BV (1992) Theory of nonsteady burning and combustion instability of solid propellants by the Zeldovich- Novozhilov method. In Nonsteady Burning and Combustion Stability of Solid Propellants, L De Luca, Price EW, and Summerfield M, (Eds.). Progr Astronautics and Aeronaut 143:pp. 601–642, American Institute of Aeronautics and Astronautics, Washington, DC, USA
28. Moore GE, Berman K (1965) Solid-liquid rocket propellant system. Jet Propul, Nov. 1965, pp 965–968
29. Dyer J, Doran E, Dunn Z, and Lohner K (2007) Design and development of a 100 km nitrous oxide/paran hybrid rocket vehicle, Joint Propul Conf Exhibit, July 2007
30. Larson DB et al (2011) Characterization of the performance of paraffin/LiAlH4 solid fuels in a hybrid rocket system, AIAA 2011–5822, 47th AIAA/ASME/SAE/ASEE Joint Propul Conf Exhibit, San Diego, CA, July 31 – August 3, 2011

Fracture Toughness Enhancement of Boron Nitride Nanosheets via Crack Edge Passivation Using Various Radicals

Bharat Bhushan Sharma and Avinash Parashar

1 Introduction

Due to their outstanding properties, BNNS have attracted a lot of attention from researchers around the world [1, 2]. Analogous to graphene, BNNS are also having sp^2 hybridized hexagonal lattice with an equal number of boron and nitrogen atoms [3, 4]. This sp^2 hybridized hexagonal structure provides BNNS outstanding thermal and mechanical properties [5, 6]. Apart from these outstanding properties, BNNS also have some distinctive properties, e.g., electrical insulation [7], thermal stability [8], chemical inertness [9] and biological compatibility [10] as compared to graphene due to their slight ionic bonds. The combination of these outstanding and distinctive properties makes BNNS as a potential material in various applications, e.g., electronic packaging [11], nanocomposite [12], biomedical [13], nano-electro-mechanical system (NEMS) [14], water desalination and nano-fluidic [15, 16]. Further, these unique and outstanding properties have also greatly modified by chemical functionalization for improved dispersion properties in nanocomposites [17, 18].

During the last few years, various researchers are trying to modify the properties of 2D nanomaterials such as graphene and BNNS by using chemical functionalization techniques to improve their dispersion properties in nanocomposites [17, 18] and further extending their applications in the field of electronic industries [19] and drug delivery [20]. Several articles had already been published on BNNS functionalized by different radicals such as hydrogen [21], fluorine [19], hydroxyl [20], CHO [19] and CH_3 [19] and reported that functionalization modulates the properties of nanosheets. Hydrogenated and hydroxyl functionalized BNNS had already been

B. B. Sharma · A. Parashar (✉)
Department of Mechanical and Industrial Engineering, Indian Institute of Technology Roorkee, Roorkee 247667, India
e-mail: avinash.parashar@me.iitr.ac.in

V. H. Saran and R. K. Misra (eds.), *Advances in Systems Engineering*, Lecture Notes in Mechanical Engineering,
https://doi.org/10.1007/978-981-15-8025-3_12

explored by the researchers using experimental- and computational-based techniques for tailoring selective materials properties [17–23]. In 2012, a chemical functionalization technique for OH exfoliated BNNS had been proposed by Sainsbury et al. [18] using solution-phase oxygen radical functionalization. In 2016, Kumar et al. [21] applied classical mechanics-based simulations to predict the mechanical behaviour of hydrogen functionalized BNNS. Apart from these fully functionalized BNNS, Huang et al. [22] also synthesized the edge-hydroxylated BNNS by a steam treatment method and investigated that edge-hydroxylated BNNS as a productive additive to enhance the thermal conductivity of hydrogels. Kumar et al. [23] also employed MD-based simulations to predict the effect of crack edge passivation on the fracture toughness of BNNS.

Based on the literatures that are available on OH and H functionalized BNNS, a lot of research has already done on synthesizing techniques, dispersion properties in nanocomposites, stability of functionalized structures, mechanical properties and thermal properties. Despite a lot of research, the literature is almost silent on the effect of crack edge passivation on the fracture properties of BNNS. This research paper explains the effects of crack edge passivation on the fracture properties of BNNS. In this paper, MD-based simulations have been employed in conjunction with ReaxFF [24], to predict the fracture behaviour of crack edge passivated BNNS.

2 Computational Details

MD-based simulations were employed using LAMMPS, whereas open visualization tool (OVITO) has been utilized for the post-processing of dump file. In this article, more accurate, but computationally intensive ReaxFF interatomic potential [24] was used in order to capture the interactions between atoms of crack edge passivated BNNS. For simulating the interaction between atoms, ReaxFF parameters given by Paupitz et al. [25] were utilized. In ReaxFF potential, total energy is the summation of various partial energies and expressed in Eq. (1).

$$\begin{aligned} E_{\text{system}} = E_{\text{bond}} + E_{\text{over}} + E_{\text{under}} + E_{\text{lp}} + E_{\text{val}} + E_{\text{pen}} + E_{\text{tors}} \\ + E_{\text{conj}} + E_{\text{vdWaals}} + E_{\text{Columb}} \end{aligned} \quad (1)$$

where E_{bond} = bond order energy; E_{over} and E_{under} are the energies because of over- and under-coordination of the atoms; E_{lp} = lone pairs energy on N atoms; E_{val} = valence angle distortion energy; E_{pen} = penalty energy; E_{tors} = energy related with the torsion angle; E_{conj} = conjugation energy; and E_{vdWaals} and E_{Columb} are the potential of energies for van der Waals and Coulomb interactions.

In this research article, size of the BNNS was taken as 190 Å × 190 Å along zigzag and armchair directions, respectively. In-plane periodic boundary conditions were applied to avoid edge effects, while, along the thickness, boundary of simulation box

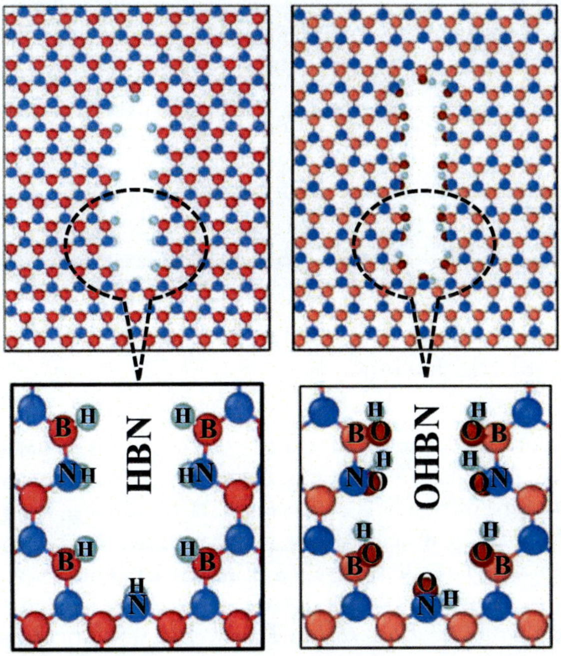

Fig. 1 Representation of various structures of crack edge functionalization in BNNS

was kept fixed. Herein, all simulations were employed at a very low temperature of 1 K to avoid the thermal disturbances. In order to predict the effect of crack edge functionalization, fully passivated configurations (H and OH radicals were attached to both *sp* hybridized N and B atoms) were considered; (i) HBN (hydrogenated crack edge passivation) and (ii) OHBN (hydroxyl crack edge passivation) as shown in Fig. 1.

After generating the atomistic structures of crack edge passivated BNNS in the simulation box, the atomic configurations were equilibrated using NPT ensemble for 20 ps with an integrating time step of 0.25 fs. To analyse the tensile behaviour of passivated and non-passivated BNNS, uniaxial tensile strain was applied on the equilibrated structures at a strain rate of 0.001 ps^{-1} with NVT ensemble. The coordinates of atoms were updated using velocity-Verlet integration scheme with an integration time step of 0.25 fs. The atomic virial stresses were calculated with the help of Eq. (2).

$$\sigma_{ij}^{\alpha} = \frac{1}{\varphi}\left(\frac{1}{2}m^{\alpha}v_i^{\alpha}v_j^{\alpha} + \sum_{\beta=1,n} r_{\alpha\beta}^{j} f_{\alpha\beta}^{i}\right) \tag{2}$$

where *i* and *j* are indices of the Cartesian coordinates system, α and β are the atomic indices, m_α and v_α = mass and velocity of atom α, $r_{\alpha\beta}$ = distance between atoms α and β, and φ = atomistic volume of BNNS. The mode-I critical stress intensity factor or fracture toughness (K_{IC}) was computed using Eq. (3) by utilizing the stress value at the fracture point (σ_f).

$$K_{IC} = Y\sigma_f\sqrt{\pi a} \tag{3}$$

where $Y =$ dimensionless geometrical parameter and its value is 1 due to larger size BNNS, $a =$ half of the crack length. Length of the BNNS was taken as 10 times the crack length of $2a = 19$Å to avoid the crack length effects on the fracture toughness.

3 Results and Discussion

Initially, the accuracy of ReaxFF parameters was validated by calculating the K_{IC} of pristine BNNS using Eq. (3). The snapshot of the crack configurations along the armchair and zigzag directions is shown in Fig. 2. Here, armchair and zigzag direction refer the loading direction which was always 90° to the crack edges to satisfy mode-I loading requirement. The estimated values of K_{IC} in armchair and zigzag directions are 3.22 MPa$\sqrt{m}$ and 3.05 MPa$\sqrt{m}$, respectively. These estimated values of K_{IC} are in agreement with the values reported by Wei et al. [26]. After estimating the K_{IC} of pristine BNNS, simulations were employed to predict the effect of H and OH crack edge passivation on the fracture toughness of BNNS. Stress–strain responses are plotted in Fig. 3 along the direction of loading for crack edge passivated BNNS. It can be seen from Fig. 3 that the fracture stresses are significantly decrease in AC loading direction due to hydroxyl crack edge passivation, whereas it increases in ZZ loading direction.

Further, the fracture toughness of these passivated configurations is determined using Eq. 3 by utilizing the value of fracture stresses, which are given in Table 1. It can

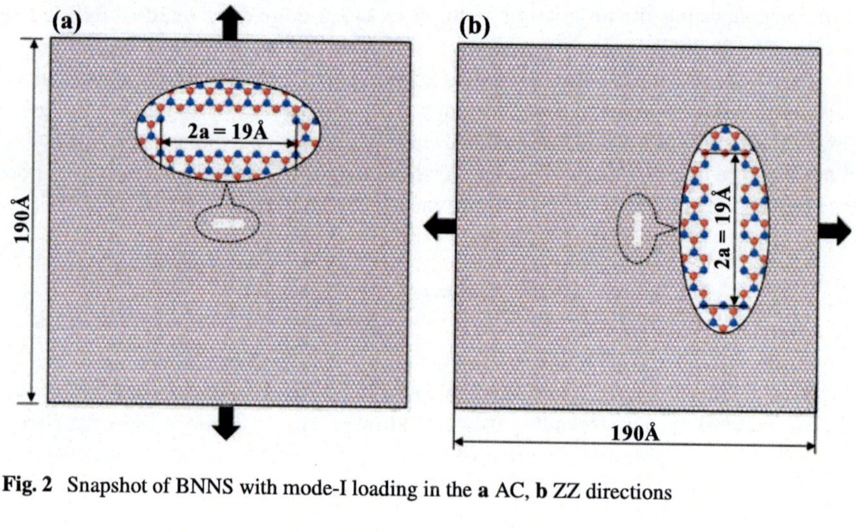

Fig. 2 Snapshot of BNNS with mode-I loading in the **a** AC, **b** ZZ directions

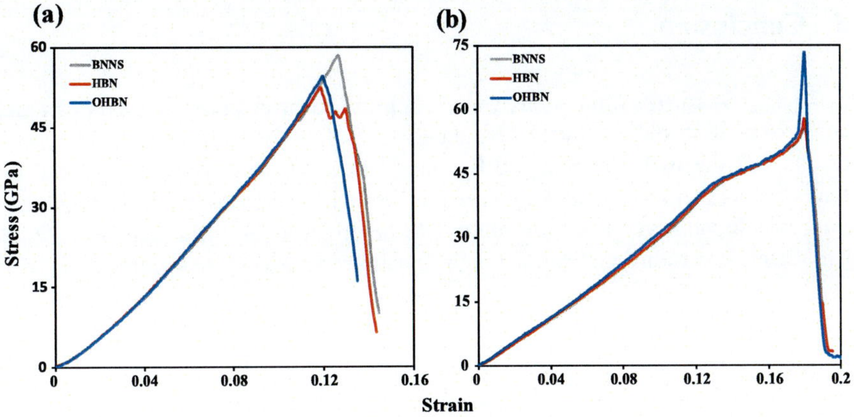

Fig. 3 Stress–strain response for various configurations of passivated BNNS in the **a** armchair, **b** zigzag loading directions

Table 1 Fracture toughness of various BNNS configurations in AC and ZZ loading directions

Directions	K_{IC} (MPa$\sqrt{m}$)		
	BNNS	HBN	OHBN
AC	3.22	2.87	2.95
ZZ	3.05	3.16	4.01

be deduced from Table 1 that the H and OH crack edge passivation adversely affected the fracture toughness of BNNS in armchair loading direction, whereas it is positively affected in zigzag loading direction. In all BNNS configurations, OHBN have maximum enhancement in ZZ loading direction, whereas it has minimum decrement in AC loading direction.

Theses contrasting trends of fracture toughness along AC and ZZ loading directions are because of different responses of edge atoms. In armchair loading direction, the upper edge of the crack has only *sp* hybridized B atoms, while the lower edge has only *sp* hybridized N atoms. In zigzag loading direction, both left and right edges of the crack have *sp* hybridized B and N atoms alternatively which is shown in Fig. 2. The hydroxyl functionalization was performed opposite to each other on both the *sp* hybridized B and N atoms; due to this, the crack edges in armchair loading direction pulled opposite to each other, while, in zigzag loading direction, no such type of pulling effect was observed because of alternate boron and nitrogen atoms on the crack edges. This opposite pulled of crack edges in AC direction leads to tearing effect at the crack tip, which ultimately reduced the fracture toughness in armchair direction.

4 Conclusion

MD-based atomistic simulations were employed to investigate the effect of crack edge passivation on the fracture toughness of BNNS. It can be deduced from the results that the hydrogenated and hydroxyl crack edge passivation improved the fracture toughness in ZZ loading direction, while its decreases in AC loading direction. Results observed from this study will be utilizing the full capacity of BNNS nanosheets in various applications such as nano-membrane for ion separation.

References

1. Sharma BB, Parashar A (2019) Mechanical and fracture behaviour of hydroxyl functionalized h-BN nanosheets. J Mater Sci 55:3228–3242
2. Verma PK, Sharma BB, Chaurasia A, Parashar A (2020) Inter-granular fracture toughness of bi-crystalline graphene nanosheets. Diam Relat Mater 102:107667
3. Sharma SS, Sharma BB, Parashar A (2019) Mechanical and fracture behavior of water submerged graphene. J Appl Phys 125:215107
4. Sharma SS, Sharma BB, Parashar A (2019) Defect formation dynamics in dry and water submerged graphene nanosheets. Mater Res Express 6:075063
5. Mortazavi B, Pereira LFC, Jiang JW, Rabczuk T (2015) Modelling heat conduction in polycrystalline hexagonal boron-nitride films. Sci Rep 5:1–11
6. Sharma BB, Parashar A (2019) A review on thermo-mechanical properties of bi-crystalline and polycrystalline 2D nanomaterials. Crit Rev Solid State Mater Sci 45:134–170
7. Watanabe K, Taniguchi T, Kanda H (2004) Direct-bandgap properties and evidence for ultraviolet lasing of hexagonal boron nitride single crystal. Nat Mater 3:404
8. Sharma BB, Parashar A (2019) Atomistic simulations to study the effect of grain boundaries and hydrogen functionalization on the fracture toughness of bi-crystalline h-BN nanosheets. Phys Chem Chem Phys 21:13116–13125
9. Zhi C, Bando Y, Tang C, Golberg D (2010) Boron nitride nanotubes. Mater Sci Eng R Rep 70:92–111
10. Chen X, Wu P, Rousseas M, Okawa D, Gartner Z (2009) Boron nitride nanotubes are noncytotoxic and can be functionalized for interaction with proteins and cells. J Am Chem Soc 131:890–891
11. Lin Z, Liu Y, Raghavan S, Moon KS, Sitaraman SK, Wong CP (2013) Magnetic alignment of hexagonal boron nitride platelets in polymer matrix: toward high performance anisotropic polymer composites for electronic encapsulation. ACS Appl Mater Interfaces 5:7633–7640
12. Chaurasia A, Verma A, Parashar A, Mulik RS (2019) Experimental and computational studies to analyze the effect of h-BN nanosheets on mechanical behavior of h-BN/polyethylene nanocomposites. J Phys Chem C 123:20059–20070
13. Roosta S, Nikkhah SJ, Sabzali M, Hashemianzadeh SM (2016) Molecular dynamics simulation study of boron-nitride nanotubes as a drug carrier: from encapsulation to releasing. RSC Adv 6:9344–9351
14. Ghorbanzadeh Ahangari M (2015) Modeling of the interaction between polypropylene and monolayer sheets: a quantum mechanical study. RSC Adv 5:80779–80785
15. Sharma BB, Parashar A (2020) Mechanical strength of a nanoporous bicrystalline h-BN nanomembrane in a water submerged state. Phys Chem Chem Phys DOI: 10.1039/D0CP03235B
16. Sharma BB, Parashar A (2019) Atomistic simulations to study the effect of water molecules on the mechanical behavior of functionalized and non-functionalized boron nitride nanosheets. Comput Mater Sci 169:109092

17. Kong D, Zhang D, Guo H, Zhao J, Wang Z, Hu H, Xu J, Fu C (2019) Functionalized boron nitride nanosheets/poly(l-lactide) nanocomposites and their crystallization behavior. Polymers (Basel) 11:440
18. Sainsbury T, Satti A, May P, Wang Z, McGovern I, Gun'ko YK, Coleman J (2012) Oxygen radical functionalization of boron nitride nanosheets. J Am Chem Soc 134:18758–18771
19. Bhattacharya A, Bhattacharya S, Das GP (2012) Band gap engineering by functionalization of BN sheet. Phys Rev B—Condens Matter Mater Phys. 85:1–9
20. Weng Q, Wang B, Wang X, Hanagata N, Li X, Liu D, Wang X, Jiang X, Bando Y and Golberg D (2014) Highly water-soluble, porous, and biocompatible boron nitrides for anticancer drug delivery. ACS Nano 8:6123–30
21. Kumar R, Mertiny P, Parashar A (2016) Effects of different hydrogenation regimes on mechanical properties of h-BN: a reactive force field study. J Phys Chem C 120:21932–21938
22. Xiao F, Naficy S, Casillas G, Khan MH, Katkus T, Jiang L, Liu H, Li H, Huang Z (2015) Edge-hydroxylated boron nitride nanosheets as an effective additive to improve the thermal response of hydrogels. Adv Mater 27:7196–7203
23. Kumar R, Parashar A (2017) Fracture toughness enhancement of h-BN monolayers via hydrogen passivation of a crack edge Nanotechnology 28
24. Van Duin ACT, Dasgupta S, Lorant F, Goddard WA (2001) ReaxFF: A reactive force field for hydrocarbons. J Phys Chem A 105:9396–9409
25. Paupitz R, Junkermeier CE, van Duin ACT, Branicio PS (2014) Fullerenes generated from porous structures. Phys Chem Chem Phys 16:25515–25522
26. Wei X, Xiao S, Li F, Tang D, Chen Q, Bando Y (2015) Comparative fracture toughness of multilayer graphenes and boronitrenes. Nano Lett 15:689–694

Measurement of Through-Thickness Residual Stresses Under Restrained Condition in Pressure Vessel Steel Weld

P. K. Taraphdar, M. M. Mahapatra, A. K. Pradhan, P. K. Singh, Kamal Sharma, and Suranjit Kumar

1 Introduction

The critical operating condition of pressure vessels in industries necessitates extra attention in the design and fabrication process. Welding-induced residual stresses severely affect the strength and load-bearing capacity of pressure vessels and other industrial components [1]. Residual stresses present in engineering machinery promote stress corrosion cracking, fatigue crack growth and impacts on fracture mechanics properties of welded joints [2]. Distortion and tensile nature of welding residual stresses affect the quality of any structural integrity under the influence of tensile loading condition during service life [3]. Therefore, detailed information regarding the distribution of residual stresses in welded joints is necessary to assess the sound service life of any component.

All the residual stress measurement techniques have their own limitations and advantages. Such as, the nondestructive techniques can be utilized to measure residual stress magnitudes at several regions of any particular specimen, whereas the destructive techniques involve complete destruction of the specimen which limits the number of possible measurements. Semi-destructive techniques engage a limited amount of material removal to obtain strains out of displacements and accommodate further space for additional measurements. Hole drilling, ring coring and deep hole drilling are well-known semi-destructive residual stress measurement techniques associated with mechanical strain relaxation [4]. Among the semi-destructive techniques, deep hole drilling (DHD) technique is capable of measuring through-thickness residual stress distribution. In the present work, through-thickness residual stress distribution

P. K. Taraphdar (✉) · M. M. Mahapatra · A. K. Pradhan
School of Mechanical Sciences, Indian Institute of Technology, Bhubaneswar, India
e-mail: pkt10@iitbbs.ac.in; p.taraphdar@gmail.com

P. K. Singh · K. Sharma · S. Kumar
Bhabha Atomic Research Centre, Mumbai, India

V. H. Saran and R. K. Misra (eds.), *Advances in Systems Engineering*,
Lecture Notes in Mechanical Engineering,
https://doi.org/10.1007/978-981-15-8025-3_13

was investigated in thick butt-welded pressure vessel steel plate of SA 516 Gr. 70 under the influence of external constraints using both numerical and experimental approaches.

2 Experimental Procedure

The base material of the present study is SA516 Gr. 70 plate of dimension 300 × 115 × 25 mm^3. It was welded using ER70S6 filler material of 1.2 mm diameter by gas metal arc welding process with 1.5 mm of root gap. The edges of the plate normal to welding torch travel path were tack welded with external links to create a constrained environment for the weld. The weld groove geometry and weld bead sequence are shown in Fig. 1. Later on, residual stresses were evaluated in the welded specimen using deep hole drilling technique. During welding, shrinkages after every pass were noted and compared with another constraint-free weld sample to find out the degree of constraint (k_s). According to Masubuchi and Ich, it can be defined as the intensity of reaction stress per unit transverse shrinkage [5]. It is expressed as below,

$$\frac{S_r}{S} = \frac{1}{1 + 0.086 \times k_s^{0.87}} \quad (1)$$

where S_r and S are the transverse shrinkages in restrained and unrestrained joints, respectively.

2.1 Measurement of residual stresses using DHD

Deep hole drilling technique comprises majorly four basic steps, as shown in Fig. 2. It is based upon the mechanical strain relaxation technique under precisely controlled material removal. In the first step, a reference hole was drilled using a CNC milling

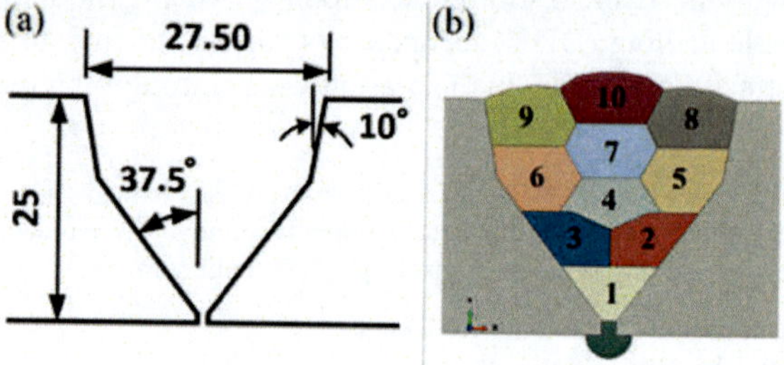

Fig. 1 **a** Weld groove geometry, **b** weld pass sequence

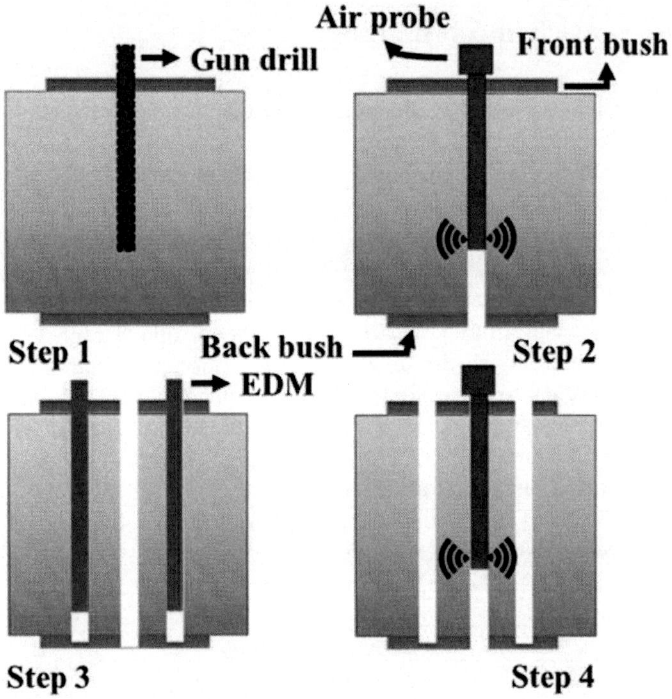

Fig. 2 Steps involved in deep hole drilling technique

machine at the desired measurement location. In the next step, the internal diameter of the drilled hole was measured at certain intervals through the thickness in both the longitudinal and transverse directions. After that, the reference hole was trepanned coaxially using the EDM process. In the last step, the internal diameter of the reference hole was remeasured at the previous locations and directions following predefined depth intervals. The following equations were used to calculate the stress magnitudes,

$$\varepsilon_x = \frac{\Delta \mathrm{d}_x}{\mathrm{d}_x} = -\frac{1}{E}\left[3\sigma_x - \sigma_y\right] \tag{2}$$

$$\varepsilon_y = \frac{\Delta \mathrm{d}_y}{\mathrm{d}_y} = -\frac{1}{E}\left[3\sigma_y - \sigma_x\right] \tag{3}$$

where ε_x and ε_y are the longitudinal and transverse strains, $\Delta \mathrm{d}_x$ and $\Delta \mathrm{d}_y$ are the diametral changes in longitudinal and transverse directions, σ_x and σ_y are the magnitude of longitudinal and transverse stresses, respectively.

3 Thermo-mechanical Analysis

A three-dimensional-coupled thermal elastic–plastic finite element model was developed to simulate the present welding operation with practical dimensions, and welding parameters were taken into consideration as input. Based on ABAQUS code, C3D8T: 8-node thermally coupled brick, trilinear displacement and temperature elements were used in the present analysis. The weld zone was assigned fine meshing grids and gradually coarsening-biased meshing order was defined to the base material in the direction normal to the fusion boundary to reduce total computational time. Element birth and death technique were followed for the deposition of weld filler material using the 'model change' option. The bottom corners and the plate edges normal to the weld axis were restrained in displacement to control the rigid body motion and replicate the clamping effect on the model. The ambient temperature and predefined temperature field of the base plates were defined as 28 °C. Heat loss in terms of radiation for the weld zone and convection for the base plates were taken into account as boundary conditions. A uniform volumetric moving heat source was developed to simulate the welding arc. The heat source was applied by user subroutine FORTRAN code, DFLUX. It can be expressed as below,

$$\text{DFLUX} = VI\eta/\text{Vol}_h \quad (4)$$

where V is voltage, I is current, η is arc efficiency assumed as 0.8 and Vol_h is the volume of the heat source. The travel speed of the moving heat source is controlled by the user subroutine code.

4 Results

The approximate degree of constraint was calculated as 9.78 kg/mm^2/mm using Eq. 1. The shrinkage and distortion were found to be less in constrained plate compared to the constraint-free weld sample. The accuracy of a finite element welding simulation majorly depends upon the obtained thermal profile during the thermal analysis. It should replicate the real-time temperature history experienced by the material during the practical welding operation. Figure 3 shows a cross-section of welding fusion zone where it can be observed that the fusion boundary reached the melting point satisfactorily. The grey region depicts the molten weld pool.

Figure 4 represents the contour plots of induced residual stresses distributed through the cross-section of the weldment. Tensile residual stresses were observed at the centre of the weld zone, and it slowly changed into compressive stresses in the direction normal to the weld axis.

The comparisons of residual stress magnitudes evaluated through DHD technique and numerical modelling are given in Fig. 5. Residual stresses induced initially in

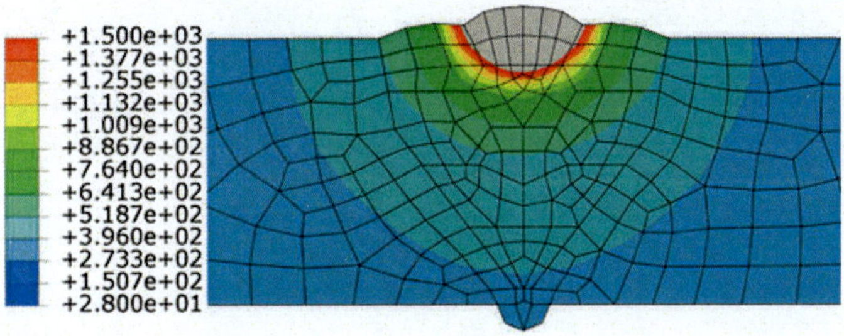

Fig. 3 Temperature distribution

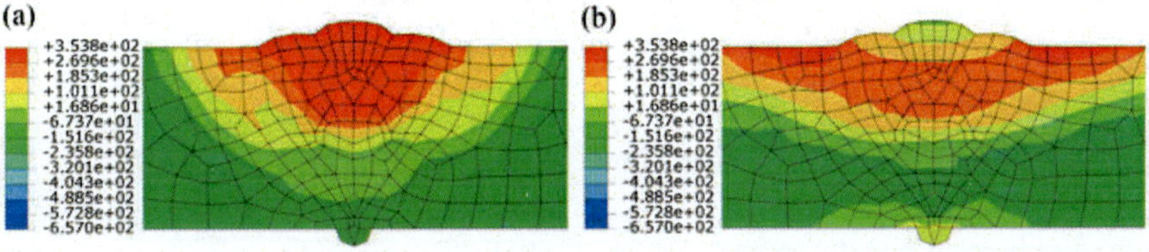

Fig. 4 Stress contour plots **a** Longitudinal, **b** Transverse

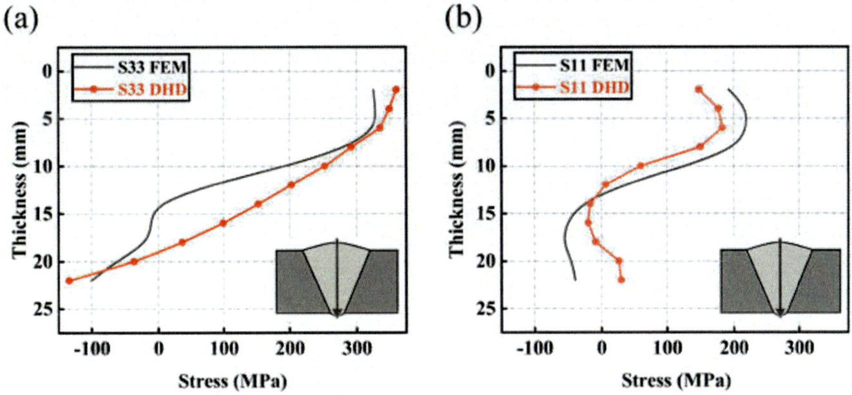

Fig. 5 Residual stress distribution through the thickness **a** Longitudinal, **b** Transverse

the base material due to other manufacturing processes were not taken into consideration during the numerical analysis. This can be the reason behind the differences observed in the comparison of results obtained experimentally and numerically. The longitudinal residual stress (tensile) reached a peak magnitude of 360 MPa at the top surface of the weld zone. The trend of longitudinal residual stress gradually decreased in the depth direction and transformed into compressive stresses. In the transverse direction, a maximum of 182 MPa tensile residual stress was observed at a depth of 6 mm from the top surface. The transverse residual stress magnitude was

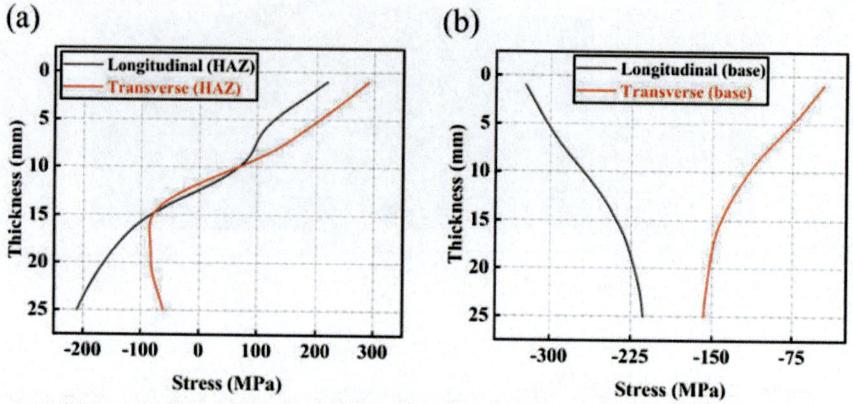

Fig. 6 Residual stress distribution through the thickness **a** HAZ, **b** base material (30 mm away from fusion boundary)

found to be significantly lower compared to longitudinal residual stress at the weld zone.

Figure 6 shows the through-thickness residual stress distribution at the heat-affected zone (HAZ) and base material (30 mm away from the fusion boundary). Unlike the weld centre, the peak residual stress at the HAZ was higher in transverse direction compared to the longitudinal direction. Both the longitudinal and transverse residual stresses present in the base material were found to be compressive through the thickness. Thus, the equilibrium state of residual stresses induced in welded joints is balanced.

5 Conclusion

The maximum residual stress magnitude observed is 84% of the yield strength of the weld material under a restrained condition. The results obtained from the experiments and numerical modelling are in a reasonably close match. The distribution of residual stresses does not remain uniform through the thickness of the specimen. The trend of longitudinal and transverse residual stresses through the thickness is notably different from each other for flat butt joints. The maximum longitudinal residual stress magnitude in a butt-welded joint is significantly higher compared to transverse residual stress at the weld centre.

References

1. Zondi MC (2014) Factors that affect welding-induced residual stress and distortions in pressure vessel steels and their mitigation techniques: a review. J Press Vessel Technol 136(4):040801
2. Jiang W, Gong JM, Woo W, Wang YF, Tu ST (2012) Control of welding residual stress and deformation of the butt welded ultrathick tube-sheet: effect of applied load. J Press Vessel Technol 134(6):061406
3. Javadi Y (2014) Investigation of clamping effect on the welding residual stress and deformation of Monel plates by using the ultrasonic stress measurement and finite element method. J Press Vessel Technol 137(1):011501
4. Giri A, Mahapatra MM, Sharma K, Singh PK (2017) A study on the effect of weld groove designs on residual stresses in SS 304LN thick multipass pipe welds. Int J Steel Struct 17(1):65–75
5. Masubuchi K, Ich NT (1970) Computer analysis of degree of constraint of practical butt joints. Weld J 49(4)

Enhancement in Mechanical Properties of Polyethylene Using h-BN Nanofiller

Ankur Chaurasia, Avinash Parashar, and Rahul S. Mulik

1 Introduction

During the last decades, research community focused on exploring the full capability of 2D nanomaterials (i.e., Graphene and h-BN) as a reinforcement in polymer matrix subjected to high mechanical loading. These 2D nanofiller exhibits a broad scope of thermal and mechanical properties [1, 2]. BNNS is structurally analogous to the monolayer graphene sheet and shows similar mechanical properties [3–5]. BNNS is electrically insulative and highly thermally stable as compared to graphene [6–12]. Because of lightweight and simple fabrication process, high-density polyethylene (HDPE) and other thermoplastics are usually utilized for synthesizing nanocomposite subjected to mechanical loading [13, 14]. Polyethylene molecular chain contains semi-crystalline molecular structures, which help in enhancing the young's moduli and flexibility [15–17]. Polyethylene is commonly used in the fields of defense, aerospace, automobile, and thermal packaging [18–20].

To capture the reinforcement capability of 2D nanofiller on mechanical properties of nanocomposites. Researchers have opted for computational modeling and experimental approach. Rahman and Haque [21] presented the effect of graphene nanoplatelets (GNP) loading, aspect ratios (AR), and dispersion on the mechanical behavior of GNP/epoxy nanocomposites. They reported in their work that the optimum concentration of graphene (ranges 1% to 3 wt%) greatly enhances the mechanical behavior of GNP/epoxy nanocomposites. Lv et al. [22] also predicted the effect of graphene functionalization on the interfacial behavior of graphene/PE nanocomposite. They reported that graphene functionalization significantly enhances the bonding energy and interfacial shear stress of nanocomposite system. Li et al.

A. Chaurasia · A. Parashar (✉) · R. S. Mulik
Department of Mechanical and Industrial Engineering, Indian Institute of Technology Roorkee, Roorkee 247667, India
e-mail: avinash.parashar@me.iitr.ac.in

V. H. Saran and R. K. Misra (eds.), *Advances in Systems Engineering*, Lecture Notes in Mechanical Engineering,
https://doi.org/10.1007/978-981-15-8025-3_14

[23] studied the effect of defective graphene (Stone–Wales,single vacancy, double vacancy) on interfacial and thermal properties of the graphene/epoxy nanocomposites systems. They revealed that Stone–Wales defects significantly enhance the interfacial shear strength and thermal conductivity, whereas single vacancy and double vacancy degrade the interfacial properties of the nanocomposites system. Ramanathan et al. [24] experimentally validated the enhancement in mechanical and thermal properties of poly (methyl methacrylate) nanocomposites reinforced with fictionalized graphene sheet. Nasrabadi and Foroutan [25] studied the interaction energy associated with polystyrene (PS), poly [m-phenylenevinylene-co-(2,5 dioctyloxy-p-phenylenevinylene)] (PmPV) and polythiophene (PT) polymer reinforced with boron nitride nanotubes (BNNTs). Sainsbury et al. [26] experimentally reported that 0.1 wt% hydroxyl functionalized BNNSs can achieve a 186% increase in the elastic modulus of a BNNS/polyvinyl-alcohol nanocomposite as to 20% when reinforced with pristine BNNSs. Lee et al. [27] revealed that adding functionalized boron nitride nanoplatelets (OH-BNNP) in PE matrix enhances the young's modulus and yield strength by 127% and 69%, respectively.

The last couple of years, BNNS is emerging as a potential candidate to replace graphene and CNTs in developing future nanocomposites [18]. In present study, MD-based simulations were performed to investigate the reinforcing capabilities of BNNS for PE-based nanocomposites.

2 Modelling Details

MD-based simulations were performed in open-source code Large-scale Atomic/Molecular Massively Parallel Simulator (LAMMPS), whereas the postprocessing of the dump files was performed with the help of open visualization tool (OVITO). So far, researchers have utilized different types of interatomic potential Tersoff, dreiding, universal force field, and reactive force field for simulating the mechanical behavior of h-BN nanosheets. Among all these interatomic potentials, reactive force field (ReaxFF) is considered as more accurate. Energy breakup of the reaction force field outlined by Eq. 1;

$$E_{\text{system}} = E_{\text{angle}} + E_{\text{tors}} + E_{\text{bond}} + E_{\text{over}} + E_{\text{vdwalll}} + E_{\text{coulomb}} + E_{\text{specific}} \quad (1)$$

E_{over} = Penalty energy ; E_{bond} = Bond formation energy; E_{angle} and E_{tors} = Three- and four-body torsional angle strain energy, respectively. E_{vdwalls} and E_{coulomb} = Van der Walls and coulombic interaction energy, respectively; E_{specific}= Energy due to bond correction, lone pair, and hydrogen binding.

However, non-bonded interaction between BNNS and PE was modelled with *L–J* potentials [28]. In order to simulate uniaxial tensile stress–strain response of nanocomposites that constitutes HDPE (approx. 0.89–0.93 g/cm^3) reinforced with BNNS, a cubical box size of 70 Å × 70 Å × 70 Å was used to pack polyethylene

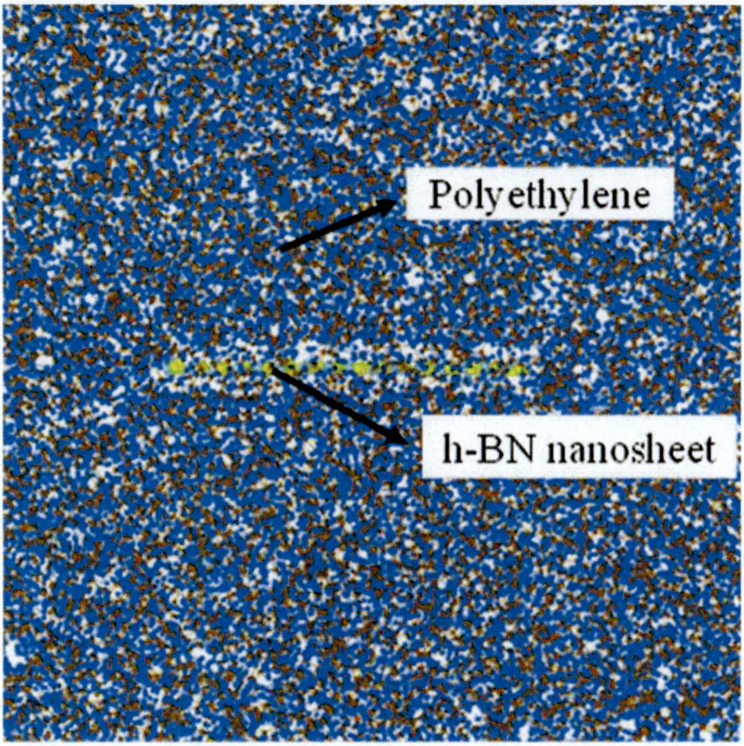

Fig.1 Atomistic model of h-BN/PE nanocomposite

linear chains. In the simulation box, rectangular BNNS of dimension 35 × 35 Å was incorporated at the middle of polyethylene matrix as shown in Fig. 1.

In this study, rectangular BNNS was considered for the study. In order to capture reinforcing effect of BNNS, authors have performed the simulation for uniaxial tensile testing of BNNS and BNNS/PE nanocomposites. During all the simulations with BNNS, the weight concentration of BNNS was kept constant at ~3.0%. In order to avoid the surface effects, periodic boundary conditions were applied in all the three mutual principal directions.

All the simulations were carried out at 100 K (below glass liquid transition temperature of polyethylene) [16]. After assigning the atomic coordinates for rectangular hexagonal boron nitride nanosheet and PE molecular chains in the simulation box, energy of the system was relaxed using conjugate gradient algorithm. After obtaining the minimum energy of the BNNS/PE atomic configuration, relaxation of the box was performed in multiple stages. During the first stage, BNNS/PE system was equilibrated at 500 K under the effect of NVE ensemble for 10^5 timesteps, whereas the temperature of the simulation box was maintained during the simulation with the help of Langevin thermostat [18]. Furthermore, the configuration was again relaxed for 40 picoseconds under the effect of NPT ensemble. In the next phase, system temperature was bought down to 100 K using isothermal–isobaric NPT ensemble, and the system was maintained at the same temperature for 25 picoseconds. With the constant timestep of 0.15 femtoseconds after equilibrating the atomic configuration, tensile deformation was applied in the desired direction at strain rates $10^9\ s^{-1}$. In all the simulations, stress was estimated with the help of virial stress [29–31] Eq. 2;

$$\sigma_{ij}^{\alpha} = \frac{1}{\varphi^{\alpha}}\left(\frac{1}{2}m^{\alpha}v_i^{\alpha}v_j^{\alpha} + \sum_{\beta=1,n} r_{\alpha\beta}^{j} f_{\alpha\beta}^{i}\right) \quad (2)$$

where $r_{\alpha\beta}$ = Distance between the atoms α and β; v^{α} and m^{α} = Velocity and mass of atom α,, respectively; φ^{α} = Volume of system; α and β = Atomic indices; i and j = Indices in cartesian coordinate system.

3 Results and Discussion

In this article, initial simulations were carried out to validate the fidelity of reaction force field (ReaxFF) potential parameters provided by Han et al. [30] to capture the stress–strain response of pure polyethylene, BNNS, and BNNS/PE nanocomposite. The uniaxial tensile response of neat PE and a pristine sheet of BNNS were calculated using ReaxFF potentials. It was predicted from the results that fracture stress for pure PE (223.30 MPa), BNNS (AC-133.4 GPa, ZZ-117.58 GPa) is in line with the previous literature.

3.1 *Effect of h-BN Nanosheets (BNNS) on Stress–Strain Response of Polyethylene*

In this section, the authors have performed the simulations for BNNS-reinforced PE subjected to tensile loading. Pure polyethylene reinforced with rectangular BNNS (3wt%) was simulated and stress–strain response was plotted in Fig. 2.

It was predicted from the results that, fracture stress of BNNS/PE increased by ~22% as compared to that of pure polyethylene. It can also be observed from Fig. 2 that reinforcing BNNS improves the energy absorption capability of the PE matrix. Enhancement in mechanical properties is due to a stable interface between BNNS and PE matrix. The interface plays a vital role in nanocomposites; strong interfacial allows higher stress transfer capabilities. In the present study, electrostatic interaction, van der walls, and π–π interactions also account for the load transfer between PE matrix and BNNS (in the interface region). In addition to this, strong interface corresponds to higher interaction energy between BNNS and PE matrix. Moreover, as the interaction energy increases, which eventually improves the non-bonded interaction between the

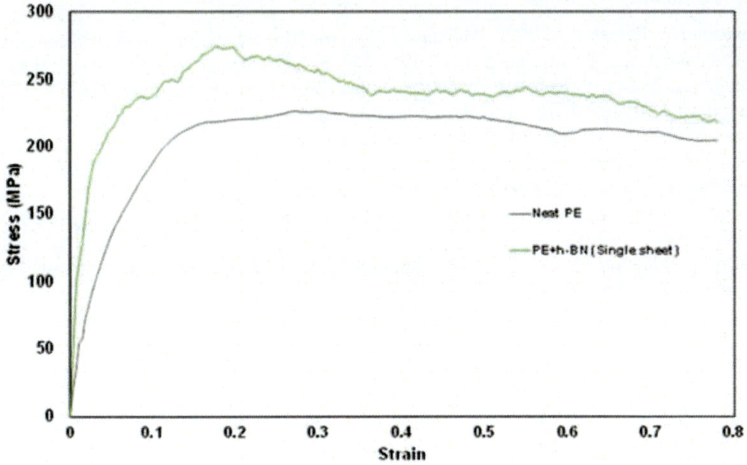

Fig. 2 Stress–strain response of neat PE and h-BN/PE nanocomposites

BNNS and PE matrix and resulting the overall improvement in tensile strength of the BNNS/PE nanocomposite.

4 Conclusion

It can be concluded from the simulations that reinforcing PE with BNNS nanofiller has significant improvement in the mechanical strength PE matrix. It was predicted from the MD simulations that tensile behavior of pure PE can be improved by 22% by reinforcing with 3wt% of BNNS. Additionally, simulations have also predicted the overall improvement in toughness PE matrix reinforced with h-BN nanosheets. It can be concluded that BNNS exhibit the superior reinforcing capabilities in PE matrix.

References

1. Li X et al (2013) Exfoliation of hexagonal boron nitride by molten hydroxides. Adv Mater 25(15):2200–2204. https://doi.org/10.1002/adma.201204031
2. Sharma BB, Parashar A (2019)A review on thermo-mechanical properties of bi-crystalline and polycrystalline 2D nanomaterialsCrit Rev Solid State Mater Sci 0(0):1–37 https://doi.org/10.1080/10408436.2019.1582003
3. Sharma SS, Sharma BB, Parashar (2019) Mechanical and fracture behavior of water submerged grapheme. J Appl Phys 125(21). https://doi.org/10.1063/1.5088884
4. Sharma SS, Sharma BB, Parashar A (2019) Defect formation dynamics in dry and water submerged graphene nanosheets. Mater. Res. Express 6(7). https://doi.org/10.1088/2053-1591/ab19fc

5. Sharma BB, Parashar A (2020) Mechanical and fracture behaviour of hydroxyl functionalized h-BN nanosheets. J Mater Sci 55(8):3228–3242. https://doi.org/10.1007/s10853-019-04163-7
6. Zunger A, Katzir A, Halperin A (1976) Optical properties of hexagonal boron nitride. Phys Rev B 13(12):5560–5573. https://doi.org/10.1103/PhysRevB.13.5560
7. Baowan D, Cox BJ, Hill JM (2008) Junctions between a boron nitride nanotube and a boron nitride sheet. Nanotechnology 19(7). https://doi.org/10.1088/0957-4484/19/7/075704.
8. Slotman GJ, Fasolino A (2013) Structure, stability and defects of single layer hexagonal BN in comparison to grapheme. J Phys Condens Matter 25(4). https://doi.org/10.1088/0953-8984/25/4/045009
9. Meng J, Tajaddod N, Cranford SW, Minus ML (2015) Polyethylene-assisted exfoliation of hexagonal boron nitride in composite fibers: a combined experimental and computational study. Macromol Chem Phys 216(8):847–855. https://doi.org/10.1002/macp.201400585
10. Sharma BB, Parashar A (2019) Atomistic simulations to study the effect of grain boundaries and hydrogen functionalization on the fracture toughness of bi-crystalline h-BN nanosheets. Phys Chem Chem Phys 21(24):13116–13125. https://doi.org/10.1039/c9cp01661a
11. Sharma BB, Parashar A (2019) Atomistic simulations to study the effect of water molecules on the mechanical behavior of functionalized and non-functionalized boron nitride nanosheets. Comput Mater Sci 169:109092. https://doi.org/10.1016/j.commatsci.2019.109092
12. Kumar R, Parashar A (2017) "Fracture toughness enhancement of h-BN monolayers via hydrogen passivation of a crack edge. Nanotechnology 28(16). https://doi.org/10.1088/1361-6528/aa6294
13. Sahputra IH, Echtermeyer AT (2013) Effects of temperature and strain rate on the deformation of amorphous polyethylene: a comparison between molecular dynamics simulations and experimental results. Model Simul Mater Sci Eng 21(6). https://doi.org/10.1088/0965-0393/21/6/065016
14. Chaurasia A, Verma A, Parashar A, Mulik RS (2019) Experimental and computational studies to analyze the effect of h-BN nanosheets on mechanical behavior of h-BN/polyethylene nanocomposites. J Phys Chem C 123(32):20059–20070. https://doi.org/10.1021/acs.jpcc.9b05965
15. Rahman R, Foster JT (2014) Deformation mechanism of graphene in amorphous polyethylene: a molecular dynamics based study. Comput Mater Sci 87:232–240. https://doi.org/10.1016/j.commatsci.2014.02.023
16. Hossain D, Tschopp MA, Ward DK, Bouvard JL, Wang P, Horstemeyer MF (2010) Molecular dynamics simulations of deformation mechanisms of amorphous polyethylene. Polymer (Guildf) 51(25):6071–6083. https://doi.org/10.1016/j.polymer.2010.10.009
17. Chawla R, Sharma S (2017) Molecular dynamics simulation of carbon nanotube pull-out from polyethylene matrix. Compos Sci Technol 144:169–177. https://doi.org/10.1016/j.compscitech.2017.03.029
18. Kumar R, Parashar A (2018) Effect of geometrical defects and functionalization on the interfacial strength of h-BN/polyethylene based nanocomposite. Polymer (Guildf) 146:82–90. https://doi.org/10.1016/j.polymer.2018.05.041
19. Chabba S et al (2007) Accelerated aging study of ultra high molecular weight polyethylene yarn and unidirectional composites for ballistic applications. J. Mater Sci 42(8):2891–2893. https://doi.org/10.1007/s10853-007-1617-7
20. Xu T, Farris RJ (2007) Comparative studies of ultra high molecular weight polyethylene fiber reinforced composites. Polym Eng Sci 47(10):1544–1553. https://doi.org/10.1002/pen.20876
21. Rahman R, Haque A (2013) Molecular modeling of crosslinked graphene-epoxy nanocomposites for characterization of elastic constants and interfacial properties. Compos Part B Eng 54(1):353–364. https://doi.org/10.1016/j.compositesb.2013.05.034
22. Lv C, Xue Q, Xia D, Ma M (2012) Effect of chemisorption structure on the interfacial bonding characteristics of graphene-polymer composites. Appl Surf Sci 258(6):2077–2082. https://doi.org/10.1016/j.apsusc.2011.04.056
23. Li M, Zhou H, Zhang Y, Liao Y, Zhou H (2017) The effect of defects on the interfacial mechanical properties of graphene/epoxy composites. RSC Adv 7(73):46101–46108. https://doi.org/10.1039/c7ra08243f

24. Ramanathan T et al (2008) Functionalized graphene sheets for polymer nanocomposites. Nat Nanotechnol 3(6):327–331. https://doi.org/10.1038/nnano.2008.96
25. Nasrabadi AT, Foroutan M (2010) Interactions between polymers and single-walled boron nitride nanotubes: a molecular dynamics simulation approach. J Phys Chem B 114(47):15429–15436. https://doi.org/10.1021/jp106330c
26. Sainsbury T et al (2012) Oxygen radical functionalization of boron nitride nanosheets. J Am Chem Soc 134(45):18758–18771. https://doi.org/10.1021/ja3080665
27. Lee D, Lee B, Park KH, Ryu HJ, Jeon S, Hong SH (2015) Scalable exfoliation process for highly soluble boron nitride nanoplatelets by hydroxide-assisted ball milling. https://doi.org/10.1021/nl504397h.
28. Ahadi Z, Shadman M, Yeganegi S, Asgari F (2012) Hydrogen adsorption capacities of multi-walled boron nitride nanotubes and nanotube arrays: a grand canonical Monte Carlo study. J Mol Model 18(7):2981–2991. https://doi.org/10.1007/s00894-011-1316-9
29. Kumar R, Mertiny P, Parashar A (2016) Effects of different hydrogenation regimes on mechanical properties of h-BN: a reactive force field study. J Phys Chem C 120(38):21932–21938. https://doi.org/10.1021/acs.jpcc.6b05812
30. Meng F, Chen C, Song J (2017) Lattice trapping and crack decohesion in graphene. Carbon N Y 116:33–39. https://doi.org/10.1016/j.carbon.2017.01.091
31. Tsai DH (1979) The virial theorem and stress calculation in molecular dynamics. J Chem Phys 70(3):1375–1382. https://doi.org/10.1063/1.437577

Dynamical Systems

Spring Tuning for Spacecraft Separation Dynamics Using Genetic Algorithm

Partha Ajit Surve, D. Vamsi Krishna, K. R. Madhavi, K. K. Raveendra Babu, L. Sowmianarayanan, and G. Ayyappan

1 Introduction

The improvement in launch vehicle technology has made it possible to launch a large number of spacecrafts simultaneously in the same launch vehicle. The collision-free separation of these spacecrafts is a critical part of any mission. The jettisoning of these spacecrafts is achieved using springs located on the adapter.

The body rate imparted to the spacecraft during the separation process should be near-zero because extra fuel will have to be consumed to kill the rates, which in turn will reduce the spacecraft life. Similarly, from collision avoidance point of view, a minimum relative velocity has to be maintained between the spacecraft and the final stage of the launch vehicle.

In this paper, an algorithm is described which provides the optimal combination of spring locations and the spacer value at each location which results in near-zero body rate to the spacecraft at separation. To achieve this, genetic algorithm was used due to its capability of handling highly nonlinear objective functions, capability to explore the entire design space, and the ability to provide a global minimum. The subsequent sections explain the methodology and formulation.

2 Methodology

The algorithm consists of two major components: the objective function evaluator and the optimization routine. The evaluator generates the function value ($f(x_i)$) with

P. A. Surve (✉) · D. Vamsi Krishna · K. R. Madhavi · K. K. Raveendra Babu · L. Sowmianarayanan · G. Ayyappan
Vikram Sarabhai Space Centre, Thiruvananthapuram, Kerala 695022, India
e-mail: parthasurve1@gmail.com

V. H. Saran and R. K. Misra (eds.), *Advances in Systems Engineering*, Lecture Notes in Mechanical Engineering,
https://doi.org/10.1007/978-981-15-8025-3_15

the given set of input design variables (x_i). The optimization routine takes these design variables and objective function values as the input and generates a new set of design variables (x_{i+1}) for the next iteration based on various constraints.

2.1 *Objective Function Evaluator*

In the present work, separation dynamics of the spacecraft needs to be simulated at each design iteration. A six-degree of freedom separation dynamics analysis is performed using an in-house developed general-purpose software 'The Stage Separation Dynamics Package' (SEPPACK). Rigid body equations of motions are solved in the body frame to obtain velocity (V) and body rates (ω) [1]. The body rates and velocities are transformed in the predefined local inertial frame.

The rigid body equations for each of the separating body are integrated by the modified Euler method with a suitable integration step size and by applying respective initial conditions.

The input to the evaluator consists of constant properties like mass, CG, and moment of inertia of the spacecraft and the final stage. The separation system specifications such as the spring constant, stroke length and the pitch circle diameter (PCD) of the spring locations have to be provided. The spring locations and the spacer values (X_i and spacer_i) are the variables that need to perturbed for the spring tuning purpose in the input. The output consists of the rates imparted on the spacecraft at the instant of separation which can be converted into the net body rate rate_{net}. Thus, the inputs for the objective function evaluator are the design variables, i.e., X_i and spaceri. The output generated is the rate_{net} on the spacecraft.

2.2 *Optimization Routine Optimization Routine*

For real-world problems such as spring tuning, the closed-form solution cannot be determined. Thus, the conventional methods become computationally expensive since the gradient or hessian has to be evaluated using methods like finite-difference. The design space itself is very nonlinear and multimodal; thus, the conventional methods won't be able to effectively determine the global minimum. To circumvent these problems, application of genetic algorithm [2] has been proposed in this study. Evolutionary algorithms are able to explore the entire design space and provide with a globally optimum solution. The in-built GA toolbox in MATLAB was used in the optimization routine. The two blocks are put together and used for spring tuning studies; the problem formulation is posed in the next section.

3 Problem Formulation

During the separation process, due to the inherent center of gravity offset present in the system, spacecraft experiences a body rate about the pitch and yaw axes.

Thus, the objective is to minimize the net rate (rate_{net}) at separation instance.

The schematic of spacecraft adapter is shown in Fig. 1 which has a total of 16 possible spring locations (X_i) ranging from A to P. The spring energy level at each of the locations can be varied by adding spacers (spacer_i) of varying sizes ranging from 0 to 6 mm. The highest spring energy is achieved by 6 mm and the least by 0 mm. Near-zero body rate during separation is achieved by tuning of these values along with the location. The springs are characterized using a universal testing machine, and the energy value is evaluated from the force displacement curve. Constraints are placed in order to make the solution physically viable and realistic. The separation velocity achieved must be within the required bounds.

Similarly, the worst-case rate during separation after including all the parametric perturbations must be within the specifications.

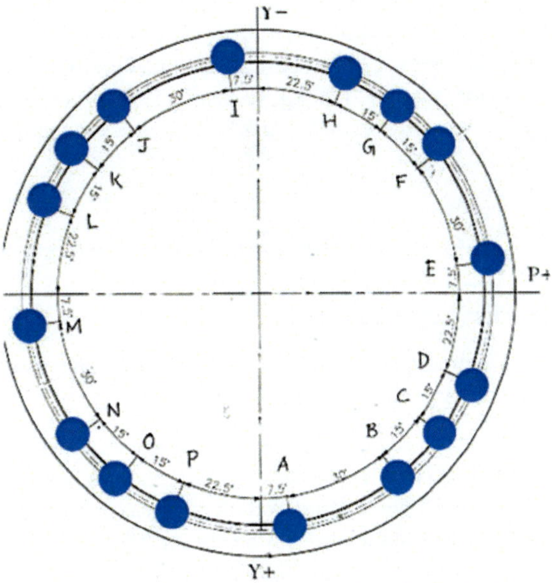

Fig. 1 Spring thruster locations on the spacecraft adapter

$$\textbf{Objective :}$$
$$\text{rate}_{\text{net}} = \sqrt{\text{rate}^2_{\text{pitch}} + \text{rate}^2_{\text{yaw}}}$$
$$\text{Minimize : } \text{rate}_{\text{net}}| \text{ @ separation instant}$$
$$\textbf{w.r.t :}$$
$$X_i \in [\mathbb{Z} : 1, 2 \ldots 16], \text{ where } i = 1, 2 \ldots n$$
$$\text{spacer}_i \in [\mathbb{Z} : 0, 1, 2, 3, 4, 5, 6], \text{ where } i = 1, 2 \ldots n$$
$$\textbf{Constraints :}$$
$$\text{rate}_{\text{worst}} \leq \text{rate}_{\text{spec}}$$
$$V_{\text{lower}} \leq V_{\text{rel}} \leq V_{\text{upper}}$$
$$X_i \neq X_j \ \forall i (i \in [1, 2 \ldots, n] \wedge i \neq j) \text{ where } j = 1, 2 \ldots n$$
$$n_{\text{quad}_i} \geq 1 \ \forall \ i (i \in [1, 2, 3, 4])$$

4 Results

In the previous section, the methodology used for this study has been briefed. This methodology has been demonstrated and verified for two different classes of spacecrafts 600 kg and 1600 kg, respectively. The configuration finalized based on these studies was used to design hardware that was used in ISRO launch vehicle missions.

4.1 *Case 1: 600 kg Class Spacecraft*

Mission Requirements:

1. Maximum resultant lateral rate on spacecraft under worst case $\leq 2°/s$
2. Relative velocity: 1.2 m/s $\pm$ 0.1 m/s.

The mass, center of gravity, and moment of inertia properties of the spacecraft are detailed in Table 1. These were used as constant values while generating the input for the objective function evaluator. It was proposed to use 6 number of springs (out of 16) based on the mass of the spacecraft and the relative velocity requirement.

A population size of 10,000 was used with a crossover and mutation rate of 0.8 and 0.2, respectively. The convergence criteria were met by 18th generation.

With the optimum spring configuration, a normalized nominal rate of 0.0008°/s was achieved. The worst-case rate generated from additive perturbation of all of the sensitive parameters was 1.3°/s which is within the specification of $\leq 2°/s$.

In Fig. 2, the first plot compares the spacecraft rates during separation with and without spring tuning. In the second plot, the red curve depicts the averaged body rate per generation. The advantage of the spring tuning algorithm is clearly visible. The second plot shows the convergence results of the genetic algorithm long with

Table 1 600 kg class spacecraft properties

Spacecraft properties in spacecraft co-ordinates									
Mass (kg)	CG (m)			MI (kg m^2)			PMI (kg m^2)		
	Roll	Pitch	Yaw	Roll	Pitch	Yaw	Roll Pitch	Roll Yaw	Pitch Yaw
600	0.02	0.04	0.95	300	350	150	7	−8	−10

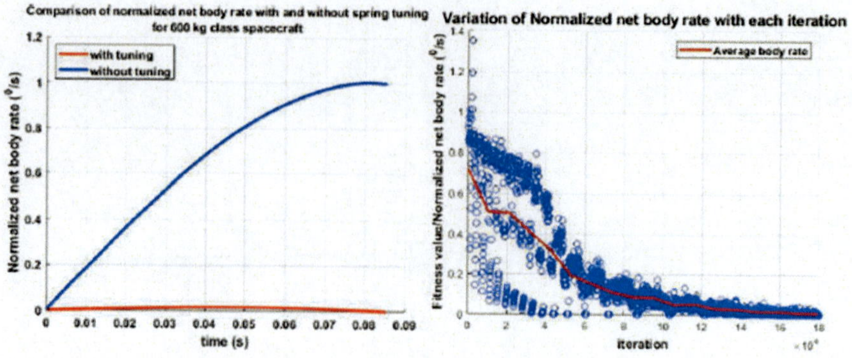

Fig. 2 Spring tuning results for 600 kg class spacecraft

the average fitness value for each generation which converges to near-zero by 18th generation.

4.2 *Case 1: 1600 kg Class Spacecraft*

Mission Requirements:

1. Maximum resultant lateral rate on spacecraft under worst case ≤5°/s
2. Relative velocity: 0.9 m/s ± 0.1 m/s.

The mass, center of gravity, and MI properties of the 1600 kg class spacecraft is detailed in Table 2. These were used as constant values while generating the input for the objective function evaluator. It was proposed to use 12 number of springs due to the heavier mass and the relative velocity requirement.

Similar to the previous case, the spring tuning algorithm was applied here.

All of the GA parameters were kept the same except the population size which was increased to 20,000 individuals to explore the design space effectively since a number of design variables have been doubled. The tolerance-based criterion showed that the solution had converged by 17th generation. The optimum spring location and spacer configuration produced a normalized nominal rate of 0.0024°/s and the worst case for this combination are 3.0°/s which is within the specification of ≤5°/s. Similar to the previous case, the results are summarized in Fig. 3 the average fitness value for each generation converges to near-zero by 17th generation. This case again demonstrates the advantage of the spring tuning algorithm.

Table 2 1600 kg class spacecraft properties

Spacecraft properties in spacecraft co-ordinates

Mass (kg)	CG (m)			MI (kg m^2)			PMI (kg m^2)		
	Roll	Pitch	Yaw	Roll	Pitch	Yaw	Roll Pitch	Roll Yaw	Pitch Yaw
1600	0.05	0.02	0.95	1250	1200	1250	10	−10	−10

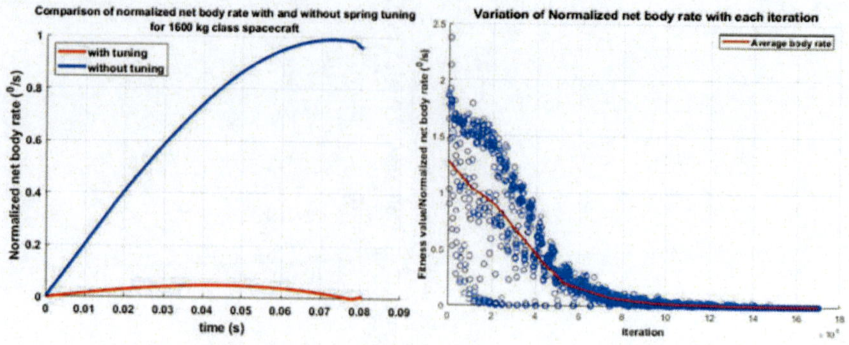

Fig. 3 Spring tuning results for 1600 kg class spacecraft

5 Conclusion and Flight Experience

Spring tuning for spacecraft separation is one of the critical parts of launch vehicle mission design. A genetic optimization-based algorithm was proposed to give optimum results. The working of the algorithm has been verified with two different test cases. Near-zero separation rates were achieved via optimum spring location and energy level (tuned using spacers) combinations. The above spacecrafts were own in ISRO launch vehicle missions, and the rates measured during flight were near-zero and matched with the preflight predictions. A significant reduction in design time for spacecraft spring tuning is observed compared to the general methods.

References

1. Goldstein H, Poole CP, Saftko JL (2001) Classical mechanics, 3rd edn. Addison-Wesley
2. Goldberg DE (1989) Genetic algorithms in search, optimization and machine learning

Health Monitoring and Assessment of the Cultural Monument Through Unmanned Aerial Vehicle (UAV) Image Processing

Ekta Baranwal, S. Raghvendra, Poonam Seth Tiwari, and Hina Pande

1 Introduction

Monitoring the health of a physical structure is necessary to validate its design assumptions for developing a safe and reliable structure to use. Day by day infrastructure is expanding to facilitate us with homes to live in, transport ways for smooth connection between two places, and service buildings such as schools, hospitals, and commercial places. The expanding infrastructure demands to be taken care of by monitoring it time to time and taking actions according to their states. For example, a crack in the overbridge needs to be fixed timely so that no accident can take place and no one can be harmed due to that. Since most of the land is covered with different types of infrastructures and human spent their time beneath it, so timely maintenance of these infrastructures is always required for a safe living which is only possible when it is monitored and assessed consistently and even after a hazard. When it comes to save the infrastructure, it also contains the buildings from our history, i.e., our architectural and cultural heritage. The architectural heritages are much sensitive to any disaster and aging. That is why for their documentation and conservation, they must be monitored very frequent as per the archaeological points.

The criterion for identification and conservancy of artistic heritage can be magnificently understood only by experts, and their endeavor is to achieve any possible

E. Baranwal (✉)
Department of Remote Sensing, Banasthali Vidyapith, Newai, Rajasthan 304022, India
e-mail: ektabaranwal01@gmail.com

S. Raghvendra · H. Pande
Photogrammetry and Remote Sensing Department, Indian Institute of Remote Sensing, ISRO, Dehradun, Uttarakhand 248001, India

P. S. Tiwari
Geoweb Services, IT & Distance Learning Department, Indian Institute of Remote Sensing, ISRO, Dehradun, Uttarakhand 248001, India

V. H. Saran and R. K. Misra (eds.), *Advances in Systems Engineering*, Lecture Notes in Mechanical Engineering,
https://doi.org/10.1007/978-981-15-8025-3_16

method for consummating this aim. The emerging technologies such as computers and digital implements have opened incipient windows and bestowed embryonic opportunities in the process of health monitoring of cultural heritage and their conservation. Cultural heritage is endangered by sundry factors such as vandalism, natural threats, aging, and expansion of cities which in a practical view, one cannot assure their eternality, and every moment, it is possible of their forfeiture. Basically, conservation or preservation process of any cultural heritage is composed of three phases: understanding the consequentiality, developing policy, and managing in accordance through the policy [1]. After the defacement of an architectural or structural heritage, an assessment is always required for reformation or reconstruction or rebuild purposes.

The trend of this era is to utilize a handy tool such as drone and our own mobile phone for the survey images and videos. In technical terms, this drone is known as unmanned aerialvehicle (UAV) or unmanned aerial system (UAS). The UAV platform can be equipped by image capturing camera, video camera, light detection and ranging (LiDAR) system, thermal camera or infrared (IR) camera systems, or a combination allied, and also with the local or global positioning system. UAV has numerous advantages over traditional methods of monitoring the site or particular object which makes it eligible for the work. Some of the major advantages which we can count are [2]:

- UAV can be utilized in the area of high-risk situations, an area where no manned aircraft is available or no flight permission is given.
- Without endangering the human life in inaccessible areas, UAV can fly at low elevation and close to the items where manned aircraft cannot be flown.
- Data acquiring in cloudy and drizzling weather conditions is not possible with manned aircraft due to a larger altitude above ground but with UAV, it is possible to acquire data by flying it below the clouds.
- Ability to operate in hostile environments (smoke, ash, an active volcano, etc.) where man held systems cannot be really put on.
- UAV is free from economic expenses and physiological limitations of human pilots.
- It has fast data acquisition and real-time capability, as it transmits the images, video and positioning data in real time to the ground controller.
- UAV has lower operating cost than manned system, but depending upon the application, the cost can be similar to manned aircraft.
- High spatial drawn to particular focal length and high temporal resolution for dynamic changes in the scene.
- Large area coverage compared to ground observations but less area coverage compared to manned aircraft.
- Long dwell times over areas of interest to focus longer on any particular spot.
- Easy and quick to deploy for data acquisition as compared to manned systems because manned systems take 1–2 days to deploy whereas UAVs take few hours.
- Variety of platforms and sensors even hybrid type by mix and match for a better platform.

- More flexible, i.e., oblique photographs at the desired location.
- The applications where two types of photogrammetry (UAV and terrestrial) are combined, the same camera setup keeping the same remoteness to the objects is used for simplifying the collective data processing.
- A mixed output (conventional aerial and terrestrial) of data acquisition can be done through UAV.

Literature shows plentiful work based on documentation, and health assessment of structural heritage through techniques comes under the image-based assessment, non-image-based assessment, and combinative methods [3]. Among all these, the most commonly used technique is image-based where photogrammetry is utilized for the purpose either through panorama photography, close-range photogrammetry or UAV photogrammetry. In case where we want to monitor and record structural health, UAV is more advantageous and portable because UAV systems are used as photogrammetric data acquisition platforms which provide various innovative applications for close-range photogrammetry, terrestrial photogrammetry, aerial photogrammetry, and even combination of two or more of them. UAV systems stand in between close-range and aerial systems in terms of the level of details in data which also provides the combined features of both with a small scene and higher data details and can be taken at low height and nearby to the objects in the area of high-risk situations where no manned aircraft is available or no flight permission is given. When any built structure is recorded via UAV images for post disaster assessment, the assessment is done through three methods: visual interpretation, classification, and comparison of pre and post disaster data either images, DEM or point cloud directly to relegate damage [4]. Many works [5–7] have been successfully done in this area where UAV images are utilized alone or in collaboration of high-resolution spaceborne images to detect damages such as cracks and apertures in roof and façade of the buildings and also for historic structures. A three-tiered measure is often utilized to relegate infrastructure damage: completely ravaged structures or those beyond repair, partially eradicated structures with a possibility of rehabilitation, and unaffected structures or those with solitary minor damage [8]. Since damage detection mainly depends on the image resolution, higher the resolution, better is the result. UAV photogrammetry provides the ultra-high-resolution images, but there are some doubts about the approaches and algorithms which is applicable to them, i.e., what are the possibilities that the algorithms utilized for satellite images also work for UAV images. In this work, UAV photogrammetry is manoeuvred to generate ultra-high dense 3D model of cultural monument for its post-earthquake documentation and defacement analysis. Textural and spectral analysis is performed on orthomosaic photo for damage detection.

2 Study Area and Data Used

For this work, the study area is Sulamani Pagoda from Bagan (nowadays also called Burmese), the ancient city of Myanmar (Fig. 1) whose archaeological belt is spread

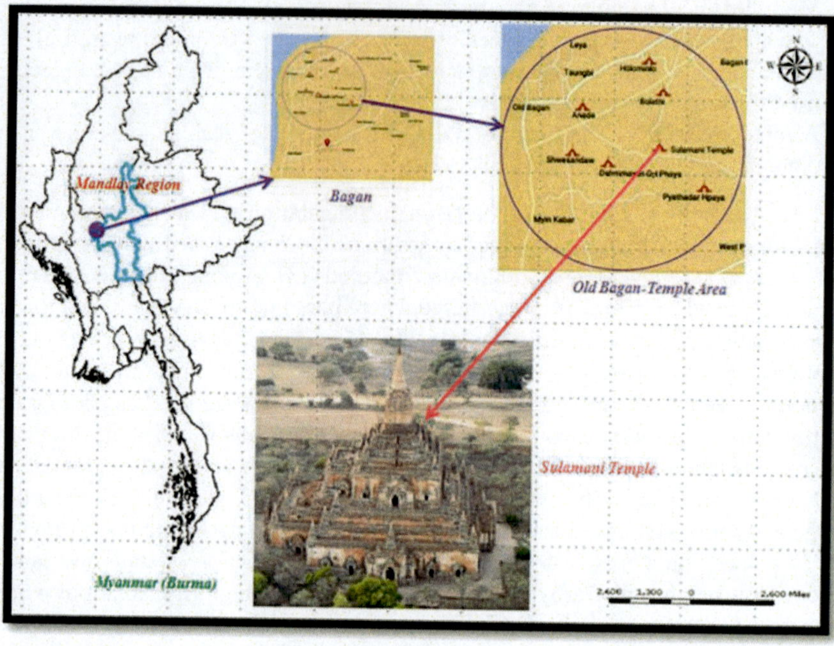

Fig. 1 Study area

over the 13 × 8 km area positioned 290 km southwest of Mandalay region and 700 km north of region of Yangon. 10,000 plus religious monuments (Buddhist temples, pagodas (stupas) and monasteries) were built and shaped in the Burmese plains in the midst of 11th and 13th eras mostly in old Bagan (Temple City), one of the among world's most noteworthy archaeological sites. These monuments are still present but only a few in numbers, because the city is settled in rustling earthquake zone, and had endured through several seismism over the dimension of time. In the year 2016 on 24 August, a hefty earthquake hit central Myanmar and did extensive damage in Bagan. About 400 temples were eradicated, in which Myauk Guni (North Guni) and Sulamani Phato were severely damaged due to this earthquake. Old Bagan has the great archaeological and religious significance which results to its preservation. Also, this city comes under substantial earth-shaking region which is the sufficient reason behind the health monitoring for documentation and conservation of the cultural monuments exist there. In this paper, the Sulamani Pagoda is focused for the purpose of monitoring and analysing its structural health because it is one of the last large and more sophisticated temples from Bagan period built by King Naraptisithu in year 1181 and decorated by impressive architecture works of art.

Total one hundred and fifty images obtained through unmanned aerial vehicle (UAV) were used for this work which were processed to generate various outputs for damage assessment of the temple. These images were obtained from the 90m of flying height with the camera model FC300X having focal length 3.61 mm, resolution 4000 × 300 (RGB), and pixel size 1.56 × 1.56 micro meter.

3 Methodology

After acquiring images, initial part of the methodology (Fig. 2) took by its processing. Processing raw UAV images does not end with a single step instead it works on a stack of steps. Firstly, all the images are aligned as per the camera positions, and then, it is checked for their overlapping pattern and percentage. After this, key points or tie points are selected to estimate depth from focus point of the camera based on its focal length. This depth calculation is used to generate 3D model of the object with its point cloud. Compactness and complexity of this point cloud depend on the size of the raw images used to create it and vary from lowest density to ultra-high dense point cloud. [9] For ultra-high dense point cloud, full size of images is taken, and it is

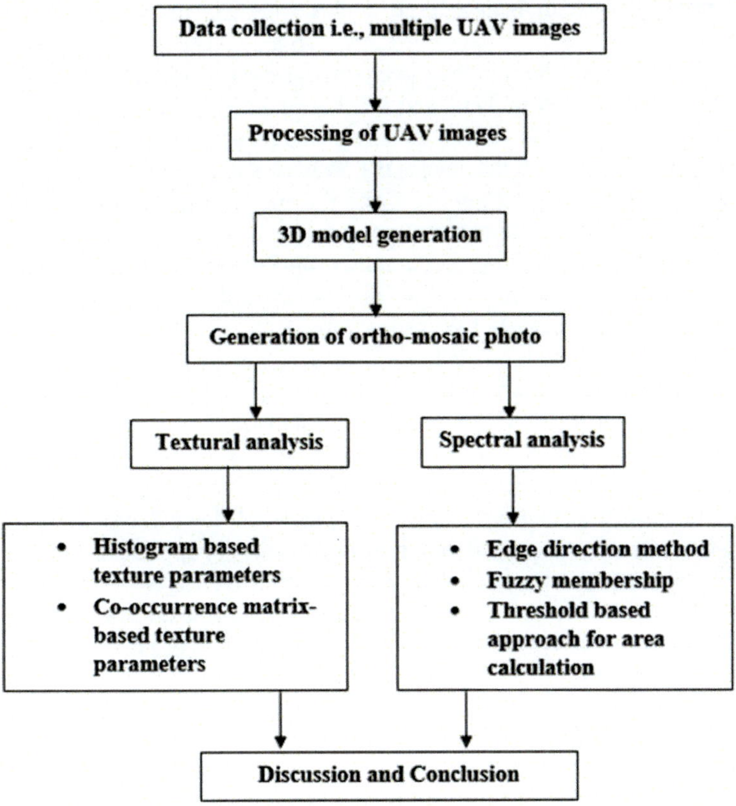

Fig. 2 Proposed workflow

downscaled to its percentage for lower density such as 6.75% for lowest dense point cloud and 25% for medium dense point cloud. The quality of 3D model is directly proportional to the excellence of the point cloud, and in this research, full size images were processed to generate point cloud of maximum density. Dense point cloud and 3D model are used to monitor the structure in a realistic view, but mostly, the point cloud is converted to digital elevation model (DEM) and orthomosaic photo of the object for assessment of its physical structure. In this research, for assessing the post-earthquake health of the Pagoda, a texture-based and spectral analysis is performed on orthomosaic photo. Many texture parameters are calculated in different position vectors, but only few parameters were succeeded to identify defaced part of the monument. On the other hand, spectral analysis is also done where many fuzzy membership, edge direction operators, and threshold-based approach are applied. All these results are shown and discussed in results section.

For texture analysis, two types of texture parameters were utilized:

- histogram-based textures, and
- co-occurrence matrix-based textures.

The histogram-based features used in this work are first-order statistics that include: Mean Euclidean distance (MED), variance, skewness and kurtosis. By definition, MED gives the middling gray level of each region by calculating mean Euclidean distance between them, and it is useful only as a rough idea of intensity not really texture, while the variance gives the amount of gray level fluctuations from the mean gray level value of the image [10]. Skewness is a quantity of the asymmetry in the gray levels around the sample mean, this means, negative skewness depicts that the data are expansive more to the left of the mean, and positive skewness shows that the data are spread out more to the right [10]. Kurtosis is an amount to describe how outlier-prone a spreading is and defines the shape of the end of the histogram to differentiate various objects in the images [10].

To analyse texture with co-occurrence matrix, we need both a direction and a spacing which is the position vector. *X* vector denotes the numeral of pixels across in co-occurrence matrices (C) to calculate the statistics and *Y* vector, the numeral of pixels up or down in co-occurrence matrices (C) to calculate the statistics. For example, with a position vector of (2, 1), each pixel in the image will be compared with the pixel located two pixels right and one pixel down. This position vector decides the track of texture. Different position vectors are applied in co-occurrence matrix-based textural analysis to find the texture difference between damage and non-damage part of the monument. The texture parameters that come under this type of texture analysis and lead to in a satisfactory assessment of post disaster state of the monument are as follows: Contrast, standard deviation (SD), dissimilarity, mean, second-order moments, and first-, second-, and third-order inverse moments. Change in position vector results in variation in output of the particular texture parameter. In most of the cases, lower value than 3 for position vector has given the coarser texture which does not provide any useful information for analysis.

In spectral analysis, three approaches (edge direction approach, fuzzy membership approach, and threshold-based approach) were utilized to determine the post-earthquake state of the cultural monument. In edge direction approach, the Sobel operator is utilized to perform a 2-D spatial gradient quantification which accentuates regions of high spatial frequency that correspond to edges. Sobel filter can be utilized in two different directions to highlight different edges. These two directions are horizontal and vertical. A 3 × 3 kernel size has different values in both directions. In fuzzy membership approach, fuzzy membership transforms the input raster into a 0 to 1 scale, denoting the vigor of a membership in a set, predicated on a designated fuzzification algorithm which is according to fuzzy membership type. A value of 1 denotes full membership in the fuzzy set, with membership decrementing to 0, denoting it is not a member of the fuzzy set. Seven fuzzy memberships: Fuzzy Gaussian, fuzzy large, fuzzylinear, fuzzy MS large, fuzzy MS small, fuzzy near, and fuzzy small are used for this fuzzy analysis of the image, but only three (Fig. 8) gave the satisfactory result in this study. Threshold approach is aimed to dissever objects utilizing a threshold for segmentation. It mainly segmented the whole image in two sections, above and below threshold based on the brightness value as a threshold. This segmented part is then utilized to calculate its area which ultimately gives the area of the damaged part in the form of two-dimensional image (Fig. 9).

4 Results and Discussion

After processing 150 UAV images, total 108,044 tie points were generated which were then used to generate a point cloud (Fig. 3) with ultra-dense quality giving more than 140 million points. Thus, the orthomosaic photo generated (Fig. 4) was also of ultra-high resolution, i.e., 3.4 cm/pixel which depict the monument with fine details.

In all of four histogram-based textures (Fig. 5), mean Euclidean distance (Fig. 5a) shows the most of the damage part in both shadow and non-shadow regions, even the area under reconstruction. Contrast is the difference of the lowest and the highest

Fig. 3 Ultra-dense point cloud

Fig. 4 Ultra-high-resolution orthomosaic photo highlighting the defaced part of the monument

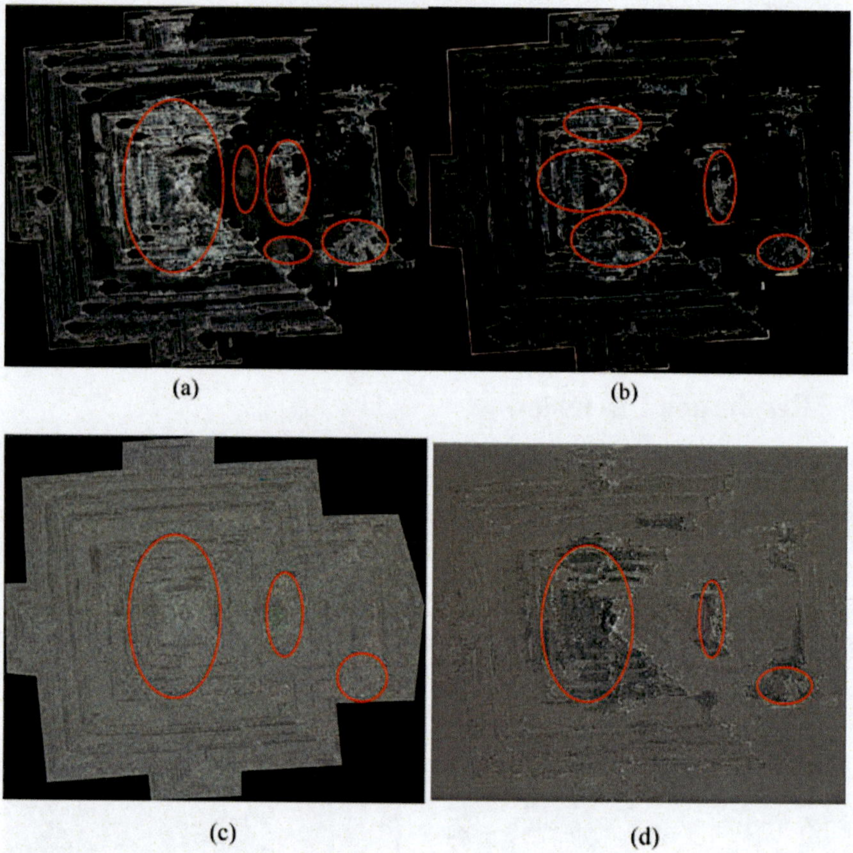

Fig. 5 Images after histogram-based texture analysis with parameters: **a** MED, **b** Variance, **c** Kurtosis, **d** Skewness

values of a connecting set of pixels which deals with the spatial occurrence of image and also measures the amount of local deviations existing in the image, while SD indicates the intensity contrast [11]. As per the result of these two parameters, debris part is clustered hence shows dark and different from other textural change in contrast image (Fig. 6a) while bright in standard deviation (SD) image (Fig. 6b). Dissimilarity measures how elements of co-occurrence matrix are different from each other. It will be the highest when the difference is large that is why the textural parameter is very much able to highlight the texture with minute differences which are dissimilar to the adjacent pixel (Fig. 6c). The mean gives the average gray level of each region [11], here (Fig. 6d), it shows the variation in the intensity with variation in position vector, and hence, the damage detection also varies [11]. As per definition the element difference, moments have a low value when pixels are similar, since the computation is least sensitive to values falling on or near the main diagonal of the co-occurrence matrix, while the inverse moments have the opposite effect. In the image of second-order moments (Fig. 6e), damage is visible not only in non-shadow region but also in shadow region. In first-order inverse moments (Fig. 6f), the texture change is uniform over the whole image. The effect of shadow is not much, and the damage is not so visible to be separated from non-damage parts. In second-order inverse moments (Fig. 6g), debris from damage area is visible and can be separable from non-damage area. It is brighter in non-shadow region and less bright in shadow region. The third-order inverse moments (Fig. 6h) produce almost same result as in second-order inverse moments.

In spectral analysis, edge direction approach is proved to be useful to analyze damage as the gradient magnitude with Sobel operator from two different directions: horizontal and vertical [10] (Fig. 7). Fuzzy logic is used to classify damage as a function of fuzzy membership which represents the degree of truth in fuzzy logic [12]. Fuzzy linear is based on the threshold of the brightness value and clearly shows the damage in non-shadow region [12] (Fig. 8b), while fuzzy MS small correctly highlights the damage in shadow region also (Fig. 8c). Threshold-based approach results in the total number of pixels segmented for damaged portion which is then used to calculate the approximate percentage of the damaged area of the monument (Fig. 9).

5 Conclusion

The use of a varied spectrum of technologies for cadence documentation of historic/cultural heritage or monument confirms the diversity of congruous alternatives for documentation of any object. Size of the entity, complexity, cost, accessibility, time, and the expertise of the surveyors, etc., depict a paramount role in picking a survey technique. Nowadays, UAV has been proved most suitable option to monitor the scene much closer and approachable than any aircraft. Unfortunately, the images used for this project have shadow which make the assessment process somewhat difficult. But, because of the ultra-high-resolution of the data, the approaches (texture and

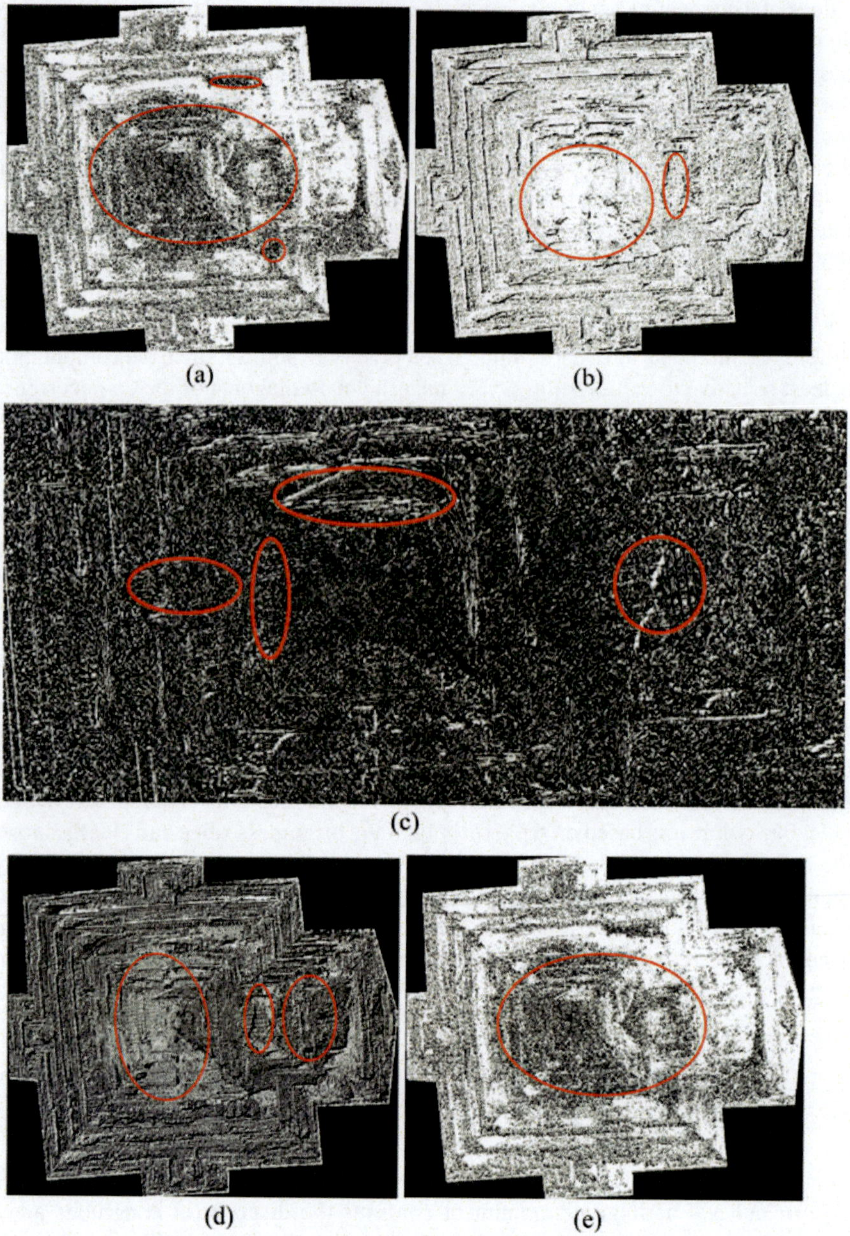

Fig. 6 Images after co-occurrence matrix-based texture analysis with parameters: **a** Contrast, **b** SD, **c** Dissimilarity, **d** Mean, **e** second-order moments, **f** firstorder inverse moments, **g** 2nd order inverse moments, **h** third-order inverse moments

(f)

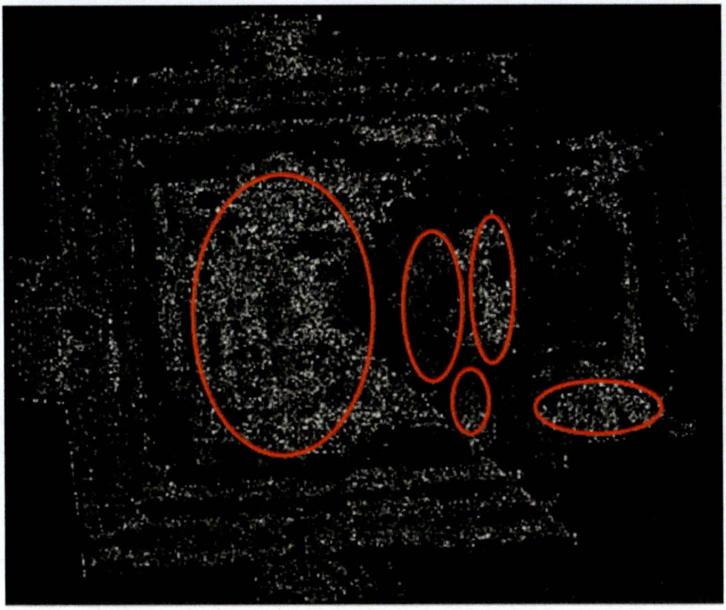

(g)

Fig. 6 (continued)

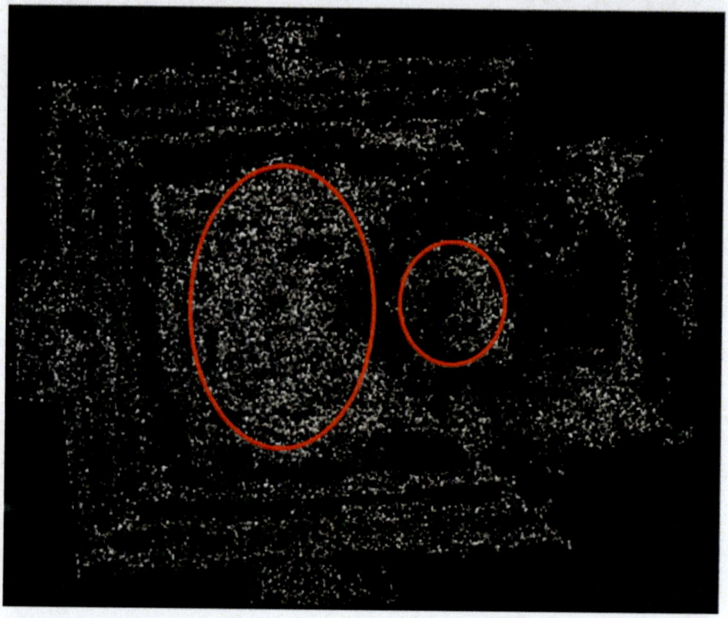

(h)

Fig. 6 (continued)

spectra-based) utilized here completely qualified the task of damage detection of the temple very satisfactory in non-shadowed and as well as in shadowed region. Besides orthophoto (generated from processing of UAV images), 3D model is also used to evaluate the volume of different part of the object. All the techniques and approaches used here worked well to detect damage in non-shadow and shadow region, although it was time consuming in comparison with process satellite images because of its resolution. UAV is proved to be most promising platform to monitor the health of an architectural structure for the purpose of documentation and conservation in terms of quality of the data and easiness to acquire these data from different angle and height. The quantified change assessment can also be done through UAV data, and it provide much closer estimate of defacement as compared to terrestrial surveys, but it needs to be validated with ground data before any final conclusion and decision making for practical approaches.

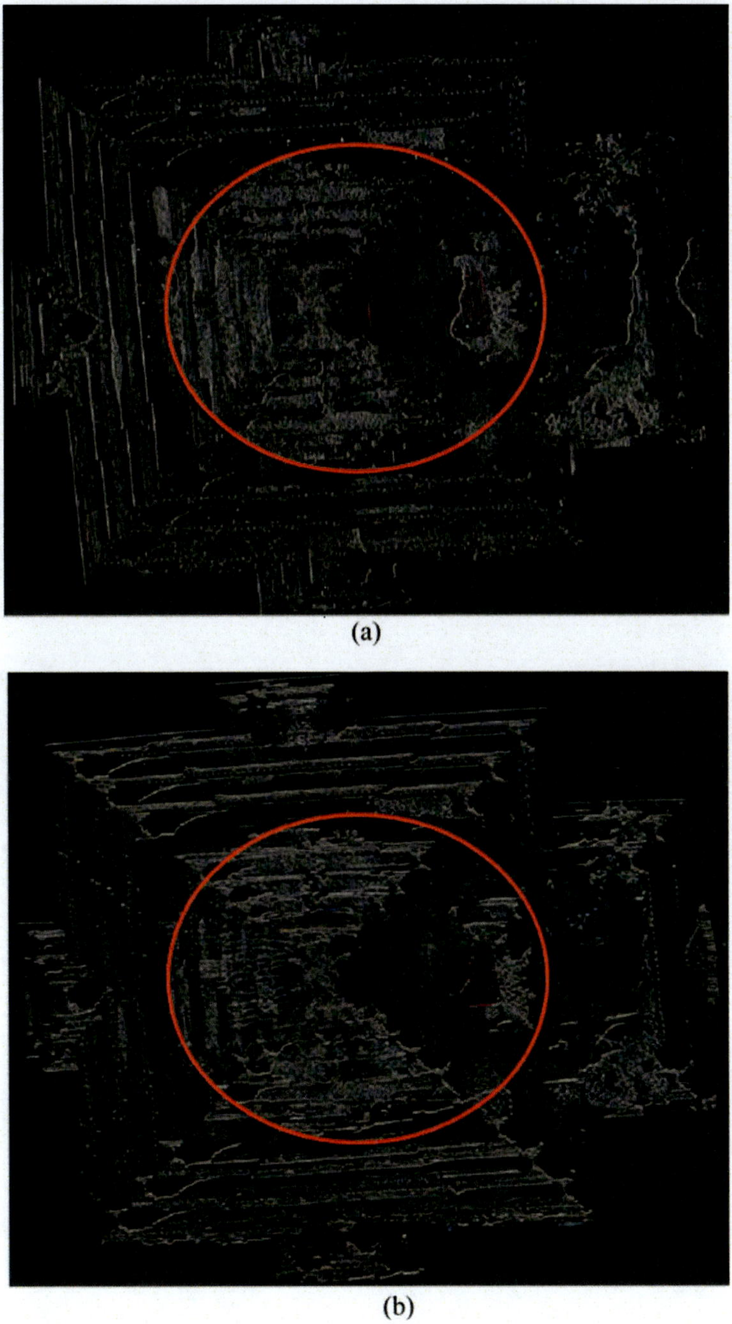

Fig. 7 Images after edge direction approach with 3 × 3 kernel size: **a** Sobel-1 **b** Sobel-2

(a)

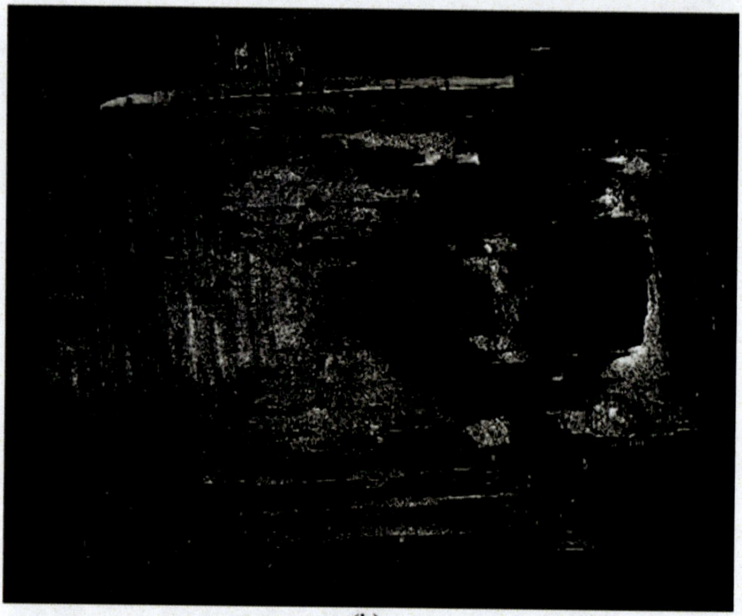

(b)

Fig. 8 Images after fuzzy logic with membership function: **a** Fuzzy Near **b** Fuzzy Linear **c** Fuzzy MS small

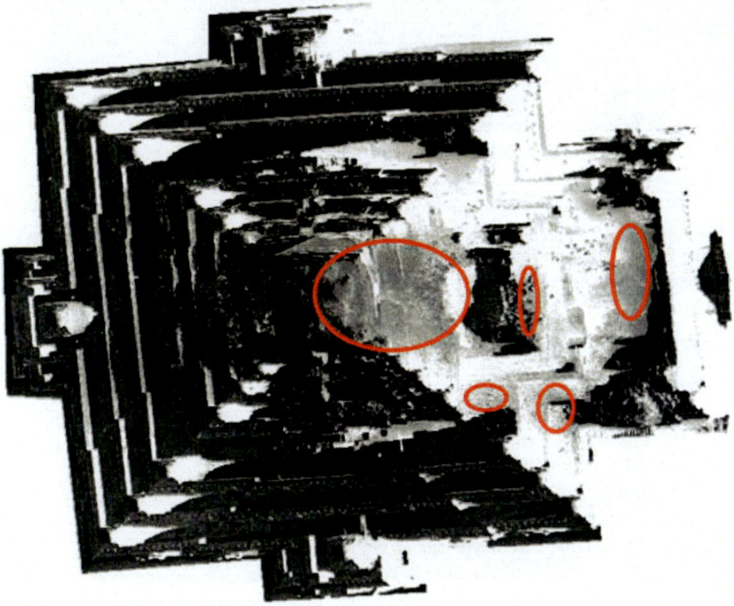

(c)

Fig. 8 (continued)

Fig. 9 Damage estimated in non-shadow region through threshold-based approach

References

1. Charter B (2013) The Australia ICOMOS charter for places of cultural significance. In: Australia ICOMOS incorporated international council on monuments and sites. Retrieved from https://portal.iphan.gov.br/uploads/ckfinder/arquivos/The-Burra-Charter-2013-Adopted-31_10_2013.pdf, on date 16th Sep, 2017
2. Eisenbeiß H (2009) UAV photogrammetry. Switzerland. https://doi.org/10.3929/ethz-a-005939264
3. Hassani F (2015) Documentation of cultural heritage techniques, potentials and constraints. The international archives of the photogrammetry, remote sensing and spatial information sciences. ISPRS Archives 40(5W7):207–214. https://doi.org/10.5194/isprsarchives-XL-5-W7-207-2015
4. Li J, Deng J, Xie W (2015) Damage detection with streamlined structural health monitoring data. Sensors 15(4):8832–8851
5. Fernandez Galarreta J, Kerle N, Gerke M (2015) UAV-based urban structural damage assessment using object-based image analysis and semantic reasoning. Nat Hazards Earth Syst Sci 15(6):1087–1101. https://doi.org/10.5194/nhess-15-1087-2015
6. Mathew G, Laura B (2017) University of Pittsburgh. Retrieved from www.pitt.edu: https://www.pitt.edu/~budny/papers/35.pdf, 27th Jan, 2018
7. Adams S et al (2010) Unmanned aerial vehicle data acquisition for damage assessment in hurricane events. In: 8th international workshop on remote sensing for disaster management, (September 2014), 1–7. Retrieved from https://www.enveng.titech.ac.jp/midorikawa/rsdm2010_pdf/13_adams_paper.pdf
8. iRevolutions (2017) Retrieved from www.irevolutions.org: https://irevolutions.org/2015/05/18/damage-from-3d-clouds/, on date 18th Nov, 2017
9. Agisoft PhotoScan User Manual (2016) Retrieved from www.agisoft.com: https://www.agisoft.com/pdf/photoscan-pro_1_2_en.pdf, on date 20th Nov, 2017
10. ERDAS Field Guide™ (1999) Retrieved from https://web.pdx.edu: https://web.pdx.edu/~emch/ip1/FieldGuide.pdf, on date 2nd Dec 2017
11. Girisha AB, Chandrashekhar MC, Kurian MZ (2013) Texture feature extraction of video frames using GLCM. Int J Eng Technol (IJETT) 4(6):2718–2721
12. ArcGIS for Desktop. Retrieved from https://desktop.arcgis.com/en/arcmap/10.3/tools/spatial-analyst-toolbox/how-fuzzy-membership-works.htm, on date 8th Jan, 2018

Assessment of Mesh-Like Surface for Water Harvesting from Moist Air

Singh Punj Lata and Sikarwar Basant Singh

1 Introduction

Our blue planet exists with finite resource and by current practise of use has led to depletion of the resource. The reason of depletion and shortage has been reported due to damage to our bionetwork and unexpected climatic change. The repetitive habits of negligence towards impact of growing population, standardizing livelihood, deforestation, modernization and development of industries, urbanization and unstoppable wastes have responded a deep effect consequence a change in ecosystem [1]. However, several steps have been invented and innovated to opt as alternative solution to formulate and solve the crisis but one of them, the water scarcity is growing speedily, has covered most of the areas of the globe beyond estimation. By 2025, the challenge of drinking water shortage will camouflage around 50–65% of the earth's surface [2]. In order to solve the issues related to the water problem, atmospheric water generating devices have seek out an attention to the researchers because water generating from the thin air is an easily available and a valuable resource of water because moist air at high relative humidity holds 20 L of water per cubic metre at 30–35 °C temperature [3, 4]. Wahlgren [4] reported diverse devices for generating water from thin air. Jensen et al. [5] explained a condensation phenomenon by the nucleation on substrate as a feasible method to obtain in numerous conditions. A conventional condensation completely depends upon the surface wettability suitably categorized into hydrophobic or hydrophilic [6, 7]. The condensation mode plays an important role while releasing energy from the condensing surface. Dropwise

S. P. Lata (✉) · S. B. Singh
Department of Mechanical Engineering, Amity School of Engineering and Technology (AUUP), Noida, India
e-mail: plsingh@amity.edu

S. B. Singh
e-mail: bssikarwar@amity.edu

V. H. Saran and R. K. Misra (eds.), *Advances in Systems Engineering*,
Lecture Notes in Mechanical Engineering,
https://doi.org/10.1007/978-981-15-8025-3_17

has high condensation coefficient as compared to others [8]. The surface wettability controls the mode of condensation, and low wettability requires for dropwise condensation. The wetting characterization of surfaces is altered by physical or chemical texturing.

When a moist air contacts a cold surface which has temperature below than the moist air's saturation temperature, the moist air condenses on the surface [9]. The condensing surface is fabricated to improve surface morphology for intensifying the condensation rate. It was reported [10] that surface has higher hydrophobicity and lower contact angle hysteresis which impels an effective and efficient dropwise condensation that has high condensation rate. Additionally, a mesh-like surfaces augment efficacy of moist air condensation process by having large contact area [10]. Although during the past decades, both natural materials and synthetic materials have been discovered and developed with principles and strategies to enhance the behaviour of water harvesting and wettability pattern. Despite the materials, methods are equally important to fabricate the framed model potentially. Therefore, the literature review directed this research work to explore the novelty, to perform assessment of copper mesh-like surface for water harvesting from moist air. While studies related to condensation of water vapour present in moist air has been reported periodically in the literature but proficient method under controlled environmental condition for extracting water from humid air for enhancing the surface condensation rate is not provided explicitly in the literature.

Against this background, for evaluation of worth efficacy, an assessment among a coated copper surface against a coated tubular copper mesh-like surface is carried out experimentally. A similar size of both outer tubular surface of mesh and bare copper surface is coated by chemical action with copper oxide (CuO)-based called "Silylation by immersion" reported elsewhere by Sharma et al. [11]. Besides it, to maintain surface temperature below than dewpoint, maintained cold water is flowed, through tubes and on opposite side of condensing surface of bare copper substrate. Furthermore, the experiments were performed ranging 60–80% relative humidity and water collected amount for bare copper surface and mesh-like surface are 2.778 and 3.472 ml/m^2-s with contact angles 150° and 155°, respectively, at relative humidity 80%.

2 Materials and Methods

The test samples selected were bare Cu surface (99.9% purity, 2 mm thickness and 40 mm × 40 mm of size) and copper mesh-like surface (99.9% purity, 3 mm outer diameter and 40 mm × 40 mm of size) which were textured for achieving different wettability. Initially to frame the mesh-like surface, the Cu tubes were cut into equally to eight pieces each of length 40 and 3 mm outer diameter. Using mixture of LPG (propane) and oxygen by a process of brazing was applied to join four pieces placed horizontally with other four vertically placed Cu tubes at maintained controlled temperature of 750–800 °C. The samples were primarily, physically cleaned for smoothing surfaces and removing small imperfections by choosing 100 to 120 grit emery paper, then etched sample is kept in HCL solution for 30 s to eliminate the oxide layer on the surface and then kept in vacuum desiccator for 30 min. Furthermore, the samples were ultrasonically cleaned with acetone for 15 min and dried at room condition. To obtain the hydrophobic surface, the bare Cu surface and mesh-like Cu surface were functionalized by immersing in the chemically silicate solution of 40 ml for the duration of 4 days and the samples were prepared by silylation process in controlled condition as reported in literature. A uniform bluish film was obtained on the surfaces of samples when pickled out and further proceeded to drying up for at least of 30 min at 100 °C to obtain the permanent layer of coating as shown in Fig. 1.

The detailed experimental set-up used for harvesting water of coated copper surface for moist air condensation is schematically shown in Fig. 2.

The processed test sample of Cu surface and copper mesh-like surface, both were placed vertically in an acrylic camber and cool water from ha thermal bath, was circulated through a pipe to cool the surfaces of the samples. In case of mesh-like surface, the tubes of the mesh were interconnected to a leak proof pipe of 3 mm internal diameter through which a maintained flow and cooling effect on surface were found. During harvesting, moist air was supplied directly to the samples inside acrylic chamber and the surface captured the nucleation of drops of water via a collision with surface following condensation phenomenon. The temperature of moist air near the test surface found was 28.5 ± 1.0 °C, the relative humidity was 70–90%, and the cold water in the tube was 14 ± 0.5°C as average temperature. In this chamber, controlled environmental condition was provided to all experiments, such as temperature and humidity (temperature accuracy of ±0.5 °C and humidity accuracy of ±3.0%), and was recorded using hygrometer and digital temperature indicator. The cooling of fluid was maintained by thermal bath and circulated using an inline pump of capacity of 5 L/min, pump pressure of 100 Psi and temperature accuracy of ±0.05 °C at 0 °C. The gliding droplet of water from the surface of the sample was collected to a circular vessel placed below the sample. The collected drops were measured using an injection syringe and a stopwatch is used to observe cyclic phenomenon of tailoring drop while dropwise condensation. The total amount of water collected is also measured for the

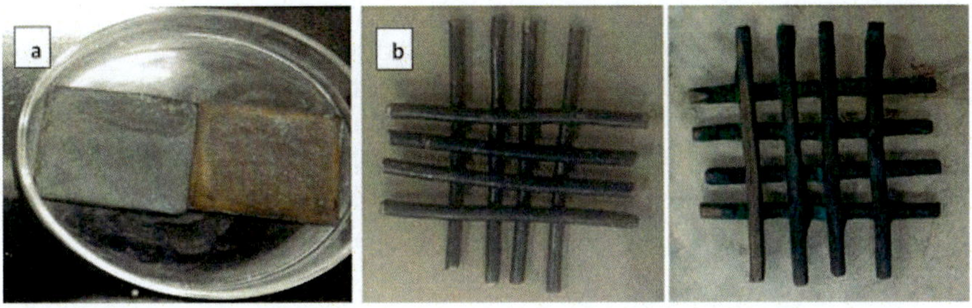

Fig. 1 **a** Image of the sample of solid copper surface showing before and after chemically treated for coating. **b** Image of the sample of mesh-like copper surface showing before and after chemically treated for coating.

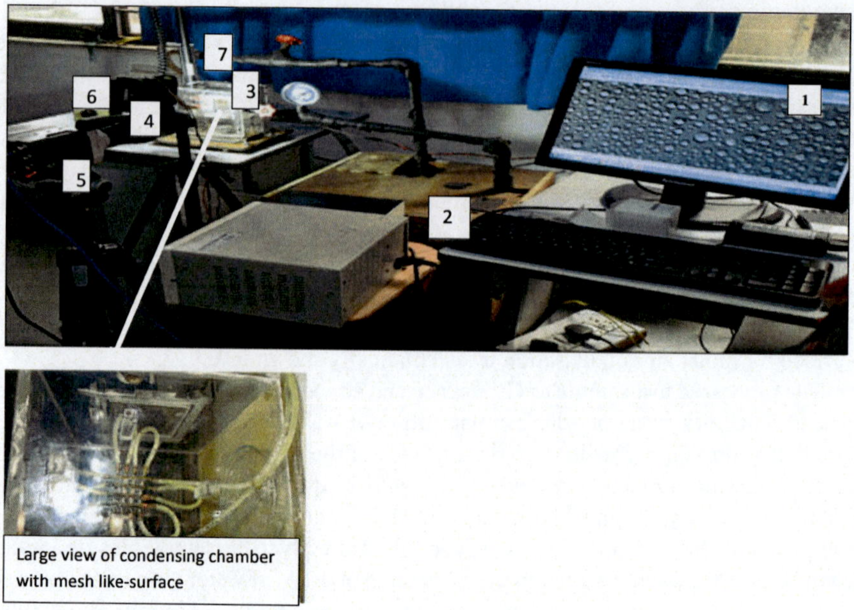

Fig. 2 Photograph of experimental set-up of coated copper surface for moist air condensation: 1—computer; 2—source for moist air; 3—coated copper surface inside Acrylic Chamber; 4—led light; 5—high-speed camera; 6—digital temperature indicator; 7—hygrometer

duration 3 h to compare the performance of water harvesting on both surface. A high-speed camera was used to capture all images while experiment in the presence of LED lights operated at 12 V.

3 Result and Discussion

The fabrication leads to the development in surface morphology as required properties of super-hydrophobicity so that repeatable and well-characterized surfaces can be obtained with a bounce, fall-off rate and high contact angles. To characterize the surface condensation at thermal equilibrium states of drop progressing from nucleation sites to slide off on the bouncing surface indicates a dropwise condensation appearing as a cyclic phenomenon was obtained by capturing images at different interval of time such as 10, 20, 30, 40, 50 min for both substrates as shown in Fig. 3. A renucleation and growth was commenced by the continued condensation of moist air and gets larger in size by the coalescence on the intimated area of the substrate. The allied heat transfer rates were also found time-dependent, larger at initiation of the cycle and attains to minimum just before to set down of the drop.

The formation of dropwise condensation pattern by as nucleation of drops and combining to grow big in size of drop throughout experiment was dependent on heat

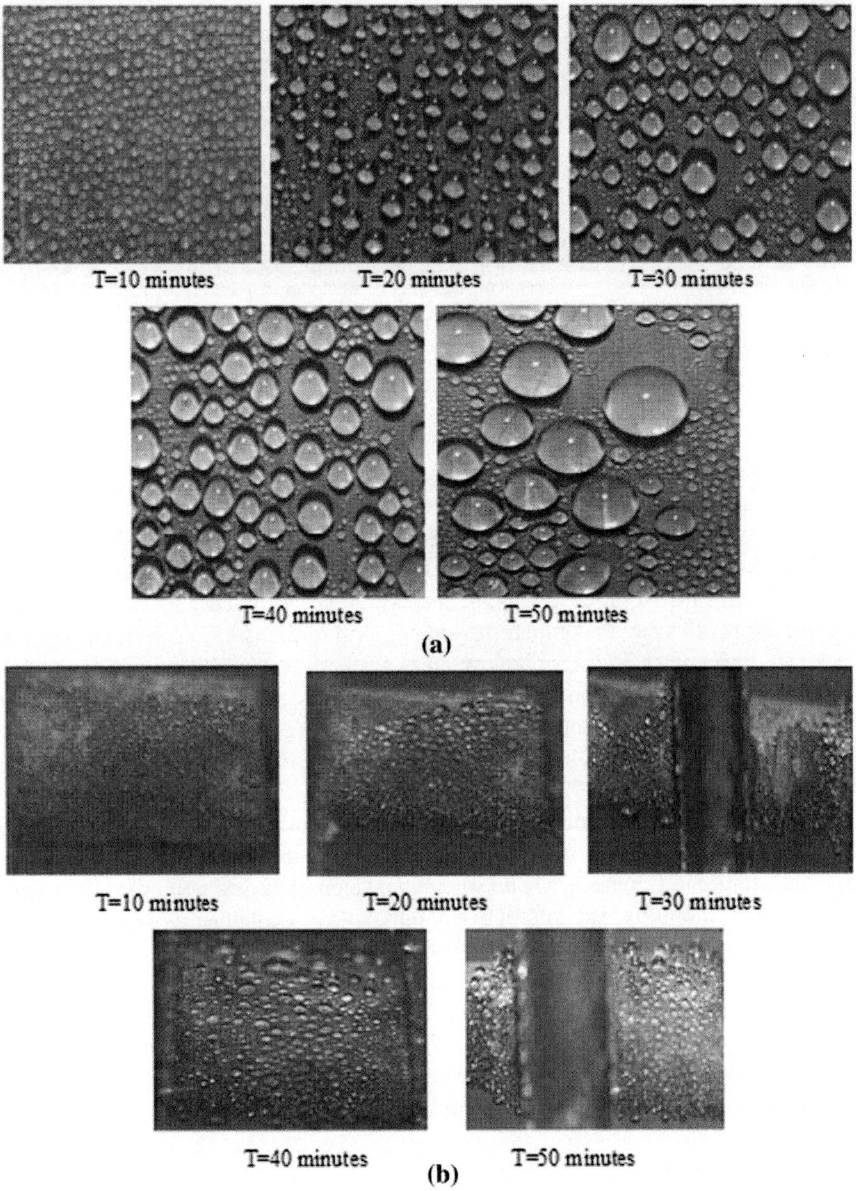

Fig. 3 **a** Condensation of moist air on copper mesh-like surface for different time interval at relative humidity (60–80%). **b** Condensation of moist air on solid copper mesh-like surface for different time interval at relative humidity (60–80%)

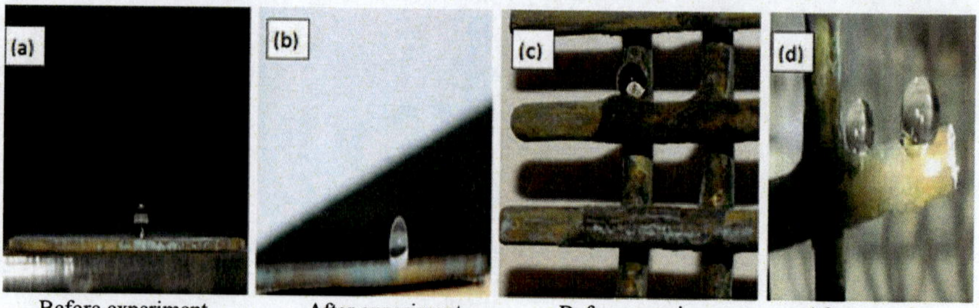

Before experiment $\theta=150^0$ After experiment $\theta=148^0$ Before experiment $\theta=155^0$ After experiment $\theta=152^0$

Fig. 4 **a** and **b** Images of drops on test surfaces of copper solid surface for measuring contact angle **c** and **d** Images of drops on test surfaces of copper mesh-like surface for measuring contact angle

transfer coefficient rate as well as surface area also related to enhance condensate rate. In such a way, the copper mesh-like surfaces were found to have larger contact area that of copper solid surface which revealed that the moist air easily flows over surface, resulted to a faster nucleation of drops to achieve effective with efficient condensation rate. In order to support the assessment of copper mesh-like surface that had more efficacies as compared to copper solid surface for water harvesting from moist air, the contact angles for both samples are as shown in Fig 4. It was found that the effect of increasing contact angle increases the heat transfer coefficient under the maintained environmental condition which can be analysed as by total water collected amount for mesh-like surface is 3.472 ml/ m^2-s approx. 20% more than the solid copper surface of 2.778 ml/m^2-s.

It is well explained from the history of literature that the increased relative humidity led to the rising dew point temperature of the moist air which signifies that temperature difference (ΔT) between moist air and its dew point decreases with rising relative humidity. This sort of behaviour could be explained, as the temperature of surfaces leads to grow more droplets when moist air temperature (28–33 °C) was kept constant and the increased relative humidity persuades condensation rate. At different relative humidity, an assessment for efficient condensation rate, a graph was plotted against time for 4 h as shown in Fig. 5 for both the cupper surfaces of solid and mesh-like. It was observed that the condensate collected time for Cu mesh-like surface was 1.3 times faster than Cu solid surface.

4 Conclusion

This research work is not only limited to express significant condensation heat transfer improvement, but this research has developed a new and inexpensive technology that produces potable water from thin air. This can resolve the problem of water stress in developing countries such as India and also scalable method to increase efficacy for applications in the field of atmospheric water harvesting and

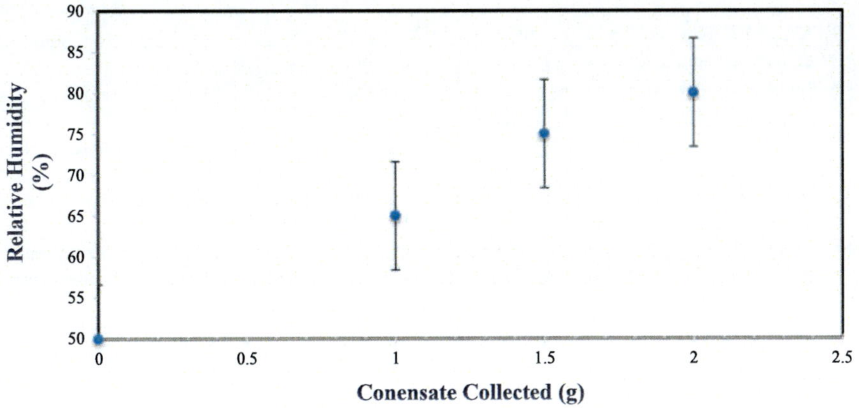

Fig. 5 Condensation rate of moist air for both on copper solid surface and mesh-like surface

dehumidification. In future, the work can be extended to perform above 80% RH under diverse days of silylation process. In order to access the water harvesting, the set-up efficiently dehumidifies, recycles and most of them are also an environment friendly so positively no wastage of water is registered and neither this system consumes any groundwater although it securely fruitages potable water. The combination of controlled environment and sustainability tracks a system which is completely autonomous and renewable energy source.

Acknowledgements The author acknowledges Mr. Deepak Kumar Sharma, Research Scholar, Department of Amity Institute of Nano Engineering and Technology, for support while Nano-texturing of surfaces and financial support from the Science and Engineering Research Board (SERB), India, established through an Act of Parliament: SERB Act 2008, Department of science & Technology (DST), India, Government of India (Project no: ECR/2016/000020).

References

1. Zanelli A, Monticelli C, Polomová B, Hanuliak P, Vargová A (2019) Application of atrium tensile structure in historic building. Case study : daylight modelling of atrium within tensile historic building. Exploring fog harvesting solution membrane structures towards sustainable development in the Italian urban context, pp 535–546
2. Liu S et al (2017) Experimental analysis of a portable atmospheric water generator by thermoelectric cooling method. Energy Procedia 142:1609–1614
3. Lawrence MG (2005) The relationship between relative humidity and the dewpoint temperature in moist air: a simple conversion and applications. Bull Am Meteorol Soc 86(2):225–234
4. Wahlgren RV (2001) Review paper atmospheric water vapour processor designs for potable water production : a review 35(1)
5. Jensen KR, Fojan P, Jensen RL, Gurevich L (2014) Water condensation: a multiscale phenomenon 14(2):1859–1871
6. Hou Y, Yu M, Chen X, Wang Z, Yao S (2015) Filmwise-to-dropwise condensation transition enabled by patterned high wetting contrast. J Heat Transfer 137(8):080907

7. Kim MH, Kim H, Lee KS, Kim DR (2017) Frosting characteristics on hydrophobic and superhydrophobic surfaces: a review. Energy Convers. Manag 138:1–11
8. Sikarwar BS, Khandekar S, Agrawal S, Kumar S, Muralidhar K (2012) Dropwise condensation studies on multiple scales. Heat Transf Eng 33(4–5):301–341
9. Aronov D, Rosenman G, Barkay Z (2007) Wettability study of modified silicon dioxide surface using environmental scanning electron microscopy. J Appl Phys 101(8)
10. Miljkovic N, Enright R, Wang EN (2013) Modeling and optimization of superhydrophobic condensation. J Heat Transfer 135(11):111004
11. Sharma DK, Baghel V, Kumar R, Avasthi DK, Sikarwar BS (2019) Recent developments in fabrication of super-hydrophobic surfaces: a review. In: Lecture notes in mechanical engineering

Mathematical Modeling and Design Optimization of an Alternative Cost-Effective Rain Sensing Wiper System

Amol M. Khatkhate, Bairagi Mina, Maurya Ramkumar, Jasdanwalla Husain, Dhamal Nitin, and Shah Varsha

1 Introduction

During to heavy rains, lot of vehicles loses their electronic capability due to water-logging which can result in loss of life of individuals. A rain sensing wiper and alert system that works on a principle other than electronics and *independent* of the microcontroller of the vehicle would be a boon in such a situation. The voltage can be made to close the windows or open the door locks/windows during heavy rain allowing passengers to escape from the vehicles. A low cost sensitive rain sensing wiper will be very useful in such a scenario.

Section 2 discusses the design of the electrochemical rain sensor (ERS) and the various parameters associated with its sensitivity. It also explains in detail the working of the ERS. The useful life of the ERS and comparison with the existing sensors in literature has been explained in detail in Sect. 2.4 and 2.5. The literature review [1–4] and working principle of various existing automatic rain sensing wiper technologies are discussed in detail in Table 2 in Sect. 2.5 of the paper. Section 3 discusses the application of electrochemical rain sensor to the wiper system. Section 4 presents the detailed mathematical model for the conventional Lucas wiper motor and for the newly proposed DC geared motor with the rack and pinion arrangement. Section 5 presents the results of the simulation of the ERS-based wiper system and shows the comparison with the conventional Lucas wiper motor arrangement for various parameters. Section 6 discusses the salient features, conclusions and future work.

A. M. Khatkhate (✉) · B. Mina · M. Ramkumar · J. Husain · S. Varsha
Rizvi College of Engineering, Off Carter Road Bandra (W), Mumbai 400050, India
e-mail: amolmk@eng.rizvi.edu.in

D. Nitin
Veermata Jijabai Technological Institute (VJTI), Matunga, Mumbai, India

V. H. Saran and R. K. Misra (eds.), *Advances in Systems Engineering*, Lecture Notes in Mechanical Engineering,
https://doi.org/10.1007/978-981-15-8025-3_18

2 Design of the In-Situ Electrochemical Rain Sensor (ERS)

The in-situ electrochemical sensor as seen in Fig. 2 is made using the conventional technique. Two electrodes of Cu and Zn are placed in their respective electrolytes. In order to complete the circuit, a salt bridge is made using the following procedure.

2.1 *Procedure for Making Accordion Joint Salt Bridge*

The salt bridge acts as a connector between the electrolytic solutions and thereby completes the circuit. Basically, the salt bridge is in the galvanic cell to regulate the charges in solution and keep them neutral. Without this, there is no flow of electrons and therefore no electrical output. It avoids voltage drop. The standard procedure of making a salt bridge is enlisted below:

1. Prepare about 150 ml of distilled water in 400 ml beaker and bring to boil.
2. Take 3 g agar-agar powder and stir the mixture as the suspension boils.
3. Remove the beaker from the heat and stir in 15 g KCL until the salt dissolves.
4. Pour the warm mixture in the bent straw salt bridge until it is completely filled. Once agar is set, store in plastic bag for preventing drying out.
5. Seal one end with cotton plug and ensure that no air bubbles are present in the salt bridge prepared.

2.2 *Working of Electrochemical Rain Sensor (ERS)*

The electrochemical alert sensor is modeled as a mass-spring-damper system as shown in Fig. 1. The parameters used in the design of the electrochemical sensor are given in Table 1. The static deflection due to the container, salt bridge and spring is 51.4 μm. In order to test the sensitivity of the sensor, raindrops were considered as a forcing input [5].

Diameter of raindrop—3 mm, terminal velocity—8 m/s, time taken to come to rest—80 ms. Force generated due to impulse = mv/t = $(1000 \times \pi \times (0.003)^3 \times 8)/(6 \times 0.08) = 1.41$ mN. Variability of the impulse force will be due to the different radii of the raindrops and will change only the magnitude of the response seen in Fig. 2 but not the time period of oscillations.

Every raindrop generates an impulse force of 1.41 mN resulting in a deflection of 3.2 μm of the assembly (container + salt bridge + spring) which slowly reaches to static deflection in about 1.2 s. This was in consideration with the value of damping of air, $\zeta air = 0.01$, while in practice, air has almost negligible damping and will continuously produce a sinusoidal oscillation of amplitude 3.2 μm.

The steady state response to raindrop force can be seen in Fig. 2. The cell will continue to generate current while it is raining and will stop when the rainwater

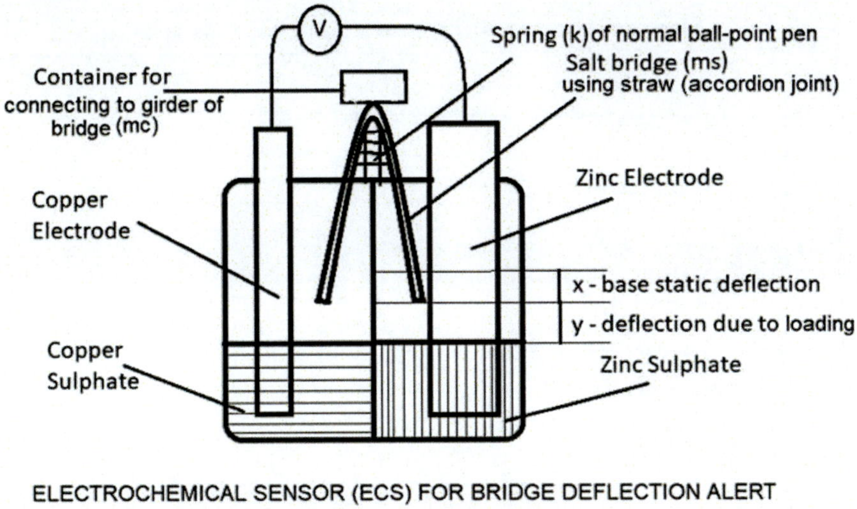

Fig. 1 Electrochemical rain sensor (ERS)

Table 1 Parameters for the electrochemical rain sensor

Parameter	Value	Unit	Availability
Mass of container (m_c)	1.47	gm	General purpose container to give medicine to children and adults (10 ml)
Mass of salt bridge (m_s)	3.5	gm	Normal straw with center bend (accordion joint)
Mass of spring (m_k)	0.17	gm	Used in normal ballpoint pen
Stiffness of spring (k)	950	N/m	NA

in the container drains out. The deflection '*y*' can be adjusted to allow for suitable amount of rainwater being collected in the container thus adjusting the sensitivity of the ERS. Sample calculation of the sensitivity of the cell is explained in the next section (Fig. 3).

2.3 *Sample Calculation of Sensitivity of Electrochemical Rain Sensor (ERS)*

Volume of 1 raindrop = πD3/6 m^3 Allowable deflection/threshold = y_{allow}

Steady state oscillation due to single raindrop = 3.2 μm Steady state response due to *N* raindrops = 3.2 N μm

Total deflection due to '*N*' raindrops = $N \times 1000\ g\pi D^3/6\ \mu m$ (Static deflection) + 3.2 N μm (oscillatory assuming all in-phase)

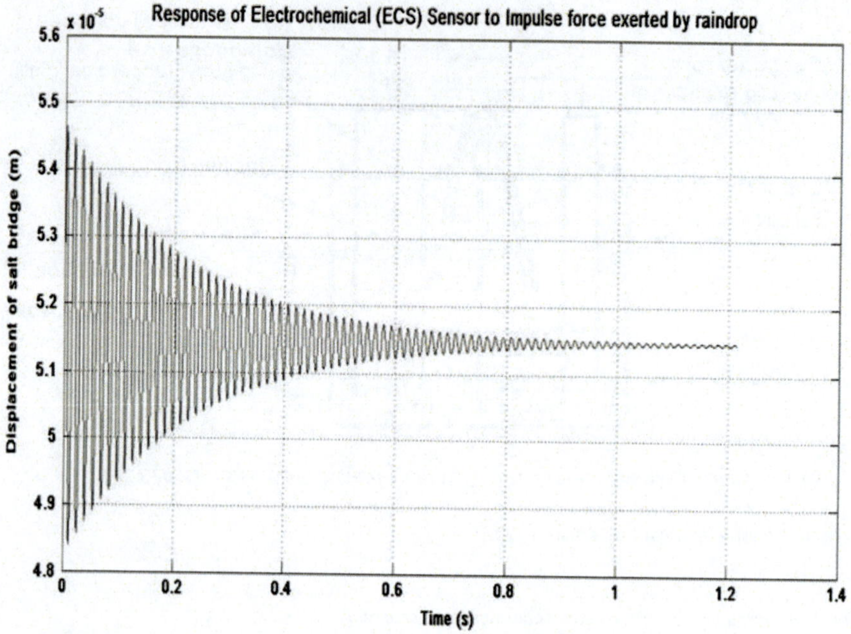

Fig. 2 Response of electrochemical rain sensor (ERS)

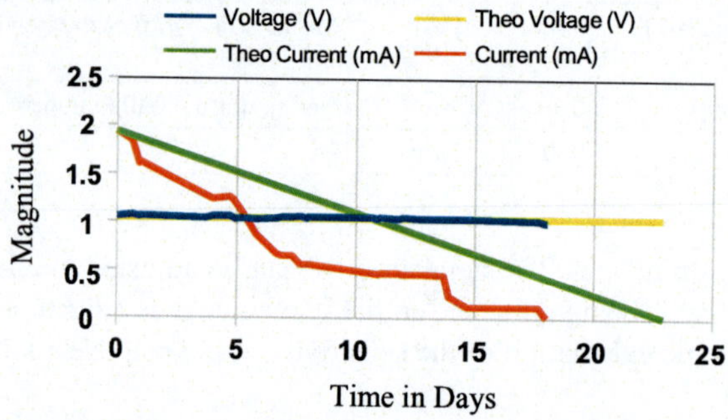

Fig. 3 Performance of ERS (Cu-Zn voltaic cell with accordion salt bridge in 0.1 M 100 ml solutions) (The data for the Cu-Zn cell can be shared upon request and proper authorization from Rizvi College of Engineering (RCOE))

Assume the time between successive raindrops be 1 s; then $y$allow $= N \times 1000\ \mathrm{g}\pi \mathrm{D}^3/6\ \mathrm{k}\ \mu\mathrm{m} + 3.2\ \mathrm{N}\ \mu\mathrm{m}$

For $D = 3$ mm; $k = 950$ N/m $y$allow $= 0.148$ N $+ 3.2$ N

So, for a threshold value equal to a thickness of an A4 paper, $y_{\mathrm{allow}} = 50\ \mu\mathrm{m}$ which gives $N = 15$ with a response time of 14 s assuming a spacing of 1 s between raindrops.

$N = ROUND(y_{allow}/3.348)$
Sensitivity of ERS, S = 3.2/Volume of 1 raindrop (ml) = 3.2/0.0141 = 227 μm/ml
Time to response = N-1 seconds

2.4 *Useful Life of the Electrochemical Rain Sensor (ERS)*

The sensor produced in the manner as described above is very light in weight and sensitive to small disturbances. Also, due to the field environment, the sensor may lose its capability of sensing. Due to changes in ambient temperature, the salt bridge characteristics may change and thereby make the sensor malfunction. Hence, in order to understand the exact behavior of the sensor, we conducted a test for determining the average lifespan of the sensor when it is continuously in the conducting state. The experiment was conducted with an accordion salt bridge over a period of 18 days using 0.1 M solution of each of the electrolytes CuSO4 and ZnSO4. The voltages and current were measured at different time intervals. The measured voltages and current were used to compute the actual power of the sensor (cell). The theoretical time for the cell life was computed as given below:

Cell life in days = (No of cells × Molarity × Volume (liters) × No of electrons/mole × 96485)/(60 × 60 × 24 × (Final current + Initial current)/2))

The theoretical current is taken to be average of initial and final current value. The theoretical cell (sensor) life turned out to be 22.9 days. The actual experiment worked for 18 days (432 h). The performance of the ERS is as seen in Fig. 4. The actual curve of the current indicates that the ERS will discharge as a battery in an exponential manner. The voltage provided by the ERS is almost constant at 1.04 V which is less than the expected value of 1.1 V due to polarization effects. The current starts at a value of 2 mA and exponentially decays to zero.

Observations:

1. The volume of electrolytes reduced to 50% after 18 days, and $CuSO_4$ became almost colorless indicating consumption of all ions.
2. Accordion salt bridge made of ***plastic*** does not allow for proper gelation of the agar-agar KCL solution, and hence, further work in this area needs to be carried out. Advantage of this method is the consumption of plastic and creating a plastic-free environment.

2.5 *Comparison of the Electrochemical Rain Sensor (ERS) with Other Sensors*

The following Table 2 explains the comparison of the ERS with various other sensors

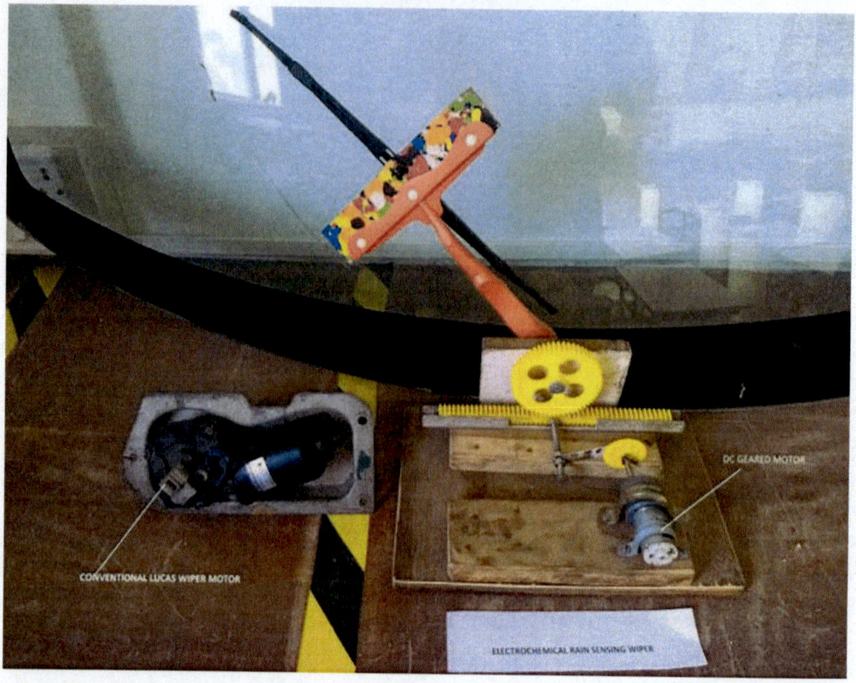

Fig. 4 **a** Conventional Lucas wiper motor. **b** Electrochemical-based rain sensing wiper system

and has been published in the literature [6].

3 Application of the ERS to the Wiper System

The following sections discuss in detail the two types of arrangements for the operation of the wiper motor using

- Conventional Lucas wiper motor arrangement [7].
- DC Geared Wiper motor with rack and pinion mechanism [8, 9].

3.1 Comparison of the Lucas Wiper Motor and the ERS-Based Wiper Motor

The conventional wiper motor assembly is as shown in Fig. 4. The wiper motor assembly developed at Mechatronics Lab, RCOE, using a DC geared motor and a rack and pinion mechanism is as shown in Fig. 4. The normal wiper motor operates at two speeds of 45 and 60 rpm (Fig. 5; Table 3).

Table 2 Comparison of various rain sensors

	Different types of rain sensors				
	Plate-based sensor [2]	Piezoelectric sensor [1]	Probe-based sensor [4]	Optical-based rain sensor [3]	Electro-chemical rain sensor [6]
Operating principle	Microcontroller-based	Microcontroller-based	External circuits connected to microcontroller	Microcontroller-based	Independent of microcontroller
Working	Small drops of water change the resistance	Water between plates decrease resistance	Contact of water with probe completes rain circuit	Change in reflection due to rainwater	Rainwater energy is converted into displacement of salt bridge
False rain detection rate	Very high	Less than 5%	Less than 2%	Less than 2%	Less than 2%
Voltage required	12–5 V DC	12–5 V DC	2–6 V DC	12–5 V DC	1–3 V per single cell
System response time	Not mentioned	500 ms after rain contact	On collection of 10 cu.cm	Very low	550 ms after rain contact
Sensor surface size	Very large	≤4 cm × 4 cm	≥2.58 cm × 2.58 cm	Not applicable	1 cm × 1 cm
Placement of sensor	On the windshield	On the windshield	Inside the font hood	Inside the car cabin	Inside the front hood
Cost of replacement	Quite high	Quite high	Around Rs. 1000	Around Rs. 7000	Around Rs. 1500
Adaptivity	Changes the aesthetics of the vehicle	Embedded on windshield	Volume of rain collected is high	Seamless integration with vehicles	Adaptable to all vehicles

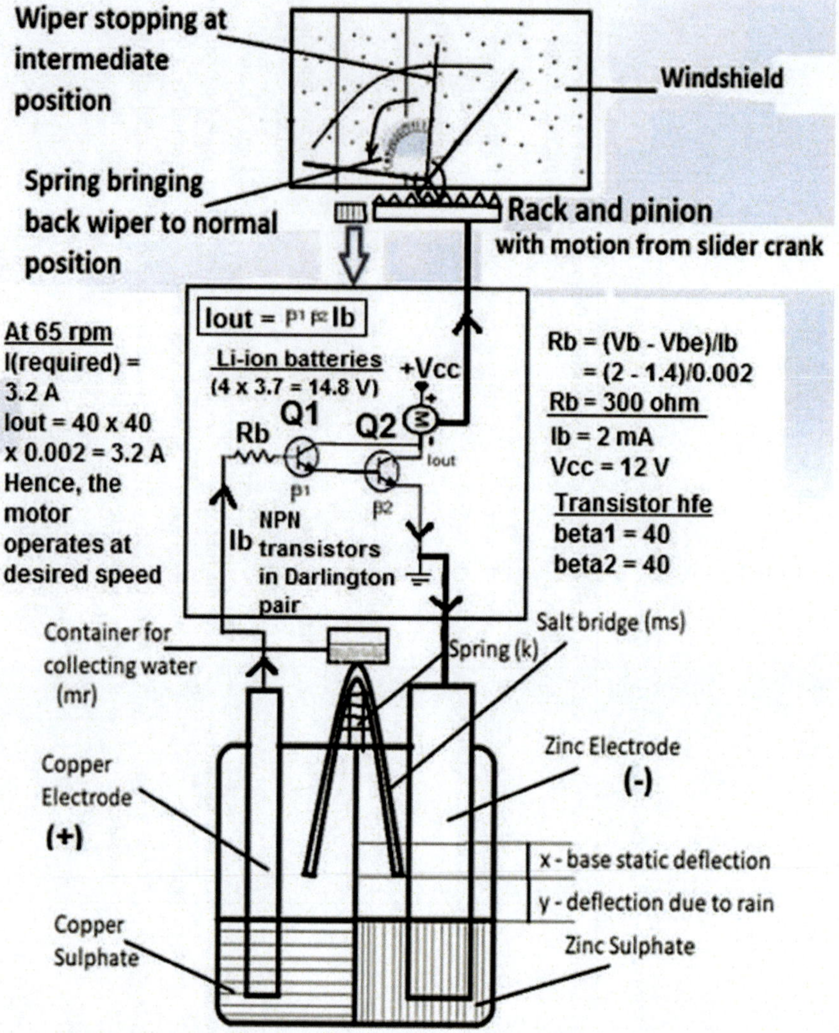

Fig. 5 Schematic of the electrochemical-based rain sensing wiper system

The torque required at 65 rpm for a wiper rod of length 450 mm with a mass of 330 gms is as shown in Fig. 6. The wiper is modeled as a rotational load along with the rack and pinion which can suitably generate the required trajectory.

Frictional force is taken as a conservative value and considered to be 0.8 N [10].

Table 3 Parameters of the two configurations

Parameters	Conventional Lucas wiper motor [8]	DC geared motor with rack and pinion [6, 7]
Weight of wiper arm	330 gms	330 gms
Length of wiper arm	450 mm	450 mm
Crank rod length	NA	25 mm
Connecting rod length	NA	50 mm
Diameter of pinion	NA	79 mm (Module 1.41, Teeth—56, Pitch—18)
Length of rack	NA	100 mm
Weight of motor	800 gms	180 gms
Stall torque of motor	180 kg-cm	80 kg-cm
Stall current	14 A	7.5 A

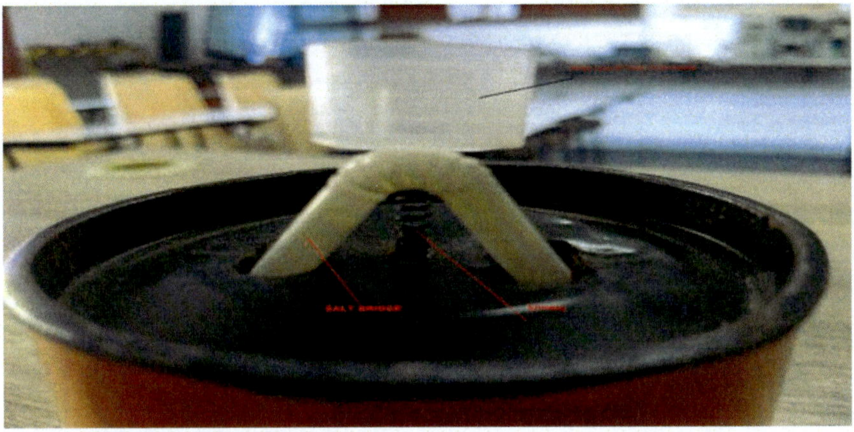

Fig. 6 Actual prototype of the electrochemical rain sensor (ERS)

3.2 *Schematic of the Proposed Electrochemical-Based Rain Sensing Wiper System*

Figure 5 shows the detailed connections for the implementation of the electrochemical-based rain sensing wiper system. As the impulse force due to the raindrops causes the salt bridge to connect with the electrolytes, the current is generated which is fed to the base of transistor with hFe = $\beta 1$. The base of the transistor is connected to the ERS which has to generate more than 1.4 V for both the transistors in pair to work in active Darlington configuration. This is achieved by a Cu-Al voltaic cell (cell voltage = 2 V) combination whose results will be similar to those mentioned for the ERS above.

The general purpose transistor BC547 has a hFe from 10 to 1000. For a BC547-A Series value of 40, the transistor can supply enough current (2 Ma × 40 × 40 = 3.2 A) to run the wiper at the speeds of 45 and 65 rpm. The voltage at 45 rpm is 9 V which is got from a voltage regulator.

The transistors in Darlington pair cause the current to be amplified by a factor of $I\text{out} = \beta 1 * \beta 2 * I$ in. This huge magnified current can be used to drive other applications like door opening/closing and windows opening, etc., at higher speeds of rainfall. The voltage supply $+Vcc$ is 14.8 V (4 Li–Ion batteries) that is charged through a solar panel. The Darlington pair require a voltage of 0.9 V across the collector–emitter junction, and hence, the voltage applied across the motor is 14.8–0.9 = 13.9 V which corresponds to a maximum speed of 70 rpm. The bases of both the transistors need to be in forward bias condition for which the voltage across the P-N junction must be greater than 0.7 V each, i.e., 1.4 V. Hence, the cell to be used is an Cu-Al cell which generates a steady voltage of 2 V instead of the Cu-Zn combination.

As seen in Table 4, the torque required by the DC geared motor arrangement is quite less as compared to the conventional Lucas motor for the three-speed arrangement. The current required is almost the same in all the cases for both the wiper arrangements.

3.3 *Sensor Location on the Vehicle*

The windshield on a passenger vehicle typically ends in the portion of the car where there is a space for the rainwater to trickle down after the wiper cleans the windshield. This area has small holes and can accommodate the sensor underneath.

The ERS kit can be very well accommodated without causing any change in the current manufacturing process and thereby provide the user the capability of automatic rain sensing for the vehicle as seen in Fig. 7a. Unlike the optical rain sensor [6] which is located behind the rearview mirror (See Fig. 7b), the ERS kit will be located in the hood. The optical sensor requires a power source and amplifier and is also connected to the microcontroller unlike the ERS kit which can function independently.

A suitable '**ERS KIT—ELECTROCHEMICAL RAIN SENSING KIT**' will be designed using the following components as shown in Table 5.

4 Mathematical Model of Existing and Proposed Wiper System

The mathematical models of both the systems are presented in this section

Table 4 Comparison of the two configurations for wiper motor systems carried out in MATLAB2014a

Motor configuration	45 RPM			65 RPM			100 RPM			Cost of system per vehicle
	*T*avg (N-m)	*I*avg (A)	*I*max (A)	*T*avg (N-m)	*I*avg (A)	*I*max (A)	*T*avg (N-m)	*I*avg (A)	*I*max (A)	
Conventional Lucas wiper motor arrangement [8]	2.4	1.71	2.18	2.44	1.71	3.29	2.44	1.71	6.2	4400/- (MIN) (TVS single wiper system) [9, 11]
DC geared wiper motor with rack and pinion mechanism [6]	2	1.68	2.13	2	1.68	3.2	2	1.68	6	1500/−1000 (DC geared motor) + 200 (Rack and pinion) + 100 (Slider crank) + 200 (ERS) [7]
% Change in parameter from conventional arrangement	**-18**	**-1.75**	**-2.75**	**-18**	**-1.75**	**-2.75**	**-18**	**-1.75**	**-3.23**	**66% reduction**

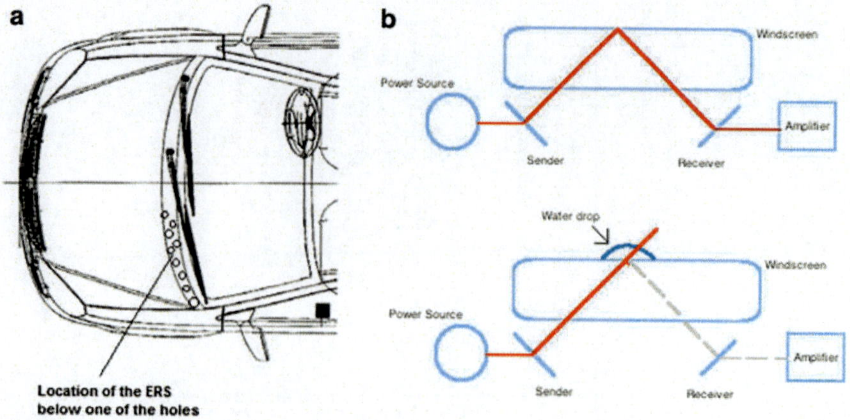

Fig. 7 **a** Location of the ERS kit in a conventional vehicle. **b** Optical rain sensor located inside the vehicle near the rearview mirror [6]

Table 5 ERS Kit components

Components	Manufacturing cost
Rain sensitive cell	200
Accordion joint salt bridge (12 QTY)	70
Rack and pinion	180
Slider crank	50
DC geared motor	1000
Total	1500

4.1 *Conventional Lucas Wiper Motor System*

The conventional Lucas motor seen in Figure 8 consists of worm and worm wheel attached to a DC motor and operating a four bar mechanism to rotate the wiper. The equation of torque produced by the Lucas motor is shown below

$$\mathbf{T = F\,\mu\, f_s\, f_l\, \mathit{l}\, (\omega_a/\omega_m)(R_h/R_c)(1/e)}$$

where:

T torque to move one wiper arm (N-m).
F force onto blade on screen $(N) = (W + 0.8)\cos(\beta + \lambda)$.
W 90 gms weight.
Frictional force, 0.8 N = maximum frictional force on blade [5].
β angle of travel by wiper (rad) = max 180–2λ.
λ relief angle on windscreen = 15.

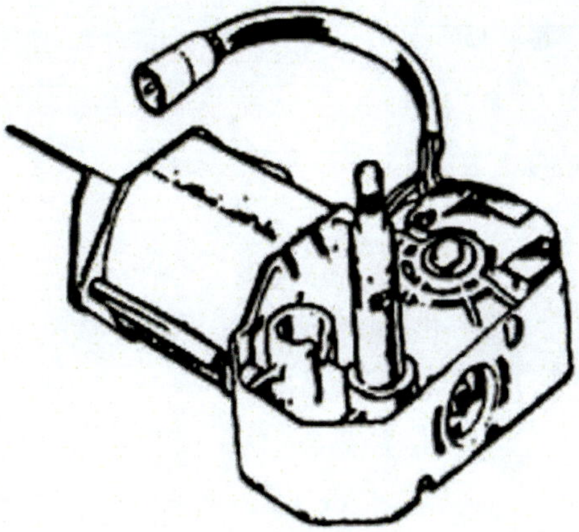

Fig. 8 Conventional Lucas wiper arrangement for single wiper [7]

μ maximum dry coefficient of friction, which is 2.5.
f_s multiplier for joint friction, which is 1.15.
f_1 tolerance factor, which is 1.12.
l wiper arm length, m = 0.24 m.
ω_a maximum angular velocity of arm, rad/s.
ω_m mean angular velocity of motor crank, rad/s.
e efficiency of motor gear unit which is 0.8.
R_h motor winding resistance, hot-Ω = 1.05.
R_c motor winding resistance, cold-Ω = 1 (Figs. 9 and 10).

4.2 DC Geared Motor Rack and Pinion Wiper System

Taking downward direction as positive

$$W + mg = N \quad (1)$$

Taking right side positive

$$-Ft + \mu N = -\text{mddot}(x) \quad (2)$$

Moment about O taking clockwise positive

$$-T - \text{WLcos}(\beta + \lambda_0)/2 + F_t R - \mu\text{NR} = \text{Iddot}(\beta) = (\text{WL}^{2/3g})\text{ddot}(\beta) \quad (3)$$

Also, for the pinion and rack arrangement,

$$\beta = -x/R$$

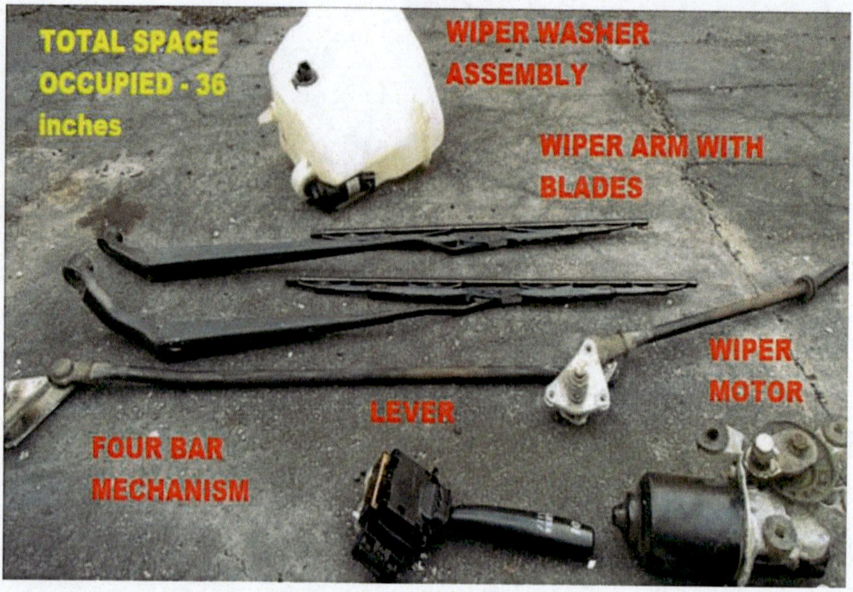

Fig. 9 Various parts of the conventional wiper assembly

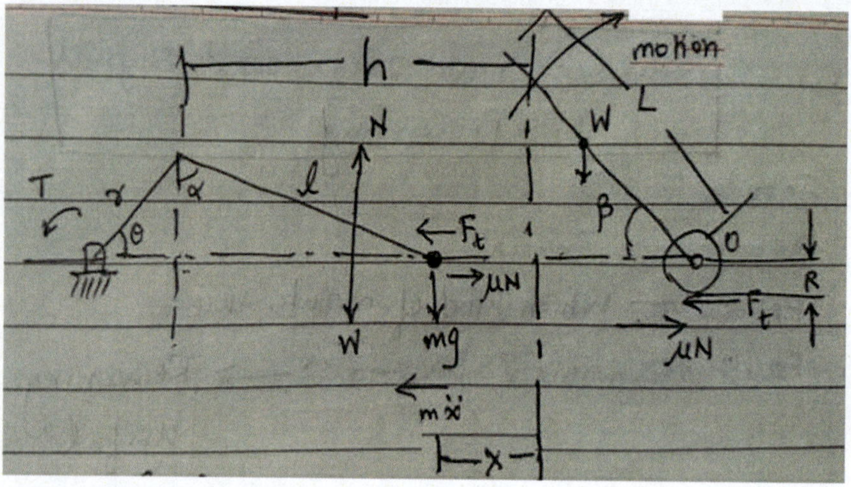

Fig. 10 Free body diagram of the ERS-based wiper system

$$\text{dot}(\beta) = +\,\text{dot}(x)/R$$
$$\text{ddot}(\beta) = -\text{ddot}(x)/R$$

Substituting in Eq. (3), we get

Fig. 11 Proposed ERS-based wiper system for automobile [9]

$$T = [(WL^2/3gR) + mR]\ddot{x} - (WL/2)\cos(\beta + \lambda_0) \tag{4}$$

Consider the rack and pinion configuration as shown in Fig. 11.

In a pinion rotation of $180–2\lambda^0$ deg in a single direction, the rack covers distance equal to ONE travel length. The relief angle on the windshield is given by $\lambda 0 = (2\pi - 2(xm/R))/4$.

5 Simulation Results and Discussion

The curves showing the displacement of rack, velocity of rack and acceleration of rack are given in Fig. 12.

The performance of the DC geared wiper motor is compared with the conventional wiper motor, and the results are shown in Fig. 13.

As seen in Fig. 13, the torque required by the DC geared motor wiper is large at the starting point and then reduces as it approaches the mean position. As seen in Fig. 14, the tangential force provided by the rack and pinion arrangement and the conventional wiper motor is shown. The force required is maximum at the extreme position and reduces toward the mean positions. The inertial torque is considered in

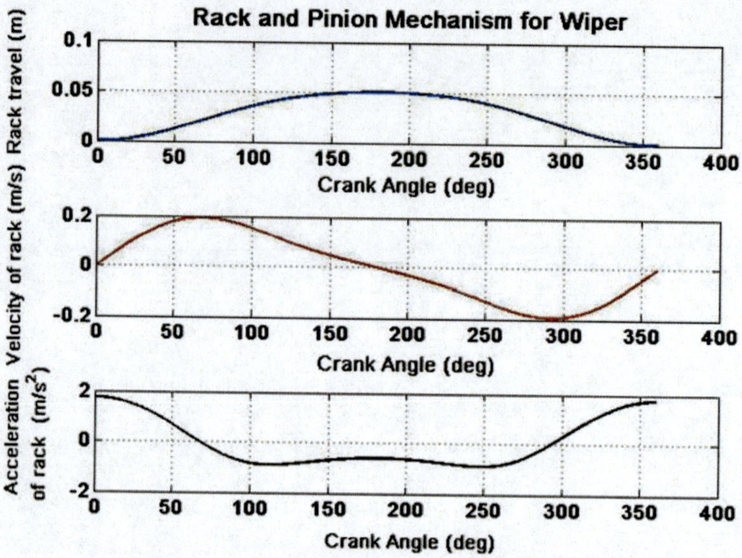

Fig. 12 Response of the slider crank mechanism actuating the rack

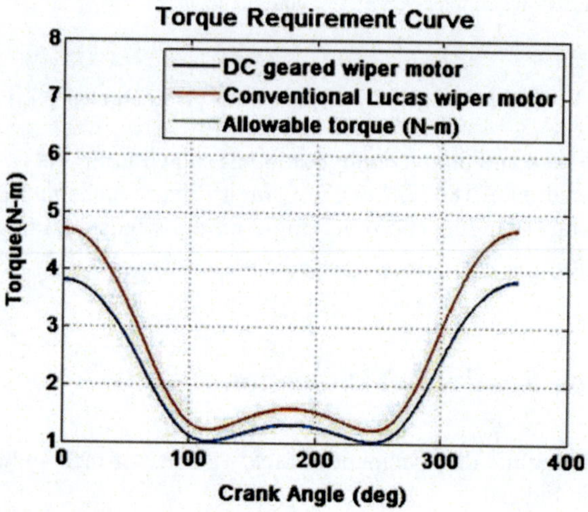

Fig. 13 Torque demand curve for the wiper systems

the analysis and is indicated by the αë both for the Lucas as well as DC geared motor arrangement as seen in mathematical derivation in Sect. 4 (Figs. 15 and 16).

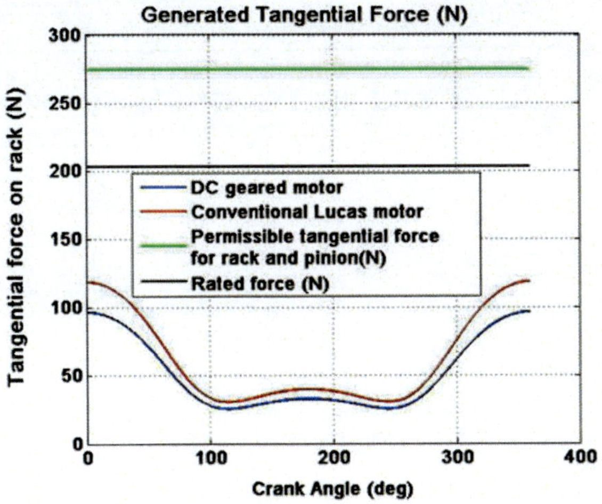

Fig. 14 Tangential force for Lucas wiper motor and DC geared motor at 65 rpm

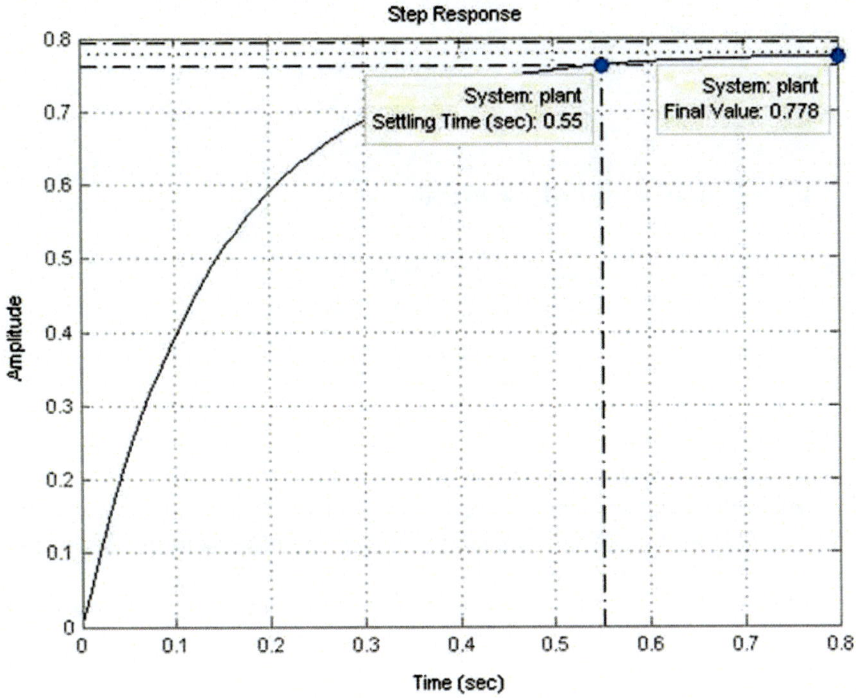

Fig. 15 Response time for rack and pinion mechanism

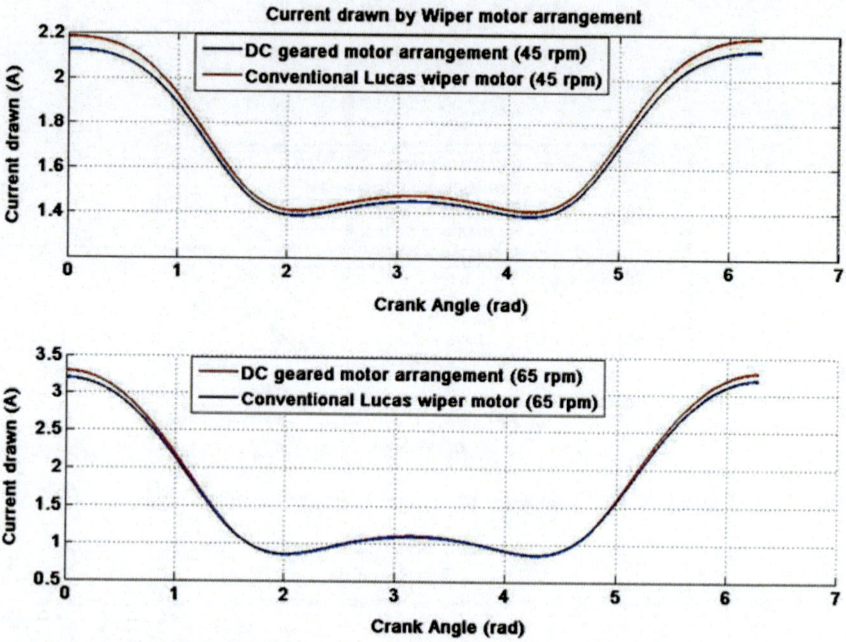

Fig. 16 Current drawn for Lucas wiper motor and DC geared motor at 65 rpm

6 Conclusions and Future Work

This section discusses the salient features of the electrochemical rain sensor (ERS)-based wiper system as compared with the existing conventional Lucas wiper system.

- Since the electrochemical reaction is spontaneous, the actuation by this kind of a cell will result in very ***less response time***. (approx 700–550 ms by rack and pinion + 150 ms by the ERS as seen in Fig. 15).
- The DC geared motor consumes almost ***18% less torque*** than the conventional Lucas motor and runs with very minimal parts.
- Also, ***80% weight reduction*** by replacement of Lucas motor with geared DC motor and the ***easy off the shelf availability of the rack and pinion*** setup are a lucrative point to choose this alternative.
- Compactness of this arrangement as seen in Fig. 6 which has a dimension of 36 inches to a reduction of 6 inches which is ***1/6th the original size*** is a great advantage for the auto industry.
- The cost reduction estimated with the alternative wiper system is around ***66% per vehicle***.
- Experimental verification of the same is a part of ongoing work.

Furthermore, a suitable ARAI standard compatible product will be designed. The sensitivity of the product will be compared with the existing products in the market.

Pilot project on an actual existing compact car will be carried out to prove the effectiveness of the technology. Also, suitable maintenance procedures and manuals will be designed so that the technology transfer to car service centers can be easily carried out. The benefits of the technology can be seen if implemented by car companies on a large scale. Furthermore, the following enhancements will be done in due course of time which will lead to value addition of the above rain sensing wiper:

- Automatic speed control of wiper based on rate of rainfall.
- Automatic mist and fog removal.
- Design of parking mechanism for wiper.
- Design of vibration isolation system for cell and effect of temperature on the performance.

The introduction of this new type of system in current cars needs to also be checked in terms of its effect on the fuel economy especially the effect on reduction in usage of air-conditioning and the gains/losses that it has as compared to an existing rain wiper system.

Acknowledgements We would like to thank Mr. Bhaskar Gareg for providing us with the windshield free of cost that motivated us in conducting this experiment. We would like to thank Mr. Milind Sanzgiri from Asha Scientific who helped us by providing us with suitable electrodes for our experiment. At last, we must express my sincere heartfelt gratitude to all the student members of Mechanical Engineering Department, Electrical Engineering Department and Chemistry/Humanities and those actively associated with Mechatronics Lab, RCOE, who helped us directly or indirectly during this course of work. We are thankful for the continuous encouragement and involvement of Dr. Sujata Rout, Head of Mechanical Department, and Dr Varsha Shah, Principal, Rizvi College of Engineering (RCOE), in this project.

References

1. Dharmadhikari S, Tamboli N, Gawali N, Lokhande NN (2014) Automatic Wiper System, Department of Instrumentation and Control Engg, Pravara Rural Engineering College, Loni. Int J Comput Technol Electron Eng (IJCTEE) 4(2) (An ISO 9001: 2008 Certified International Journal)
2. Prabhakaran N, Surendran P (2015) An efficient low cost wiper system for autonomous vehicle. School of Electronics Engineering VIT University, Vellore, Tamil Nadu, India
3. A J Eng Res (AJER) 03(04):24–34. e-ISSN: 2320-0847 p-ISSN: 2320-0936
4. Kulkarni TS, Holalad HS (2012) Semi-Automatic Rain Wiper System. Int J Emerg Technol Adv Eng 2(7). ISSN 2250-2459 (Dept of Automobile Engg, B V Bhoomaraddi College of Engg & Tech Hubli)
5. Van Boxel J (1998) Numerical model for the fall speed of raindrop. Sepsis
6. https://cecas.clemson.edu/cvel/auto/sensors/rain-sensor.html
7. http://what-when-how.com/automobile/windscreen-wipers-and-washers-automobile/
8. Chandekar P, Kelaskar S, Dhanawade A, Aiyapillai A, Khatkhate AM, Mitwa Y (2017) An electrochemical rain sensing wiper. Int J Innovative Res Eng Manage (IJIREM) 4(4). ISSN: 2350-0557
9. DC geared motor—http://www.robonist.com/Rack and pinion–VJTronics, Pune

10. Constantin-Buta A (2014) Study of friction between wiper blade and windshield. Bulletin of the *Transilvania* University of Braşov Series I: Eng Sci 7(56):1–2014
11. TVS Wiper system cost (ITEM 2107)—https://drive.google.com/file/d/1LgMucOQ875ZBWxGxtCW_o5IHHVcN498Q/view?usp=sharing

Dynamic Modelling and Payload Response Analysis of a 3-D Overhead Gantry Crane

Aakash Krishna, Ravindra Singh Bisht, and Soraj Kumar Panigrahi

1 Introduction

Crane systems are commonly used at civil construction sites, warehouses, manufacturing plants, and port facilities to handle various types of payloads. The crane system helps the industry for transporting various payloads from one place to another. Industry uses various popular cranes such as truck-mounted crane, tower crane, gantry crane, rough-terrain crane, and port crane. These cranes may be classified as per their dynamic properties according to various coordinate systems. At high-rise construction sites, generally, a tower crane is used, and a cylindrical coordinate system is used to describe the mathematical modelling for movement of the payload during the crane operation. The payload is supported by wire hoist, and the trolley moves along the boom arm in case of a gantry crane. The gantry crane uses a Cartesian space to describe the position and movement of the system. The port/boom crane is used as a spherical coordinate to describe the movement of the system. The payload is backed by the wire rope suspension. A port crane is normally placed on a base to facilitate the change of the workspace. All types of crane systems must work very quickly, safely, and precisely, because efficiency is directly related to the productivity of material handling in industries. Industry uses double-pendulum [1–11] suspension-type cranes where both the hook and the payload are subjected to dynamic excitation.

A double-pendulum crane system represents the motion of both the hook and payload. The operation of the crane system becomes extremely difficult due to the nonlinear dynamic behaviour of the double pendulum [11–15] particularly when the

A. Krishna
Quantum School of Technology, Roorkee 247667, India

R. S. Bisht (✉) · S. K. Panigrahi
CSIR-Central Building Research Institute Roorkee, Roorkee 247667, India
e-mail: rsbisht@cbri.res.in

V. H. Saran and R. K. Misra (eds.), *Advances in Systems Engineering*, Lecture Notes in Mechanical Engineering,
https://doi.org/10.1007/978-981-15-8025-3_19

crane base or rail and trolley of the crane is movable. The overhead crane system based on the assumption of a unique pendulum model is not always true. In some cases, the mass of the hook cannot be ignored, which has led to a double pendulum-type system [8, 9]. Unlike a simple pendulum that only balances the payload, a double pendulum must take into account the balancing of the payload and the balancing of the hook. It has been reported that a double-pendulum crane underwent oscillations of the payload [1–4, 10–15] and its effective control strategies [1, 2, 5–9, 12, 13, 16, 17]. Therefore, designing a precise, safe, and fast-operated double-pendulum crane system to position the load with minimum payload swing is more challenging due to the complexity and chaotic dynamic behaviour of the double pendulum.

This study presents the mathematical modelling of a three-dimensional (3-D) gantry crane with a double-pendulum dynamics. The dynamic equations are derived using the Lagrangian approach. Subsequently, the output payload response is determined in time and frequency domain by developing algorithms in MATLAB for better crane payload swing control. This paper is divided into four sections. The first section gives a brief overview of the problem. The second section deals with the physical model and parameters of the gantry crane. The 3-D gantry crane mathematical modelling is described in the third section. The fourth section presents results and discussion of the key findings. Conclusions are drawn in the final section.

2 Physical Model Representation of the Gantry Crane System

The physical model of the overhead gantry system shown in Fig. 1 incorporates its motion in 3-D space including double-pendulum motion, where an external driving force ($F \in \mathrm{R}^2$) moves the trolley and rail simultaneously. The gantry crane system comprises four subsystems. i.e., rail for X-axis movement, trolley for Y-axis movement, a combination of hook and payload where θ_1 and φ_1 are the hook rope swing angles; θ_2 and φ_2 is the payload rope swing angles; L_1 and L_2 are the wire rope length of hook and payload, respectively.

The hook ball mass moment of inertia (j_h) is assumed constant [14]; however, the payload ball mass moment (j_p) of inertia is varying according to variations in the payloads (m_p). For simplified mathematical modelling, the rail and trolley-wheel friction force and the flexibility of the wire rope are neglected. The physical parameters of the gantry crane [15, 18] are given in Table 1.

3 Mathematical Modelling

The total kinetic and potential energy associated with the gantry crane needs to be derived, and then, the Lagrangian formulation for dynamic modelling is used to

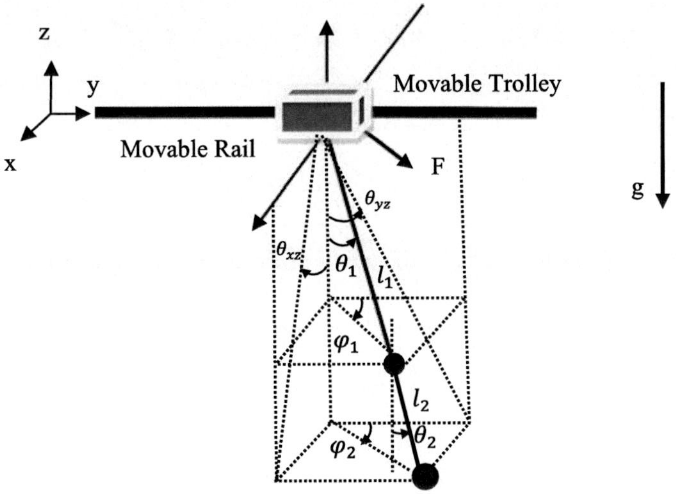

Fig. 1 Mathematical representation of the 3-D gantry crane using a double-pendulum modelling approach

Table 1 Parameters of the crane system

Parameter	Trolley mass	Rail mass	Hook ball mass	Payload ball mass	Hook moment of inertia	Payload moment of inertia	Hook rope length	Payload rope length
Symbol	m_t	m_r	m_h	m_p	j_h	j_p	l_1	l_2
Value	1.06 kg	6.4 kg	0.1 kg	Variable	0.005 kg-m^2	Variable	2.0 m	1.0 m

derive the 3-D gantry crane overall system response. Simulation is performed using a state-space model of the gantry crane by MATLAB ode45 solver.

3.1 Kinetic Energy

The rail, trolley, hook, and payload position vectors are derived using Fig. 1. The position vectors are $r_r = [x, 0, 0]^T, r_t == [x, y, 0]^T$, $r_h = [x + L_1 \sin\theta_1 \sin\varphi_1, y + L_1 \sin\theta_1 \cos\varphi_1, -L_1 \cos\theta_1]^T$, and $r_p = [x + L_1 \sin\theta_1 \sin\varphi_1 + L_2 \sin\theta_2 \sin\varphi_2, y + L_1 \sin\theta_1 \cos\varphi_1 + L_2 \sin\theta_2 \cos\varphi_2, -L_1 \cos\theta_1 - L_2 \cos\theta_2]^T$, respectively, where the trolley positions in X- and Y-directions are represented by parameters x and y, respectively. The kinetic energy, KE, of the whole crane system is given in Eq. (1),

$$\text{KE} = k_{\text{rail}} + k_{\text{trolley}} + k_{\text{hook}} + k_{\text{payload}} \tag{1}$$

where $k_{\text{rail}} = \frac{1}{2}m_r\dot{r}_r^2$, $k_{\text{trolley}} = \frac{1}{2}m_t\dot{r}_t^2$, $k_{\text{hook}} = \frac{1}{2}m_h\dot{r}_h^2 + \frac{1}{2}J_h\dot{\theta}_1^2 + \frac{1}{2}J_h\dot{\varphi}_1^2$, $k_{\text{payload}} = \frac{1}{2}m_p\dot{r}_p^2 + \frac{1}{2}J_p\dot{\theta}_2^2 + \frac{1}{2}J_p\dot{\varphi}_2^2$.

3.2 *Potential Energy*

The total potential energy, PE, of the whole crane system is given in Eq. (2),

$$\text{PE} = p_{\text{hook}} + p_{\text{payload}} \tag{2}$$

where $p_{\text{hook}} = -m_h g L_1 \cos\theta_1$, $p_{\text{payload}} = -m_p g(L_1 \cos\theta_1 + L_2 \cos\theta_2)$ and g is the gravity acceleration vector.

3.3 *Lagrangian Formulation*

The Lagrangian approach as shown in Eq. (3) is used to derive the system equations of motion for dynamic response analysis.

$$\frac{\mathrm{d}}{\mathrm{d}t}\left(\frac{\partial L}{\partial \dot{q}_i}\right) - \frac{\partial L}{\partial q_i} = F_i, \qquad i = 1, \ldots, 6 \tag{3}$$

where $L(q,\dot{q}) = \text{KE} - \text{PE}$, $q(t) = \left[x\ y\ \theta_1\varphi_1\theta_2\varphi_2\right]^T$ is the state vector and $F = \left[F_x F_y 0000\right]^T$ is the input excitation force vector, which is defined by a bang–bang control signal where F_x and F_y denote the input control force acting on the trolley along with the X- and Y-axis directions.

From Eq. (3), the dynamic model of the crane system is then expressed in the form of matrices for further conversion in the state-space representation of dynamic payload simulations.

$$M(q)\ddot{q} + V(q,\dot{q}) + G(q) = F(t) \tag{4}$$

The system response can be expressed in the following form of the equation of motions from Eq. (4).

$$\ddot{q}(t) = M(q)^{-1}(-V(q,\dot{q}) - G(q) + F(t)) \tag{5}$$

where in the above Eq. (5), the matrices $M(q) \in R^{6\times6}$, $V(q,\dot{q}) \in R^{6\times1}$, and $G(q) \in R^{6\times1}$ represent the Inertia, Centripetal–Coriolis, and gravity terms, respectively, and these matrices terms are further defined in the given appendix.

4 Results and Discussions

For swing and motion response of the payload analysis, the trolley of the gantry crane is excited using a bang–bang signal of amplitude 2 *N* force along the *X*-axis and 1 *N* of force along the *Y*-axis. The applied force signal is represented in Fig. 2. The bang–bang force signal has equal acceleration and deceleration periods. Some examples of the used bang–bang signal for payload response simulation can be found elsewhere [15, 18].

Simulation steps using developed algorithms for swing and motion responses of the payload analysis in both time and frequency domain are shown in Fig. 3.

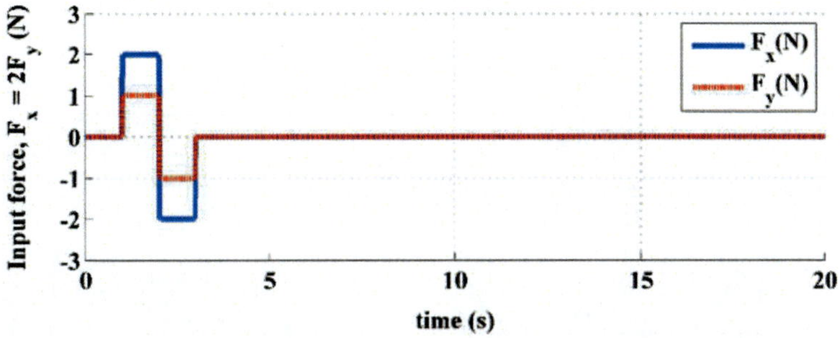

Fig. 2 Input force excitation applied on the trolley for dynamic payload response

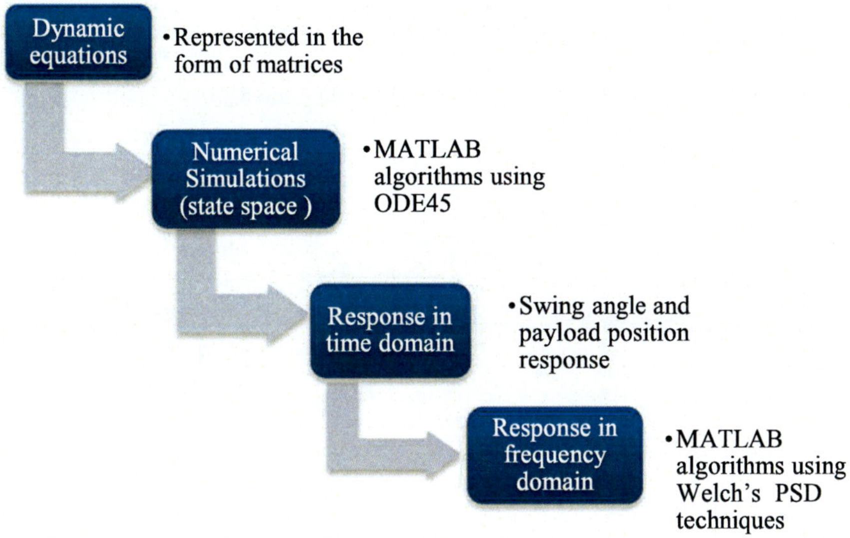

Fig. 3 Simulation steps using the developed algorithms for dynamic response analysis

The results given in Fig. 4 show the payload swing angle response (θ_{yz} and θ_{xz}) with respect to *XZ* and *YZ* planes with the various mass (m_p) of the payload. It shows that the oscillation of the swing angles decreases with increasing mass (m_p) of the payload. For other cases, in the simulation, if we increase the hook rope length (l_1), the swing angles also increase.

Simulation results in Fig. 5a and b show the variation of time response along *X*-, *Y*-, and *Z*-axis in 3D space of the payload motion with the variation of payloads mass. It shows that the amplitude of the *X*-, *Y*-, and *Z*-axis oscillation of the payload decreases with increasing mass of the payload. It is more visible when we see the 3-D phase plot of the payload amplitude oscillations shown in Fig. 5b. It is observed from Fig. 5b that when the mass of payload increases, the amplitude of the response decreases.

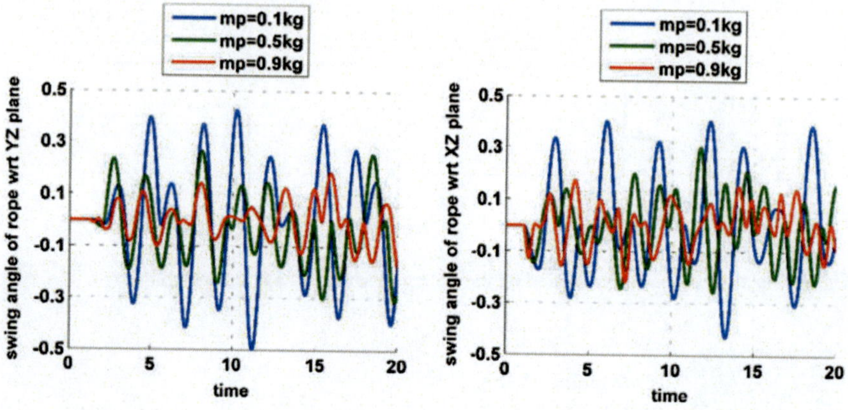

Fig. 4 Dynamic response of the payload swing angles (rad.) versus time (s)

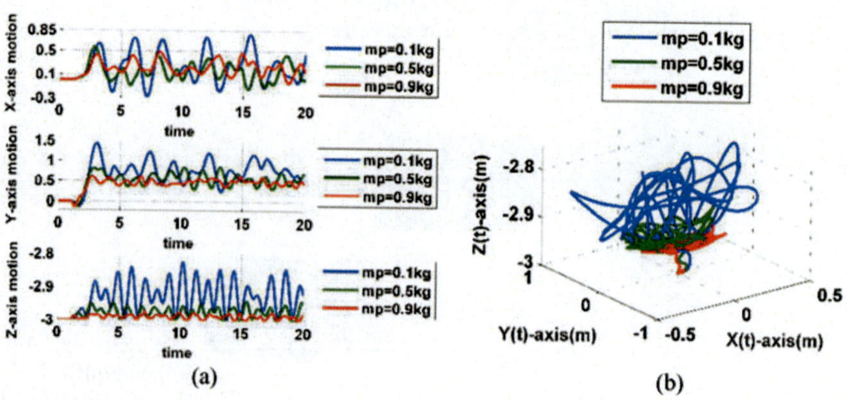

Fig. 5 Dynamic response of the payload, **a** time response plot, displacement (m) versus time (s), **b** 3-D phase response plot

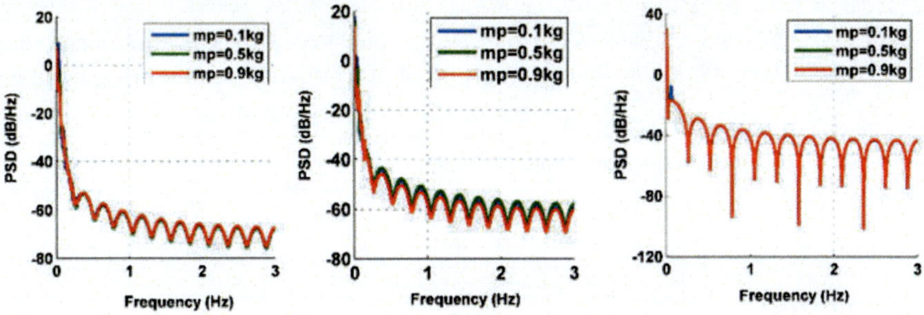

Fig. 6 Power spectral density of *X*, *Y*, and *Z* payload dynamic response

Table 2 Frequencies of payload response along the *X*-, *Y*-, and *Z*-axis for the first two modes

Payload (kg)	Resonance frequency (Hz)					
	Payload response *X*-axis		Payload response *Y*-axis		Payload response *Z*-axis	
	Mode-1	Mode-2	Mode-1	Mode-2	Mode-1	Mode-2
0.1	0.024	0.360	0.036	0.640	0.061	0.660
0.5	0.030	0.374	0.049	0.645	0.062	0.660
0.9	0.042	0.380	0.055	0.650	0.067	0.660

After knowing the dynamic response of the payload along *X*-, *Y*-, and *Z*-axis in 3-D space, the power spectral density (PSD) is then compared and evaluated using MATLAB algorithms. It is observed from Fig. 6 that the higher the payload mass, the resonance frequency is also higher, and there is not much frequency variation for higher modes of vibration. The resonance frequencies for the first two modes and payloads relation is summarized in Table 2.

5 Conclusion

In the present study, a 3-D gantry crane system considering double-pendulum effect due to hook and payload has been modelled, and payload response simulation is performed. Lagrange approach has been used to develop the system equations of motion. Bang–bang force input is used to simulate the gantry crane payload dynamic response behaviour. The crane payload position in 3-D space and swing angle responses have been obtained using the developed dynamic modelling, and further analysis is performed both in time and frequency domains by developing MATLAB algorithms. The influence of the parameters such as payload mass and hook rope length variations has been considered in the gantry crane system dynamic study and simulation. These dynamic response study may be further useful for effective payload response control of gantry cranes with varying system parameters. This study presented on the dynamics of 3-D gantry crane may also be useful for studying the dynamic behaviour and control strategies of other types of mobile crane systems.

Acknowledgements The work presented in the paper is the output of a CSIR, India, sponsored R&D project on Mass Housing Development (HCP-015).

Appendix

$$M(q) = \begin{bmatrix} m_{11} & m_{12} & m_{13} & m_{14} & m_{15} & m_{16} \\ m_{21} & m_{22} & m_{23} & m_{24} & m_{25} & m_{26} \\ m_{31} & m_{32} & m_{33} & m_{34} & m_{35} & m_{36} \\ m_{41} & m_{42} & m_{43} & m_{44} & m_{45} & m_{46} \\ m_{51} & m_{52} & m_{53} & m_{54} & m_{55} & m_{56} \\ m_{61} & m_{62} & m_{63} & m_{64} & m_{65} & m_{66} \end{bmatrix}$$

$$\begin{aligned}
m_{11} &= m_h + m_p + m_r + m_t, \\
m_{12} &= m_{21} = 0, \\
m_{13} &= m_{31} = L_1 \cos\theta_1 \sin\varphi_1 m_h + L_1 \cos\theta_1 \sin\varphi_1 m_p, \\
m_{14} &= m_{41} = L_1 \sin\theta_1 \cos\varphi_1 m_h + m_p L_1 \sin\theta_1 \cos\varphi_1 \\
m_{15} &= m_{51} = L_2 \sin\varphi_2 \cos\theta_2 m_p, \\
m_{16} &= m_{61} = L_1 \sin\theta_2 \cos\varphi_2 m_p, \\
m_{22} &= m_h + m_p + m_t, \\
m_{23} &= m_{32} = L_1 \cos\varphi_1 \cos\theta_1 m_h + L_1 \cos\varphi_1 \sin\theta_2 m_p, \\
m_{24} &= m_{42} = -L_1 \sin\theta_1 \sin\varphi_1 m_h - m_p L_1 \sin\varphi_1 \sin\theta_1, \\
m_{25} &= m_{52} = L_2 \cos\varphi_2 \cos\theta_2 m_p, \\
m_{26} &= m_{62} = -L_1 \sin\varphi_2 \sin\varphi_2 m_p, \\
m_{31} &= L_1 \cos\theta_1 \sin\varphi_1 m_h + L_1 \cos\theta_1 \sin\varphi_1 m_p, \\
m_{32} &= L_1 \cos\varphi_1 \cos\theta_1 m_h + L_1 \cos\theta_1 \cos\varphi_1 m_p, \\
m_{33} &= m_h L_1^2 + m_p L_1^2 + J_h, \\
m_{34} &= m_{43} = m_{35} = m_{53} = m_{36} = m_{63} = 0, \\
m_{44} &= -L_1^2 \cos^2\theta_1 m_h - L_1^2 \cos\theta_1^2 m_p + m_h L_1^2 + m_p L_1^2 + J_h, \\
m_{45} &= m_{54} = m_{46} = m_{64} = 0, \\
m_{55} &= m_p L_2^2 + J_p, m_{56} = m_{65} = 0, \\
m_{66} &= m_p L_2^2 + J_p - L_1^2 \cos\theta_2^2 m_p
\end{aligned}$$

$$V(q, \dot{q}) = \begin{bmatrix} V_{11} \\ V_{21} \\ V_{31} \\ V_{41} \\ V_{51} \\ V_{61} \end{bmatrix},$$

$$
\begin{aligned}
V_{11} &= -L_1 \sin\varphi_1 \dot{\theta}_1^2 \sin\theta_1 m_h - L_1 \sin\varphi_1 \dot{\theta}_1^2 \sin\theta_1 m_p \\
&\quad + 2L_1 \cos\varphi_1 \dot{\varphi}_1 \dot{\theta}_1 \cos\theta_1 m_h + 2L_1 \cos\varphi_1 \dot{\varphi}_1 \dot{\theta}_1 \cos\theta_1 m_p \\
&\quad - L_1 \sin\varphi_1 \dot{\varphi}_1^2 \sin\theta_1 m_h - L_1 \sin\varphi_1 \dot{\varphi}_1^2 \sin\theta_1 m_p - L_2 \sin\varphi_2 \dot{\theta}_2^2 \sin\theta_2 m_p \\
&\quad + 2L_2 \cos\varphi_2 \dot{\varphi}_2 \dot{\theta}_2 \cos\theta_2 m_p - L_2 \sin\varphi_2 \varphi_2^2 \sin\theta_2 m_p, \\
V_{21} &= -2L_1 \sin\varphi_1 \dot{\varphi}_1 \cos\theta_1 \dot{\theta}_1 m_h - 2L_1 \sin\varphi_1 \dot{\varphi}_1 \dot{\theta}_1 \cos\theta_1 m_p \\
&\quad - L_1 \cos\varphi_1 \dot{\theta}_1^2 \sin\theta_1 m_h \\
&\quad - L_1 \cos\varphi_1 \dot{\theta}_1^2 \sin\theta_1 m_p - L_1 \cos\varphi_1 \dot{\varphi}_1^2 \sin\theta_1 m_h \\
&\quad - L_1 \cos\varphi_1 \dot{\varphi}_1^2 \sin\theta_1 m_p \\
&\quad - 2L_2 \sin\varphi_2 \dot{\varphi}_2 \dot{\theta}_2 \cos\theta_2 m_p - L_2 \cos\varphi_2 \dot{\theta}_2^2 \sin\theta_2 m_p \\
&\quad - L_2 \cos\varphi_2 \dot{\varphi}_2^2 \sin\theta_2 m_p \\
V_{31} &= -\dot{\varphi}_1^2 L_1^2 \cos\theta_1 \sin\theta_1 \left(m_h + m_p\right), \\
V_{41} &= 2L_1^2 \cos\theta_1 m_h \sin\theta_1 \dot{\varphi}_1 \dot{\theta}_1 \left(m_h + m_p\right), \\
V_{51} &= -L_2^2 \cos\theta_2 m_p \sin\theta_2 \dot{\varphi}_2^2,\ V_{61} = 2L_2^2 \cos\theta_2 m_p \sin\theta_2 \dot{\varphi}_2 \dot{\theta}_2
\end{aligned}
$$

$$
G(q) = \begin{bmatrix} 0 \\ 0 \\ m_h g L_1 \sin\theta_1 + m_p g L_1 \sin\theta_1 \\ 0 \\ m_p g L_2 \sin\theta_2 \\ 0 \end{bmatrix}.
$$

References

1. Karajgikar A, Vaughan J, Singhose W (2011) Double-pendulum crane operator performance comparing pd-feedback control and input shaping. In: Proceedings of the ECCOMAS thematic conference on advances in computational multibody dynamics, Brussels, Belgium, pp 1–14
2. Qian D, Tong S, Yang B, Lee S (2015) Design of simultaneous input-shaping-based SIRMs fuzzy control for double pendulum-type overhead cranes. Bull Pol Acad Sci Tech Sci 63(4):887–896
3. Weiping G, Diantong L, Jianqiang Y, Dongbin Z (2004) Passivity-based-control for double-pendulum-type overhead cranes. In: IEEE region 10 conference TENCON 2004, pp 546–549
4. Kenison M, Singhose W (1999) Input shaper design for double-pendulum planar gantry cranes. In: Proceedings of the 1999 IEEE international conference on control applications, Hawaii, USA, pp 539–544
5. Vaughan J, Maleki E, Singhose W (2010) Advantages of using command shaping over feedback for crane control. In: Proceedings of the 2010 american control conference, Baltimore, USA, pp 2308–2313
6. Zhang M, Ma X, Rong M, Tian X, Li Y (2016) Adaptive tracking control for double-pendulum overhead cranes subject to tracking error limitation, parametric uncertainties and external disturbances. Mech Syst Signal Process 76:15–32
7. Vaughan J, Kim D, Singhose W (2010) Control of tower cranes with double-pendulum payload dynamics. IEEE Trans Control Syst Technol 18(6):1345–1358

8. Qian D, Tong S, Lee S (2016) Fuzzy-logic-based control of payloads subjected to double-pendulum motion in overhead cranes. Automa Constr 65:133–143
9. Kim D, Singhose W (2010) Performance studies of human operators driving double-pendulum bridge cranes. Control Eng Pract 18(6):567–576
10. Ahmad MA, Raja Ismail RMT, Ramli MS, Hambali N (2010) Investigations of NCTF with input shaping for sway control of a double-pendulum-type overhead crane. In: International conference on advanced computer control, Shenyang, China, pp 456–461
11. Huang J, Liang Z, Zang Q (2015) Dynamics and swing control of double-pendulum bridge cranes with distributed-mass beams. Mech Syst Signal Process 54–55:357–366
12. Kim D, Singhose W (2006) Reduction of double-pendulum bridge crane oscillations. In: The 8th international conference on motion and vibration control (MOVIC 2006), 27–30 August, Daejeon, Korea, pp 300–305
13. Singhose WE, Towell ST (1998) Double-pendulum gantry crane dynamics and control. In: Proceedings of the 1998 IEEE international conference on control applications, Trieste, pp 1205–1209
14. Chang CY, Chiang KH (2008) Fuzzy projection control law and its application to the overhead crane. Mechatronics 18(10):607–615
15. Ismail RMTR, Ahmad MA, Ramli MS, Rashidi FRM (2009) Nonlinear dynamic modelling and analysis of a 3-D overhead gantry crane system with payload variation. In: 2009 Third UKSim European symposium on computer modeling and simulation. IEEE, pp 350–354
16. Singhose W, Kim D, Kenison M (2008) Input shaping control of double-pendulum bridge crane oscillations. J Dyn Syst Meas Contr 130(3):034504
17. O'Connor W, Habibi H (2013) Gantry crane control of a double-pendulum, distributed-mass load, using mechanical wave concepts. Mech Sci 4:251–261
18. Ismail RMTR, Ahmad MA, Ramli MS, Ishak R, Zawawi MA (2011) Dynamic modelling of a double-pendulum gantry crane system incorporating payload. AIP Conf Proc 1337(1):118–122

Design and Development of 3-DoF Mechanism for the Exoskeleton Hip Joint

V. M. Akhil, Jobin Varghese, P. K. Rajendrakumar, and K. S. Sivanandan

1 Introduction

Robotic systems, which can make most complex movements including human, are well suited for the assistance of human beings, who have lost their ability to move their body organs. People loses their movement ability due to afflicted lower-extremity joints either by birth, diseases or by ageing. Simple affordable transportation systems are always an answer to these problems. In practice, wheel chairs have been employed as safest transportation system for afflicted and differently abled subjects [1]. This is best suited for the subjects whose upper body is healthy, as they can ride it by themselves. However, this practice of using wheel chairs will never help to get rid of those afflicted conditions. A human motion-assistive device helps to reproduce the gait cycle, which in fact produces the effects of physiotherapy. Exoskeletons are the general active assistive devices that are under research interest, which have the actuators to reproduce the motion. Re-Walk [2] is the exoskeleton with powered hip and knee joints to actively facilitate the gait cycle of below-hip disabled subjects. But it does not have a three degrees of freedom at the hip joint. These exoskeletons are not only designed for differently abled subjects, but also Raytheon XOS like exoskeleton [3] is designed for military purposes as well.

Another area where the exoskeleton finds its application is to lift heavy load. Cyber dyne's HAL 5 is the cyborg-type wearable robot which aids the wearer to lift higher loads. Some widely available commercially available exoskeletons are

V. M. Akhil (✉) · P. K. Rajendrakumar
National Institute of Technology Calicut, Calicut, Kerala, India
e-mail: akhilvmtvm@gmail.com

J. Varghese
Adi Shankara College of Engineering and Technology, Ernakulam, India

K. S. Sivanandan
Manipal Institute of Technology, Udupi, Karnataka, India

V. H. Saran and R. K. Misra (eds.), *Advances in Systems Engineering*,
Lecture Notes in Mechanical Engineering,
https://doi.org/10.1007/978-981-15-8025-3_20

Honda [4], M.I.T. Media Lab's Bio Mechatronics Group [5] legs and Parker Hannifin Indego Exoskeleton [6]. Rex [7] and IHMC [8] are providing two degrees of freedom actuation at hip joint that is. flexion/extension and abduction/adduction. However, none of these exoskeletons have more than two degrees of freedom at the hip joints although this joint gives great flexibility in movement. The major reason for the exoskeleton developers to stick on two degrees of freedom hip joints is due to the complexity in implementation of three degrees of freedom joints. In the present work, a lower-extremity exoskeleton with a 3-DoF hip joint is designed and fabricated. The developed exoskeleton is tested on subjects, and its gait pattern is compared with the gait pattern of healthy subject. Both gait patterns are found to be in good agreement.

2 Method and Materials

The developed three degrees of freedom hip joint exoskeleton was tested on 10 subjects. Training to the subjects to use the exoskeleton for walking was given beforehand to the data collection process. Hip joint rotation angle data were measured using accelerometer (ADXL 335), and motion was captured with video camera (Vicon V16). The accelerometer and video camera data were collected using National instruments Data Acquisition card (NI-USB 3636) and LabVIEW software.

2.1 Three-DOF Hip Joint Mechanism Design Steps

Method for the selection of criteria for design of 3-DoF hip joint is described Table 1 in detail. Methodology used in this work as shown in Fig. 1.

Problem identification: efficient three degrees of freedom mechanisms are not currently available for assisting hip joints of afflicted subjects. Hence, a three degrees of freedom mechanism is needed to have unconstrained walking ability for an afflicted subject.

Concept generation: different concept models of three degrees of freedom hip joint assisting mechanism are developed. The major constraints imposed during concept generation are number of components, cost, safety, comfort, manufacturability, weight, durability and quick response.

Concept screening: The concepts that are developed by strictly following the imposed constraints are screened out.

Concept selection: The best out of the screened-out concept needs to be found out. So, the concept selection is done again based on the previously selected constraints, and this time the weightage of each constraint is found out and the individual concepts score a value out of five.

Table 1 Selection of criteria for design of 3-DoF hip joint

Systems							
Concept variants							
		A		B		C	
Selection criteria	Weightage (%)	Rating	Weighted score	Rating	Weighted score	Rating	Weighted score
No. of components	10	2	0.2	4	0.4	2	0.2
Cost	20	3	0.6	4	0.8	3	0.6
Safety and comfort	15	3	0.45	2	0.3	4	0.6
Manufacturability	20	2	0.4	4	0.8	3	0.6
Weight	15	3	0.45	3	0.45	2	0.3
Durability	10	3	0.3	3	0.3	4	0.4
Quick release	10	1	0.1	2	0.2	2	0.2
Total		2.5		3.25		2.9	
Rank		3		1		2	
Continue?				Preferred			

2.2 CAD Design

The model B was finalized as the best suited one. The best suited CAD model of three degrees of freedom hip joint mechanism is drafted in SOLIDWORKS 2015 (Fig. 2) with the average hip and thigh dimensions of 10 subjects took part in the study. The further development of the design is to be done in this model and the dimensions need to be optimized for minimum material usage.

2.3 Torque Required at Joint

The major focus is on the analysis of the torque required at the joint (τ). Lagrangian method (Fig. 3) is utilized for the calculation of required torque at hip joint of the exoskeleton.

$$\tau = \frac{\mathrm{d}}{\mathrm{d}t}\left(\frac{\partial L}{\partial \dot{q}_i} - \frac{\partial L}{\partial q_i}\right) \tag{1}$$

where

$$L = K - P, \tag{2}$$

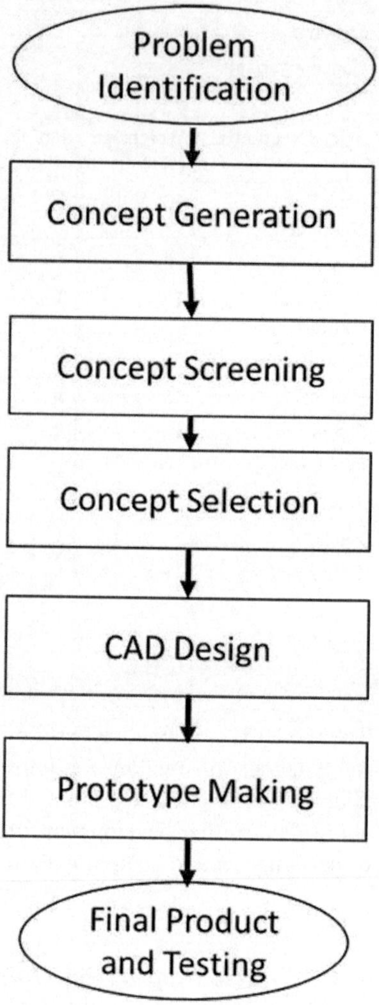

Fig. 1 Methodology used in this work

K is the kinetic energy and P is the potential energy, q_i is the angle of ith joint and $\dot{q}_i$ is the angular velocity of ith joint.

$$M(\theta)\ddot{\theta} + C\big(\theta, \dot{\theta}\big)\dot{\theta} + G(\theta) = \tau \tag{3}$$

where

L	Lagrangian term,
K	Kinetic energy,
P	Potential energy,
τ	Torque,
$M(\theta)$	Inertia matrix,

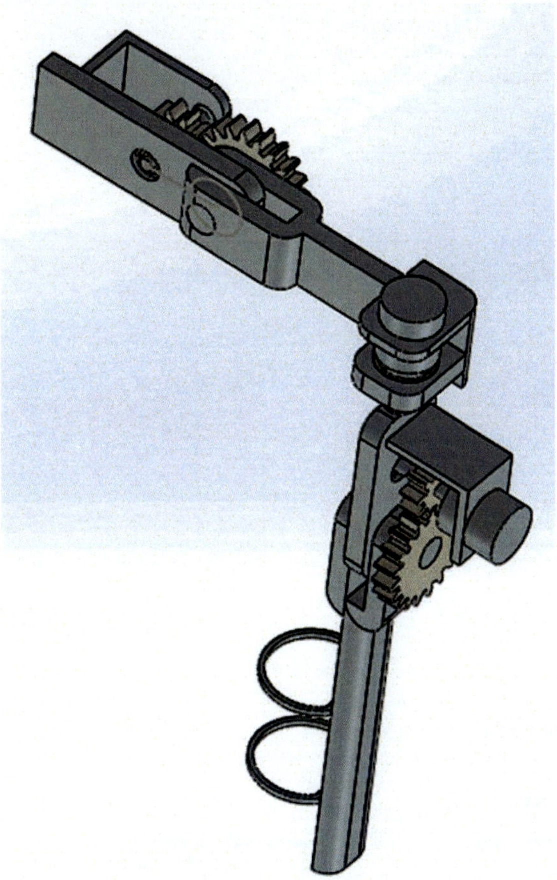

Fig. 2 CAD model of three degrees of freedom hip joint mechanism

$C(\theta, \dot{\theta})$ Coriolis and centrifugal force,
$G(\theta)$ Gravity term,
$\ddot{\theta}$ Angular acceleration,
$\dot{\theta}$ Angular velocity,
θ Angular displacement

From Fig. 4, it is observed during flexion and extension that maximum angle from −20° to 20° should be produced by an exoskeleton hip joint. This in fact requires higher torque and hence a higher power rated motor should be used.

2.4 Prototype

Prototype of the proposed hip joint actuator is 3-D printed (Fig. 5) using the PLA

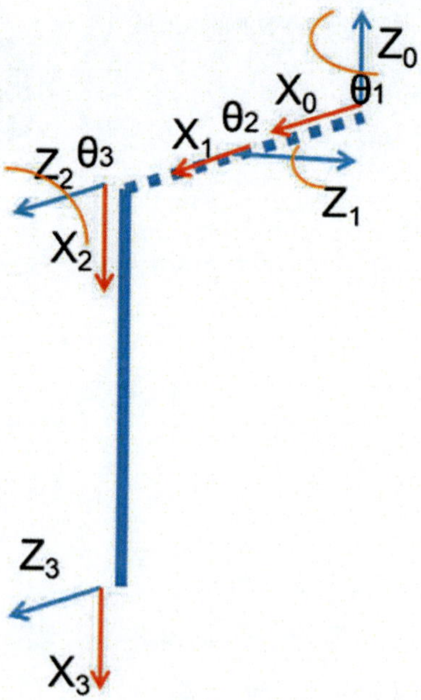

Fig. 3 D-H method to calculate the torque required to actuate the hip joint

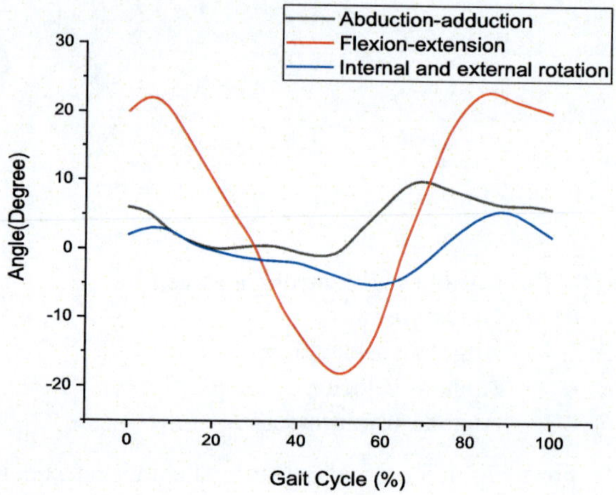

Fig. 4 Abduction–adduction, flexion–extension and internal and external rotation angles of hip plotted against gait cycle percentage

material. The individual components are initially 3-D printed and later they are assembled together to make a complete working prototype.

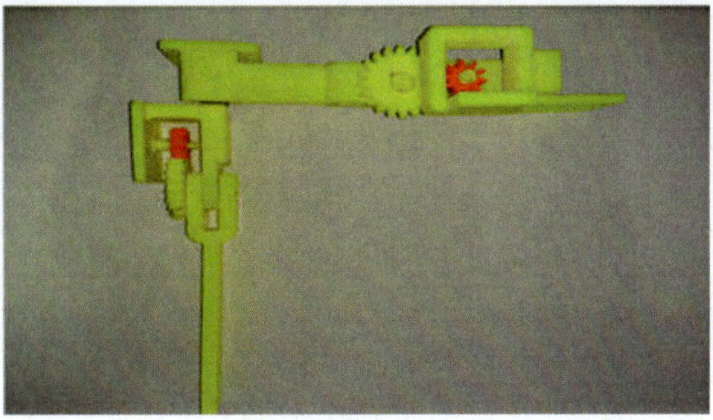

Fig. 5 Prototype of the proposed model

3 Analysis

3.1 Structural Analysis

The structural analysis was done, and the critical points with higher stress values were taken into consideration. The analysis is done using ANSYS 16.0. For surface meshing, triangular elements are used and the interior meshing is inherently program controlled. The dimensions of the CAD model are varied in during analysis until the required von Mises stress is obtained. The allowable value for von Mises stress for PLA material was considered to be 25 MPa whereas the maximum value of stress in the model was found to be 12.83 MPa at abduction gear shaft 1. Figure 6 gives the critical parts of the model under structural analysis as well as the whole model. Figure 7 shows the stress variation in the model.

Fig. 6 Stress variation in flexion gear 1 shaft at flexion pivot in structural analysis

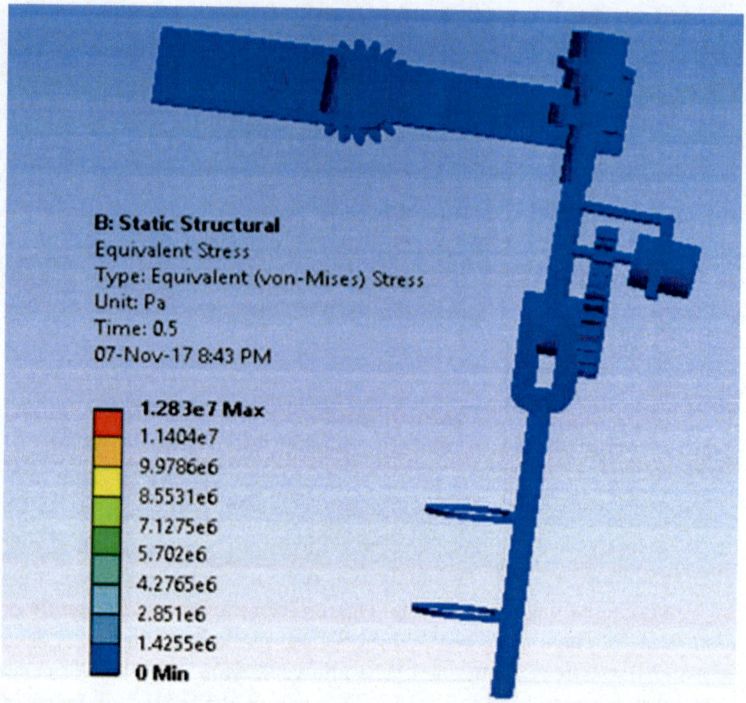

Fig. 7 Stress variation of the model in structural analysis

3.2 *Dynamic Analysis*

The dynamic analysis was conducted and the critical points of high stress values was found and necessary corrections were made in the dimensions to get to the final design of the CAD model. Three of the parts were considered to the most critical parts (Table 2) which are abduction gear 1 shaft, abduction gear 2 shaft and the fixed link. The corrections were made to get the final factor of safety (fos) value to be between 1.5 and 2.

4 Actuator Selection

As torque required at the hip joint is determined, a suitable actuator that can deliver a torque above the required torque should be selected. In addition to the torque requirement, other necessary factors for the smooth functioning of the exoskeleton to be considered are (i) size of the component, (ii) weight of the component, (iii) ease of availability and (iv) cost of the component. By adding a factor of safety to the calculated torque (14.42 N-m), required torque to be produced at the hip joint of

Table 2 Dynamic Analysis

Parts in model	Stress variation	Factor of safety
ABDUCTION GEAR 1 SHAFT	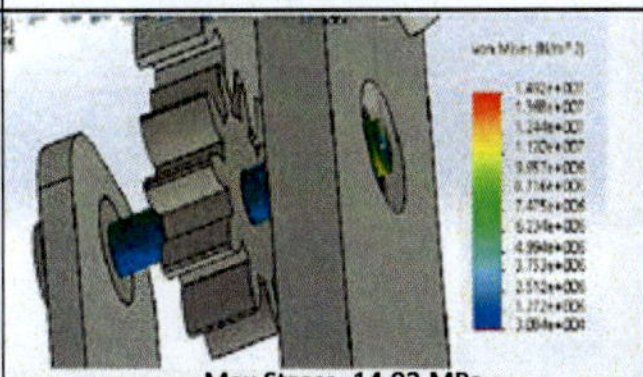 Max Stress=14.92 MPa	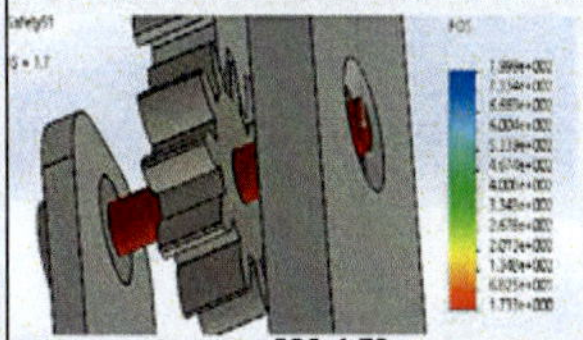 FOS=1.73
INTERNAL ROTATION LINK	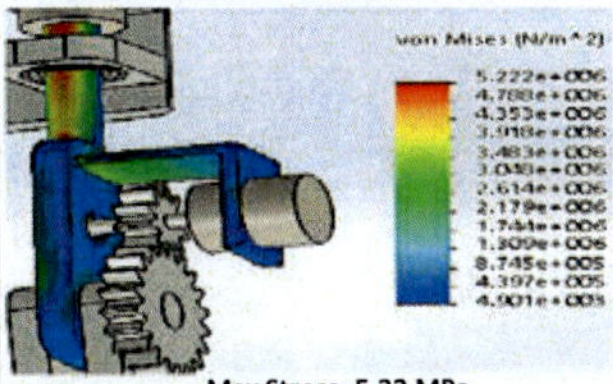 Max Stress=5.22 MPa	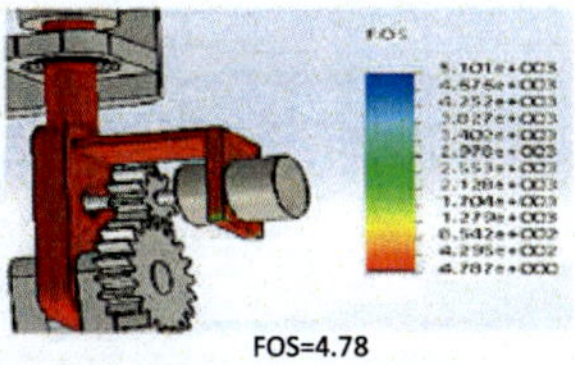 FOS=4.78
ABDUCTION LINK	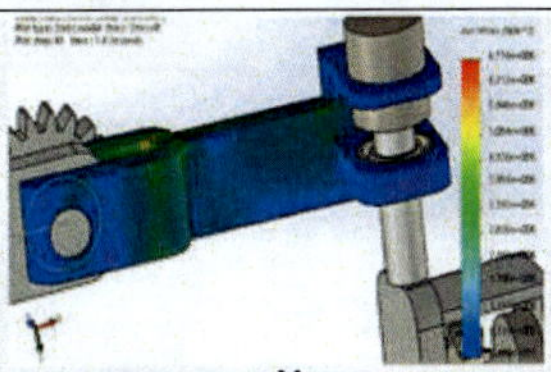 Max Stress=6.77 MPa	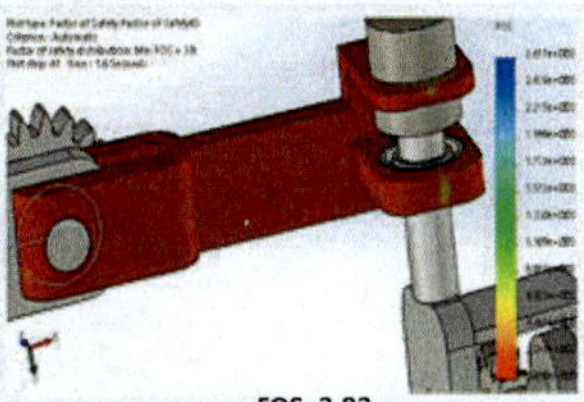 FOS=3.83

exoskeleton is decided as 16 N-m. A double acting rotary pneumatic actuator brand called AIRA which can deliver the required torque was selected. It is an actuator that converts the translator motion of a pneumatic piston into rotary motion by means of a rack and pinion mechanism. The rack is attached to the piston and pinion is placed along the axis of the output shaft from which the torque is obtained as output.

ABDUCTION GEAR 2	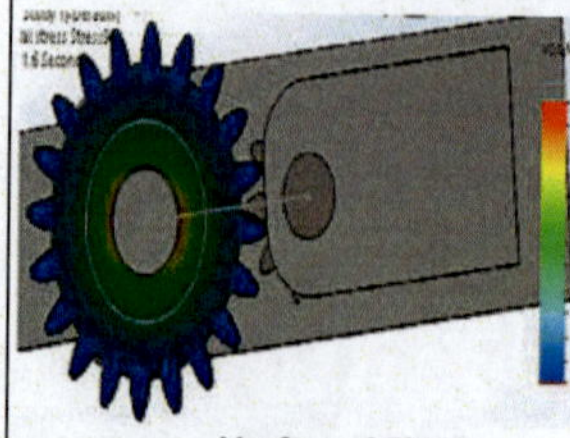 Max Stress=2.78 MPa	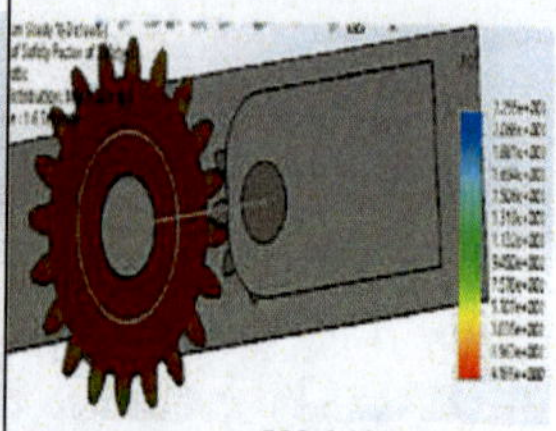 FOS=9.18
ABDUCTION GEAR 2	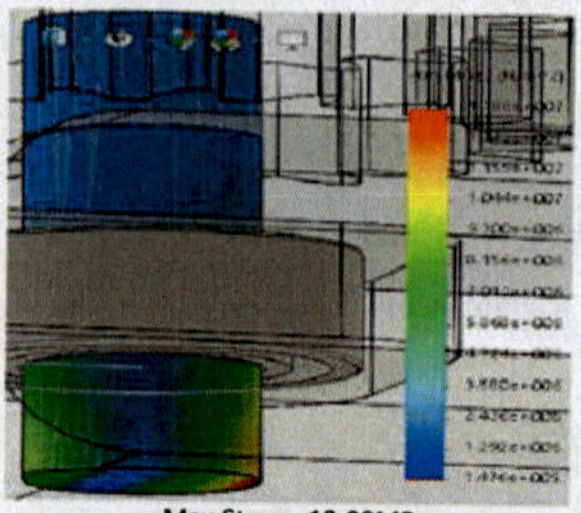 Max Stress=13.88MPa	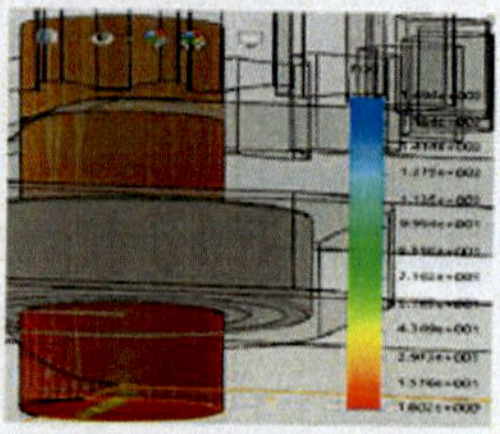 FOS= 1.8
fixed LINK	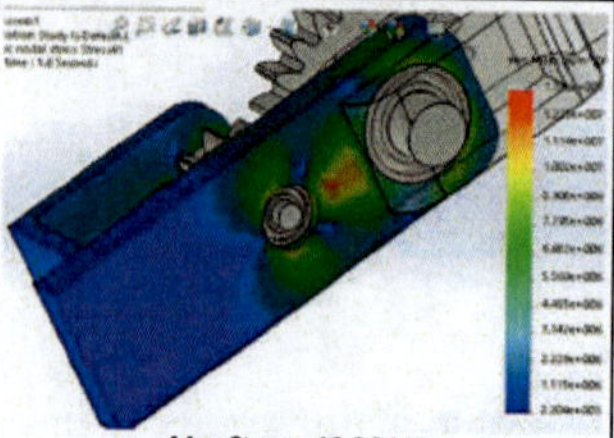 Max Stress=13.36 MPa	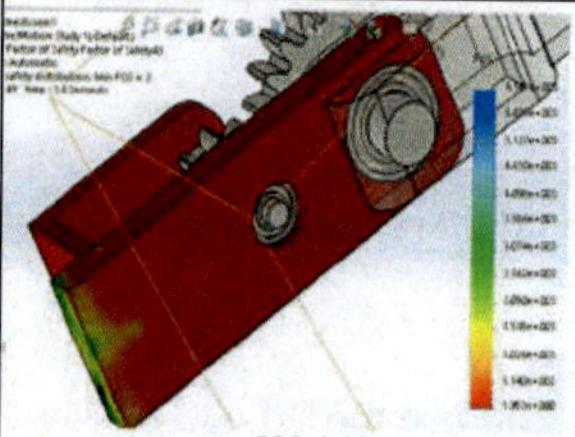 FOS=1.98

5 Fabrication of the Model

With the final CAD model being decided upon and having designed the pneumatic and electronic circuit, further step is to continue with the fabrication of the components of the mechanism and later to assemble them to generate the final mechanism (Fig. 8). The exoskeleton mechanism mainly consists of four links. The actuators are mounted on these links with its axis parallel to the axis of the links. The CAD model of the exoskeleton mechanism is made by giving a preference to easy manufacturability. The fabricated model is assembled and the rotary actuator shaft is welded to these links, so that when the actuator actuates the links can also rotate along with it.

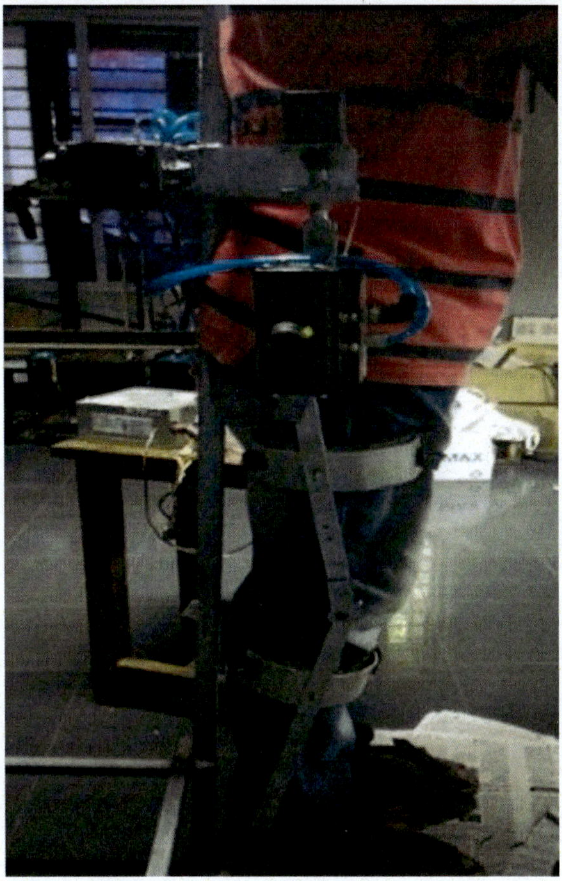

Fig. 8 A person wearing the exoskeleton with 3-DOF hip joint

6 Testing of the System

After fabrication and assembly, the mechanism was tested to see its performance characteristics. Data were collected during testing with the help of accelerometers as well as a video camera. Accelerometers are mounted (Fig. 9) on the mechanism, and camera is placed in the sagittal plane to record the video. Video analysis is done with the help of KINOVEA software. Through the video analysis using KINOVEA software changes in all values of position, velocity and acceleration during the link rotation are determined. The changes in position are determined with respect to a general stationary reference in the video itself. The variation in acceleration of the abduction–adduction motion is used to compare the performance of the mechanism and if found satisfactory the acceleration motions of all other motions are also used as the characteristic performance curve of the mechanism.

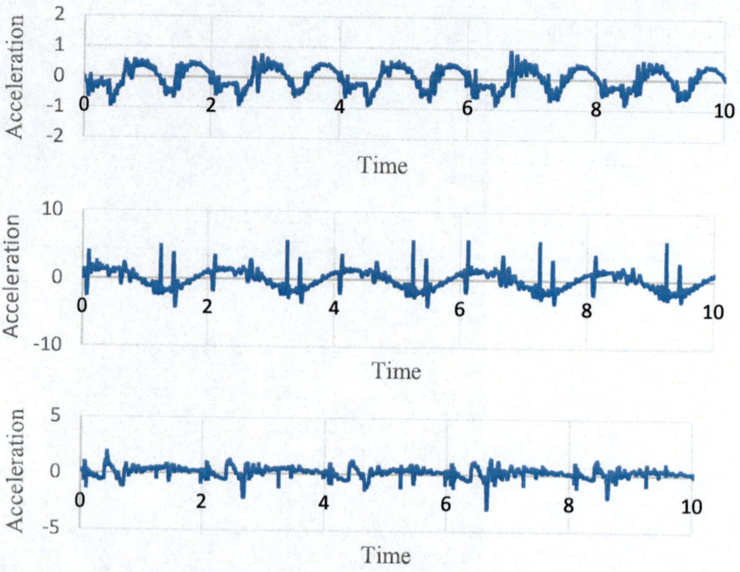

Fig. 9 Accelerometer data for abduction–adduction, flexion–extension and internal and external rotation of developed 3-DOF hip joint are plotted against time

7 Results

A mechanism which is able to provide all three degrees of freedom to a human hip joint was designed and the mechanism was analysed in static and dynamic conditions to check for its ability to work as a hip joint in a lower body exoskeleton mechanism. A 3-D printed prototype of the mechanism was made from final CAD model initially and then the model was fabricated in aluminium alloy and assembled the parts. With the model fully fabricated and the control systems incorporated into the mechanism, the testing of the mechanism was done. It was seen that the gait pattern of the mechanism resembles that of a normal human. The response characteristics of the mechanism were obtained in terms of their acceleration and velocity. The advantage of the whole mechanism lies in the fact that the performance of the mechanism can be varied by changing the Arduino program and by controlling the air flow rate by means of the valves. Thus, in short, the whole mechanism was designed, analysed, fabricated and tested to verify if the mechanism is able to replicate a normal gait cycle values and it was found that the response is in terms with the standard values.

8 Conclusion

A mechanism for incorporating all degrees of freedom that are expected in a human hip joint was designed as planned. Initially, different concepts were produced and from them the best that suits the needs was selected and a CAD model of it was made. This was followed by optimizing the dimensions of the model. Motion and dynamic analyses were done on the CAD model, and the final design was obtained by optimizing each stages of analysis. Final CAD model was made by considering the ease of manufacturability. With the final CAD model, a prototype of the model was made. The actuators are mounted on the prototype. The gait patterns developed from the actuation of the actuator are compared with that of normal human gait patterns and found to be in good agreement. An appropriate rotary pneumatic actuator which could deliver the required torque was mounted on the exoskeleton to impart the motions to the joints to produce abduction–adduction and flexion–extension movements. Aluminium is used as the material for the fabrication of the mechanism. The various operations such as welding, drilling, turning and sheet metal operations are performed for fabricating the mechanism. The developed exoskeleton mechanism is tested for any gait pattern mismatch. Although the model possesses some disadvantages, the methods to improve the design are discussed. Variations in exoskeleton mechanism are possible for accommodating different gender dimensions for their assistance.

References

1. For A, Walking F (1890) Apparatus for facilitating walking, running and jumping. No 420, 179 'Patented Jan. 28, 1890.v (420)
2. Esquenazi A, Talaty M, Packel A, Saulino M (2012) The rewalk powered exoskeleton to restore ambulatory function to individuals with thoracic-level motor-complete spinal cord injury. Am J Phys Med Rehabil 91(11):911–921
3. Raytheon XOS 2 Exoskeleton, second-generation robotics suit 2010 [Online]. Available: http://www.army-technology.com/projects/raytheon-xos-2-exoskeleton-us
4. Kusuda Y (2009) In quest of mobility—honda to develop walking assist devices. Ind Robot Int J 36:537–539
5. Herr Hugh (2009) Exoskeletons and orthoses: classification, design challenges and future directions. J Neuroengineering Rehabil 6(1):21
6. Aliman N, Ramli R, Mohamed Haris S (2017) Design and development of lower limb exoskeletons: a survey. Robot Auton Syst 95:102–116
7. Viteckova S, Kutilek P, de Boisboissel G, Krupicka R, Galajdova A, Kauler J, Lhotska L, Szabo Z (2018) Empowering lower limbs exoskeletons: state-of-the-art. Robotica 36(11):1743–1756
8. Young AJ, Ferris DP (2016) State of the art and future directions for lower limb robotic exoskeletons. IEEE Trans Neural Syst Rehabil Eng 25(2):171–182

Crack Detection in Rolling Element Bearings Using the Wavelet Transform

Pravajyoti Patra, V. H. Saran, and S. P. Harsha

1 Introduction

Roller bearings are one of the vital and frequently used components, which need constant monitoring to avoid malfunction or catastrophic failure. The bearing vibration signal broadly classified as time domain, frequency domain and time-frequency domain. In the case of time and frequency domain, it is very difficult to identify the defect due to low signal-to-noise ratio. Hence, a better signal processing technique is very much essential to predict the fault in the rotor-bearing system. Recently, many researchers tried the combination of various signal processing methods to enhance the signal-to-noise ratio to improve accuracy in measuring the actual condition of the machinery components. Mishra et al. [1] used envelope analysis and wavelet transform for the fault detection in a rotor-bearing at a slow speed. Kankar et al. [2] studied the combined effect of cyclic auto-correlation and wavelet transform to improve the fault classification using the statistical parameters. Kankar et al. [3] proposed a method using the combination of the auto-correlation and continuous wavelet transform for fault detection. Peng et al. [4] compared the results between improved Hilbert–Huang Transform and wavelet transform for fault detection for the rolling bearing system. Quan et al. [5] compared the results based on the wavelet packet transform and ensemble empirical mode decomposition for the precision of diagnostics methods. Kulkarni [6] presents the methodology to compare the results of wavelet packet transform and discrete wavelet transform for defective bearings. Khanam et al. [7] estimated the fault size using the Symlet-5 mother wavelet to study the sudden change in the vibration signal. Prabhakar et al. [8] monitored the vibration level for the non-stationary signal to detect the crack in the rotor-bearing

P. Patra (✉) · V. H. Saran · S. P. Harsha
Department of Mechanical and Industrial Engineering, Indian Institute of Technology Roorkee, Roorkee, Uttarakhand 247667, India
e-mail: ppatra@me.iitr.ac.in

V. H. Saran and R. K. Misra (eds.), *Advances in Systems Engineering*, Lecture Notes in Mechanical Engineering,
https://doi.org/10.1007/978-981-15-8025-3_21

Table 1 Comparison of mother wavelets

Wavelet type	Maximum relative wavelet energy
Coif 5	0.375736
Db 10	0.374996
Db 44	0.374673
Rbio 5.5	0.366732
Sym 5	0.377276

system. Pandya et al. [9] compared various AI techniques for fault detection in rolling element bearings using RBIO 5.5 mother wavelet. Bhavaraju et al. [10] compared the various machine learning techniques like artificial neural network (ANN) and self-organizing maps (SOM) extracting features from the continuous wavelet transform (CWT). Wang et al. [11] proposed an improved combined method of Hilbert–Hung and wavelet transform to examine the fault criteria in railway bearing.

Wavelet transform: Fault identification of defect component is a complex task as the signal acquired from a data acquisition system is the combination of noise, vibrations from other components of machine and fault signal. Hence, a signal processing technique with improved signal-to-noise ratio is very much essential. Wavelet denoising analyzed the signal (image) by multiplying it by a window function and performing an orthogonal expansion, analogously to other linear integral transformations. Formally, a wavelet series represents a square-integrable function with respect to a complete, orthonormal set of basis functions called wavelets, meaning a small wave. Proper selection of mother wavelets from the wavelet family provides the edge for preserving the low-frequency fault components of the signal. Based on the review of previous researchers, five real wavelets (db10, db 44, sym 5, coif 5, rbio 5.5) are compared (Table 1) using "maximum relative wavelet energy" criteria.

There are two types of wavelet transform one is continuous wavelet transform (CWT) and the other one is the discrete wavelet transform (DWT). The main applications of CWT are time-frequency analysis and filtering of time localized frequency component where DWT is used for denoising and compression of the signal. The two transformations defer based on how they discretize the scale and translational parameters.

CWT of a signal $s(t)$, in terms of scale (a) and shifted (b) parameters can be represented as:

$$W(a,b) = \int_R s(t)\frac{1}{\sqrt{a}}\psi^*\left(\frac{t-b}{a}\right)\mathrm{d}t \quad a \in R^+,\ b \in R \tag{1}$$

where ψ = mother wavelet and ψ^* = complex conjugate of ψ.

In the case of DWT, the signal is decomposed into low and high pass frequencies named "approximation" and "detail," respectively, and the approximation part is further decomposed into next level approximation and detail as shown in Fig. 1.

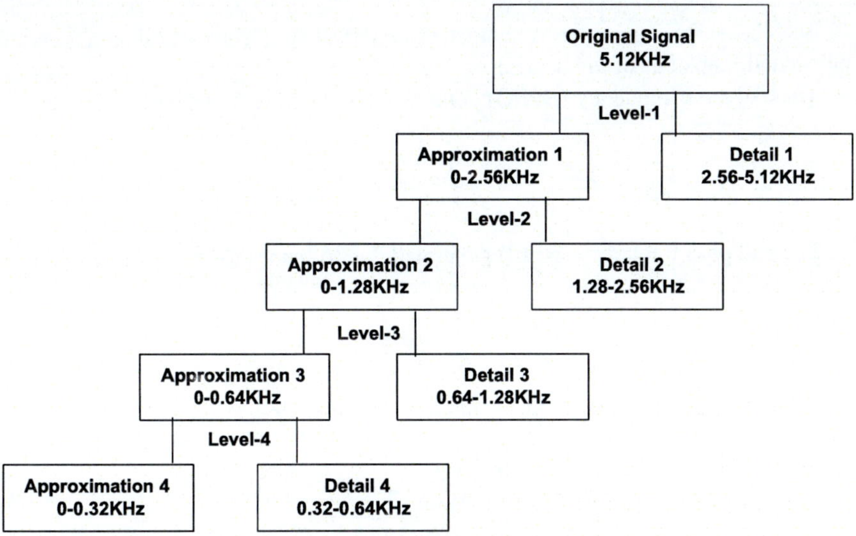

Fig. 1 Wavelet transform frequency bandwidth decomposition

The mathematical formulation used by DWT transformation can be expressed as:

$$\mathrm{DWT}(j,k) = \frac{1}{\sqrt{2^j}} \int_{-\infty}^{\infty} s(t)\psi^*\left(\frac{t-2^i k}{2^j}\right)\mathrm{d}t \tag{2}$$

where j, k are integers, 2^j and 2^j k represent the scale and translation parameter, respectively.

ψ represents the "mother" wavelet and ψ^* is the complex conjugate of ψ, $s(t)$ is the original signal.

2 Experimental Setup

The experiments are performed in bearing test rig (Fig. 2) in loaded condition with motor and shaft connecting with a flexible shaft coupling. The NU-205 cylindrical roller bearing is used for the vibration signature analysis of both healthy and faulty conditions at a speed of 30 Hz. The details of the roller bearing dimensions are given in Table 2. The loader is set in the experiment in a way to provide some amount of unbalance to the rotor-bearing system which makes the analysis more realistic. Four types of case studies, i.e., 1-healthy bearing and 3-defects at the inner race, outer race and roller with 100 microns square size defect throughout the length are conducted to observe the vibration response of the rotor-bearing system.

Each bearing element has a characteristic defect frequency that depends on the physical dimensions of the bearing.

The ball pass frequency (BPFO) generates due to the defect at the outer race is

$$\mathrm{BPFO} = \frac{N_r}{2}\omega_{\mathrm{shaft}}\left(1 - \frac{R_r}{R_p}\right)$$

The ball pass frequency (BPFI) generated due to inner race defect is

$$\mathrm{BPFI} = \frac{N_r}{2}\omega_{\mathrm{shaft}}\left(1 + \frac{R_r}{R_p}\right)$$

The ball spin frequency (BSF) generated due to defect at roller is

$$\mathrm{BSF} = \frac{R_p}{2R_r}\omega_{\mathrm{shaft}}\left(1 - \left(\frac{R_r}{R_p}\right)^2\right)$$

Fig. 2 Bearing test rig

Table 2 NU 205 cylindrical roller bearing dimensions

Outer ring outside diameter (D)	52 mm
Inner bore diameter (d)	25 mm
Pitch circle diameter (P_{d})	38.5 mm
Roller radius (ρ_{R})	3.25 mm
Width (B)	15 mm
Effective roller length (l_{eff})	6 mm
Number of rollers (N_R)	12
Contact angle	0°

Table 3 Theoretical frequencies for NU 205 bearings at 30 Hz

Ball pass frequency outer race (BPFO)	150 Hz
Ball pass frequency inner race (BPFI)	210 Hz
Ball spin frequency (BSF)	82.5 Hz
Fundamental train frequency (FTF)	12.5 Hz

The fundamental train frequency (FTF) due to the rotation rate of the cage is

$$\mathrm{FTF} = \frac{\omega_{\mathrm{shaft}}}{2}\left(1 - \frac{R_r}{R_p}\right)$$

The theoretical frequencies for the above set of bearings at 30 Hz are given in Table 3.

3 Results and Discussion

The effectiveness of the Symlet-5 mother wavelet decomposition has been validated with experimental signals. Vibration signal analysis for both healthy and defective bearing using the fourth-order wavelet decomposition, each level approximations and details information and frequency spectrum of fourth level approximation has been shown below. The noise level that is present in the original signal has been successfully removed using the four-level decomposition, and the defect frequencies can be clearly observed from the FFT spectrums after the decomposition. Figure 3 shows the vibration response and four levels of decomposition of a healthy bearing at 30 Hz, in the form of approximations (Fig. 3c) and details (Fig. 3d) of the raw signal (time data—Fig. 3a and frequency data—Fig. 3d) captured from the MFS. The frequency spectrum (Fig. 3e) clearly indicates the varying compliance frequency (VC, i.e., fundamental train frequency multiplied with the number of rollers) at 150 Hz and the harmonics of rotational frequency (X) at 30 and 60 Hz. Figure 3c clearly indicates the reduction of noise level as we increase the level of approximations. This technique helps in getting better signal-to-noise ratio to study the signal characteristics. Figure 4 shows four-level decomposition for the inner race defect bearing and frequency spectrum after the denoising. The defect frequency (BPFI) at 210 Hz and rotational frequencies can be observed. Similarly, Figs. 5 and 6 characterize the vibration signal and decomposition levels of outer race defect and roller defect, respectively. The outer race defect frequency and the ball spin frequency from the theoretical frequencies are 150 Hz and 82.5 Hz, respectively, which is well validated from the experimental results. The approximation at the fourth-level bears clear information regarding the defect frequencies.

The above analysis shows the sensitivity of wavelet transform to each case of faults present in the bearings. In the case of FFT spectrum, defect frequencies are surrounded by the frequencies related to noise which creates difficulties in identifying

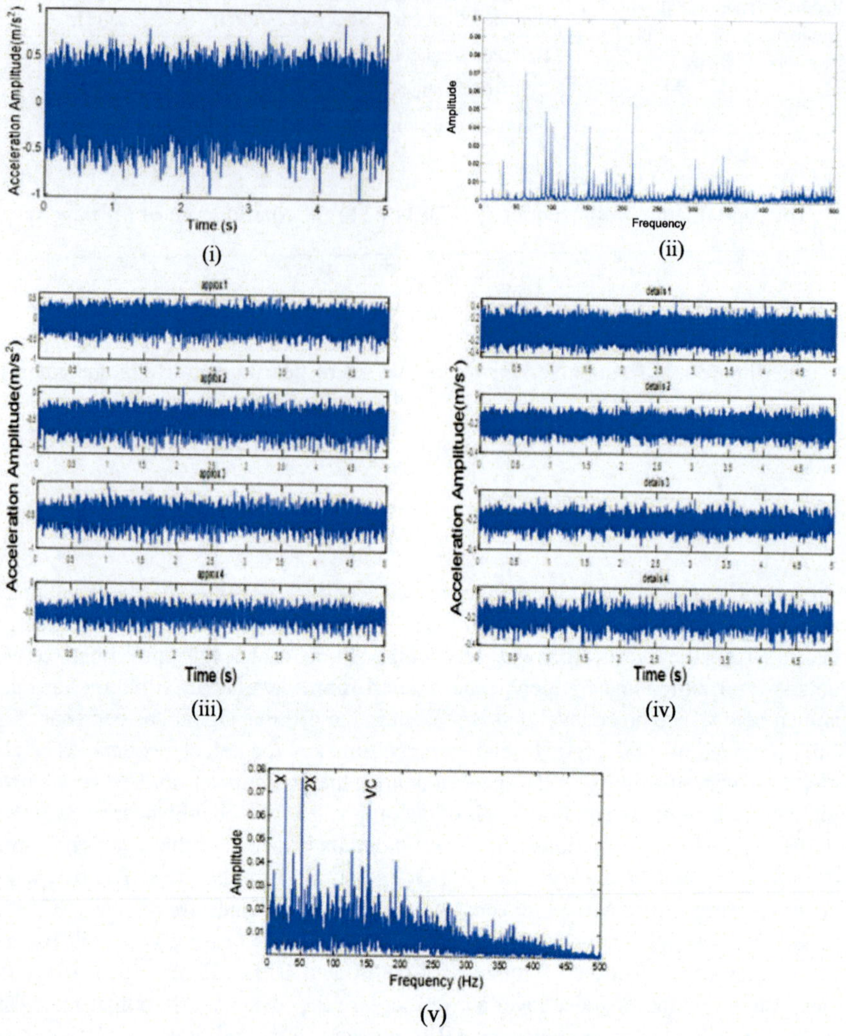

Fig. 3 Healthy bearing signal at 30 Hz. **a** Time-domain signal, **b** frequency domain signal, **c** approximations of the signal, **d** details of the signal, **e** FFT of approximation level 4

the required frequencies. So, a selection of mother wavelet using proper wavelet selection criteria improves the performance to remove noise and extract the key features of a signal.

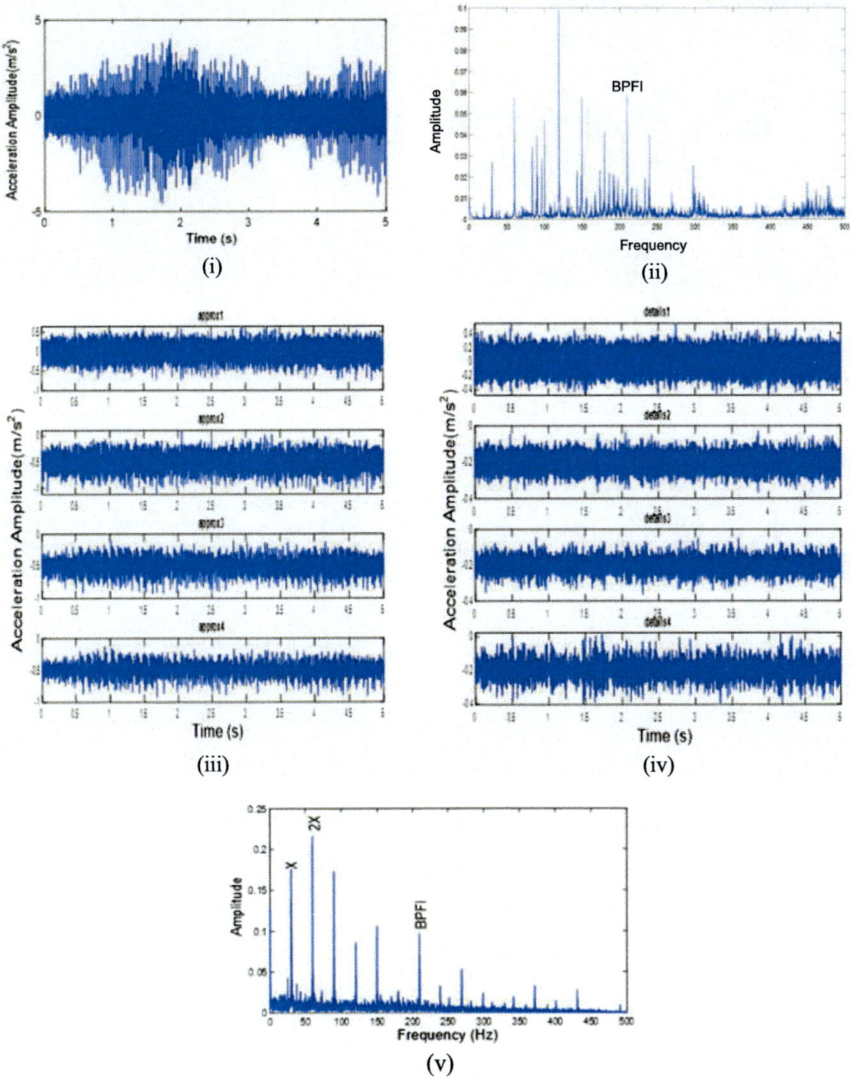

Fig. 4 Inner race defect signal at 30 Hz. **a** Time-domain signal, **b** frequency domain signal, **c** approximations of the signal, **d** details of the signal, **e** FFT of approximation level 4

4 Conclusions

The time-frequency analysis of bearing signals performs a better result than the frequency domain method. Decomposed data using the wavelet transform indicates the better signal-to-noise ratio than the frequency domain data. Among various

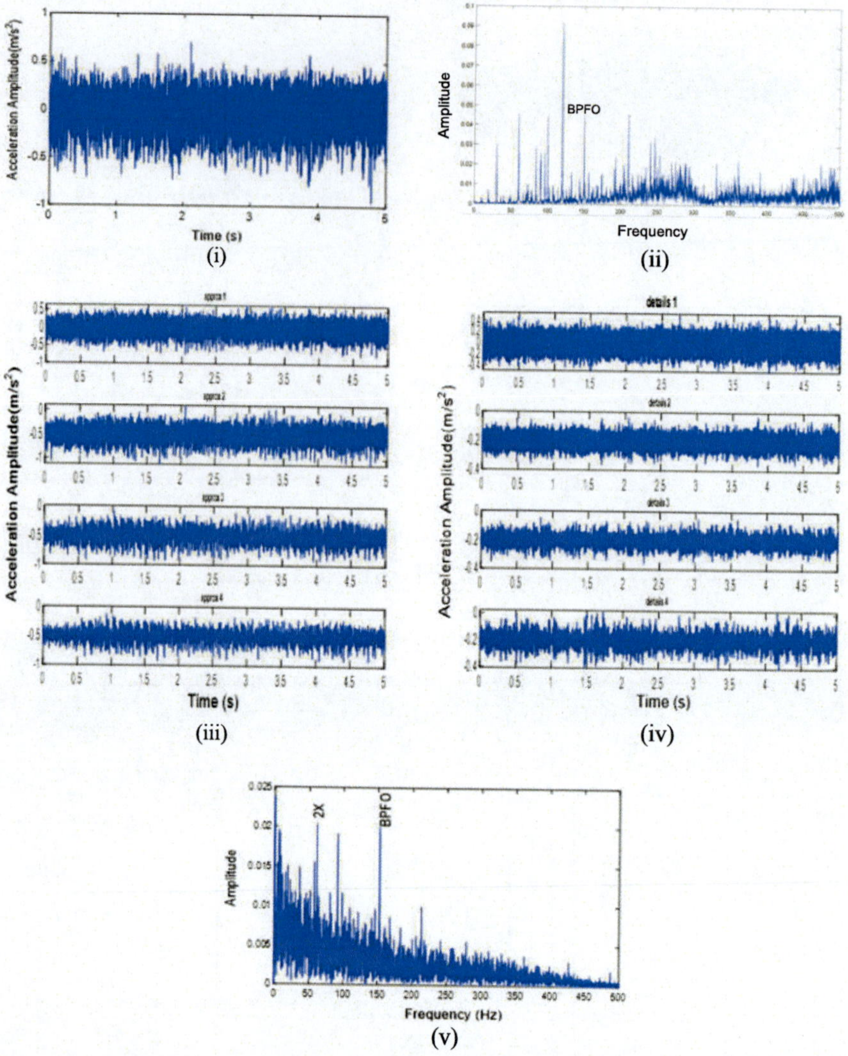

Fig. 5 Outer race defect signal at 30 Hz. **a** Time-domain signal, **b** frequency domain signal, **c** approximations of the signal, **d** details of the signal, **e** FFT of approximation level 4

mother wavelet families; the "Symlet-5" mother wavelet has a better effect considering the maximum energy-to-Shannon entropy-ratio criteria. The decomposition of vibration signal using the Symlet-5 mother wavelet clearly points out the defect frequencies.

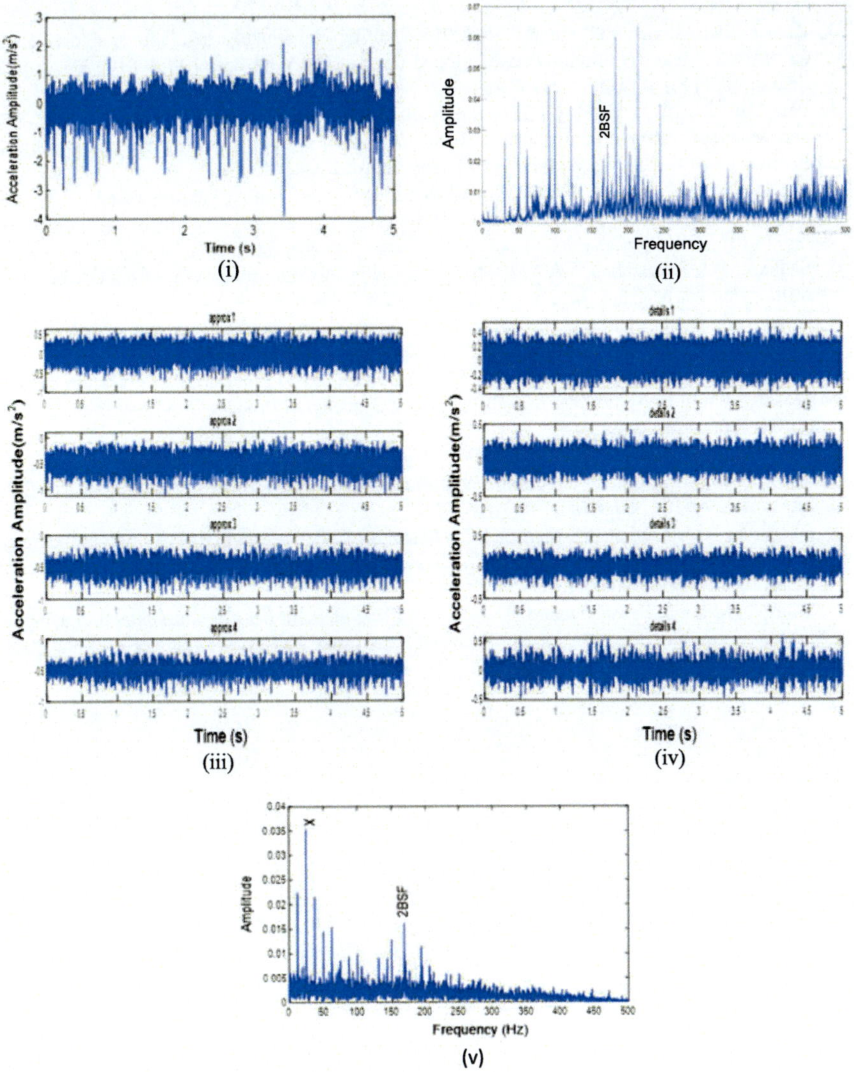

Fig. 6 Roller defect signal at 30 Hz. **a** Time-domain signal, **b** frequency domain signal, **c** approximations of the signal, **d** details of the signal, **d** FFT of approximation level 4

References

1. Mishra C, Samantaray AK, Chakraborty G (2017) Rolling element bearing fault diagnosis under slow speed operation using wavelet de-noising. Meas J Int Meas Confed 103:77–86
2. Kankar PK, Sharma SC, Harsha SP (2013) Fault diagnosis of rolling element bearing using cyclic autocorrelation and wavelet transform. Neurocomputing 110:9–17

3. Kankar PK, Sharma SC, Harsha SP (2011) Rolling element bearing fault diagnosis using autocorrelation and continuous wavelet transform. Journal Vib Control 17:2081–2094. https://doi.org/10.1177/1077546310395970
4. Peng ZK, Tse PW, Chu FL (2005) A comparison study of improved hilbert-huang transform and wavelet transform: application to fault diagnosis for rolling bearing. Mech Syst Signal Process 19:974–988. https://doi.org/10.1016/j.ymssp.2004.01.006
5. Liu Q, Chen F, Zhou Z, Wei Q (2013) Fault diagnosis of rolling bearing based on wavelet package transform and ensemble empirical mode decomposition. Adv Mech Eng. https://doi.org/10.1155/2013/792584
6. Kulkarni PG, Sahasrabudhe AD (2013) Application of wavelet transform for fault diagnosisof rolling element bearings. Int J Sci Technol Res 2(4):138–148
7. Khanam S, Tandon N, Dutt JK (2014) Fault size estimation in the outer race of ball bearing using discrete wavelet transform of the vibration signal. Procedia Technol 14:12–19
8. Prabhakar S, Sekhar AS, Mohanty AR (2001) Detection and monitoring of cracks in a rotor-bearing system using wavelet transforms. Mech Syst Signal Process 15:447–450. https://doi.org/10.1006/mssp.2000.1381
9. Pandya DH, Upadhyay SH, Harsha SP (2014) Fault diagnosis of rolling element bearing by using multinomial logistic regression and wavelet packet transform. Soft Comput 18:255–266. https://doi.org/10.1007/s00500-013-1055-1
10. Bhavaraju KM, Kankar PK, Sharma SC, Harsha SP (2010) A comparative study on bearings faults classification by artificial neural networks and self-organizing maps using wavelets. Int J Eng Sci Technol 2:1001–1008
11. Wang D, Miao Q, Fan X, Huang HZ (2010) Rolling element bearing fault detection using an improved combination of Hilbert and wavelet transforms. J Mech Sci Technol 23:3292–3301. https://doi.org/10.1007/s12206-009-0807-4

Manufacturing Systems

Structural Impact of Presence of Metallic Spacers Between Shims in a Flex Nozzle System

Mohit Agarwal, B. Kiran Kumar, A. Yezhil Arasu, Thomas Kurian, and G. Levin

1 Introduction

Launch vehicles generally work with active control system and there are several ways to achieve that. Flex nozzle control (FNC) is one of these systems through which this can be achieved effectively. This vehicle flight control system is necessary to guide the rocket through the desired trajectory as well as to maintain the flight stability by countering the disturbance forces like external winds, thrust misalignment, differential thrusts produced between pair motors, gusts, shift in the vehicle center of gravity, etc. The deviations created due to these factors need to be corrected in real time during flight and to keep the vehicle under control as well as to maintain the structural integrity and to minimize the structural and thermal loads on to the rocket. The control system induces a lateral force acting on to the rocket at some distance from center of gravity causing a couple to rotate the rocket about this point, and hence causing the desirable change in its attitude. In flex nozzle control (FNC) system, nozzle orientation is manipulated to control the thrust direction and hence the attitude of the rocket. Flex nozzle system comprises of an alternate stack up of layers of elastomeric material and metallic or composite rings sandwiched together to get flexibility as well as strength and stiffness, which can be on real time controlled with the help of electrohydraulic or electromechanical actuators. These rings are generally sandwiched such that they have a common center of pivoting. However, other configurations also exist with considerations of the manufacturing easiness, cost and space constraints. Based on these parameters, the main types of FNC systems can have spherical, conical and cylindrical configurations [1].

M. Agarwal (✉) · B. Kiran Kumar · A. Yezhil Arasu · T. Kurian · G. Levin
Vikram Sarabhai Space Center (VSSC), Indian Space Research Organization (ISRO), Trivandrum, Kerala 695022, India
e-mail: mohit.agarwal.iit@gmail.com

V. H. Saran and R. K. Misra (eds.), *Advances in Systems Engineering*, Lecture Notes in Mechanical Engineering,
https://doi.org/10.1007/978-981-15-8025-3_22

ISRO's heaviest rocket GSLV Mk-III comprises of 2 nos. of S200 boosters as the strap-on motors. S200 is the biggest solid motor ever made by ISRO. This booster uses flex nozzle system with cylindrical body as the means of thrust vector control.

2 S200 FNC Configuration

S200 flex nozzle system comprises of eight numbers of elastomer pads made from natural rubber and seven numbers of metallic rings known as shims made up of 15CDV6 material. These pads and shims are sandwiched together in an alternate pattern. This system is further sandwiched between two end rings known as fore end ring or FER and aft end ring or AER also made of 15CDV6 material, to interface the flex seal with the other subsystems of nozzle.

This subsystem is manufactured through compression molding process in which the sand blasted metallic shims are assembled with uncured rubber pad in between after applying the bonding solution (rubber solution) at bonded interfaces. Curing of the system is done under the simultaneous application of pressure and temperature. For ensuring fixed thickness for the elastomer layers, small metallic spacers are used which are to be removed after the completion of curing.

For a configuration of S200 FNC system, a total of 256 nos. of such spacers are used during molding of S200 flex seal, i.e., at every 22.5° location in the elastomer pad. The study is carried out to analyze the impact of the presence of at least one of these spacers (between aft end ring and shim1) during the actuation of flex nozzle system. Figure 1 shows the location of spacer.

Flex seal before induction into space program go through several acceptance tests to assess on the structural margins and to confirm its functional performance after the full manufacturing process. The detailed testing scheme includes the simulation of axial loading in the flex seal along with actuation and also its ability to withstand the hoop loads. The detailed testing schemes are finalized and fixed after assessing the structural characterization requirements as reported in [1–6]. For simulation of hoop loads into the flex seal system, the flex seal is subjected to an internal pressure corresponding to 1.1 times of maximum expected operating pressure (MEOP) with axial relief to avoid over testing of flex seal in axial direction. This is the proof pressure testing condition of the flex seal and associated nozzle component. For demonstrating the structural adequacy and design margins of flex seal and associated nozzle components, the test pressure is adjusted to have a closed system condition on and nozzle is actuated to the required angle using actuators. This test procedure is known as pressure vectoring test.

Finite element static structural analyses were carried out to assess the structural impact of the spacer on the flex seal while conducting the above acceptance tests, viz., proof pressure test (PPT), null position test (NPT) and pressure vectoring test (PVT). This paper summarizes on the simulation methodology and the results of the analyses. A comparison of these results obtained from analyses with experimental results is also provided.

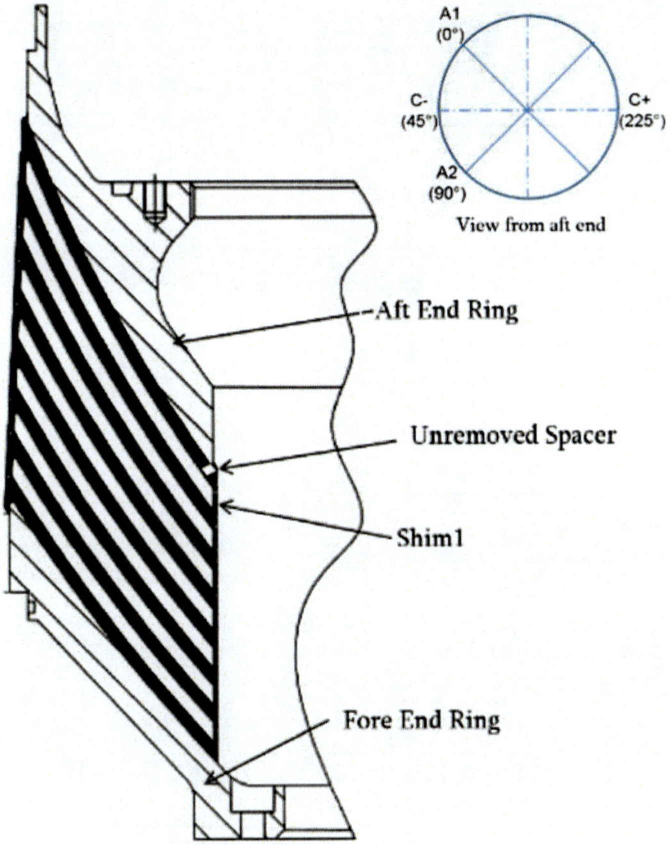

Fig. 1 Location of 'unremoved spacer'

3 Finite Element Model

Extensive and comprehensive static structural analyses of the S200 flex nozzle system toward assessment of the structural impact of left out spacer are carried out using ANSYS®. Different analyses were carried out for simulating each load cases. The results of the analyses are further compared with the ground actuation test and a close match is found out between the predicted and measured parameters.

One metallic spacer is simulated in the 225° resultant plane to take the advantage of symmetry of FE model. Figure 2 shows the finite element model used for the simulation along with the actuator. To reduce the computational effort for the problem, half model is simulated with symmetric boundary conditions. Shims, aft end ring and fore end ring are modeled with the nonlinear material properties of 15CDV6. Appropriate material properties are also chosen for relevant hardware and two parameter Mooney–Rivlin material model is used for the elastomer. Extensive tests were

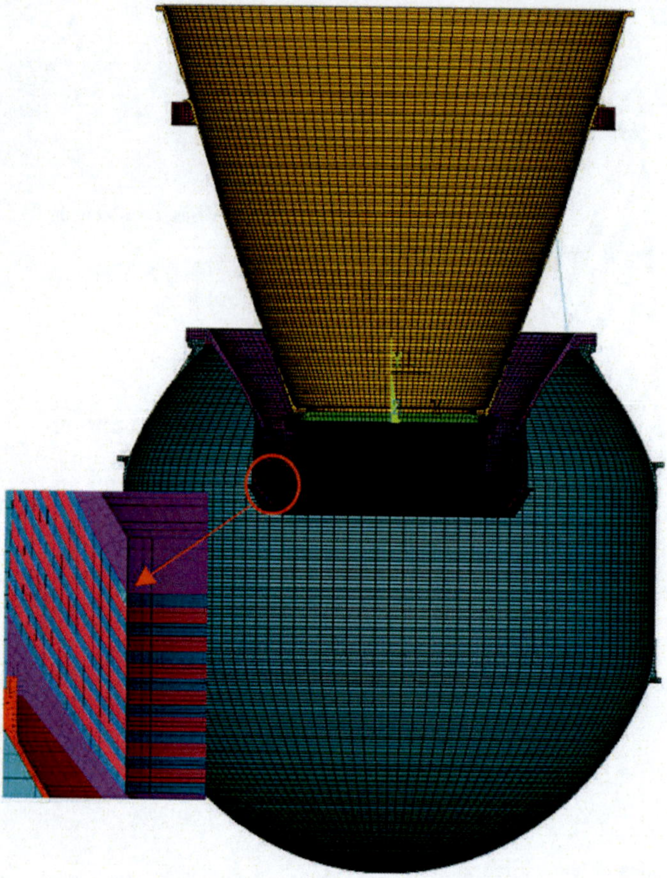

Fig. 2 Finite element model considered for the analysis

performed for the evaluation of these characterizing parameters. The model considered is having a degree of freedom of 12 lac (approximately) to properly capture the stresses and strains in the shims and other components while vectoring the flex seal. Selectively finer mesh is used for the concerned areas. The increased number of degree of freedom is on account of the finer mesh requirement in the model to achieve a converged solution. Convergence checks are carried out to confirm on the convergence of solution. Furthermore, the result of analysis, viz., strain on salient location on the hardware is compared against the measured values during the development tests. Proper contact conditions are simulated at interfaces to address the movement tendencies of spacer.

4 Results and Discussion

Analyses were carried out for PPT (6.475 MPa) condition along with floating piston for axial load relief, NPT (internal pressure of 3.776 MPa) and PVT (internal pressure along with 5.5° simultaneous vectoring in actuator plane for both push and pull). Displacements and maximum effective stresses for all seven shims along with end rings were extracted from the analysis results and tabulated. For comparison sake, results with generic condition (no left out spacer) also were included in the tables.

4.1 Results for PPT Condition

From the analysis results for PPT condition, it is observed that the spacer is not getting dislocated from its position, i.e., spacer does not have a tendency to come out from its position. Displacements and maximum effective stresses from FE analyses extracted for this load case are tabulated in Table 1. Strains from FE analysis along with measured strains during the test are tabulated in Table 2.

- Due to the presence of spacer, there is a reduction in axial displacements of shims when compared to the generic case (maximum of 0.92 mm in shim1). However, this reduction trend got almost nullified when one sees the displacement in shim7.
- Maximum effective stress of 757.7 MPa is observed (68.4 MPa increase over the value in generic case) in AER. However, this augmentation in effective stress is observed locally near the spacer region for a circumferential distance of 222 mm and for a meridional length of 30 mm from ID toward OD.
- Maximum compressive hoop strain of 3324 μs noted in shim4 (an increase of 124 μs when compared to generic case).

Table 1 Displacements and max. eff. stresses for PPT at 6.475 MPa

Component	Displacement (mm)			Eff. stress (MPa)	
	Generic	With spacer		Generic	With spacer
	Max	At spacer	Max		
AER	6.81	**6.94**	**6.94**	689.3	**757.7**
Shim1	10.17	9.25	10.34	571.8	604.5
Shim2	11.58	11.09	11.77	651.7	657.6
Shim3	**12.75**	12.42	12.92	**690.3**	693.6
Shim4	13.82	13.58	13.96	687.9	691.3
Shim5	14.94	14.61	14.89	656.4	658.8
Shim6	15.66	15.51	15.72	606.08	610.1
Shim7	16.53	16.4	16.56	528	530.6

Table 2 Strains at 225° plane for PPT at 6.475 MPa

Component	Hoop/axial	FEA		Measured
		Generic	With spacer	Generic
Shim1	Hoop	−2400	−2078	
	Axial	542	212	
Shim2	Hoop	−2900	−2959	−3001
	Axial	779	706	
Shim3	Hoop	−3219	−3299	−2964
	Axial	900	888	625
Shim4	Hoop	−3200	**−3324**	−3369
	Axial	928	934	836
Shim5	Hoop	−3050	−3149	−3150
	Axial	894	902	816
Shim6	Hoop	−2820	−2856	−3033
	Axial	820	826	
Shim7	Hoop	−2395	−2414	
	Axial	697	701	

4.2 Results for NPT Condition

Like the case of PPT, for NPT condition also spacer dislocation was not observed. Results for the NPT loads are tabulated in Tables 3 and 4.

- Due to the presence of spacer, there is a reduction in axial displacements of shims when compared to the generic case. Maximum reduction of 1.04 mm is noted in shim1.

Table 3 Displacements and max eff. stresses for NPT at 3.776 MPa

Component	Displacement (mm)			Eff. stress (MPa)	
	Generic	With spacer		Generic	With spacer
	Max	At spacer	Max		
AER	6.17	**6.32**	**6.32**	**688.6**	**756.6**
Shim1	9.96	8.92	10.98	553.6	591.6
Shim2	11.54	10.91	11.83	632.6	638
Shim3	**12.68**	12.4	13.17	673.7	674.6
Shim4	14.08	13.71	14.4	677.6	679.6
Shim5	15.22	14.83	15.55	655.5	657.7
Shim6	16.27	15.87	16.6	616.2	617.7
Shim7	17.38	16.92	17.66	561.3	563.2

Table 4 Strains at 225° plane for NPT at 3.776 MPa

Component	Hoop/axial	FEA		Measured
		Generic	With spacer	generic
Shim1	Hoop	−2570	−2139	
	Axial	541	179	
Shim2	Hoop	−3128	−3087	−3054
	Axial	810	718	
Shim3	Hoop	−3450	−3486	−2986
	Axial	952	926	573
Shim4	Hoop	−3517	**−3563**	−3388
	Axial	997	991	827
Shim5	Hoop	−3403	−3437	−3161
	Axial	977	977	807
Shim6	Hoop	−3166	−3188	−3062
	Axial	915	916	
Shim7	Hoop	−2739	−2749	
	Axial	814	795	

- Maximum effective stress of 756.6 MPa is observed (67.9 MPa increase when compared to generic) in AER. However, this augmentation in effective stress is noted locally near the spacer region for a circumferential distance of 222 mm and for a meridional length of 30 mm from ID toward OD.
- Maximum compressive hoop strain of 3563 μs noted in shim4 (increase of 46 μs when compared to generic).

4.3 *Results for VT Condition*

As mentioned already, analyses were carried out for an internal pressure of 3.432 MPa with simultaneous actuation load conditions at actuator plane. While vectoring the nozzle, it is observed that the spacer is dislocating from its original position and moving toward ID of flex seal. From the analysis results, it is noted that, while vectoring the nozzle to an angle of 2° under pull-pull case, spacer can slip out from the flex seal. However, as a conservative case, to see the effect of spacer on nozzle vectoring (if it is not slipping out from flex seal), the analysis was repeated by constraining the spacer not to fall. Results for these analyses are tabulated in Tables 5 and 6.

- Maximum effective stress of 870.1 MPa is observed (3.7 MPa increase over the stress in generic case) in shim3 at 225° location (where spacer was present) for both actuators in simultaneous push condition. This augmentation in effective stress is noted locally near the spacer region for a circumferential distance of 185 mm and for a length of 4 mm along the shim from ID toward OD.

- Maximum increase in effective stress (16.8 MPa compared to generic case) is noted in shim1 due to the presence of spacer, which means that, had a spacer been present between shim2 and shim3 or shim3 and shim4 during the vectoring test, shim3 or shim4 would have undergone significant yielding.

Table 5 Max. effective stresses for VT at 3.432 MPa +5.5° simultaneous vectoring

Component	Push		Pull	
	Generic	With spacer	Generic	With spacer
AER	803.4	817.5	754.1	753.5
Shim1	818.4	835.2	776.2	774.8
Shim2	854.3	856.2	823.1	822.8
Shim3	**866.4**	**870.1**	**835**	**835.6**
Shim4	862.5	865.1	829.1	829.9
Shim5	850.2	850	810.8	811.1
Shim6	840.5	840.4	798.6	798.9
Shim7	805.5	805.5	743.9	744.5

Table 6 Strains at 225° plane for VT at 3.432 MPa +5.5° simultaneous vectoring

Component	Push			Pull		
	FEA		Measured	FEA		Measured
	Generic	With spacer	Generic (flex seal)	Generic	With spacer	Generic (flex seal)
Shim1	−4375	−4898		−648	−610	
	892	609		181	155	
Shim2	−5884	−6066	−5282	−503	−569	−40
	1590	1598		142	130	
Shim3	**−6211**	**−6496**	−5149	−476	−521	364
	1820	1878	1508	137	134	−336
Shim4	−6010	−6189	−6458	−538	−573	516
	1800	1849	1866	155	155	−89
Shim5	−5237	−5322	−5637	−725	−750	359
	1620	1623	1661	106	206	−46
Shim6	−4099	−4118	−4588	−1036	−1054	−140
	1255	1263		290	263	
Shim7	−2949	−2945		−1334	−1346	
	884	882		377	375	

- Maximum compressive hoop strain of 6496 μs noted in shim3 (increase of 283 μs compared to generic) for both actuators under simultaneous push condition.
- No significant change in maximum effective stress (compared to generic case) is noted in case of simultaneous pull condition at both actuation planes because the maximum effective stress is observed at 45° location, where the spacer effect is not felt.
- Unlike actuators in push condition, in case of actuators under pull condition, both hoop and axial strains are smaller at 225° location. However, a slight increase in strains is noted with the presence of spacer.

5 Conclusion

Structural impact of a spacer present in a booster flex seal during the acceptance tests is studied through a set of detailed finite element analyses. Analyses were carried out for PPT, NPT and VT load conditions using 3D models. Implication of the presence of spacer is studied for all the three test conditions.

It is observed from the analyses that, during PPT and NPT spacer is not getting dislocated from its position. However, during the vectoring test with pull-pull condition, at a resultant plane angle of 2°, spacer can slip out from its position and can fall out of the flex seal assembly.

An increase in effective stress of 68.4 MPa is observed for AER for PPT and NPT conditions. Margin of safety over minimum guaranteed 0.2% PS of the material is 0.10. The maximum effective stress in shim3 in VT for both actuators under push condition is increased only by 3.7 MPa (from 866.4 to 870.1 MPa) as compared to the generic case (no spacer). This augmentation in maximum effective stress is seen very locally near the spacer region for a circumferential distance of 185 and 4 mm along shim from ID toward OD.

It was observed that had the spacer been present between shim2 and shim3 or between shim3 and shim4 it can lead to the significant yielding in shim3 or shim4 and could cause the permanent deformation in the shims.

Hence, it is recommended to ensure that all of the 256 spacers from the flex seals are removed prior to acceptance tests to safe the flex seal from undergoing any structural damage. Therefore, it has to be made as a mandatory inspection to confirm it after the completion of molding process and before subjecting to any kind of structural test.

References

1. Woodberry RFH, Zeamer RJ (1974) Solid Rocket thrust vector control. In: Keller RJ (ed) Tr. NASA SP- 8114
2. Graves SR, Finnedberg JE, Roth RB (1989) Development and certification of the space shuttle redesigned solid rocket motor nozzle AIAA paper 89–2418

3. Le Merror O (1980) ARIANE 5 solid motor nozzle design. AIAA paper 90–2083
4. Prins WS, Meyer SA, Cox PD (1992) Advanced solid rocket motor nozzle thrust vector control flexseal development status. AIAA paper 92-3553
5. Shani S, Patter S, Pretez A (1995) Development of high-performance flexible joint for thrust vector control. AIAA 95-3017
6. Arasu Y, Kurian T, Kiran Kumar B, Venkitaraman S (2010) Advances structural analysis of flex nozzle for solid rocket booster. IAC-10-C2.1.6

Microstructure and Hardness Evaluation of Functionally Graded A356-20 wt.% SiC$_P$ Composite

J. Prem Kumar, V. R. Rajeev, K. Krishnakumar, and M. Hashim

1 Introduction

Automobile and aerospace sectors demand for more incipient materials considering the consequentiality of reliability, strength, wear resistance, and thermal conductivity. Reduction in the weight of components amends fuel economy. Weight reduction of the piston in IC engines, for example, reduces the unbalancing force due to reciprocating mass and thereby vibration. The development of innovative materials like aluminium matrix composites seems an auspicious task under these circumstances. Wear resistance of this composite material is very similar to that of steel even though the weight is one-third of steel [1]. Functionally graded materials belong to the advanced class of inhomogeneous materials characterized by a gradual change of material properties with the location. The gradation of reinforcement materials in FGMMC causes a smooth transition from one material to another reduces the distinctive interfaces/stress risers, persisting in conventional composites where failure is initiated. Hence, these classes of materials replaced conventional monolithic materials for automotive components such as piston ring, cylinder liner, and brake disc. A significant advantage of FGMMC is the possibility of tailoring its gradation to amend its performance [2]. Hence, the region other than surface and subsurface locale where reinforcement is not as pivotal, machining would be simpler without the reinforcement. Thus, the utilization of selective reinforcement, whereby the hard ceramic particles are utilized where high strength and wear resistance are required,

J. P. Kumar (✉) · V. R. Rajeev
College of Engineering, Trivandrum, Kerala 695016, India
e-mail: prem24970@yahoo.in

K. Krishnakumar
Sree Vellappally Natesan College of Engineering, Kayamkulam, Kerala 690503, India

M. Hashim
Government Engineering College, Wayanad, Kerala 670644, India

V. H. Saran and R. K. Misra (eds.), *Advances in Systems Engineering*,
Lecture Notes in Mechanical Engineering,
https://doi.org/10.1007/978-981-15-8025-3_23

is very appealing [3]. Powder metallurgy, vapour deposition, solid freeform fabrication, centrifugal slurry, or centrifugal casting are the different techniques utilized to fabricate graded structures by the various researchers [4]. Centrifugal casting is one of the most flourishing and economical ways for the preparation of FGMMC made of a liquid aluminium metal matrix reinforced with SiC ceramic particulate composite. In the present study, attempts were made to prepare and cast FGMMC through vertical centrifugal casting technique and further to correlate the hardness with the microstructural features.

2 Materials and Methods

Commercial A356 aluminium alloy has been used for the preparation of composite, procured from M/S Fenfe Metallurgicals, Bangalore, India. Table 1 shows the elemental composition of the alloy selected for the study. Silicon enhances casting features such as fluidity, hot tear resistance, and feeding. Besides, the amount of Si present in this alloy eases to decrease reactivity with SiC particles while casting [5]. Silicon carbide (SiC_p) ceramic particulates with a mass density of 3.2 g/cc were chosen for the reinforcement since it is inexpensive and readily available. As received SiC_p from M/S Carborundum Universal Ltd. Kochi, India, with an average particle size of 28 μm, were cleaned in water and dried. Figure 1 shows the pictorial view of the casting set up used for the preparation of FGMMC.

The composite was prepared through a liquid metallurgy route. The A356 alloy was initially melted and maintained to a temperature of 750 °C in the steel retort of the 5kg capacity bottom pouring electric stir casting furnace supplied by SwamEquip Private Ltd, Chennai, India. 2 wt.% of magnesium cubes were dipped into the melt before the addition of SiC_p to improve wettability between matrix and reinforcement particles [6]. SiC powder was heated at 500^0C in the powder heating furnace and was fed to the molten alloy by controlling the valve of the powder heating furnace along with mechanized mixing at 400–500 rpm using graphite coated twin-blade stirrer[7]. SiC particle heating improves the wettability of SiC with matrix alloy and aids in eliminating surface impurities. Following the powder adding up, the stirring was extended for 5 minutes. The molten mixture of processed metal matrix composite then transferred through a metal pouring cup into the graphite coated rotating metallic mould, mounted on the SwamEquip, make vertical centrifugal casting machine by operating the motorized valve located at the bottom of the stir casting furnace. The rotational speed and preheating temperature of mould were kept at 1300rpm and 350 °C, respectively. The cast FGMMC ring removed from the mould and cleaned.

Table 1 Elemental composition of the A356 alloy

Element	Si	Cu	Mg	Zn	Ti	Fe	Mn	Al
wt. %	7.00	0.02	0.40	0.04	0.18	0.11	0.0014	Balance

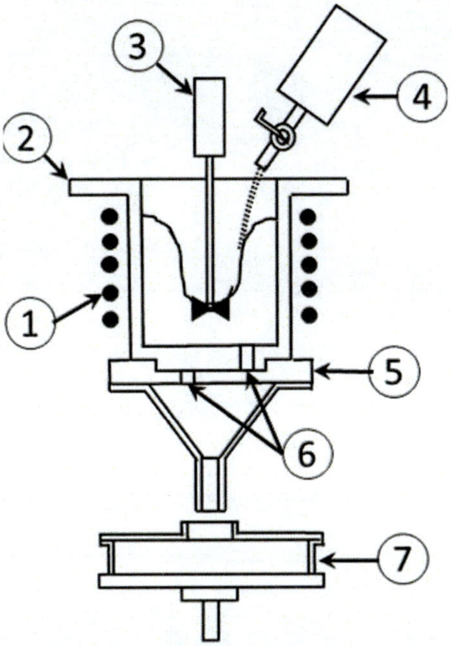

Fig. 1 Schematic diagram of bottom pouring stir casting furnace with rotating mould. (1) Heater, (2) steel retort, (3) stirrer, (4) powder heating furnace, (5) cone assembly, (6) port holes for pouring melt, and (7) heated and rotating mould for centrifugal casting

The weight of the prepared A356-20 wt.% SiC_p FGMMC ring was 2.4 kg. Outer diameter, average inner diameter, and thickness of the as-cast ring were 240 mm, 128 mm, and 28 mm, respectively.

As-cast test samples of identical-shaped rectangular blocks (Fig. 2) for microstructural characterization and hardness evaluation were cut from prepared FGMMC ring using water jet cutting. The standard metallographic procedures were used for sample preparation of microstructural evaluation. The microstructural test sample was polished using emery papers of the rough, medium, and fine grit sizes ranging from 80 to 1200. Further, the polishing was done using a micro-cloth fixed disc polishing machine at a rotational speed of 350rpm. Medium grade (3 μm) and fine grade (0.25 μm) diamond abrasive pastes were used for polishing along with kerosene as a lubricant. Subsequently, the sample was cleaned in soapy water, followed by distilled water. Then, the sample was dried in hot air and etched with Keller's reagent. Microstructures were captured through the Olympus-LC30 optical microscope at an interval of 2.5 mm from outer to inner edge along the radial path of the FGMMC ring ample (Fig. 2) and images were captured.

Bulk hardness testing was carried out (Brinell hardness tester) to study the hardness variation along the radial direction from outer to the inner edge of the FGMMC composite ring. Hardness indentations were taken (Fig. 3) by applying a load (P), of

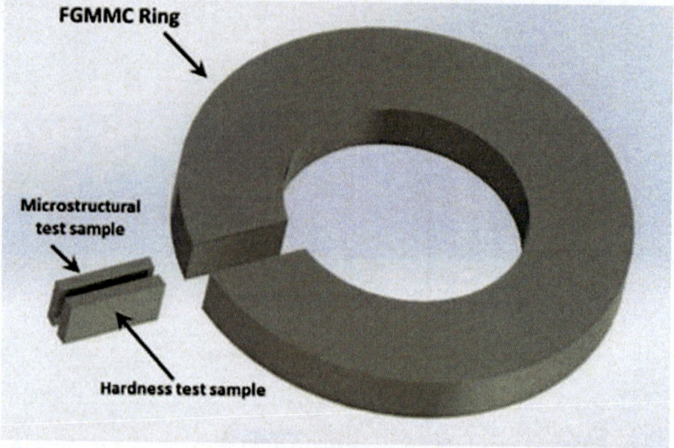

Fig. 2 Schematic diagram of cast FGMMC ring and test samples for microstructural analysis and hardness evaluation

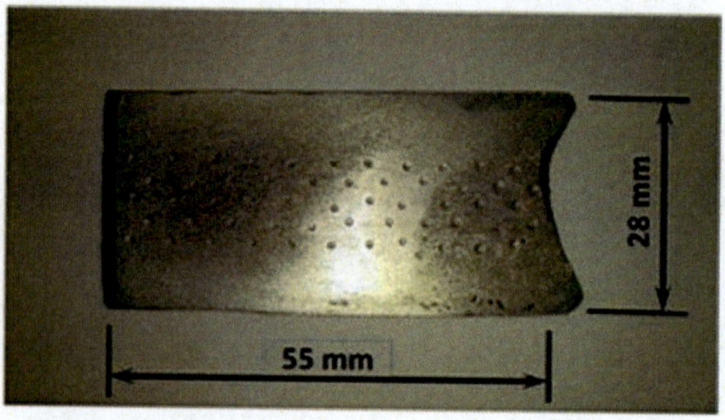

Fig. 3 Hardness test sample with ball indentations

60 kgf with a 2.5 mm cast iron ball indenter (D), for 20 s. The indentation images were captured using the Olympus optical microscope, and the diameter (d) of indentations was measured with the image analysis software of the microscope. Further, hardness values were computed in Brinell hardness number (BHN) using Eq. (1) and plotted against distance from outer to the inner edge of the FGMMC ring.

$$\mathrm{BHN} = \frac{2P}{\pi D\left(D - \sqrt{(D^2 - d^2)}\right)} \tag{1}$$

3 Results and Discussion

3.1 Microstructure

Based on the SiC particle concentration along the radial direction in the prepared FGMMC ring, different regions could be identified through optical microscopic image analysis. The outermost chilled, particle rich, transition, matrix rich, and particle agglomerated/gas porosities regions were the five distinct regions. The concentration of reinforcement particles was more in the outer region spanning from 5.0 to 20.0 mm (Fig. 4) along the radial direction. In this particle-rich zone, maximum particle concentration was observed at 7.5 mm from the outer periphery (Fig. 4a), whereas at distances 10.0, 12.5, and 15.0 mm, Fig. 4b–d, almost same particle concentration were observed. Figure 5a–d showed the microstructure of transition zone spanning from 20.0 to 27.5 mm, wherein a gradual decrease in particle concentration was observed and further from 30.0 to 40.0 mm, matrix-rich/particle-free zone, no particles were seen, Fig. 6a, b. Particle agglomeration and gas porosities were observed, Fig. 7a, b, beyond 40 mm towards the inner periphery. The combined effect of centrifugal action and difference in the solidification rate of composite melt

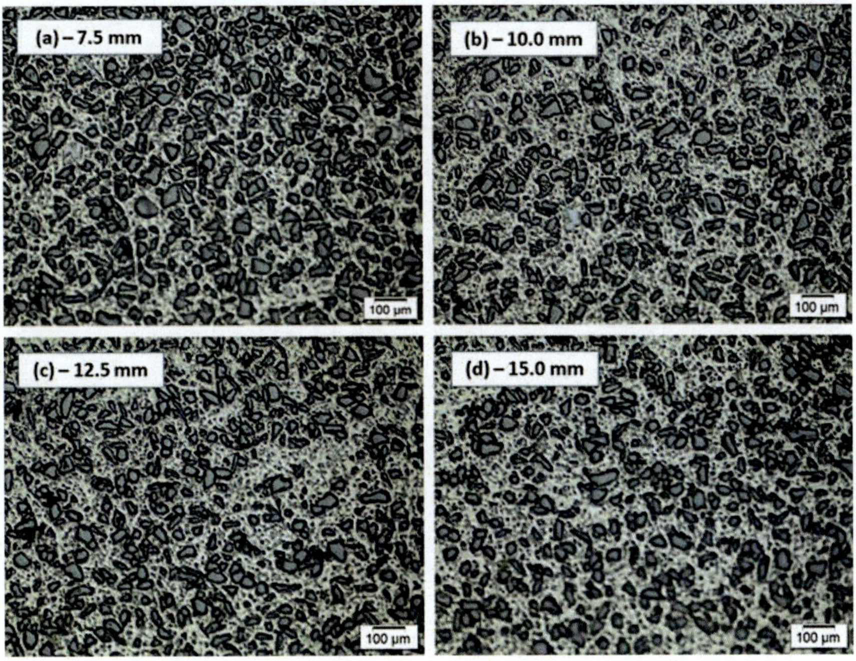

Fig. 4 a–d Microstructure images of particle-rich outer zone at locations 7.5, 10.0, 12.5, and 15.0 mm

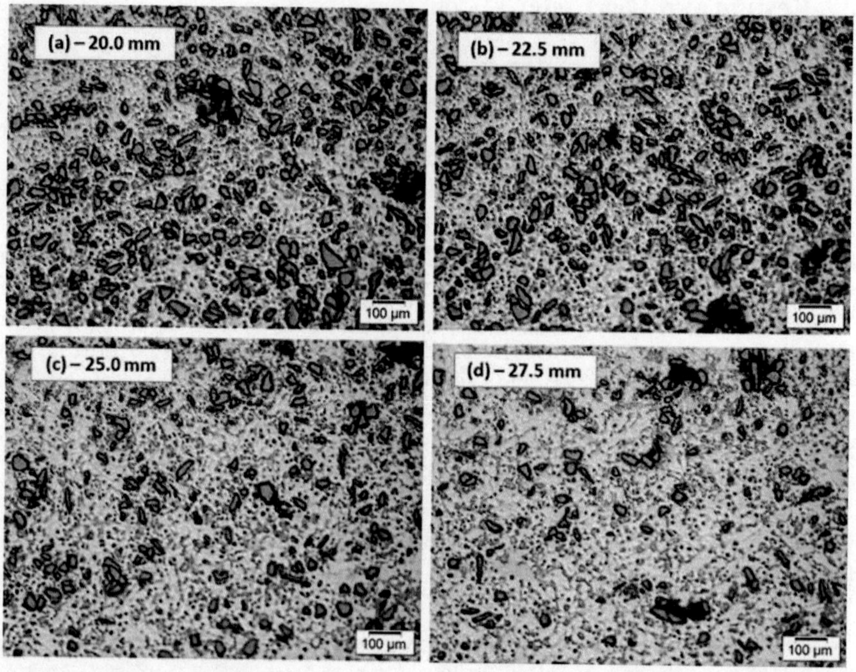

Fig. 5 **a–d** Microstructure images of transition zone at locations 20.0, 22.5, 25.0, and 27.5 mm

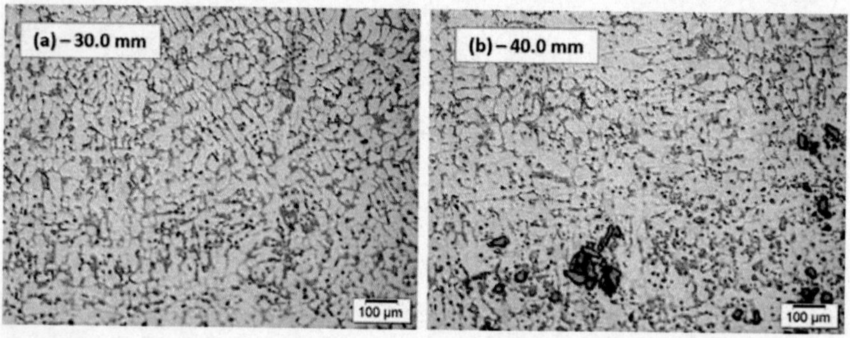

Fig. 6 **a** and **b** Microstructure images of matrix-rich zone at locations 30.0 and 40.0 mm

in the rotating mould during casting causes the reinforced particle gradation in the FGMMC ring.

Grain size variations were also observed along the radial direction of the fabricated ring. This grain size variation must be due to the difference in the solidification rate of melt in the mould where the melt solidifies rapidly once it comes into contact with colder mould surface, forming a coverage of chilled zone. The growth of dendritic

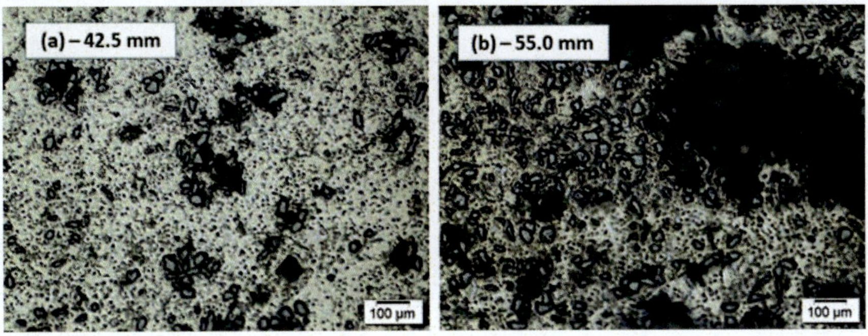

Fig. 7 **a** and **b** Microstructure images of inner zone with porosities and particle agglomeration at locations 42.5 and 55.0 mm

primary α-aluminium prevents due to the pressure created by centrifugal action and the high concentration of SiC_p in the outer region.

3.2 *Hardness*

Figure 8 shows the panoramic view of microstructure and hardness values of the as-cast sample at different locations at an interval of 2.5 mm along the radial path from the outer to the inner edge of the casting. It was noticed from the microstructure images and hardness plots that hardness values improved with an increase in SiC particle concentration. The hardness value of 62 and 72 BHN was observed in the prepared FGMMC ring at locations 2.5 and 5.0 mm away from the outer edge of the ring, and a maximum value of 101 BHN was found at 7.5 mm from the outer edge of the ring where particle concentration was maximum. In the chilled zone region (0–2.5 mm), because of sudden solidification due to the difference in mould wall and melt temperatures, particles were unable to reach this region under the action of centrifugal force. The outer edge of the ring, chilled zone, showed hardness equivalent to matrix alloy. The solidification front moving radially inward direction slowed down the particle movement in outward direction caused gradient distribution of particles in the outermost zone ranging from 2.5 to 7.5 mm and a corresponding increase in hardness from 62 to 101 BHN was noticed. It was also clear from microstructural images that the higher particle concentration (particle-rich) region was between 7.5 and 20.0 mm, and the hardness values decreased from 101 to 80 BHN in this region (Fig. 8). Particle concentration further decreased steeply in the transition region (20.0–27.5 mm from outer), and the corresponding hardness values were obtained in the range from 80 to 62 BHN. The hardness value at the matrix-rich inner region varied from 62 to 60 BHN. Further, towards the inner periphery, the average hardness dropped to 40 BHN. Because of the drop in centrifugal force magnitude in the

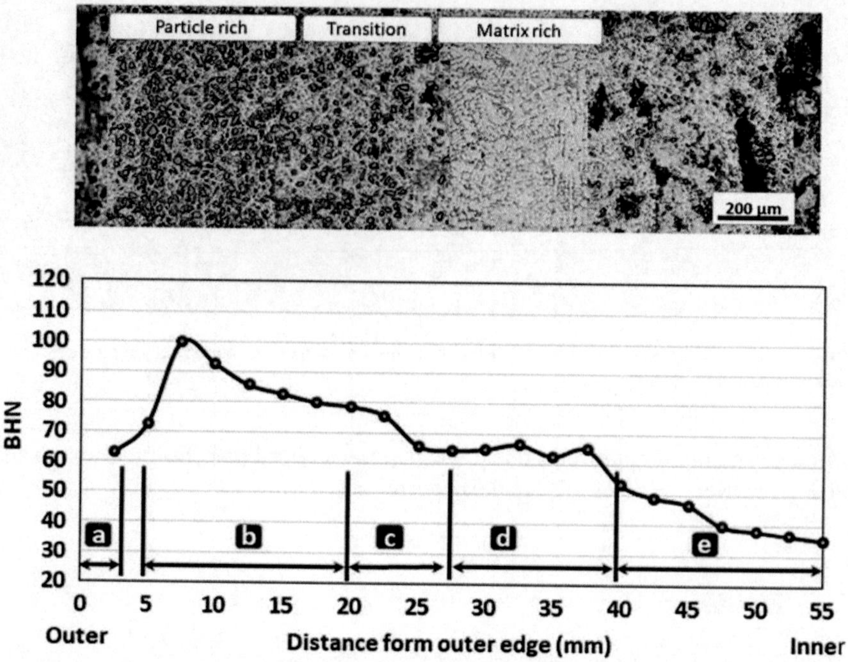

Fig. 8 Panoramic view of microstructure image and BHN plot against distance measured along radial distance from outer to inner periphery of the FGMMC ring. (a) Chilled zone, (b) particle-rich zone, (c) transition zone, (d) matrix-rich zone, and (e) porosities and particle agglomerated zone

inner region, the presence of gas porosities and particle agglomeration inner region exhibited lower hardness.

4 Conclusion

In this work, attempts were made to prepare A356 aluminium alloy reinforced with SiC_p functionally graded metal matrix composite using liquid metallurgy route accompanied by centrifugal casting. Based on the microstructural images and hardness test results, the following conclusions were drawn.

- A356-20 wt.% SiC_p FGMMC was successfully processed by the vertical centrifugal casting technique. Based on the concentration of SiC_p reinforcement, chilled zone (0–3 mm), particle rich (5–20 mm), transition (20–27.5 mm), matrix rich (27.5–40.0 mm), and particle agglomerated weaker inner (40.0–55.0 mm) regions were observed through microstructure evaluation.
- Microstructure and hardness correlation of the prepared FGMMC revealed that the highest value of hardness, i.e., 101 BHN occurred at 7.5 mm from the outside edge

of the ring. As a result of radial gradient distribution of SiC particles, towards the inner edge of the ring, noticed a decrease in hardness from 101 BHN (particle-rich region) to 60 BHN (matrix-rich region).

Acknowledgements The first author gratefully acknowledges the QIP-Ph.D scholarship received from AICTE, Government of India, for doing the doctoral research work. The authors acknowledge the fund received from the Centre for Engineering Research and Development (CERD), Govt. of Kerala, India, vide letter no. KTU/RESEARCH-3/2647, dated: 14.11.2016 for the smooth conduct of the work.

References

1. Bonollo F, Moret A, Gallo S, Mus C (2003) Experimental techniques and design in composite materials, 6th International seminar 18th to 20th June, Vicenza
2. Watanabe Y, Yamanaka N, Fukui Y (1998) Control of composition gradient in a metal-ceramic functionally graded material manufactured by the centrifugal method. Compos Part A: Appl Sci Manufact 29:595–601
3. Rajan TPD, Pai BC (2014) Developments in processing of functionally gradient metals and metal–ceramic composites: a review. Acta Metall Sin (Engl Lett) 27(5):825–838
4. El-Galy IM, Ahmed MH, Bassiouny BI (2017) Characterization of functionally graded Al-SiC_p metal matrix composites manufactured by centrifugal casting. Alexandria Eng J 56:371–381
5. Rodnguez-Castro R, Wetherhold RC, Kelestemur MH (2002) Microstructure and mechanical behaviour of functionally graded Al-A359/SiC_p composite. Mater Sci Eng A 323:445–456
6. Hashim J, Looney L, Hashmi MSJ (2001) The enhancement of wettability of SiC particles in cast aluminium matrix composites. J Mater Process Technol 119:329–335
7. Aniban N, Pillai RM, Pai BC (2002) An analysis of impeller parameters for aluminium metal matrix composites synthesis. Mater Des 23:553–556

Optimum Design of Magnetic Field Assisted Finishing Tool: A Finite Element Study

Yogendra Kumar and Harpreet Singh

1 Introduction

In the past recent years, the finishing technologies have growing faster rate with the immense effect on the developed new material and product. The manufacturing personals and associated researchers are meeting to challenges of finishing these materials to fulfill the functional requirements due to complex geometry. Traditional finishing processes like grinding, lapping, and honing have lack of tool flexibility during finishing of sculptured surface, harder material, and shape variation [1]. To overcome with these limitations, scientist and researchers have developed various advanced non-traditional finishing processes. These advanced non-traditional processes can be utilized flexible abrading medium for finishing of complex/sculptured surfaces. The flexible abrading medium is prepared by the mixing of abrasives in the base medium like liquid or air. For finishing, these processes which uses magnetically influenced smart fluid such as in magnetorheological finishing (MRF), magnetorheological polishing (MRP), magnetic abrasive finishing (MAF), and magnetic field assisted finishing (MFAF), ball end magnetorheological finishing (BEMRF) [2–5]. In all these finishing processes, the force induced in abrasive particles is regulated externally by varying the dependent parameter of finishing like magnetic field density, working gap, abrasives, magnetic particles, and base medium. To interpret the finishing processes correctly, it is also required to understand various rheological properties like viscosity and yield stress of magnetorheological fluids inhabitancy of external magnetic range. Figure 1 shows the CIPs chain formation on account of externally applied magnetic field [6]. It were observed during the study that magnetic flux was the highest contributed parameter in the yield stress and viscosity of the magnetorheological fluid, with 92.72%, 49.95%,

Y. Kumar (✉) · H. Singh
Mechanical Engineering Department, PDPM IIITDM Jabalpur, Jabalpur 482005, India
e-mail: ykd.gni@gmail.com

V. H. Saran and R. K. Misra (eds.), *Advances in Systems Engineering*,
Lecture Notes in Mechanical Engineering,
https://doi.org/10.1007/978-981-15-8025-3_24

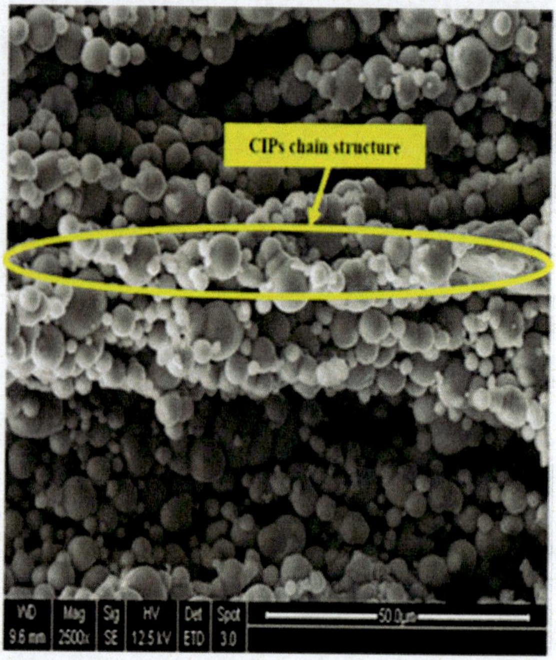

Fig. 1 CIPs chains formation with due presence of magnetic field [6]

respectively, among all the main factor and their interactions term with varying concentration of CIPs and abrasives as shown in Fig. 2 [7].

In magnetic field assisted finishing, the applied magnetic field with the use of permanent magnet/electromagnet and abrasive particles is aligned in magnetic flux lines to form chain-like structure for strong magnetic abrasive brush [8]. Comparatively abrasive particles are smaller in size as compared to magnetic particle (CIPs). The forces enforce on magnetic particles that lead to alignment of abrasive particle results great value of hardness (Al_2O_3, SiC) to access on work surface. The rotary

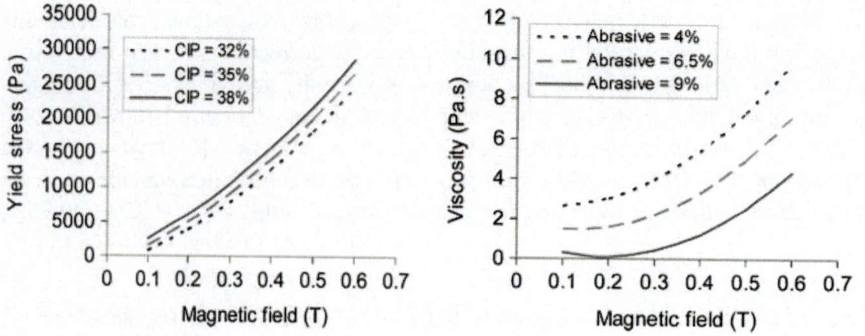

Fig. 2 Magnetic field effect on yield stress and viscosity [7]

Table 1 Literature work on permanent magnet used by various researchers

S. No.	Author	Work material	Permanent magnet	B_{max} (T)	Ra (nm)
1	Das et al. [10]	Brass	Nd-Fe-B	0.2	50
2	Kim et al. [11]	Silicon	Nd-Fe-B	–	88.4
3	Sidpara et al. [12]	EN-31	N50	0.3–0.05	28
4	Sidpara et al. [13]	Silicon	N48	0.46	5.36
5	Barman et al. [14]	Titanium	N48	0.4	50–70

movement in between magnetized abrasive brush and workpiece during finishing due to magnet tool movement, it becomes feasible to remove the micro-/nanomaterial. Any kind of workpiece surface with intricate/complex shapes can be finished due to flexibility by these abrasives. The magnetic field assured the certain uniformity of the magnetic force which imposed on the work surface.

Kim et al. [9] reported the improvement in magnetic flux value during his study in available working gap with maximum value of 0.2 mT. It is also found that permanent magnet enhances the surface roughness when compared with electromagnet during finishing. Various researchers have reported their work on use of permanent magnet on magnetic field assisted finishing as mentioned in Table 1.

The study aimed to assess the effect of permanent magnet during selection of appropriate one while finishing. The objective of the study to find the possible location/design of the permanent magnet to get uniform and maximum value of magnetic field for the formation of strong finishing brush.

2 Design and FEA Analysis of Finishing Tool

To investigate the simulated effect of all possible location of permanent magnets, 2D Ansoft Maxwell® is employed during study. Permanent magnet provides strong magnetic field in the finishing zone. Different grades permanent magnet which were available and worked by various researchers are enlisted in Table 2.

Circular size of fix size magnet fixture with multiple small permanent magnet of different diameter (2–4 mm) attached internally are considered during simulation

Table 2 Different grades of permanent magnet with their properties

S. No.	Magnet grade	Br (G)	H_c (kA/m)	BH_{max} (MGOe)
1	N30	10,800	796	28
2	N35	11,700	867	33
3	N48 [13]	13,800	836	46
4	N50 [14]	14,000	796	48
5	N52 [6]	14,300	796	50

as shown in Fig. 3. The permanent magnet of N52 is considered for the study. The following assumptions were considered during finite element simulation [2].

- Maxwell 2D solver is considered
- All object during simulation is stationary
- Source of magnetic field used with permanent magnet (N52).

In FEA simulation, cylindrical permanent magnet tool for the analysis is as shown in Fig. 3. During simulation, the magnet fixture material is mu metal for constant diameter magnet tool of 20 mm diameter in all the three different design combinations. Magnetic field distributions were analyzed for three possible designs as shown in Fig. 3, which makes the basis for magnet optimum dimension. In magnetic field distribution plots as shown in Fig. 3, it is observed for 2-mm-magnet diameter (17 magnet) magnetic flux distribution is non-uniform as compared to other design set of magnet.

It is observed from Fig. 4 that the 3-mm-diameter magnets have produce more

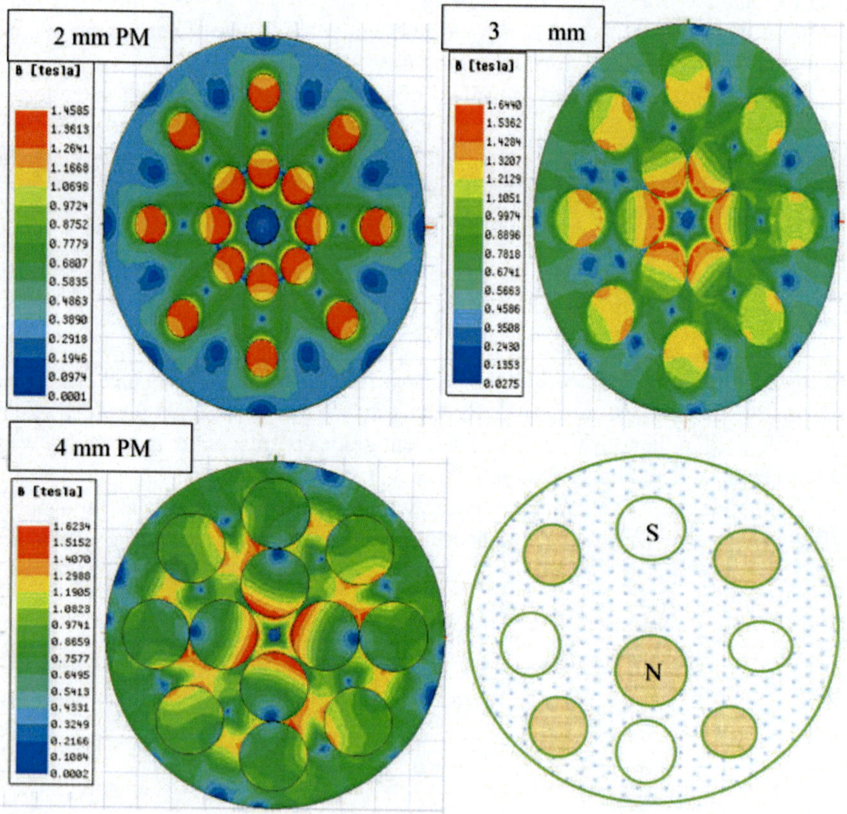

Fig. 3 Simulated magnetic flux density value for different size permanent magnet (N52 grade) and sample magnet arrangement used during simulation

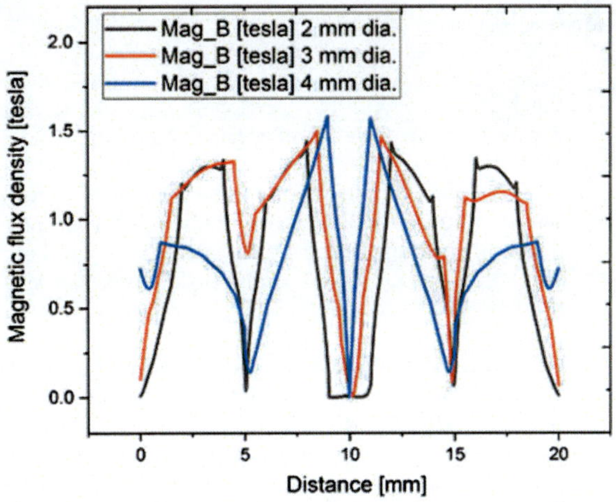

Fig. 4 Magnetic flux density plot along with reference line for various diameters of the permanent magnets

Table 3 Obtained simulated magnetic flux density value for different design

S. No.	Simulation study	outer fixture dia (mm)	Magnet dia (mm)	Magnet grade	No. of magnet	Magnetic flux density
1	I	20	2	N52	17	1.437–0.2
2	II	20	3	N52	14	1.495–0.2
3	III	20	4	N52	12	1.579–0.2

uniform magnetic flux in comparison with 4-mm-diameter magnet. The II and III type designs produce approximately same value of minimum magnetic flux value (Table 3), but the area under the magnetic flux value plot is more for II (3 mm) type of design.

3 Conclusions

A simulation study via FEA-based Ansoft Maxwell® was carried out to find the best possible design of finishing tool on the basis of uniform magnetic flux value with incorporating three different diameters of permanent magnet. Based on the simulation results, the following conclusions can be derived.

- The II-type simulation study is found best among all three types of simulated design.
- The II-type of design magnetic flux density plot is consistent and stable. A steeper change in plot of magnetic flux density in III type of design.
- The minimum achieved constant magnetic flux density value in all three types of probable design is 0.2 Tesla to carry the further finishing.

References

1. Singh H, Jain PK (2016) Developments in electro-chemical honing (ECH): a review on experimental investigation of precision finishing of mechanical parts. Int J Precis Technol 6(1):17–41
2. Barman A, Das M (2017) Design and fabrication of a novel polishing tool for finishing freeform surfaces in magnetic field assisted finishing (MFAF) process. Precisi Eng 49:61–68. https://doi.org/10.1016/j.precisioneng.2017.01.010
3. Jain VK, Singh P, Kumar P, Sidpara A, Das M, Suri VK, Balasubramaniam R (2009) Some investigations into magnetorheological finishing (MRF) of hard materials. ASME 2009 Int Manufact Sci Eng Conf 2:261–270
4. Nagdeve L, Jain VK, Ramkumar J (2018) Nanofinishing of freeform/sculptured surfaces: state-of-the-art. Manufact Rev 5(6). https://doi.org/10.1051/mfreview/2018005
5. Aravind T, Arunachalam N, Xavier K (2018) Physical insights about magnetic flux distribution and its effect on surface roughness in MR fluid based finishing process. Mater Res Express 6(1)
6. Nagdeve L, Jain VK, Ramkumar J (2017) Preliminary investigations into nano-finishing of freeform surface (femoral) using inverse replica fixture. Int J Adv Manuf Technol. https://doi.org/10.1007/s00170-017-1459-7
7. Sidpara A, Das M, Jain VK (2009) Rheological characterization of magnetorheological finishing fluid. Mater Manufact Processes 24:1467–1478
8. Das M, Jain VK, Ghoshdastidar PS (2015) A 2D CFD simulation of MR polishing medium in magnetic field-assisted finishing process using electromagnet. Int J Adv Manufact Technol 76:173–187. https://doi.org/10.1007/s00170-014-5847-y
9. Mun SD (2010) Micro machining of high-hardness materials using magnetic abrasive grains. Int J Precis Eng Manufact 11(5):763–770
10. Das M, Jain VK, Ghoshdastidar PS (2012) Nanofinishing of flat workpieces using rotational–magnetorheological abrasive flow finishing (R-MRAFF) process. Int J Adv Manuf Technol 62:405–420. https://doi.org/10.1007/s00170-011-3808-2
11. Kim WB, Lee SH, Min BK (2004) Surface finishing and evaluation of three-dimensional silicon microchannel using magnetorheological fluid. Trans ASME 126:772–778. https://doi.org/10.1115/1.1811113
12. Sidpara A, Jain VK (2012) Nanofinishing of freeform surfaces of prosthetic knee joint implant. Proc IMechE Part B: J Eng Manufact 226(11):1833–1846. https://doi.org/10.1177/0954405412460452
13. Sidpara A, Jain VK (2011) Effect of fluid composition on nanofinishing of single-crystal silicon by magnetic field-assisted finishing process. Int J Adv Manuf Technol 55:243–252. https://doi.org/10.1007/s00170-010-3032-5
14. Barman A, Das M (2018) Simulation and experimental investigation of finishing forces in magnetic field assisted finishing process. Mater Manufact Processes 33(11):1223–1232. https://doi.org/10.1080/10426914.2018.1453157

Mechanical Properties of AA2014 Matrix Composites Reinforced with TiO_2 Particles Through Multi-step Stir Casting

J. S. Khalkho, CH. S. Vidyasagar, and D. B. Karunakar

1 Introduction

Recently, it has become very necessary to achieve the composites which are lighter, stronger and cheaper over conventional monolithic materials. The metal matrix composites symbolize new category of materials that possess combined behavior of properties, i.e., tensile-wear, hardness–ductility, etc., which have not been manifested by other conventional monolithic materials [1]. These types of metal matrix composites are utilized for applications requiring superior torque including aero transportation, automobile, marine and other transportation industries, etc. [2]. Aluminum matrix composites fall under the group of MMCs in which aluminums is the matrix material. Currently, AMCs have been the center of focus of researchers all over the world as they have the potential to meet the current demand of above-mentioned materials. Aluminum composites exhibit superior mechanical and physical abilities, for example, high specific strength, excellent ductility, superior tensile, compression, shear strength, better elasticity, bulk and shear modulus, less coefficient of thermal expansion, good wear properties, extraordinary resistance to corrosion, creep resistance at high temperature and good fatigue limit [3]. Moreover, an important factor of AMCs is that these materials have higher specific strength than steel, and also, their properties can be tailored to fit the various demands of the engineering and structural applications [4]. AMCs could be reinforced with a variety of reinforcements such as (a) whisker or particulates, (b) continuous fibers, (c) discontinuous fibers and (d) natural fibers. These varieties of reinforcements could be incorporated either in weight or volume percentage in varying amounts [5]. Generally, hard particulate

J. S. Khalkho (✉) · D. B. Karunakar
Mechanical and Industrial Engineering Department, IIT Roorkee, Roorkee 247667, India
e-mail: johnsk.dme2017@iitr.ac.in

CH. S. Vidyasagar
Metallurgical and Materials Engineering Department, IIT Roorkee, Roorkee 247667, India

V. H. Saran and R. K. Misra (eds.), *Advances in Systems Engineering*,
Lecture Notes in Mechanical Engineering,
https://doi.org/10.1007/978-981-15-8025-3_25

ceramics are used as reinforcement to these AMCs. However, other materials have also shown potential to be used as reinforcement. The fabrication method plays a vital role in defining the properties of the end product.

Classical liquid metallurgy technique is the most utilized fabrication route due to its simplicity, flexibility, low cost and commercial viability of all the various methods for aluminum composite fraction. Generally, the fabrication methods can be classified into some major categories such as casting, in situ processing, pressure infiltration and spray deposition [6]. Compared to the other processing methods, casting process with the introduction of stirring into the molten melt has some important advantages. Examples include production ability of various materials, formability of complex shape, good matrix–particle interaction, matrix structure can be controlled easily, production of large size parts, low cost processing, suitable for mass production and extraordinary productivity of near-net shaped components [7]. Even though there are some negative issues concerning stir casting of AMCs like poor wettability and non-uniform reinforcement distribution in the matrix [8]. However, even liquid metallurgy route suffers defects and drawbacks. The following are few drawbacks that are addressed in this paper.

Basically, the meaning of improper wettability related to the reinforcing particles in the molten matrix melt implies that the liquid metal does not completely adsorb into the reinforcement particles. Henceforth, even after mixing the molten metal through stirring, the reinforcement particles float on the surface of the melt. This phenomenon is caused due to surface tension of the molten melt [9]. During mechanical stirring, usually the particles are mixed into the entire molten melt, but when stirring stops, the particles tend to return to the surface due to gravity effect [10]. Literature gives us the application of some techniques to improve the wettability of the reinforcing powder or fiber particles into the molten metal matrix. Examples include preheating the reinforcing particles to a temperature equal to that of the molten melt before adding them to the matrix melt, addition of magnesium to the melt to absorb the gases induced on the surface of the particle during stirring, etc. [11].

Another important problem that raises question in casting is the uniform distribution of the reinforcing powder particles into the matrix melt. As the reinforcing particles get wet when introduced into the molten liquid, these particles will tend to immerse or remain on the surface of the liquid metal because of their difference in densities between the matrix alloy melt and reinforcement, as a result, the distribution of hard whiskers or fibers is found to be non-uniform, and at that stage, reinforcements exhibit very high possibility of clustering and agglomeration [12]. Moreover, apart from stirring, other methods are available to introduce particles into the liquid matrix. For example, directly injecting the reinforcements with carrier in an inert atmosphere like Argon, etc., into the liquid matrix. It is established that this method is useful in increasing the uniform dispersion of the particles in the liquid matrix.

Hence, the present research is emphasized on the influence of multi-step semisolid stir casting in improving the wettability and distribution. The materials used are AA2014 matrix and ceramic material TiO_2 as reinforcement. The effect of this processing method is evaluated by relating the microstructure by the mechanical properties [13].

2 Materials and Methods

The materials chosen for the present research work were AA2014 ingot and TiO_2 powder. The AA2014 ingots were procured from Ampco Metal India. Pvt. Ltd., and the TiO_2 powder was procured from Alfa Aesar, USA, with average particle size of 10 μm and 99.96 purity. The chemical composition of AA2014 is given in Table 1, and the properties of TiO_2 is given in Table 2.

The as-received ingots were cut into several pieces weighing 300 g. The reinforcing TiO_2 powder was weighted according to the intended compositions as shown in Table 3. An as-received sample was also characterized as benchmark to evaluate the performance and effect of multi-step stir casting. Table 3 shows the intended compositions of the composite samples developed.

Reinforcement was weighted according to Table 3. The as-received ingots were cut into 300 g pieces. These 300 g AA2014 pieces were further cut into several pieces for conveniently placing then in the furnace. When all the pieces were placed in the furnace, the temperature was gradually raised to a preset temperature of 800 °C (above liquidus temperature). Once the temperature has reached to the set value, it was checked whether the metal pieces were melted and had become molten liquid. Once the metal pieces were turned into molten liquid, proportions of TiO_2 powder particles as shown in Table. 3 which were preheated at 300 °C were put into the molten melt wrapped in a thin aluminum foil in ex-situ method. After this, mechanical stirring was gradually increased to 100 RPM and was performed for 10 min. A graphite stirrer was used for stirring. The furnace top plate was opened and exposed to atmosphere, and as a result, the furnace temperature started to drop, making the molten metal semisolid, and during this, the stirring speed was also gradually decreased. Once the molten metal became semisolid, next proportion of reinforcement was sprinkled, and gradually, the stirring speed and temperature were increased by replacing the top plate of the furnace until the semisolid melt became molten metal. Again the top plate of the furnace is removed and the process is repeated the third time. Finally, after three steps of semisolid stirring, the molten metal is poured into a preheated steel die of dimensions 15 × 30 × 250 mm. The reinforcement for all the samples was added in three steps by following the same procedure discussed above. The entire experimentation was done in normal atmosphere, i.e., no special atmosphere was used. The stir casting furnace used in this study is shown in Fig. 1.

Table 1 Chemical composition of the as-received AA2014 ingot

Element	Cu	Si	Mg	Zn	Fe	Ti	Cr	Mn	Al
Weight %	4.67	1.1	0.8	0.25	0.47	0.15	0.001	0.80	Balance

Table 2 Properties of TiO_2 used as reinforcement in the present study

Molecular formula	Molar mass	Density	Melting point	Boiling point
TiO_2	79.866 g/mol	4.23 g/cm3	1843 °C	2972 °C

Table 3 Intended proportions of the composite samples

S.No	AA2014 weight (wt. %)	TiO_2 weight (wt. %)
1	100	0.0
2	99	1.0
3	98	2.0
4	97	3.0
5	96	4.0

Fig. 1 Aluminum stir casting furnace used in the present study

After cooling the samples at normal atmospheric temperature, the solidified pieces were taken out from the die and were cut into various pieces for metallurgical characterization like optical microscopy and the mechanical characterization like hardness, tensile, etc. Density analysis was also done for various developed composite samples. The developed composite cast samples are shown in Fig. 2.

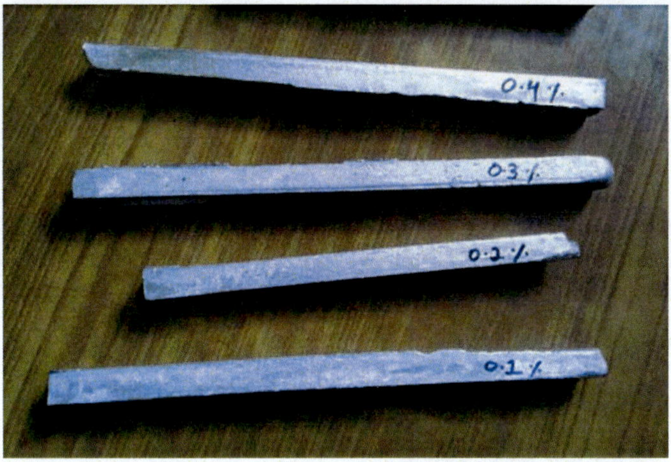

Fig. 2 Multi-step stir cast AA2014 composite samples with varying wt.% of TiO_2

3 Results and Discussion

3.1 Density Analysis

Some part of the developed composite samples are cut for performing experimental density calculation by Archimedes principle. Ten readings were taken, and the average is used for the calculation of relative density of the samples. The theoretical density was estimated by calculating using the standard "rule of mixtures" formulae. Figure 3 reveals the relative density values of the composites samples with varying amount of TiO_2 reinforcement.

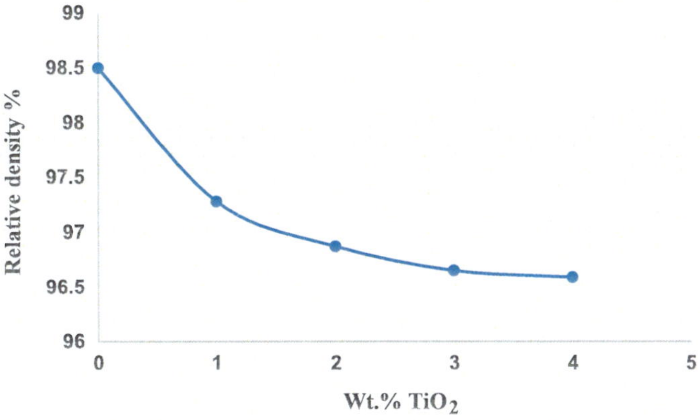

Fig. 3 Relative density values of the composites with varying amount of TiO_2 addition

It is well known that the density of TiO_2 is higher than that of the density of aluminum. Therefore, the theoretical density of the composites increases with an increase in the weight % of TiO_2. However, the experimental density does not show same kind of variation due to the presence of porosity. Therefore, there is a variation in the relative densities of the developed composites. During stir casting, the majority time of stirring was done when the liquid melt is solidifying, so the possibility of the formation of pores became less, and hence, the increase in density of all the composites above 96.5% was achieved. Further, no degassing agent was used which would have helped the trapped gasses to escape and thus the decrease in the relative densities of the composite samples. However, the density of the composite samples has decreased with the addition of TiO_2 compared to pure AA2014 sample.

3.2 *Mechanical Properties*

3.2.1 Hardness Analysis

Some pieces were cut from the developed cast samples, polished, and their hardness was calculated using a computerized Vickers hardness testing machine (Model: FIE VM 50 PC). Ten readings at different places all around the sample were taken, and the average value is reported with error bars showing maximum and minimum values. The variation in hardness with the addition of different incremental amounts of TiO_2 is given in Fig. 4.

From Fig. 4, it can be observed that the hardness of the composite samples increases with the increase in the reinforcement of TiO_2. There is a clear trend of increase as the wt.% TiO_2 increases compared to pure AA2014 sample. This increase in the hardness can be attributed to the factors like.

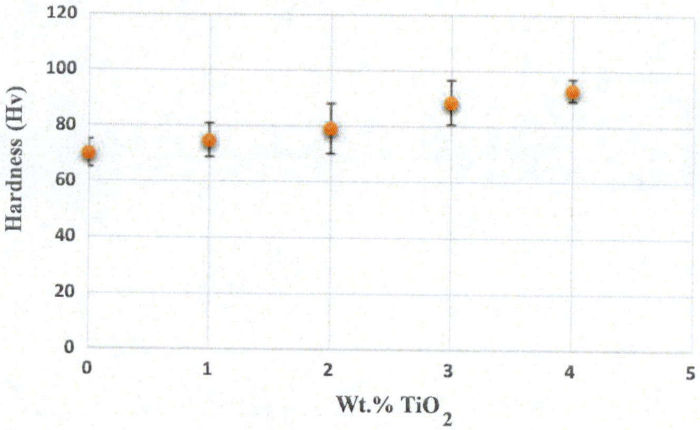

Fig. 4 Change in the hardness of the composites with the addition of incremental amounts of TiO_2

a) The addition of hard ceramic TiO_2 and its proper distribution in the matrix which can effectively hinder the dislocation motion by dispersion strengthening. As the TiO_2 content increases, the dispersion strengthening also increases.
b) Another factor contributing to the hardness is that the grain size decrease as TiO_2 wt.% reinforcement increases.

3.2.2 Tensile Properties

As discussed above, three tensile samples were cut from each developed sample through Electric discharge wire cut machining. Three readings were taken from each sample, and the average values are reported with error bars showing maximum and minimum values. The tests were conducted on a universal testing machine (UTM) (Model: H 75 KS) at a strain rate of 0.003/second. The variation in the tensile properties related to the developed cast composites with the variation in TiO_2 wt% is shown in Fig. 5.

From Fig. 5, it can be observed that the properties like UTS and YS increase with the increase in TiO_2 addition. There is a clear trend of increment in both YS and UTS. However, the improvement in YS is not quite obvious and remarkable. On the other hand, the UTS increases considerably with the increase in TiO_2 addition. Also, the YS and UTS follow the identical increasing variation like the hardness. The improvement in the YS and UTS can also be explained in the same lines of hardness as dispersion strengthening achieved by the hard ceramic TiO_2 and the decrease in the grain size as predicted by Hall–Petch [14]. The elongation of the developed composite samples is shown in Fig. 6. In the present investigation, elongation at breaking of the sample is taken into account.

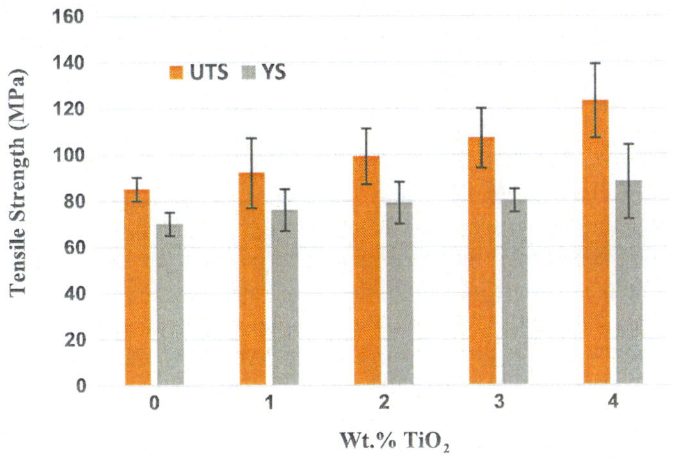

Fig. 5 Variation in the tensile properties of cast composites with the variation in TiO_2 wt%

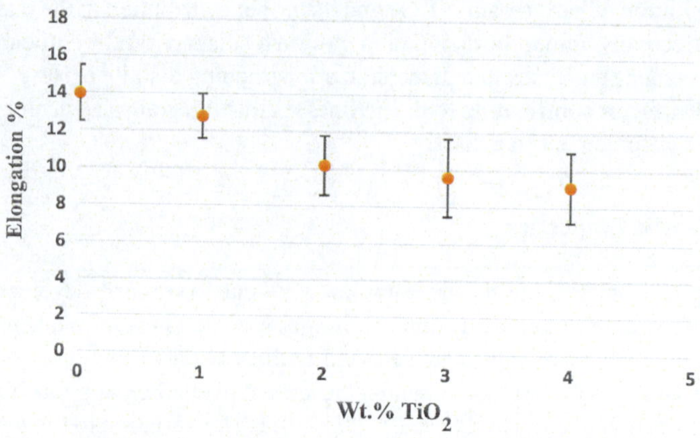

Fig. 6 Elongation % of cast composites with incremental addition of TiO_2 wt%

From Fig. 6, it can be seen that there is a steep drop in the elongation with an increase in TiO_2. It is obvious that addition of ceramic reinforcement has reduced the elongation % as the ceramic material makes the composites brittle.

3.3 *Microstructure*

The developed composite samples have been cut in to pieces for microstructure characterization. The optical microscope used in this study was LEICA DMI 5000 M (Model: Leica Microsystems, Baffalo Grove, IL) microscope. The microstructures were captured at 100×, and the average grain sizes were determined by linear intercept method taking 20 images per each sample and IMAGE J software. The microstructure-property relationship was determined for all composite samples. The composite samples were further cut into smaller pieces, ground and after that polished on 1/0, 2/0, 3/0 and 4/0 silicon carbide emery papers. The polished samples were further polished on a mechanical rotating coarse cloth disk with fine Al_2O_3 powder particles until the visible scratches were removed. Finally, the coarse cloth on the rotating disk was replaced by a fine cloth, and the samples were polished with a nano size diamond particle paste until the fine scratches were removed and a mirror like finish was obtained. After that, the specimens were thoroughly cleaned with water, and the moisture is dried using a hot air dryer. Now, the dried samples were etched with modified Keller's reagent for 30–40 s. The optical microstructures of the developed composite samples are shown in Fig. 7.

From the microstructures in Fig. 7, it can be observed that with the addition of TiO_2, grain refinement took place, and the refinement increased with an increase in TiO_2 addition. The mechanism of grain refinement can be explained on the basis of increase in the dislocations which increase with an increase in TiO_2 and these

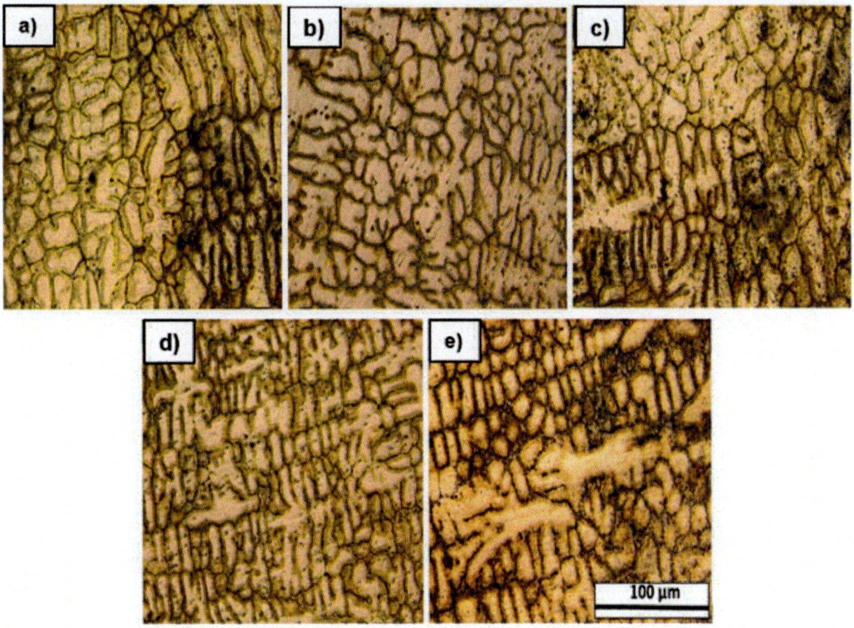

Fig. 7 Microstructure of the composite samples reinforced with **a** 0.0 wt.% TiO_2, **b** 1 wt.% TiO_2, **c** 2 wt.% TiO_2, **d** 3 wt.% TiO_2, **e** 4 wt.% TiO_2

nucleating sites leading to the formation of grains. The precipitates which have an approximate chemical composition of Al_2Cu phase (not shown here) are usually present at the grain boundaries of the aluminum matrix which leads to a uniform distribution of the reinforcement in the aluminum matrix [14, 15]. A secondary phase is also visible all the microstructures in the midst of dendrite arms. It has proposed that under shear force, the dendrite arms bend plastically and brought to large misorientations into the arm and forming dislocation. The dendrite arm spacing also decreased with an increase in TiO_2 addition. Other than grain size, the reason for dislocation movement is dendrite arm which act as a barrier. Therefore, the higher the number of arms, the more the resistance offered to dislocation movements [16–18]. The grain sizes of the cast composites calculated through "Image J" software are 46, 43, 32, 26, 22 μm, respectively, for the samples starting from without TiO_2 addition to 4wt.% TiO_2 addition. Also, very little porosity was observed in the microstructures.

4 Conclusions

Based on the present work, the given under conclusions can be drawn.

- The reinforcement of TiO_2 improved the mechanical properties of all the multi-step stir cast composite samples.
- The composite sample with 4 wt.% TiO_2 showed highest mechanical properties like 93 Hv of hardness, 123 MPa of UTS, 88 MPa of YS, respectively, which are 75%, 69% and 79% higher compared to that of the un-reinforced sample.
- Microstructure revealed that there was grain refinement and it increased with an increase in TiO_2. Dendrite arm spacing also reduced with TiO_2 addition.
- The grain sizes of the composite samples calculated from the microstructures are 46, 43, 32, 26, 22 μm, respectively.
- The main strengthening mechanisms were found to be dispersion strengthening and grain boundary strengthening.

Acknowledgements Authors would like to thank The Mechanical and Industrial Department of IIT Roorkee for providing the necessary equipment required for this study.

References

1. Haghshenas M (2016) Metal–matrix composites. Ref Modul Mater Sci Mater Eng 0–28 (January). https://doi.org/10.1016/B978-0-12-803581-8.03950-3
2. Nturanabo F, Masu Land Kirabira JB (2016) We are IntechOpen , the world ' s leading publisher of Open Access books Built by scientists , for scientists TOP 1 %, Intech, vol i, no. tourism, p 13. https://doi.org/10.5772/intechopen.86225
3. Mavhungu ST, Akinlabi ET, Onitiri MA, Varachia FM (2017) Aluminum Matrix Composites for Industrial Use: Advances and Trends. Proc Manuf 7:178–182. https://doi.org/10.1016/j.promfg.2016.12.045
4. Bodunrin MO, Alaneme KK, Chown LH (2015) Aluminium matrix hybrid composites: a review of reinforcement philosophies; mechanical, corrosion and tribological characteristics. J. Mater. Res. Technol. 4(4):434–445. https://doi.org/10.1016/j.jmrt.2015.05.003
5. Surappa MK (2003) Aluminium matrix composites: challenges and opportunities. Sadhana Acad Proc Eng Sci 28(1–2):319–334. https://doi.org/10.1007/BF02717141
6. Ervina EMN, Siti SN, Abdullah MMAB (2016) Fabrication method of aluminum matrix composite (Amcs): a review. Key Eng Mater 700:102–110. https://doi.org/10.4028/www.scientific.net/KEM.700.102
7. Garg P, Jamwal A, Kumar D, Sadasivuni KK, Hussain CM, Gupta P (2019) Advance research progresses in aluminium matrix composites: manufacturing & applications. J Mater Res Technol 8(5):4924–4939. https://doi.org/10.1016/j.jmrt.2019.06.028
8. Mazahery A, Shabani MO (2013) Plasticity and microstructure of A356 matrix nano composites. J King Saud Univ-Eng Sci 25(1):41–48. https://doi.org/10.1016/j.jkssues.2011.11.001
9. Almadhoni K, Khan S (2016) Review of effective parameters of stir casting process on metallurgical properties of ceramics particulate Al composites review of effective parameters of stir casting process on metallurgical properties of ceramics particulate Al composites. Rev Eff Parameters Stir Cast Process Metall Prop Ceram Part Al Compos 6:21–40. https://doi.org/10.9790/1684-12642240
10. Kandpal CB, Kumar J, Singh H (2018) Manufacturing and technological challenges in Stir casting of metal matrix composites—a review. Mater Today Proc 5(1):5–10. https://doi.org/10.1016/j.matpr.2017.11.046

11. Sobczak N, Stobierski L, Radziwill W, Ksiazek M, Warmuzek M (2004) Wettability and interfacial reactions in Al/TiO2. Surf Interface Anal 36(8):1067–1070. https://doi.org/10.1002/sia.1838
12. Hashim J, Looney L, Hashmi MSJ (2002) Particle distribution in cast metal matrix composites—Part II. J Mater Process Technol 123(2):258–263. https://doi.org/10.1016/S0924-0136(02)00099-7
13. Girish KB, Shobha BN (2018) Synthesis and mechanical properties of Zirconium nano-reinforced with Aluminium alloy matrix composites. Mater Today Proc 5(1):3008–3013. https://doi.org/10.1016/j.matpr.2018.01.100
14. Sajjadi SA, Ezatpour HR, Torabi PM (2012) Comparison of microstructure and mechanical properties of A356 aluminum alloy/Al_2O_3 composites fabricated by stir and compo-casting processes. Mater Des 34:106–111. https://doi.org/10.1016/j.matdes.2011.07.037
15. Murthy KVS, Girish DP, Keshavamurthy R, Varol T (2017) Progress in natural science: materials international mechanical and thermal properties of AA7075/TiO_2/Fly ash hybrid composites obtained by hot forging. Prog Nat Sci Mater Int 27(4):474–481. https://doi.org/10.1016/j.pnsc.2017.08.005
16. Pola A, Tocci M, Kapranos P (2018) Microstructure and properties of semi-solid aluminum alloys: a literature review. Metals (Basel) 8(3). https://doi.org/10.3390/met8030181
17. Ezatpour HR, Sajjadi SA, Sabzevar MH, Huang Y (2014) Investigation of microstructure and mechanical properties of Al6061-nanocomposite fabricated by stir casting. Mater Des 55:921–928. https://doi.org/10.1016/j.matdes.2013.10.060
18. Radhakrishna K, Seshan S (1989) Dendrite arm spacing and mechanical properties of Aluminium alloy castings. Cast Met 2(1):34–38. https://doi.org/10.1080/09534962.1989.11818980

An Investigation of Behaviour of Mild Steel Under Tensile Test for Industrial Applications

Nerkar Pravin, Tawale Sanket, Shrikhande Prachit, and Pophale Prajwal

1 Introduction

Stainless steel is defined as iron-based alloy having up to 1.2% carbon content, whereas mild steel (MS) or low carbon steel has approximately 0.05% to 0.25% carbon content with up to 0.4% manganese. It also contains varying proportions of phosphorus and sulphur. The surface hardness of mild steel can be increased through heat treatment processes like carburizing. This material has moderate strength but as it is cheap and easy to shape, hence, most of the products are of mild steel. Material testing is necessary to determine its mechanical properties and behaviour. Tensile test is the most significant mechanical testing technique in order to select appropriate materials. Tensile test is also helpful for validation and fulfilment of minimum strength and elongation requirements. It also determines elastic and plastic material properties like Young's modulus, ultimate strength, yield strength, plastic deformation, and strength coefficient of various materials. The results obtained from tensile test is in the form of stress–strain curve which is used to calculate strength of materials in various regions.

The theoretical stress–strain curve is shown in above Fig. 1. When the material is subjected to axial tensile load the elongation starts. This elongation varies as a function of axial load. The stress varies directly with elongation at an initial stage and follows Hooke's law of elasticity, i.e. the stress is proportional to the strain until its elastic limit is reached, and this limit is called the proportionality limit. When the elastic limit is reached, the yielding of material starts followed by strain hardening. In this region, the material shifts to the plastic region and gets permanently deformed. From this point, the slope of stress–strain curve increases in opposite direction leading

N. Pravin (✉) · T. Sanket · S. Prachit · P. Prajwal
Mechanical Engineering Department, St Vincent Pallotti College of Engineering & Technology, Nagpur 441108, Maharashtra, India
e-mail: psnerkar@yahoo.co.in

V. H. Saran and R. K. Misra (eds.), *Advances in Systems Engineering*,
Lecture Notes in Mechanical Engineering,
https://doi.org/10.1007/978-981-15-8025-3_26

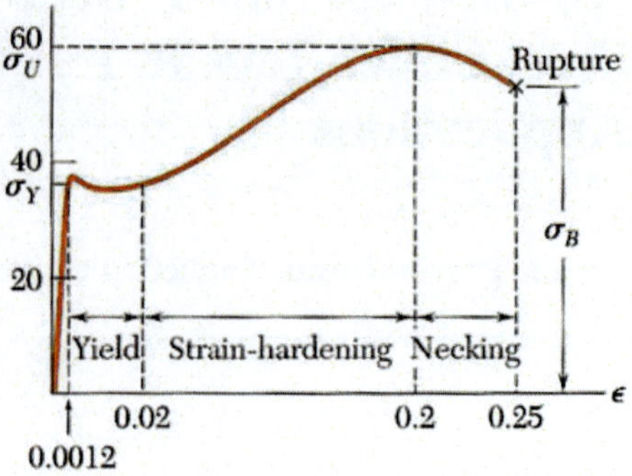

Fig. 1 Theoretical stress–strain curve

to formation of neck. The material further deforms under same load and finally leads to rupture. The value of stress at peak of the stress–strain curve is known as the ultimate tensile strength of the material.

Various experiments performed by different researchers were studied. Based on a review, an experiment is performed using a specimen of mild steel. The results plotted on stress–strain curve indicate that the mechanical properties can be improved by selecting proper design forces. The results obtained can help designers to predict life of material in their intended applications.

2 Experimental Overview

Finite element analysis (FEA) can stimulate the stiffness and visualization of component strength and can predict the expected way of behaving under operating conditions [1]. The authors' purpose of using the results obtained after tensile test is to use this in FEA of MS (mild steel) samples bearing a constant tensile load. For this, author also compare existing manual results with finite element test. In an comparison with actual test data, finite element analysis (FEA) provides more realistic results and reduces product development time. A tension test of mild steel using Abaqus CAE software was performed [2]. The result obtained from stress–strain diagram shows mild steel follows Hooke's law. It was found that in graph, at a particular point, linear necking in a sample occurs and finally leads to fracture. Thus, it is quite easy to analyse tensile test of mild steel using finite element analysis. With an increase in nickel content and reducing manganese content improves the impact strength but slightly reduces the tensile strength and hardness of mild steel in compare with EN 8 steel [3]. The fine grain-structure of steel has been shown to increase tensile strength [4]. It was also found that mild steel specimen with smaller notch diameter and larger notch angle results in higher K_{IC} values. The effects on different temperature on 316 and 201 types of stainless steel were examined [5]. The test is performed using trapezium software called universal testing machine. It was found that with an increasing test temperature on tensile properties indicates an increase in uniform elongation, and the yield strength increases with increasing test temperature. The mechanical properties show that the ductility of the sample increases with increasing yield strength. It was concluded that as the test temperature increases, the expansion rate increases. Heat treatment effect on properties of mild steel due to quenching media was checked [6]. Heat treatment was carried out at 900 °C for 4 h, and each of six specimens were then quenched by various media used as a major source of strength improvement. A universal testing machine (UTM) was used for various tests. A change in a material properties of mild steel material in a term of high tensile strength toughness, ductility, and hardness has been obtained from the analysis of test results.

Thus, the author concludes that water quenching should be used for heat treatment of mild steel components which is used to connect agricultural machinery. Mechanical properties such as tensile strength, elongation in mild steel, and stainless steel are tested on UTM to determine the effect of various heat treatments such as annealing, quenching, and normalizing on material properties [7]. Heat treatment increases tensile strength, as well as hardness of sample. A uni-axial tensile test is often used to determine the mechanical properties of isotropic materials. Normalization of heat treatment in mild steel samples gives better results than other heat treatment methods, while for stainless steel annealing is prefer. Friction welded joints of various metals are produced to evaluate the mechanical properties of mild steel of aluminium alloy. The tensile strength and microhardness of both materials and fusion joint are in good condition, but the ductility was reasonable. Tensile strength and strain gradually increase until the welding current reaches 160A and then decreases in all the grades of mild steel [8, 9]. As the welding current increases, the compressive strength of all the steel grades decreases, similarly, its opposite condition occurs in compressive strain. A new test database for a modern metal at various temperature and strain rate was developed [10]. The aim is to suggest new coefficient for complex equation based on new and existing test databases and to develop statistical analysis method for predicting material properties. Seven types of different steel used in marine application were tested. In general, the yield strength and tensile strength increases with increasing load speed, while fracture strain trends to decrease. The Cowper-Symonds equation for measuring the effect of strain rates and low temperature shows satisfactory results when compared with existing test database. It is important to understand that incorrect definition of materials properties is the main source of uncertainly associated with structural design.

A series of experiments and tests on mild-carbon steel grade S275 and S355 was performed under monotonic and cyclic loading [11]. For a purpose of accurate and constitutive model of steel under cyclic loading, chatoche hardening model was used for calibration. It was found that hysteresis stress–strain curve is much higher than monotonic stress curve between the yield strength and ultimate strength of loaded specimen. The increased in cyclic loop and strain amplitude directly affect the ductility of material. The obtained hardening parameter from this test can be used in engineering practices and FEM analysis of cyclic behaviour of steel structure under seismic conditions. A problem related to the issue of accuracy (e.g. elongation and necking) in stress–strain measurements (e.g. tensile strength) during static testing of

Table 1 Dimension of specimen

Specimen	Gauge length	Diameter	Cross-head speed
Mild steel	60 mm	12 mm	1 mm/min

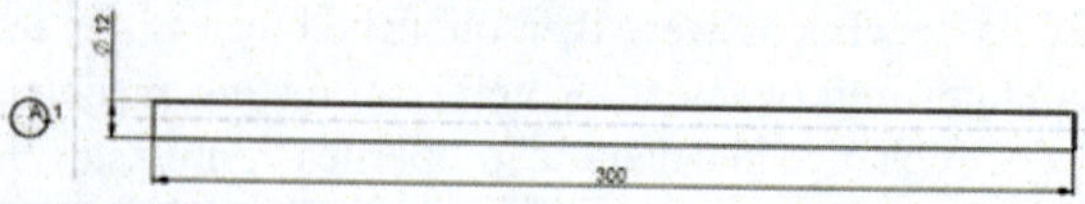

Fig. 2 Geometry of specimen

materials at room temperature were discussed [12]. This study mainly focuses on determining the limiting error of quantities which are obtained from tensile test of a rectangular specimen.

3 Methods and Materials

3.1 Specimen Details and Test Setup

A material used in a current study is mild steel consist of 0.05–0.25% of carbon, 0.4% of manganese, and small amounts of phosphorous and sulphur. The specimen is polished with silicon carbide paper because removing the dirt layers from the surface of specimen is very important which is done by grinding and polishing of specimens. The grinding and polishing were carried on both sides of specimen. The dimensions of specimen are shown in Table 1 and geometry at Fig. 2.

A tensile test is performed in an universal testing machine (UTM) of 400 KN capacity machine which is operated by software. A universal testing machine has both ability and efficiency to perform standard tensile as well as compressive tests on specimen. Figure 3 shows the detailed description of UTM used to perform the test. The loading unit is hydraulic in operation with piston inside which moves up and down. The specimen is mounted in the jaw of upper cross-head while lower cross-head moves down. After mounting the specimen, axial tensile force is subjected, until the specimen breaks. The tensile force was observed to be proportional to the increase in gauge length. The tensile test is carried out at room temperature. The actual force at which specimen breaks was recorded.

3.2 Result and Discussion

The result obtained from the test is plotted on stress–strain curve as shown in Fig. 5. When the applied forces are small enough, the deformation is entirely elastic and it will return to its original state after removal of load with further increase in force,

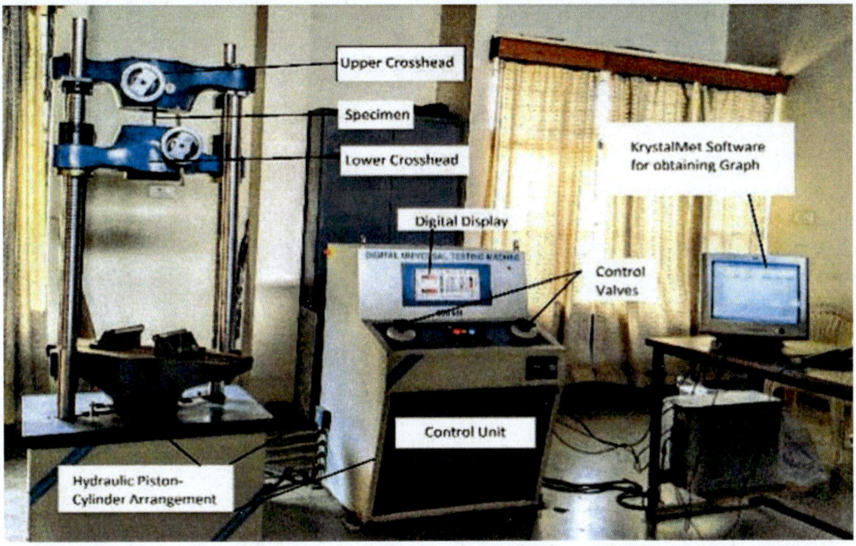

Fig. 3 Detailed of UTM

and the material tends to be deformed permanently. For a certain material, the initialization of plastic deformation is expressed as a sudden drop in load indicating both an upper and a lower yield point. This stage is the ultimate strength point which is defined as the ratio of maximum load on a specimen to original cross-section area. From a graph, it was observed that material deformed for a longer period. At a certain point, this deformation is concentrated and its cross-section area decreases, which results in neck formation and leads to rupture from that point (Fig. 4a). The dimension of specimen at neck formation is shown in Fig. 4b. The neck point can be prevented by knowing actual fracture forces. The maximum tensile force and strength of materials in various regions calculated from graph are shown in Table 2 (Fig. 5).

4 Conclusion

The objective of study is to predict the life of materials by increasing their mechanical properties in industrial applications. Tensile test is performed using specimen of mild steel. The specimen is subjected to tensile force in universal testing machine. The results plotted on stress–strain curve indicate that the mechanical properties can be improved by selecting proper design forces.

1. Maximum stress and strain points were obtained from graph and the percentage of elongation was measured. Heat treatment can improve material properties of

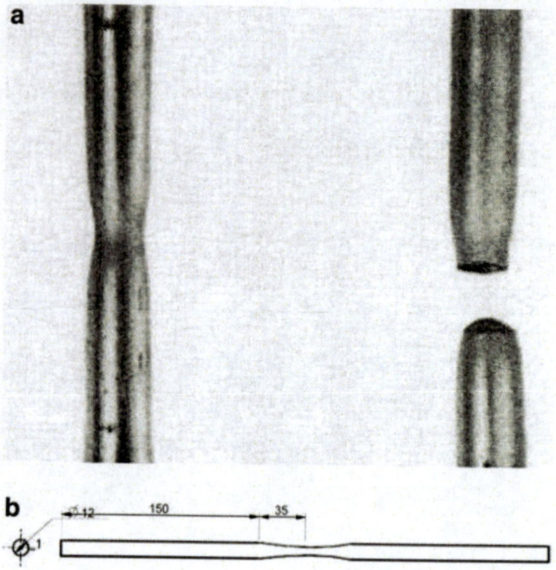

Fig. 4 **a** Neck formation and fracture specimen, **b** geometry of specimen with neck formation

Table 2 Yield calculation by graph method

Maximum force	35.1 kN
Tensile strength (UTS)	310.353 MPa
% Elongation	16.667%
% Area reduction	75%
Yield load	21.72 kN
Stress at yield (YS)	192.048 MPa
Strain at yield	0.133
UTS/YS	1.616
YS/UTS	0.619

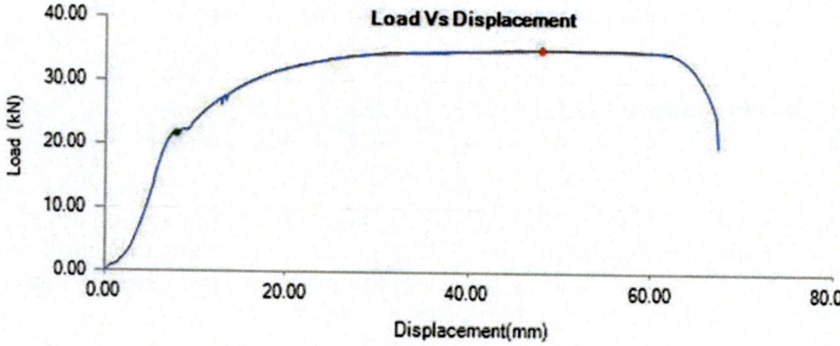

Fig. 5 Stress–strain curve obtained from test

mild steel, hence, mild steel component must undergo heat treatment before its use. Normalization of heat treatment is preferred.
2. Finite element analysis (FEA) provides more realistic results and reduces product development time, hence, the specimen must undergo finite element analysis before tensile test.
3. The accuracy of specimen preparation is most importance as it directly affects the test quality. Evaluation of errors and to eliminate is essential during measurement.
4. It was observed that the percentage elongation at the fracture point of the mild steel specimens is 15–30% at various strain rates, which provides an indication of ductile fracture. Hence, mild steel is sensitive to strain rate.

The values obtained from the experiment may not accurate for practical applications and for complex specimen as these values are obtained by approximation of plot generation with help of UTM and associated software and the experiment was performed on ideal specimen.

References

1. Um-E-Azra, Ramesh Nayak B, Appaiah M (2016) Design of Mild Steel specimen using FEA and tensile test data, materials today. In: Proceeding, International Conference on Advanced Material and Application (ICAMA 2016), pp 1–4
2. Jose E, Anto T (2015) Analysis of tensile test of mild steel using finite element method. Int J Innov Eng Technol (IJIET) 5(4):247–251
3. Jha SK (2017) Investigation of micro-structure and mechanical properties of three steel alloys. Int J Autom Mech Eng 14(2):4315–4331
4. Nath SK, Das UK (2006) Effect of micro structure and notches on the fracture toughness of medium Carbon steel. J Naval Archit Marine Eng 3(2006):15–22
5. Zafar H, Haider SK (2013) A comparative study of tensile properties of two different grades of stainless steel at different temperatures. J Nat Sci 1(1):23–28
6. Ahaneku IE, Kamal AR, Ogunjirin OA (2012) Effect of heat treatment on the properties of mild steel using different quenchants. Front Sci 2(6):153–158
7. Tanwer AK (2014) Effect of various heat treatment processes on mechanical properties of mild steel and stainless steel. Am Int J Res Sci Technol Eng Math 8(1):57–61
8. Kumar S, Kumar R, Singla YK (2012) To study the mechnaical behaviour of friction welding and aluminium alloy and steel. Int J Mech Eng Robot Res 1(3):44–50
9. Sanjay KN, Ageshabanu, Baig A, Krishnamurthy CV, Taj Pasha R (2015) Effect of welding current on mechanical properties of different grade of mild steel. Int J Eng Technol 2(5):406–410
10. Paik JK, Kim KJ, Lee JH, Gyeong B, Kim SJ (2017) Test database of mechanical properties of mild, high-tensile and stainless steel and Aluminium alloy associated with cold temperature and strain rates, pp 1–34
11. Krolo P, Grandic D, Smolcic Z (2016) Experimental and numerical study of mild steel behaviour under cyclic loading with variable strain ranges. In: Advances in materials science and engineering, pp 1–13
12. Adamczak S, Bochnia J, Kundera C (2012) Stress and strain measurements in static tensile tests. Meteorol Measure Syst XIX(3):531–540

Mathematical Systems

Predator–Prey System: Two Predators Feed on the Different Stages of a Single Prey

Rajat Kaushik and Sandip Banerjee

1 Introduction

In recent years, stage-structured predator–prey models have been widely studied by many researchers [1–6]. Stage-structured modeling makes the predator–prey ecology more realistic as well as complicated. In [6], Rui, Xu and his co-authors investigated the predator–prey ecology in which predator kills only adult preys. Maiti et al. [7] studied a predator–prey model with Crowley-Martin type functional response in which predators only depend on the adult preys. On the other hand, Hastings (1983) has outlined a mathematical model where predator eats only juvenile preys [8, 9], which are ecologically justified if adult preys are not vulnerable to predators.

Motivated by the previously mentioned work, in this paper, we consider a predator–prey system with two predators and one prey. Prey has been divided into juvenile and adult class. We consider the case when strong predators feed on the adult preys while the weak predator feed on the small size consumers, that is, the juvenile preys. We assume that adult preys are not easily accessible to weaker and smaller predators; therefore, they search and hunt exclusively the juvenile preys. On the other hand, stronger predators go in the search of the large adult preys and often ignore the juvenile preys, concealed in the mountain caves [6].

We now formulate the model. Here, x_J is the population density of juvenile preys, and x_A is the population density adult preys, y and z are the population densities of weak and strong predators, respectively. We assume that growth of juvenile preys depends on adults (rx_A, r being the reproduction rate of juvenile preys). The growth

R. Kaushik (✉) · S. Banerjee
Department of Mathematics, Indian Institute of Technology Roorkee, Roorkee, Uttarakhand 247667, India
e-mail: rkaushik@ma.iitr.ac.in

S. Banerjee
e-mail: sandip.banerjee@ma.iitr.ac.in

V. H. Saran and R. K. Misra (eds.), *Advances in Systems Engineering*, Lecture Notes in Mechanical Engineering,
https://doi.org/10.1007/978-981-15-8025-3_27

of adult preys depends on conversion of juvenile to adult (mx_J in Eqs. (1) and (2)). K, γ, β and δ are natural mortalities of juvenile preys, adult preys, first predator (weak predators), and second predators (strong predators), respectively. ψ is the intra-specific competitive rate of adult preys. We assume that weak predators feed on the small size consumers (juvenile preys), which explains $\alpha_1 x_J y$ in Eq. (1). This predation increases the population density of weak predators with conversion efficiency η. Also, we assume that strong predators feed exclusively on the adult preys as small size juvenile preys are not good enough to satisfy their hunger. The term $\alpha_2 x_A z$ represents the predation by strong predators that increases the population density of strong predators with efficiency ν. With above assumptions, the proposed model is

$$\frac{dx_J}{dt} = rx_A - \alpha_1 x_J y - Kx_J - mx_J \tag{1}$$

$$\frac{dx_A}{dt} = mx_J - \alpha_2 x_A z - \gamma x_A - \psi x_A^2 \tag{2}$$

$$\frac{dy}{dt} = \eta\alpha_1 x_J y - \beta y \tag{3}$$

$$\frac{dz}{dt} = \nu\alpha_2 x_A z - \delta z \tag{4}$$

1.1 Positivity and Boundedness

For positivity and boundedness, we can use classical theorem proposed by Nagumo [10] and method used by Birkhoff and Rota [11], respectively. Now, we state the following theorems:

Theorem 1 *Solutions of the system (1–4) with positive initial conditions remains positive for all* $t \geq 0$.

Theorem 2 *All solutions of the proposed system (1–4) are ultimately bounded.*

2 Equilibrium Point and Stability Analysis

1. Trivial equilibrium point: : $E_0 = \left(x_J^0, x_A^0, y^0, z^0\right) = (0, 0, 0, 0)$.

2. Complete predator-free equilibrium point: $E_1 = (\overline{x}_J, \overline{x}_A, 0, 0)$, where

$$\overline{x}_J = \frac{m(r^2 - r\gamma) - rK\gamma}{(m+K)^2\psi}, \quad \overline{x}_A = \frac{m(r-\gamma) - K\gamma}{(m+K)\psi}.$$

E_1 exists when $mr > (K+m)\gamma$.

3. Weaker predator-free equilibrium point: $E_2 = \left(x_J^{'}, x_A^{'}, 0, z'\right)$, where

$$x_J^{'} = \frac{r\delta}{(m\alpha_2\nu + K\alpha_2\nu)}, x_A^{'} = \frac{\delta}{\alpha_2\nu}, z' = \frac{rm\nu\alpha_2 - (K+m)(\psi\delta + \alpha_2\gamma\nu)}{(m\alpha_2\nu + K\alpha_2\nu)\alpha_2}.$$

The existence condition of E_2 is $rm\nu\alpha_2 > (K+m)(\psi\delta + \alpha_2\gamma\nu)$.
4. Stronger predator-free equilibrium point: $E_3 = \left(x_J^{''}, x_A^{''}, y'', 0\right)$, where

$$x_J^{''} = \frac{\beta}{\alpha_1\eta}, x_A^{''} = \frac{-\alpha_1\gamma\eta + \sqrt{4\alpha_1 m\psi\eta\beta + \alpha_1^2\gamma^2\eta^2}}{2\alpha_1\eta\psi},$$

$$y'' = \frac{r\sqrt{4\alpha_1 m\psi\eta\beta + \alpha_1^2\gamma^2\eta^2} - 2m\psi\beta - 2K\psi\beta - r\alpha_1\gamma\eta}{2\alpha_1\psi\beta}.$$

E_3 exists if $r\sqrt{4\alpha_1 m\psi\eta\beta + \alpha_1^2\gamma^2\eta^2} > 2m\psi\beta + 2K\psi\beta + r\alpha_1\gamma\eta$.

5. Interior equilibrium point: $E^* = \left(x_J^*, x_A^*, y^*, z^*\right)$ where

$$x_J^* = \frac{\beta}{\alpha_1\eta},$$

$$x_A^* = \frac{\delta}{\alpha_2\nu},$$

$$y^* = \frac{r\alpha_1\eta\delta - m\alpha_2\beta\nu - K\alpha_2\beta\nu}{\alpha_1\alpha_2\beta\nu},$$

$$z^* = \frac{\alpha_1(\psi\eta\delta^2 - \alpha_2\gamma\eta\nu\delta + m\alpha_2^2\beta\nu^2}{\alpha_1\alpha_2^2\eta\nu\delta},$$

E^* exists when $r\alpha_1\eta\delta > m\alpha_2\beta\nu + K\alpha_2\beta\nu$ and $m\alpha_2^2\beta\nu^2 > \alpha_1\left(\psi\eta\delta^2 + \alpha_2\gamma\eta\nu\delta\right)$.

2.1 Stability Analysis

Theorem 3 *If* $(m+K)\gamma > mr$ *holds, then the system (1–4) is locally asymptotically stable around* E_0*; otherwise unstable.*

Proof The Jacobian matrix at E_0 is given by

$$[J]_{E_0} = \begin{bmatrix} -(K+m) & r & 0 & 0 \\ m & -\gamma & 0 & 0 \\ 0 & 0 & -\beta & 0 \\ 0 & 0 & 0 & -\delta \end{bmatrix}$$

The corresponding characteristic equation is given by

$$(\lambda+\delta)(\lambda+\beta)\left[\lambda^2 + (\gamma+m+K)\lambda + \gamma(m+K) - mr\right] = 0.$$

The first two eigenvalues are $(-\delta)$ and $(-\beta)$, which are negative. The other two eigenvalues will be negative if $(m+K)\gamma > mr$. This proves the theorem.

Theorem 4 *If the conditions* $\nu\alpha_2 mr < (m+K)(\delta\psi + \gamma\nu\alpha_2)$, $\eta\alpha_1 mr^2 < (m+K)\left[\beta(m+K)\psi + r\eta\alpha_1\gamma\right]$ *and* $\gamma < r$ *hold, then the system (1–4) is locally asymptotically stable around* E_1*; otherwise unstable.*

Proof The Jacobian matrix at E_1 is given by

$$[J]_{E_1} = \begin{bmatrix} -(K+m) & r & -\alpha_1\overline{x}_J & 0 \\ m & -(\gamma+2\psi\overline{x}_A) & 0 & -\alpha_2\overline{x}_A \\ 0 & 0 & \eta\alpha_1\overline{x}_J - \beta & 0 \\ 0 & 0 & 0 & \nu\alpha_2\overline{x}_A - \delta \end{bmatrix},$$

and the corresponding characteristic equation is given as follows:

$$[\lambda - (\nu\alpha_2\overline{x}_A - \delta)][\lambda - (\eta\alpha_1\overline{x}_J - \beta)]\left[(\lambda+K+m)(\lambda+\gamma+2\psi\overline{x}_A) - mr\right] = 0.$$

The first root of above equation is negative when $\nu\alpha_2\overline{x}_A - \delta < 0$, i.e., $\nu\,\alpha_2\,mr < (m+K)\,(\delta\psi + \gamma\nu\,\alpha_2$. The second root will be negative if $\eta\alpha_1\overline{x}_J - \beta < 0$, which is equivalent to $\eta\alpha_1 mr^2 < (m+K)\left[\beta(m+K)\psi + r\eta\alpha_1\gamma\right]$. The other two roots will be negative if $(K+m)(\gamma+2\psi\overline{x}_A) > mr$, which implies $r > \gamma$. This completes the proof of the theorem

Theorem 5 *If* $r\alpha_1\eta\delta < m\alpha_2\beta\nu + K\alpha_2\beta\nu$ *and* $A_1A_2 > A_3$ *hold true, then system (1–4) is locally asymptotically stable around* E_2*; otherwise unstable.*

Proof Let us define Jacobian matrix at E_2

$$[J]_{E_2} = \begin{bmatrix} -(K+m) & r & -\alpha_1 x'_J & 0 \\ m & -\left(\alpha_2 z' + \gamma + 2\psi x'_A\right) & 0 & -\alpha_2 x'_A \\ 0 & 0 & \eta\alpha_1 x'_J - \beta & 0 \\ 0 & \nu\alpha_2 z' & 0 & 0 \end{bmatrix}.$$

The corresponding characteristic equation is given by

$$\left[\lambda - \left(\eta\alpha_1 x'_J - \beta\right)\right]\left[\lambda^3 + A_1\lambda^2 + A_2\lambda + A_3\right] = 0, \tag{5}$$

where $A_1 = \alpha_2 z' + \gamma + 2\psi x'_A + K + m > 0$,

$$A_1 = \nu\alpha_2^2 z' x'_A + (K+m)\left(\alpha_2 z' + \gamma + 2\psi x'_A\right) - rm,$$

$$A_3 = (K+m)\left(\nu\alpha_2^2 z' x'_A\right) > 0$$

The first root of equation Eq. (5) is negative if $\eta\alpha_1 x'_J < \beta$, which is equivalent to $r\alpha_1\eta\delta < m\alpha_2\beta\nu + K\alpha_2\beta\nu$. By Routh–Hurwitz criterion, the other cubic equation will have negative roots or roots with negative real parts if $A_1 A_2 > A_3$. This completes the proof of the theorem.

Theorem 6 *Interior equilibrium point E^* is locally asymptotically stable if $PQR > R^2 + P^2 S$; otherwise unstable.*

Proof The Jacobian matrix at E^* is given by

$$[J]_{E^*} = \begin{bmatrix} -\frac{rx_A^*}{x_J^*} & r & -\alpha_1 x_J^* & 0 \\ m & -\left(\alpha_2 z^* + \gamma + 2\psi x_A^*\right) & 0 & -\alpha_2 x_A^* \\ -\eta\alpha_1 y^* & 0 & 0 & 0 \\ 0 & \nu\alpha_2 z^* & 0 & 0 \end{bmatrix}.$$

The corresponding characteristic equation is given by

$$\lambda^4 + P\lambda^3 + Q\lambda^2 + R\lambda + S = 0, \tag{6}$$

where $P = \alpha_2 z^* + \gamma + 2\psi x_A^* + \frac{rx_A^*}{x_J^*} > 0$,

$$Q = \alpha_2^2 x_A^* \nu z^* + \frac{rx_A^*}{x_J^*}\left(\alpha_2 z^* + \gamma + 2\psi x_A^*\right) - mr + \alpha_1^2 \eta x_J^* y^*,$$

$$R = \alpha_1^2 \eta x_J^* y^* \left(\alpha_2 z^* + \gamma + 2\psi x_A^*\right) + \frac{r x_A^{*2} \alpha_2^2 \nu z^*}{x_J^*} > 0,$$

$S = \alpha_2^2\alpha_1^2 \nu\eta x_A^* x_J^* y^* z^* > 0$. Since $P > 0$, $Q > 0$ and $S > 0$; therefore, by Routh–Hurwitz criterion, condition for stability of E^* will be $PQR > R^2 + P^2 S$. This completes the proof of the theorem.

Remark 1 Since existence condition of stronger predator-free equilibrium point E_3 is not easily conceivable, we skip the stability analysis of E_3.

2.2 *Transcritical Bifurcation*

Existence condition of E_1 and stability condition of E_0; (Theorem 3) are just opposite to each other. The system (1–4) undergoes a transcritical bifurcation involving E_0 and E_1 when $(m + K)\gamma = mr$. The existence condition of E_2 and first stability condition of E_1; (Theorem 4) are just opposite to each other. Therefore, the system (1–4) undergoes a transcritical bifurcation involving E_1 and E_2 is when $rm\nu\alpha_2 = (K + m)(\psi\delta + \alpha_2\gamma\nu)$. Also, first existence condition of E^* and first stability condition of E_2 are just reverse of each other. Therefore, system (1–4) experiences a transcritical bifurcation involving E_2 and E^*, provided, $r\alpha_1\eta\delta = m\alpha_2\beta\nu + K\alpha_2\beta\eta$.

3 Numerical Simulation

We consider the set of values of the parameter $\alpha_1 = 0.9, \alpha_2 = 1, \eta = 8/9, \nu = 0.9, K = 0.3, m = 0.6, \gamma = 0.3, \psi = 0.4, \beta = 0.4, \delta = 0.3$. We vary the parameter r (reproduction of the prey), which is the key parameter for the model. For $r = 0.4$ (Fig. 1a), reproduction of preys is not good enough for their survival which results in extinction of all species. At $r = 0.5$, (Fig. 1b) prey survives in the habitat but prey abundance is not sufficient for the survival, of any predator and complete predator-free equilibrium point is obtained meanwhile we get the transcritical bifurcation at $r = 0.45$(see TB-1 in Fig. 1e). At $r = 0.7$, strong predator also survives in the habitat along with the preys and weak predator-free equilibrium point is obtained. Notice that $r = 0.65$ (TB-2 in Fig. 1e) is the point of transcritical bifurcation. Finally, at $r = 1.75$, all species co-exist successfully, and interior equilibrium point is obtained. Here also we get a transcritical bifurcation point at $r = 1.35$ (see Fig. 1e).

4 Conclusion

In this paper, we mainly focus on the predator–prey ecology with two kinds of the predators. Our research shows that strong predator-free steady-state is very unlikely to occur which is biologically reasonable because environment is always more favorable for strong predators. Also, in the survival of all species, strong predator's population density is greater than the weaker predator's population density which also justifies the fact that environment is appropriate only for the faster and the strong species. Lot of transcritical bifurcation points enriches the dynamics of the predator–prey system. The future work in this dimension can be predator–prey ecology in which strong predator also feed on the weaker predators as an alternative option.

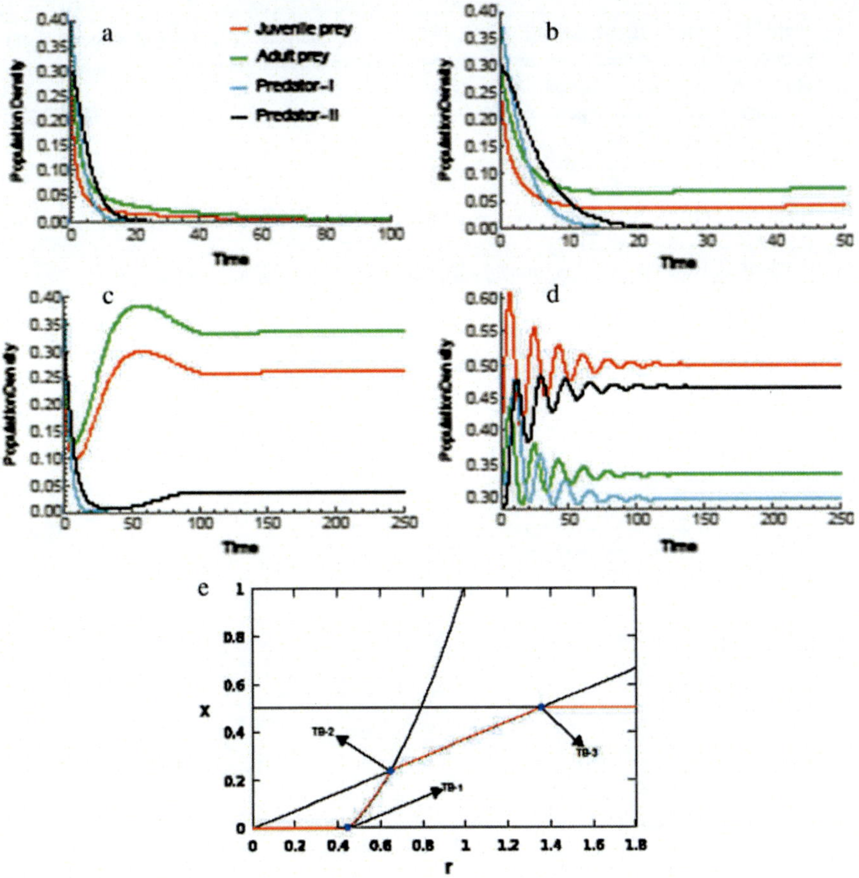

Fig. 1 Time evolution and bifurcation diagram of all the species for the system (1–3)

References

1. Agarwal M, Devi S (2010) Persistence in a ratio-dependent predator-prey resource model with stage-structure for prey. Int J Biomaths 3:313–336. https://doi.org/10.1142/s17935245100010x
2. Liu M, Wang K (2011) Global stability of stage-structured predator-prey models with beddington de angelis functional response. Commun Nonlinear Sci Numer Simul 16:3792–3797
3. Liu C, Zhang Q, Jinna L, Yue W (2014) Stability analysis in a delayed prey-predator resource model with harvest effort and stage structure. Math Comput 238:177–192
4. Ma X, Shao Y, Wang Z, Luo X, Fang JuZ (2016) An impulsive two-stage predatorprey model with stage structure and square root functional response. Math Comput Simulation 119:91–107
5. Wei F, Fu Q (2016) Global asymptotic stability of a predator-prey model with stage structure incorporating prey refuge. Int J Biomaths 9. https://doi.org/10.1142/s1793524516500583
6. Xu R, Chaplain MAJ, Davidson FA (2016) Persistence and global stability of a ratio dependent predator-prey model with stage structure. Appl Math Comput 9:729–744

7. Maiti AP, Dubey B, Chakraborty A (2019) Global analysis of a delayed stage structure prey-predator model with CrowleyMartin type functional response. Math Comput Simulation 162:58–84
8. Hastings A (1983) Age-dependent predation is not a simple process. I. Continuous time models. Theor Popul Biol 23:347–362
9. Hallam TG, Levin SA (1986) Mathematical ecology: an introduction. Springer, Heidelberg, p 293
10. Nagumo N (1942) Ber die lage der integralkurven gewhnlicher differentialgleichungen. Proc Phys-Math Soc Jpn 24:551
11. Birkhoff G, Rota GC (1982) Ordinary differential equations. Ginn, Boston

Development of Techno-Feasible Mobile App for Process Optimization in Textile Industry

Tushar Patil, Bhushan Chaudhari, Yatin Patale, Tushar Shinde, Rajendra Parsi, Sujit Gulhane, and P. P. Raichurkar

1 Introduction

In the existing modern techno-scientific era, Internet of things (IoT) is emerged as one of the key tools which can be made available easily anywhere at any time. Similarly, connectivity within short time period allows peoples can get many different benefits from devices like computer, laptop and mobile. Several mobile apps have been developed for the industrial applications to cope up with quality and productivity of manufacturing processes. In textile industry, segments like spinning, weaving, chemical processing and garment manufacturing, process performance and quality of product are governed by the three M's, i.e. machine, material and man. Whereas in each of these sectors, efficiency and quality targets are the heartburn issue faced by the management and technical personnel. Quality and efficiency enhancement through downtime reduction in this case will give boost to the textile industry. Use of ICT tools either in terms of mobile app will make the data available to resolve the respective issue on the production floor within the short time spam [1, 2]. Mobile app can be used as master key solutions through providing effective technical data-based optimized process parameters on production floor and will help to resolve respective arise difficulty related to quality and production aspects. With same concern, in textile industry, particularly fabric manufacturing sector (weaving) is high focused area where production and quality will directly contribute to gross sale value of product and ultimately the profitability of the organization [3–5]. Hence, weaving of fault-free fabrics with higher production level is the prime concern. Various fabric defects generated during weaving are likely to deteriorate the fabric quality, but start mark contributes at higher level towards rejecting the fabric produced with required specifications. Start mark is appearing as dense or gap in the pick spacing

T. Patil (✉) · B. Chaudhari · Y. Patale · T. Shinde · R. Parsi · S. Gulhane · P. P. Raichurkar
Centre for Textile Functions, SVKM's NMIMS, MPSTME, MPTP Campus, Shirpur, India
e-mail: tushar.patil@nmims.edu

V. H. Saran and R. K. Misra (eds.), *Advances in Systems Engineering*,
Lecture Notes in Mechanical Engineering,
https://doi.org/10.1007/978-981-15-8025-3_28

and being generated during restarting the loom after mending warp or weft breakage. The variation could be either an increase or decrease of pick spacing, which takes the appearance of a thick or thin place, respectively. Such faults become prominent when the variation exceeds a certain threshold. Generally, spun yarns, due to their inherent hairiness, mask the fault to some extent, while the fault is more apparent when continuous filament yarns are used. Basic cause of start marks generation is displacement of fell of cloth from its original position during stoppage of weaving machine, and this problem arrives due to tension and relaxation of warp yarns and woven fabrics. Displacement of cloth fell position before restarting the loom is one of the major causes of start mark generation [6, 7]. However, methodology to prevent starting marks should also be adopted as same as measurement of starting marking is been done according to their size or denseness in the fabrics [8]. This will help to help to analyse the root cause of concern start mark for the ongoing fabric quality in loom shed. Work has been done in the core shed area to find the various solutions for the prevention of starting marks. An effective work on cloth fell position and its displacement with laser-based technique is conducted to prevent the start mark generation during weaving [9]. Results of various studies conducted in the respective cause analysis to avoid the starting marks show the track where they have concluded on the shed geometry as well as the stress on warp shed at different machine positions (degree) during stop and restarting the loom. This study focuses on the effect of loom settings such as backrest position, drop box settings and electronic settings include automatic machine positioning, automatic start offset on starting marks. Whereas in loom shed, fabric sorts those are highly sensitive to start marks defects, its very difficult to recover it within short time spam because conducting trails during troubleshooting will increase the downtime and decreases the productivity. Henceforth, optimization of respective loom settings and formulation of techno-feasible mobile app based on it will help to reduce the downtime during conduction of trials to avoid the starting marks.

2 Technical Review

Variation in pick spacing during restarting the loom leaves the cloth fell with prominent gap or higher picks caused the start mark defect. Warp tension during weaving cycle and relaxation during stoppage time in concern with cloth fell is the basic cause of generation of start mark [10]. Dense mark and open mark (crack) generate due to increase and decrease in consecutive pick spacing after restarting the weaving machine after recovering the respective breakage (Fig. 1).

Reasons of starting marks Improper ratio of take up and let off, sometimes take up is higher than let off, and sometimes, let off is higher than take up. Usually, this leads to improper warp tension.

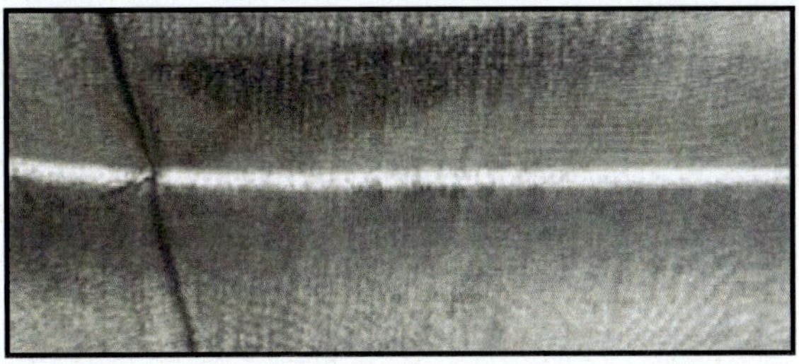

Fig. 1 Start mark defect

- Machine stoppage for long time.
- Backrest roller position.
- Different settings of microprocessor.
- Position of reed according to weaving cycle.
- Due to different pick spacing.
- Beam collars are loose, cracked, broken or bent beam.

The reed striking of the fell during this turning over of the loom has a take up is higher than let off, and sometimes, let off is higher than take up. Usually, this has come due to improper warp tension. When machine is stopped, tension on warp yarn is relaxed, and then, difference between take up and let off may occur. Prolonged stoppage of weaving machine (loom) also causes the displacement of clothfell from the beat-up position. This occurs due to relaxation of warp yarns.

2.1 Methods for Prevention of Starting Mark

Asymmetric shedding is troughing off of shed and creates tension differences between top and bottom layers at time of beat up. This can be achieved by either raising or lowering the backrest position. When the backrest position is changed with respect to the shed geometry, tension on top and bottom layers will be changed according to their position from the hypothetical warp line. As a result, when the backrest is raised, the tension on top layer of warp reduced effectively when the shed is fully open. Whereas tension of bottom shed layer is increased, this will help to hold the newly inserted pick in position during beat up. This methodology has been used to improve cloth cover and reduce the generation of start marks (Fig. 2).

Staggering of Healds During the crossing of the warp threads, considerable friction and therefore abrasion take place amongst end. This abrasion and obstructed movement can be reduced by using split shedding in which the warp is made to cross in separate layer instead of all the threads of one shed line crossing all the threads of the other sheds line at the same time. By split shedding, the total threads and healds crossing is reduced is called staggering of healds.

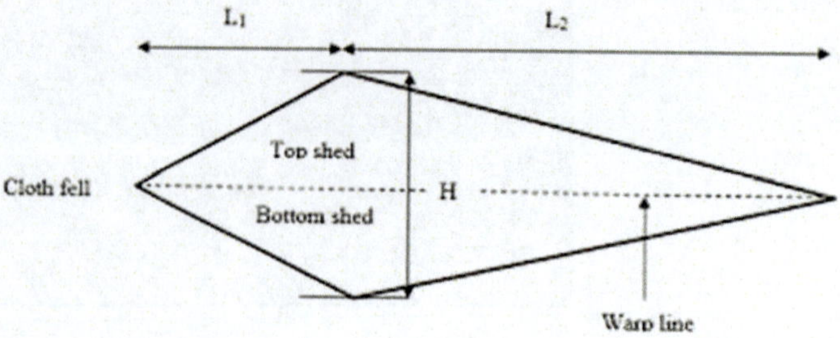

Fig. 2 Geometry of warp shed

2.2 *Electronic Settings*

Automatic start mark prevention (ASP) The simple function of ASP with absolutely reproducible function stands for quality reliability in event of machine stoppage and restarting. Starting behaviour can be set with dynamic reed beat up, and the starting process is individually programmable. According to different weaving material and different kind of start marks, halting and starting angle of the weaving machine can be set freely. According to downtime and specific reasons duration and reason of machine stopping, cloth fell will be adjusted appropriate position when weaving machine is restarted to prevent stop mark and ensure high quality weaving. Cloth fell following functions makes sure that the cloth fell moves forward at the movement of halt to prevent the slow back run reed from touching the cloth fell which eliminates the risk of activating the stop mark. When the weaving machine is started again, the cloth fell will be repositioned by synchronizing the warp tension through let-off motor and thus it prevents the variation in weft density during beat-up (Fig. 3).

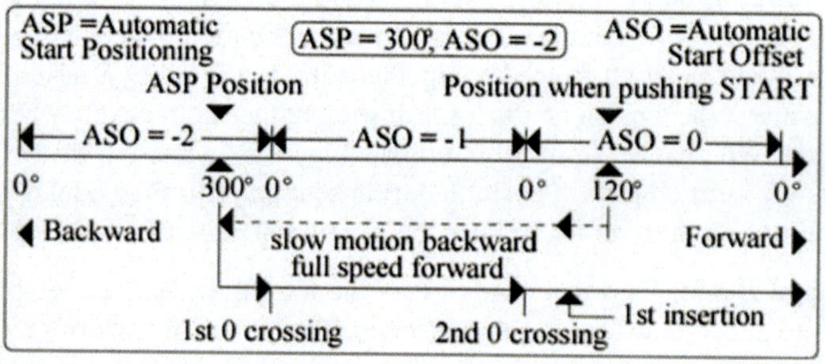

Fig. 3 Automatic start mark prevention

2.3 Mechanical Settings

Back Rest setting With reference to shed geometry, backrest roller positions affect the variations in the warp tension level during the weaving cycle. Increasing the backrest height makes the top shed layer slack and bottom layer tight and vice versa if backrest roll is lowered. Whereas if the backrest moved forward towards the heald frames, warp tension increases due to higher pull of warp during shed formation and decreases as backrest roll set backwards (Figs. 4 and 5).

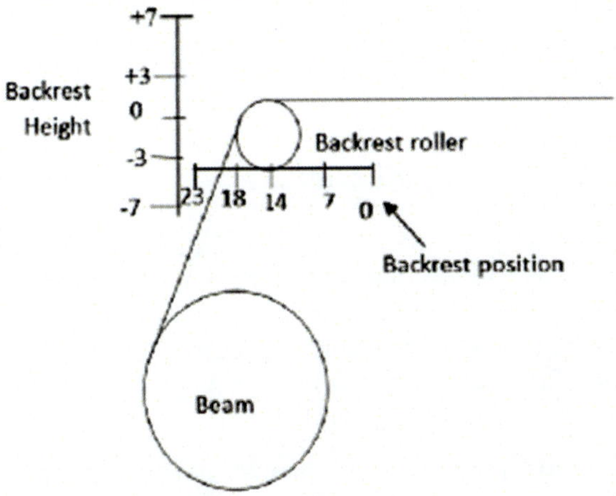

Fig. 4 Backrest setting. *Source* ASRJETS

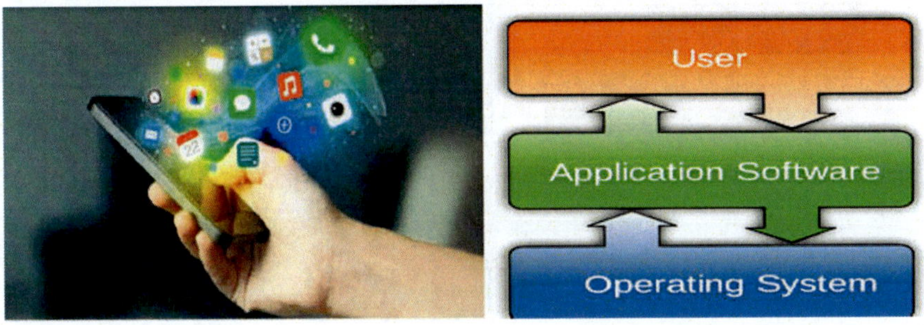

Fig. 5 Mobile App Working Concept

3 Concept of Design Development of Mobile App

Designing of mobile app initially requires the effective data input which will help to filter out the test results and will scrutinize and gives the best effective outputs. In case of weaving industry, conduction of trails for loom settings lays more time consumption and increases the downtime and deteriorates the process efficiency as well as the quality of the fabric. Data collected while conducting of trials during troubleshooting will be optimized through the mobile app and to be used on the production floor whenever required. This will give the resource available, i.e. recommended machine settings to recover the concern quality or production issue within the short time period (Fig. 6).

3.1 *Basics for Mobile App Designing*

Data Analysis and its specification: The purpose of the data analysis and specification step is to know the core needs of the project on which app will be developed, and this includes the scrutiny of problems and various ways to provide solutions to the concern inputs. Specification of data inputs helps to frame out the credentials to be given to get the best possible outputs [11].

Gathering and analysis of input data Initially, all the requirements based on the software are collected and then analysed. The aim of the analysis phase is to eliminate and inconsistencies [11].

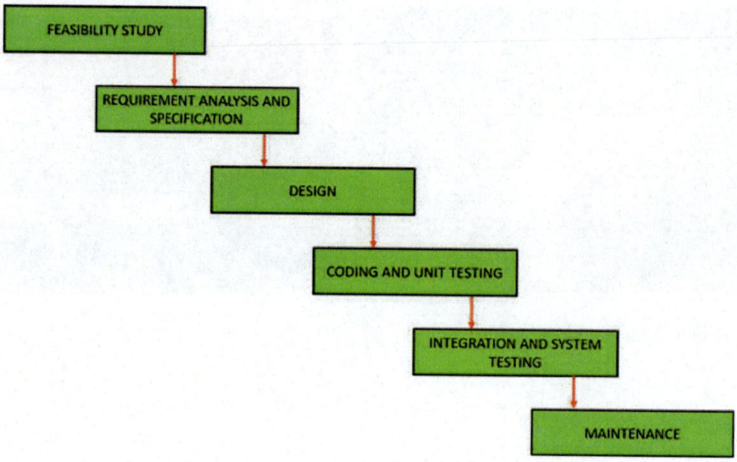

Fig. 6 Phases of app manufacturing. *Source* GreeksforGreeks Computer Science Portal

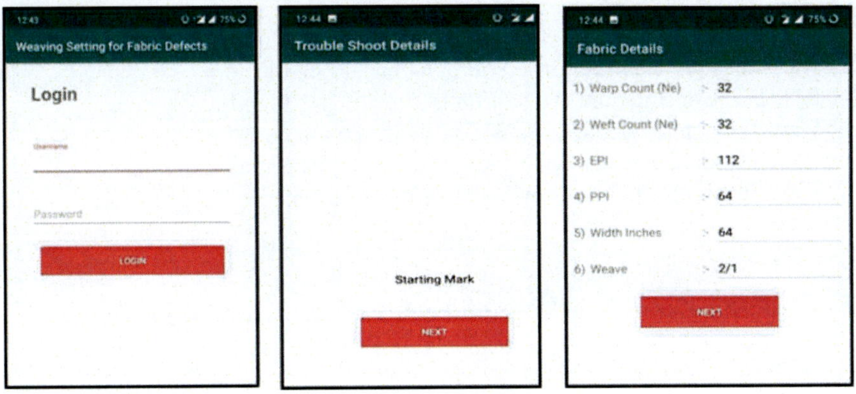

Fig. 7 Guidelines/Steps for input credentials in mobile app

DesignThis phase framing out the requirements specified in the SRS document into a structure which is suitable for input of data or values in some programming language [11].

Coding of Input Credentials This phase consists of translation of design of software into concern code by using programming language. As a result, coding of every module is done. After this phase, working of module is been tested for its proper functioning [11] (Fig. 7).

Integration and System testing After modules have been coded, integration of different modules has been conducted till the formulation of resultant system. Each module section is added in the system and finally is tested. As all modules have been integrated, the complete system is obtained and will be processed for testing [11].

4 Experimental Work

4.1 Optimization of Loom Settings

- Fabric sorts those are sensitive to generation of starting marks are taken under consideration. Air-jet weaving machines are considered for conduction of trails.
- Three fabric sorts with different fabric parameters are taken under trails. Starting marks were assessed in terms of penalty points, during conduction of study trails.
- In case of mechanical settings, backrest height, depth and drop box settings were taken under trial.
- Whereas in case of electronic settings ASO, AMP and ASP offsets were considered.

A] Fabric Construction: (32 × 32)/ (112 × 74) [3/1 Twill].
Warp Tension: 2.85 kN, Grey Width: 64 inch (Table 1).
B] Fabric Construction: (40 × 40)/(132 × 60) [1/1 Plain].
Warp Tension: 2.50 kN, Grey Width: 63.2 inch (Table 2).
C] Fabric Construction: (60 × 60)/(176 × 98) [4/1 Satin].
Warp Tension: 3.75 kN, Grey Width: 120 inch (Table 3).

5 Designing and Working of Techno-Feasible Mobile App

See Fig. 8.

6 Conclusion

Prevention of any fabric defect at weaving stage leads to significant increase in fabric quality as well as productivity. Based on the technology upgradation in weaving sectors, it is required to reduce the downtime on the production floor while conducting trails for quality setting or defect elimination specially when its about starting marks. Trials conducted in above project work show that mechanical (backrest, drop box and shed crossing timing) and electronic settings such as ASO, AMP and ASP influences on starting marks.

Trials were conducted on the air-jet loom for wider width fabric sorts, and data collection shows significant reduction in generation of starting marks. Also, the loom settings for various fabric sorts have been optimized and formulated through mobile app. This project can help weaving industry to improve the productivity and efficiency of loom shed department, and it will help to reduce the downtime of machine. It will directly give the e-solutions on mobile app. This will help to reduce the effort of Tackler by providing e-solutions on mobile. With the help of IT tools or e-solutions, we can work smartly and effectively.

Table 1 Mechanical & Electronic Setting—Twill weave

Trial	Mechanical settings			Electronic settings						Grade/penalty point
	Backrest height	Backrest depth	Drop box height	ASO		AMP		ASP		
				Weft stop	Warp stop	Weft stop	Warp stop	Weft stop	Warp stop	
1	+ 6	3	2	0	1	270°	290°	299°	299°	2.5
2	+ 4	5	−1	−1	−1	299°	310°	299°	299°	0.5
3	+ 3	2	3	1	1	320°	320°	310°	310°	4

Table 2 Mechanical & Electronic Settings—Plain Weave

Trial	Mechanical settings			Electronic settings						Grade/penalty point
	Backrest height	Backrest depth	Drop box height	ASO		AMP		ASP		
				Weft stop	Warp stop	Weft stop	Warp stop	Weft stop	Warp stop	
1	+ 8	3	5	0	−1	310°	300°	35°	45°	4
2	+ 6	3	3	−1	0	315°	315°	40°	40°	2
3	+ 4	4	0	−1	1	320°	315°	45°	45°	1

Table 3 Mechanical & Electronic Settings—Satin Weave

Trial	Mechanical settings			Electronic settings						Grade/penalty point
	Backrest height	Backrest depth	Drop box height	ASO		AMP		ASP		
				Weft stop	Warp stop	Weft stop	Warp stop	Weft stop	Warp stop	
1	+ 5	4	0	0	−1	299°	310°	295°	295°	3
2	+ 4	3	−1	0	−1	320°	320°	305°	305°	2
3	+ 4	4	−1	−1	−1	310°	320°	310°	320°	1

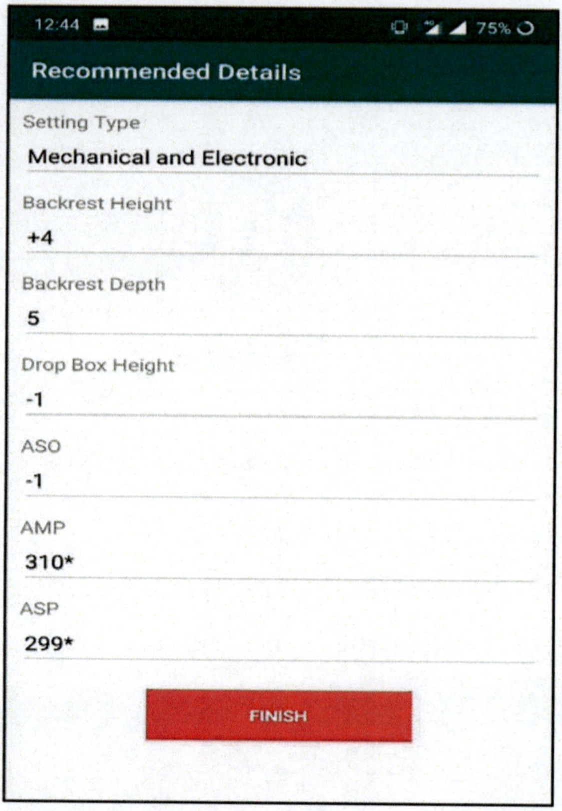

Fig. 8 Guidelines/Steps for input credentials in mobile app

References

1. A review paper on cross platform mobile application development IDE 2015
2. Mobile application development feasibility studies: a case study in https://www.academic.edu
3. Dorugade VA, Patil TC, Mali D (2013) Enhancement of loom shed efficiency. Melliand International, pp 173–175
4. Patil V, Patil T (2011) Hard waste control in loom shed. Textile Mag 52(8):16–21
5. Patil LC, Patil T (2016) Denim weaving-control of fabric defects. Int J Textile Eng Process 20–27
6. Greenwood K (1957) Relaxation phenomena in weaving. J Text Inst 48:206–213
7. Greenwood K (1975) Weaving: control of fabric structure. Merrow, Durham
8. Islam SA (2000) Prevention of set marks in weaving. J Inst Textile Eng Tech ITET 8:42–47
9. Islam A (1996) Prevention of start-up marks in fabrics woven on power looms
10. Vangheluwe L, Kiekens P (1997) Simulation of procedures to avoid set marks in weaving caused by relaxation. Textile Res J
11. Software Engineering|Classical Waterfall Model, Greeksforgreeks, Computer Science portal for Greeks
12. Android tutorial-Tutorials point of PD https.www.tutotrialspoint.com android

Re-engineering Infill Density System for 3D Printing

Dheeraj Kumar Angajala, Ankit Sahai, and Rahul Swarup Sharma

1 Introduction

3D printing or additive manufacturing is the process of making an object by addition of material layer by layer [1]. It enables the fabrication of geometrically complex objects which cannot be achieved through conventional manufacturing processes [2]. Among various techniques of 3D printing, fused filament fabrication (FFF) is widely popular among hobbyists, educational institutes, and industries for rapid prototyping using thermoplastics. In FFF, a heated nozzle extrudes the material layer by layer [1]. Slicing softwares are used to convert CAD models in to 2D layers. We can change various parameters in slicing software to optimize the process and produce better quality products. In general, main parameters which affect the mechanical properties and process parameters are build orientation, raster angle, raster-to-raster air gap, infill density, and layer thickness [3].

The internal structure of the 3D printed object is known as infill. The density and pattern of infill is often controlled in the slicing software. The pattern and density percentage significantly affect the printing and physical properties. In general, higher infill density leads to higher print time and more material consumption while improving resistance to external loads [4]. The optimization of infill pattern which goes beyond regular patterns is necessary to build lightweight and strong objects and is achieved through topology optimization.

Topology optimization (TO) is a mathematical tool which is used to optimize the performance of the component by optimizing the material layout based on the design and loading boundary conditions. TO has wide range of applications in various fields like civil, aerospace, mechanical, bio-chemical engineering. Engineers widely

D. K. Angajala (✉) · A. Sahai · R. S. Sharma
Department of Mechanical Engineering, Dayalbagh Educational Institute, Agra, Uttar Pradesh 282005, India
e-mail: dheerajkumar.dei@gmail.com

V. H. Saran and R. K. Misra (eds.), *Advances in Systems Engineering*, Lecture Notes in Mechanical Engineering,
https://doi.org/10.1007/978-981-15-8025-3_29

use it to create lightweight components for aerospace and automobile sectors [5]. Using TO, free-form designs naturally occur which is difficult to manufacture using conventional subtractive manufacturing, which makes it and ideal fit for TO and 3D printing. In the next section, we discuss different optimization techniques for topology optimization.

2 Analysis of Infill Optimization Techniques

Fernandez-Vicente [6] had evaluated the influence of two controllable variables in fabrication using AM—pattern and density of the infill. They used open-source 3D printer and printed objects with different infill density and patterns. The results reflected that the different patterns can cause a variation in tensile strength up to a maximum of 5%. It was observed that the tensile strength is also affected by change in infill density. The highest tensile strength was observed in rectilinear infill pattern. With density parameter as constant, honeycomb pattern shows better tensile strength than rectilinear pattern possibly due to small variations in amount of plastic deposited for each pattern. Further research might be required to understand the relation between the three variables and to understand the relation between crystallinity volume fraction and tensile strength. Below, we discuss certain studies which have demonstrated the effect of variation of density patterns on mechanical properties of the object.

In this section, we analyze different recent infill optimization techniques reported by researchers. Wu [4] reported an optimization approach for generation of bone-like porous structures using voxel optimization scheme. Given a design domain and set of external loads, this technique optimizes the distribution of material thereby maximizing the mechanical stiffness. They have proposed a formulation to measure the local volume fractions and imposed constraints on these values thereby regulating local material distribution and allowing generation of porous structures. The objective function is to maximize the stiffness. The manufacturability of bone-like infills had been tested by fabricating femur model using selective laser sintering (SLS) process and three more models using FDM printers. The bone-like infills had been tested for robustness and manufacturability. It is observed that by using this technique, robust structures can be achieved, thereby making suitable structures for specific application using optimized infills for AM.

Clausen [7] had demonstrated that the buckling load-carrying capacity of an infill structure is superior when compared with a solid structure of same mass. In this paper, projection-based minimum compliance approach and coating approach were compared. The standard projection-based minimum compliance approach produced perfectly black and white structure while the other approach produced structure composed of solid shell with porous infill. Compliance and buckling load were used to compare the optimized structures. Coating approach allowed exploitation of AM infill to create components. As a result, it was observed that buckling properties of the structure fabricated using minimum compliance approach improved as compared to the structure generated using other approach. The samples were subjected to specific MBB beam test case, and it was observed that the buckling load-carrying capacity of a porous structure is increased by 500% compared to the solid structure at a cost of decrease in stiffness of about 20–25%. This study is an example of how TO can be adapted for improvements in structural performance and AM-specific manufacturing opportunities.

In context of additive manufacturing applications, non-traditional 3D infill patterns were characterized as an alternative to classical 2D infill patterns by Podroužek [8]. The investigated patterns are biologically inspired, emerged from numerical topology optimization, and selected based on their mechanical properties such as double curvature. Mechanical properties of 2D and 3D infills like stiffness, load capacity, and infill density were tested, and results were compared with Gyroid and Schwarz D. Linear scaling law was observed with infill density and the stiffness and load capacity results. In contrast to the 2D hexagonal infills, 3D infill specimens exhibited ductile behavior and showed smooth strain periodicity without stress concentrations. The dog-bone tensile test was used to evaluate sensitivity to slicing strategy and build orientation. Nonlinear FEM analysis calibrated values show a good agreement with the experimental results in terms of peak load, periodic strain, and stiffness fields. Based on the findings, it is observed that custom 3D infills can be an alternative to 2D infills. 3D infills can cause general stress states to occur in case of geometric shapes, and they can also achieve large weight reductions even without introducing macroscopic anisotropy in the object while manufacturing using AM.

Wu [9] had presented a TO method for non-uniform and shell infill. Objects printed using this method achieve prescribed shell thickness as the method allows filling the interior of the object with non-uniform structures till a required local volume fraction is obtained. Numerical analysis was used to test the effectiveness of the shell-infill interpolation model for 2D objects. The results demonstrate that in uniaxially loaded bars and the joints connecting these bars, elongated and crossing infills are found, respectively. The shell-infill formula maybe extended for 3D structures. In powder-based fabrication, closed-wall infill structures provide stiffness but might trap unsintered powders. Closed-wall structures might turn into truss-like structures if the local volume bound is restricted or the allowable volume is insufficient to create closed-wall structures. Despite complex optimization, this model is robust to parameter and geometric variations.

Gebisa and Lemu [10] had studied topology optimization in additive manufacturing to maximize the stiffness and minimize the weight of the specimen. They have performed a case study on jet engine bracket and found that the weight of the component is reduced 65% without compromising the structural and functional performance of the component.

Table 1 summarizes the different infill optimization methods.

In the next section, we summarize different techniques in topology optimization and discuss future trends.

Table 1 Summary of recent work in topology optimization of infill for 3D printing

References	Structure generated	Optimization technique	Objective function	Combinations of infill	Test performed	Outcome
[4]	Bone-like porus strucutre	Local volume-constrained topology optimization	Analyze the optimality, robustness, and manufacturability of the obtained structures	Honeycomb structure, structures generated by TO with a total volume constraint and local volume constraints	FEM for stress analysis and 3D printing with SLS and FFF to check manufacturability	Increased stiffness, larger bending stiffness-to-weight ratio
[7]	Messerschmitt-Bölkow-Blohm(MBB) beam	Coating approach to topology optimization	Designing infill-based components that possess a strongly improved buckling load and, as a result, improved structural stability	Optimized trainagular infill and solid structure	Comparison of the standard and coating approaches to topology optimization for the MBB beam benchmark case	The buckling load for the porous structure was more than five times that of the solid structure, for a decrease in stiffness of about 20 to 25%

(continued)

Table 1 (continued)

References	Structure generated	Optimization technique	Objective function	Combinations of infill	Test performed	Outcome
[8]	Compression test cylinders (200 mm height, 100 m dia), ISO tensile test specimen	Topology optimization	Analyze the qualitative and quantitative differences in mechanical response between the investigated 3D infill patterns (3 types) in comparison with the reference 2D pattern (honeycomb)	Gyroid, Schwarz D, and Schwarz P, honeycomb	Compression and tensile test	Gyroid and Schwarz D. infills showed similar stifness and load-carrying capacity, and Schwarz P showed least performance
[9]	Messerschmitt-Bölkow-Blohm (MBB) beam	Topology optimization	Topology optimization method for evolving the shell and non-uniform infill concurrently	Optimized infill	Numerical tests	Optimized, non-uniform infill performs better than uniform infill for the same material volume
[10]	Case study of rocket engine bracket	Topology optimization	To minimize weight and maximize stiffness	Optimized infill	FEM analysis	65% Weight reduction

3 Conclusion and Future Trends

Recent developments in optimization of infill for 3D printing are discussed in this paper. It is observed that researchers are trying to use TO to optimize infill density and patterns to achieve better strength, robustness, and lightweight structures without increasing the amount of material used. Biologically inspired and nature-inspired patterns are being increasingly used for infill and are being optimized using TO.

However, research in this field needs more work, and further investigation is required to better tune the existing algorithms. Many algorithms like the robust TO approach, heterogeneous two-scale TO, have not yet been closely associated and tested for AM. There are many open challenges like—concurrently optimizing the build direction when performing overhang-free topology optimization, establishing relationship between the variables—deposition trajectories, interlayer bonding zones and elastic modulus, residual stress-constrained topology optimization for metal AM, and so on.

Another research problem that could be explored is free material optimization in which instead of the macro-density of usual TO, the macro-material physical properties are directly optimized. We forecast an affluent and stimulating future in TO for AM.

References

1. Jiménez M, Romero L, Domínguez IA, Espinosa MDM, Domínguez M (2019) Additive manufacturing technologies: an overview about 3D printing methods and future prospects. Complexity
2. Huang Y, Leu MC, Mazumder J, Donmez A (2015) Additive manufacturing: current state, future potential, gaps and needs, and recommendations. J Manuf Sci Eng Trans ASME 137(1):1–11
3. Popescu D, Zapciu A, Amza C, Baciu F, Marinescu R (2018) FDM process parameters influence over the mechanical properties of polymer specimens: a review. Polym Test 69:157–166
4. Wu J, Aage N, Westermann R, Sigmund O (2018) Infill optimization for additive manufacturing-approaching bone-like porous structures. IEEE Trans Vis Comput Graph 24(2):1127–1140
5. Liu J et al (2018) Current and future trends in topology optimization for additive manufacturing. Struct Multidiscip Optim 57(6):2457–2483
6. Fernandez-Vicente M, Calle W, Ferrandiz S, Conejero A (2016) "Effect of infill parameters on tensile mechanical behavior in desktop 3D printing, 3D Print. Addit Manuf 3(3):183–192
7. Clausen A, Aage N, Sigmund O (2016) Exploiting additive manufacturing infill in topology optimization for improved buckling load. Engineering 2(2):250–257
8. Podroužek J, Marcon M, Ninčević K, Wan-Wendner R (2019) Bio-inspired 3D infill patterns for additive manufacturing and structural applications. Materials (Basel) 12(3):1–12
9. Wu J, Clausen A, Sigmund O (2017) Minimum compliance topology optimization of shell–infill composites for additive manufacturing. Comput Methods Appl Mech Eng 326:358–375
10. Gebisa AW, Lemu HG (2017) A case study on topology optimized design for additive manufacturing. In: IOP conference series: materials science and engineering, vol 276, no 1.

Experimental and Numerical Studies of Heat Transfer Through Heat Sinks in Electronic Devices

Md Adil, Asif Khan, Abhishek Gupta, Sohail Ahmad, and M. Altamush Siddiqui

1 Introduction

Extended surfaces are employed for cooling electronic systems by proper regulation of its temperature up to allowable limit. Research and development on heat sink has a long history and is still continuing with efforts to improve its design and performance by bringing innovations in modelling and analytical techniques. The development of various heat sinks with various fin geometries has revolutionized the heat sink industry. The very first experiment conducted by Starner and McManus [1] presents free convection data for four rectangular fin arrays in three positions including the vertical position from the fin base. An experimental investigation was conducted by Welling and Wooldridge [2] where free convection heat transfer takes place from rectangular fins attached to a vertical plate. They found that there is an optimum fin height for maximum heat transfer with particular fin spacing. Harahap and McManus [3] have reported the pattern of flow associated with free convection heat transfer from horizontal fin arrays. Computational fluid dynamics (CFD) codes are widely used as a tool for thermal analysis. However, CFD-based thermal analysis is not necessarily easy to perform where the object of analysis is geometrically complex. With the advent of CFD in recent years, computation of flow and heat transfer has become quite possible. Arularasan and Velraj [4] optimized the geometric parameters for the rectangular heat sink by conducting CFD modelling and simulation in Ansys Fluent and found certain specific values of the same. Lee [5] concluded that the cooling performance of a heat sink depends on different types of parameters including thermal conduction resistance, dimensions, location and concentration of heat sources as well as the airflow bypass conditions. Air cooling investigation on heat sink with square

Md Adil (✉) · Asif Khan · Abhishek Gupta · Sohail Ahmad · M. Altamush Siddiqui
Department of Mechanical Engineering, ZHCET, Aligarh Muslim University, Aligarh 202002, India
e-mail: mdadil@zhcet.ac.in

V. H. Saran and R. K. Misra (eds.), *Advances in Systems Engineering*,
Lecture Notes in Mechanical Engineering,
https://doi.org/10.1007/978-981-15-8025-3_30

module array in electronic devices reported by Mohammad [6]. Yuncu and Anbar [7] reported experimental results on free convection heat transfer on rectangular fin arrays taking base plate horizontal and by varying parameters like fin height, fin spacing and power supply 8–50 W. Paisarn et al. [8] based on real PC conditions performed numerical simulation on mini rectangular fin with the deionized water as cooling medium.

In present work, experimental and numerical heat transfer studies with air have been carried out on a heat sink of a CPU. The experiment was conducted for different heat flux by means of a nichrome wire heater. And for the same configuration, numerical simulation was carried out by a computational fluid dynamics (CFD) model developed using the commercial software package ANSYS Workbench. Different values of the heat transfer coefficient of the surrounding media, varied from $h = 5$ W/m^{2}K to 20 W/m^{2}K, were studied in order to find its optimal value. The contour plots are presented for different heat fluxes and heat transfer coefficients and compared with the experimental observations. Experimental and numerical values are also validated.

2 Experiments

2.1 Experimental Setup

The experimental arrangements are shown in Figs. 1 and 2. It has an aluminium heat sink with base dimension of 34.17 × 60.5 mm and 12 fins in which 2 fins are 2.36 mm thick, and 10 fins are 1.25 mm thick. Such heat sinks are generally employed on a processor in a CPU. A heating element made from nichrome wire embedded on an iron plate is used for providing heat flux at the base of the heat sink. A data logger is used for recording temperatures. The asbestos sheet is used for restricting heat loss at the base of the heat sink. Eighteen T-type (Copper and constantan) thermocouples are used for recording the temperatures. A voltmeter and an ammeter are used for

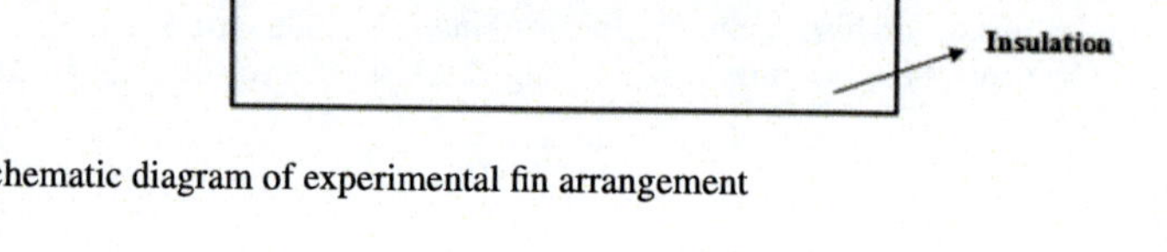

Fig. 1 Schematic diagram of experimental fin arrangement

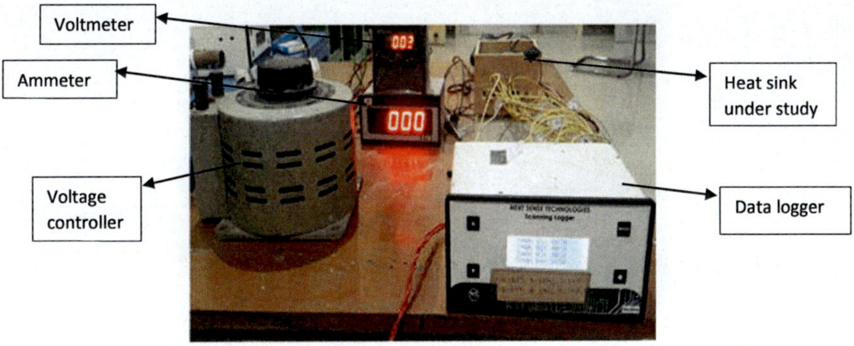

Fig. 2 View of experimental setup

measuring heat supply. A dimmerstat is used for controlling the power provided in the experiment.

2.2 *Data Collection*

The experiments are conducted for the heat flux of magnitudes 4.8 W, 7.3 W, 9.9 W. Heat was supplied to the heat sink by means of a nichrome wire heater. Here, copper-constantan thermocouples are used in the experiment, which is also called T-type thermocouples. Figure 3 shows various thermocouples attached to different fins, namely A1, A2 and A3. At the central part, two thermocouples are attached to fin, namely B1 and B2. Three thermocouples are attached to the right end fin named C1, C2 and C3. And one thermocouple is attached to the base of the fin named as BB. The

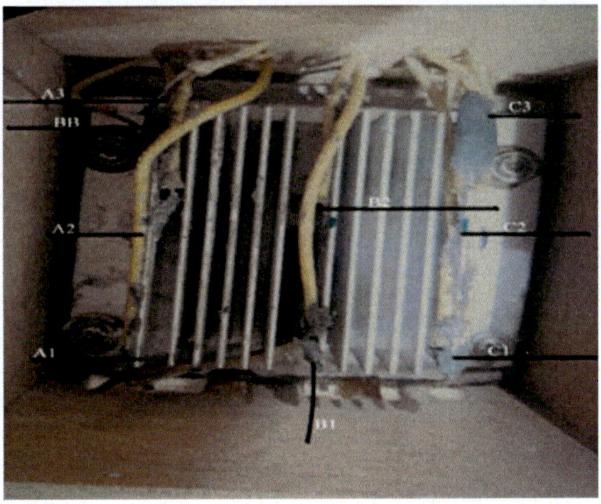

Fig. 3 Top view of the heat sink with attached thermocouples

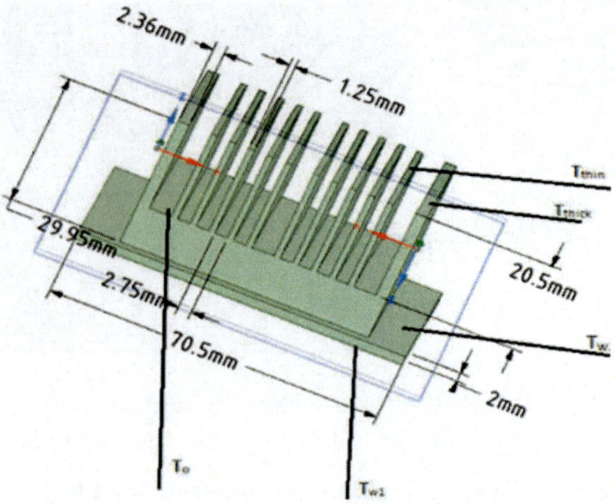

Fig. 4 Heat sink model developed

temperature was recorded with the help of data logger (Next Sense Technologies, 16 channels) for the base and tips of various fins at different air velocities including free convection. Heat supplied was estimated by means of voltage and current to the heater.

3 Numerical Scheme

3.1 Modelling and Design of Heat Sink

Generally heat sinks are made of copper and aluminium, because they have a desirable value of thermal conductivity. But mostly aluminium is chosen for the heat sinks as it is easily available and cheaper than copper. Therefore, material for a heat sink in the commercial package ANSYS FLUENT and experiment aluminium having thermal conductivity as 202.5 W/mK [9] is chosen. The detailed specification of the heat sink model developed in ANSYS is shown in Fig. 4.

3.2 Meshing and Simulation

The numerical simulation was carried out by a CFD model developed using the commercial software package ANSYS Fluent. The dimensions for analysis are taken same as it is for the actual heat sink. The structured mesh was generated using the Hexa-dominatrix method. The number of nodes and elements is 804,730 and 177,270, respectively. The heat transfer coefficient (HTC) of the surrounding medium varied from $h = 5$ W/m^{2}K to 20 W/m^{2}K.

3.3 Governing Equations

The general form of fluid flow and heat transfer equations of compressible Newtonian fluid with time independency used in solver execution is given below, where ρ is the density, *u*, *v* and *w* are velocity components, vector, *p* is the pressure, *V* is the velocity vector, *k* is the thermal conductivity and h_o is the total enthalpy.

- Conservation of mass

$$\frac{\partial(\rho u)}{\partial x}+\frac{\partial(\rho v)}{\partial x}+\frac{\partial(\rho w)}{\partial x}=0 \tag{1}$$

- Momentum equation

  X momentum:

$$\frac{\partial(\rho u^2)}{\partial x}+\frac{\partial(\rho uv)}{\partial x}+\frac{\partial(\rho uw)}{\partial x}=-\frac{\partial(P_m)}{\partial x}+\rho v(\frac{\partial^2 u}{\partial x^2}+\frac{\partial^2 u}{\partial y^2}+\frac{\partial^2 u}{\partial z^2}) \tag{2}$$

  Y momentum:

$$\frac{\partial(\rho vu)}{\partial x}+\frac{\partial(\rho v^2)}{\partial x}+\frac{\partial(\rho vw)}{\partial x}=-\frac{\partial(P_m)}{\partial y}+\rho v\left(\frac{\partial^2 v}{\partial x^2}+\frac{\partial^2 v}{\partial y^2}+\frac{\partial^2 v}{\partial z^2}\right)+g(\rho-\rho a) \tag{3}$$

  Z momentum:

$$\frac{\partial(\rho wu)}{\partial x}+\frac{\partial(\rho wv)}{\partial x}+\frac{\partial(\rho w^2)}{\partial x}=-\frac{\partial(P_m)}{\partial z}+\rho v\left(\frac{\partial^2 w}{\partial x^2}+\frac{\partial^2 w}{\partial y^2}+\frac{\partial^2 w}{\partial z^2}\right) \tag{4}$$

- Conversation of energy:

$$\nabla\cdot(\rho h_0\overline{V})=-p\nabla\cdot\overline{V}+\nabla\cdot(k\nabla T) \tag{5}$$

4 Results and Discussions

4.1 Experimental

For specified power supply to the heater, the velocity of air across the heat sink was varied. This created different heat transfer coefficient over the heat sink. Steady-state temperatures at different locations over the heat sink as specified in Fig. 3 were recorded using a data logger. Temperatures recorded are shown in Table 1. The tip temperatures of all the fins which are in observations are found to have minimum temperatures and they increases with the increase in the heat flux, and temperature

Table 1 Observation table for experimental data

Power supply (W)	HTC (W/m^2K)	Measured temperatures (Celsius)													
		A1	A2	A3	Ab	B1	B2	Bb	C2	C1	C3	Cb	BB	T1	T2
4.8	6.9	55	55	54	56	55	54	55	54	55	54	56	57	45	36
4.8	9.2	44	44	43	45	44	43	45	44	44	44	45	47	37	32
4.8	13.01	40	39	39	41	40	40	41	40	40	40	41	45	35	31
4.8	18.54	37	36	36	38	37	36	37	37	36	37	38	40	33	31
7.32	7.4	60	60	59	61	61	59	61	60	61	60	62	69	48	35
7.32	9.76	52	52	51	54	52	52	53	52	52	52	54	58	41	34
7.32	12.6	46	46	45	48	47	46	48	47	46	47	48	52	39	34
7.32	18.84	45	43	43	46	44	43	45	43	43	44	45	49	36	32
9.9	8.4	81	80	78	82	81	79	81	78	81	79	82	85	54	35
9.9	9.2	63	63	62	65	63	62	64	62	63	63	65	69	53	36
9.9	13.6	49	50	50	52	51	50	52	51	50	50	53	58	40	34
9.9	18.84	45	45	45	48	46	45	47	45	45	45	47	53	37	33

decreases with the increase in the fan speed and found minimum temperature at the maximum HTC.

4.2 Numerical Results

Temperature contour plots obtained numerically are presented in Fig. 5a–h in which heating load is varied from 4.8 W to 9.9 W at the base of the heat sink. The heat transfer coefficient is changed across the heat sink by varying air velocity. The temperature along the fin length varies from base to the tip. The lowest temperature is obtained at the tip of each fin. The highest temperature is seen at the base of the fin, which is a very good agreement with the literature. For obtaining the best performance of electronic devices, we need to dissipate more heat away from the devices [10]. At heat load of 4.8 W and heat transfer coefficient of 18.4 W/m^2K, the maximum temperature is just 40.6 °C which is quite low as compared to those at higher heat fluxes. From Fig. 5a–h, the maximum temperatures at different loads and heat transfer coefficient are selected and plotted in Fig. 6 which shows that maximum temperature decreases gradually with increase in the HTC. It can also be seen in Fig. 6 that the rate of decrement of maximum temperature is lowered at larger values of HTC. This indicates faster cooling rate at high values of HTC of the coolant (air). However, at around HTC $\geq$ 16 W/m^2K, the change in maximum temperature on the sink becomes almost constant. Therefore, there is no need to further increase the HTC because this will result in consumption of more electricity and hence become costly.

Faster rate of cooling is observed near ends of the heat sink where fresh air enters as compared to the middle part of the heat sink where air movement lowers down. Also

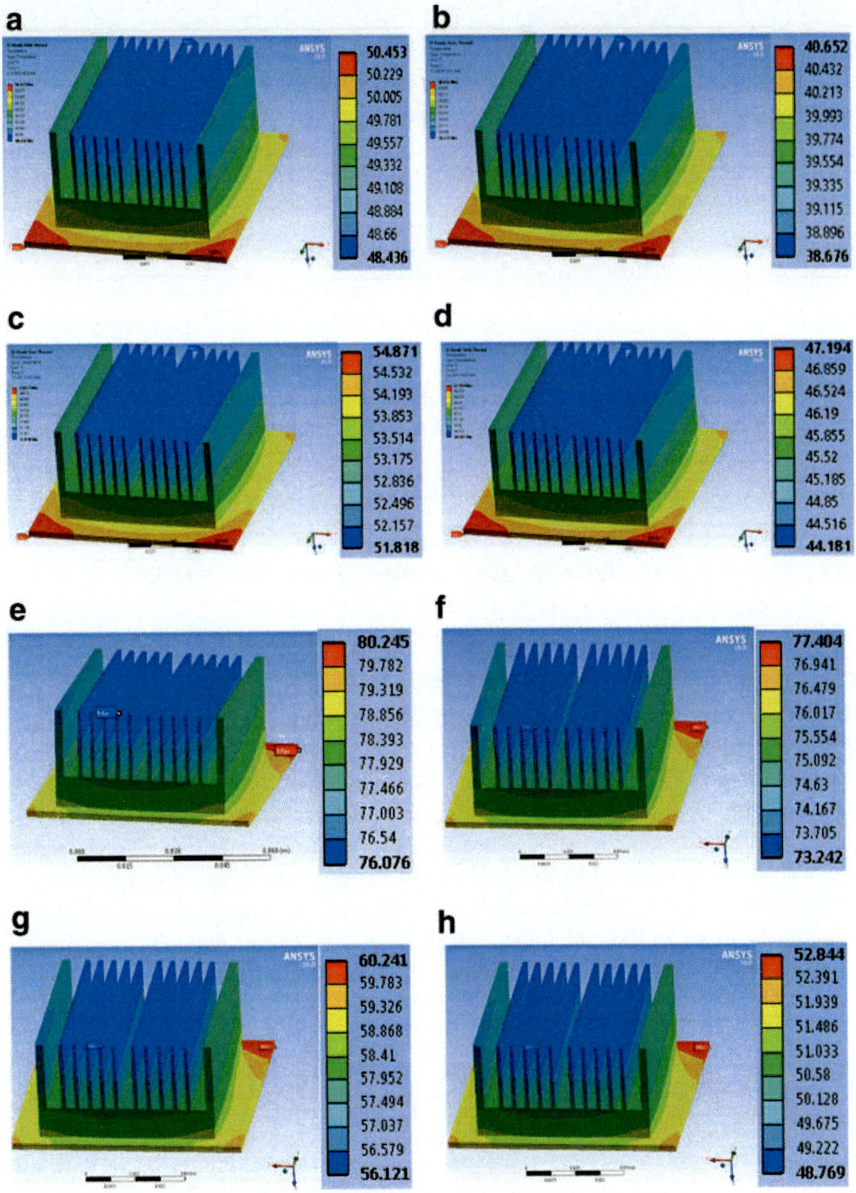

Fig. 5 **a** Temperature contour for heat flux 4.8 W and HTC 9.2 W/m^2k, **b** Temperature contour for heat flux 4.8 W and HTC 18.54 W/m^2k, **c** Temperature contour for heat flux 7.32 W and HTC 12.6 W/m^2k, **d** Temperature contour for heat flux 7.32 W and HTC 18.54 W/m^2k, **e** Temperature contour for heat flux 9.9 W and HTC 8.4 W/m^2k, **f** Temperature contour for heat flux 9.9 W and HTC 9.2 W/m^2k, **g** Temperature contour for heat flux 9.9 W and HTC 13.6 W/m^2k, **h** Temperature contour for heat flux 9.9 W and HTC 18.54 W/m^2k

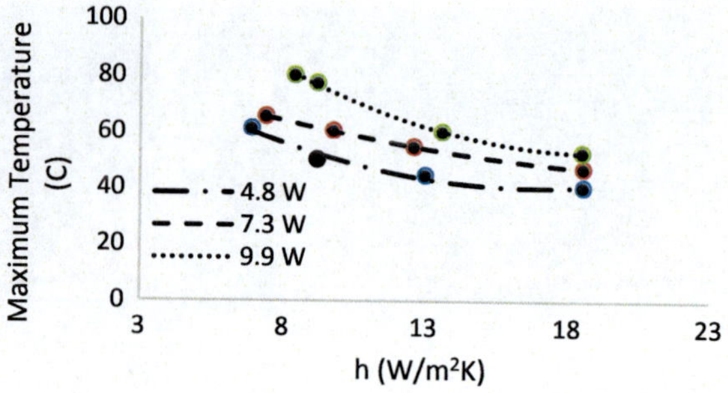

Fig. 6 Variation of maximum temperature on heat sink with varying heat transfer coefficient at various load

Fig. 5a–h shows that the temperature across the fin is increasing with the increase in the heat flux. Thin fins (i.e.1.25 mm) show lower temperatures as compared to the thick fins (i.e. 2.36 mm).

4.3 Comparison Between Experimental and Numerical Results

Comparison of tip temperatures for one of the fins was made. Errors with the analytical and experimental values were calculated and found that maximum deviation is $\pm$ 8%. This error might be due to HTC, which may not be uniform throughout the fin's surfaces during the experimentation due to the channelized flow of air, as we have taken constant HTC in numerical analysis. As there is space constraint for mounting thermocouples, we have used M-Seal to fix these thermocouples, and its usage changes the mode of heat transfer from convection to conduction in the vicinity of the thermocouples. This may also lead to increase in temperature. It may also be due to limitations of the data logger, ammeter and voltmeter readings in giving the exact value. Least count of data logger was limited to one degree Celsius (Table 2).

5 Conclusion

In the present work, numerical and experimental work for aluminium heat sink was carried. It is obvious that for obtaining best performance of electronic devices we need to dissipate more heat from the devices. We therefore need high air velocity for high heating of electronic devices. At the same time, a very large HTC may be useless and unnecessarily increase the electricity cost. Therefore, there should be some optimal value for HTC, which in the present study comes out to be around 16 W/m^{2}k, for better performance as well as economy. The numerical values of temperature on the fins are compared with the experimental values of the temperatures and found agreement with the same.

Table 2 Comparison of tip temperatures for one of the fin in experimental and numerical studies

S. no.	Heat source (W)	HTC (W/m^2K)	Type of convection	Experimental (°C)	Numerical (°C)	%Error b/w EandN
1	4.8	6.9	Free	54.75	59.92	8.63
2	4.8	9.2	Forced	43.75	48.88	−4.43
3	4.8	13.01	Forced	39.25	43.16	−3.51
4	4.8	18.54	Forced	36.25	39.12	−2.69
5	7.32	7.4	Free	59.75	63.29	2.18
6	7.32	9.76	Forced	51.75	58.17	−5.34
7	7.32	12.6	Forced	45.75	52.5	−4.55
8	7.32	18.54	Forced	43.5	44.85	−3.57
9	9.9	8.4	Free	79.75	77.47	−6.92
10	9.9	9.2	Forced	62.5	74.63	−6.6
11	9.9	13.6	Forced	49.75	57.49	−5.99
12	9.9	18.54	Forced	45	50.13	−5.23

References

1. Starner KE, McManus HN (1963) An experimental investigation of free convection heat transfer from rectangular fin-arrays. J Heat Trans Trans ASME Ser C 85:273–278
2. Welling JR, Wooldridge CB (1965) Free convection heat transfer coefficient from rectangular vertical fins. J Heat Trans Trans ASME Ser C 87:439–444
3. Harahap F, McManus HN (1967) Natural convection heat transfer from horizontal rectangular fin array. J Heat Trans Trans ASME Ser C 89:32–38
4. Arularasan R, Velraj R (2008) CFD analysis in a heat sink for cooling of electronic devices. Int J Comput Internet Manage 16(3):1–11
5. Mohamed MM (2006) Air cooling characteristics of a uniform square modules array for electronic device heat sink. Appl Thermal Eng 26:486–493
6. Yuncu H, Anbar G (1998) An experimental investigation on performance of rectangular fins on a horizontal base in free convection. Heat Mass Transf 33:507–514 (Springer)
7. Naphon P, Klangchart S, Wongwises S (2009) Numerical investigation on the heat transfer and flow in the mini-fin heat sink for CPU. Int Commun Heat Mass Transf 36:834–840
8. Lee S (1995) Optimum design and selection of heat sinks. In: IEEE semiconductor thermal measurement and management symposium, San Jose, USA, February, pp 48–54
9. Ch. Srinivasa Rao V, Himachandra Raju CFD (2018) Modeling and analysis of a heat sink with rectangular pin fin. Int J Res Appl Sci Eng Technol (IJRASET) 6(III). ISSN: 2321-9653 (IC Value: 45.98; SJ Impact Factor: 6.887)
10. Sam Sukumar R, Sriharsha G, Bala Arun S, Dilip Kumar P, Sanyasi Naidu Ch (2013) Modelling and analysis of heat sink with rectangular fins having through holes. Int J Eng Res Appl (IJERA) 3(2):1557–1561

Epileptic Seizures Detection

Ayushi Sharma and Sandeep Paul

1 Introduction

Epilepsy is a dysfunctionality of brain that affects people of all age. Globally, about 70 million people around the world suffer from it. About 84% epileptic patients belong to weak financial background. Usually, epileptic patients respond slowly to corresponding treatment. About 75% people suffer from the problem and are not able to get right treatment due to financial problems. Sometimes, patients and their families face inequity [1]. Anyone can be affected by epilepsy, at any time of the life. In children and in people over 65, epilepsy is the most commonly diagnosed. Machine learning proves its superiority over human capabilities many times in many of the tasks and applications like audio recognition [2], picture recognition [3], detection of Alzheimer's disease [4], sales prediction [5], estimation strength [6], truth detection, etc. It is also a promising field of computer science for designing an automated system for finding and curing any disease. Traditionally, neurologists consider direct observation to detect epileptic form abnormalities. This technique is time-consuming and limited in detecting abnormalities. So, it is essential to create an automated or computer-based system [7] for categorizing unwillingly classes of these EEG signals using machine learning approaches.

A. Sharma · S. Paul (✉)
Dayalbagh Educational Institute, Agra, India
e-mail: spaul@dei.ac.in

A. Sharma
e-mail: ayushi.sharma515@gmail.com

V. H. Saran and R. K. Misra (eds.), *Advances in Systems Engineering*,
Lecture Notes in Mechanical Engineering,
https://doi.org/10.1007/978-981-15-8025-3_31

2 Proposed Study

There are many methods that are used to detect and cure epileptic seizures. EEG test is the most popular among all of them [1]. EEG signals can be used to study the behavior of epileptic seizures. EEG signals are generally non-stationary and nonaligned which make the study of signals complex [8]. In this work, an application has been developed using neural networks. Multiple experiments have been performed using machine learning techniques like *K*-nearest neighbors (KNN) [9], logistic regression [10], random forest [11], gradient boosting (SGD), extreme random forest, extreme gradient boost [12], artificial neural network (ANN) [13], convolutional neural network (CNN) [14], and others in this study to identify seizure and non-seizure classes. And based on the performances of the different classifiers, the best one is chosen for development of the application.

The *K*-nearest neighbor algorithm is very simple. It works on the basis of training samples at a minimum distance from the query example to determine the nearest neighbors.

Artificial neural networks (ANN) are large networks using neurons, same as the brain, to learn certain information and predict on the basis of that learning. ANN consists of three types of layers, viz. the input layer (IL), hidden layers (HL), and the output layer (OL) [15]. As the name suggests, the input layer (IL) is used to provide the input to the net, i.e., it consists as many neurons as there are inputs. Now, the output of the input layer is passed on the next layer. This layer is called the hidden layer (HL). The last layer is always the output Layer, going by the name predicts the class label or a certain output. The number of nodes or neurons is the same as the number of classes in output layer. For binary class problem, we have only one node at output layer, whereas these neurons use various activation function, according to which net gets simulated. Some of the common activation functions are linear, reactifier linear unit (Relu), sigmoid, Gaussian, etc. The neural network is broadly classified into two categories feed forward and recurrent. Feed forward is further divided into single layer perceptron, multilayer perceptron, etc. Initially, network works in forward manner to achieve desired output from given input, so we call it as feed forward neural network. Finding error or loss in generation of output is calculated and balanced by Backpropogation Algorithm.

CNN works similar to a basic neural network. The CNN has three-dimensional input such as height (*h*), depth (*d*) and width (*w*). There are three main layers in CNN such as: fully connected layer (FC), convolutional layer (CONV), and pooling layer (POOL). There are some other layers are also used such as a rectifier linear unit (RELU) and dense. Each layer may also have some parameters and hyperparameters such as zero padding and stride [16]. This work is done focusing majorly on neural networks. The other classifiers were used as an experiental purpose only.

3 Experiments

We have used the UCI's seizure detection EEG dataset [17] which is labeled and preprocessed. The dataset is the combination of five different sets namely A, B, C, D and E. Each comprises 100 signal channel EEG signals of nearly 23 s duration. The recordings were recorded in the following manner: Set A, B: Five healthy patients' recording using 20 electrodes, while patients were in relaxed position with eyes open and closed, respectively. Set C: Intracranial EEG signals recording of five different epileptic patients during non-seizure time from brain's epileptogenic zone. Set D: Intracranial EEG signals recording of five different epileptic patients during non-seizure time from brain's different zone during hippocampal formation. Set E: Seizure recording of five different epileptic patients signals is sampled at 174 Hz and digitized using 12 bit ADC. Here, the dataset we have does not have artifacts. The artifacts had been handled by band pass filter with cutoff frequencies of 1–40 Hz in the raw EEG segments. For experiments, the dataset is split into training, validation and testing in 70:15:15 ratio, and the labels are converted into binary class such that for 1 for seizures (E) and 0 for others (A B C D). The proposed study has been implemented using Python AIML libararies.

Seizure detection is based on supervised learning classes, viz. seizure and non-seizure. The algorithms determine the best performance of the dataset. The main objective here is to detect seizures. Multiple classification techniques have been used during experiments such as: K-nearest neighbors, logistic regression, stochastic gradient descent, Gaussian Naïve Bayes, decision tree, random forest, gradient boosting, extreme random forest, extreme gradient (XG) boost, artificial neural network, and convolutional neural network. And on the basis of metric values such as area under curve (AUC), accuracy (Acc), recall (R), precision (P), specificity (Spec), and prevalence (Prev), best one is selected.

We have used neural network as a main classifier to test data, as the accuracy of NN is higher than any other classifier. The final architecture of CNN includes four convolutional layers (Conv), three max pooling layers (Pool), and one fully connected layer (FC). The binary cross entropy has been used as a loss function with Adam as an optimizer. The final architecture of CNN is as follows: Input-Conv1(500)-BN-ReLU-Pool1-Conv2(500)-BN-ReLU-Pool2-Conv3(500)-BN-ReLU-Pool3-Conv4(500)-BN-ReLU-FC1(500)-DR(0.2)-Output. Here, BN, ReLU, and DR represent batch normalization, rectifier linear unit, and dropout, respectively. The dropout technique with rate of 0.2 has been applied after the fully connected layer. There are 500 filters in each convolutional layer. The fully connected layer FC1 has 500 neurons. The output layer has two neurons which denote the target classes as seizure and non-seizure. This achieves 99.39% cross validation and 99.39% test accuracy. Other classifiers metric values are shown in Table 1.

Normally, it is difficult to keep track of abnormalities of patient. Patients are not well versed with manual treatment which doctors normally use for tracking the health of patient [18]. To overcome these problems, a patient health monitoring app

Table 1 Multiple classifiers metric values

S.No.	Classifiers	AUC	Acc	*R*	*P*	Spec	Prev
1	KNN	0.932	0.922	0.666	0.965	0.993	0.219
2	Logistic regression	0.536	0.696	0.435	0.345	0.769	0.219
3	SGD	0.523	0.608	0.477	0.273	0.644	0.219
4	Gaussian NB	0.981	0.950	0.886	0.884	0.967	0.219
5	Decision tree	0.852	0.894	0.849	0.717	0.907	0.219
6	Random forest	0.981	0.941	0.881	0.856	0.958	0.219
7	Gradient boosting	0.985	0.950	0.936	0.851	0.954	0.219
8	Extra tree	0.994	0.961	0.973	0.864	0.957	0.219
9	XG boost	0.991	0.958	0.931	0.884	0.966	0.219
10	ANN	0.991	0.965	0.925	0.902	0.975	0.201
11	CNN	0.978	0.9939	0.997	0.989	0.959	0.219

is used to manage the daily records of patient. The application can be accessed from everywhere to check on patient's health by doctor.

The detection method integrates at backend of the application and gets the results at user (front) end. The working of the application is shown in Fig. 1. This application aims to record user's health condition and informs user at the time of risk. It also gives some general prediction to the user for maintaining and improving his or her health. This is a prototype application which has some limitations such as: capturing

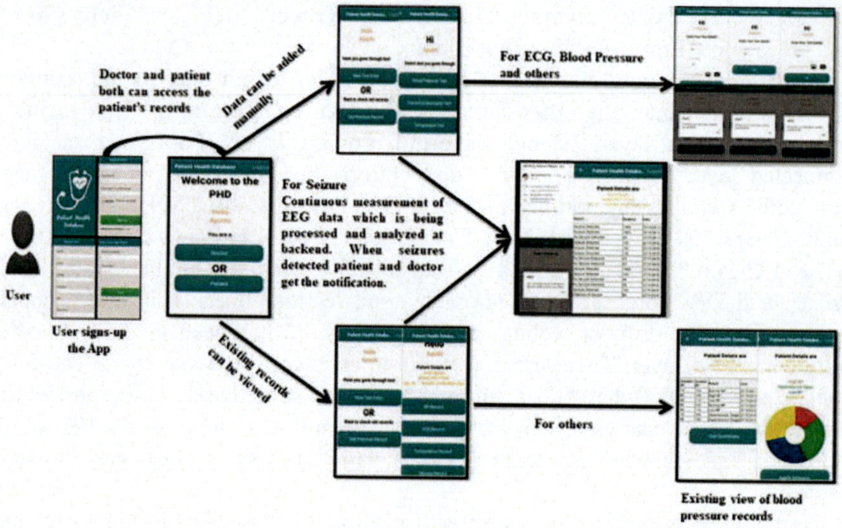

Fig. 1 Flow diagram of working of the mobile application. User can either be a doctor or patient

Table 2 Comparison of selected studies in the seizure detection using UCI's dataset

S.No.	Author	Year	Classifier	Data ratio	Accuracy
1	Guo et al. [14]	2010	ANN	50:50	98.27
2	Rivero et al. [9]	2011	KNN	Variable	98.40
3	Fu et al. [19]	2015	SVM	80:20	98.40
4	Peker et al. [13]	2016	ANN	60:40	99.33
5	Ullah et al. [21]	2018	CNN	80:20	99.7
6	Hussein et al. [20]	2018	LSTM	80:20	100
7	Proposed CNN	2019	CNN	70:15:15	99.39

EEG data, improper notification method. These limitations will be overcome in the near future.

4 Results and Discussions

The comparative study has been performed with other existing approaches in the literature as shown in Table 2. Based on the experimental results, we obeserved that the accuracy of CNN is higher than any other classifiers which is 99.39%. The obtained results using the proposed approach are better than the workdone in [9, 13, 14, 19]. Moreover, these results are comparable with [20, 21] which are obtained using the CNN.

The workdone in [21] used ensemble of CNN to handle different segments of EEG signal. Similarly, in [20], multiple LSTM modules were used for capturing different segments of input signal. The used CNN in our work showed the comparable results with much simple network.

5 Conclusion

The proposed study helps the patient's doctor to take appropriate action at the appropriate time. Machine learning approach-based application will prevent accidents. It keeps track of patient's condition after every interval of time and maintains health records. Also at the time of risk, it notifies patient and respective doctor. The other attributes of the application help doctor to take more precise decision.

References

1. Epilepsy. Accessed: 22 Apr 2019 [Online]. Available: https://www.who.int/news-room/fact-sheets/detail/epilepsy
2. Lecun Y, Bengio Y (2015) Deep learning. Nature 521:436–444
3. Krizhevsky A, Sutskevar I, Hinton GE (2012) ImageNet classification with deep convolutional neural networks. In: Neural ınformation processing system conference
4. Ortiz A, Munilla J, Górriz JM, Ramírez J (2016) Ensembles of deep learning architectures for the early diagnosis of the Alzheimer's disease. Int J Neural Syst 26(07):1650025
5. Rafiei MH, Adeli H (2016) A novel machine learning model for estimation of sale prices of real estate units. J Constr Eng Manage 142(2):04015066
6. Baldominos A, Blanco I, Moreno AJ, Iturrarte R, Bernárdez O, Afonso C (2018) Identifying real estate opportunities using machine learning. Appl Sci 8:2321
7. Faust O, Acharya UR, Adeli H, Adeli A (2015) Wavelet-based EEG processing for computer-aided seizure detection and epilepsy diagnosis. Seizure 26:56–64
8. Acharya UR, Vinitha Sree S, Swapna G, Martis RJ, Suri JS (2013) Automated EEG analysis of epilepsy: a review. Knowl Based Syst 45:147–165
9. Rivero D, Fernandez-Blanco E, Dorado J, Pazos A (2011) A new signal classification technique by means of genetic algorithms and kNN in Evolutionary Computation (CEC). IEEE Congress on. IEEE, pp 581–586
10. Cizek GJ, Fitzgerald SM (1999) Methods plainly speaking: an introduction to logistic regression. Measur Eval Couns Dev 31(4):223
11. Andy L, Matthew W (2002) Classification and regression by random. Forest 2/3
12. https://xgboost.readthedocs.io/en/latest/tutorials/model
13. Peker M, Sen B, Delen D (2016) A novel method for automated diagnosis of epilepsy using complex-valued classifiers. IEEE J Biomed Health Inf 20(1):108–118
14. Guo L, Rivero D, Pazos A (2010) Epileptic seizure detection using multiwavelet transform based approximate entropy and artificial neural networks. J Neurosci Methods 193(1):156–163
15. Kumar S (2004) Neural networks: a classroom approach. Tata McGraw-Hill, New York
16. LeCun Y, Bengio Y (1995) Convolutional networks for images, speech, and time-series
17. https://archive.ics.uci.edu/ml/datasets/Epileptic+Seizure+Recognition
18. Ramgopal S et al (2014) Epilepsy & behavior seizure detection, seizure prediction, and closed-loop warning systems in epilepsy. Epilepsy Behav 37:291–307
19. Fu K, Qu J, Chai Y, Zou T (2015) Hilbert marginal spectrum analysis for automatic seizure detection in EEG signals. Biomed Signal Process Control 18:179–185
20. Hussein R, Palangi H, Ward R, Wang ZJ (2018) Epileptic seizure detection: a deep learning approach. Computing Research Repository (CoRR), pp 1–12. arXiv/abs/1803.09848v1
21. Ullah I, Hussain M, Qazi E, Aboalsamh H (2018) An automated system for epilepsy detection using EEG brain signals based on deep learning approach. Computing Research Repository (CoRR). arXiv/abs/1801.05412v1

Numerical Simulation on Heat Transfer Improvement in Wavy Micro Channel

Khursheed Afroz Ansari, Shahid Raza, Asif Anwar, Md. Adil, Mohsin Ali, and S. M. Yahya

1 Introduction

In recent years, miniaturisation of semiconductor devices has led to serious heat dissipation problems in the electronics industry. The speed, capacity and performance of the chips had improved enormously over the last fifty years. These devices contain a large number of transistors. Although individual heat dissipation by the transistors is very small, but their overall effect is significant and requires considerable attention for proper functioning. Also, thermal management is a critical bottleneck for researchers to develop advanced electronic products. Nowadays, hot spots in an advanced processor are approaching the power density of one kilowatt per square centimetre which is much higher than the heat inside a rocket nozzle. All transistors in advanced microprocessors are still not operated as they would generate too much heat that is not in the safe limit of the devices. Further, a processor designer has not many ways to prevent the chips from generating too much heat. So, there is a need to develop a system that can help in dissipating heat to assist in the future development of advanced electronics products.

Conventionally, a silicon die is attached to a copper or aluminium block carved with elaborated fins and ridges. A plastic fan is used to blow air across the metal to dissipate heat, but these systems are bulky, noisy and power-hungry. One idea proposed is to place something cold few micrometres away from the heat-generating transistors. It could be achieved by carving a microfluidic channel into a chip or substrate and pumping coolant through them. Although relevant studies include both single-phase as well as two-phase cooling, but usually single-phase cooling

K. A. Ansari (✉) · S. Raza · A. Anwar · Md. Adil · M. Ali · S. M. Yahya
Department of Mechanical Engineering, ZHCET, Aligarh Muslim University, Aligarh 202002, India
e-mail: khursheed.afroz@gmail.com

V. H. Saran and R. K. Misra (eds.), *Advances in Systems Engineering*,
Lecture Notes in Mechanical Engineering,
https://doi.org/10.1007/978-981-15-8025-3_32

is preferred as it can avoid critical issues like critical heat flux, condensation, saturation temperature, etc. The use of micro channel cooling for cooling of high power density applications was first proposed in the classical paper by Tuckerman and Pease [1]. The progress in the development of micro channel heat sinks was given by Kandlikar and Grande [2]. Conventionally, a straight micro channel is employed in which streamline of coolant is nearly straight resulting in poor fluid mixing, and thereby, inefficient heat transfer. Further, straight channels are inefficient in managing the thermal load at the hot spot regions on the chip.

Kowalewski et al. [3] studied the wavy channel with crest and troughs facing each other for a wide range of Reynolds number and observed the onset of instability for low Reynolds number. Sui et al. [4] studied the flow and heat transfer in a three-dimensional wavy micro channel and investigated the fluid mixing through dynamic system technique (Poincare section). The micro channel heat sinks have gained much attention due to high heat transfer coefficient, small coolant requirement, small sizes and volume per heat load. Further, wavy channels are planner and easy to fabricate. Since then, different types of materials and channel dimensions have been explored.

Two different types of wavy channels were employed one with crests and troughs facing each other (i.e. racoon channel) and other with crests and trough facing each other with a phase difference of 180° (i.e. serpentine channel) as shown in Fig. 1. Both the configurations were investigated for different wave amplitudes, wavelengths and Reynolds number to recognise the flow conditions and geometry that leads to improved overall thermal performance. The channel considered was of 20 mm in length and 200 μm in width with enterance length L_e and exit length L_{ex} are taken to be 3 mm and 5 mm, respectively, with wavy wall length L of 12 mm length in the middle of the channel. To study the individual impact of factors on overall performance, the variables were varied one at a time at various Reynolds number (viz. 50, 100 and 150).

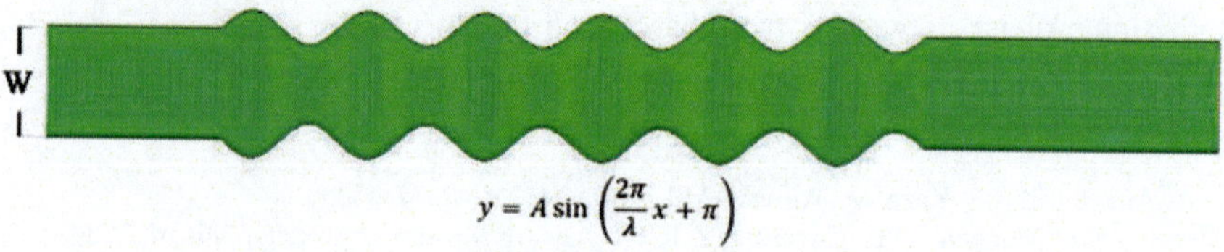

Fig. 1 Physical model of serpentine and raccoon channel

2 Numerical Simulation

A two-dimensional forced convective flow and heat transfer was modelled using a commercial CFD solver ANSYS Fluent. To simplify the analysis, the following assumptions are made

1. Steady-state, Laminar flow;
2. Incompressible fluid;
3. Constant fluid properties;
4. Negligible viscous dissipation;
5. Negligible radiative and natural convective heat transfer from the micro channel heat sink.

The conservative form of steady-state mass, momentum and energy equations is given by:

$$\nabla.(\rho u) = 0 \tag{1}$$

$$(u.\nabla)\rho u = -\nabla p + \mu\nabla^2 u \tag{2}$$

$$u.\nabla T = \frac{K}{\rho C_{\mathrm{p}}}\nabla^2 T \tag{3}$$

Energy equation in the solid domain is given by:

$$\nabla^2 T = 0 \tag{4}$$

2.1 *Validation*

To ensure the accuracy of the results obtained in this work, local Nusselt number values as obtained by Wang and Chen [5] were reproduced for steady Newtonian fluid flow with Pr = 6.93 at a constant wavy wall temperature of 301 K. The results were reproduced for the wavy channel with A = 0 mm and 200 mm at Re = 500. Figure 2 shows a close agreement between the two results and ensures the accuracy of present study.

2.2 *Boundary Conditions*

For simulation, a velocity-inlet condition was specified according to the Reynolds number at 300 K. It was given a pressure-outlet boundary condition, assuming that

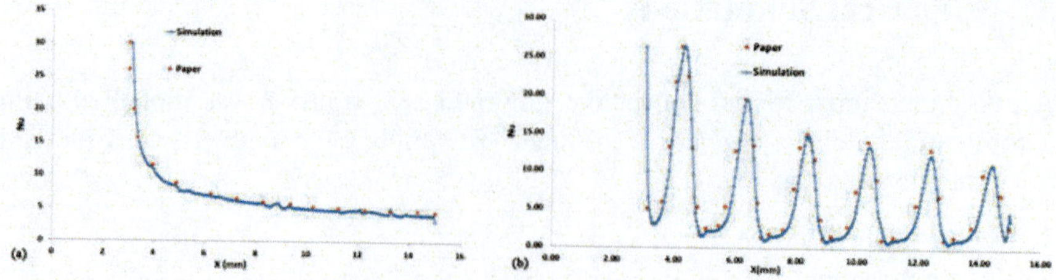

Fig. 2 Variation of Nu along channel length x for **a** straight channel **b** wavy channel

flow at the outlet is fully developed and 100% of fluid is discharged from the outlet. The walls except the wavy section were assumed to be insulated. The upper and lower wavy wall is given a pre-specified heat flux of 20 W/m^2. The fluid is assumed to be liquid water having $Pr = 7$, and solid surfaces of the channel were assumed to be made up of aluminium.

2.3 *Grid Independency*

An All-Quad mapped mesh as illustrated in Fig. 3 was used for all the simulations. Grid sizes ranging from 10,000 to 2 million nodes were employed to ensure the accuracy of the results, independent of the mesh size. A difference of 0.9 and 1.1% was found in the flow pressure drop and Nusselt number between the solutions with 0.4 million nodes and 2 million nodes, respectively. To save the computing time, grid size of around 0.4 million nodes was employed. A SIMPLEC method was used to achieve velocity–pressure coupling. A relative error ϵ of around 10^{-8} was assumed as the convergence criteria for velocity and pressure variables, while $\epsilon < 10^{-12}$ was assumed for temperature to stop the iterating procedure.

Fig. 3 Illustration of meshing in serpentine and racoon channel

2.4 *Performance Factor (PF)*

A performance factor was considered to compare the overall performance of the wavy channel with that of a straight channel with constant pumping power condition as defined in Ref. [6]. It describes the overall performance by simultaneously considering a rise in both the pressure drop and heat transfer. The wall averaged Nusselt number and pressure drop were used to evaluate the PF as given by:

$$\text{P.F} = \frac{\left(\frac{\text{Nu}_\text{w}}{\text{Nu}_\text{s}}\right)}{\left(\frac{\Delta P_\text{w}}{\Delta P_\text{s}}\right)^{1/3}} \tag{5}$$

3 Results and Discussions

The parametric analysis was carried for the configurations with different range of variables as given in Table 1.

3.1 *Effect of Wave Amplitude with Reynolds Number*

Serpentine Channel Figure 4 shows the velocity vectors of serpentine channel at different amplitudes. It can be seen that on increasing the amplitude, more vortices are

Table 1 Range of variables for parametric study

S. No.	Variables	Constant	Channel
1.	A (μm) = 0, 300, 500, 700	For λ = 1 mm	Serpentine and racoon channel
2.	λ (mm) = 1, 2, 3	For A = 300 μm	Serpentine channel

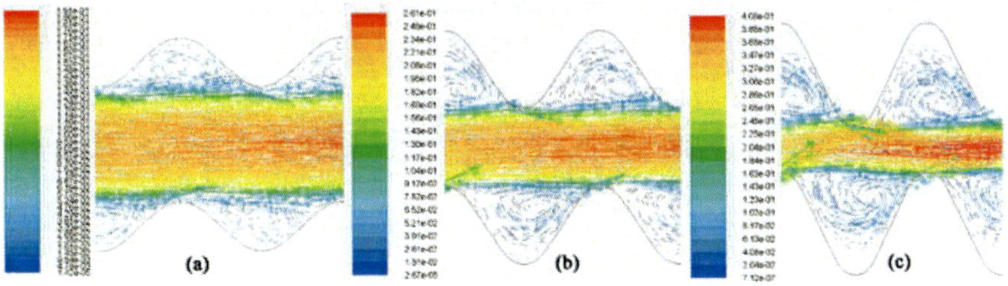

Fig. 4 Velocity vectors at Re = 100 and **a** A = 300 mm **b** A = 500 mm **c** A = 700 mm

formed near the wavy wall which enhances fluid mixing and hence the heat transfer. The increase in heat transfer through the wall is characterised by an increase in Nusselt number as shown in Fig. 5a. It can also be seen that the effect of increasing amplitude is significant after 300 μm. Further, on increasing the Reynolds number for a given amplitude, the density of vortices increases and therefore Nusselt number. However, on increasing the amplitude as well as Reynolds number, there is a simultaneous increase in pressure drop as shown in Fig. 5b. The adverse effect of an increase in pressure drop for higher amplitude and Reynolds number is nullified by an improved heat transfer as the PF is still greater than unity as shown in Fig. 6. Figure 7 shows that local Nusselt number for higher amplitude is greater than the lower one at the same axial positions due to which maximum wall temperature will also decrease. A fully developed parabolic Poiseuille flow type behaviour was observed with recirculation and appears to be dominated by the viscous forces. The hydrodynamic boundary layer and thereby thermal boundary (as Pr = constant) were thinned as a consequence of an increase in the amplitude leading to significant improvement in thermal performance as shown in Fig. 8.

Racoon Channel Racoon channel shows a similar behaviour as shown by the serpentine channel for heat transfer and pressure drop. Figure 9 shows the velocity vectors of racoon channel at different amplitude. It can be seen that dense vortices

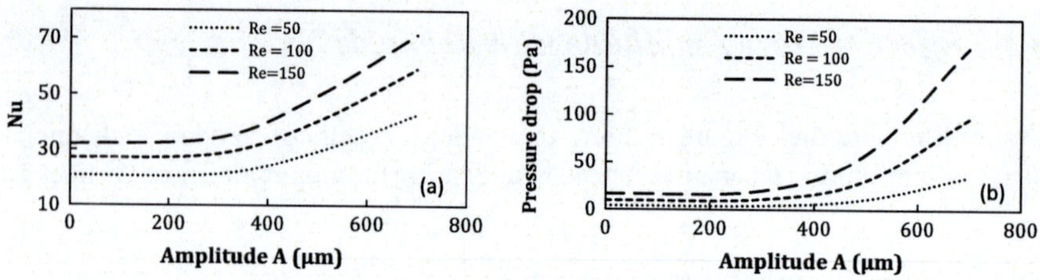

Fig. 5 Variation in **a** average Nusselt number and **b** pressure drop with amplitude for serpentine channel

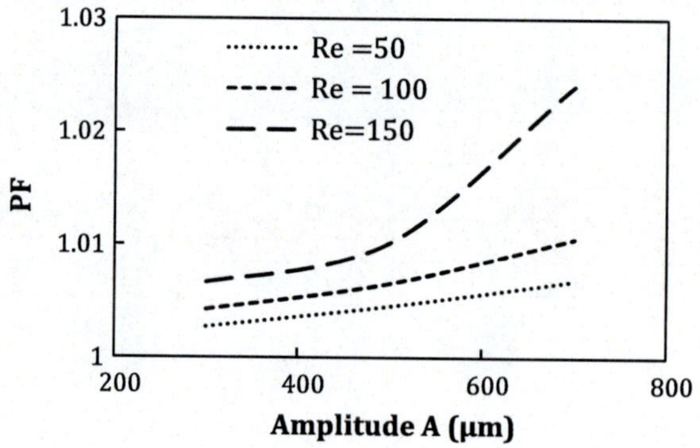

Fig. 6 Variation of PF with amplitude *A*

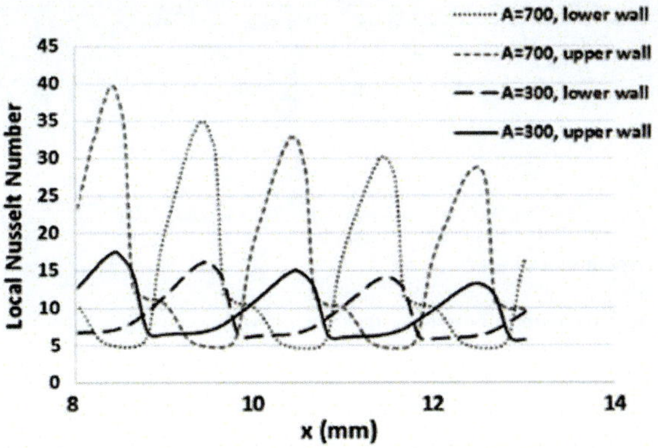

Fig. 7 Local Nusselt number plot

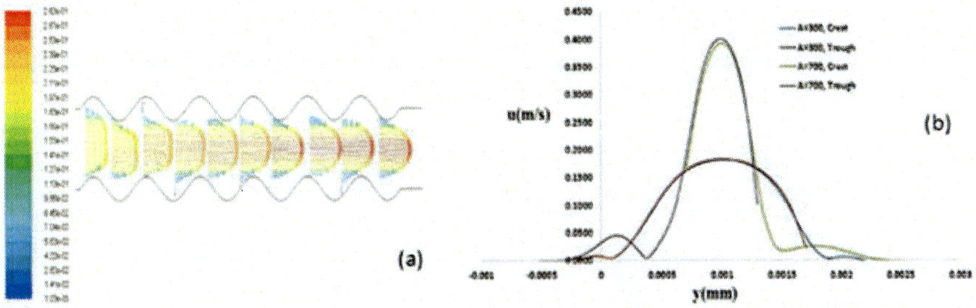

Fig. 8 Velocity vectors for **a** $A = 300$ mm **b** velocity profile at Re = 150

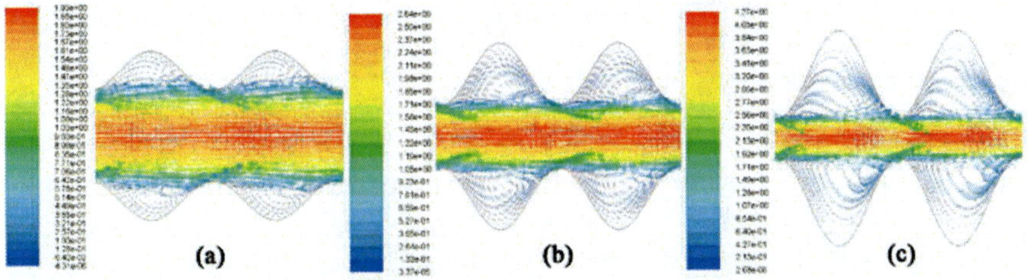

Fig. 9 Velocity vectors at Re = 100 and **a** $A = 300$ mm **b** $A = 500$ mm **c** $A = 700$ mm

were formed near the wavy wall and therefore enhances the heat transfer which can also be observed in Fig. 10a. However, improvement in the heat transfer by racoon channel as compared to the serpentine channel was quite marginal. Also, there was a quite sharp increase in pressure drop for the channels as the amplitude is increased beyond 400 μm as the narrow contacting sections of the channel contribute to a significant increase in the pressure drop as shown in Fig. 10b. Figure 11 confirms the existence of secondary flow by negative values of flow directional velocity for

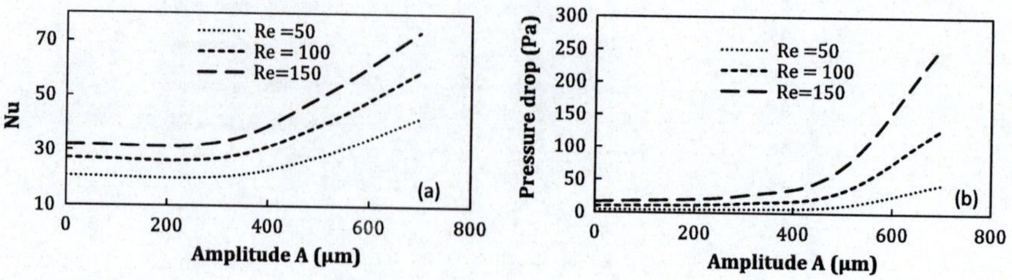

Fig. 10 Variation in **a** average Nusselt number and **b** pressure drop with amplitude for racoon channel

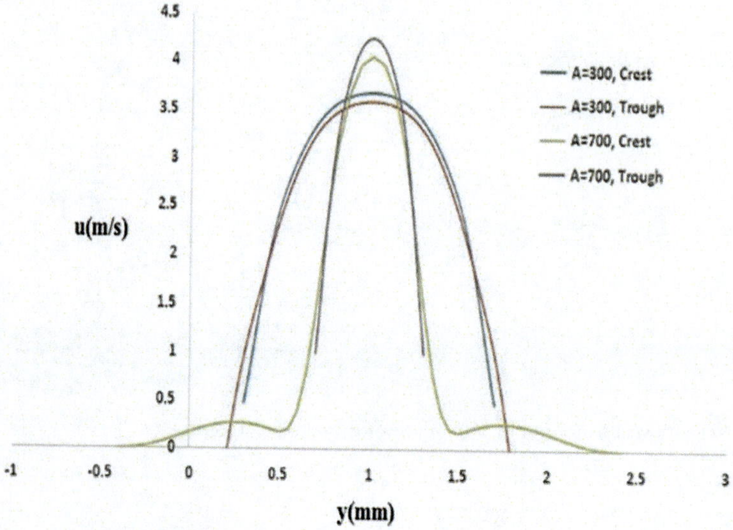

Fig. 11 Velocity profile at different amplitude

high amplitudes only as there were no significant geometry variations in contracting and expanding sections. Figure 12 shows the local Nusselt number to be greater for higher amplitude. However, the overall performance was still less than that of the straight channel as PF is below unity because the adverse effect of pressure drop overshadows the positive effect of heat transfer enhancement as observed in Fig. 13. So, it can be inferred that serpentine channel provides better overall performance than the racoon channel.

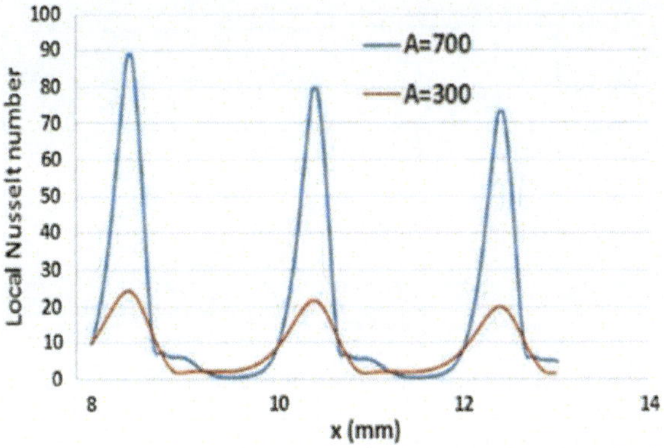

Fig. 12 Local Nusselt number plot

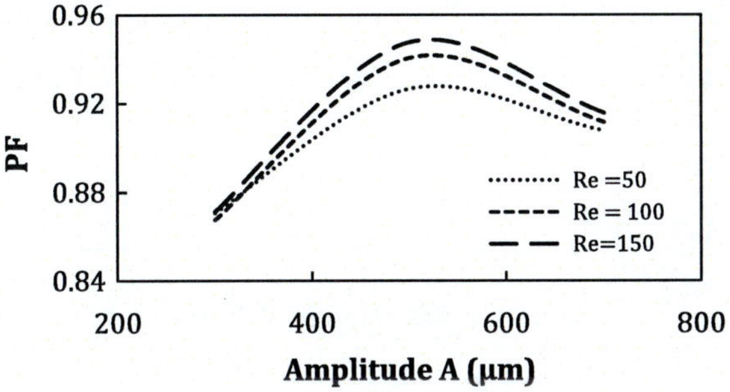

Fig. 13 Variation of PF with amplitude *A*

3.2 *Effect of Wavelength with Reynolds Number*

The effect of wavelength was analysed for serpentine channel at 300 μm wave amplitude. It was observed that on increasing the wavelength, both the heat transfer and pressure drop show a decreasing trend as observed in Figs. 14 and 15, respectively. The effect of a decrease in wavelength is analogous to an increase in amplitude of the wavy channel. A decrease in wavelength leads to the formation of vortices near the wavy wall resulting in thinning of the velocity boundary layer and consequently the thermal boundary layer. Due to the enhanced heat transfer, wall temperature will also decrease consequently for a decrease in wavelength and increase in Re. The adverse effect of a rise in pressure drop overshadows the improvement in the heat transfer for channels with higher wavelength as the PF is less than unity. From Fig. 16, it can be observed that beyond 1 mm wavelength, the overall performance decreases for the channel.

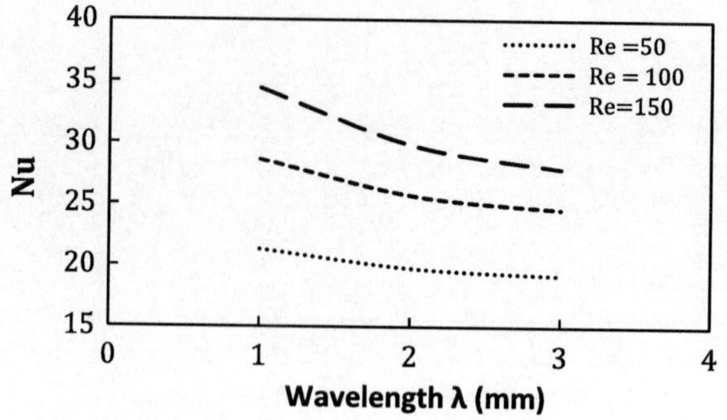

Fig. 14 Variation of average Nusselt number with wavelength

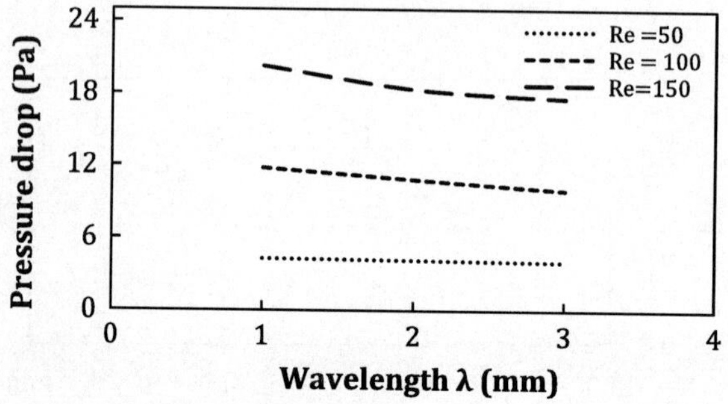

Fig. 15 Variation of pressure drop with wavelength

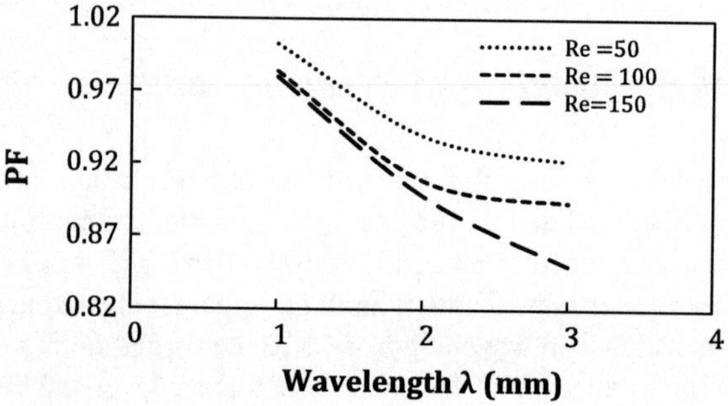

Fig. 16 Variation of PF with wavelength

4 Conclusion

This work is concerned with the improvement in the heat transfer from the wavy walls of the micro channel employed as a heat sink in electronic devices for chip cooling. The study was made for low Reynolds number flow using a two-dimensional commercial CFD software using ANSYS Fluent. Rectangular cross section channels with a constant hydraulic diameter of 1 mm were considered at different amplitude and wavelength. It was found that wavy surfaces with proper geometry and flow parameters can be employed as a potential alternate for performance improvement in low Re laminar flow regime without employing any external mixing devices. At 1 mm wavelength, the rate of heat transfer is almost the same for both channels for different wave amplitude at a given Re. However, the pressure drop in the racoon channel is greater than the serpentine channel, and the difference is magnified with increasing amplitude. Also, PF for serpentine channel is greater than unity and less than unity for racoon channel. Thus, serpentine channel performs better than raccoon channel as the improvement in heat transfer takes place at the cost of pressure drop in the latter case. Further, at 300 μm wave amplitude of the serpentine channel, rate of heat transfer decreases with an increase in wavelength at a given Re, whereas the pressure drop remains almost constant with wavelength. However, beyond 1 mm wavelength, PF is less than unity.

Therefore, it was found that the serpentine channels with lower wavelength and higher amplitude were the ideal geometry for overall performance improvement as the configuration leads to thinning of both the boundary layer at the wave peaks as well as fluid mixing and thereby enhancing the overall performance. With respect to straight channel, the best configuration was obtained for serpentine channels with wavelength = 1 mm and amplitude = 700 μm with an improved PF of 2.34% at 150 Reynolds number. It also suggests that channels with high relative waviness can also be designed at high heat flux regions for hot spot mitigation purposes.

References

1. Tuckerman DB, Pease RFW (1981) High-performance heat-sinking for VLSI. IEEE Electr Device Lett 2(5):126–129
2. Kandlikar SG, Grande WJ (2004) Evaluation of single phase flow in micro channels for high heat flux chip cooling—thermohydraulic performance enhancement and fabrication technology. Heat Transfer Eng 25(8):5–16
3. Kowalewski TA, Szumbarski J, Blonski S (2008) Low-Reynolds-number instability of the laminar flow between wavy walls. In: Proceedings of the sixth international ASME conference on nanochannels, microchannels and minichannels, Darmstadt, Germany, 23–25 June
4. Sui Y, Teo CJ, Lee PS, Chew YT (2010) Fluid flow and heat transfer in wavy microchannel. Int J Heat Mass Transf 53:2760–2772
5. Wang CC, Chen C-K (2002) Forced convection in wavy-wall channel. Int J Heat Mass Transf 45:2587–2595
6. Gee DL, Webb RL (1980) Forced convection heat transfer in helically rib-roughened tubes. Int J Heat Mass Transf 23:1127–1136

A Novel Heuristic for Estimating Two-Colorability for Conflict-Free Coloring in Simple Graphs

Rishabh Singhal, Ayushi Sengar, V. Prem Prakash, and Mayank Agrawal

1 Introduction

Graph coloring is a special case of graph labeling, wherein labels conventionally referred to as 'colors' are assigned to graph elements, subject to some constraints. Basically, graph coloring aims to assign colors to graph vertices in such a manner that adjacent vertices are not assigned the same color. In this context, the chromatic number refers to the smallest number of colors that can be used to cover all the vertices of the graph.

This process of simple graph coloring has a serious drawback, especially when the problem of assigning different frequencies to the n base stations shows up, involving a number of colors to be assigned to each vertex. The traditional approach to graph coloring needs large values of chromatic number, which is not desirable due to the high costs associated [1]. Hence, in order to further bring down the number of frequency assignments such that there is no interference of frequencies, conflict-free coloring comes into use.

Conflict-free coloring is the process of coloring vertices of a graph such that every vertex has a unique color when compared to other vertices in its neighborhood. The main purpose is to reduce the chromatic number of a graph. We consider two cases:

R. Singhal (✉) · A. Sengar · V. Prem Prakash · M. Agrawal
Dayalbagh Educational Institute, Agra, India
e-mail: r.singhal.com@gmail.com

A. Sengar
e-mail: ayushisengar48@gmail.com

V. Prem Prakash
e-mail: vpremprakash@dei.ac.in

M. Agrawal
e-mail: dei.mayank.agrawal@gmail.com

V. H. Saran and R. K. Misra (eds.), *Advances in Systems Engineering*, Lecture Notes in Mechanical Engineering,
https://doi.org/10.1007/978-981-15-8025-3_33

1. Open Neighborhood, in which the concerned vertex does not belong to its own neighborhood, so each vertex is colored such that exactly one vertex in its neighborhood is uniquely colored.
2. Closed Neighborhood, in which the concerned vertex also belongs to its own neighborhood, thus requiring that each vertex is colored such that there is a vertex with a unique color including itself and its neighborhood.

Conflict-free closed neighborhood two-coloring was shown to be NP-complete by Gargano and Rescigno [2]. The recent work in this field has focused on coloring special graphs generated by specific geometric objects such as intervals [3], rectangles [4] and unit disks [5]. A heuristic for conflict-free closed neighborhood coloring has also been proposed for an interval graph with at most four colors [6].

This paper proposes a novel polynomial-time heuristic to determine, for any given graph $G = (V, E)$, if it has a conflict-free closed neighborhood chromatic number of 2. The rest of the paper is organized as follows. The proposed heuristic is described in Sect. 2, and experimental results are given in Sect. 3, and concluding remarks made in Sect. 4.

2 Proposed Heuristic

2.1 Overview

The novel heuristic proposed in this work is described in this section. In this context, some relevant definitions and terminology are given as follows.

Given a $G = (V, E)$ be a graph, for a vertex $v \in V(G)$, $N(v)$ denotes the set consisting of all vertices which are adjacent to v, called open neighborhood of v [6].

- The set $N[v] = N(v) \cup \{v\}$ is called the closed neighborhood of v.
- A coloring C_G of a graph G is called conflict-free closed neighborhood (CF-CN) coloring if for every vertex $v \in V(G)$, the set $N[v]$ has a unique color.
- A coloring C_G of a graph G is called conflict-free open neighborhood (CF-ON) coloring if for every vertex $v \in V(G)$, the set $N(v)$ has a unique color.
- The minimum value k for which there is a CF-ON (respective CF-CN) coloring of G with k colors is called the CF-ON (resp. CF-CN) chromatic number of G and is denoted as $\chi_{cf}(G)$ (respective $\chi_{cf}[G]$).

The heuristic makes use of the following ideas. Consider a closed neighborhood for a vertex, say 'X'. For 'X', we can have three possible coloring conditions:

- There are some vertices which are colored in a specific color, say color 2
- There are some vertices which are colored in a specific color, say color 3
- The remaining vertices are not assigned any color as yet.

Our objective is to now color these uncolored vertices on the same principles using just two colors—color 2 and color 3 in this case. Assume the count of color 2 be 'α'

and count of color 3 be 'β'. The relationship between 'α' and 'β' can physically be interpreted in different ways, which eventually will help us to color those uncolored vertices.

- 'α' = 'β' = 0: No vertex has been colored yet in the closed neighborhood of 'X'. Our strategy here would be to color the vertex 'X' with color 2 and remaining vertices in its neighborhood with color 3. In this way, color 2 will be unique in the closed neighborhood of 'X'. Now, if 'X' has exactly one neighbor, then it cannot be colored using only two colors. Hence, $\chi_{cf}[G] \neq 2$.
- 'α' = 1 and 'β' = 0: Color 2 is already unique, and no vertex has been assigned color 3. Thus, our task here is to assign the color 3 to all the remaining vertices in the neighborhood of 'X'. Again, if 'X' has exactly one neighbor, then it cannot be colored using only two colors. Hence, $\chi_{cf}[G] \neq 2$.
- 'α' = 0 and 'β' = 1: Color 3 is already unique, and no vertex has been assigned color 2. Thus, our task is now to assign the color 2 to all the remaining vertices in the neighborhood of 'X'. Again, if 'X' has exactly one neighbor, then it cannot be colored using only two colors. Hence, $\chi_{cf}[G] \neq 2$.
- 'α' = 0 and 'β' > 1: If we have at least one uncolored vertex, then assign one vertex color 2 and remaining vertex with color 3. If we have no uncolored vertex with this condition, then the given graph cannot be colored using two colors. Hence, $\chi_{cf}[G] \neq 2$.
- 'α' > 1 and 'β' = 0: If we have at least one uncolored vertex, then assign one vertex color 3 and remaining vertex with color 2. If we have no uncolored vertex with this condition, then the given graph cannot be colored using two colors. Hence, $\chi_{cf}[G] \neq 2$.
- 'α' = 1 and 'β' > 1: This condition is already satisfying the definition. Our job is to color the remaining vertices with color 3.
- 'α' > 1 and 'β' = 1: This case also satisfies the definition. Our job is to color the remaining vertices with color 2.
- 'α' = 1 and 'β' = 1: If there is no uncolored vertex left, then $\chi_{cf}[G] \neq 2$. But if there are some uncolored vertex left, choice has to be made between color 2 and color 3. Assign color 2 and proceed, if the graph cannot be colored in this fashion, go back to the last occurring case and then assign color 3 to validate the result.
- 'α' > 1 and 'β' > 1: In this case, graph cannot be colored using just two colors. Hence, $\chi_{cf}[G] \neq 2$.

The proposed heuristic operates using the above simple rules. The closed neighborhood is accessed by the rows of the adjacency matrix of the generated graph, taken one at a time. Before actually accessing the adjacency matrix, replace all the diagonal elements with '1' so as to include that vertex also in the conflict-free closed neighborhood coloring. The neighborhood of 'X' is simply defined as the non-zero columns of the matrix for the row 'X'. For each row taken under operation, assign two counters to count color 2 and color 3 assigned with nodes in row 'X' and an array to hold the nodes position, which is simply the column number, that are not yet been assigned any color and are neighbor of 'X'. Whenever we assign any node a color, replace '1' in the columns of the adjacency matrix for the corresponding row with that color—'2' or '3', in our case. In next step, replace all the row elements with entry '1' for that specific column whose color has been changed, with the color assigned. It will help us to calculate number of color 2 and color 3 assigned to nodes when we change the neighborhood. Repeat the above steps '*N*' times, where '*N*' is the number of nodes. Pseudo-code for the proposed heuristic is given in listing 1.

2.2 Erdos–Renyi Graph Model

The test graphs used in the experimental work in this paper were generated using the Erdos–Renyi graph model [7]. Specifically, the model used is the *G*(*n*, *p*) model, wherein a graph is constructed by connecting nodes randomly. Each edge is included in the graph with probability *p* independent from every other edge. Equivalently, all graphs with *n* nodes and *M* edges have equal probability of

$$p^M(1-p)^{\binom{n}{2}C - M} \tag{1}$$

The parameter *p* in this model can be thought of as a weighting function; as *p* increases from 0 to 1, the model becomes more likely to include graphs with more edges and relatively less likely to include graphs with fewer edges. In particular, the case $p = 0.5$ corresponds to the case where all $2^{\binom{n}{2}C}$ graphs on *n* vertices are chosen with equal probability.

Listing 1

```
Function conflictfree(Adjacency Matrix amat, Nodes)
{
Array: nodecolor[nodes] = {0}
amat[i][i] = 1
Repeat from 0 to nodes-1
{
Array: nocol [nodes] = {0} // position ofuncolored
nodes
int col2 = 0, col3 = 0 // count of nodes with color
2 & 3
int flag = 0 // avoid repetitive coloring of same
node
IF (col2 = 1), u2 = 1
ELSE IF (col3 = 1 and u2 ≠ 1), u3 = 1.

  Repeat from 0 to length(nocol) -1
  {
```

```
{
IF (u2 ≠ 1 and flag = 0); color of first uncolored
node = 2, update nodecolor, change all the values of
corresponding rows to 2 of that column and set flag
to 1.
IF (col2 > 1 and col3 ≠ 0), u3 = 1.
IF (col2 = 1 or (col2 ≠ 1 and u3 ≠ 1)), u2 = 1
}
{
IF (u3 ≠ 1 and flag = 0); color of first uncolored
node = 3, update nodecolor, change all the values of
corresponding rows to 3 of that column and set flag
to 1.
IF (col3 > 1 and col2 ≠ 0), u2 = 1.
IF (col2 = 1 or (col2 ≠ 1 and u2 ≠ 1)), u3 = 1
}
Flag = 0
}

  Col2 = 0, col3 = 0

}
Check the conditions that could violate conflict
free coloring. If found return "Not possible", else
return the array of colors with elements at diagonal
of adjacency matrix.
}
```

3 Experimental Work

The proposed heuristic was implemented in Python 3.6 environment: idle (Python 3.6 64-bit), OS: Windows 10 Home, Processor: Intel (R) Core (TM) i5-7200U CPU @ 2.5 GHz, RAM: 8.00 GB, System Type: 64-bit OS.

For $n = 5, p = 0.7$ the resultant colored graph is shown in Fig. 1.

Likewise, different set of probabilities, number of nodes and the computation time, several graphs were plotted.

The result of graphs generated using Erdos–Renyi is summarized in Table 1. The first column indicates the number of nodes in given graph and the second indicates probability of generating edge between two pair of nodes. Column three represents the time taken (seconds) to compute the result and the last column indicates the computational result: ‘positive’ indicates that the graph can be colored and ‘negative’ means that graph cannot be colored with two colors.

The heuristic took 0.00157 s to determine that a graph with 10 nodes and edge probability of 0.33 can be two-colored (Fig. 2). Next, it took 0.01562 s to evaluate a graph of 50 nodes with edge probability of 0.45. Similarly, various other instances were also evaluated likewise. Lastly, a graph of 1000 nodes with 2,499,500 edges was tested in 0.46362 s.

Some cyclic graphs were generated and their computational result is summarized in Table 2. The first column indicates number of nodes followed by computational

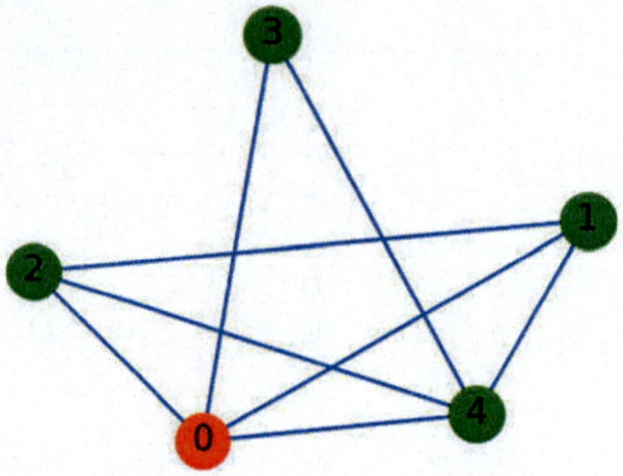

Fig. 1 Conflict-free closed neighborhood-colored five-node graph

Table 1 Test cases for Erdos–Renyi graphs

Nodes	Probability	Computation time (s)	Result
10	$p = 0.33$	0.0015784	Positive
50	$p = 0.45$	0.01562	Negative
500	$p = 0.7$	0.11591	Negative
846	$p = 0.8$	0.2853	Negative
500	$p = 1$	0.11269	Positive
1000	$p = 1$	0.46362	Positive

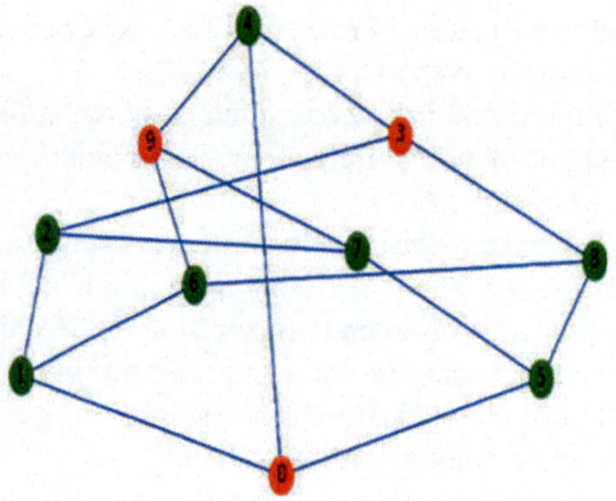

Fig. 2 Conflict-free closed neighborhood-colored 10-node graph

Table 2 Test cases for cyclic graphs

Nodes	Computation time (s)	Result
20	0.00565	Positive
400	0.10025	Positive
555	0.15371	Positive
1009	0.4168	Positive
10,000	46.5499	Positive

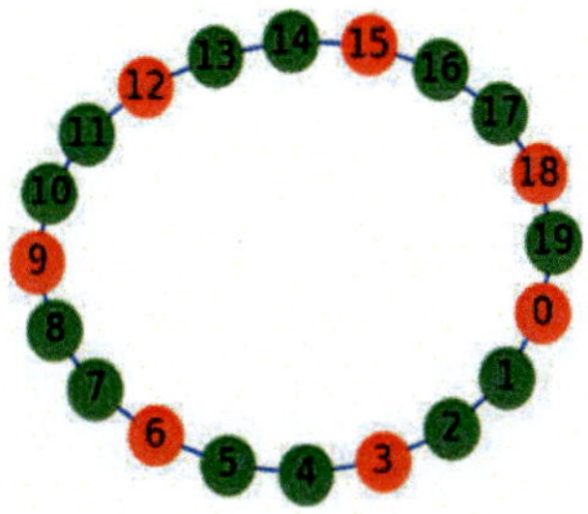

Fig. 3 Cyclic graph with 20 nodes

time (seconds) and the last column gives the result.

- Cyclic graph of 20 nodes was generated, and heuristic gave positive result in 0.00565 secs. Coloring pattern is shown in Fig. 3.
- Similarly, odd cycles of 555 nodes and 1009 nodes and even cycles of 400 and 10,000 nodes were generated and were tested.

Further, cyclic graphs with nodes ranging from 1000 to 15,000 were plotted against computational time in seconds (Fig. 4).

Graphs with nodes ranging from 100 to 5000 were generated using Erdos-Renyi graph with $p = 1$ (Fig. 5) and $p = 0.5$ (Fig. 6). Working of heuristic on those nodes were plotted against computational time.

4 Conclusions

Conflict-free coloring in closed neighborhood as presented in this paper is one possible generalization of the traditional approach to graph coloring. Conflict-free coloring of graphs finds practical applications in the cellular network domain for problems ranging from radio frequency identification to frequency assignment. The

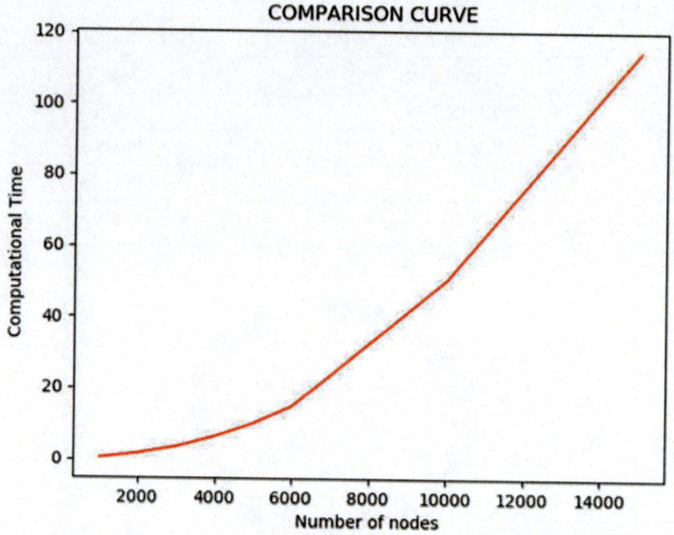

Fig. 4 Compare size of graph against computational time (seconds) for cyclic graph

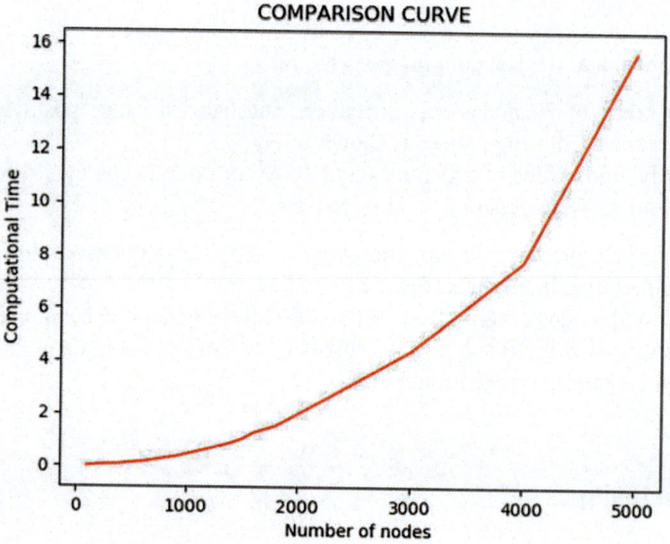

Fig. 5 Compare size of graph against computational time (seconds) for Erdos–Renyi graph with edge probability 1

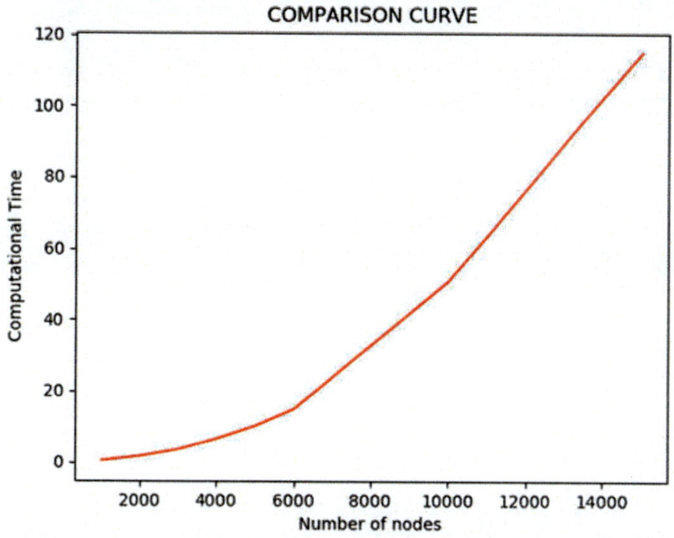

Fig. 6 Compare size of graph against computational time (seconds) for Erdos–Renyi graph with edge probability 0.5

problem of two-coloring a conflict-free closed neighborhood is known to be NP-complete. The novel heuristic presented in this work solves this problem in polynomial time with big O complexity of $O(n^3)$, and the results are given on a wide range of random graphs instances.

References

1. Cheilaris P (2009) Conflict-free coloring. Dissertation, City University of New York, p 5
2. Gargano L, Rescigno AA (2015) Complexity of conflict-free colorings of graphs. Theoret Comput Sci 566:39–49
3. Bar-Noy A, Cheilaris P, Smorodinsky S (2008) Deterministic conflict-free coloring for intervals: from offline to online. ACM Trans Algorithms (TALG) 4(4):44
4. Ajwani D, Elbassion K, Govindarajan S, Ray S (2007) Conflict-free coloring for rectangle ranges using $O(n^{0.382})$ colors. In: Proceedings of the nineteenth annual ACM symposium on parallel algorithms and architectures. ACM, pp 181–187
5. Lev-Tov N, Peleg D (2009) Conflict-free coloring of unit disks. Discrete Appl Math 157(7):1521–1532
6. Reddy V (2017) Parameterized algorithms for conflict-free colorings of graphs. arXiv:1710.00223v1 [cs.DS], 30 Sept 2017
7. Erdős P, Rényi A (1959) On random graphs. I (PDF). Publ Math 6:290–297

Exploring Antecendents of FinTech Adoption Using Adapted Technology Acceptance Model

Shubhangi Singh, Marshal M. Sahni, and Raj K. Kovid

1 Introduction

Owing to multiple technological innovations and their ubiquity, financial service industry is witnessing a disruptive structural change. FinTech is understood as a technological innovation that is challenging the traditional banking and financial companies [1]. Consumers are drifting away from conventional payment methods to FinTech which offers an enhanced and efficient client experience. FinTech has become an important dimension of financial services industry because of constant innovations. Although there is a growth in number of FinTech users, but there is no conclusive research findings which determine attributes affecting FinTech adoption and usage. However, both academic researches and global evidence support that FinTech services provide increased personalization, flexibility and ease of delivery of financial services [2], which in turn lead to higher productivity, profitability and reach of financial services [3]. Fintech services have improved efficiency, reduced risk and contributed to inclusive growth [4]. These technological innovations have the potential to substantially influence traditional business models of highly regulated financial services industry [5].

However, we see selective adoption of FinTech services. To put things in perspective, at present, though there are numerous FinTech services available, yet only few of them have been proven successful. Success and large-scale adoption of any technology is largely determined by the willingness of potential users to adopt technological innovation [6]. For this, actual Usage is the necessary precursor and is thus adopted as measure of performance for technology products or services.

S. Singh (✉) · M. M. Sahni
Amity University, Noida, India
e-mail: shubhangi.singh@hotmail.com

R. K. Kovid
Sharda University, Noida, India

V. H. Saran and R. K. Misra (eds.), *Advances in Systems Engineering*,
Lecture Notes in Mechanical Engineering,
https://doi.org/10.1007/978-981-15-8025-3_34

The previous researchers investigate FinTech adoption through acceptance of technology. Most of the existing literature on technology acceptance uses the most tested and established model, the technology acceptance model (TAM) [7, 8]. It states that perception of ease of use and usefulness of technology predicts attitude and subsequent acceptance and use of technology [7]. TAM 2 (a revision of TAM) [9] and unified theory of technology acceptance and use of technology (UTAUT) [10] are also other popular technology acceptance models. TAM and UTAUT models are adapted from theory of reasoned action (TRA) [11] and theory of planned behavior (TPB) [12, 13] and thus, share several features with them. Some studies have also noticed its link with theory of diffusion of innovation [6]. The TAM model [7] is proposed based on these variables using perceived ease of use (PEU) and perceived usefulness (PU), factors assessed through a few short questions. It proposed that attitude helps in evaluating intention to perform a behavior, thereby resulting in use of innovation.

In many studies, additional variables have been introduced to increase the predictive power of TAM variables. Research suggests that security, responsiveness and social influence also significantly influence acceptance of technology [14–16]. For FinTech service providers, the pertinent question is: what attributes enhanced consumer perceptions that make FinTech services easy, useful, secure and responsive? This study examines FinTech adoption and usage from technology acceptance perspective adding constructs for security, responsiveness and social influence are added as antecedents of behavior intention and FinTech usage. The antecedents of FinTech usage intention and actual usage are investigated through a comprehensive framework weaving sub-constructs of TAM, UTAUT and ServPerf. In Sect. 2, review of literature is presented on technology adoption and usage explaining various factors affecting it. Based on this review, the study proposes a model of customer intention to use FinTech and formulate associated research hypothesis. Research methodology is discussed in Sect. 3 and Sect. 4 discusses the research findings of the study derived from the analysis of empirical data. Section 5 concludes the paper with discussion, limitation of the study, managerial implications and future research directions.

The conclusions generated by this study will be useful in many ways. First, it adds to the technology acceptance literature by incorporating relevant adoption beliefs and investigating the antecedents of FinTech usage. Second, a thorough investigation of the impact of user reaction to FinTech usage and intention would help FinTech companies learn to entice users for frequent use and attract non-users to do their first online financial transaction. Lastly, by knowing which attributes influence user belief to improve FinTech services.

2 Review of Literature and Proposed Hypotheses

With the number of FinTech service providers offering and improving FinTech services, it is imperative to study the factors affecting adoption and use of these services. For this study, TAM [7, 8] provides the conceptual framework and is discussed with two main objectives: first, to investigate antecedents of FinTech usage and second, to investigate user perception on adoption and usage of FinTech services. TAM suggests that two beliefs, i.e., perceived ease of use and perceived usefulness, determine a user's initiative toward using a new technology, which in turn determine their intention to use it. TAM is validated by various studies as a robust framework to understand user's adoption of technology in varied contexts, such as banking technology, m-commerce, email, online games, lotus notes, desktop video conferencing, telemedicine technology and so on. Since FinTech is a retail format of innovation, TAM provides framework for this study to investigate consumers' acceptance of FinTech.

FinTech helps in assessing the rapid development of the financial system and financial institutions. It has made consumption of financial services convenient by doing the technological advancement in fundamental services and building new applications for delivery such as making payments, saving, borrowing, managing risk and seeking financial advice [17]. With digital transformations in other industries, there is an increased demand of technology-based financial solutions by consumers [18]. FinTech companies are meeting these consumer demands with convenient and less costly ways to transfer, borrow or invest money [19]. Now, FinTech is not limited to banking services and investment fund but is adopted by retail groups and telecom operators who are innovating to offer financial services through their existing networks.

2.1 Behavior Intention to Use FinTech

With the advent of FinTech and its integration with brick-and-mortar financial services industry, behavior intention for use has become an important dimension to indicate the possibility of use and adoption of FinTech services by the users [20]. The original TAM model [7] purports that behavioral intention affects usage pattern. Pace of technological advancement in financial services vis-à-vis level of consumer awareness has significant influence on a person's behavioral intention. Also, if the technology advancement is at a higher rate than consumer awareness and use, then FinTech companies may not be able to reap the benefits of the innovation or the gestation time to earn profits will increase for them [21]. Hence, technology use and adoption have gained researchers' attention, and several theories and models have been proposed to study behavior intention [22].

Various studies have identified that behavioral intention is influenced by many factors. Accessibility of technology, information about its utility and usage, and then the direct experience of usage of technology enable users to form stable behavior intention for continued use in the future [23]. Other factors like performance expectancy, effort expectancy, social influences and facilitating conditions are important factors which influence behavior intention [10]. Behavior intention was also assessed through behavioral and technological factors with two main constructs—ease of use and usefulness [7]. In this study, actual usage is the dependent variable affected by behavior intention. Actual usage is the frequency and approximate number of times FinTech services is used in a given period. As the relationship between intention and actual usage in consumer perspective is not tested in the previous literature, thus, the first hypothesis is proposed as:

H1: Behavior intention positively affects use of FinTech services.

2.2 *Perceived Usefulness*

According to TAM, perceived usefulness is the degree to which one believes that using the technology will enhance his/her performance [8]. There are many studies proving the significant impact of perceived usefulness on behavior intention [9, 10, 24]. Perceived usefulness is a crucial construct in determining adoption of technology [25, 26]. Because of ubiquity of FinTech services coupled with convenience, higher the perceived usefulness, higher is the behavior intention for adoption and usage of FinTech services. Therefore, the third hypothesis is proposed as:

H2: Perceived usefulness positively affects behavior intention (H2a) and use (H2b) of FinTech services.

2.3 *Perceived Ease of Use*

Perceived ease of use is the degree to which one believes that using the technology will be free of effort [8]. It is further defined as the extent of effort a technology needs to make it easy to use [9]. Also, TAM further asserts that perceived ease of use is the key factor explaining the variance in perceived usefulness. It is posited that ease of use has a positive and direct effect on behavior intention to reuse technology services [9, 15, 27]. This, in long run, influences users to adopt technological services. It is also found that perceived ease of use enhances the prediction of use of technology services including Internet banking [28]. As there are numerous variants of FinTech services, it can be tested that degree of ease of use will affect behavior intention. Based on the previous literature and considering this as one of the major behavioral factors affecting behavior intention and use FinTech service, the second hypothesis is proposed as:

H3: Perceived ease of use positively affects perceived usefulness (H3a), behavior intention (H3b) and use (H3c) of FinTech services.

2.4 Social Influence

Social influence is the extent of influence by others to use a specific technology [29]. It is assumed that person consults his social circle about new technologies and can be influenced by them. Perception about characteristics of technology, requirement to use technology and attitude toward technology is influenced by social norms, actions and statements of significant peers, friends and family, and by their past attitude and usage. Among the existing literature, social influence is the most hypothesized and tested construct of UTAUT and there is significant relation established between social influence and its effect on behavior intention to use. While some of the studies state contradictory results [30], this may be partly due to differents mechanism of application of social influence in one's social context. Thus, it will be interesting to test the relationship between social influence and behavior intention and use, as users modify their actions according to actions of others to gain social acceptance. Thus, the fourth hypothesis is proposed as:

> H4: Social influence positively affects perceived ease of use (H4a), perceived usefulness (H4b) and use (H4c) of FinTech services.

2.5 Responsiveness

Responsiveness refers to user perception of the effectiveness and excellence of service offered online. It refers to the accuracy of the services delivered and clear, current and complete information available on the Web site. It is largely affected by Web site or interface performance about prompt service and assistance provided while doing financial transactions. Buyer perceptions about nature of interaction with the service provider are largely affected by responsiveness. It also helps customer in absence of awareness of technology use or any unforeseen event requiring effective information. Thus, effective communication is the key factor determining the perceived quality of the online service, and thus, responsiveness is also a major element of service quality [14, 31]. It influenced user assessment for perceived usefulness and ease of use of the technology and indirectly affects the user adoption [32]. It has been empirically shown to have a significant impact on quality of service and the second strongest predictor of behavior intention and attitude toward an online interface [33]. Considering its effectiveness in predicting ease of use and usefulness, sixth hypothesis is proposed as:

> H5: Responsiveness positively affects perceived ease of use (H5a) and perceived usefulness (H5b) of FinTech services.

2.6 Security

One key reason for not doing online financial transactions by users regardless of being Internet users is because of the belief about safety of the transactions [32]. Security of online transactions and reputation of the service provider are the key factors influencing belief about safety while doing financial transactions [34]. As compared with brick-and-mortar financial services providers, security is more critical while doing online financial transactions [35, 36]. Because of the absence of face-to-face interaction while using FinTech services, users feel a greater risk and uncertainty. Security is the most effective tool for reducing this risk and uncertainty [34, 37]. Many prior empirical studies included security into TAM [38–40]. Results show security to be an antecedent of ease of use [34] and usefulness [34, 41]. The enhanced security features also increase the responsiveness of services at the time of crisis or problem. Considering its effectiveness in explaining various belief behaviors, fifth hypothesis is proposed as:

H6: Security positively affects perceived ease of use (H6a), perceived usefulness (H6b) and responsiveness (H6c) of FinTech services.

3 Method

3.1 Stage I: Qualitative Assessment at Pre-test Stage

Prior to the main survey, the study did pre-test to (1) identify FinTech adoption drivers that emerge when doing financial transactions and (2) to refine and finalize main survey instruments. As security, responsiveness and social influence are key factors in many of the adoption studies for e-services [9, 14], they were selected for finalization of factors through iterative discussion and then pre-test.

Sample 79 FinTech users, which included domain experts from industry and academia, participated in an iterative discussion. This led to redefining factors and identification of various dimensions of constructs.

Instrument The iterative discussion was conducted on four broad dimensions: (1) number of FinTech services used and their usage frequency; (2) behavior intention to use FinTech service; (3) belief variables for ease of use, perceived usefulness, security, responsiveness, social influence and attitude toward FinTech usage; and (4) demographic information. All variables except for demographic information and FinTech service usage frequency were assessed on a five-point Likert scale (5 = strongly agree and 1 = strongly disagree). To assess FinTech service usage frequency, the experts were requested to give their usage frequency for each service, namely 'more than once a week,' 'weekly,' 'monthly,' 'less than once a month' or 'never.' They were also asked to name the FinTech service which is most preferred or frequently used. Davis [7, 8, 24] scale was used to measure usefulness with seven items, ease of use with eight items and behavior intention to use with three items. Five items from WebQual 4.0 [16] and three items from ServQual [14] were used to assess security, three items from UTAUT [10] were used to assess social influence and three items form ServQual [14] were used to assess responsiveness.

Findings Principal component analysis (PCA) of all the constructs was conducted with varimax rotation to identify underlying factors for FinTech adoption. The analysis identified 29 items under five factors, namely ease of use, usefulness, security, social influence and responsiveness. These factors accounted for total variance of 71.376%. A minimum Eigenvalue of 1.0 and scree plot criteria was used to decide number of factors. The scale was then refined and reworded considering PCA findings and iterative discussion with domain experts for further data collection and empirical analysis.

3.2 Stage II: Empirical Validation

The main data collection survey used the same measures refined and reworded in the pre-test. A total of 850 users were invited for online survey, which were chosen based on the past usage and/or experience of the listed FinTech service. The survey was conducted by electronic distribution of the survey questions to potential participants. The target respondents were customers of leading public and private sector banks, academicians and staff working in university or college, and post-graduate students of private universities in Mumbai and Delhi. A combination of convenient and random sampling method was used, as it is difficult to obtain the sampling frame containing the details of all the FinTech users. The universities and banks were conveniently selected, and users were randomly chosen. Response rate was 51.6% with collection of 439 responses.

4 Result

4.1 Sample Characteristics

Out of 439 respondents, 148 respondents (33.7%) were in the age group of less than 30 years. Approximately, 48% respondents were between 30 and 50 years of age who are also incidentally the major user of FinTech services [3]. Approximately, 70% of the respondents were male, and 30% were female respondents.

One hundred and ninety five (44.4%) of the respondents have more than 10 years of Internet experience. The preferred FinTech service for the respondents was Internet banking with 56% of responses. The second preferred service was payment wallet service. Except for four respondents, all respondents have used at least one of the FinTech services in the last month.

4.2 Measurement Model Results

Exploratory factor analysis (EFA) of 29 measures of FinTech adoption was again conducted with maximum likelihood extraction for further confirmatory factor analysis. Presuming correlation among FinTech adoption factors, oblique rotation method (promax) was used. Three items, one item from construct ease of use and two items from construct usefulness, were omitted for further analysis as they showed poor psychometric properties and significant cross loading. The Cronbach's alpha of the measurement model was conducted to ensure reliability. This analysis gives 26 items under the five factors extracted in PCA. These factors accounted for total explained variance of 69.965%.

Confirmatory factor analysis was done using AMOS to test the measurement model on the items remained after the EFA. One item each from ease of use and usefulness was removed from the final scale due to large standardized residual. The final set of 21 items exhibit construct reliabilities, average variance extracted (AVE) and Cronbach's alpha exceeding the recommended standard of reliability and unidimensionality (Table 1). If AVE is greater than 5.0 of the total variance, convergent validity can be established. Discriminant validity denotes that each factor is different than others, and it is established if average variance extracted (AVE) is greater than the squared correlation coefficient between factors [42] (Table 2). The goodness of fit statistics is: $\chi^2(177) = 429.256$, CMIN = 2.425, CFI = 0.967, GFI = 0.914, AGFI = 0.888, SRMR = 0.0397, RMSEA = 0.057, PCLOSE = 0.045.

4.3 Structural Model Results

The structural model suggested in this study was tested through maximum likelihood method with AMOS. The result shows an acceptable fit for the proposed model with $\chi^2(209) = 490.831$, CMIN = 2.348, CFI = 0.965, GFI = 0.910, AGFI = 0.882, SRMR = 0.0379, RMSEA = 0.055, PCLOSE = 0.077. Hypotheses test support all postulated paths except H1, H2b, H4b. (see Fig. 1 for standardized path coefficients and significance level).

Hypotheses 1 examine the impact of behavior intention on actual usage of FinTech services. The factor loading on intention—usage—(Table 3) was −0.03 (with CR = −0.331, $p = 0.741$), which was not significant, and thus, H1 is not supported. The factor loading on usefulness—usage—was −0.04 (with CR = −0.402, $p = 0.688$) and did not support hypothesis H2b. Factor loading for social influence—usefulness—was 0.03 (with CR = 1.005, $p = 0.315$) not supporting H4b. The

Table 1 Exploratory factor analysis and confirmatory factor analysis

Factor/Items	EFA					CFA		
	Factor 1	Factor 2	Factor 3	Factor 4	Factor 5	Std. factor loading	Construct reliability	AVE
Factor 1: Ease of use							0.921	*0.661*
Easy to learn	0.904	0.173	0.017	−0.146	−0.122	0.85		
Easy to browse	0.929	−0.105	−0.004	−0.056	0.050	0.84		
Easy to use	0.894	0.015	−0.068	0.020	−0.068	0.86		
Create positive experience	0.679	−0.086	0.010	0.163	0.077	0.77		
Clear and comprehensible interaction	0.723	0.073	0.025	−0.020	0.037	0.78		
Has attractive layout and appearance	0.640	−0.025	0.014	0.063	0.070	NA		
Easy to become skillful for use	0.723	−0.039	0.015	0.071	0.056	0.77		
Factor 2: Security							0.928	*0.720*
Has good reputation	0.127	0.687	−0.027	0.093	−0.075	0.78		
Safe to complete financial transaction	−0.058	0.948	−0.028	−0.045	0.006	0.88		
Gives the feeling of safety for personal information	−0.036	0.870	−0.028	−0.070	0.067	0.81		
Service is trustworthy	0.120	0.670	0.002	0.071	0.064	0.89		
Has adequate security features	−0.018	0.816	0.048	0.053	0.029	0.87		
Factor 3: Social influence							0.952	*0.868*
Family and friends insist to use	−0.020	−0.052	0.919	0.012	0.011	0.90		
Colleagues and peers insist to use	0.017	−0.021	0.987	0.006	−0.011	0.99		

(continued)

Table 1 (continued)

Factor/Items	EFA					CFA		
	Factor 1	Factor 2	Factor 3	Factor 4	Factor 5	Std. factor loading	Construct reliability	AVE
Spouse insist to use	−0.001	0.045	0.909	−0.046	0.002	0.90		
Factor 4: Usefulness							0.893	*0.676*
Useful in daily life for doing financial transactions	0.193	0.171	−0.031	0.515	0.021	0.81		
Enhance independence for doing financial transactions	−0.074	0.078	−0.015	0.731	0.057	0.76		
Helps to do financial transactions quickly	0.055	−0.029	−0.053	0.908	−0.028	0.87		
Increase productivity	0.031	−0.062	0.005	0.872	−0.013	0.84		
Have necessary knowledge to do fin. transactions	0.094	0.188	0.177	0.416	−0.060	NA		
Factor 5: Responsiveness							*0.863*	*0.679*
Give info about time taken to complete financial transaction	0.067	−0.010	0.020	0.089	0.694	0.82		
Give prompt service as promised	0.000	0.024	−0.027	−0.063	0.980	0.89		
Helps at the time of problem or issue	−0.004	0.191	0.027	0.010	0.581	0.76		
Eigen value	11.269	2.398	1.704	1.067	1.005			
Variance %	35.815	21.185	5.733	4.020	3.212			
Cronbach alpha	0.923	0.922	0.950	0.890	0.858			

These are the factors (constructs) used in study affecting the FinTech Adoption. These factors are explained in literature review section.

Table 2 Convergent and discriminant validity

	Social influence	Usefulness	Ease of use	Security	Responsiveness
Social influence	**0.932**				
Usefulness	0.316	**0.822**			
Ease of use	0.349	0.758	**0.813**		
Security	0.295	0.733	0.672	**0.849**	
Responsiveness	0.225	0.656	0.595	0.718	**0.824**

Note The numbers in the diagonal is AVE of each construct. Numbers below diagonal are squared correlation coefficients between constructs

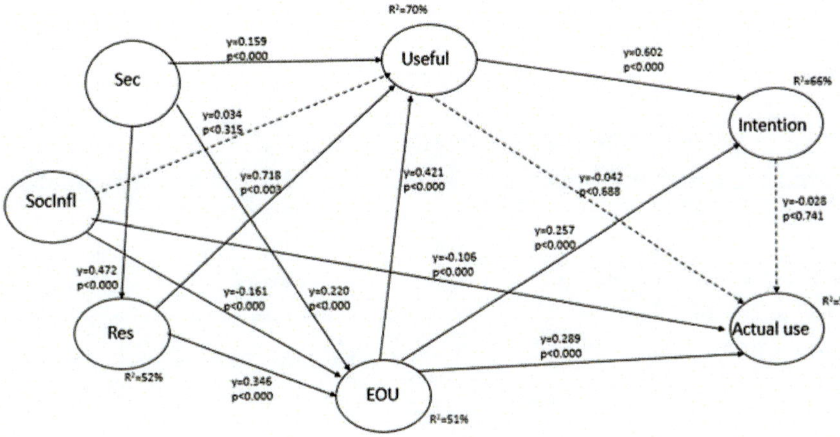

Fig. 1 Research model for antecedent of usage and behavior intention TAM. $\chi^2(209) = 490.831$, CMIN = 2.348, CFI = 0.965, GFI = 0.910, AGFI = 0.882, SRMR = 0.0379, RMSEA = 0.055, PCLOSE = 0.077. *Note* Solid paths are significant. Dotted paths are insignificant

result for having no significant impact of behavior intention on actual usage is not surprising [43, 44] as using behavior intention does not predict the actual usage. Hypotheses 2a–2b examine the influence of perceived usefulness on behavior intention and actual usage. Perceived usefulness has significant impact on intention ($\gamma = 0.60$, CR = 10.481), but not on the actual usage ($\gamma = -0.04$, CR = −0.402) for FinTech services. So, while hypothesis 2a is supported, hypothesis 2b is not supported.

Hypotheses 3a–3c examine the influence of ease of use on three variables: perceived usefulness, behavior intention and actual use of FinTech services. Ease of use has significant influence on perceived usefulness ($\gamma = 0.42$, CR = 8.004), behavior intention ($\gamma = 0.26$, CR = 4.751), and actual usage ($\gamma = 0.29$, CR = 3.312), supporting all the hypotheses.

Hypotheses 4a–4c examine the impact of social influence on ease of use, usefulness and actual usage. Social influence has significant positive impact on ease of use ($\gamma = 0.16$, CR = 4.075) and negative impact on actual usage ($\gamma = -0.11$, CR

Table 3 Standardized estimates (hypothesized model)

Hypothesis	Hypothesized path	Standardized estimate	*Z*-value	*P*-value	Hypotheses support
H1	Intention → SerUsage	−0.028	−0.331	0.741	Not supported
H2a	Useful → Intention	0.602	10.481	***	Supported
H2b	Useful → SerUsage	−0.042	−0.402	0.688	Not supported
H3a	EOU → useful	0.421	8.004	***	Supported
H3b	EOU → Intention	0.257	3.312	***	Supported
H3c	EOU → SerUsage	0.289	3.213	***	Supported
H4a	SocInf → EOU	0.161	4.075	***	Supported
H4b	SocInf → Useful	0.034	1.005	0.315	Not supported
H4c	SocInf → SerUsage	−0.106	−2.082	0.037	Supported
H5a	Rel → EOU	0.220	3.521	***	Supported
H5b	Rel → Useful	0.159	2.936	0.003	Supported
H6a	Sec → EOU	0.472	7.323	***	Supported
H6b	Sec → Useful	0.346	5.852	***	Supported
H6c	Sec → Rel	0.718	12.743	***	Supported

Notes CMIN = 2.348, CFI = 0.965, GFI = 0.910, AGFI = 0.882, SRMR = 0.0379, RMSEA = 0.055

= −2.082). But social influence has no significant impact on perceived usefulness ($\gamma = 0.03$, CR = 1.005) of FinTech services, rejecting hypothesis 4b.
Hypotheses 5a–5b examine the influence of responsiveness on perceived ease of use and usefulness. Responsiveness has significant impact on ease of use ($\gamma = 0.22$, CR = 3.521) and usefulness ($\gamma = 0.16$, CR = 2.936), supporting both the hypotheses.
Hypotheses 6a–6c examine the impact of security on ease of use, perceived usefulness and responsiveness. Security has significant impact on ease of use ($\gamma = 0.47$, CR = 7.323), perceived usefulness ($\gamma = 0.35$, CR = 5.852) and responsiveness ($\gamma = 0.72$, CR = 12.743). Thus, the three hypotheses are supported.

The hypotheses test explained 66% variance in behavior intention to use for FinTech services explained by two significant belief factors, namely perceived ease of use and perceived usefulness. Only 5% variance was explained for actual usage of FinTech services explained by ease of use and social influence. Perceived ease of use, responsiveness and security explained 70% variance in perceived usefulness. 51% of variance in perceived ease of use is explained through three antecedents, namely responsiveness, security and social influence.

Furthermore, given that security, responsiveness and social influence may have direct impact on behavior intention and actual usage of FinTech services, a modified structural model allowing these paths was estimated. All three paths are not significant, and thus, the finding supports the proposed hypotheses.

5 Discussion and Conclusion

The study explores drivers for FinTech adoption. It proposes social influence to have a direct impact on FinTech adoption in addition to perceived ease of use and perceived usefulness. Also, responsiveness and security are proposed as antecedents of ease of use and usefulness. The study tests the impact of these factors by modifying TAM. The result of this study gives support to the added antecedents in TAM. The original TAM asserts that usage behavior is significantly predicted by three belief constructs, namely perceived ease of use, perceived usefulness and behavior intention [8]. But, in this study, actual use of FinTech services is not significantly predicted by the behavior intention to use. Positive perception of security and responsiveness predicts user perception for both ease of use and usefulness, while social influence has strong negative impact on actual usage.

The study found no significant influence of behavior intention on actual use, which differs from prior study findings. It may be attributed to difference in user reaction when asked for intended use and actual use. Also, behavior beliefs and behavior intention are more related to intended use then with actual use [43, 44]. Also, the responses for utilitarian services may differ as the user agrees with the features but is still comfortable in the existing technology.

This study confirms the robustness of TAM explaining the behavior intention to use for FinTech services. Usefulness emerges as the most powerful indicator of intention to use supporting the prior TAM research, while ease of use [8, 24] and social influence are secondary determinants. The secondary impact of ease of use can be explained with increase user friendliness and less difficulty in using online systems. Also, it can be because of omnipresence of online technology, thereby increasing the competence of users to use them. Suggesting an extended TAM with addition of social influence as antecedent of ease of use, usefulness and actual use, this study finds that social influence has significant negative impact on actual usage of FinTech services. This can be attributed to higher age group of most of the respondents as with increase in age and experience in use of technology, users are not much impacted by peer pressure. Also, it is related to financial transactions, the user must be more individualistic in their approach while decision making and thus, are more concerned about usefulness and ease of use of the services offered.

This study also revealed that security and responsiveness have significant positive influence on actual usage and were mediated through ease of use and usefulness. Security and responsiveness are not yet considered in the previous TAM research in context of FinTech services. In comparing path coefficients of antecedents of ease of use and usefulness, security emerges as the powerful indicator of perceived ease of use and perceived usefulness toward FinTech usage in comparison with responsiveness. Therefore, this study suggests improving the security features of FinTech services and to stress on responsiveness of their system and/or Web site. With these features, FinTech service providers can reassure their users that the services are safe to do financial transactions. They can also influence perception of security and responsiveness of the FinTech services by emphasizing the newly added positive features in marketing campaign. Proper training should be offered to the users on safe usage of FinTech services. This can also improve security and enhance user confidence, thereby enabling users to feel secure and comfortable in using FinTech services. However, it may be possible that with properly incorporated enhanced security features, security and responsiveness might not remain a key concern in the future.

In conclusion, these results provide several key insights for the drivers of FinTech usage. First, perceived usefulness is the key factor positively influencing behavior intention to use FinTech services, while perceived ease of use is the second significant factor. Second, social influence is significant negative determinant of intention to actual usage, while ease of use positively impacts the actual usage. Thirdly, security and responsiveness were also a significant determinant for intention but are mediated through usefulness and ease of use. Lastly, security highly influences responsiveness of FinTech services. All hypotheses were supported except H1 (intention—actual usage), H3b (usefulness—actual usage) and H4c (social influence—ease of use). The model testing gives set of fit indices indicating model fitted well with the data. The results of hypotheses test through SEM analysis provide satisfactory support for the modified TAM.

6 Limitations and Future Research

FinTech industry is seeing the rapid changes, and everyday, newer technologies are introduced in the market. The result of the present set of FinTech services may not be perfectly generalizable to the newer set of FinTech services. A longitudinal study investigating different stages of launch and the mass usage of FinTech service will be a significant contribution to TAM literature. Relationship between users' perception about behavior beliefs, behavior intention and actual usage should be investigated further with objective measurement of actual usage at different times of the technology implementation process. This is will enhance the predictive power of TAM. The study only covers the construct of security, responsiveness and social influence. Future research may investigate issues related to information quality and Web site quality and its impact on behavior intention and actual usage. In addition, the present measurement instrument can be further refined in further studies to enhance its validity. Future research can be done in other product categories for generalization of result.

References

1. McWaters J, Bruno G, Lee A, Blake M (2015) The future of financial services—how disruptive innovations are reshaping the way financial services are structured, provisioned and consumed. In: World economic forum, pp 1–178
2. PWC (2017) FinTech trends report. Retrieved from https://www.pwc.in/assets/pdfs/publications/2017/fintech-india-report-2017.pdf on 27th Aug 2018
3. EY (2016) EY FinTech Adoption Index 2016. Ey, pp 1–44. Retrieved from https://www.ey.com/GL/en/Industries/Financial-Services/ey-fintech-adoption-index on 25th Mar 2018
4. RBI (2017) FinTech and Digital Banking, November, pp 1–84
5. Leong C, Tan B, Xiao X, Tan FTC, Sun Y (2017) Nurturing a FinTech ecosystem: the case of a youth microloan startup in China. Int J Inf Manage 37(2):92–97
6. Rogers E (1983) Diffusion of innovations. Free Press, New York
7. Davis FD (1986) A technology acceptance model for empirically testing new end-user information systems. Cambridge, MA
8. Davis FD, Bagozzi RP, Warshaw PR (1989) User acceptance of computer technology: a comparison of two theoretical models. Manage Sci 35(8):982–1003
9. Venkatesh V, Davis FD (2000) A theoretical extension of the technology acceptance model: four longitudinal field studies. Manage Sci 46(2):186–204
10. Venkatesh V, Morris MG, Davis GB, Davis F (2003) User acceptance of information technology: toward a unified view. MIS Q 425–478
11. Fishbein M, Ajzen I (1975) Belief, attitude, intention and behavior: an introduction to theory and research, pp 181–202
12. Ajzen I (1985) From intentions to actions: a theory of planned behaviour. In: Action control: from cognition to behavior, pp 11–39
13. Ajzen I (1991) The theory of planned behavior. Organ Behav Hum Decis Process 50(2):179–211
14. Cronin JJ Jr, Taylor SA (1992) Measuring service quality: a reexamination and extension. J Mark 56(3):55–68
15. Venkatesh V, Davis FD (1996) A model of the antecedents of perceived ease of use: development and test. Decis Sci 27(3):451–481
16. Barnes SJ, Vidgen RT (2002) An integrative approach to the assessment of e-commerce quality. J Electron Commer Res 3(3):114–127
17. He MD, Leckow MRB, Haksar MV, Griffoli MTM, Jenkinson N, Kashima MM, Tourpe H (2017) Fintech and financial services: initial considerations. Int Monetary Fund 49. Retrieved from https://www.imf.org/~/media/Files/Publications/SDN/2017/sdn1705.ashx on 21st Mar 2018
18. Saal M, Starnes S, Rehermann T (2017) Digital financial services: challenges and opportunities for emerging market banks. EM Compass Note 42, IFC
19. Manyika J, Lund S, Singer M, White O, Berry C (2016) Digital finance for all: powering inclusive growth in emerging economies. McKinsey Global Institute
20. Feng TT, Tien C, Feng ZY, Lai PJ (2014) Web site quality and online trading influences on customer acceptance of securities brokers. Asia Pac Manage Rev 19(1)
21. Abbasi T, Weigand H (2017) The impact of digital financial services on firm's performance: a literature review. arXiv preprint arXiv:1705.10294
22. Slade EL, Dwivedi YK, Piercy NC, Williams MD (2015) Modeling consumers' adoption intentions of remote mobile payments in the United Kingdom: extending UTAUT with innovativeness, risk, and trust. Psychol Mark 32(8):860–873
23. Kaba B, Touré B (2014) Understanding information and communication technology behavioral intention to use: applying the UTAUT model to social networking site adoption by young people in a least developed country. J Assoc Inf Sci Technol 65(8):1662–1674
24. Davis FD, Bagozzi RP, Warshaw PR (1992) Extrinsic and intrinsic motivation to use computers in the workplace 1. J Appl Soc Psychol 22(14):1111–1132
25. Venkatesh V, Morris MG (2000) Why don't men ever stop to ask for directions? Gender, social influence, and their role in technology acceptance and usage behavior. MIS Q 115–139

26. Chen YH, Barnes S (2007) Initial trust and online buyer behaviour. Ind Manage Data Syst 107(1):21–36
27. Amin H (2007) An analysis of mobile credit card usage intentions. Inf Manage Comput Secur 15(4):260–269
28. Gounaris S, Koritos C (2008) Investigating the drivers of internet banking adoption decision: a comparison of three alternative frameworks. Int J Bank Mark 26(5):282–304
29. Venkatesh V, Thong JY, Xu X (2012) Consumer acceptance and use of information technology: extending the unified theory of acceptance and use of technology. MIS Q 36(1):157–178
30. Shin DH (2010) Modeling the interaction of users and mobile payment system: conceptual framework. Int J Hum Comput Inter 26(10):917–940
31. Parasuraman A, Zeithaml VA, Berry LL (1985) A conceptual model of service quality and its implications for future research. J Mark 49(4):41–50
32. Gefen D, Straub D (2003) Managing user trust in B2C e-services. e-Service 2(2):7–24
33. Wolfinbarger M, Gilly MC (2003) eTailQ: dimensionalizing, measuring and predicting etail quality. J Retail 79(3):183–198
34. Pavlou PA (2003) Consumer acceptance of electronic commerce: integrating trust and risk with the technology acceptance model. Int J Electron Commer 7(3):101–134
35. Reichheld FF, Schefter P (2000) E-loyalty: your secret weapon on the web. Harvard Bus Rev 78(4):105–113
36. Grewal D, Lindsey-Mullikin J, Munger J (2004) Loyalty in e-tailing: a conceptual framework. J Relat Mark 2(3–4):31–49
37. Suh B, Han I (2002) Effect of trust on customer acceptance of internet banking. Electron Commer Res Appl 1(3–4):247–263
38. Gerrard P, Barton Cunningham J, Devlin JF (2006) Why consumers are not using internet banking: a qualitative study. J Serv Mark 20(3):160–168
39. Laukkanen P, Sinkkonen S, Laukkanen T (2008) Consumer resistance to internet banking: postponers, opponents and rejectors. Int J Bank Mark 26(6):440–455
40. Sujana A (2008) Adoption of internet shopping: cultural considerations in India and Australia. J Internet Banking Commer 13(2):1–17. Retrieved from https://www.icommercecentral.com/open-access/adoption-of-internet-shopping-cultural-considerations-in-india-and-australia.php?aid=38428
41. Dahlberg T, Mallat N, Öörni A (2003) Trust enhanced technology acceptance model consumer acceptance of mobile payment solutions: tentative evidence. Stockholm Mobility Roundtable 22(1)
42. Hair JF, Black WC, Babin BJ, Anderson RE, Tatham RL (2006) Multivariate data analysis, 6th edn. Pearson Prentice Hall, Upper Saddle River
43. Adams DA, Nelson RR, Todd PA (1992) Perceived usefulness, ease of use, and usage of information technology: a replication. MIS Q 227–247
44. Straub D, Limayem M, Karahanna-Evaristo E (1995) Measuring system usage: implications for IS theory testing. Manage Sci 41(8):1328–1342

Power Quality Enhancement in a Stand-Alone WECS Fed Nonlinear Load Through UPQC

M. Hari Babu, M. Hari, R. S. Shivakumara Aradhya, Dharmendra Kumar Singh, and Anurag Verma

1 Introduction

Nowadays, due to day-by-day increment in the load, lack of sources and global warming restricts the usage of coal, diesel and other thermal power generation which leads to new sources and other ways of generation. In mean time, the nonconventional energy sources are of more importance. Due to the usage of nonconventional or renewable sources such as solar, wind and geothermal, the output may be less or undermining (which is unable to tell as a constant one) and has less power quality. When these types of sources are operated as stand-alone or connected to grid to supply, the surplus power has any one or more power quality issue like under voltage, over voltage, voltage sag, voltage swell, noise, distortion and harmonic problems including sub-harmonics and waveform distortions due to the presence

M. Hari Babu (✉) · R. S. Shivakumara Aradhya
Department of Electrical and Electronics Engineering, Acharya Institute of Technology, Bangalore, India
e-mail: haribabu3e@gmail.com

R. S. Shivakumara Aradhya
e-mail: rssaradhya@gmail.com

M. Hari
Department of Electrical and Electronics Engineering, CMR Institute of Technology, Bengaluru, India
e-mail: harimaaperuman@gmail.com

D. K. Singh
Department of Electrical and Electronics Engineering, Noida Institute of Engineering & Technology, Greater Noida, India
e-mail: er.dksingh1994@gmail.com

A. Verma
Faculty of Engineering, Dayalbagh Educational Institute, Agra, India
e-mail: rsaverma@gmail.com

V. H. Saran and R. K. Misra (eds.), *Advances in Systems Engineering*, Lecture Notes in Mechanical Engineering,
https://doi.org/10.1007/978-981-15-8025-3_35

of nonlinear loads in such a system. As the power electronics introduced various controlled electronic switches, the reactive power management is more possible with multilevel inverter concepts. From various reasons, the same FACTS devices can be used in distribution side and the same device can be operated for improving the power quality issues. There are various power quality problems nullified or compensated, namely power quality conditioners such as DSTATCOM, active power filters, UPQC and dynamic voltage restorer (DVR), which are found from [1–8]. UPQC uses two inverters. Among various methods of triggering inverters, PWM gives a simple control structure and has good control capability [9, 10].

As the distributed generation, DC microgrid, smart grid, etc., are few systems which emphasize the usage of stand-alone systems, when compared to grid-connected systems, as stand-alone systems are cheaper, has simple control methods, no transmission losses, no grid connected problems, easy way of detection of fault conditions and maintenance than the grid-connected systems [8, 11–13]. The wind energy is more popular in the hilly areas and places where more wind energy is available so that the wind energy conversion system (WECS) is quite fit for such conditions as mentioned. The WECS contains the wind turbine with rear system and yaw control to control the direction of rotation and speed, which is coupled with either permanent magnet synchronous generator (PMSG) or doubly fed induction generator (DFIG). When it is stand-alone, the load is connected directly or indirectly to the load through a converter for the purpose of control [11–16].

As mentioned earlier, when the load is nonlinear in nature, the input current is affected by the load which leads to a power quality issue. When any load is sensitive such as a DSO in a hospital or electronic equipment which show a wrong result, it leads to a big problem which is to be reduced and it is to be noted that in such times, the power quality necessitates the usage of power quality conditioners. UPQC is the combination of series and shunt active power filters which is capable of serving the necessary service what this article expects. Section 2 shows the principle operation of UPQC, Sect. 3 represents simulation result and its discussion, and at the end, Sect. 4 indicates the conclusion of this article.

2 Principle of UPQC

The use of nonlinear loads to a greater extension further can increase the harmonics in voltage as well as current. As wind energy and, solar energy, and fuel cell-related grid-connected systems will increase the harmonics due to its fundamental issues both at high and low voltage transmission and distribution levels. Electrical power industry is facing new challenges to accommodate the increased integration of emerging distributed generation systems [2] and is creating problems in maintaining the power quality according to regulations; it is general procedure to do power conditioning/smoothing at various power levels [15–17]. UPQC is found to be one of such cost effective solutions to reduce or eliminate few or most major power quality issues at the distribution level [15–19].

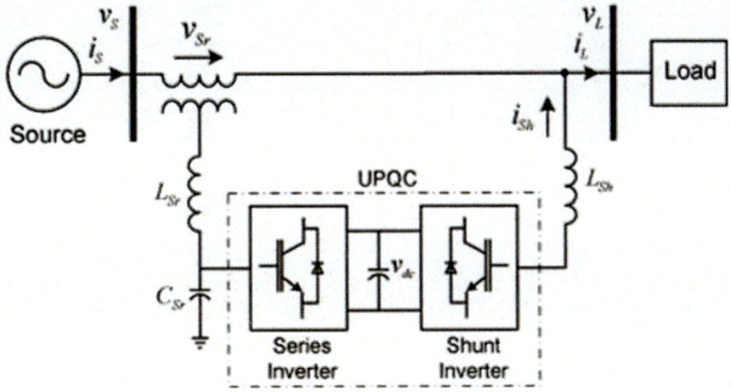

Fig. 1 Schematic representation of UPQC, [19]

Figure 1 shows the schematic of general UPQC, and it contains two inverters (one shunt and one series separated by dc link capacitor), and it may use transformer to reduce the level of voltage and current. v_{sr} and i_{sh} are the voltage and current injected by the UPQC for the specified application based on the conduction period of series and shunt inverters, respectively, with the inclusion on exclusion of inductance L_{Sr} and capacitance C_{Sr}. The impact of load (dominated nonlinear one) degrades the quality of v_L and i_L which indirectly affects the v_s and i_s, respectively, in terms of any of the power quality issue as mentioned earlier.

2.1 Power Angle Control (PAC)—Principle

In controlling voltage sag/swell and THD in the system, the UPQC is one of the most distinct solutions. Control of UPQC's rating is governred by compensating the % of maximum voltage sasg/swell [19]. These voltage variations such as sag/swell are of very short duration power quality issues, so a dedicated series inverter of UPQC connected to the line is not utilized to its full extent or true capacity under normal operating conditions. The proper power angle control in UPQC between the source voltage and load voltage, and required reacative power demand can be met by connecting both shunt and series inverters without creating any troble to the overall rating of the UPQC [19]. The principle of PAC is dipicted in Fig. 2 phasor representation under a rated steady-state condition [19, 20]. As per the phasor diagram, when a voltage vector $\overline{V}_{Sr}$ with proper magnitude V_{Sr} and phase angle ϕ_{Sr} is injected through series inverter, it leads to an increased power angle δ between the source voltage V_S and resultant load voltage V_L voltages being maintained at the same magnitudes. The shift in power angle will try to modify the relative phase advancement between the supply voltage and resultant load current I_L, denoted as angle β. So that with the help of PAC principle, the load reactive power demand can be supported by the series inverter, and hence, the reactive power demand shared by

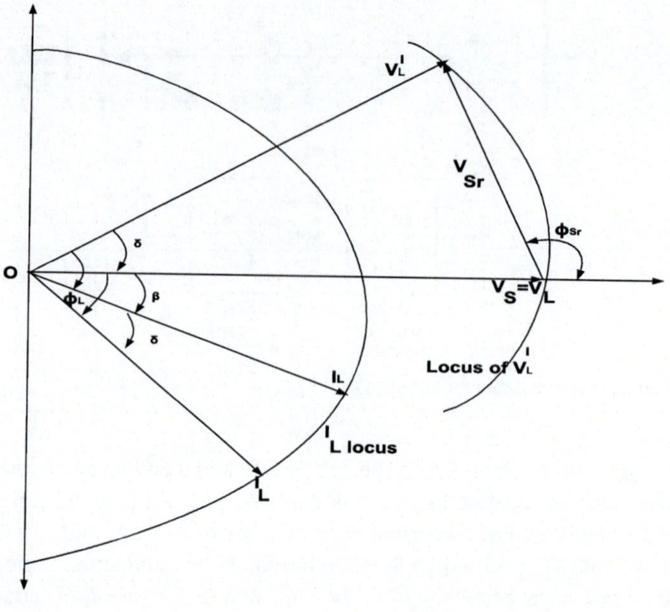

Fig. 2 Principle of operation of UPQC in terms of power angle control (PAC) [19]

the shunt inverter will be reduced. For a given rated steady-state condition

$$|V_S| = |V_L| = |V_L^*| = |V_L^*| = k \tag{1}$$

From Fig. 2,

$$\overline{V}_{Sr} = |V_{Sr}|\angle\phi_{Sr} = \left\{k\sqrt{2}\sqrt{(1-\cos\delta)}\right\}\angle\left\{\frac{90^\circ+\delta}{2}\right\} \tag{2}$$

$$\delta = \sin^{-1}\left(\frac{Q_{Sr}}{P_L}\right) \tag{3}$$

From Eqs. 1 and 2, it can be concluded that the phase angle (δ) injected by UPQC controls the reactive power, thereby sending end voltage. So, ultimately, the power quality will be improved. The block diagram in brief is shown in Fig. 3. The control circuit is taking various inputs such as currents and voltages at the input side of the nonlinear load (three phase), and it will convert into *d*–*q* transformation coordinates, and then, through sinusoidal pulse width modulation (SPWM), the switching pulses are producing. Based on the conduction period of each switch in VSC, the respective amount of inductance and capacitance is controlling, which can control the reactive power to be injected into the system to modify the quality of the voltage and current at output side of WECS.

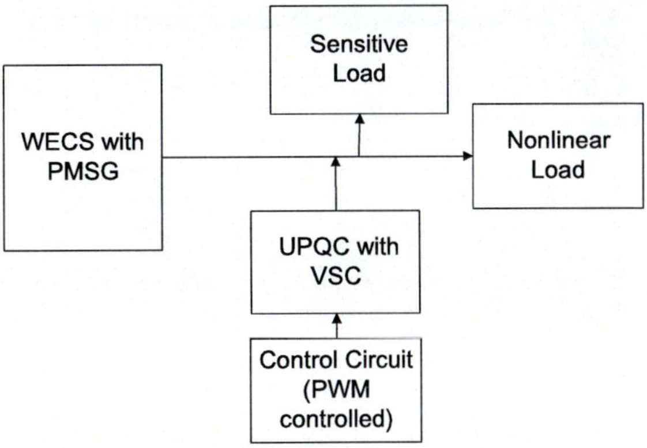

Fig. 3 Block diagram of UPQC

3 Simulation Results and Discussion

The proposed system is consisting of WECS whose output is connected with a load which can have both sensitive and nonlinear loads. The WECS has PMSG whose rating is of 5 kVA and 415 V, and the wind turbine has 3.5 kW nominal input and 40 m/maximum wind speed. From starting of the simulation the sensitive load and base nonlinear load of the values 1 kΩ, rectifier based *R–L* load of 1 mH and 1 kΩ, 30 μH respectively is applied. PWM-based VSC-controlled UPQC contains two star/star with grounded transformers which will be used. In this research PWM, carrier frequency is of 5 kHz is considered. The value of capacitor C_{Sr} is of 100 μF and inductance L_{Sr} is of 50 mH. The simulation is made through MATLAB/SIMULINK software. In the process of simulation, it is assumed that the WECS input (wind speed) as almost constant just for the effective study of the proposed system.

Three major changes are happening in various waveforms during the simulation. One is due to base sensitive load and base load of diode rectified nonlinear load, and the second is due to occurrence of three-phase line to ground fault (L-L-L-G) at $t = 0.3$ s and 0.4 s, and the third is due to the dominant nonlinear load which is being added at $t = 0.5$ s which effect is to be minimized at the output side of WECS ultimately by using UPQC. The results are divided into to two cases. Case-1 is the system without UPQC and case-2 is the system with UPQC.

3.1 Case-1: Without UPQC

Figures 4 and 5 show voltage across the nonlinear load, and it is harmonic analyze window (FFT) without UPQC in the system. It is understood from Fig. 5 that the voltage fundamental component is of 539.3 V (amplitude) and its %THD is 45.91% which is due to the impact of nonlinear load presence in the system considered.

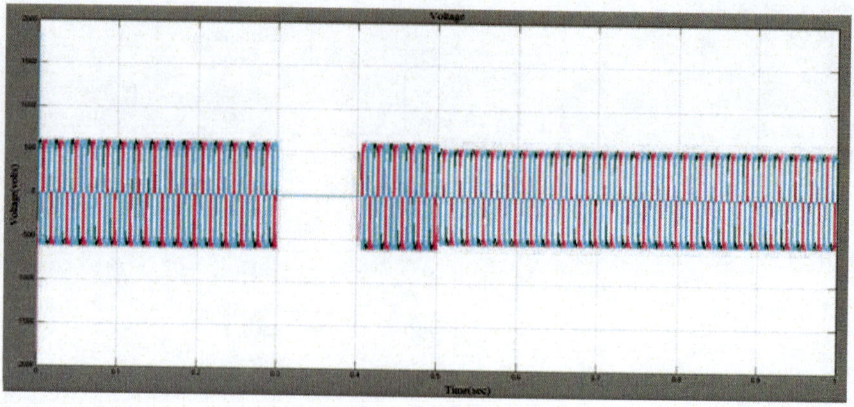

Fig. 4 Voltage across the nonlinear load without UPQC

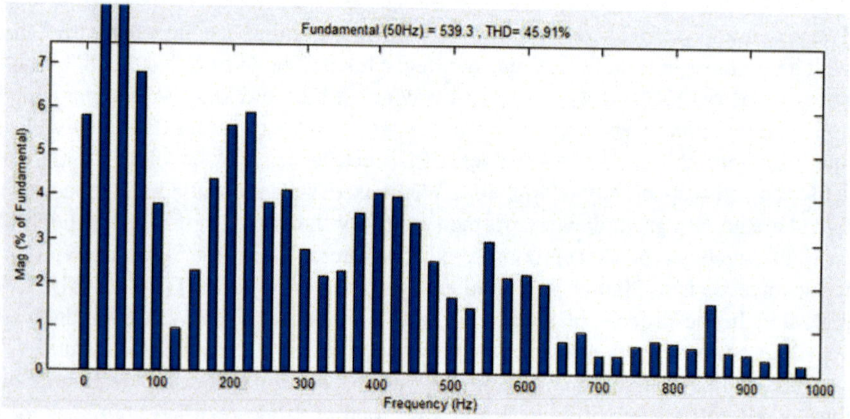

Fig. 5 Harmonic analyze window through FFT of voltage across nonlinear load without UPQC

The main thing to be observed here in Fig. 4 is sudden switching of the load which leads to voltage spike at the start to around 2000 V (peak), and it comes back to a normal value of 560 V (peak). When three-phase line to ground fault occurred at $t = 0.3$ s, the voltage will become zero, and again, it will persist its previous value. Once the nonlinear load suddenly applied $t = 0.5$ s, it makes the voltage sag (due to drop) and will be reduced to a value of 530 V (peak) and it will be same as it is till the end of the simulation time.

Figures 6 and 7 show the current through the nonlinear load, and it is harmonic analyze window (FFT), respectively, without UPQC in the system. It is understood from Fig. 5 that the voltage fundamental component is 0.5292 A (amplitude) and its %THD is 44.63% which is due to the impact of nonlinear load presence in the system considered. It is to be observed here from Fig. 6 that sudden switching of the load will make to a value of 0.6 A (peak) at the starting which may be to supply the power

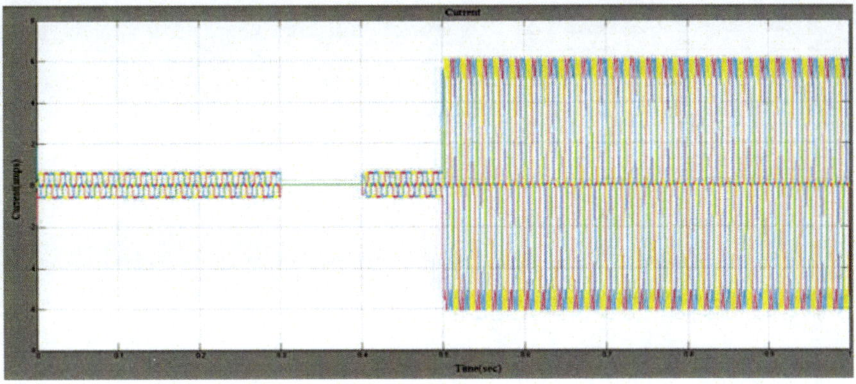

Fig. 6 Current through the nonlinear load without UPQC

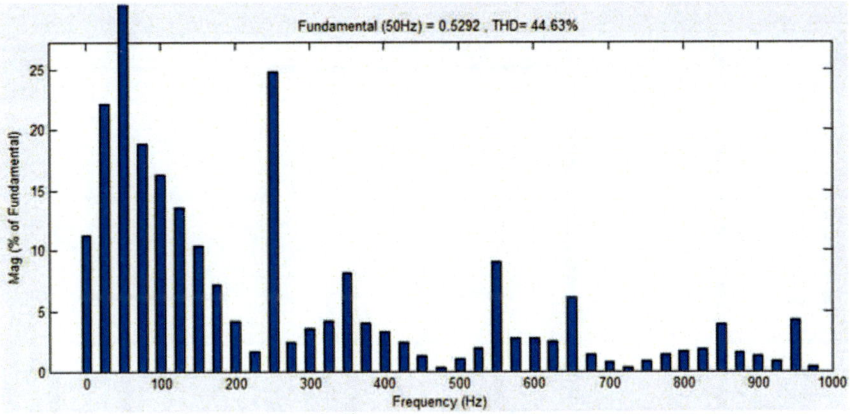

Fig. 7 Harmonic analyze window by FFT of current through the linear load without UPQC

to the load. When three-phase line to ground fault occurred at $t = 0.3$ s, the current at the load side will become zero (as the fault is of symmetrical in nature), and again, it will persist its previous value once after the fault is exhausted. The nonlinear load suddenly applied at $t = 0.5$ s makes the current to raise (due to drop) to a value of 6 A (peak), and it will be same as it is till the end of the simulation time.

Figures 8 and 9 show the output voltage, and it is harmonic analyze window (FFT) of WECS without UPQC in the system. It emphasizes that the voltage fundamental component is 561.4 V (amplitude) and its %THD is 32.96% which is due to the impact of nonlinear load presence in the system that considered.

Figures 10 and 11 show the output current, and it is harmonic analyze window (FFT) of WECS without UPQC in the system. It emphasizes that the fundamental component of current is 6.567 A (amplitude) and of it's %THD is 20.01% is again due to the impact of nonlinear load presence in the system that considered. Figure 12

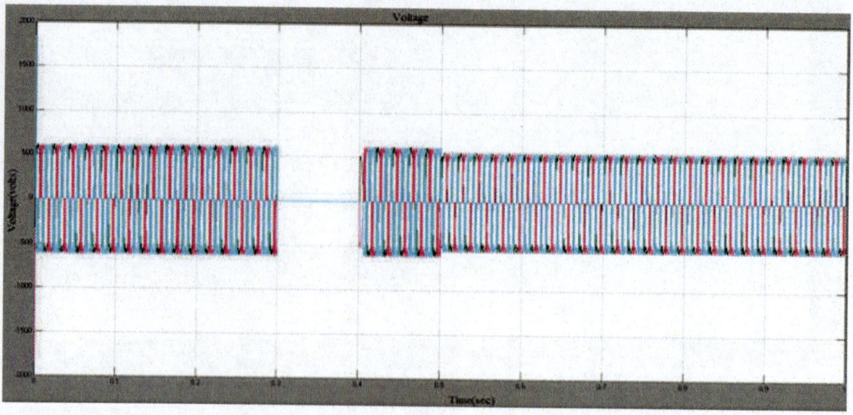

Fig. 8 Output voltage of WECS without UPQC

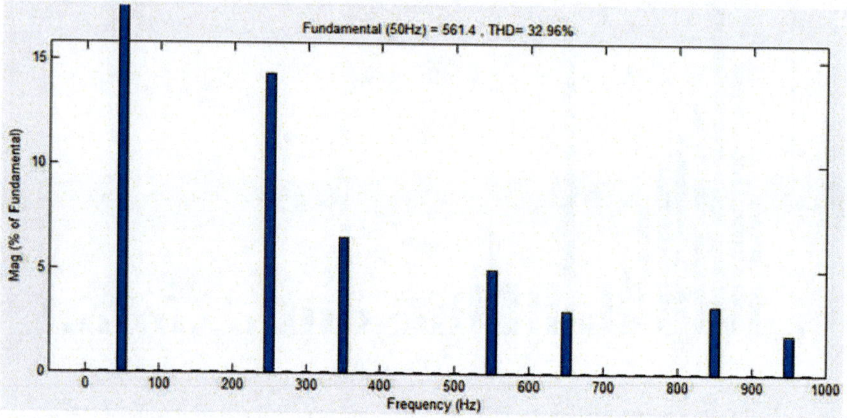

Fig. 9 Harmonic analyze window through FFT of output voltage of WECS without UPQC

shows the output active power and reactive powers of WECS without UPQC.It is to be observed here in Fig. 10 is sudden switching of the load will make the WECS to supply the power with a current of 0.65 A (peak) at the starting.When three-phase line to ground fault occurred at $t = 0.3$ s, the current at WECS output side will raise to a high value of 80 A (as the fault is of symmetrical in nature) and it will reduce slowly as already the circuit breaker operates and cut down the healthy part of the network. Again in this article it is expecting with a worst case of forcibly applying closing the circuit breaker to supply the load, the current is trying to come back to a safe value of 0.65 A as at the initial base load levels. The nonlinear load suddenly applied at $t = 0.5$ s which makes the current to raise (due to drop) to a value of 6.62 A (peak), it will be same as it is till the end of the simulation time.

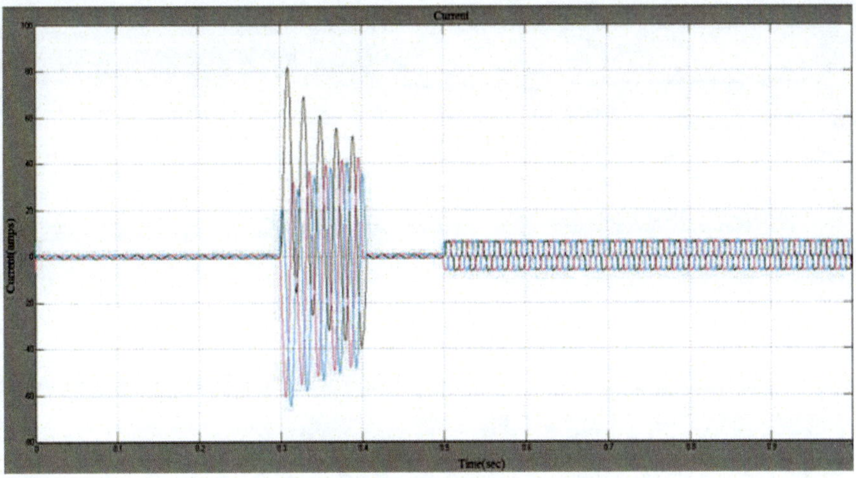

Fig. 10 Output current of WECS without UPQC

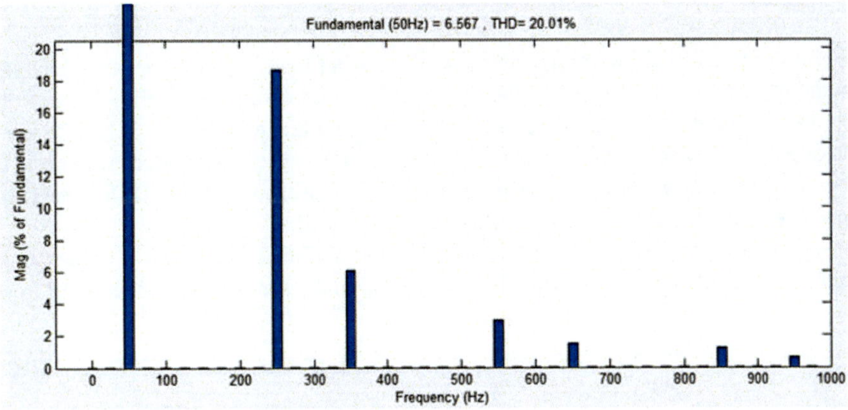

Fig. 11 Harmonic analyze window through FFT of output current of WECS without UPQC

From Fig. 12 it is evident that initially when the base load (both sensitive and nonlinear) is applied the output of WECS, active power (P) is at its normal values of 400 W and have a small fluctuations and reactive power (Q) of 10 kVar. Once the faultoccurred, no power transfer will happen due to opening of circuit breaker (another reason is due to three-phase symmetrical fault). When the fault exhausted at $t = 0.4$ s, P and Q will persist its previous values. Once the nonlinear load is switched on at $t = 0.5$ s, P and Q values increases which are making the changes in current and voltage levels as well as the shape of the waveforms (is one of the reason). P is at 2.5 kW and Q is at 6.5 kVar with more fluctuations when the case is without UPQC.

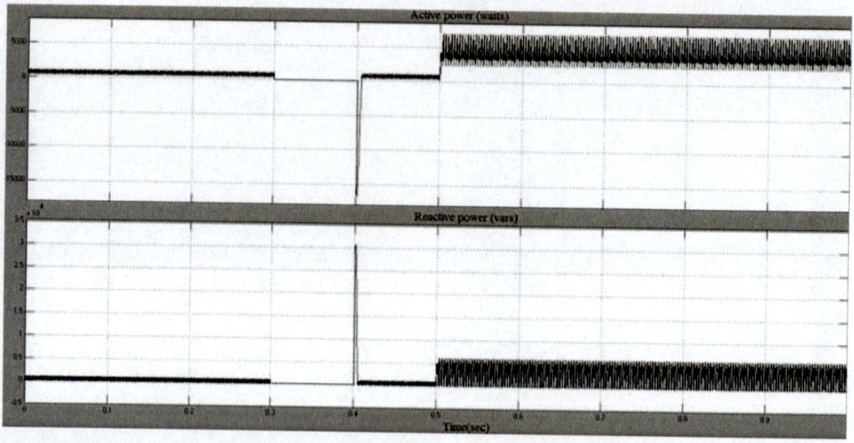

Fig. 12 Output active power and reactive powers of WECS without UPQC

3.2 *Case-2: With UPQC*

Figures 13 and 14 show voltage across the nonlinear load, and it is harmonic analyze window (FFT) with UPQC in the system. It is understood from Fig. 14 that the voltage fundamental component is 559.3 V (amplitude) and of it's %THD is 20.04% is due to the impact of nonlinear load presence in the system considered. But it is to be noted that when compared to the results with respect to without UPQC, the THD level is better and impact of nonlinear load is reduced.

Figures 15 and 16 show current through the nonlinear load, and it is harmonic analyze window (FFT) respectively with UPQC in the system. It is understood from

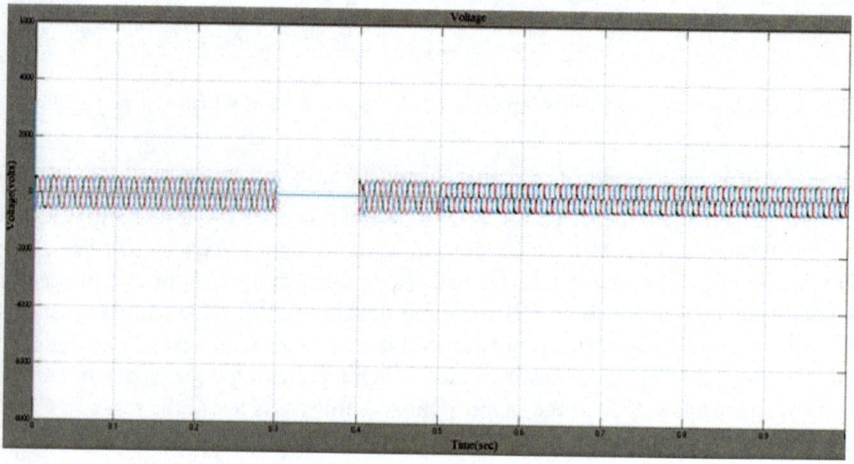

Fig. 13 Voltage across the nonlinear load with UPQC

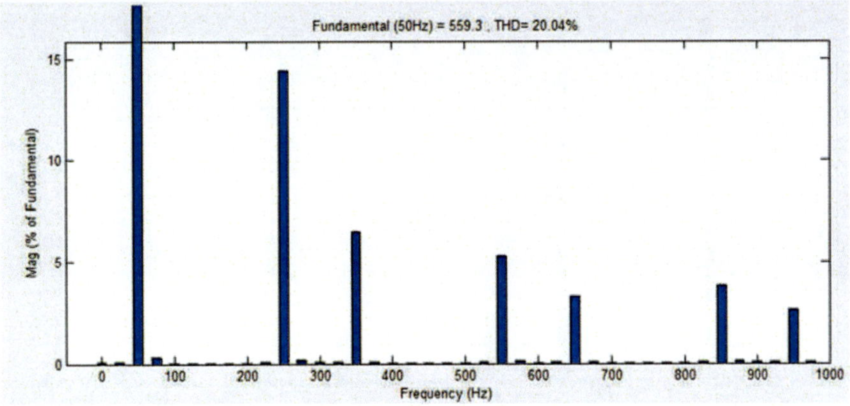

Fig. 14 Harmonic analyze window through FFT of voltage across nonlinear load with UPQC

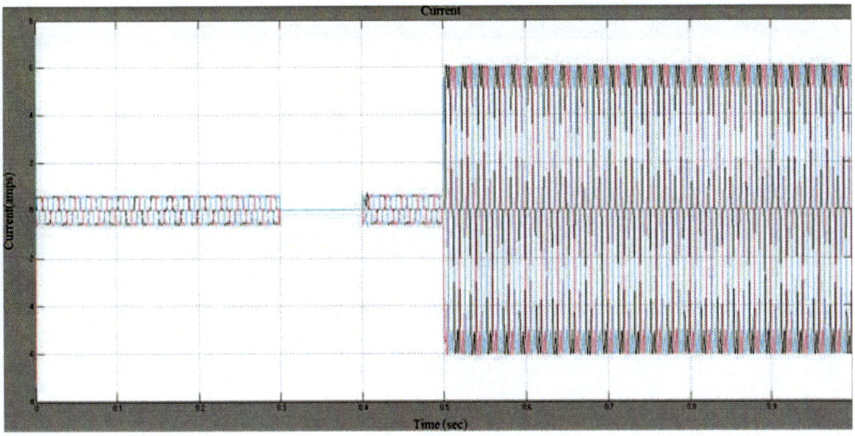

Fig. 15 Current through the nonlinear load with UPQC

Fig. 16 that the voltage fundamental component is 6.233 A (amplitude) and of it's %THD is 21.02% is due to the impact of nonlinear load presence in the system considered. Again it is to be noted that when compared to the results with respect to without UPQC, the THD level is better and impact of nonlinear load is reduced.

Figures 17 and 18 show the output voltage, and it is harmonic analyze window (FFT) of WECS with UPQC in the system. It emphasizes that the voltage fundamental component is 539.2 V (amplitude) and its %THD is 18.95% which is due to the impact of nonlinear load presence in the system that considered. Again, it is to be noted that when compared to the results with respect to without UPQC, the THD level is better and impact of nonlinear load is reduced. And the THD is reduced almost 50% of the previous one when the system is without UPQC. It can be aid here that the power quality is improved almost 50% than the earlier in terms of voltage. It is to be

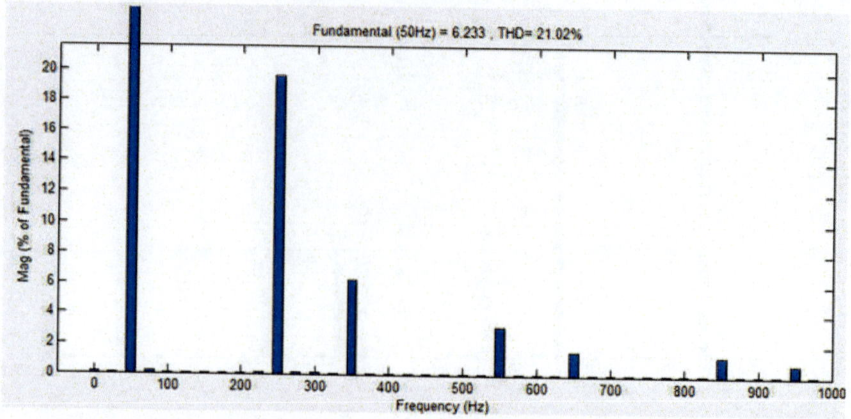

Fig. 16 Harmonic analyze window by FFT of current through the linear load with UPQC

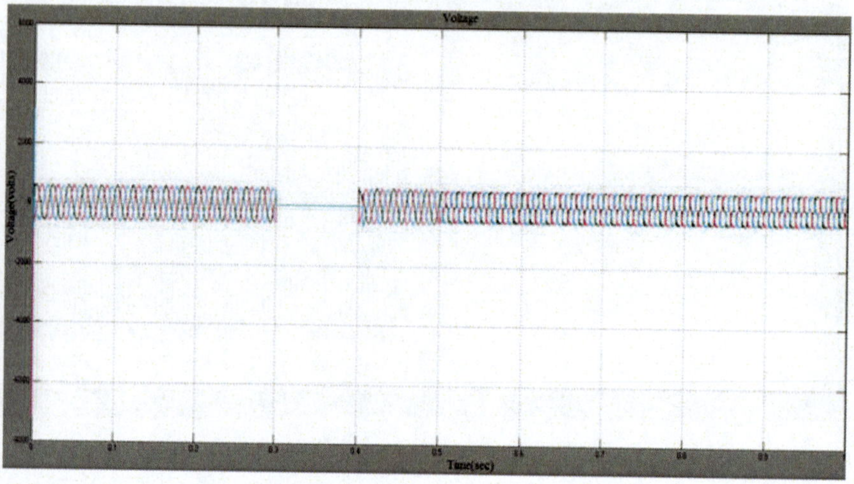

Fig. 17 Output voltage of WECS with UPQC

observed here in Fig. 4 that sudden switching of the load leads to voltage spike at the start to around 2000 V (peak), and it comes back to a normal value of 560 V (peak). When three-phase line to ground fault occurred at $t = 0.3$ s, the voltage will become zero, and again, it will persist its previous value. Once the nonlinear load suddenly applied $t = 0.5$ s, it makes the voltage sag (due to drop) and will be again at a value of 560 V (peak) (and is not be reduced as the UPQC support the voltage and reduces the voltage sag) and it will be same as it is till the end of the simulation time.

Figures 19 and 20 show the output current, and it is harmonic analyze window (FFT) of WECS with UPQC in the system. It emphasizes that the voltage fundamental component is 1.038 A (amplitude) and its %THD is 17.54% which is due to the impact

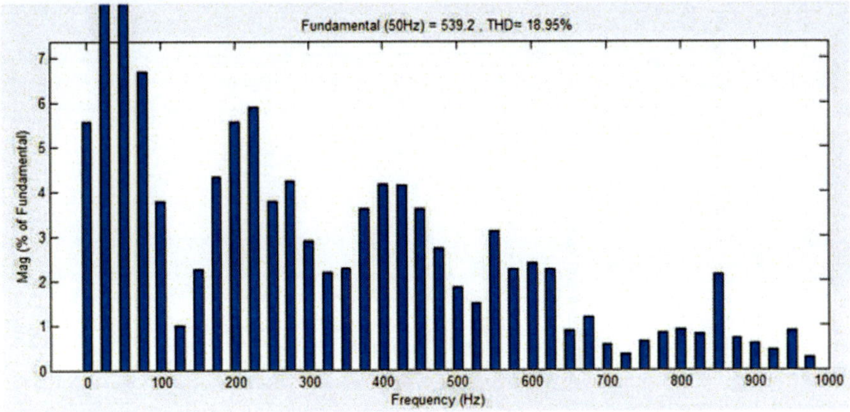

Fig. 18 Harmonic analyze window through FFT of output voltage of WECS with UPQC

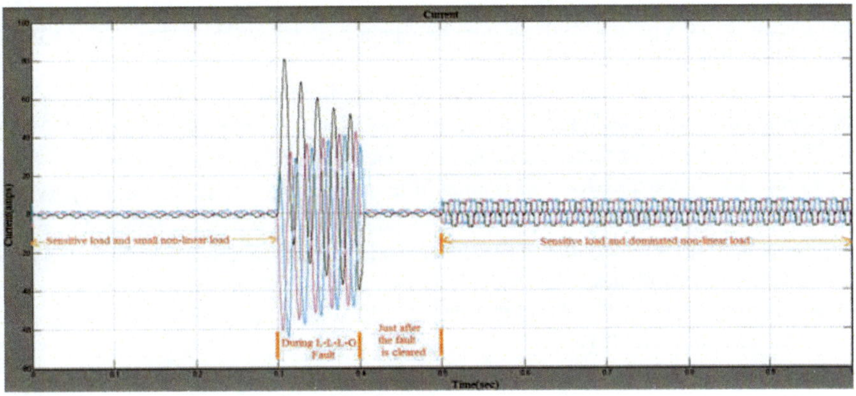

Fig.19 Output current of WECS with UPQC

of nonlinear load presence in the system that considered. Again, it is to be noted that when compared to the results with respect to without UPQC, the THD level is better and impact of nonlinear load is reduced. And the THD is reduced almost 2% with the previous one when the system is without UPQC. It can be aid here that the power quality is improved than the earlier in terms of current also. It is to be observed here in Fig. 19 that sudden switching of the load will make the WECS to supply the power with a current of 0.646 A (peak) at the starting. When three-phase line to ground fault occurred at $t = 0.3$ s, the current at WECS output side will raise to a high value of 79.8 A (as the fault is of symmetrical in nature) and it will reduce slowly as already the circuit breaker operates and cuts down the healthy part of the network. The current is trying to come back to a safe value of 0.646 A as it is at the initial base load levels. The nonlinear load suddenly applied at $t = 0.5$ s makes the current

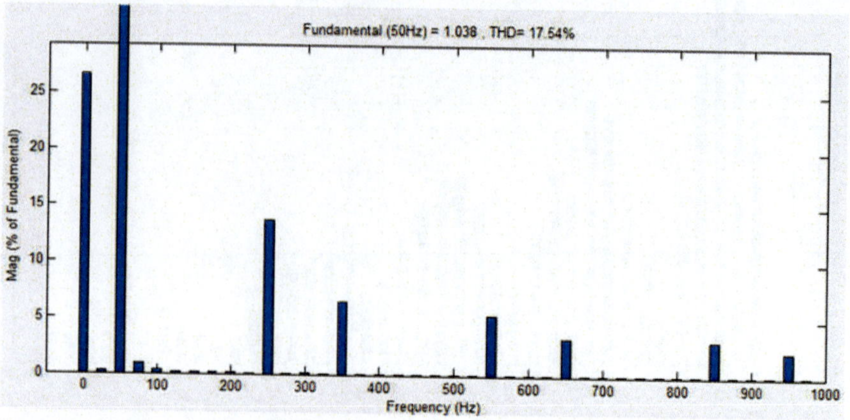

Fig. 20 Harmonic analyze window through FFT of output current of WECS with UPQC

to raise (due to drop) to a value of 6.615 A (peak), and it will be same as it is till the end of the simulation time.

From Fig. 21, it is evident that initially, when the base load (both sensitive and nonlinear) the output of WECS, active power (*P*) is at its normal values of 1.2 kW and has very much less fluctuations and reactive power (*Q*) of 9 kVar (Once the fault occurred, no power transfer will happen due to opening of circuit breaker). When the fault exhausted at $t = 0.4$ s, *P* and *Q* will persist its previous values. Once the nonlinear load is switched on at $t = 0.5$ s, *P* and *Q* values increase which are making the changes in current and voltage levels as well as the shape of the waveforms (is one of the reason). *P* is at 4.5 kW and *Q* is at 5.2 kVar with very much lesser fluctuations when the case is with UPQC.

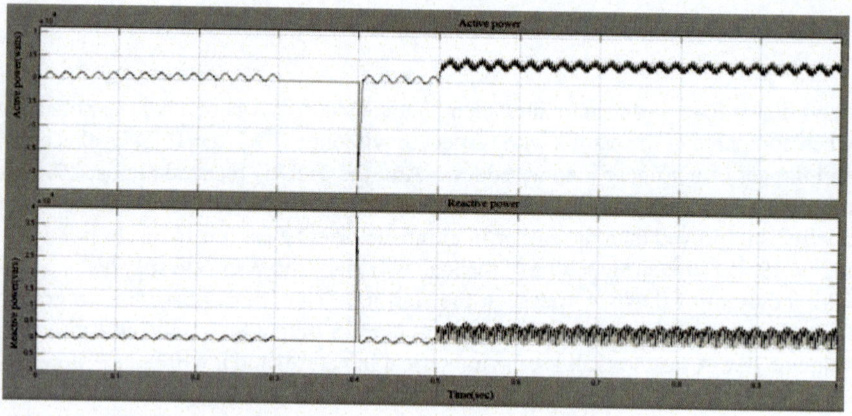

Fig. 21 Output active power and reactive powers of WECS with UPQC

It is to be mentioned here that %THD of current in this article has better results in current THD of 17.54%, when compared to 20.6% [11] and 17.6% [13]. It is also witnessed that the utilization of active power is more and reactive power drawn by various loads from WECS is lesser in this case (with UPQC).

4 Conclusion

This paper is attempted on the WECS whose output is being connected to a nonlinear load. Various simulation results were shown and compared with both the cases that considered. The power quality of the WECS output which is degraded by the nonlinear load is being improved with the help of UPQC. The utilization of active power is more with UPQC, and at the same time, the reactive power drawn by various loads from WECS is reduced with the help of UPQC. As the UPQC is supporting the reactive power to load, the voltage sag at the output side of WECS is also reduced. The result is better in both the voltage and current of WECS, and the impact of nonlinearity of the load is reduced on the output side of the WECS.

References

1. Hingorani NG, Gyugyi L (2000) Understanding FACTS: concepts and technology of flexible AC transmission systems. IEEE Press, New York, p 432. https://doi.org/10.1109/MEI.2002.981326
2. Walling RA, Saint R, Dugan RC, Burke J, Kojovic LA (2008) Summary of distributed resources impact on power delivery systems. J IEEE Trans Power Delivery 23(3):1636–1644. https://doi.org/10.1109/TPWRD.2007.909115
3. Bhim S, Kasal GK (2008) Voltage & frequency controller for a 3 phase 4 wire autonomous wind energy conversion system. J IEEE Trans Energy Convers 23(2)
4. Sood VK (2004) HVDC and FACTS controllers—applications of static converters in power systems. Kluwer, Boston, MA, p 295. https://doi.org/10.1007/b117759
5. Ghosh A, Ledwich G (2002) Power quality enhancement using custom power devices. Kluwer, Boston, MA, p 460. https://doi.org/10.1007/978-1-4615-1153-3
6. Benlamoudi A, Rachid A (2012) Autonomous SEIG in a small wind power plant with voltage and frequency control. J Serb J Electr Eng 9(3):343–359
7. Deepika G, Elakkiya M, Mohamed Iqbal M (2017) Harmonics reduction using multilevel based shunt active filter with SMES. Int J Eng Technol (IJET) 9(2):1001–1011. https://doi.org//10.21817/ijet/2017/v9i2/170902270
8. Singh B, Al-Haddad K, Chandra A (1999) A review of active power filters for power quality improvement. J IEEE Trans Ind Electron 45(5):960–971. https://doi.org/10.1109/41.793345
9. Hari M, Verma A, Halakurki RR, Ravela RR, Kumar P (2018) A dynamic analysis of SVM based three-level NPC for a 3-phase ınduction motor. In: 1st International conference on power energy, environment and ıntelligent control (PEEIC), pp 162–167. IEEE. https://doi.org/10.1109/PEEIC.2018.8665611
10. Ramesh P, Sharmeela C, Prabakaran B, Rajkumar S (2016) ANN controlled sepic cascaded with multi level inverter fed induction motor drive. In: International conference on computation of power, energy information and commuincation (ICCPEIC), Chennai, pp 817–822. https://doi.org/10.1109/ICCPEIC.2016.7557330

11. Nithara PV, Lakshmi Prabha MR (2015) Active and reactive power controller for an autonomous wind system using 12 pulse PWM voltage source converter. In: International conference on emerging trends in engineering
12. Manyonge AW, Ochieng RM, Onyango FN, Shichikha JM (2012) Mathematical modelling of wind turbine in a wind energy conversion system: power coefficient analysis. Appl Math Sci 6(91):4527–4536. https://doi.org/10.1088/1757-899X/322/7/072061
13. Nithara PV, Hari M, Sreelakshmi CS, Eldho RP (2019) V-f controlled autonomous wind energy conversion system using Z2 transformer connected DSTATCOM. Int J Eng Adv Technol 8(6):1492–1496. https://doi.org/10.35940/ijeat.F8135.088619
14. Kasal GK, Bhim S (2009) Three-phase 4-wire ısolated wind energy conversion system employing VSC with a T-connected transformer for neutral current compensation. J Electr Eng Technol 4(2):211–218. https://doi.org/10.1016/j.epsr.2011.12.014
15. Khadkikar V, Chandra A (2008) A new control philosophy for a unified power quality conditioner (UPQC) to coordinate load-reactive power demand between shunt and series inverters. J IEEE Trans Power Delivery 23(4):2522–2534. https://doi.org/10.1109/TPWRD.2008.921146
16. Vilathgamuwa M, Zhang ZH, Choi SS (1998) Modeling, analysis and control of unified power quality conditioner. In: Proceedings of IEEE Harmon. Quality power, pp 1035–1040. https://doi.org/10.9790/1676-1106016469
17. Gon M, Liu H, Gu H, Xu D (2002) Active voltage regulator based on novel synchronization method for unbalance and fluctuation compensation. In: Proceedings of IEEE ındustrial electronics society (IECON), pp 1374–1379. https://doi.org/10.1109/IECON.2002.1185476
18. Khadkikar V, Chandra A, Barry AO, Nguyen TD (2006) Analysis of power flow in UPQC during voltage sag and swell conditions for selection of device ratings. In: Proceedings of IEEE electrical and computer engineering (CCECE), pp 867–872. https://doi.org/10.1109/CCECE.2006.277324
19. Khadkikar V, Chandra A (2006) A novel control approach for unified power quality conditioner Q without active power injection for voltage sag compensation. In: Proceedings of IEEE international conference on industrial technology (ICIT), pp 779–784. https://doi.org/10.1109/ICIT.2006.372356
20. Khadkikar V, Chandra A (2011) UPQC-S: a novel concept of simultaneous voltage sag/swell and load reactive power compensations utilizing series ınverter of UPQC. J IEEE Trans. Power Electron 26(9):2414–2424. https://doi.org/10.1109/TPEL.2011.2106222

Comparative Analysis of Metaheuristic Algorithms for the Implementation of Photovoltaic Solar Panel Model in MATLAB/Simulink

Ashish Kumar and D. Bhagwan Das

1 Introduction

In photovoltaic research, estimation of equivalent circuit parameters of a solar panel is an open research problem [1]. A solution to which can be used to carry out the detailed simulation of solar panel behaviour and performance optimization of the solar panels [2–4]. In order to model solar panels in an adequate manner, two things are required firstly, an appropriate circuit model to represent the solar panel characteristics and secondly, a method to accurately determine solar panel circuit parameters [1]. There are several circuit models considered by researchers to represent a solar panel, viz. single-diode [5], double-diode [6] and triple-diode model [7]. The primary reason given to choose a multiple-diode model is to consider the effect of low illumination conditions, under which multiple-diode models have provided better accuracy than the single-diode model. But as the scope of present work is limited to perfectly illuminated conditions, the model considered here is single-diode model; it is shown in Fig. 1.

In single-diode model of a solar panel, there are five unknown parameters, viz. I_{ph}, I_{sat}, A, R_{se} and R_{sh}. So in order to simulate an accurate model of a solar panel, estimation of these five unknown equivalent circuit parameters is needed to be carried out first. The basic equation which relates these five unknown parameters to the output voltage (V_{out}) and output current (I_{out}) of solar panel is given by Kirchoff's voltage law in Eq. 1 (Table 1)

A. Kumar (✉) · D. Bhagwan Das
Department of Electrical Engineering, Dayalbagh Educational Institute, Deemed University, Agra, Uttar Pradesh, India
e-mail: ashish.postgrad.dei@gmail.com

D. Bhagwan Das
e-mail: dbhagwandas@gmail.com

V. H. Saran and R. K. Misra (eds.), *Advances in Systems Engineering*, Lecture Notes in Mechanical Engineering,
https://doi.org/10.1007/978-981-15-8025-3_36

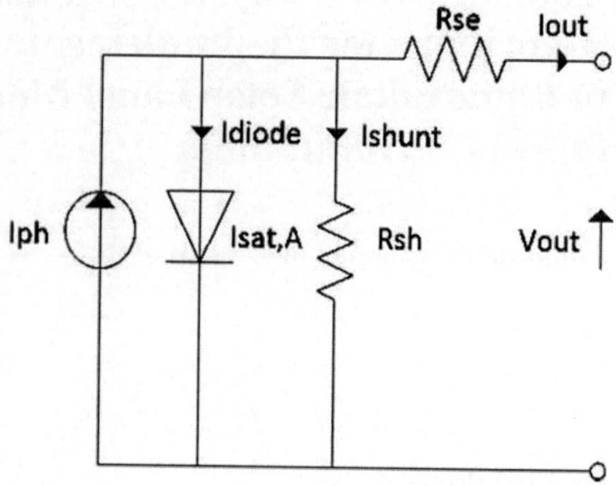

Fig. 1 Single-diode circuit model of a solar panel

Table 1 Symbols used

Symbol	Meaning
I_{ph}	Photocurrent
I_{sat}	Dark Saturation Current
A	Ideality factor
R_{se}	Series resistance
R_{sh}	Shunt resistance
V_{out}	Output voltage
I_{out}	Output current
I_{sc}	Short circuit current
V_{oc}	Open circuit voltage
V_{m}	Voltage at maximum power point
I_{m}	Current at maximum power point
V_{t}	Equivalent voltage to temperature at p–n junction of one solar cell
n_{s}	Number of solar cells connected in series in the panel

$$f(I_{\text{out}}, V_{\text{out}}) = I_{\text{ph}} - I_{\text{sat}}\left(exp\frac{V_{\text{out}} + I_{\text{out}} \cdot R_{\text{se}}}{n_{s.} \cdot A \cdot V_t} - 1\right) - \left(\frac{V_{\text{out}} + I_{\text{out}} \cdot R_{se}}{R_{\text{sh}}}\right) - I_{\text{out}} = 0 \quad (1)$$

Here V_t is given by:

$$V_t = \frac{K_b.T}{q} \quad (2)$$

where K_b is Boltzmann's constant $K_b = 1.38064852 \times 10^{-23}$ m^2 kgs^{-2} K^{-1}, q is elementary charge $q = 1.60217662 \times 10^{-19}$C and T is the module temperature. Now by using Eq. 1, a system of five equations has to be formed to formulate the solar panel parameter estimation problem for single-diode solar cell model. There are two prevalent approaches reported in literature for this; these are three-point and five-point parameter estimation problems [8, 9]. The difference between these two approaches is that the accurate modelling using three-point method is hard to be carried out without using a detailed manufacturer data sheet because the three-point approach requires a beforehand knowledge of the numerical value of ($\mathrm{d}I/\mathrm{d}V$) at short circuit point to develop one of the five equations, which is either provided directly in the detailed data sheet of solar panel or needed to be calculated from the standard solar panel i-v curves provided in such detailed data sheets. But usually the manufacturer data sheets, even if available, are not that much detailed especially in laboratory settings. The five-point parameter estimation technique on the other hand is free from this requirement of a detailed manufacturer data sheet as it is based only on five experimentally acquirable (I, V) data points. So five-point method has been used for parameter estimation in this work.

2 Five-Point Method

To formulate the five required equations, five points from the solar panel *i-v* curve and Eq. 1 have been used. The five points used are shown in Fig. 2. The five equations are as follows.

- **First Equation** For Point 1 ($V_{\text{out}} = 0$, $I_{\text{out}} = I_{\text{sc}}$),

$$f_1(I_{\text{sc}}, 0) = I_{\text{ph}} - I_{\text{sat}}\left(\exp\frac{I_{\text{sc}} \cdot R_{\text{se}}}{n_{s.} \cdot A \cdot V_t} - 1\right) - \left(\frac{I_{\text{sc}} \cdot R_{\text{se}}}{R_{\text{sh}}}\right) - I_{\text{sc}} \tag{3}$$

- **Second Equation** For Point 2 ($V_{\text{out}} = V_{0.25}$, $I_{\text{out}} = I_{0.75}$),

$$\begin{aligned} f_2(I_{0.75}, V_{0.25}) = I_{\text{ph}} - I_{\text{sat}}\left(exp\frac{V_{0.25} + I_{0.75} \cdot R_{\text{se}}}{n_{s.} \cdot A \cdot V_t} - 1\right) \\ - \left(\frac{V_{0.25} + I_{0.75} \cdot R_{\text{se}}}{R_{\text{sh}}}\right) - I_{0.75} \end{aligned} \tag{4}$$

- **Third Equation** For Point 3 ($V_{\text{out}} = V_{\text{m}}$, $I_{\text{out}} = I_{\text{m}}$),

$$f_3(I_m, V_m) = I_{\text{ph}} - I_{\text{sat}}\left(exp\frac{V_m + I_m \cdot R_{\text{se}}}{n_{s.} \cdot A \cdot V_t} - 1\right) - \left(\frac{V_m + I_m \cdot R_{\text{se}}}{R_{\text{sh}}}\right) - I_m \tag{5}$$

- **Forth Equation** For Point 4 ($V_{\text{out}} = V_{0.75}$, $I_{\text{out}} = I_{0.25}$),

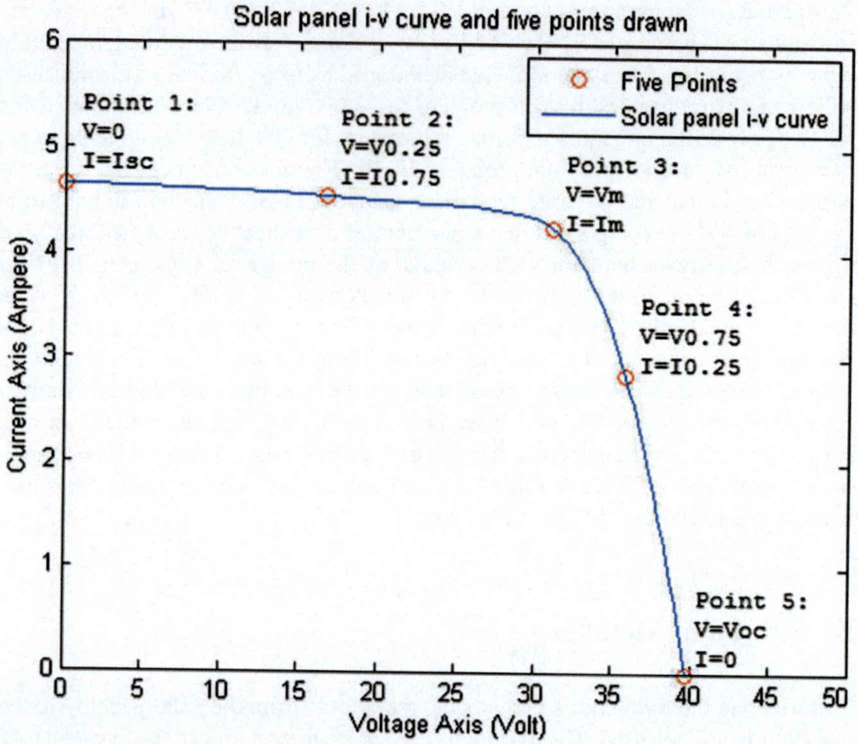

Fig. 2 Solar panel *i-v* curve and five points

$$f_4(I_{0.25}, V_{0.75}) = I_{\text{ph}} - I_{\text{sat}}\left(\exp\frac{V_{0.75} + I_{0.25} \cdot R_{\text{se}}}{n_{s.} \cdot A \cdot V_t} - 1\right) - \left(\frac{V_{0.75} + I_{0.25} \cdot R_{\text{se}}}{R_{\text{sh}}}\right) - I_{0.25} \tag{6}$$

- **Fifth Equation** For Point 5 ($V_{out} = V_{oc}$, $I_{out} = 0$),

$$f_5(0, V_{\text{oc}}) = I_{\text{ph}} - I_{\text{sat}}\left(\exp\frac{V_{\text{oc}}}{n_{s.} \cdot A \cdot V_t} - 1\right) - \left(\frac{V_{\text{oc}}}{R_{\text{sh}}}\right) \tag{7}$$

This system of five equations f1, f2, f3, f4 and f5 represents the single-diode model of the solar panel and the cost function defined using them is given as:

$$\text{Cost function} = \sqrt{\frac{f_1^2 + f_2^2 + f_3^2 + f_4^2 + f_5^2}{5}} \tag{8}$$

Fig. 3 The experimental setup

Hence, the parameter estimation problem gets translated to an unconstrained optimization problem. A solution to which can be obtained by using a numerical or a metaheuristic method. In this work, metaheuristic methods are used to solve the optimization problem.

Now, in order to capture these five data points at different ambient conditions, the experimental setup as shown in Fig. 3 is used to trace nine complete i-v curves of the solar panel under different ambient conditions. In each curve, 52 distinct *i-v* points have been captured by the curve tracer. Out of these 52 *i-v* points, 5 *i-v* points as required are taken out. The solar panel curve tracing has been carried out three times a day on 8, 9 and 10 January 2019. For *i-v* curve tracing, Seaward's solar *i-v* curve tracing kit has been used. In Fig. 3, it is shown connected to the solar panel. The five points fetched from the nine solar panel i-v curves are shown in Table 2.

3 Comparative Analysis of Meta-Heuristic Algorithms

The five metaheuristic algorithms, viz. Real-coded Simulated Annealing, Real-coded Genetic Algorithm, Differential Evolution, Constriction Coefficient-based Particle Swarm Optimization and Levy-flight-based Cuckoo Search are compared for their performance in solving the parameter estimation problem on the basis of two criteria,

Minimum value of cost function obtained, while estimating the solar panel circuit parameters.

Root mean square error in the output current estimation, with respect to the experimentally obtained output current.

Table 2 Data collection of five *i-v* curve points at different temperature and irradiance

Day	Time	Temp (°C)	Irradiance (W/m^2)	(V_{sc}, I_{sc})	($V_{0.25}$,$I_{0.75}$)	(V_m, I_m)	($V_{0.75}$, $I_{0.25}$)	(V_{oc}, I_{oc})
1	10:35	34	651	(0, 3.61)	(16.71, 3.54)	(33.55, 3.31)	(37.65, 2.23)	(41.02, 0)
	13:30	45	798	(0, 4.23)	(08.90, 4.13)	(31.39, 3.80)	(36.60, 2.33)	(39.92, 0)
	15:47	31	425	(0, 1.98)	(17.69, 1.93)	(34.28, 1.81)	(37.92, 1.28)	(40.88, 0)
2	10:39	35	635	(0, 3.60)	(22.43, 3.45)	(33.50, 3.24)	(37.74, 2.12)	(40.93, 0)
	12:47	48	917	(0, 4.80)	(16.52, 4.71)	(31.23, 4.37)	(35.92, 2.88)	(39.73, 0)
	15:35	37	487	(0, 2.34)	(15.44, 2.29)	(33.42, 2.13)	(37.21, 1.48)	(40.31, 0)
3	10:40	36	660	(0, 3.71)	(17.32, 3.64)	(33.07, 3.41)	(37.46, 2.27)	(40.86, 0)
	12:31	47	877	(0, 4.64)	(16.78, 4.54)	(31.06, 4.24)	(35.82, 2.83)	(39.69, 0)
	15:17	34	551	(0, 2.64)	(16.23, 2.59)	(33.22, 2.46)	(37.20, 1.82)	(40.85, 0)

To do so first, the parameters are estimated by each algorithm using the five points obtained from each experimentally captured curve at different ambient conditions. Once estimated, the circuit parameters are used to further estimate the output current of the solar panel at the same voltages at which the experimental solar panel *i-v* curve points are obtained. Then the root mean sqaure error between the estimated and experimentally obtained current values is calculated for each data instances. For the purpose of output current estimation because of its fast convergence and easy implementation, the univariate Newton–Raphson method has been used to solve the single-diode model equation for the output current, given by Eq. 1, once the other unknown parameters are estimated by a metaheuristic method. The number of iterations performed by each metaheuristic algorithm is 10,000 with a population size of 50. Now, Figure 4 presents the minimum value of the cost function in semilogy scale reached by each algorithm in all nine different data instances, after performing all the iterations. On the other hand, Figure 5 presents the root mean square error in output current estimation in semilogy scale reached by the different algorithm for all the different nine data instances.

Based on the best cost and root mean square error plots in Fig. 4 and Fig. 5, it could be concluded that the performance of Cuckoo Search algorithm is better than other competitive algorithms. The Particle Swarm Optimization can be considered as the second best. So using the Cuckoo search algorithm, a Simulink model has been developed. Then to demonstrate the functioning of the developed solar panel model in Simulink, the *i-v* and *p–v* curves corresponding to different temperature and irradiance on 8-Jan-2019 (day 1) at three different time instances have been obtained and are shown in Fig. 6 and Fig. 7, respectively. Here, the dots (o) represent the experimentally acquired points and the coloured lines (-) represent the estimated curve obtained from the solar panel MATLAB/Simulink simulation. From the given figures, it could be observed that the estimated curves satisfy the experimental data to a good degree.

4 Conclusion

In this paper, comparison between several metaheuristic algorithms to solve the solar panel parameter estimation problem has been carried out. Minimum value of cost function obtained by a particular algorithm and the root mean square error in the solar panel output current estimation are considered as two scales for comparing their performance. The metaheuristic algorithms among which the comparison has been carried out are: Simulated Annealing, Genetic Algorithm, Differential Evolution, Particle Swarm Optimization and Cuckoo Search. In terms of obtaining the minimum value of the cost function, Cuckoo Search outperforms all the other algorithms while in terms of root mean square error in output current estimation, it also performed better. Based on these outcomes, Cuckoo Search has been used to implement the model of the solar panel in Simulink.

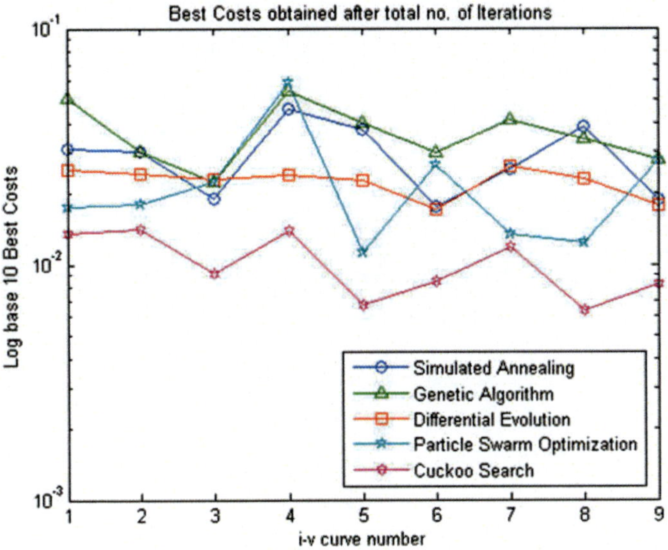

Fig. 4 Minimum values of the cost function

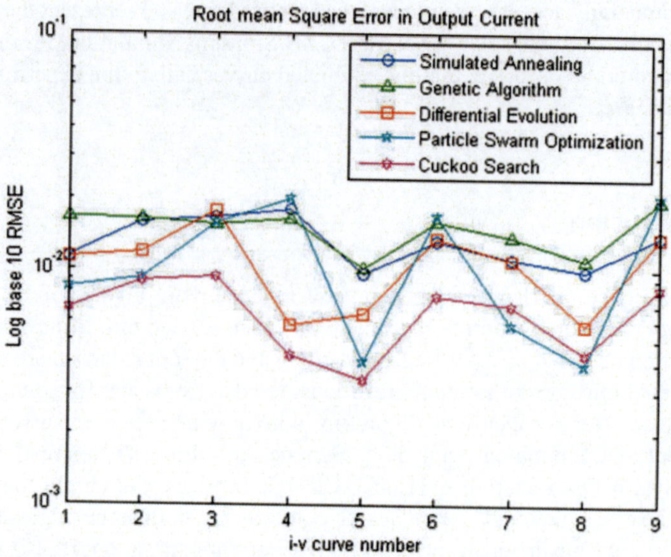

Fig. 5 Root mean square error in output current estimation

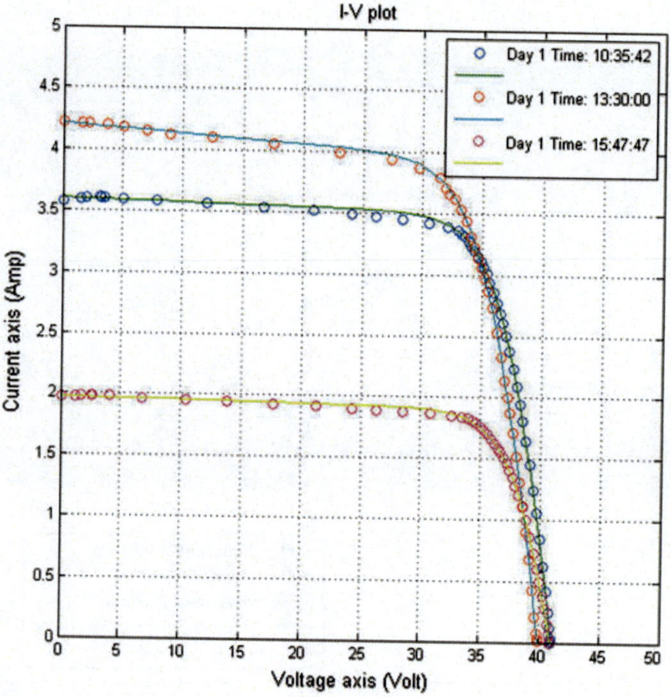

Fig. 6 I-V plot for day 1 data

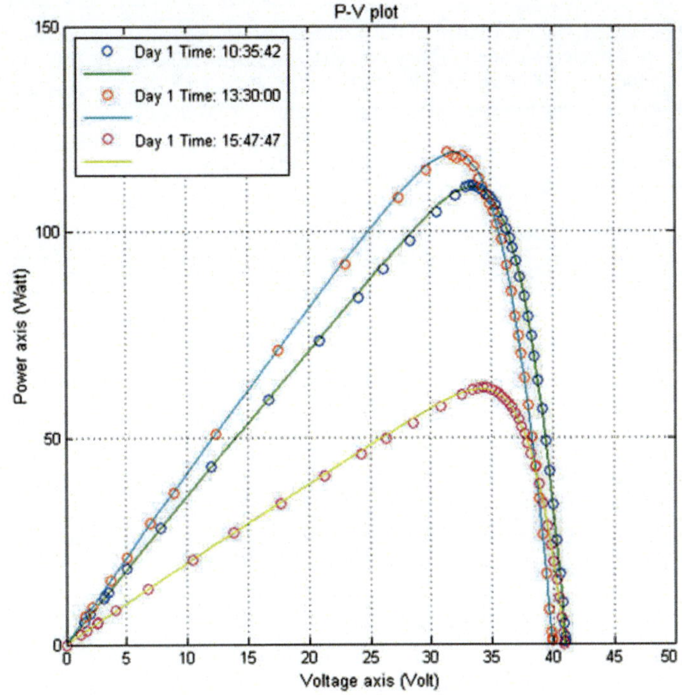

Fig. 7 P–V plot for day 1 data

Acknowledgements This work is supported by a DST Project titled "DST MI Smart Grid R&D Project" no. DST/CERI/MI/SG/2017/060.

References

1. Rezaee Jordehi A (2016) Parameter estimation of solar photovoltaic (PV) cells: a review. Renew Sustain Energy Rev 61:354–371 (Elsevier)
2. Bellia H, Youcef R, Fatima M (2014) A detailed modeling of photovoltaic module using MATLAB. J Astronomy Geophys NRIAG 3:53–61 (Elsevier)
3. Salmi T, Bouzguenda M, Gastli A, Masmoudi A (2012) PV panel model based on datasheet values. Int J Renew Energy Res IJRER 2(2)
4. Argyrou MC, Christodoulides P, Kalogirou SA (2018) Modeling of a photovoltaic system with different MPPT techniques using MATLAB/Simulink. In: IEEE International energy conference, IEEE, pp 1–6
5. Etienne S, Alberto T, Mikhal S (2011) Explicit model of photovoltaic panels to determine voltages and currents at the maximum power point. Solar Energy, Elsevier 85(5):713–722
6. Vandana Khanna, Das BK, Dinesh Bisht, Vandana, Singh PK (2011) Simple, fast and accurate two diode model for photovoltaic modules. Solar Energy Mater Solar Cells 95(2):586–594 (Elsevier)

7. Kashif I, Zainal S, Hamed T (2015) A three diode model for industrial solar cells and estimation of solar cell parameters using PSO algorithm. Renew Energy, Elsevier 78:105–113
8. Sera D, Teodorescu R, Rodriguez P (2007) PV panel model based on datasheet values. In: Proceedings of the international symposium on industrial electronics, Spain, IEEE
9. Reis LRD, Camacho JR, Novack DF (2017) The Newton Raphson method in the extraction of parameters of PV modules. In: Proceedings of the international conference on renewable energies and power quality, Spain

Data Analytics Systems

Emotion Recognition from Facial Biometric System Using Deep Convolution Neural Network (D-CNN)

Chevella Anil kumar and Kancharla Anitha Sheela

1 Introduction

Since computers were invented, people desired to build systems that are mentally and physically equal to humans, that is, having artificial intelligence (AI). In decades, the increase in computational power that is available provided a helping hand in the development of fast learning machines, while the online sources offered a huge amount of training data. Among the lots of modern machine learning technologies that have been built to this stage, adequate thought is given to each other by the most promising technologies for creating AI machines that approach human-level intelligence. Recognition of expression is the method of recognition of human emotions based on facial expressions. Children who are only 36 h of age can interpret some basic facial emotions.

A person produces identical muscle movements that belong to a specific state of mind despite birth, race, education, etc. Therefore, in human–machine interaction, if properly modelled, this can be a convenient feature. A well-trained system, independent of who the subject is, can understand emotions. Automated FER has numerous practical applications for machine interactions such as robots, medical applications, neuromarketing, e-learning, monitoring, entertainment, law and games in human. If robots are able to predict human emotions, they can react to it and behave accordingly. We introduce deep learning techniques in this paper and suggest active deep convolution neural network architectures to solve the problem of recognition of facial expression.

C. Anil kumar (✉) · K. A. Sheela (✉)
JNTUH College of Engineering, Hyderabad 500085, India
e-mail: chevellaanilkumar@gmail.com

K. A. Sheela
e-mail: kanithasheela@gmail.com

V. H. Saran and R. K. Misra (eds.), *Advances in Systems Engineering*, Lecture Notes in Mechanical Engineering,
https://doi.org/10.1007/978-981-15-8025-3_37

2 Literature Survey

Paul Ekman has been conducting extensive studies on human facial expressions [1] since 1970. He found that emotional facial expressions are not culturally determined, but are universal to human culture and therefore biological in origin. He developed the Facial Action Coding System [2] where a set of action units (Aus) [2] is described as movement on the face. Computer-assisted facial expression recognition began in the 1990s. Usually, the features used are based on the local spatial location of the face (eye top, eyebrows and mouth) different points and regions.

One of the first researchers to classify facial expressions using image processing techniques is Mase (1991). K-nearest neighbours (KNN) [3] rule was used with an 80 per cent accuracy for the classification task of emotions. Tian et al. (2000) investigated recognition using facial features (lips and wrinkles) that are permanent and temporary. For finding their shapes and features, geometric models were used. With a neural network (machine learning)-based classification, i.e. SVM [4, 5], they achieved 96 per cent accuracy. Cohen (2003, 2004) implemented the static description of the Bayesian network.

Previous research focused on different techniques for image processing to match features that describe facial geometry. While these methods worked under very restricted settings, they showed that computers can be used to automatically recognize faces. Thereafter, statistical methods, principal component analysis (PCA) and linear discriminate analysis (LDA) [6, 7], are used. Feature-based emotion recognition approaches consist of matching local image-wide features. Holistic methods were developed and combined with feature-based methods into hybrid methods. Up until recently, facial recognition systems based on hybrid methods remained state of the art. Deep learning [8] emerges [9] as the leading approach to most applications for computer vision, including face recognition. The deep learning methods for face recognition are convolution neural networks (CNNs) [10–13] and deep convolution neural networks (D-CNNs) [14, 15].

3 Deep Learning

Deep learning techniques are machine learning methods that teach computers to do what comes naturally to humans; deep learning is a crucial technology behind facial emotion recognition to understand human expressions for different applications, driverless cars, and also alerts them to recognize a stop sign. Computer learns to perform tasks of classification from images in deep learning. Deep learning achieves high accuracy than ever before in independent word accuracy recognition. It helps consumer electronics meet the expectations of users, and it is important for many applications in real time.

Table 1 Evaluation of CNN architecture

Year	CNN (layers)	Place	Top-5 error rate
1998	LeNet(8)		
2012	AlexNet(7)	1st	15.396
2013	ZFNet()	1st	14.896
2014	GoogLeNet(19)	1st	6.6796
2014	VGG Net (16)	2nd	7.396
2015	ResNet (152)	1st	3.696

3.1 CNN Evolution

LeNet, AlexNet, ZFNet, GoogLeNet, VGG and ResNet (https:/iq.opengenus.org/evolution-of-cnn-architectures/) It all began in 1998 with LeNet and eventually, almost fifteen years later, resulted in groundbreaking models winning the ImageNet Massive Scale Visual Recognition Challenge incorporating AlexNet in 2012 with GoogLeNet in 2014 with ResNet in 2015. There has been no significant progress over the last two years, and the new models are an ensemble of earlier groundbreaking models (Table 1).

4 Convolutional Neural Network

A convolutional neural network is a multilayer neural network designed to recognize visual patterns with minimal pre-processing directly from pixel images. These are aligned with ordinary neural networks in the sense that these consist of neurons with biases and weights. Growing neuron receives few inputs, performs a dot product and follows it with optional nonlinearity.

4.1 Architecture Overview

Neural networks took an input and transformed it into a sequence of invisible layers. Every secret layer is created from a community of neurons, wherever every nerve cell is fully connected to all or any neurons within the previous layer, and wherever neurons perform entirely severely during a single layer and have no connections with each other. The last layer that is totally connected with previous layer is called the "output layer", and it represents the category scores in the classification settings.

In JAFFE, pictures are only 192 × 192 × 1 (192 wide, 192 high, one channel), a single neuron of a neural network having 192 * 192 * 1 = 36,864 nodes in the first layer. This amount appears to be manageable, but this fully connected structure

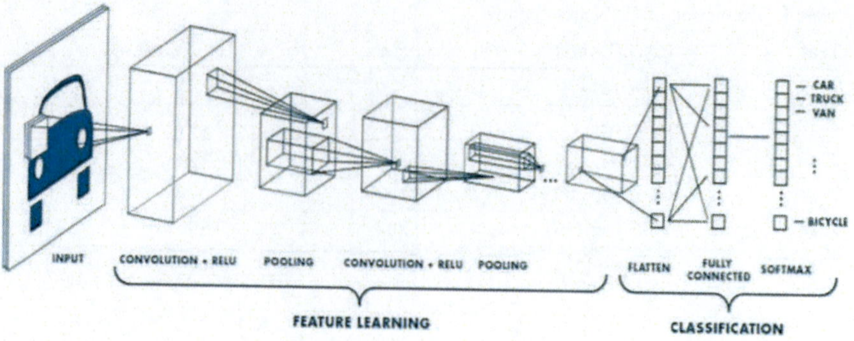

Fig. 1 Structure of convolution neural network

clearly does not scale up to larger images. This complete connectivity is wasteful, and the huge number of parameters leads to over-fitting.

CNNs take benefit of the fact that the input is made up of objects and they are more receptive to the architecture. In particular, unlike a neural network, a ConvNet's layers have three-dimensional neurons arranged: height, width and depth. The input images in JAFFE, for example, are an activation input volume, and the size has 192 × 192 × 3 dimensions (width, height, depth). Before it, neurons in a layer will be connected to a particular area of the layer. The output layer would have dimensions 1 × 1 × 7 for JAFFE, because the full image will be reduced by the end of the ConvNet architecture into a single class scoring vector, organized along the *d* length (Fig. 1).

CNN is very useful for studying usability and high abstraction modelling. This comprises seven input layer elements, convolution layer, rectified linear unit (ReLU), pooling layer (subsampling layer), fully connected layer, output layer and softmax layer.

Input layer—Input image takes the raw data.

Convolution layer—This layer calculates the convolutions in the input between the neurons and different patches as determined by the size of the kernel. For the feature map, the kernel is moved over the input image. The output can be calculated as follows:

$$\text{Conv}_{x,y} = \sum w_i * v_i \tag{1}$$

w_i = kernel weights, v_i = values of the spatially corresponding elements in the input maps

$$Z_{x,y} = \sum w_i \; * \; v_i + b_i \tag{2}$$

$z_{x,y}$ is the output of the convolution layer with adding bias value b_i

With the convolution, there are other tricks we may apply, one of which is padding and the other is striding.

When using a CNN, we need to know the relationship between the size of the filter, image and padding during convolution. Without padding of the image, the pixels at the edge of the image are processed only partially and the result of the convolution will be smaller than the actual image size.

Rectified linear unit (ReLU)—It applies an activation function to the previous layer output. It adds nonlinearity to the network so that it can generalize well to any function and it simply threshold at zero and can be calculated as follows:

$$R_{x,y} = \max\left(0, Z_{x,y}\right) \tag{3}$$

Pooling layer—Pooling helps to keep only the important parts of the network as we progress. The pooling layer works on each input deep slice and resizes it in space. It includes many types of max pooling, average pooling and L2 pooling. There are two hyperparameters in pooling—filter(L) and stride(S). Generally, if we have $A_1 \times B_1 \times C_1$ input dimensions, then $A_2 = (A_1 - L)/S + 1$, $B_2 = (B_1 - L)/S + 1$, $C_2 = C_1$, where A_2, B_2, C_2 are the width, height and depth of output.

Flattening: Flattening converts the data into a one-dimensional array to be added to the next layer. To create a single long feature vector, flatten the output of convolution layers.

Fully connected layer: Fully connected layers in general neural networks are similar to neurons in which their neurons in the previous layer are fully connected to each neuron in the next layer.

$F(X, W) = \phi(W * X)$, where ϕ is nonlinear activation function preferably bipolar sigmoid.

Output layer—Layer calculates the last layer output scores. The resulting output is $1 \times N \times N$, where $N =$ number of classes in the training data set.

Softmax Layer: The network error is propagated by a softmax layer backwards. Softmax can calculate a mapping, and the calculation of each component of the softmax layer is as follows:

$$S(x) = \frac{e^{x_i}}{\sum_{j=0}^{k} e^{x_j}} \quad i = 0, 1, 2 \ldots k, \tag{4}$$

CNN learning works by determining neuron weights for the strongest synapses. CNN includes raw images as input, unlike the general neural network that consists of handcrafted features. During the training phase, the network is fed by training data that includes greyscale images with their respective tags, and a set of images is selected for the validation set to determine the best weight set. The network receives a greyscale image during the test phase, and network outputs are used to predict a given image's category.

Adam optimization: Adam optimization is one that works well across a wide variety of architectures of deep learning for optimizing the weights of neurons.

Some advantages of Adam include:

1. Relatively low memory requirements.
2. Works well even a little tuning of hyperparameters.

Categorical cross-entropy loss, or log loss, tests the efficiency of the classification system whose output is between 0 and 1. Cross-entropy decreases as a sample's expected likelihood varies from the actual value.

$$-(y_i \log(p_i) + (1 - y_i) \log(1 - p_i) \tag{5}$$

Here, y_i is the binary indicator (0 and 1) denoting the class for the sample i, while p_i indicates the predicted probability between 0 and 1 for that sample.

Alternatively, we define a term known as categorical cross-entropy if there are more than two classes. The following equation is given mathematically:

$$-\sum_{C=1}^{M} y_{i,c} \log(p_{i,c}) \tag{6}$$

Here, M denotes the number of classes and$y_{i,c}$ is a binary indicator 0 or 1 that indicates whether C is the correct class or not.

5 JAFFE Database [16]

The database contains 1596 static (Table 2) images of emotional facial expressions posed by 10 female Japanese models of 192 × 192 size. Each image was rated by 60 Japanese subjects on seven emotional adjectives (Fig. 2).

6 Proposed Work

In this proposed approach, in order to define seven basic emotions of JAFFE and my own data set, we tend to develop a three-layer convolution neural network structure with ReLU activation function, maxpooling and a fully connected softmax layer in accordance with the Adam optimization algorithm with categorical cross-entropy loss function.

Table 2 JAFFE database information

	Disgust	Happy	Anger	Surprise	Sad	Fear		Neutral
	16	30	24	30	30	20		24
2	22	28	24	20	22	32		24
3	12	24	24	22	24	24		22
4	22	24	18	24	24	26		20
5	20	16	24	24	24	24		22
6	30	24	24	22	24	22		22
7	22	24	22	24	24	22		24
8	22	20	22	22	20	24		20
9	24	20	24	24	22	24		22
10	22	22	22	20	22	22		22

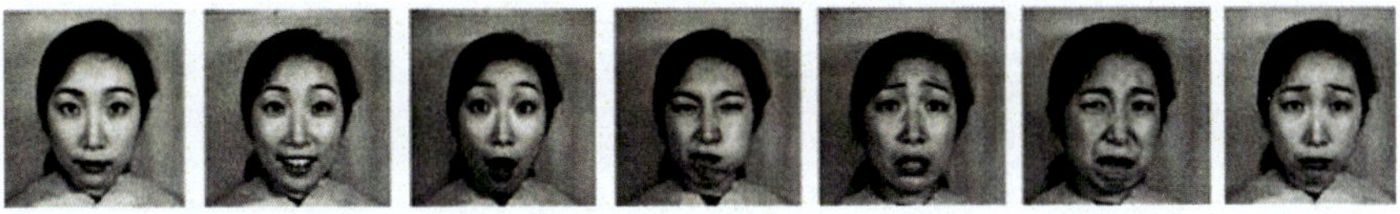

Fig. 2 Different emotions of JAFFE data set

Data set was labelled as seven different emotions (anger, fear, happiness, neutrality, disgust, surprise and sadness), and 80 per cent was used for training and 20 per cent for testing. 85% was used for training and 15% for validation from training data set to find the exact termination point (early stop point) for training. All the modules were developed using the Keras deep learning library's TensorFlow framework.

7 Testing

Testing for a group of images for different epochs and confusion matrices generated for labelled versus predicted images are tabulated (Table 3).

7.1 Confusion Matrix

Total number of images used for testing = 324.

Table 3 Different confusion matrix for actual versus predicted emotions

Confusion matrix for 5 epochs average accuracy = 0.447

	Neutral	Happy	Surprise	Anger	Fear	Disgust	Sad
Neutral	18	5	5	2	2	3	1
Happy	10	12	9	2	1	0	2
Surprise	2	1	35	4	0	0	0
Anger	18	5	2	15	1	3	4
Fear	15	7	15	3	16	2	2
Disgust	0	4	1	1	4	38	0
Sad	8	18	5	8	0	4	11

Confusion matrix for 50 epochs average accuracy = 0.975

	Neutral	Happy	Surprise	Anger	Fear	Disgust	Sad
Neutral	36	0	0	0	0	0	0
Happy	0	36	0	0	0	0	0
Surprise	0	0	42	0	0	0	0
Anger	0	0	0	48	0	0	0
Fear	0	0	0	0	52	2	6
Disgust	0	0	0	0	0	48	0
Sad	0	0	0	0	0	0	54

Confusion matrix for 100 epochs average accuracy = 0.895

	Neutral	Happy	Surprise	Anger	Fear	Disgust	Sad
Neutral	36	0	0	0	0	0	0
Happy	0	36	0	0	0	0	0
Surprise	0	0	42	0	0	0	0
Anger	0	0	0	48	0	0	0
Fear	0	0	0	0	52	2	6
Disgust	0	0	0	0	0	48	0
Sad	0	0	0	0	0	0	54

8 Result

From the confusion matrices, it is observed that 44.7% accuracy is achieved with five epochs, 97.5% accuracy with 50 epochs and 89.5% with 100 epochs, which indicates under-fitting with five epochs and over-fitting with 100 epochs of data. Hence, training was stopped at 50 epochs. A sample of image labelled probabilities and the different emotions recognized is shown in Figs. 3 and Fig. 4, respectively.

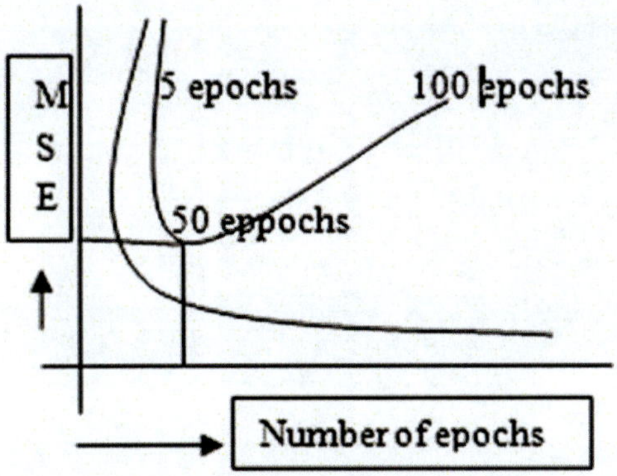

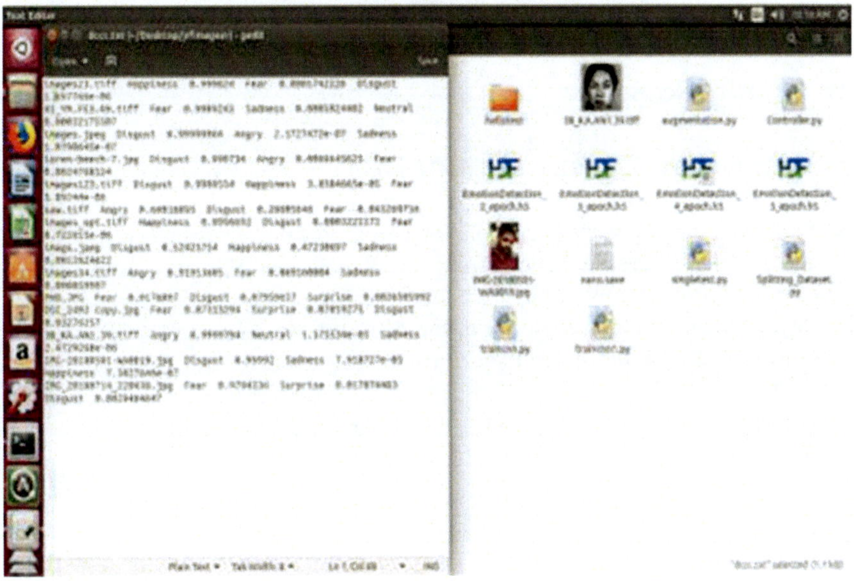

Fig. 3 Each image labelled probabilities

9 Conclusion

This paper presented an emotion recognition system for the JAFFE database and real-time images for the group as well as individual images using deep convolution neural network with three hidden layers for multiple epochs to avoid over-fitting and under-fitting problems and generated different confusion matrices along with an accuracy of each epoch. Proposed convolution neural network (CNN) model is giving an accuracy of 97.5%. It was given a better accuracy compared to the machine learning algorithms by Cheng et al. [16] which is 93.43% on the same JAFFE data set.

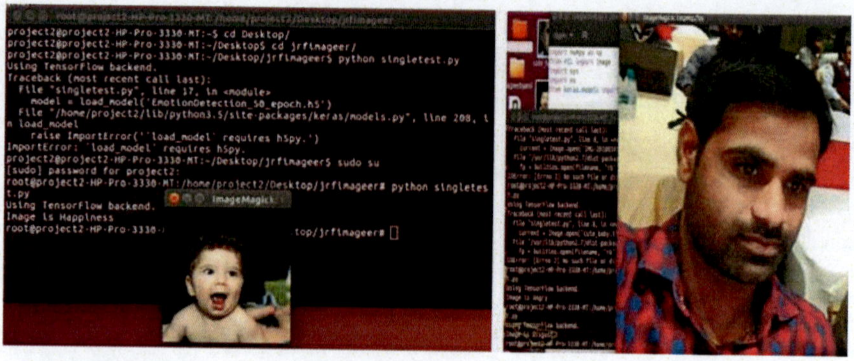

Fig. 4 Different emotions recognition

10 Future Scope

As the technology is growing in today's world, emotion recognition from facial expressions plays a vital role in real-time applications and human-to-machine interface for communication. Deep convolution neural network structure can also be used for recognizing different emotions from speech data sets. However, deep convolution neural networks need a large amount of database for training and validation. Further, to increase the availability of the database, we have created an Anechoic Chamber recently in our department for recording ground truth facial and speech signal database with various emotions, which helps us to create the needed database for recognition of different emotions from multiple databases.

References

1. Cohen KT, Tian Y (2000) Comprehensive database for facial expression analysis. In: International conference on automatic face and gesture recognition, pp 46–53
2. Lien JJ et al (1988) Automated facial expression recognition based on FACS Action Units. In: 3rd international conference, 14–16 April 1988 at Japan
3. Reney D et al (2015) An efficient method to face and emotion detection. In: 2015 fifth international conference on communication systems and network technologies, 978-1-4799-1797-6/15 $31.00 © 2015 IEEE https://doi.org/10.1109/CSNT.2015.155
4. Bottou L, Cortes C, Denker J, Drucker H, Guyon I, Jackel L, LeCun Y, Muller U, Sackinger E, Simard P, Vapnik V (1994) Comparison of classifier methods: a case study in handwriting digit recognition. In: Proceedings of international conference pattern recognition, pp 77–87
5. Kotsia I, Pitas I (2007) Facial expression recognition in image sequences using geometric deformation features and support vector machines. IEEE Trans Image Process 16(1)
6. Belhumeur P, Hespanha J, Kriegman D (1997) Eigenfaces vs. Fisherfaces: recognition using class specific linear projection. IEEE Trans Pattern Anal Mach Intell 19:711–720
7. Turk MA, Pentland AP (1991) Face recognition using Eigenfaces. In: Proceedings of the IEEE conference computer vision and pattern recognition, pp 586–591
8. Real Time Facial Expression Recognition with Deep Learning. https://github.com/a514514772

9. https://github.com/mbadry1/DeepLearning.ai-Summary/tree/master/4-20Convolutional%20Neural%20Networks.
10. Convolutional Neural Networks for Visual Recognition, Spring 2019. https://cs231n.stanford.edu/
11. Tumen V et al (2017) Facial emotion recognition on a dataset using convolution neural network. 978-1-5386-1880-6/17/$31.00
12. Lawrence S, Lee Giles C, Ah Ch (1997) Face recognition. A convolutional neural-network approach. IEEE Trans Neural Netw 8(1)
13. Wang T, Wu DJ, Coates A, Ng AY (2012) End-to-end text recognition with convolutional neural networks. In: 21st international conference on pattern recognition (ICPR 2012), Stanford University, 353 Serra Mall, Stanford, CA 94305, November 11–15, 2012. Tsukuba, Japan
14. Pour MM et al (2017) Facial emotion recognition using deep convolutional networks. In: IEEE 4th international conference on knowledge-based engineering and innovation (KBEI), pp 0017–0021
15. Krizhevsky A et al (https://code.google.com/p/cuda-convnet/) ImageNet classification with deep convolution neural networks
16. Cheng F et al (2010) Facial expression recognition in JAFFE dataset based on Gaussian process classification. IEEE Trans Neural Netw 21(10)
17. Lyons M, Akamatsu A, Kamachi M, Gyoba J (1998) Coding facial expressions with Gabor wavelets. In: IEEE international conference o automatic face and gesture recognition, pp 200–205

Tracking of Football During Set Piece Scenarios Using Variants of Unscented Kalman Filter

Pritam Pathak, Sarbamoy Mallick, and Abhik Mukherjee

1 Introduction

Set piece scenario in football refers to the kick or throw that restarts the game after foul or line crossing interruption. Statistics show that about 40% of the goals are scored from set piece situation. Tracking of set piece scenarios could be immensely helpful for coaches to plan about engaging players with right height at right positions running in the right direction during set piece. Tracking here implies estimation of the position coordinates of the ball during flight. To compute the motion of the ball, the kinematic equations of the ball are needed. The forces acting on the ball include the thrust on the ball at the time of kick, aerodynamic drag acting on the ball and the force of gravity [1–3].

For linear system dynamics, the system states can be predicted as apriori estimate with assumed noise statistics. This apriori state estimate is then refined using available observations, which too are contaminated with measurement noise. The proportion of belief in the apriori estimate and measurement can be optimized based on assumed noise statistics for the system and measurement, giving estimation convergence. This concept of estimator for linear systems, known as conventional Kalman filter (CKF), serves the purpose of estimation with linear system as well as measurement models [4, 5]. When system or measurement equations becomes nonlinear, extended Kalman filter (EKF) and unscented Kalman filter (UKF) can be more suitable for estimator design.

For the present problem, the state variables to be estimated are the position, velocity and acceleration along the three axes. Since aerodynamic drag depends on one of the state variables, i.e. velocity, the system model becomes nonlinear. The measurements considered are range, azimuth and elevation angles—synthesized

P. Pathak · S. Mallick · A. Mukherjee (✉)
CST Department, IIEST (BESU), Shibpur, India
e-mail: abhik@cs.becs.ac.in

V. H. Saran and R. K. Misra (eds.), *Advances in Systems Engineering*,
Lecture Notes in Mechanical Engineering,
https://doi.org/10.1007/978-981-15-8025-3_38

as measurement of electromagnetic radiation intensity from the ball detected by a sensor placed at the centre of the bar post of the defending goal. Taking the line of sight between the sensor and the ball as boresight, the measurement states can be computed from the received electromagnetic energy intensity. However, since the state variables are modelled in Cartesian coordinates and measurement is available in polar coordinates, the conversion of measurement vector to state vector is nonlinear. Due to weakness in system model, the apriori estimate is noise prone and due to the ambient noise, the measured intensity is also noisy. All this requires the design of a nonlinear estimator.

Proper tuning of system and measurement noise statistics is the key to good target tracking [6, 7]. EKF tackles nonlinearity in each time step by forming Jacobian of the state transition matrix. UKF reduces the computational complexity in this respect [8–11]. This work is a continuation of the earlier work [12] that compared UKF and EKF design but considers only range and elevation angle in 2D plane with fixed azimuth. In Sect. 2, the system model, measurement model and the filter are described in details. This is followed by the details of implementation in Sect. 3. The results obtained by running the proposed algorithm for a typical case study are shown in Sect. 4. Finally, the limitations and scope for further work are discussed in the concluding section.

2 Methodology

The translation and rotation of a football in flight in 3D space with the direction of velocity and angles made with the respective axes are shown in Fig. 1. Angular velocity (ω) calculation uses positive z-direction and negative y-direction. θ is the angle of elevation of the ball, α is the azimuth angle, $\hat{i}$, $\hat{j}$ and $\hat{k}$ being the unit vectors along the x, y and z axes, respectively.

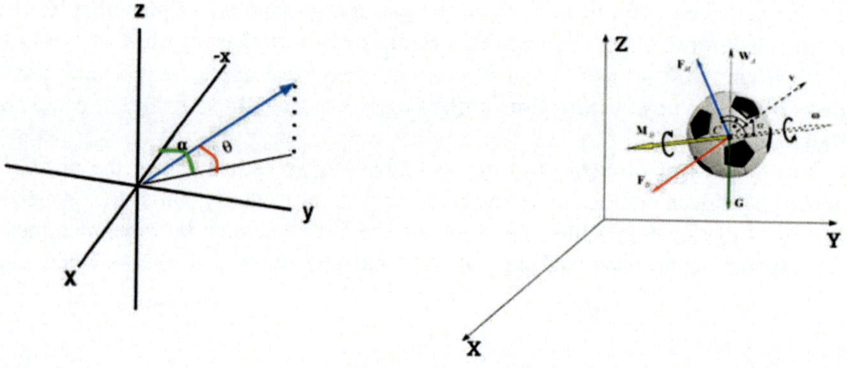

Fig. 1 Angle representation in 3D space (L) free body of a football in 3D space (R) tracking football with UKF 3

$$\text{Velocity, } v = v \cdot \cos\theta\left(-\cos\alpha\hat{i} + \sin\alpha\hat{j}\right) + v \cdot \sin\theta\hat{k} \tag{1}$$

$$\text{Angular velocity, } \omega = \sin\theta\left(\cos\alpha\hat{i} - \sin\alpha\hat{j}\right) + \cos\theta\hat{k} \tag{2}$$

2.1 System Description

The forces acting on the ball during its flight in 3D space are as follows [1, 2]: **Gravitational force**: the force that attracts any two objects with mass

$$G = m \cdot g(-\hat{k}) \tag{3}$$

where m is the mass of the football and g is the acceleration due to gravity.

- **Archimedes Force**: The buoyant force that lifts the ball by acting directly opposite to the force of gravity and is given by:

$$W_A = -(0.40)\pi R^3 \rho_a g\hat{k} \tag{4}$$

where R is mean radius of ball, ρ_a the air density at ambient temperature.
- **Drag force**: Force acts opposite to the direction of ball's velocity, given by:

$$F_D = \frac{1}{2} C_D \rho_a A v^2 \cos\theta(-\cos\alpha\hat{i} + \sin\alpha\hat{j}) + \sin\theta\hat{k} \tag{5}$$

with $A = \pi R^2$, middle cross-sectional area of the ball and drag coefficient is C_d..
- **Magnus force**: The force is generated by a spinning object travelling through a viscous fluid. It is perpendicular to the velocity vector of the object. Direction of spin dictates orientation of the Magnus force on the object. C_L is the lift coefficient depending on Reynolds number (R_e). It is given by:

$$FM_D = \frac{1}{2} C_D^{(m)} \rho_a A v^2 \frac{\omega \times v}{\|\omega \times v\|} \tag{6}$$

$$\omega \times v = -\sin\alpha\hat{i} - \cos\alpha\hat{j} + 0\hat{k} \tag{7}$$

Drag moment: The force acts opposite to angular velocity with the presumption of the ball being an ideal sphere and is given by:

$$M_D = \frac{1}{2} C_D^{(m)} \rho_a A v^2 \frac{\omega}{\|\omega\|} \tag{8}$$

2.2 System and Measurement Model

In 3D model, the state variables are represented as: $X = [x\ y\ z\ \dot{x}\ \dot{y}\ \dot{z}\ \ddot{x}\ \ddot{y}\ \ddot{z}]^T$.

The state equations are: $X_{k+1} = FX_k + w$ where X_k is the state at the kth instant of time, F is the model and w is the Gaussian noise of the process, dT being the time interval between two measurements.

Based on the choice of F, the filters are henceforth referred to as standard UKF (SUKF) and adaptive UKF (AUKF). SUKF uses the Newtonian laws of motion to formulate the state model of the moving ball. The adaptive filter, however, imparts the knowledge of all the forces that are acting on the ball, as described in the earlier section, to formulate the state model.

The measurement can be taken from any position on the field. To simplify calculations, it can be placed at the initial position as the ball itself. But practically and without loss of generality, the sensor can be assumed to be placed above the goal post. Radar principle of operation yields the range, r and the angle of elevation, θ of the football, α as shown in Fig. 1 and this forms the measurement space $Z = [r\ \theta\ \alpha]^T$ with $r = px^2 + y^2 + z^2$, $\alpha = \tan^{-1}\frac{y}{x}$ and $\theta = \tan^{-1}\frac{z}{\sqrt{(x^2+y^2)}}$. Residual Y can be calculated as: $Y = Z - H\hat{X}^-$ where H is the Jacobian of the nonlinear function which transforms the apriori estimate of the process, $\hat{X}^-$ from state space to measurement space.

2.3 Filter Design

UKF design involves a deterministic sampling technique called unscented transformation to tackle the nonlinearity, bypassing Jacobian computation. For UKF, the nonlinear function describing the system dynamics using sigma points is directly used to compute the apriori state vector and required noise statistics. These sigma points, generated using sigma function along with mean and variance weights, are used for time and measurement update [8–10]. However, although computational overhead for Jacobian computations is absent, due to sigma point computation the complexity is not reduced much.

The adaptive UKF adapts to the dynamics of the process model without changing any conventional steps of the filter. In order to refine the system equations with proper system parameters, specific playing conditions, atmospheric conditions like wind and temperature, as well as knowledge about thrust, angle and curving skills of the striker may be incorporated.

3 Implementation

According to the standards, height of the inner post is 2.44 m above the ground. Off-the-mark kicks are excluded from the study as these are irrelevant from tracking perspective. The engagement variables considered for the realistic set-up are as follows:

- Initial velocity, $v = 28$ m/s
- Initial angle of elevation of the kick, $\theta = 18°$
- Initial azimuth angle, $\alpha = 2°$
- Distance of the spot from goal = 30 m
- Interval of $\mathrm{d}T = 0.0005$ s between measurements, total 3000 s
- Mass and radius of ball, $m = 0.45$ kg, $R = 0.11$ m
- Convergence threshold of error = 10^{-1}.

Air density $\rho_a = 1.205$ kg/m^3 at 20 °C, middle cross-sectional area of the ball is $A = \pi R^2$ and C_D the drag coefficient, for the case of translation without spin, depends mainly on Reynolds number R_e which ranges from 2.93×10^5 to 4.4×10^5 for velocity in range 20–30 m/s. From the wind tunnel data for C_D as a function of R_e and v, according to [3] $C_D \approx 0.20$ in the absence of rotation. Standard deviation of measurement = [0.1 m, 0.01 rad, 0.01 rad] and that of process = [0.1 m, 0.1 m, 0.1 m, 1 m/s, 1 m/s, 1 m/s, 2 m/s^2, 2 m/s^2, 2 m/s^2].

4 Results

Present study removes assumptions like known azimuth angle and considers 3D trajectory equation, together with adaptations like player knowledge and ambient noise statistics. The filter is tested in simulation environment through Monte Carlo runs. Convergence of error covariance for the tracked trajectory and histogram analysis of the residual are satisfactory. All performance metrics are presented here as the mean of 100 Monte Carlo simulation runs unless stated otherwise. This is done so as to study the behaviour of the models under multiple randomly generated inputs. Standard UKF and adaptive UKF are compared to each other as shown in Table 1. The analysis of the residual mean and variance of the filters is tabulated.

Table 1 Performance comparison of different filters for 3D motion

Filter	Maximum overshoot (m)	Fall time (s)	Time to converge (s)	Steady-state MSE	Residual mean (m)	Residual Variance (m^2)
SUKF	68.509	0.003	Does not converge		2.49	92.15
AUKF	0.696	0.0335	0.5835	0.0997	−0.45	79.74

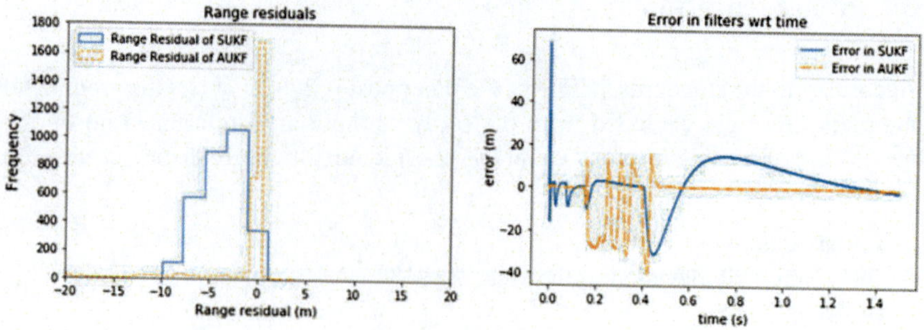

Fig. 2 Histogram of the range residual compared for AUKF and SUKF for 3D motion (L) comparison of estimation error propagation in 3D motion (R)

The difference in performance for AUKF and SUKF can further be appreciated from Fig. 2 (Left) where it can be readily seen that the residual for AUKF becomes zero mean Gaussian with small variance whereas the distribution for SUKF has nonzero mean and high variance. The trajectory residual error is compared for the filter estimates and the zero tracking by the filters for a typical run is shown in Fig. 2 (Right).

5 Conclusion

In this work, a full-fledged three-dimensional tracking of football during set piece scenario has been implemented in the simulation environment. Both system and measurement noise are nonlinear. The noise associated with observations of the ball involves polar coordinates. Errors creep in the system model due to uncertainties of the ambient environment parameters. Overall, this calls for design of a nonlinear filter, for which performance of UKF has been studied.

This work is an important first step that may lead to more scientific defending of set piece scenarios. This simulation set-up can serve useful purpose for studying video archives of key players involved in set piece movements. However, since multiple cameras get involved during the video shoot, correspondence problem can arise during such study. Another limitation of the simulation study for real-life implementation is the availability of signals from the ball. In spite of such limitations, the main hurdle of designing an estimation algorithm with time and space complexity suitable for real-time implementation has been achieved.

References

1. Javorova J, Ivanov A (2018) Study of soccer ball flight trajectory. MATECWeb Conf 145:01002
2. Goff JE, Carr MJ (2009) Trajectory analysis of a soccer ball. Am J Phys 77(11):1020–1027

3. Asai T, Seo K, Kobayashi O, Sakashita R (2007) Fundamental aerodynamics of the soccer ball. Sports Eng 10:101110
4. Zarchan P, Musoff H. (2009) Fundamentals of Kalman filtering: a practical approach, Third edition. American Institute of Aeronautics and Astronautics, Inc
5. Gelb A (1974) Applied optimal estimation. MIT, Cambridge, MA
6. Bar-Shalom Y, Li XR, Kirubarajan T (2001) Estimation with applications totracking and navigation: theory algorithms and software. Wiley, New York (ISBNs 0-471-41655-X (Hardback) 0-4s71-22127-9 (Electronic))
7. Li XR, Jilkov VP (2003) Survey of maneuvering target tracking—Part I: dynamic models. IEEE Trans AES 39(4):1333–1364
8. Julier SJ, Uhlmann JK (2004) Unscented filtering and nonlinear estimation. Proc IEEE 92(3):401–422
9. Wan EA, Van Der Merwe R (2002) The unscented Kalman filter, Chapter 7. In: Simon Haykin (ed) Kalman filtering and neural networks, Wiley Online Library
10. Wan EA, Van der Merwe R (2000) The unscented Kalman filter for nonlinear estimation. Proceedings of the IEEE 2000 Adaptive Systems for Signal Processing, Communications, and Control Symposium, pp 153–158
11. Konatowski S, Kaniewski P, Matuszewski J (2017) Comparison of estimation accuracy of EKF, UKF and PF filters. Annual of Navigation 23(1):69–87
12. Mallick S, Pathak P, Mukherjee A (2019) Design of an adaptive estimator for tracking football in flight. TenSymp 2019, pp 631–636

Reduction of Pre-Kiln Rejections for Improving Productivity in Ceramic Industry

Shyam Sunder Suthar

1 Introduction

Ceramics products are one of the major inputs of the construction sector. The structure of Indian ceramics industry is highly split with very few large players and a large number of SMEs who are facing the problems of poor productivity due to demand not met by present production line. A look at the profit margin of all the ceramics companies in India shows that the profit margin is low [1]. Thus, industry required ways of productivity improvement.

Rapid development in ceramic industries noticed in India due to development in construction, space research sector, automotive sector, health sector, and implementation of the Svatchh Bharat Abhiyaan. The sanitary ware market is growing steadily with a growth rate of 13% with organized segment of 55% [2]. Unorganized segment is facing the problem of a traditional method, labor intensive, low-volume and low-quality product, irrational structure, and high rejection.

The production process of ceramic manufacturing is divided into three stages like pre-kiln stage, kiln stage, and post-kiln stage [3]. A post-kiln stage cannot be recycled, but a pre-kiln process can be recycled. However, the pre-kiln stage takes more time as compared to post-kiln process which increases the cost of reusing. The increased pre-kiln rejection leads to low productivity and high product cost. This provides a base of research to reduce pre-kiln rejections with the help of root cause method. Kaizens are introduced for reducing pre-kiln rejections and improving productivity.

S. S. Suthar (✉)
Mechanical Engineering, Government Engineering College Bikaner, Bikaner, Rajasthan 334004, India
e-mail: sthr_shym@yahoo.co.in

V. H. Saran and R. K. Misra (eds.), *Advances in Systems Engineering*,
Lecture Notes in Mechanical Engineering,
https://doi.org/10.1007/978-981-15-8025-3_39

2 Problem Definition

2.1 Company Profile

The *XYZ* is a renowned company in the area of producing sanitary ware product required for construction and housing sector. It was started in 1996 as micro small and medium enterprises (MSME). The total cost of production plant is 10 Crores. Presently, company is manufacturing 18,000 pieces per month of various designs as per demand with an annual turnover of Rs. 4.6 Crores and 38 employees. It manufactures various designs sanitary ware. It is renowned for product quality and customer satisfaction.

2.2 Production Flow

The company is engaged in manufacturing of sanitary ware product. It is made of ceramic material. The various steps used for producing sanitary ware from raw material are shown in Fig. 1.

A sanitary ware produced by the use of ball clay, china clay, silica, feldspar, and water as raw material. A mold prepared as per the design requirement of a product by the use of plaster of Paris in the mold section. A sanitary ware manufacturing process consists of the following three stages

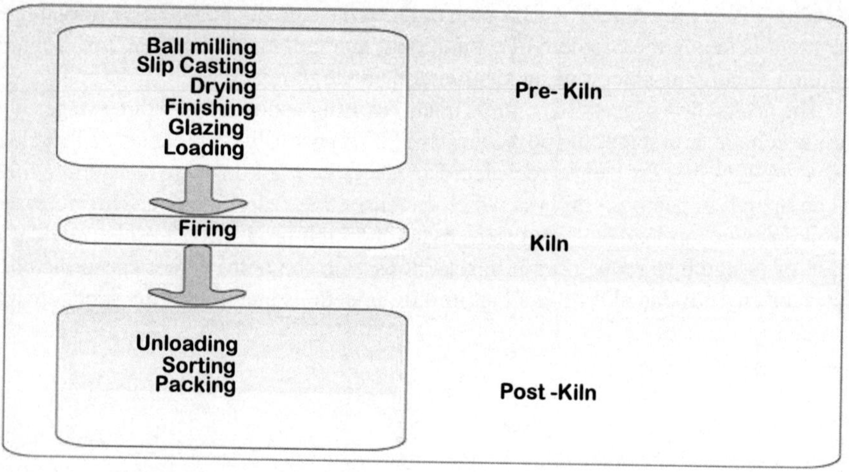

Fig. 1 Stages of sanitary ware manufacturing process [1]

Stage I: Pre-Kiln Stage

Ball Milling: It is the first step of sanitary production of slip which consists of ball clay, china clay, and feldspar. The triaxial body has three components, namely former, filler, and flux. The former provides strength to greenware and is in the form of plastic material. They are mixed in a ball mill, and slip is prepared. Produced slip having property of fluidity, thixotropic, and liter weight.

Glaze Preparation: It is parallel to ball milling process in which different color glaze prepared according to the need.

Slip Casting: Produced slip is poured into a cavity know as mold and allowed to form a thick layer. Extra slip is drained out from mold cavity and usually reused in the agitator tank. Slip is allowed to dry in the mold and is removed after some time. Produced product is called greenware

Drying: For complete drying, the greenware allowed for drying in atmospheric temperature for one to three day. During the process of drying greenware lost his weight and size shrink. Rejected greenware supplied to scrap mixing tank.

Finishing: In this section, greenware inspected for defects. Defected ware repaired or rejected depends upon the type of defects. Kerosene applied to find the defects in the green wares and after checking greenware sent to glazing and rejected ware supplied to scrap mixing tank.

Glazing: The glaze formed by proper mixing of china clay, calcite, and color in ball mill. The formed glaze slurry is then poured in an agitator tank. The glaze is used for different color on the green ware as per production requirement. The glaze is sprayed manually.

Phase II: Kiln Stage

Firing: After vacuum cleaning or air blowing greenware is loaded into the kiln cart. The maximum temperature of the kiln is 1250 °C. There are two types of kiln which are named shuttle kiln and tunnel kiln. Tunnel kiln is mostly used for firing. There are mainly three zones in kiln which are named pre-heating, firing, and cooling zones. During pre-heating of glazed ware moisture is removed from the glazed ware, and then, it is fused in firing zone. Glossy surface is created in the cooling zone. It is supplied to the sorting area after firing.

Phase III: Post-Kiln Stage

Sorting: In the sorting area, the defect in the ware is checked, and packing is done when corrected. Minor defect is fired again. It is rejected when cracked. Standard quality tests are done which are flush test, smoke test, leak test, and load test.

Packing: It this final stage of production, quality standard pass product packed and shipped to the warehouse.

2.3 *Problem Identification*

Information pertaining to demand and supply has been collected. The company reported that a large amount of finished products were being rejected due to mishandling, and the company is not able to supply as per demand due to high pre-kiln rejection rate. In this view of the above problem being noticed that demand is not fulfilled as per current state production line. Thus, the company required a detailed study of production process and rejection analysis to improve the productivity to satisfy the demand of the customer.

3 Root Cause Analysis

Root cause analysis is a set of approaches that can be applied to determine the underlying cause(s) of a problem or failure in equipment design and production quality as well as supporting management processes [4, 5]. Each of the root cause analysis approaches is valuable to inform a broader system-level understanding of recurring challenges. Root cause analysis has previously been applied to many manufacturing issues, including problems with refrigerators [6, 7]. RCA is an approach which can be used to enhance productivity in many manufacturing industries. Root cause analysis (RCA) is also called Ishikawa diagram. RCA is the simplest technique to implement, and to find the cause behind any defect to decrease defect leads to increase productivity. The concept of RCA was originally developed by Sakichi Toyoda in 1958, which developed a process called the "Five Whys" to understand the potential cause of problems [3].

4 Methodology

Methodology adopted for improvement is shown in Fig. 2.

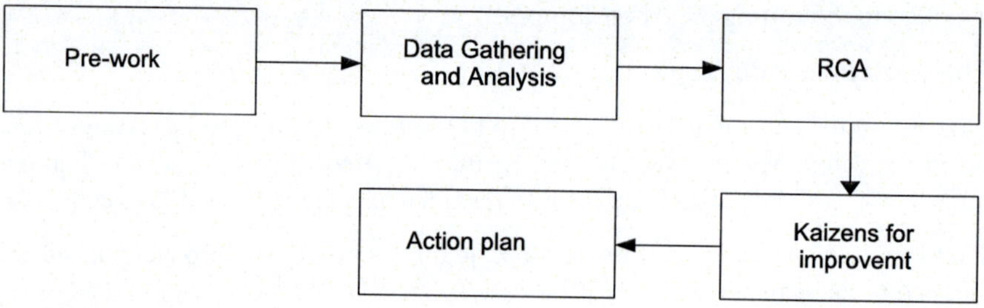

Fig. 2 Methodology

Table 1 Rejection analysis in pre-kiln stage before RCA

Pre-kiln stage	Monthly average rejections/defect			
	Handling	Broken	Surface cracks (SFC)	Cavity
Drying	500	200	100	100
Finishing	1200	500	400	NIL
Glazing	250	50	NIL	NIL
Total	1950	750	500	100

4.1 Pre-work

Plant visited and thorough study of production line carried out. Literature reviewed related to work. A team was formed and trained to analyze the complete production process. After a complete analysis done by team main rejection were found. Suggested root cause and corrective action suggested by team member noted down.

4.2 Data Gathering and Analysis

Data pertaining to rejection in each process of the pre-kiln stage was collected and summarized as shown in Table 1. To find the trend of different defect in the pre-kiln stage a bar chart analysis was drawn which is shown in Fig. 4.

4.3 Root Cause Analysis (RCA)

A Ishiwaka diagram was drawn to find the cause of problem in pre-kiln stage and corrective action found for each problem. Figure 3 shows the cause effect diagram for rejections, and Table 2 summarized the action plan to reduce pre-kiln rejections.

Drying, finishing, and glazing are pre-kiln processes. Various defects are caused by reasons as mentioned in Table 2 which leads to reduce quantity of final products. Reduction in final product quantity leads to decrease in productivity as it measured in terms of final product produce

4.4 Kaizen Introduced

The various actions were taken as per suggestion in RCA diagram. Some Kaizens were implemented to decrease the rejection and enhance productivity. Some are as under:

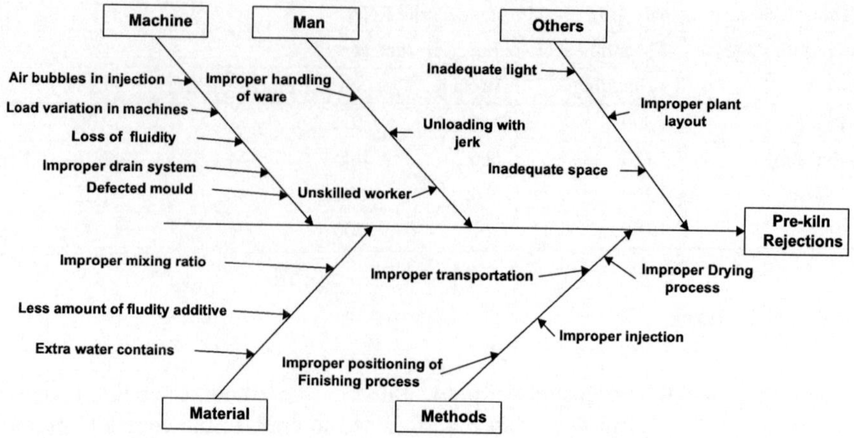

Fig. 3 Cause effect diagram for rejection

Table 2 Action plan to reduce rejections

Reasons		Actions	Defects
Man	Improper handling of ware	Foam-based trolley designed	Handling
	Unskilled worker	Worker trained	Broken
	Unloading with jerk	Rubber sheet used	SFC
Machine	Air bubble in injection	Bleed of circuit designed	Cavity
	Loss of fluidity	Stirrer used	SFC
	Improper drain system	Effective drain system	SFC
	Defected mold	Replacement of mold	Broken/SFC
Material	Improper mixing ratio	Proper mixing ratio used	SFC/Cavity
	Less fluidity additive	Proper fluidity additive added	Broken
	Extra water contains	Water contains reduced	Handling/SFC
Methods	Improper transportation	Improved by introducing trolley	SFC/Handling
	Improper finishing process	Improved by reducing travel time	Handling/SFC/Broken
	Improper drying process	Improved by introducing hot air	Handling/Cavity/SFC/Broken
Others	Inadequate light	Lighting improved	Broken
	Inadequate space	Space utilized	Broken
	Improper layout	Improved by shifting finishing area near drying area	Handling/SFC

Table 3 Rejection analysis in pre-kiln phase after RCA

Pre-kiln processes	Monthly average rejections/defect			
	Handling	Broken	Surface cracks (SFC)	Cavity
Drying	210	95	20	10
Finishing	450	75	25	NIL
Glazing	50	10	NIL	NIL
Total	710	180	45	10

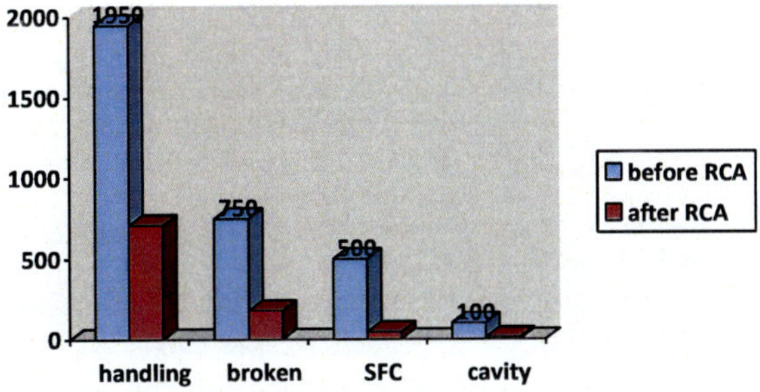

Fig. 4 comparison of rejections

- Pumps are used to collect drained slip into an agitator tank which is removed, and a gravity-based channel system introduced with filter.
- Pumps are used to collect drained glaze into a tank which is removed, and a gravity-based system introduced with filter.
- Manual transportation is using which was replaced by introducing a new designed trolley.
- Hot air introduced for drying process which was purely natural.
- Using a polyethene sheet during glazing for collecting extra spillage of paint.
- Supervisor trained to take corrective action.

After the implementation of the action plan and Kaizens data pertaining to rejection collected and are summarized as shown in Table 3 and Fig. 4.

By incorporating actions as mentioned in Table 2 and Kaizens, it was found that pre-kiln rejections are reduced as mentioned in Table 3 which leads to improve productivity as the final product quantity increased.

5 Result

In this paper, a case study method of research was used to implement RCA in a ceramic industry. Pre-kiln stage rejection reduced. By using RCA, root cause of defects found. On the basis of corrective action and some Kaizens, pre-kiln rejection minimized by 70% and productivity increased by 19.62%. RCA also helped to enhance skill of employees and minimization of environmental issues in ceramic industry.

References

1. Prowess, CMIE
2. Ceramic and Sanitary-ware Market Investment Report: Focus Morbi 2016. Source: www.vibrantceramics.com
3. Bhamu J, Sangwan KS (2015) Reduction of Post-kiln rejections for improving sustainability in ceramic industry: a case study. In: 12th Global Conference on Sustainable Manufacturing, Procedia CIRP, vol 26, pp 618–623 (Science Direct)
4. Rooney JJ, Vanden Heuvel LN (2004) Root cause analysis for beginners. Quality Progress 45–53
5. Tomic B, Spasojevic Brkic V (2011) Effective root cause analysis and corrective action process. J Eng Manage Competit 1(1/2):16–20 (15)
6. Kalantri R, Chandrawat S (2013) Root cause assessment for a manufacturing industry: a case study. J Eng Sci Technol Rev 6(1):62–67
7. Suk Han H, Bong Jeong W, Seong Kim M, Hoon Kim T (2009) Analysis of the root causes of refrigerant-induced noise in refrigerators. J Mech Sci Technol 23(12):3245–3256 (18)

Application of Emotion Detection Using Facial Expression Recognition

Yadu Meemansha

1 Introduction and Literature Review

Emotion of a person is his/her response toward one's mood, thoughts, external situations, ethics, control, interactions, etc. These emotions can be identified by their facial expression [1] as well as their verbal expression. But it is always easier to know a person's emotion by recognizing their facial expression, rather than verbal expression. These emotions can be happiness, sadness, surprise, anger, fear, calmness, disgust, irritation, uneasy, and so on. The behavior of any person can be well-defined by the facial expression that the person shows unknowingly in response to some situations. Researches from 1990s [2, 3] have already proved that emotion of a person can be recognized by his/her facial expression. Many features of the face are observed which help to identify the emotion of the person [4]. These features can be eyebrows corners, eyes, nose, lip corners, etc.

The automatic detection of the emotion in a video [5] was developed when human–computer interaction was an emerging field in computer science. Initially, the architecture was designed, which was later implemented using various techniques like eigenface recognizer (using PCA) [6], and fisherface recognizer (using LDA) [7]. Gradually, it was realized that videos are better way to immediately and accurately identify the human activities, and hence, real-time applications were developed to predict the emotion expressed by any individual on the video being played or the face being captured in real life using web camera [8]. Hence, the teachers of preschools were evaluated based on their recording in class for the application defined in the paper.

There are many applications of emotion recognition [9] like finding student engagement during class or video lectures [10], e-learning [11], response of children

Y. Meemansha (✉)
Dayalbagh Educational Institute, Agra, Uttar Pradesh, India
e-mail: meemansha1995@gmail.com

V. H. Saran and R. K. Misra (eds.), *Advances in Systems Engineering*, Lecture Notes in Mechanical Engineering,
https://doi.org/10.1007/978-981-15-8025-3_40

with syndrome [12], and many more. This paper proposes an innovative application of emotion detection to analyze person for a job based on their emotions and expected emotions in that job. It can be used in many professions and jobs like astronomer going to space must have controlled emotions and be calm in any situation, administrator fulfilling public needs must not get angry and must take right decisions, economist should not have extreme fear of losses or extreme joy of profits and must remain neural and work, parents should not show extreme emotions to children which can affect their minds and behavior, doctor's emotion should not overpower their daily work and many more.

This paper is organized as follows: Sections 2 and 3 deal with dataset used and implementation, respectively. They provide all the steps and calculations involved in the application that is introduced in the paper. Sections 4 and 5 talk about the results, conclusion, and future scopes.

2 Dataset

For facial expression recognition, many datasets have been designed. Initially, data was collected by setting up an environment that contains good lighting and were posed. For example, Kaggle dataset for facial expression recognition [13] which have around 13,700 images, but all are posed and are captured in a controlled environment. This might give good accuracy, but cannot be used in the application where we want to evaluate person in surveillance camera. In this project, we wanted dataset in which the images are captured in unconstrained environment.

Other criteria of choosing the dataset was the cultural aspect. Since every country or different part of the world have different culture, so is the way of expressing emotion. Since we are evaluating Indian employees at the initial stage and there can be different ways of expressing the same emotions by Indians and non-Indians due to differences in culture, so we have used the Indian Movie Face Database (IMFD) [14] for training the emotions. The dataset contains 34,512 images of 100 Indian actors from about 103 Indian movies.

All the faces were first detected in each image, cropped to same size, and converted into grayscale using dlib face recognizer in python before the training phase. Some samples are shown in Fig. 1. The dataset was divided into 80:20 ratio to achieve training and testing sets. Local binary pattern histogram (LBPH) face recognizer was used to train the model because it learns each pixel as well as its surrounding, instead of the whole image. It is also not affected by light contrast in the image. This makes LBPH appropriate method for working on real-world images. Its performance is presented in Table 1.

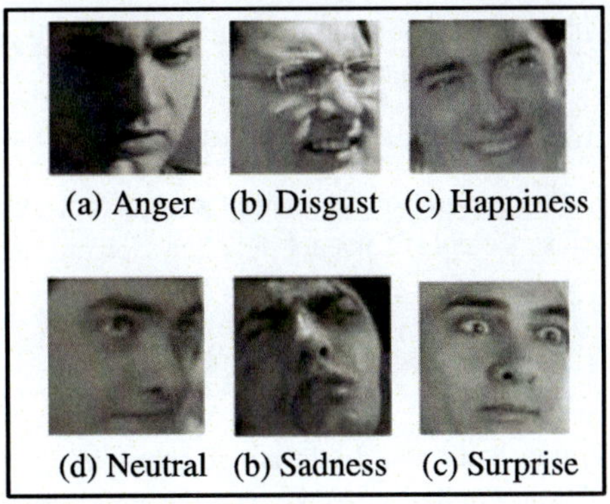

Fig. 1 Amir Khan emotions in IMFD refined dataset

Table 1 Performance results of LBPH model on IMFD dataset

Accuracy (%)	Precision (%)	Recall (%)	F_Measure (%)
87.539	87.437	87.460	87.433

3 Implementation

3.1 Score Generation

Studies have already been done on performing emotion detection on videos [15]. These studies can be used to achieve a realistic and better results and analysis of a person in a job. For the application proposed in the paper, faces were detected from surveillance camera or digital camera video recordings of each classroom in a preschool. Teacher's faces were detected, and emotions were predicted for them using the trained model. Frequency of each emotion for a teacher is calculated, and then according to the weightages given to the emotion for an ideal teacher, rank as well as percentage of each emotion shown by the teacher is calculated. Percentage of emotion also tells where the teachers lack and what needs to be improved with respect to the emotions to increases their position in the job. Weightage given to each emotion for the purpose of teacher evaluation is given in Table 2.

These weights are given according to the required emotions of a teacher. Negative weightages are given to the negative emotions like anger, disgust, sadness with anger given maximum negative weightage as it will have very negative effect on child's mind during the youngest age. Positive weightages are given to the positive emotions

Table 2 Emotions and corresponding weightage allocated

Neutral	Anger	Disgust	Happy	Sadness	Surprise
20	−20	−15	25	−10	15

like neutral, happy, surprise with happily given maximum positive weightage. Thus, a vector is created of the weightages:

```
Weightage = [20, − 20, − 15,25, − 10,15]
```

After prediction is made, we get a vector of frequency of each emotion in the class video(s) for each teacher. For one teacher, for example, we get the following frequency array:

```
predCount = [3,7,4,9,5,2]
```

where neutral expression is shown 3 times in given video(s), anger is expressed 7 times, and so on. In the above example, a total number of images are 30. The equation defined in the paper for calculating score of each teacher is shown below.

$$\text{Score} = \sum_{i=1}^{e} (\text{predCount}[i] * \text{weightage}[i]) \tag{1}$$

where e = number of emotions.

This score is normalized to get the value in the range [0, 1]. The Teacher ID, score, and emotion percentages are stored in text files. Then scores are arranged in ascending order and according to that corresponding Teacher ID and emotion percentages are arranged to display the final ranking of the teachers.

3.2 *Automatic System for Employee Evaluation*

A case study of employee evaluation application of emotion detection was performed on a preschool. Figure 2 clearly shows the flowchart for the evaluation system proposed in the paper, particularly for the teachers in preschools. System developed for this application follows the steps shown in the flowchart below. Teacher mentioned in the flowchart can be replaced by any type of employee that needs to be evaluated.

4 Results and Discussions

After training and obtaining model for emotions, we implemented them on data collected from Euro Kids Preschool branch at Agra. It had four classes, with 2 teachers in each class. For emotion evaluation, video recording was done using hidden cameras for 1 week for each class. Each video was of 2 h, i.e., the time before lunch break. This period of time was recorded because it was the time when children comes to school crying, and it is the teacher's duty to handle them properly and keep

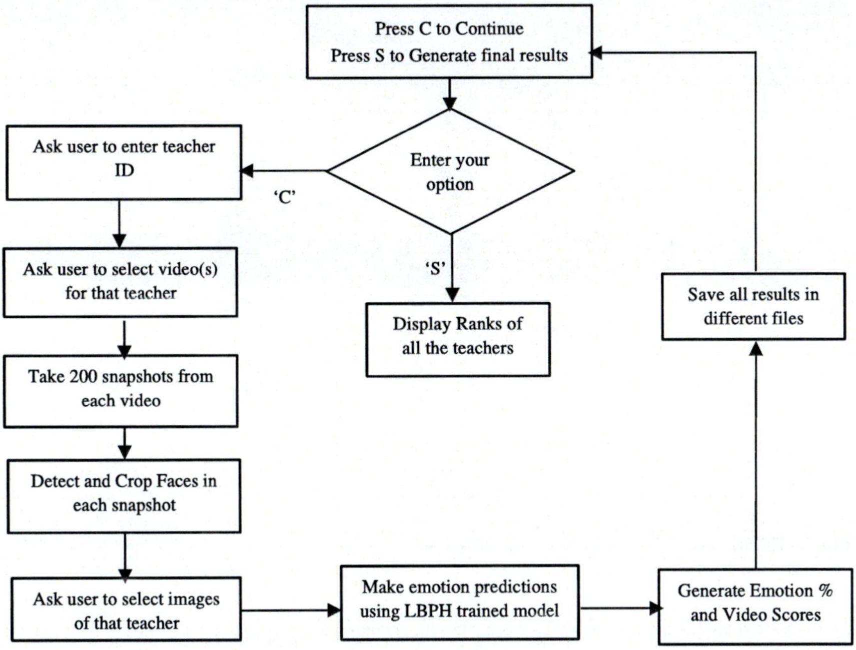

Fig. 2 Automatic preschool teacher evaluation system

them happy and involved by showing some positive emotions, instead of scolding and showing anger.

An automatic system was designed for the purpose of taking 200 snapshots from each video at regular interval. In each snapshot, face was detected, cropped, and saved in each teacher's folder. Teacher ID was allocated as ET001, ET002..., ET008 for 8 teachers.

4.1 Results of Highest Ranked Teacher (ET003) at Euro Kids Preschool

For demonstrating the results obtained after implementing the system on preschool teachers, the results for teacher ET003 are shown below. The results for other teachers obtained are like the results demonstrated with their respective percentage of emotions expressed.

Figure 3 shows a pie chart of percentage of each emotion expressed by teacher ET003 in 1 week. As it can be seen that most positive emotion (i.e., happiness) has highest percentage, even more than cumulative of all the other emotions. Negative emotions (i.e., disgust, anger, and sadness) have very low percentage as compared to happiness. Since happiness was allocated highest weightage, the score of teacher

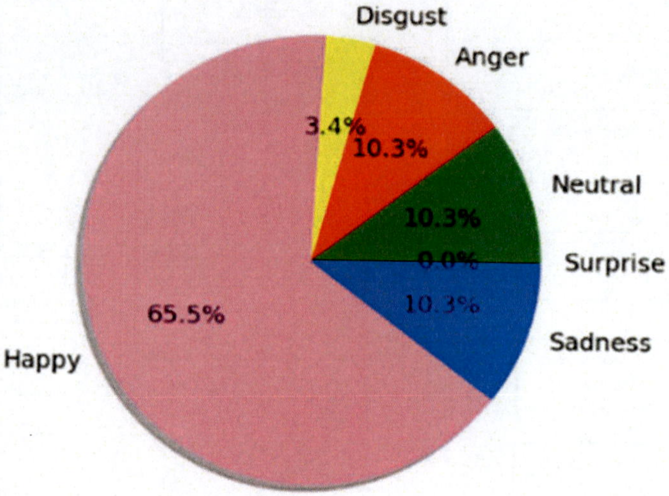

Fig. 3 Emotion percentage for teacher ET003

Table 3 Teachers ranked according to their score (ET003 ranked highest). Emotion percentages are also displayed in the table for each teacher

Rank	TeacherID	Score	Neutral(%)	Anger(%)	Disgust(%)	Happy(%)	Sadness(%)	Surprise(%)
1	ET003	0.774	10.345	10.345	3.448	65.517	10.345	0.0
2	ET008	0.724	35.484	9.677	3.226	25.806	9.677	16.129
3	ET002	0.712	12.5	6.25	6.25	43.75	15.625	15.625
4	ET004	0.572	13.333	23.333	6.667	40.0	15.0	1.667
5	ET001	0.487	29.412	17.647	26.471	2.941	2.941	20.588
6	ET007	0.472	18.367	14.286	0.0	10.204	42.857	14.286
7	ET006	0.467	13.793	31.034	10.345	27.586	13.793	3.448
8	ET005	0.349	24.138	58.621	3.448	10.345	0.0	3.448

ET003 is highest among all the teachers as can been seen from Table 3. To visualize and compare all the teachers based on each emotion and scores corresponding to results in Table 3, bar graph is also generated for emotion percentage (Fig. 4) as well as for scores (Fig. 5).

After the results were generated, these results were shown to the principal/in-charge of the preschool and a feedback was taken from them about the result. They told if the ranks generated are the correct ranks or not according to the best of their knowledge about the teachers employed under him/her in the preschools. As per them, the method of using emotion for employee evaluation is unique and innovative way. It will lead to proper, reasonable, and valid way to evaluate and compare the employees, which is not possible by any other method. It is not possible to keep a watch on every teacher all the time even by using surveillance cameras. But this emotion detection method will do the job for them. This will thus improve the performance and outputs of the institute.

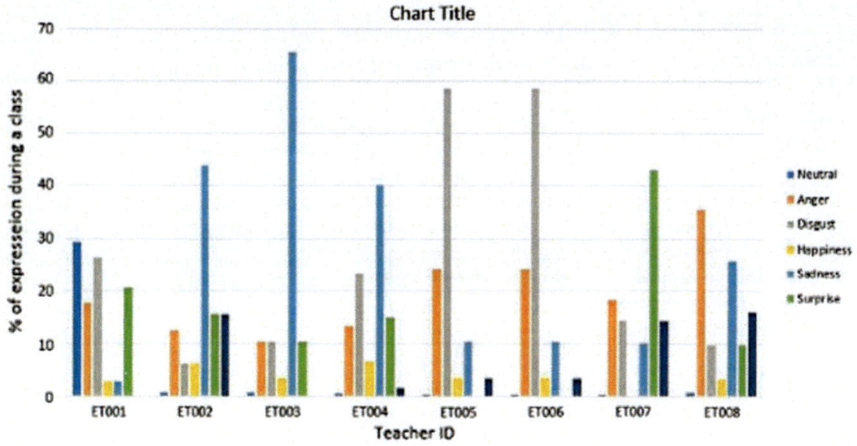

Fig. 4 Percentage of emotion expressed by teachers during the class

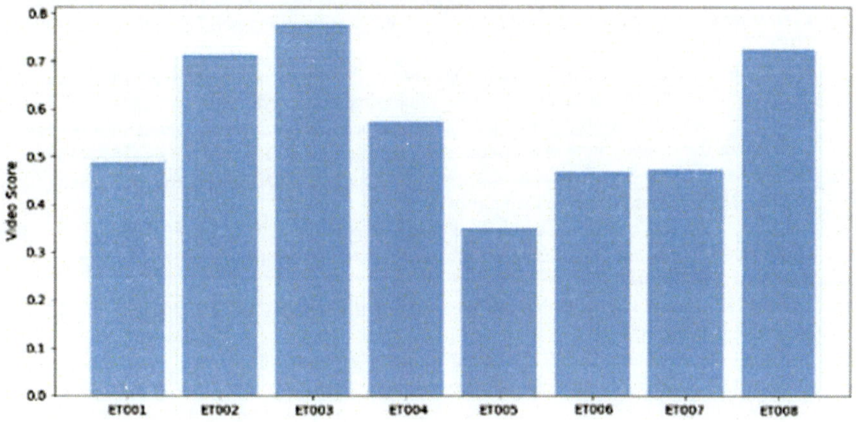

Fig. 5 Emotion score of the teachers

5 Conclusion and Future Scopes

In this paper, the emotion detection of people in video [16–18] is performed for a very useful application. The condition of the society is dependent on the performance of various institutes, companies, etc., which in turn are solemnly dependent on the performance of the employees recruited there. This system will enable the employees to be evaluated based on their performance while doing their jobs. The proposed application will thus help in increasing the outputs of the companies, better care given to juvenile at detention center, special care given to child with syndrome at health centers, etc., and help to improve any individual based on the feedbacks obtained.

Although the proposed system works on a case study of preschool teachers, it can further be expanded and used for other organizations also in future.

Acknowledgements I am highly pleased to express my sincere gratitude to my highly respected and esteemed guide Dr. Lotika Singh for her valuable guidance, encouragement, and help for completing this work. Her useful suggestions for this whole work and cooperative behavior are sincerely acknowledged. At the outset, I would like to thank all my faculty members for their extra effort to help me throughout the project. I also extend my gratitude toward the Head of Euro Kids preschool at Agra for cooperating with me, providing me with the recordings of the classes and all the possible help needed for my project, without which the project would not be completed. At the end, I would like to express my sincere thanks to all my friends and family members who helped me directly or indirectly during this project work.

References

1. Tarnowski P, Kołodziej M, Majkowski A, Ra RJ (2017) Emotion recognition using facial expressions. In: International Conference on Computational Science, ICCS 2017, 12–14 June 2017, Zurich, Switzerland
2. Russell JA (1994) Is there universal recognition of emotion from facial expression? A review of the cross-cultural studies. American Psychological Association Inc., vol 115
3. Strand PS, Downs A, Barbosa-Leiker C (2016) Does facial expression recognition provide a toehold for the development of emotion understanding?. Dev Psychol 52(8):1182–1191
4. Happy SL (2015) Automatic facial expression recognition using features of salient facial patches. IEEE Trans Affective Comput 6:1–12
5. Ehsan S, Clark AF, ur Rehman N, McDonald-Maier KD (2015) Integral images: efficient algorithms for their computation and storage in resource-constrained embedded vision systems. Sensors 15:16804–16830
6. Dauda AK, Bhoi N (2014) Facial expression recognition using PCA & distance classifier. Int J Sci Eng Res 5, May-2014
7. Wang Z, Ruan Q, An G (2015) Facial expression recognition using sparse local Fisher discriminant analysis. Neurocomputing, September 2015
8. Salih H, Kulkarni L (2017) Study of video based facial expression and emotions recognition method. In: 2017 International Conference on I-SMAC (IoT in Social, Mobile, Analytics and Cloud) (I-SMAC), IEEE, Palladam, India, 05 Oct 2017
9. Szwoch LM, Szwoch W, Wróbel MR, Kołakowska A (2014) Emotion recognition and its applications, Chapter from book Human-computer systems interaction: backgrounds and applications, vol 3, pp 51–62, July 2014
10. Whitehill J, Serpell Z, Lin Y-C, Foster A, Movellan JR (2014) The faces of engagement: automatic recognition of student engagement from facial expressions. IEEE Trans Affective Comput 5 (January–March 2014)
11. Bahreini K, Nadolski R, Westera W (2014) Towards multimodal emotion recognition in e-learning environments. Interactive Learn Environ 24:590–605
12. Cebula KR, Wishart JG, Willis DS, Pitcairn TK (2017) Emotion recognition in children with down syndrome: influence of emotion label and expression intensity. Am J Intellectual Dev Disabilities 122(2):138–155
13. Goodfellow IJ, Erhan D, Carrier PL, Courville A, Mirza M, Hamner B, Cukierski W, Tang Y, Thaler D, Lee D-H, Zhou Y, Ramaiah C, Feng F, Li R, Wang X, Athanasakis D, Shawe-Taylor J, Milakov M, Park J, Ionescu R, Popescu M, Grozea C, Bergstra J, Xie J, Romaszko L, Xu B, Chuang Z, Bengio Y (2015) Challenges in representation learning: a report on three machine learning contests. Neural Netw 64:59–63, April 2015

14. Setty S, Husain M, Beham P, Gudavalli J, Kandasamy M, Vaddi R, Hemadri V, Karure JC, Raja Raju R, Vijay Kumar, Jawahar CV (2013) Indian movie face database: a benchmark for face recognition under wide variations. In: IEEE National Conference on Computer Vision, Pattern Recognition, Image Processing and Graphics (NCVPRIPG), Bombay
15. Mishra P, Ratnaparkhi S (2018) HMM based emotion detection in games. In: IEEE 3rd international conference for Convergence in Technology (I2CT), Pune, India, April 06–08, 2018
16. Yoon J, Pohlmeyer A, Desmet P (2017) EmotionPrism: a design tool that communicates 25 pleasurable human-product interaction. J Des Res 15:174–196
17. Tyng CM, Amin HU, Saad MNM, Malik AS (2017) The influence of emotion on learning and memory. Front Psychol J 8, Article 1454, August 2017
18. Kalaiselvi R, Kavitha P, Shunmuganathan KL (2014) Automatic emotion recognition in video. In: 2014 international conference on green computing communication and electrical engineering (ICGCCEE), IEEE, Coimbatore, India, 6–8 March 2014

Interactive Learning System for Children

Binathi Bingi, Harsh Vardhan, and Lotika Singh

1 Introduction and Literature Review

The advent of technology in education sector has led to the development of AI-based tools and educational robots [1]. An example of AI-based tool is Quillionz, question generation software. The input to the software is the content and keywords, following which the questions are generated. Multiple choice questions (MCQ) are used as an instrument for active learning and assessment at various levels of education [2]. MCQs are generated using rule-based, template-based and ontology-based approaches [3]. It is proved by several researchers that conceptual understanding of students on a topic can be improved by asking MCQ on the topic immediately after teaching [4, 5]. But generation of MCQ requires domain expertise and is time consuming. Consequently, several MCQ-based systems have been developed [2, 3, 6–8].

Generating MCQ involves three steps: (i) Selection of sentences from the passage or input text, (ii) identification of keyword and generation of MCQ and (iii) generating distractors which are the incorrect answers [7]. In general, the approaches for sentence selection are either rule-based, feature-based or template-based. The quality of sentences selected depends on the type of rules, features or templates and is extremely domain-specific. From the selected text, keywords are identified using techniques like conditional random field (CRF) algorithm [8], dependency trees [6], noun/phrase extraction [9], firefly-based phrase extraction [10] and so on, while distractors are identified using set of rules, hypernyms and hyponyms of key [9], similarity distance and gazetteer list-based approaches [3].

B. Bingi (✉) · H. Vardhan · L. Singh
Department of Physics and Computer Science, Dayalbagh Educational Institute, Agra, Uttar Pradesh, India
e-mail: b.binathi@gmail.com

V. H. Saran and R. K. Misra (eds.), *Advances in Systems Engineering*,
Lecture Notes in Mechanical Engineering,
https://doi.org/10.1007/978-981-15-8025-3_41

Table 1 RoSAS 18 items [15]

Warmth	Competence	Discomfort
Organic	Reliable	Awkward
Social	Competent	Scary
Emotional	Knowledgeable	Strange
Compassionate	Interactive	Awful
Happy	Responsive	Dangerous
Feeling	Capable	Aggressive

In this research, an interactive learning system is developed and interfaced onto humanoid (Nao) which acts as a teaching assistant. Currently, in China and Singapore, robots are deployed as teaching assistants in schools [10–13]. Chin-Wei Chang and his colleagues observed that robots have the ability to create interactive and engaging learning experience for students [14]. C. Angeli and N. Valanides explained in their study how young children's computational thinking can be developed with educational robotics [14]. Humanoid Nao was programmed to explain a story/concept by exploiting its speech synthesis and body gestures features. The interaction and learning experience of students with Nao was evaluated using the RoSAS [15]. It is an 18-item scale adopted from Godspeed questionnaire [16] and social perception studies. The items are further categorized into three dimensions listed in Table 1. These dimensions determine the quality of interaction with the robot.

The goal of the paper is to facilitate effective learning among students by using AI-based education tools, humanoid as a teaching aid and to evaluate its effectiveness in Indian school scenario. This paper is further divided into MCQ generation, interactive system using humanoid as an interface, evaluation of human–robot interaction and conclusion.

2 Question Generation

The text considered for explanation by humanoid was stories from NCERT textbooks and environmental science topics from Wikipedia. The text explained by Nao is considered as input for MCQ generation. The input text is subjected to coreference resolution. According to Stanford NLP group, "Coreference resolution is the task of finding all expressions that refer to the same entity in a text". Coreference in the input text was found using the Python library neural coref which is integrated into spaCy NLP pipeline. Word cloud library from Python was used to generate the most frequently used words in the text. These words are stored and passed as input to Nao, to explain the meaning of the most important words in a concept that it is explaining. The second step is topic modeling [17] in which latent Dirichlet allocation algorithm

(LDA) [18] was used to create topic and extract keywords under each topic. The third step is summary extraction, which is accomplished using text rank [19] derived from Google page rank algorithm [20].

Text rank algorithm uses the intuition behind page rank algorithm. The sentences in the input text are treated as analog to the web pages in page rank algorithm, and the probability of going from one sentence to the other is calculated as the similarity of the two sentences. The similarity score is computed using cosine similarity. The sentences that are similar to most of the other sentences in the text are used to generate the summary. This is given as input to the question generator.

For question generation, key needs to be identified for each sentence followed by distractor generation for every key. The key has to be a word in the sentence such that its absence does not make the context of the sentence unclear and the remaining part of the sentence should facilitate to find the key. The keyword must not be easy enough to be predicted by persons who do not have enough knowledge about the domain. For finding the keyword, the sentence is POS tagged. Proper nouns or common nouns with describing adjective in the start or end of the sentence, adjectives or nouns at any part of the sentence are chosen as keywords. If none of the above cases are identified, then the sentence is termed as unfit for MCQ generation and is discarded. A random key is chosen from the list of the keywords and is identified as the key/blank. The summary was further refined by removing the sentences that did not contain any keywords.

Goodness and toughness of an MCQ depend on how similar the distractors are to the key [10]. In this paper, the similarity distance between the distractors and key is found through cosine similarity. For a given key, the distractors are the list of synonyms and antonyms generated through cosine similarity from WordNet. If no synonyms or antonyms are found for a key, the key is discarded and a new key is selected from the list of keywords. Three options for the MCQ are randomly selected from the list of distractors, and the fourth option will be the key. The order of the options of the MCQ is shuffled before presenting to the student.

The readability of generated MCQ is calculated using the readability indexes. Readability of text depends on the complexity of vocabulary, syntax and typographical aspects (font size, line height and length). There are four different readability indexes, which are used here: Flesch reading ease test, Gunning fog index, SMOG grade and Dale–Chall readability formula. Flesch reading ease test is designed to indicate the difficulty level in understanding the English passage. Higher scores above 100 indicate that the text is easier to read while lower scores below 60 indicate the text is difficult to read. The second readability test is Gunning fog index, which estimates the years of formal education a person needs to understand the text. For a wide audience, the index must be less than 12. For near universal understanding, texts need to have index less than 8.

The third readability test used is simple measure of Gobbledygook (SMOG) grade, which determines the reading level of the text. Words with more than three syllables are identified, and SMOG conversion tables compiled by Harold C. McGraw are used to find the grade level. The fourth readability index used is Dale–Chall readability formula. It provides a numeric gauge of the comprehension difficulty level of the text. For score of 4.9 or lower, the text can be understood by grade 4 and below, and for a score above 9, text is designed for university level students. As the MCQs are targeted for students between grade 2 and grade 4, the questions are interfaced onto Nao only if the readability values of the generated questions satisfy any of the following conditions: Flesch reading ease test score—greater than 100, fog index—less than 8, SMOG index—less than 8 and Dale–Chall index less than 4.9. In the next section, interfacing of the words, questions and narration by Nao is discussed.

3 Interactive System Using Humanoid as an Interface

Nao which has been deployed as teaching assistant has the following advantages when used for learning by teaching: The height of Nao v6 allows to establish proximity and resemblance to a student [21]. This humanoid can perform human-like movements due to its 25 degrees of freedom, and each joint is equipped with a sensor. It can perform gestures useful for learning [15, 16] and also mimic gestures [22]. It has speech synthesis module which allows it to speak and listen [18]. It can be programmed using Choregraphe software (Aldebaran, Softbank Robotics). It is a graphical environment that allows simple programming and also programming in Python [23] to implement any desired objective. Its features can be seen in Fig. 1.

A database of ten stories from NCERT textbook and ten science articles from Wikipedia targeting the students between grade 2 and grade 4 was created. Nao has been programmed to narrate the stories and perform gestures during narration. For the purpose of narration with emotion, an audio file is created for every text. The voice file contains the narration of the text along with the sounds. Based on the audio at a given moment, movements for humanoid are coded into the program. The

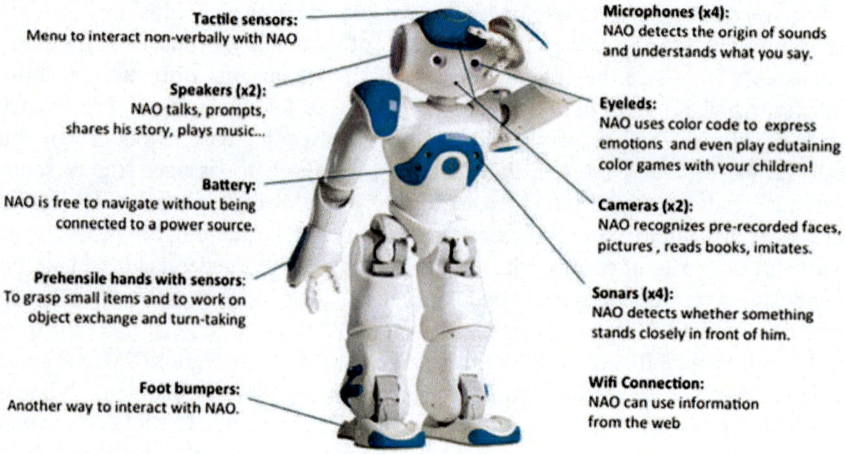

Fig. 1 Nao features [24]

movements can be defined by setting the position of the robot or its body part and saving it in the motion keyframe. The motion keyframe will save the angle values of the joints at that particular moment of time. The interpolation of joint values is performed to obtain the intermediate position between two motion keyframes. This facilitates to program Nao to perform two synchronous tasks—to tell a story or explain concept and perform corresponding actions.

Nao asks the user to choose from the list of stories and articles available. It captures the title of the text to narrate from the voice of the user and executes the corresponding program. Consider the following story from NCERT second grade English textbook. “This is Mohan. He lives in Kuppam village. He is a potter. He uses clay to make pots…. People like his pots and jugs very much.” Coreference resolution is applied, and word cloud is generated to find out the most frequently used words in the text. Example of words in word cloud is Mohan, drinking, lamps and potter. Nao explains the meaning of the top ten most frequently used words in the text by retrieving their meanings from the PyDictionary, an inbuilt dictionary in Python. It also makes the student repeat the important words generated through word cloud for three times to help the students learn the pronunciation. Then, it executes the corresponding program for narration.

While Nao is performing narration of text, the corresponding text is retrieved from the database and MCQ generation is performed using the steps in Sect. 2. An example of generated MCQ is as follows:

Mohan uses a potter’s ___________ to shape the clay and make the pots.

A. roulette_wheel B. bike C. pedal D. wheel

The readability index values generated for the above MCQ are as follows: flesch = 144.58, gunning_fog = 16.900305810397555, smog index = 3.1, dale_chall = 1.95. After the narration, the questions are given as input to the text to speech model in Nao. For every question, Nao identifies a student through facial recognition and poses the question to the student and provides the student with the option. The student has to give the correct option or the correct answer. In case, if the student fails, Nao gives a second chance with a waiting period of three minutes for the student to reply. It converts the reply into text by speech to text module and compares with the correct option. If the student fails for second time, it explains the question and the correct answer to the class. The working of interactive learning system is shown in Fig. 2.

4 Evaluation of Human–Robot Interaction

The user experience of Nao as teaching assistant was measured using RoSAS [15] on a sample size of 54 school students. After interacting with Nao, the students filled out a questionnaire where they had to rate how well the 18 items mentioned in Table 1 describe the robot. A rating system of one to five was used where one represents “not at all” and five represents “very much”. The mean values [25] for the items were calculated and used to calculate mean values for the dimensions warmth, competence and discomfort. The values indicated in Table 2 show how the robot

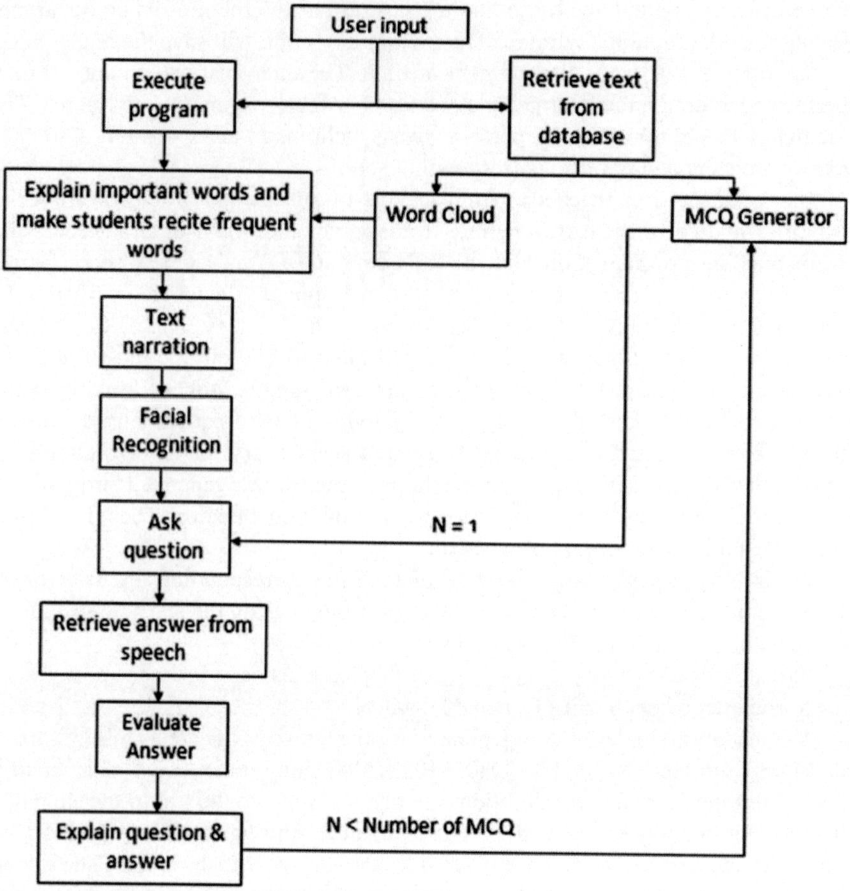

Fig. 2 Block diagram of interactive system

Table 2 Result RoSAS factors

Warmth		Competence		Discomfort	
Organic	3.509434	Reliable	1.641509	Awkward	1.188679
Social	4.415094	Competent	4.433962	Scary	1.283019
Emotional	3.396226	Knowledgeable	4.584906	Strange	1.528302
Compassionate	3.773585	Interactive	4.415094	Awful	2.54717
Happy	4.981132	Responsive	4.358491	Dangerous	1.09434
Feeling	4.811321	Capable	4.301887	Aggressive	1.301887
Mean value	4.147799		3.955975		1.490566

was perceived by the students. The robot has achieved a good competence score but the low reliability score indicates the lack of trust on robot. It is observed that the students feel very warmth in the presence of a humanoid teaching assistant and not at all experience discomfort. The results in Table 2 indicate that narration of text by humanoid draws curiosity, calls student's attention and simulates interest to study.

5 Conclusion

An innovative interactive learning system was developed for the benefit of primary school children that narrates concepts and interacts with students. The results in Table 2 indicate the acceptability of humanoid teaching assistant by students. Explanation and recitation of words before narration of the text facilitated better understanding among students. The MCQ generation was enhanced with the addition of coreference resolution and topic modeling generating more useful and readable questions. Identification of student by facial recognition and posing the question to an individual motivated the student to learn and understand the concepts. As a test case, this system was deployed at primary schools in Agra and is achieving satisfiable results. This model could be further enhanced by synchronizing visual representations on screen with narration. Along with MCQ, generation of Wh question may also be useful. To understand the impact of humanoid on students, further studies may be carried out using different human–robot interaction scales like negative attitudes toward robots scale [26], ethical acceptability scale [27] and Frankenstein syndrome questionnaire [28].

References

1. Wikipedia (219) Educational robotics. Wikipedia. [Online]. Available: https://en.wikipedia.org/wiki/Educational_robotics. Accessed 25 May 2019
2. Majumder M, Saha SK (2015) A system for generating multiple choice questions: with a novel approach for sentence selection. In: Proceedings of the 2nd workshop on natural language processing techniques for educational applications, pp 64–72
3. Papasalouros A, Kanaris K, Kotis K (2008) Automatic generation of multiple choice questions from domain ontologies. In: Proceedings of e-learning, pp 427–434
4. Mazur E (1997) Peer Instruction: a user's manual series in educational innovation. In: The changing role of physics departments in modern universities, part two. In: Redish EF, John RS (eds) American Institute of Physics, pp 981–988
5. Nicol D (2015) Keynote Paper principles of good assessment and feedback: theory and practice
6. Afzal N, Mitkov R (2014) Automatic generation of multiple choice questions using dependency-based semantic relations. Soft Comput 18(7):1269–1281
7. Bernhard D (2010) Educational applications of natural language processing. In: NATAL, pp 1–123
8. Sang EFTK, Veenstra J (1999) Representing text chunks. In: Proc of EACL'99, pp 173–179
9. Sathiyamurthy K, Geetha TV (2012) Automatic question generation from documents for e-learning. Int J Metadata Semant Ontol 7(1):16–24

10. Kumar SV, Prasad V, Santhanavijayan A, Balasundaram SR, Narayanan SH (2017) Automatic generation of multiple choice questions for e-assessment. Int J Signal Imaging Syst Eng 10:54–62
11. Low A (2018) China is using adorable robot teachers in kindergartens. CNET
12. Chin C (2016) Exclusive: Singapore is using robots to teach social skills. GovInsider
13. Masson O, Baratgin J, Jamet F, Ruggieri F, Filatova D (2016) Use a robot to serve experimental psychology: Some examples of methods with children and adults. In: International conference on information and digital technologies (IDT), pp 190–197
14. Angeli C, Valanides N (2019) Developing young children's computational thinking with educational robotics: An interaction effect between gender and scaffolding strategy. Comput Human Behav
15. Carpinella CM, Wyman AB, Perez MA, Stroessner SJ (2017) The robotic social attributes scale (RoSAS): Development and Validation. In: International conference on human-robot interaction—HRI '17, pp 254–262
16. Bartneck C, Kulić D, Croft E, Zoghbi S (2009) Measuring the anthropomorphism, animacy, likeability, perceived intelligence and perceived safety of robots. Int J Soc Robot 1:71–81
17. Yadollahi E, Johal W, Paiva A, Dillenbourg P (2018) When deictic gestures in a robot can harm child-robot collaboration. In: Proceedings of the 17th ACM conference on interaction design and children pages, pp 195–206
18. van der Drift E, Beun RJ, Looije R, Blanson Henkemans O, Neerincx M (2014) A remote social robot to motivate and support diabetic children in keeping a diary. In: International conference on human-robot interaction, pp 463–470
19. Mihalcea R, Tarau P (2004) TextRank. bringing order into texts. In: Proceedings of the conference on empirical methods in natural language processing
20. Brin S, Page L (1998) The anatomy of a large-scale hypertextual Web search engine. Comput. Networks ISDN Syst 30(1–7):107–117
21. Mwangi E, Diaz M, Barakova E, Catala A, Rauterberg M (2017) Can children take advantage of nao gaze-based hints during GamePlay? In: Proceedings of the 5th international conference on human agent interaction, pp 421–424
22. Akalin N, Kose H (2013) Humanoid robots communicating with children using sign language. In 2013 IEEE workshop on advanced robotics and its social impacts, pp 181–186
23. Pot E, Monceaux J, Gelin R, Maisonnier B (2009) Choregraphe: a graphical tool for humanoid robot programming. In: RO-MAN 2009—the 18th IEEE international symposium on robot and human interactive communication
24. Gigabotics Nao Features. [Online]. Available: http://gigabotics.com/robotics/features-of-nao-robot/
25. Pan MKXJ, Croft EA, Niemeyer G (2018) Evaluating social perception of human-to-robot handovers using the robot social attributes scale (RoSAS). In 2018 ACM/IEEE international conference, pp 443–451
26. Syrdal K, Dautenhahn DS, Koay K, Walters M (2009) The negative attitudes towards robots scale and reactions to robot behaviour in a live human-robot interaction study. In: Proceedings of AISB09
27. Peca A (2016) Robot enhanced therapy for children with autism disorders: measuring ethical acceptability. IEEE Technol Soc Mag 35(2):54–66
28. Syrdal DS, Dautenhahn TNK (2013) The frankenstein syndrome questionnaire—results from a quantitative cross-cultural survey. In: International conference on social robotics pp 270–279

Monthly Averaged All Sky Solar Irradiance Prediction Using Artificial Neural Networks for Chandigarh Region

Isha Arora, Jaimala Gambhir, and Tarlochan Kaur

1 Introduction

Renewable energy is the energy extracted from naturally existing sources that can be replenished such as solar energy, wind energy, tidal energy and biomass energy. Non-renewable fossil fuels have been utilized over many decenniums for fulfilling global energy needs, which are rising day by day. These resources are not only on the verge of exhaustion, but are also a major contribution towards global warming and harmful greenhouse gases emissions that has now brought the focus on the RESs [1]. India being a tropical nation has huge solar energy potential, as it lies north to the Equator, between 8°4′ and 37°6′ N latitude and 68°7′ and 98°5′ E longitude. Incoming solar energy in India lies in the range of 4–7 kWh/m^2-day. MNRE, India has also goal of reaching 40% of power from non-fossil-based sources, 175 GW of installed capacity from RESs by year 2022, out of which 100 GW is to be met by solar power [2]. Sun's radiation is the pivotal driving force of the Earth. Sun is a clean and green energy source that has huge ability to cater to the day-by-day increasing energy needs worldwide. The amount of solar radiation received on the Earth's surface can be measured by employing various devices such as pyrheliometer, pyranometer, photoelectric sunshine recorder and quantum sensors with the data logging system.

It is not feasible to place these measuring devices at every location, due to economic constraints, and/or technical issues (maintenance and calibration of device being the significant ones), that leads to inconvenience in solar radiation data accessibility for the largest portion of the world. Forecasting of solar radiation intensity in remote, far-flung and secluded regions is considered as tedious task due to no

I. Arora (✉) · J. Gambhir · T. Kaur
Punjab Engineering College (Deemed to Be University), Chandigarh, India
e-mail: arora.isha02@gmail.com

V. H. Saran and R. K. Misra (eds.), *Advances in Systems Engineering*, Lecture Notes in Mechanical Engineering,
https://doi.org/10.1007/978-981-15-8025-3_42

meteorological stations established in these regions. Solar radiation forecasting has gathered interest recently due to its utilization in various RESs related researches and applications [3].

Various statistical and artificial intelligent approaches have been applied for solar irradiance prediction that includes linear regression models, mathematical modelling, time series, fuzzy systems, neural networks, hybrid systems, etc. [4].

Global solar irradiation in Zhejiang has been forecasted based on horizontal direct and diffused irradiance from 6SV LUT, Digital Elevation Model (DEM) and MODIS water vapour (MOD05) system [5]. Solar radiation intensity has been forecasted ranging from 30 min to 5 h, by utilizing geostationary satellite [6].

Autoregression moving average (ARMA) model has been used to deliver an apt fit for sub-hourly solar radiation values, correspondent to global irradiance records of radiometric stations in south Spain. Time series analysis and system identification MATLAB toolboxes have been utilized for forecasting. [7]. Comparative study between physical and statistical means has been conducted to forecast PV plant power. Physical method comprises mathematical analysis of PV system that is incorporation of Sun position model and diode model. Forecasted weather parameters that have been derived from NWP are taken as model input. Humidity and position of Sun can be considered to enhance the accuracy of NN method [8].

Fuzzy and NN integrated model has been designed to predict hourly and day-to-day solar radiation. Sky and temperature information is categorized as different fuzzy sets based on fuzzy rules [9]. Fuzzy prediction interval model has been presented which considers uncertainties related to renewable energy sources and loads. Model has then been employed for one day ahead forecast using solar and wind generation and regional load data of microgrid in Chile [10].

The paper is compiled as follows: idea about ANN is given in Sect. 2, networks specifically; multi-layer feed forward and layer recurrent networks are discussed. Section 3 describes the database, technique employed, result comparison for ANN structure learning. Results for amount of solar radiation in numerous towns of H.P. are presented in Sect. 4 and conclusion in Sect. 5.

2 Artificial Neural Networks

Neuron is the fundamental component of an ANN, which operates on the framework inspired by the inference forming learning of brain. ANNs have humongous number of immensely interlinked nodes or neurons, which undergo computational analysis of networks in parallel pattern. These links transfer the signals from one node to other. Neural networks, via training procedure, discover the relation between model input variables and output [11]. Output of each ANN is calculated by some nonlinear function of its addition of inputs. The linkages between the neurons usually have a weight that updates as learning progresses. ANNs are simple, flexible, powerful

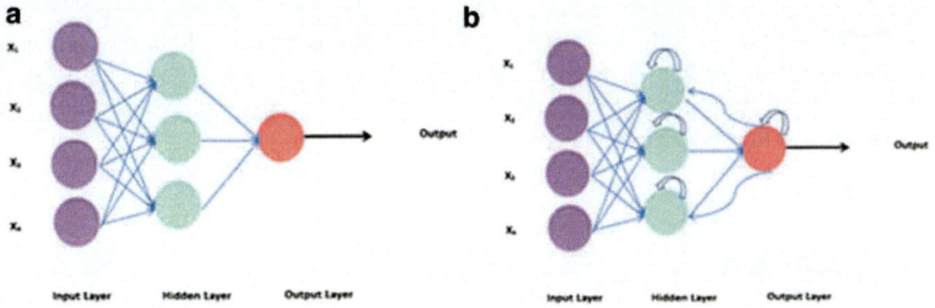

Fig. 1 **a** Multi-layer feed forward network structure **b** Layer recurrent network structure

means used for training of large amount of database, solving complex and greatly nonlinear problems and have comparatively lesser processing time due to its parallel architecture [12].

ANNs are constituted of some main elements such as weights, summation unit, activation function, output function and error function. After summation, each node has to yield a value as a function of its net through activation or transfer function. Weight enhances or reduces the strength of a signal at any linkage. Artificial neurons have a threshold, and a signal is only transmitted if the composite signal crosses the threshold. ANN has different architectures [13]. ANN is the assembly of different number of layers, viz. input, hidden and output layer; each of the layers is composed of specific number of neurons [14]. Signals travel from the foremost layer (input layer) to the rearmost layer (output layer), after crossing numbers of hidden layers [15].

2.1 *Multi-layer Feed Forward Networks*

Multi-layer feed forward networks (FFD) differentiate themselves from single layer feed forward networks by the existence of one or more number of hidden layers as shown in Fig. 1a. The nodal units of these layers are termed as hidden neurons. The role of hidden neurons is to intercede between the extrinsic input parameters and the model output in some effective manner [16].

2.2 *Layer Recurrent Networks (RNNs)*

When outputs are fed back to the neurons in the same or previous layers, networks are classified as feedback networks. A recurrent network is composed of minimum one feedback loop. A recurrent network might contain single layer of nodes with each nodal unit directing its output signal back to the inputs of all the other units, as shown in Fig. 1b. Self-feedback is termed as structures in which the output of a node is given back into its own input. Unalike FFDs, RNNs can employ their internal memory to process random series of inputs. In such networks, self-loops and rearward interconnections between the nodal units are present [17].

3 Simulation

Solar radiation intensity prediction is conducted in the following steps:

1. Optimal network architecture selection.
2. Training of patterns.
3. Validation and testing.

3.1 Network Architecture Selection

The work presented seeks to evaluate the influence of different network topologies, number of hidden layers and number of hidden nodes on the prediction errors.

The number of nodes in hidden layer has an effect on the learning potential of the system, and the complexity of neural network differs with it. Number of nodes in intervening layers should be thoroughly selected, as too many nodal units cause the issue of overfitting, which yields in loss of generalizing ability [18]. With excessive trainable parameters, the structure does not succeed in learning the training database and results in poor performance. But, if the number of neurons in hidden layer is insufficient, it may be inconvenient for the structure to train according to the historical data.

Activation Function

For different network architectures assessed, *tansig* has been opted for as transfer or activation, to calculate output of a layer from its net input. Bipolar sigmoidal transfer function is used that gives the value between −1 and +1:

$$f(\mathrm{net}) = \frac{2}{1 + \mathrm{e}^{-\lambda \mathrm{net}}} - 1 \tag{1}$$

where *f*(net) denotes activation function,

λ denotes steepness of curve, and

net denotes net additive value obtained after multiplication of input patterns with respective weights.

The performance of the forecasting model is evaluated by different accuracy parameters. An effective forecasting network should attain the requirements of being good in accuracy with lesser MSE, faster in speed, that is with lesser training time and with higher R-value. MSE, training time and R-values are considered as measures for assessment.

Mean Squared Error (MSE) The basic performance function for networks is MSE—the averaged square difference between network outputs and the target output. It is given by the formula:

$$\text{MSE} = \frac{1}{n}\sum_{i=1}^{n}(d_i - y_i)^2 \tag{2}$$

where d_i is target output of the ith pattern,
y_i is actual output of the ith pattern, and
n is total input pattern count.

Training Time
It is the time taken for the network to undergo training as per the input parameters set for learning. Its value is considered in seconds.

Correlation Coefficient (*R*-values)
Correlation coefficient (R-values) depicts the relation between the model outputs and the target outputs of the ANN model. The perfect fit exhibits the slope close to 1, which is, data is falling across the 45° line, which means the model outputs are equal to target outputs. R-value of 1 means a strong; whereas, 0 means a random relation between actual and desired outputs [19].

A number of hidden layers and number of neurons in a hidden layer are estimated by employing hit and trial means. The selection of number of neurons in the hidden layer is started of as a fraction, which is $2/3$ the number of neurons in the input layer [20]. Structure of the prediction model is varied by changing the number of hidden neurons and/or by changing the number of hidden layers.

4 ANN-Based Prediction—Results and Discussion

The prime aim of this work is to predict the monthly averaged all sky solar radiation intensity (kW-hr/m^2/day) over the mountainous areas of H.P., India, from attributes such as geographical parameters and meteorological parameters employing the ANN model and comparing it with actual values to verify the accuracy of the model.

Data utilized to carry out training of ANN models is collected from January 2016 to December 2017 for 25 different cities all over India, with widely ranging climatic conditions. The climatic data values of these regions are taken from National Aeronautics and Space Administration (NASA) [21, 22]. The input database is normalized using min–max normalization technique. Following variables are taken as input neurons:

Latitude of the region (°), longitude of the region (°), altitude of the region (m), monthly averaged temperature of the region (°C), monthly averaged maximum temperature of the region (°C), monthly averaged minimum temperature of the region (°C), monthly averaged relative humidity of the region (%), monthly averaged wind speed of the region (m/s), monthly averaged sunshine hours of the region (min) and month of the year.

So, to begin with choosing optimal network architecture, a comparative analysis has been carried out on two different configurations that are FFD and RNN. ANN model developed has one input layer that is composed of ten input neurons same as number of input variables. As the target outcome is monthly averaged solar radiation intensity, model has one output that presents the amount of solar radiation for that particular month. Month variable denotes the month of the year as the solar radiation intensity varies from one month to another. Topography modifications start with five nodal units in hidden layer. The model is trained over 1000 epochs. Epoch is one entire stretch of training, testing and validation.

Data has been split into 60%, 20% and 20% for training, testing and validation, respectively. Testing has been conducted to check if the network is generalizing well, but it does not have impact on the training of network.

The results of the different structure combinations evaluated on the basis of performance indicators are tabularized in Tables 1 and 2, where L denotes number of hidden layers and N depicts number of neurons in a hidden layer.

It is observed that as the number of neurons in hidden layer exceeds thrice the number of input neurons, system begins to overfit.

Figure 2a and b shows the performance of different network topographies considered, as MSE and training time when trained over 1000 epochs. The training is carried out on the basis of gradient descent algorithm. The connection weights between neurons and biases of the network are updated in the direction in which the performance function reduces more swiftly that is negative of the gradient [23].

So, from the above outcomes, it is inferred that FFD networks have higher R index, which undergoes training in lesser time and that too with higher accuracy.

Table 1 MSE, training time and *R*-values for different FFD structure combinations

S.No.	Architecture	MSE	Training time (s)	R-value (all)	Structure
A1	L=1, N=5	0.0828	23	0.9196	
A2	L=2, N=5	0.0238	35	0.9384	
A3	L=1, N=10	0.0652	28	0.9268	
A4	L=2, N=10	0.0675	41	0.8966	
A5	L=1, N=15	0.0509	37	0.9278	
A6	L=2, N=15	0.0778	107	0.9334	
A7	L=1, N=20	0.1149	71	0.9531	
A8	L=1, N=25	0.0609	72	0.9511	
A9	L=1, N=30	0.043	76	0.9243	
A10	L=1, N=35	0.0436	81	0.9221	Overfit
A11	L=1, N=40	0.0526	123	0.9324	Overfit

Table 2 MSE and training time for different RNN structure combinations

S. No.	Architecture	MSE	Training time (s)	Structure
A12	L = 1, N = 5	0.0668	16	
A13	L = 2, N = 5	0.0714	36	
A14	L = 1, N = 10	0.0255	67	
A15	L = 2, N = 10	0.0704	232	
A16	L = 1, N = 15	0.0248	181	
A17	L = 1, N = 20	0.0556	372	Overfit
A18	L = 1, N = 25	0.1208	865	Overfit
A19	L = 1, N = 30	0.07	2006	Overfit

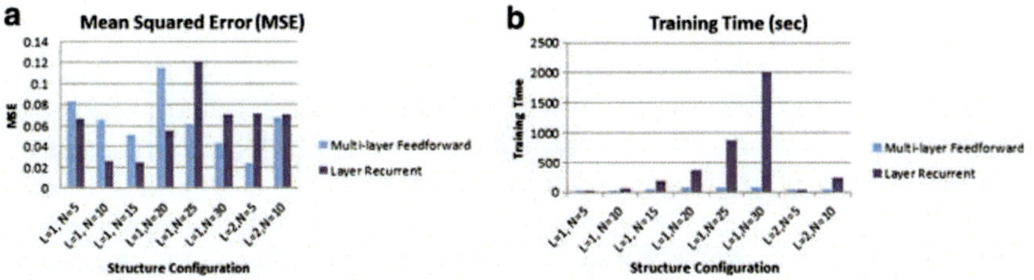

Fig. 2 **a** MSE for different structure combinations and **b** Training time for different structure combinations (for H.P.)

Selected optimal network configuration in the previous section, that is FFD network with two hidden layers composed of five hidden neurons in each, is employed for solar irradiance prediction. Forecasting is carried out from January 2017 to December 2017. Monthly averaged all sky solar radiation intensity for ten districts is predicted by ANN model, resulting in total MSE of 2.385%.

In comparison with the actual values taken from pyranometer readings for Hamirpur region, prediction error comes out to be nearly 1.26%. The selected feed forward configuration comprising two hidden layers is then applied on Chandigarh region (30.73° N, 76.77° E), to predict its monthly averaged solar radiation intensity over year 2018 using monthly averaged input attributes (geographical and climatic) for year 2016 and 2017. Figure 3a and 3b depicts MSE and *R*-values for different network configurations when trained over 1000 epochs for Chandigarh. Least error is obtained for feed forward configuration with two hidden layers. The forecasting yields in about 1.009% MSE. The results indicate good solar potential that can be utilized for various solar applications including PV power production in the region.

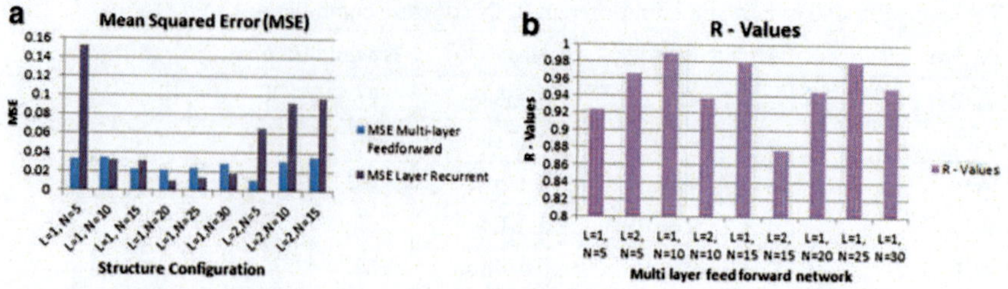

Fig. 3 **a** MSE for different structure combinations and **b** *R*-values for different structure combinations (for Chandigarh)

5 Conclusion

This work has given ANN-based technique to estimate monthly averaged solar radiation intensity. The FFD network with two hidden layers has been observed as the optimal structure amongst the different architectures considered and compared. It can be successfully employed over any designed stand-alone solar energy transformation application, or the regions where installation of meteorological stations is infeasible due to technical or economic constraints. The results have depicted the forecasting ability of the ANN model and its capability for any location with changing weather conditions. The work can be further extended to sizing of isolated PV systems and estimating the power production by the panels.

References

1. Sumithira TR, Kumar AN (2012) Prediction of monthly global solar radiation using adaptive neuro fuzzy inference system (ANFIS) technique over the state of Tamilnadu (India): a comparative study. Appl Solar Energy 48:140–145
2. http://pib.nic.in/newsite/PrintRelease.aspx?relid=133220
3. Yadav AK, Malik H, Chandel SS (2014) Selection of most relevant input parameters using WEKA for artificial neural network based solar radiation prediction models. Renew Sustain Energy Rev pp 509–519 (Elsevier)
4. Zhang X, Li Y, Lu S, Hamann HF, Hodge BM, Brad Lehman B (2019) A solar time based analog ensemble method for regional solar power forecasting. IEEE Trans Sustainable Energy 10(1):268–279
5. Su G, Xin X, Liu Q (2008) Estimation of global solar irradiance at the complicate terrains in Zhejiang Province, China. In: IEEE international geoscience and remote sensing symposium (IGARSS) pp 732–735
6. Peng Z, Yoo S, Yu D, Huang D (2013) Solar irradiance forecast system based on geostationary satellite. In: IEEE SmartGridComm symposium—support for storage, renewable resources and micro-grids, pp 708–713
7. Munoz AM, Rosa JG, Posadillo R, Bellido F (2008) Very short term forecasting of solar radiation. In: IEEE photovoltaic specialists conference
8. Huang Y, Lu J, Liu C, Xu X, Wang W, Zhou X (2010) Comparative study of power forecasting methods for PV stations. In: IEEE international conference on power system technology

9. Chen SX, Gooi HB, Wang MQ (2013) Solar radiation forecast based on fuzzy logic and neural networks. Renew Energy 60:195–201 (Elsevier)
10. Sáez D, Ávila F, Olivares D, Cañizares C, Marín L (2015) Fuzzy prediction interval models for forecasting renewable resources and loads in microgrids. IEEE Trans Smart Grid 6(2):548–556
11. Arora I, Kaur M (2016) Unit commitment scheduling by employing artificial neural network based load forecasting. In: IEEE 7th India international conference on power electronics (IICPE)
12. Zarezadeh M, Naghavi A, Ghaderi SF (2008) Electricity price forecasting in Iranian electricity market applying artificial neural networks. IEEE electrical power and energy conference
13. Muhammad A, Lee JM, Hong SW, Lee SJ, Lee EH (2019) Deep learning application in power system with a case study on solar irradiance forecasting. In: ICAIIC IEEE, pp 275–279
14. Kaur T, Kumar S, Kaur R, Gera A (2018) ANN based global solar radiation prediction: a case study. Kaav Int J Sci Eng Technol 5(4):54–63
15. Kumar S, Kaur T (2016) Development of ANN based model for solar potential assessment using various meteorological parameters. Energy Procedia 90:587–592
16. Shahat AE (2014) Artificial neural network (ANN): smart and energy systems applications. Scholar's Press
17. Lin CT, Lee CSG (1996) Neural-Fuzzy Systems: a neuro-fuzzy synergism to intelligent systems. Prentice Hall PTR
18. Adepoju GA, Ogunjuyigbe SOA, Alawode KO (2007) Application of neural network to load forecasting in Nigerian electrical power system, Pacific J Sci Technol 8(1)
19. http://in.mathworks.com
20. Kandil N, Wamkeue R, Saad M, Georges S (2006) An efficient approach for short term load forecasting using artificial neural networks, IEEE ISIE
21. http://www.timeanddate.com/
22. https://power.larc.nasa.gov/data-access-viewer/
23. Leholo S, Owolawi P, Akindeji K (2019) Solar energy potential forecasting and optimization using artificial neural network: South Africa case study. Amity Int Conf Artif Intell, IEEE, pp 533–536

Performance Comparison of Soft Computing Algorithms for Voiced and Unvoiced Speech Detection

Anurag Verma, Sitaram Jana, Ratna Raju Ravela, Himanshu Bansal, and Kishore Kumar Nune

1 Introduction to Voiced–Unvoiced Detection

The speech signal classification as voiced and unvoiced is the base for speech processing. Unvoiced part of speech is not periodic signal and looks like random signal as consonants are spoken. Voiced part of speech has constant frequency components of letters—a, e, i, o and u. Hence, periodicity is one of the features to classify voiced and unvoiced speech. Qi and Hunt [1] applied multi-layer feed forward ANN for the classification of voiced and unvoiced audio signals using zero-crossing rate(ZCR) and short-term energy (STE) of acoustic signal. Bachu [2] proposes methods of voiced–unvoiced distinction which applied in long window lengths of 40 ms and tested in silence. Voiced–unvoiced distinction under noisy environment is a difficult work as it can suppress low-energy unvoiced speech signals called fricatives (*f* and th) and hide letters—b and d [2].

A. Verma (✉)
Faculty of Engineering, Dayalbagh Educational Institute, Agra, Uttar Pradesh 282005, India
e-mail: rsaverma@gmail.com

S. Jana · R. R. Ravela · H. Bansal · K. K. Nune
Department of Electrical and Electronics Engineering, Noida Institute of Engineering and Technology, Greater Noida, India
e-mail: sijana2014@gmail.com

R. R. Ravela
e-mail: ratnamwait@gmail.com

H. Bansal
e-mail: bansal617@gmail.com

K. K. Nune
e-mail: kishorekumar.noon@gmail.com

V. H. Saran and R. K. Misra (eds.), *Advances in Systems Engineering*,
Lecture Notes in Mechanical Engineering,
https://doi.org/10.1007/978-981-15-8025-3_43

2 Critical Issues in Voiced–Unvoiced Detection (VUD)

The issue of detecting unvoiced speech part has many difficulties in low SNR situations common today. This requires getting the start and end point of any speech word uttered as shown in Fig. 1. Every speech word has a fundamental frequency called pitch which is different in voiced from unvoiced portion of speech as shown in Fig. 2 [3]. Some methods are using energy and ZCR of the speech signal for voiced, unvoiced and/or silence portion of speech detection. BPFF-neural networks are mostly used for pattern recognition and classification of speech [4]. Mostly voiced sounds are:

- Periodic in time and frequency,
- Peaking power at smaller frequencies and
- Strong energy content.

But unvoiced sounds exhibit higher power at high frequencies. An important feature for voiced–unvoiced distinction is the ratio of signal power at high frequency to signal power at lower frequency. ZCR is used for VUD when noise level is very low [5]. The block diagram for VUD decision shown in Fig. 3 below for a speech part needs values of ZCR and STE. As they are not precise, it will be good to use fuzzy logic. Hence, we can apply fuzzy logic to classify the speech [6]. A false alarm in the VUD problem of periodicity can produce classification errors due to silence or noise portion of speech signal [7]. Hence, it is required to remove silence in speech signals to detect clear speech parts. This will be useful in speech derived HMI applications, surveillance systems based on speech and all automatic speech recognition systems

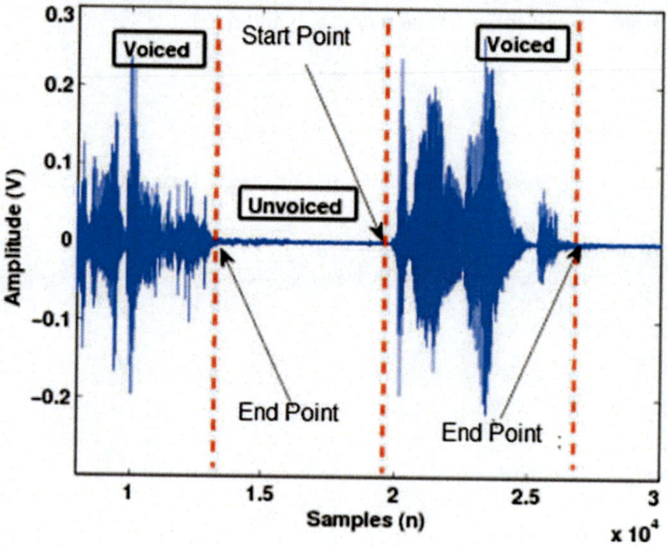

Fig. 1 Start and end point of speech signal acquired [3]

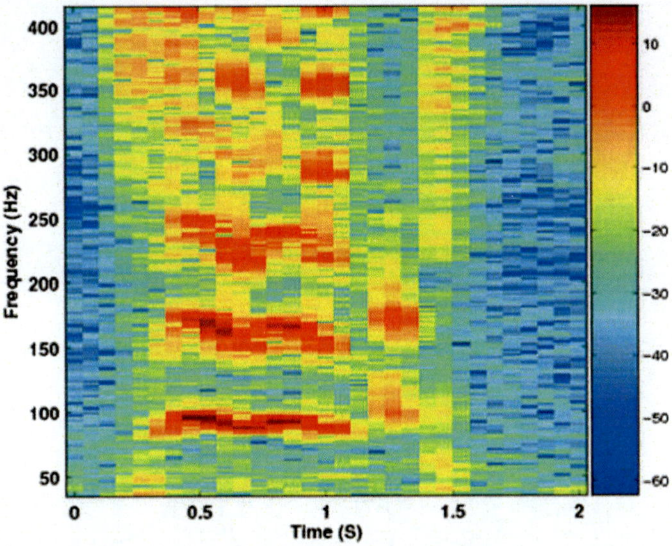

Fig. 2 Spectrogram of a speech signal acquired [3]

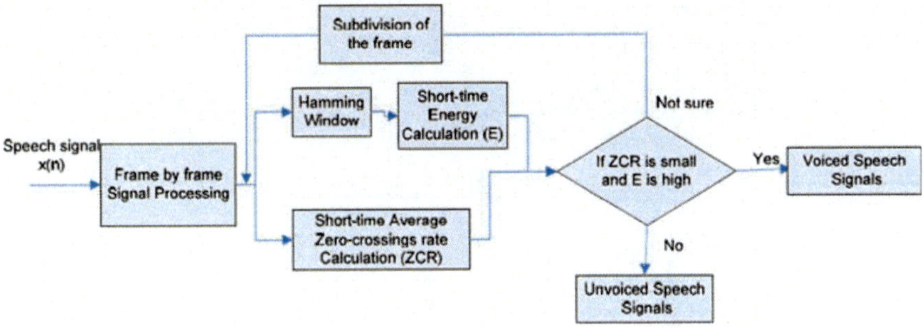

Fig. 3 Block diagram of VUD [8]

[8]."Voiced speech (which is 66.67% part of speech) has constant frequency signals, it is produced when vowels are spoken, and it is more important for intelligibility". On the other hand, unvoiced speech is random and non-periodic. This forms the basis of VUD.

Unvoiced and voiced speech signals overlap with each other and making it cumbersome to identify the start and end points of speech signals. Low amplitude voiced speech can be interpreted as unvoiced speech. Also it is possible for plosives can be confused with voiced or unvoiced signals [9]. In frame-based VUD, the time of the segment should be short enough to let the speech features to be stationary B20. Following are the features applied for VUD as pitch frequency, ZCR, cepstral peaks, STE and so on (as shown in Figs. 4, 5, and 6). Algorithms that use more

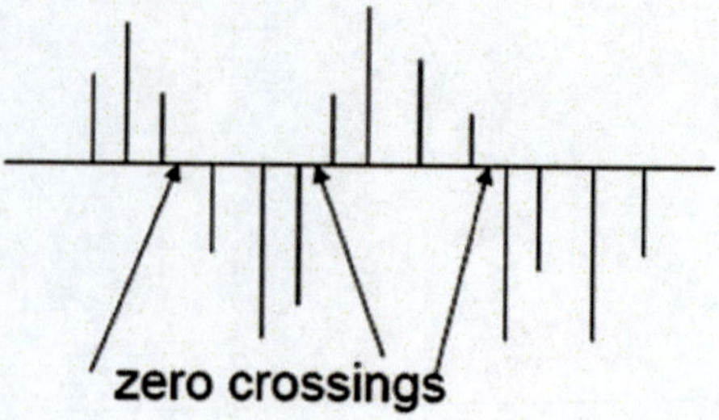

Fig. 4 ZCR [8]

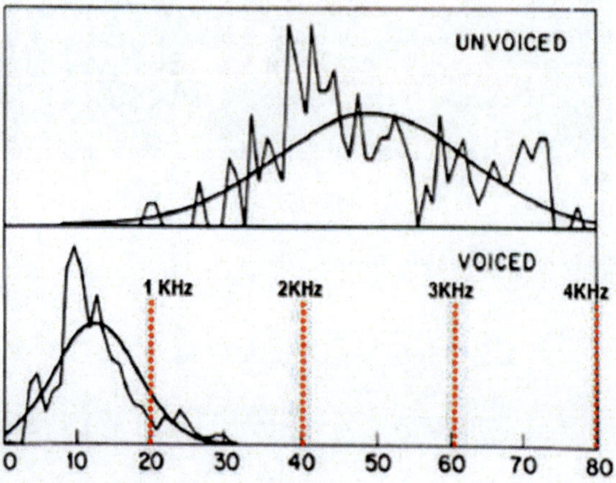

Fig. 5 ZCR for voiced and unvoiced speech [8]

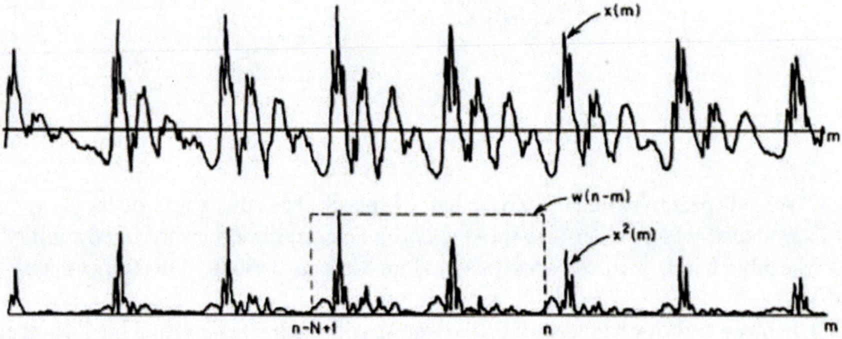

Fig. 6 STE computations [8]

than one feature give better accuracy than single-feature techniques. Speech signals are stationary in time frames of 10–30 ms. Some of the features are defined below mathematically [10]:

ZCR: It indicates the frequency content in the signal as follows:

$$\mathrm{ZCR} = \sum_{n=0}^{N-1} |\mathrm{sgn}[x_i(n)] - \mathrm{sgn}[x_i(n-1)]| \tag{1}$$

STE: It helps in understanding the variation of energy with time. Energy associated with short portions of speech signal is given by

$$E_i = \left(\sum_{n=0}^{N-1} \left|x_i(n)^2\right|\right) \Big/ N \tag{2}$$

3 Algorithm Description

In this work, the following steps are followed:

3.1 Pre-processing of speech words was performed using Audacity. First, the noise reduction was done by 40 dB and then normalization was done by 5 dB. The speech words were saved as .wav files. An example of word 'ZEBRA' is shown in Fig. 7.

3.2 Single-sided amplitude spectrum and spectrogram of speech word—zebra—are illustrated in Figs. 8 and 9. Three features—ZCR, STE and RMS—of 21 speech words were extracted using MATLAB software as shown in Fig. 10 (example of speech word—zebra) uttered by different speakers (all five authors of this paper)

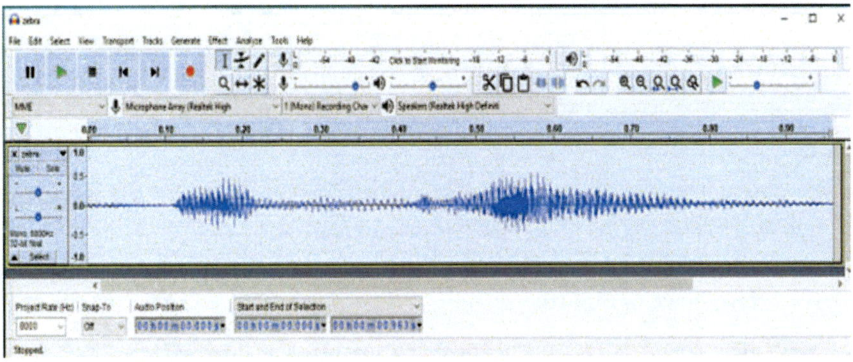

Fig. 7 Audacity software

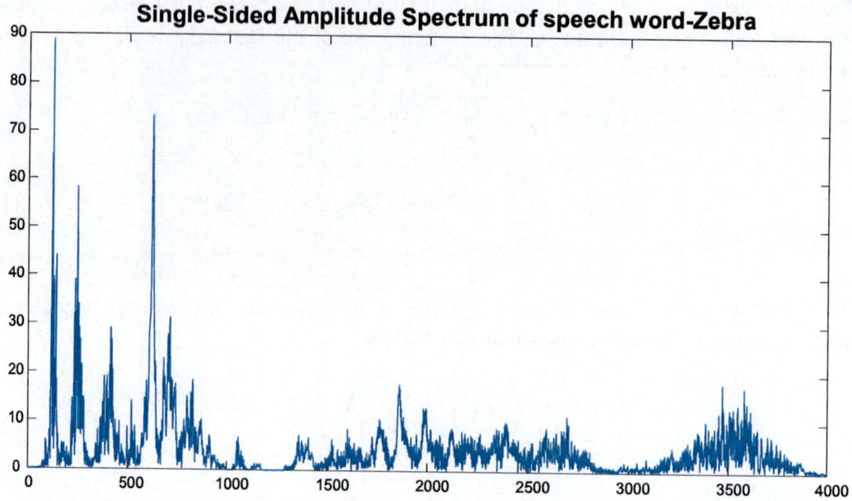

Fig. 8 Single-sided amplitude spectrum of speech word—zebra

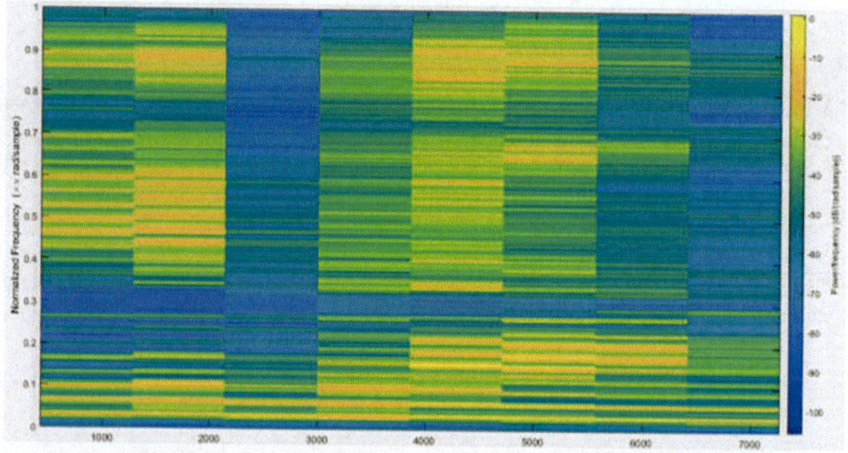

Fig. 9 Spectrogram of speech word—zebra

and recorded using Audacity software (sampling frequency of 8000 Hz and mono recording settings) as shown in Table 1.

3.3 For each word, three thresholds are estimated with respect to three features in MATLAB as shown in Fig. 10.

3.4 Labelling of words for voiced–unvoiced markings is done in MS-EXCEL by visual inspection of graphs of data plotted generated from MATLAB. This provides data for training and testing of soft computing algorithms.

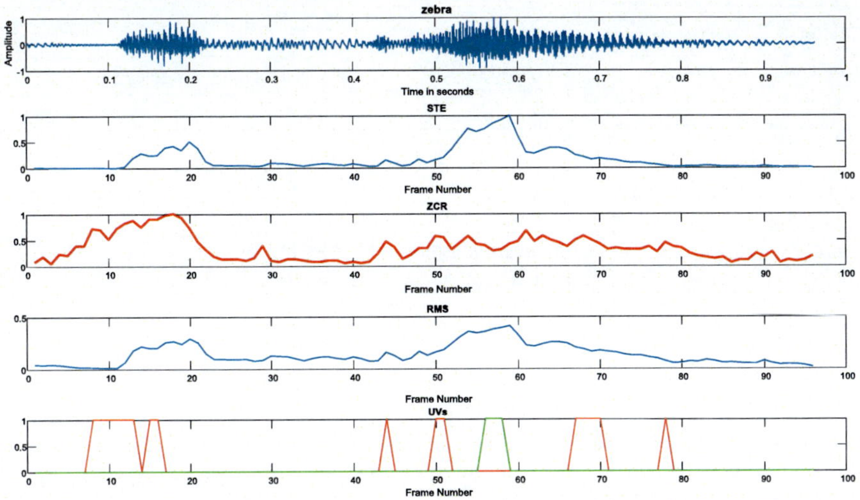

Fig. 10 Features of speech word—zebra

Table 1 Speech words recorded

S. No.	Word uttered	Author no.	Time duration
1	Acorn	1	1.30
2	Apple	1	1.35
3	Cieling	4	1
4	Eight	3	1.0
5	Eleven	3	1.0
6	Happy	4	1.1
7	Home	1	0.8
8	Igloo	4	0.94
9	Moon	5	1
10	Octopus	4	1.25
11	Oval	5	0.8
12	Paint	1	1.0
13	Safety	1	1.1
14	Safron	1	1.4
15	Shield	3	1
16	Steak	3	1
17	Tissue	5	1
18	Tree	2	1.0
19	Umbrella	2	1.40
20	Unicorn	5	1.20
21	Zebra	2	0.95

Table 2 Comparison of soft computing algorithms

S. No.	Algorithm	Classification accuracy (%)
1	MLP_NN	76.2
2	Decision trees	75.0
3	SVM	76.2
4	k-NN	75
5	Random forest	85.12
6	AdaBoost	79.76
7	GaussianNB	74.4

3.5 The labelled data is given as input to Jupyter Notebook in Python programming environment for evaluating segmentation performance of voiced and unvoiced portion of speech using soft computing algorithms like SVM, neural networks, fuzzy logic, decision trees, k-NN, random forest, AdaBoost and GaussianNB, etc.

4 Analytic Comparison of Performance for Soft Computing Algorithms

Performance of soft computing techniques is compared with respect to classification accuracy in Table 2 and is shown in Fig. 11.

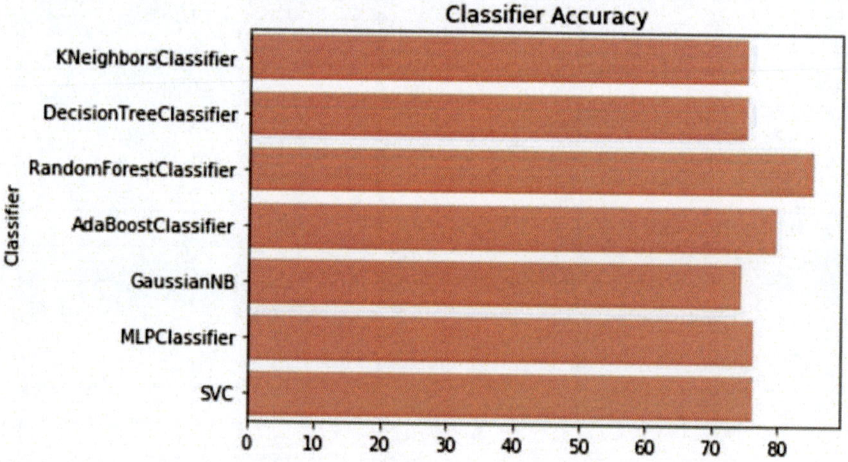

Fig. 11 Classification accuracy of different classifiers

5 Conclusion and Future Work

The experimental results in this work show that among the different classifiers like SVM, neural networks, fuzzy logic, decision trees, k-NN, random forest, AdaBoost and GaussianNB, etc. Random forest classifier is performed better with 85.12% accuracy of classification for voiced–unvoiced detection. In our future research, we will consider fuzzy classifiers and wavelet-based features.

References

1. Qi Y, Hunt BR (1993) Voiced-unvoiced-silence classifications of speech using hybrid features and a network classifier. İn: IEEE transactıons on speech and audıo processıng, vol 1
2. Bachu RG, Kopparthi S, Adapa B, Barkana BD (2008) Separation of voiced and unvoiced using zero crossing rate and energy of the speech signal
3. Zenteno E, Sotomayor M, Robust voice activity detection algorithm using spectrum estimation and dynamic thresholding. İn: IEEE Latin-American conference on communications
4. Tuan PV, Kubin G (2005) DWT-based phonetic groups classification using neural networks. In: Proceedings-ICASSP '05
5. Deng H, O'shaughnessy D (2007) Voiced-unvoiced-silence speech sound classification based on Unsupervised learning. In: IEEE ınternational conference on multimedia and expo
6. Algabri M, Alsulaiman M, Muhammad G, Zakariah M, Bencherif M, Ali Z (1997) Voice and unvoiced classification using fuzzy logic. In: IEEE workshop on speech coding for telecommunications proceedings
7. Ahmadi S, Spanias AS (1999) Cepstrum-based pitch detection using a new statistical V/UV classification algorithm. In: IEEE transactions on speech and audio processing, vol 7, issue 3
8. Giannakopoulos T (1999) A method for silence removal and segmentation of speech signals, implemented in Matlab. IEEE Trans Speech and Audio Process
9. Marks JA (1988) Real time speech classification and pitch detection. In: Southern African conference on communications and signal processing. Proceedings
10. Cai RS, Zhu YT, Guo YM (2007) Wavelet-based multi-feature voiced unvoiced speech classification algorithm. In: IET conference on wireless, mobile and sensor networks

Modal Analysis of Vibratory Finishing Machine Using ANSYS Software

Harshita Khangarot, Vikas Kumar, Rohit Sharma, Ambrish Maurya, and Jaspreet Hira

1 Introduction

Mass finishing is a standard explanation for vibrating and flowing of an abrasive media around loose work pieces, moving randomly with the abrasive media, compound, water and sometimes abrasive powder. *Mass finishing* is a process that automates the mechanical and chemical finishing of various shaped parts that permits for the mechanical finishing of a number of parts simultaneously which are small and complex in shape. Along with deburring, mass finishing may also be used to de-scale, brighten and polish parts. Mass deburring is considered a 'loose-abrasive' method in which batches of parts are vibrated or tumbled in especially designed bowls, tubs or barrels along with a loose abrasive media.

Vibratory finishing has been used for the commercial purpose for over the last 40 years. Empiricism and trial and error method is practiced in most of the industries today. The present paper is directed towards the designing of a bowl-type vibratory finishing machine and deburring and finishing of the parts. Vibratory finishing is the most common mass finishing process. Range of frequency of the vibration can be

H. Khangarot (✉) · V. Kumar · R. Sharma · A. Maurya · J. Hira
Department of Mechanical Engineering, Amity School of Engineering and Technology, Amity University Noida, Noida, Uttar Pradesh, India
e-mail: hkhangarot@gmail.com

V. Kumar
e-mail: viksmied@gmail.com

R. Sharma
e-mail: rsharma19@amity.edu

A. Maurya
e-mail: ambrish.iitr@gmail.com

J. Hira
e-mail: Jhira@amity.edu

V. H. Saran and R. K. Misra (eds.), *Advances in Systems Engineering*, Lecture Notes in Mechanical Engineering,
https://doi.org/10.1007/978-981-15-8025-3_44

from 800 to 3600 vibrations per minute (VPM). Vibratory Finishing Machine has several components of mechanical vibrations i.e. horizontal, vertical and gyratory, which may be tuned individually to suit any particular job for the optimum finish of work pieces. The parts to be machined are loaded into the bowl, along with particular ratio of water, media and abrasive compound. Then optimum vibration level is set, the components, abrasive media and finishing compound roll over and rub rigorously against each other due to these vibrations, resulting in deburring, finishing, descaling, de-coloring, burnishing, smoothening and brightening of the work pieces. The media operate continuously in multiple directions. This replaces hand deburring or manual refurbishment of components individually. Mohajerani et al. [1] numerically modeled the edge wear of borosilicate glass in vibratory finishing process in their paper. The model predicted the evolution of the cross-sectional edge profile through an iterative process of edge chipping. The chips produced of order of 10 μm in length. Pramanik et al. [2] presents the mechanisms of elementary challenges, like disparity of chip thickness, high heat stress, high pressure loads, spring back and residual stress. Latest techniques like implementation of high pressure coolant, cryogenic cooling, tap testing, thermally enhanced machining have also been concluded. Yabuki et al. [3] presented in their paper that the largest normal impact forces were found to be associated with the mode of contact in which a part of abrasive media is static on the surface while adjacent media rub over it. Hashimoto et al. [4] proposed a model of a mechanical vibratory system in a vibratory finishing machine. The consequences of key criteria in the vibratory system on the performance of the machine were refined and ideal set up constraints were given. The model was validated in their paper on the basis of experimental readings. Hashimoto et al. [5] evolved a mathematical modeling which makes the unambiguous prediction of surface roughness and material dismissal of machined feasible units. The ideal machining operation was explained in terms of least operation time and cost to attain the objective and the formula to design the ideal operation was proposed. Domblesky et al. [6] explained that for aluminium, brass and steel, MRR was constant over time and has shown sensitivity to hardness and bowl acceleration. Surface roughness did not show any sign of sensitivity to bowl acceleration. Domblesky et al. [7] developed a model to explain the MRR in vibratory finishing as a function of bowl acceleration, work pieces mass, velocity and specific energy. It was investigated that MRR essentially constant overtime when using media that was ‘broken in’. MRR is directly related to the acceleration of bowl. Steel demonstrated an intermediate MRR where brass showed the most inflated whereas aluminum had the minimum. Bortoli et al. [8] states that the lateral motion of the suspension springs during the compressor start/stop has an important influence on the spring’s fatigue life. For each start/stop occurence, more cycles could take place causing damage to the spring life. Pranitbelhekar et al. [9] states that there are some factors to be considered for material selections are formability, availability, cost, stress relaxation, corrosion resistance, fatigue strength and conductivity. Belapurkar et al. [10] proposed that for each spring, stiffness is the most important factor. Springs are accepted or rejected on precise data analysis. The results obtained were on average 93.01% accurate with reference to Universal Testing Machine. Mediratta et al. [11] investigates that the two main mechanisms play an important role in a vibratory finishing operation-plastic deformation by abrasive media collisions and metal dismissal by relative movement of abrasive media and workpiece. Media selection and quantity of compound play a crucial role in getting about the correct surface finish. Material Removal Rate for a specific material remain persistent over time and are known to be more when the work piece is particular to and deeper in the flow of media. Harak et al. [12] states that Modal Analysis of any vibratory system determines the natural frequencies and mode shapes of a continuous structure. The material chosen as the vibratory bowl for vibratory finishing machine is 316 stainless steel. Vakte et al. [13] proposes that to improve the life of the component the material used is very high strength steels and standard manufacturing process is followed which gives the fatigue failure of mechanical component depends on loading condition. The machine is incomplete without springs and the material of the spring should be of correct properties and use. Carbon-steel valve-spring-quality wire (ASTM A 230) is chosen as the material of spring. Rączka et al. [14] investigated that the stiffness coefficient of SMA springs can be controlled and the properties of a spring considered can be controlled by electric methods. Tests revealed that spring characteristics are non-linear. It was found that the NTC01 have poorer fatigue endurance than spring 65G. Tian et al. [15] developed a metal dismissal mechanism of vibratory machining and it was unveiled through kinematic model and simulation. It was found that after machining for 2.5 hours, initial surface roughness value approached from 1.5 μm to 0.13 μm. It was concluded that the relative velocity and metal removal intensity would gain the maximum value when test component was attached far from the rolling center.

Many variables and parameters can affect the machining behavior of the vibratory surface finishing machine. The modal analysis and natural frequencies are not known in the literature; however, it can affect the machining operation. Therefore, in the present work, the mode shapes and natural frequencies of the vibratory finishing machine have been investigated.

2 Dynamic analysis of vibratory finishing machine

2.1 FE Analysis

The general equation of motion for a dynamic system is $m\ddot{x} + c\dot{x} + kx = f(t)$.

Modal analysis is an efficient method in determining vibration properties of any mechanical system.

2.2 *Model for Modal Analysis*

Failure of any mechanical component depends on the loading conditions, material, component design and its manufacturing [2]. The model is always required for the modal analysis. FEM is used to determine the mode shapes and natural frequency of a dynamic system such as vibratory finishing machine. To do the numerical analysis, the design of the vibratory finishing machine was made in *Solidworks* software and then it was analyzed in *ANSYS*.

2.2.1 Vibratory Finishing Machine

Figure 1 shows the vibratory finishing machine designed in *Solidworks*. The vibratory finishing machine is utilized for the finishing of metals using media and compound. A 30 L machine used for the present work was manufactured by Delhi Metco, New Delhi, India. The bowl of the machine is made up of stainless steel 316 with a Polyurethane lining to protect the fabrication of the bowl chamber and PU lining is highly durable for coating of vibratory finishing bowl, it also has high abrasion resistance. The bowl sits above nine compression springs with stiffness 50.34 N/mm (each 5.59 N/mm). The whole system runs due to the vibratory mechanism which comprises of 0.5 HP motor connected to a shaft which in turn connected to eccentric mass. These three components make a complete vibratory mechanism of the machine. The base in the bottom supports the whole equipment.

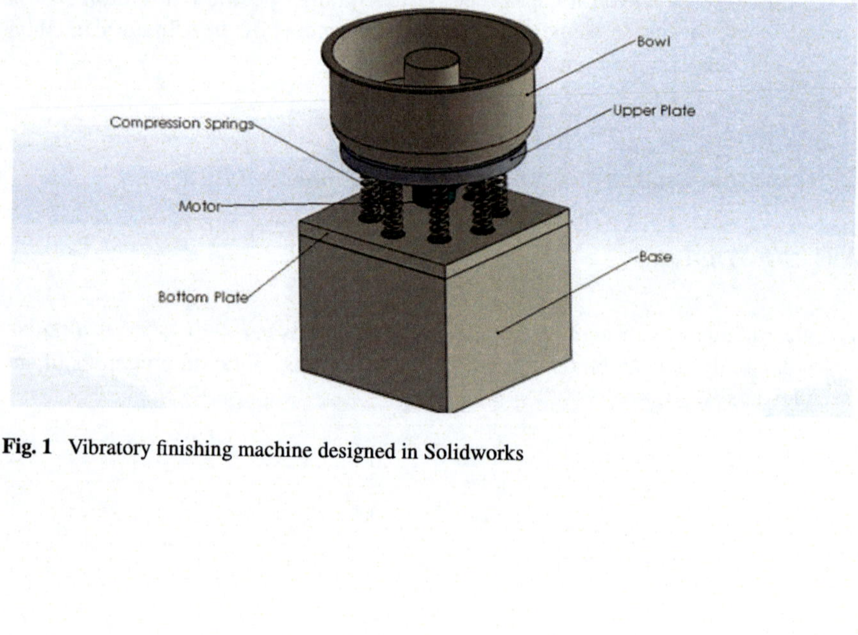

Fig. 1 Vibratory finishing machine designed in Solidworks

2.2.2 Spring Rate or Stiffness of the Spring

The spring material chosen for the present work is *carbon-steel valve- spring-quality wire ASTM A 230.* It is suitable for cyclic applications [3].

Considering the type of load application, the assumed working stress is checked:

$$\begin{aligned} k &= \left[79.3 \times 10^3 \times (6.2)^4\right]/\left[8 \times (34.64)^3 \times 7\right] \\ &= 117,176,344.5/232,7671.435 \\ &= 50.34\ \text{N/mm} \\ \text{Stiffness of each spring} &= 50.34/9 = 5.59\ \text{N/mm.} \end{aligned} \quad (1)$$

2.2.3 Deflection of the Spring

$$\begin{aligned} \partial &= 578.97/50.34 \\ &= 11.50\ \text{mm} \\ \text{Deflection on each spring} &= 11.50/9 = 1.27\ \text{mm} \end{aligned} \quad (2)$$

2.2.4 Maximum Shear Stress

$$\begin{aligned} \tau(\max) &= (4C - 1)/(4C - 4) + 0.615/C \\ &= [4(5.58) - 1]/[4(5.58) - 4] + 0.615/5.58 \\ &= 1.27 \\ \tau(\max) &= 8\text{FDKw}/\pi \text{d}^3 \\ &= (8 \times 578.97 \times 34.64 \times 1.27)/3.14 \times 6.2^3 \\ &= 203,764.09/748.34 \\ &= 272.28\ \text{N/mm}^2 \\ \text{Stress on each spring} &= 272.28/9 \\ &= 30.25\ \text{N/mm}^2 \\ \text{Shear limit} &= 1700 \times 0.5 = 850\ \text{MPa} \end{aligned} \quad (3)$$

The maximum shear stress is a function of the spring geometry and the load on it [4].

2.3 Properties of Bowl Material 316 Stainless Steel

- Resistance to corrosion and pitting crevices

Table 1 Calculated values of different properties of the spring

		Value	Unit
Stiffness of each spring	k	5.59	N/mm
Deflection on each spring	δ	1.27	mm
Maximum shear stress on each spring	τ (max)	30.25	N/mm

Table 2 Physical and mechanical properties of 316 stainless steel

S. No.	Property	Value
1.	Melting range	1390–1440 °C
2.	Density	8.027 g/cm^3
3.	Modulus of elasticity in tension	200 GPa
4.	Modulus of shear	82 GPa
5.	Yield strength 0.2% offset	205 MPa
6.	Ultimate tensile strength	515 MPa
7.	Hardness, Brinell	217

- Provide higher creep, stress-to-rupture and tensile strength at elevated temperature
- Can be formed and drawn easily.

Tables 1 and 2 presents some of the physical-mechanical properties and chemical composition of bowl material 316 stainless steel respectively.

3 Results

The first six frequencies are dominating the total frequencies. The natural frequencies of *vibratory finishing machine* as illustrated in Fig. 2 for six modes which are achieved through ANSYS analysis and the value of natural frequencies are depicted in Table 3. The vibratory finishing machine contains six parts viz. working bowl, electric motor, upper plate, lower plate, springs and base. All of them behave as a single unit when they are connected with each other. The lower plate is fixed to the base hence the motion will be provided only by the springs, upper plate and bowl.

Following mode shapes of vibratory finishing machine were obtained:

Accordingly, the first mode shape obtains a natural frequency of 11.094 Hz while 0.11547 is the highest value and 0 is the least value. The second mode shape attains a natural frequency of 4.9862 Hz with 0.11633 is the highest value and 0 is the least.

The above six figures shows the six different mode shapes and frequencies of the vibratory machine. The mode shapes were obtained by designing the machine in *Solidworks* software and exported to ANSYS for numerical analysis. From the above figures we can see the following frequencies in Table 3.

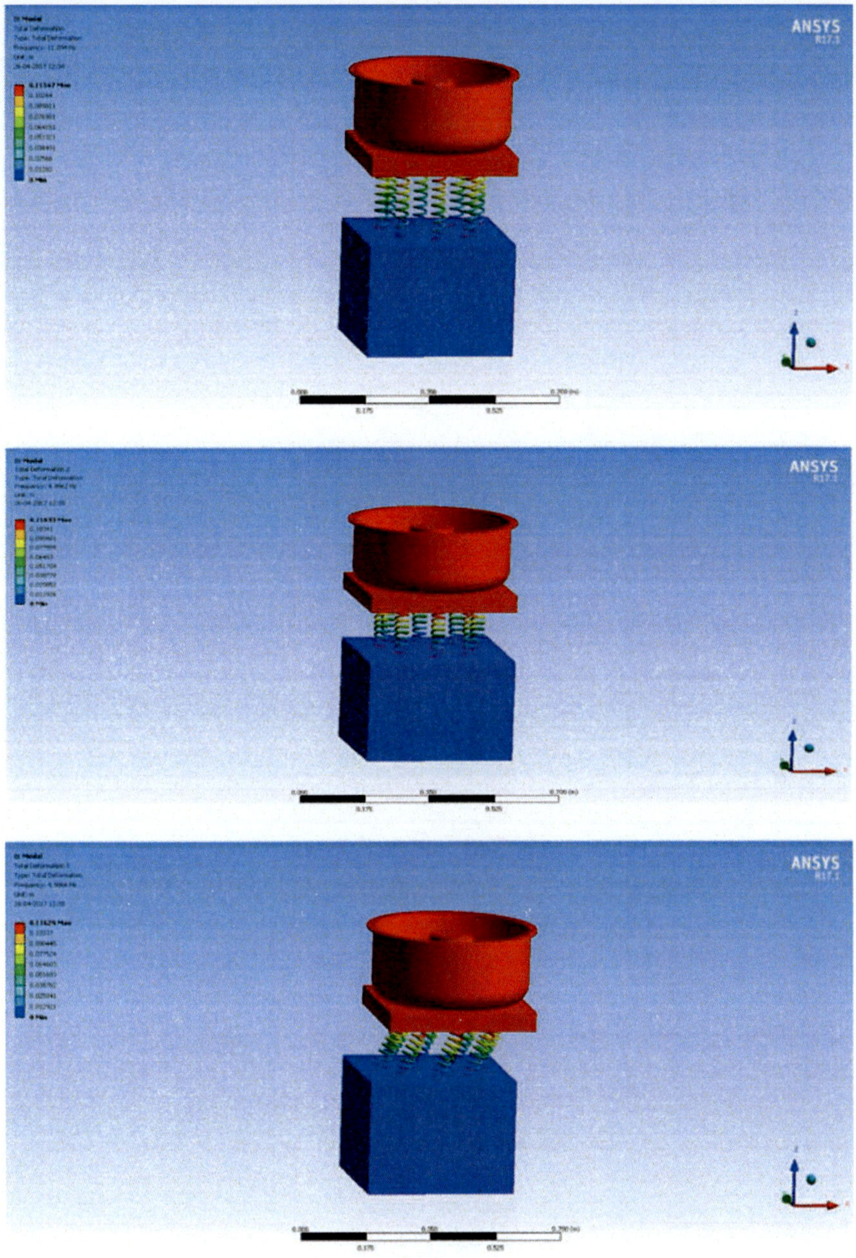

Fig. 2 Mode shapes with frequencies of vibratory finishing machine

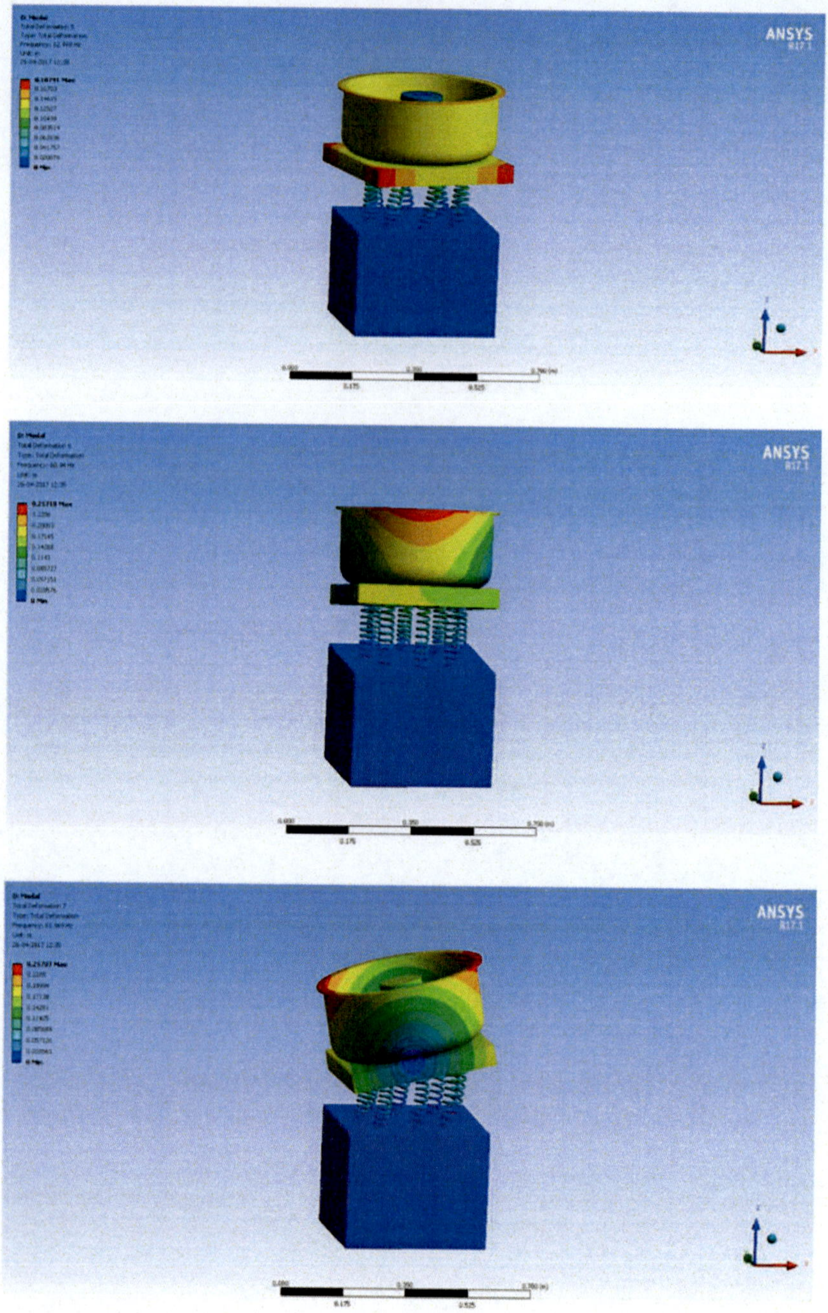

Fig. 2 (continued)

Table 3 Natural frequencies of vibratory finishing machine

Modes	Frequency (Hz)
1	11.094
2	4.9862
3	4.9866
4	12.498
5	60.94
6	61.969

4 Conclusion

A working model of three-dimensional vibratory finishing machine was designed using *solid works software* considering all the important aspects of design such as material of the working bowl and the spring, stiffness and deflection in the spring, type of media and compound used during the machining and the value of frequency of vibration obtained.

Some of the conclusions which can be drawn out from this dissertation work are as follows:

1. The stiffness of the spring was found to be 50.34 N/mm and for each spring it was 5.59 N/mm.
2. Six different mode shapes were found, and the six natural frequencies were found to be in the range 4.9862–61.969 Hz.

References

1. Mohajerani A, Spelt JK (2010) Numerical modeling of the edge rounding of brittle materials by vibratory finishing. Elsevier, Amsterdam, pp 1002–1012
2. Pramanik A (2014) Problems and solutions in machining of titanium alloys. Int J Adv Manuf Technol 919–928
3. A. Yabuki, M. R. Baghbanan, J. K. Spelt (2002) Contact forces and mechanisms in vibratory finisher. Elsevier, Amsterdam, pp 635-643.
4. Hashimoto F (1996) Modeling and optimization of vibratory finishing process. Ann CIRP 303–306. https://doi.org/10.1016/S0007-8506(07)63068-6
5. Hashimoto F, Johnson SP (2015) Modelling of vibratory finishing machines. CIRP Annu Manuf Technol 345–348
6. Domblesky J, Cariapa V, Evans R (2010) Investigation of vibratory bowl finishing. Int J Prod Res 3943–3953
7. Domblesky J, Evans R, Cariapa V (2007) Material removal model for vibratory finishing. Int J Prod Res 1029–1041
8. de Bortoli MGD, Bosco R Jr, Puff R (2010) Fatigue analysis of helical suspension springs for reciprocating compressors. In: International compressor engineering conference
9. Pranitbelhekar, Mate M, Jagpat G, Bhekare P (2016) Design, analysis and comparison of helical spring used in tractor seat application. Int J Sci Eng Technol Res 1645–1648

10. Belapurkar PD, Jadhav SS (2015) A new methodology for testing spring stiffness. Int J Sci Res 867–869
11. Mediratta R, Ahluwalia K, Yeo SH (2015) State-of-the-art on vibratory finishing in the aviation industry; an industrial and academic perspective. Int J Adv Manuf Technol. https://doi.org/10.1007/s00170-015-7942-0
12. Harak SS, Sharma SC, Harsha SP (2015) Modal analysis of prestressed draft pad of freight wagons using finite element method. J Modern Transp 43–49
13. Vakte V, Dhamande LS (2015) Analysis and comparison of helical springs used in tractor seat application. Int Eng Res J 4400–4403
14. Rączka W (2006) Testing of a spring with controllable stiffness. Mechanics 79–86
15. Tian YB, Zhong ZW, Tan SJ (2016) Kinematic analysis and experimental investigation on vibratory finishing. Int J Adv Manuf Technol

Heuristics for Shortest Path Problem with Intuitionistic Fuzzy Arc Weights

Stuti Srivastava and Richa Bansal

1 Introduction

Shortest path problem (SPP) is a well known problem where path with least cost is found in a given graph (network) from one vertex to another where cost of the path is equal to the sum of the weights of all edges (arcs) consisting in the path. These weights may be travelling time, travelling cost, capacity, demand etc. This is a fundamental problem in graph theory which have been drawn attention of many scientists from beginning as transportation problem, routing problem, communication, supply chain management, wire length minimization problems etc. can be converted into this problem.

Although, in traditional shortest path problem, the arc weights are crisp values but in real life situations arc weights may represent time, cost, capacity, demand etc. Due to variations in different variables, the arc weights cannot be represented as a single value. In this condition, fuzzy numbers may better represent the arc weights and the shortest path problem where the arc weights are fuzzy numbers is generally known as fuzzy shortest path problem (FSPP). But the concept of fuzzy numbers is very much different from crisp numbers as one cannot simply add and compare two fuzzy numbers. To deal with fuzzy numbers the concept of fuzzy set theory has been introduced by Zadeh [21]. Dubois and Prade [4] were first who analyzed fuzzy shortest path problem.

Later Atanassov in 1983 [1] extended fuzzy sets to intuitionistic fuzzy sets to solve fuzzy shortest path problem more precisely with the help of degree of membership, degree of non-membership and degree of hesitation. Since last two decades it has been

S. Srivastava (✉) · R. Bansal
Dayalbagh Educational Institute, Agra, Uttar Pradesh 282005, India
e-mail: stutisrivastava664@gmail.com

R. Bansal
e-mail: richabansal2007@gmail.com

V. H. Saran and R. K. Misra (eds.), *Advances in Systems Engineering*,
Lecture Notes in Mechanical Engineering,
https://doi.org/10.1007/978-981-15-8025-3_45

become topic of great interest for many scientists. Intuitionistic fuzzy set theory has successfully used to solve problems in many areas like logical programming, medical diagnosis, artificial intelligence, decision making problems etc. Since comparison of intuitionistic fuzzy numbers (IFNs) is a difficult task, a lot of attention has been paid to develop ranking methods for comparison of IFNs. Although ranking of intuitionistic fuzzy numbers has been started since late seventies of last century but complete ranking method is developed in 2014 by Wang and Westmant [19]. Since then few other ranking methods have appeared in the literature [2, 11, 15–18]. In this paper, we have used ranking methods given in [15] to compare the costs of paths in a given network with trapezoidal intuitionistic fuzzy arc weights.

Yu and Wei [20] proposed a fuzzy linear programming approach to find fuzzy shortest path in a graph. Liu and Kao [10] calculated Yager ranking indices for fuzzy arcs to change fuzzy formulation in crisp formulation. A fuzzy algorithm based on multiple labeling methods was proposed by Okada and Soper [14] to extract shortest paths in a given graph. Nayeem and Pal [13] also proposed an algorithm to deal with fuzzy shortest path problem with two different types of intuitionistic numbers. A dynamic programing algorithm for solving FSPP has been proposed in [17]. Later they proposed a genetic algorithm [7] to solve FSPP with mixed fuzzy arc weights in 2013. A particle swarm optimization algorithm to solve FSPP is proposed in [5]. For other papers on fuzzy shortest path problems, see [6, 8, 9]. It is worth to note that no metaheuristic has been proposed for SPP with intuitionistic fuzzy arc weights.

In this paper, we proposed two metaheuristic approaches: a genetic algorithm (GAIFSP) and a tabu search based algorithm (TSIFSP) to solve intuitionistic fuzzy shortest path problem. As there is no benchmark graph available for this problem, we performed experiments for both the proposed algorithms on randomly generated graphs. Furthermore, there is no previously proposed metaheuristic approach available for FSPP to compare the results with our proposed algorithms so we have compared the performance of our proposed GAIFSP with that of TSIFSP.

Organization of the paper is as follows: Some useful basic definitions are given in Sect. 2 to understand the problem and proposed algorithms. A genetic algorithm is described briefly in Sect. 3 followed by explanation of its application on shortest path problem step by step. Basic structure of tabu search algorithm is explained in Sect. 4 followed by explanation of proposed TSIFSP step by step. Experimentation and results are discussed in Sect. 5. The work is concluded in Sect. 6.

2 Some Basic Definitions

In this section some basic definitions regarding graph theory and fuzzy theory are presented as follows:

Graph: A graph $G(V, E)$ is a combination of a set of n vertices $V = \{v_1, v_2, \ldots, v_n\}$ together with a set of m edges $E = \{e_1, e_2, \ldots, e_m\}$, such that each edge e_k is represented as a pair of vertices (v_i, v_j) and is denoted by $v_i v_j$ for convenience. Cardinality of the graph is denoted as $|V| = n$.

Directed Graph: A simple graph where each edge has a direction i.e. each edge is represented by an ordered pair of vertex is called directed graph. In other words if an edge (v_i, v_j) connects vertex v_i to v_j then it is not necessary that it also connects vertex v_j to vertex v_i. In this paper, we have considered only directed networks and call them simply a network.

Path: A $u - v$ path is defined as a sequence of vertices starting at u and ending at v, where consecutive vertices in the sequence are adjacent vertices in the graph and no vertex is repeated.

Intuitionistic Fuzzy Sets: An intuitionistic fuzzy set X in universal set Y is defined as $X = \{y, \mu'_X(y), \nu'_X(y) \in Y\}$, where $\mu'_X(y) : Y \to [0, 1]$ and $\nu'_X(y) : Y \to [0, 1]$ are called the membership and non-membership grades respectively and $\forall y \in Y, 0 \leq \mu'_X(y) + \nu'_X(y) \leq 1$. Pair $(\mu'_X(y), \nu'_X(y))$ is called an intuitionistic fuzzy number. For any intuitionistic fuzzy number B let eight numbers $a'_1, a'_2, a'_3, a'_4, b'_1, b'_2, b'_3, b'_4 \in \mathbb{R}$ such that $b'_1 \leq a'_1 \leq b'_2 \leq a'_2 \leq a'_3 \leq b'_3 \leq a'_4 \leq b'_4$ and four functions $f_B, g_B, h_B, k_B : \mathbb{R} \to [0, 1]$, such that f_B and k_B are non decreasing and g_B and h_B are non increasing then the membership function μ'_B can be described as

$$\mu'_B(y) = \begin{cases} 0, \text{ if } y < a'_1 \\ f_B(y), \text{ if } a'_1 \leq y \leq a'_2 \\ 1, \text{ if } a'_2 \leq y \leq a'_3 \\ g_B(y), \text{ if } a'_3 \leq y \leq a'_4 \\ 0, \text{ if } a'_4 < y \end{cases}$$

and non-membership function ν'_B as

$$\nu'_B(y) = \begin{cases} 0, \text{ if } y < b'_1 \\ h_B(y), \text{ if } b'_1 \leq y \leq b'_2 \\ 1, \text{ if } b'_2 \leq y \leq b'_3 \\ k_B(y), \text{ if } b'_3 \leq y \leq b'_4 \\ 0, \text{ if } b'_4 < y \end{cases}$$

Graphical representation of the intuitionistic fuzzy number is shown in Fig. 1.

2.1 Trapezoidal Intuitionistic Fuzzy Number

A trapezoidal intuitionistic fuzzy number with parameters $b'_1 \leq a'_1 \leq b'_2 = a'_2 \leq a'_3 = b'_3 \leq a'_4 \leq b'_4$ is denoted by $X = (< a'_1, a'_2, a'_3, a'_4 >, < b'_1, b'_2, b'_3, b'_4 >)$ where, membership function is defined as

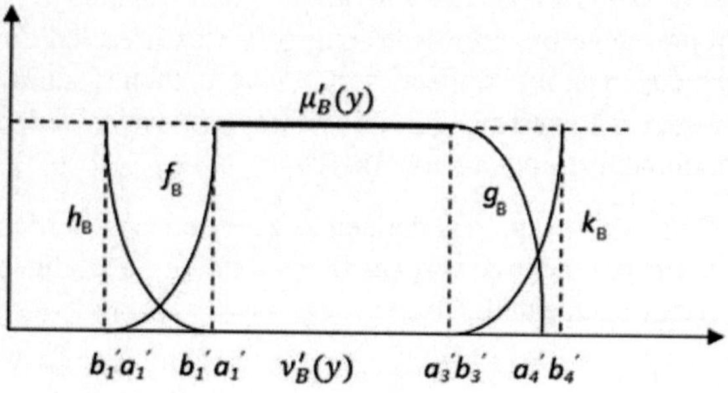

Fig. 1 Graphical representation of the intuitionistic fuzzy number

$$\mu'_X(y) = \begin{cases} 0, \text{ if } y < a'_1 \\ \frac{y-a'_1}{a'_2-a'_1}, \text{ if } a'_1 \le y \le a'_2 \\ 1, \text{ if } a'_2 \le y \le a'_3 \\ \frac{y-a'_4}{a'_3-a'_4}, \text{ if } a'_3 \le y \le a'_4 \\ 0, \text{ if } a'_4 < y \end{cases}$$

and the non-membership function is given by

$$\nu'_X(y) = \begin{cases} 0, \text{ if } y < b'_1 \\ \frac{y-b'_1}{b'_1-b'_2}, \text{ if } b'_1 \le y \le b'_2 \\ 1, \text{ if } b'_2 \le y \le b'_3 \\ \frac{y-b'_4}{b'_3-b'_4}, \text{ if } b'_3 \le y \le b'_4 \\ 0, \text{ if } b'_4 < y \end{cases}$$

Figure 2 shows the graphical representation of trapezoidal intuitionistic fuzzy number.

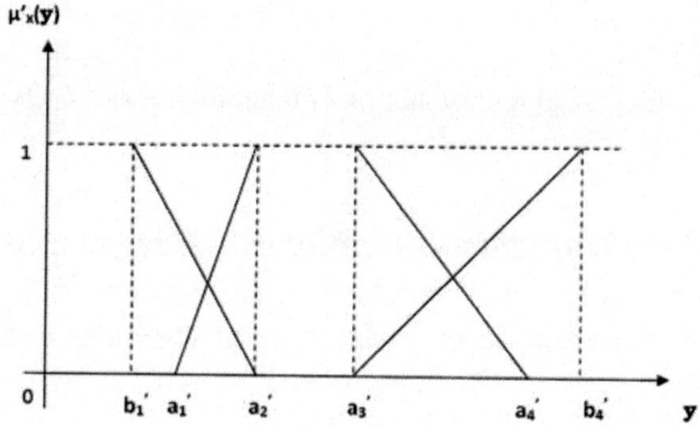

Fig. 2 Graphical representation of the trapezoidal intuitionistic fuzzy number

2.2 Addition of Two Trapezoidal Intuitionistic Fuzzy Numbers

Let $X = (< a'_1, a'_2, a'_3, a'_4 >, < a'_5, a'_6, a'_7, a'_8 >)$ and $Y = (< b'_1, b'_2, b'_3, b'_4 >, < b'_5, b'_6, b'_7, b'_8 >)$ be two trapezoidal intuitionistic fuzzy numbers then addition of X and Y, denoted by $X \bigoplus Y$, is defined as $X \bigoplus Y = (< a'_1 + b'_1, a'_2 + b'_2, a'_3 + b'_3, a'_4 + b'_4 >, < a'_5 + b'_5, a'_6 + b'_6, a'_7 + b'_7, a'_8 + b'_8 >)$.

2.3 Centroid Ranking Technique

Let $X = (\langle a'_1, a'_2, a'_3, a'_4 \rangle, \langle b'_1, b'_2, b'_3, b'_4 \rangle)$, where $b'_1 \leq a'_1 \leq b'_2 = a'_2 \leq a'_3 = b'_3 \leq a'_4 \leq b'_4$ and $a'_1, a'_2, a'_3, a'_4, b'_1, b'_2, b'_3, b'_4 \in \mathbb{R}^+$ be a trapezoidal intuitionistic fuzzy number. Then centroid of X is defined by four relations given as follows [15]:

$$p_{\mu'}(X) = \frac{1}{3}\left[\frac{a_3'^2 + a_4'^2 - a_1'^2 - a_2'^2 - a'_1 a'_2 + a'_3 a'_4}{a'_4 + a'_3 - a'_2 - a'_1}\right]$$

$$p_{\nu'}(X) = \frac{1}{3}\left[\frac{2b_4'^2 - 2b_1'^2 + 2b_2'^2 + 2b_3'^2 + b'_1 b'_2 - b'_3 b'_4}{b'_3 + b'_4 - b'_1 - b'_2}\right]$$

$$q_{\mu'}(X) = \frac{1}{3}\left[\frac{a'_1 + 2a'_2 - 2a'_3 - a'_4}{a'_1 + a'_2 - a'_3 - a'_4}\right]$$

$$q_{\nu'}(X) = \frac{1}{3}[\frac{2b'_1 + b'_2 - b'_3 - 2b'_4}{b'_1 + b'_2 - b'_3 - b'_4}]$$

The ranking function of the trapezoidal intuitionistic fuzzy number A is defined by

$$R(X) = \sqrt{\frac{1}{2}([p_{\mu'}(X) - q_{\mu'}(X)]^2 + [p_{\nu'}(X) - q_{\nu'}(X)]^2)}$$

3 Genetic Algorithm

Genetic algorithm (GA) is a heuristic search procedure and an optimization technique which is inspired from the process of natural evolution and generally used to generate high quality solution to a problem. It is a type of evolutionary algorithm based on the principle of evolution via natural selection. This process begins with initialization of a population of randomly generated solutions. It is an iterative process where each iteration generates a population of new generation. A fitness function is

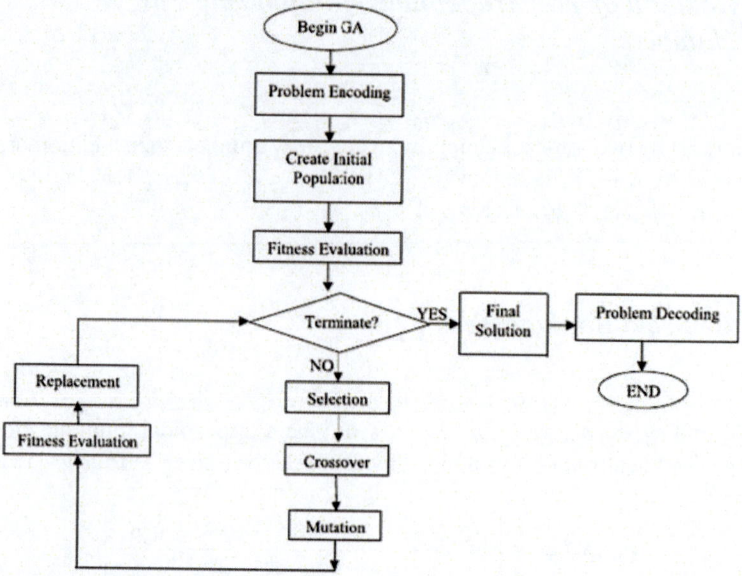

Fig. 3 A flow chart of genetic algorithm

concerned with the problem is defined to evaluate each solution in each generation and a comparison takes place between each two solutions of the current population which selects the solution with better fitness value and rejects the other one keeping the population size constant. Population obtained after this process is modified by reproduction (crossover) and mutation operators to generate the population of next generation. The new generated population is then used in the next iteration of the algorithm. In general, the procedure continued until there is no change in the best observed fitness value or number of iterations are less than the pre-defined fixed value. The flow chart of genetic algorithm is shown in Fig. 3. To know more about genetic algorithm see [3].

3.1 Genetic Algorithm for Solving Intuitionistic Fuzzy Shortest Path Problem

In this section, the proposed genetic algorithm for shortest path problem with intuitionistic fuzzy arc weights, is explained with description of genetic operators such as initialization of the parent population, selection, crossover, and mutation operator etc.

3.1.1 Initial Population

In general, a solution in the population of genetic algorithm is represented as a chromosome of binary numbers. But in case of FSPP, solution is a path in a given network. So it is not convenient to generate the solution randomly as the random sequence of vertices may not correspond to a path in a given network. Also the length of the path is not a fixed number as it may vary from two to $n - 1$ in a given network with n vertices. So, there is a necessity to use some strategy to generate a solution. Here, an approach given in [7] is used to obtain various paths in a given network. The algorithm starts with first vertex of the network and ends with the last vertex or terminating vertex by applying specific steps in between. This algorithm is called repeatedly to generate desired population of paths. The framework of the algorithm is given below [7]:

Algorithm 1 Algorithm to generate an initial population

1. Find adjacency matrix A of a directed graph $G = (V, E)$ and initialize population matrix *par_pop*.
2. Set *pop_size* (population size) and $k = 1$.
3. Set $i = l = p(l) = 1$.
4. Define $a^1(i) = \{j|(i, j) \in A, A(i, j) = 1\}$ and select a member of it, say j. Let $l = l + 1$ and $p(l) = j$.
5. If $j \neq n$ then $i = j$ and go to 4.
6. Save the produced path in par_pop[k]. let $k = k + 1$.
7. If $k <$ pop_size then go to 3 else stop.

At the end of execution of this algorithm the population of paths is stored in a matrix namely par_pop. The algorithm is explained below with the help of a network shown in Fig. 4 where A is the adjacency matrix of this network.

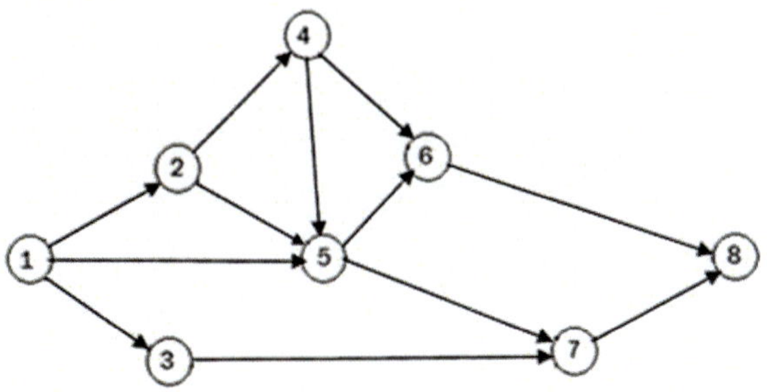

Fig. 4 An example of a network

$$A = \begin{bmatrix} 0\,1\,1\,0\,1\,0\,0\,0 \\ 0\,0\,0\,1\,1\,0\,0\,0 \\ 0\,0\,0\,0\,0\,0\,1\,0 \\ 0\,0\,0\,0\,1\,1\,0\,0 \\ 0\,0\,0\,0\,0\,1\,1\,0 \\ 0\,0\,0\,0\,0\,0\,0\,1 \\ 0\,0\,0\,0\,0\,0\,0\,1 \\ 0\,0\,0\,0\,0\,0\,0\,0 \end{bmatrix}$$

Here $n = 8$. Initially set $l = i = p(l) = 1$. Then $a^1(1) = \{2, 3, 5\}$. Select an element of this set randomly say $j = 5$. Set $l = l+1$ i.e. $l = 2$ and $p(2) = 5$. Since $5 \neq 8$, set $i = 5$ and $a^1(5) = \{6, 7\}$. Now let $j = 6$. Set $l = 3$ and $p(3) = 6$. Since $6 \neq 8$, set $i = 6$ and $a^1(6) = \{8\}$. Now $j = 8$, $l = 4$ and $p(4) = 8$. Since $p(4) = 8$, procedure stops. The generated path is 1-5-6-8. In this paper, we call it as a solution.

3.1.2 Fitness of a Solution

For selection of good solutions, it is necessary to define their fitness which we call as the cost in this paper. To find the fitness (cost) of each solution in par_pop, a centroid ranking technique defined in Sect. 2 is used. To understand it, let us consider one of the solution 1-5-6-8 of network shown in Fig. 4. The trapezoidal intuitionistic fuzzy weights of all the arcs of this network are generated randomly and shown in Table 1. Using this table, fitness of the solution 1-5-6-8 is calculated as follows:

First add all the arc weights corresponding to each edge of the solution, by applying addition procedure defined in Sect. 2. So total arc weight of the solution 1-5-6-8 is (⟨72, 87, 107, 120⟩, ⟨55, 87, 107, 133⟩). Fitness of the solution 1-5-6-8 calculated

Table 1 Edges of a network (Fig. 4) and their corresponding weights

Edge	Intuitionistic trapezoidal weight
1–2	(⟨24, 35, 43, 52⟩, ⟨16, 35, 43, 58⟩)
1–3	(⟨24, 28, 29, 37⟩, ⟨18, 28, 29, 48⟩)
1–5	(⟨27, 31, 39, 43⟩, ⟨19, 31, 39, 48⟩)
2–4	(⟨23, 32, 36, 40⟩, ⟨19, 32, 36, 43⟩)
2–5	(⟨19, 22, 25, 33⟩, ⟨12, 22, 25, 41⟩)
3–7	(⟨18, 28, 37, 39⟩, ⟨11, 28, 37, 45⟩)
4–5	(⟨20, 31, 35, 43⟩, ⟨11, 31, 35, 48⟩)
4–6	(⟨16, 26, 29, 31⟩, ⟨12, 26, 29, 32⟩)
5–6	(⟨19, 26, 33, 36⟩, ⟨17, 26, 33, 39⟩)
5–7	(⟨18, 21, 30, 35⟩, ⟨11, 21, 30, 37⟩)
6–8	(⟨26, 30, 35, 41⟩, ⟨19, 30, 35, 46⟩)
7–8	(⟨20, 31, 32, 39⟩, ⟨15, 31, 32, 44⟩)

by centroid ranking technique mentioned in Sect. 2 is 154.579. So the cost of the solution 1-5-6-8 is 154.579. In context of the problem presented here, the solution with lower cost is more fit than the solution with greater cost. Note that the minimum cost is maintained at each iteration of the algorithm which is treated as the Best_cost and the corresponding path is treated as the Best_solution at the end of the algorithm. Note that the Best_cost obtained at the end of the execution of the algorithm is treated as final_cost.

3.1.3 Selection

This process aims to keep good solutions in the population by omitting the poor ones without reducing the population size. In this work, tournament operator is used for selection procedure wherein a comparison has been made between each two consecutive solutions of par_pop and the best one (with lower cost) is placed in a dummy matrix namely new_pop with same size as that of par_pop. It is worthy to note that each solution in par_pop participates two times in a tournament and the best solution is placed in new_pop. In this way new_pop may have one, two or zero copy of a solution in par_pop.

3.1.4 Crossover

Crossover combines properties of two parent solutions and generate two child solutions having some common properties with parent solutions. In this step two new solutions are produced by using any two randomly selected solutions from new_pop. Crossover rate (p_c) is used to determine the frequency of this job. In our proposed GAIFSP, one point crossover is used. This is a very common crossover technique used in genetic algorithms where two random paths (parent solutions) are selected from the population obtained after selection and if any vertex is common between them except from the first and the last vertex, left or right parts of the common vertex of both parent solutions are exchanged.

Example: Consider two solutions 1-5-7-8 and 1-2-4-5-6-8 of network shown in Fig. 4. Here vertex 5 is common in both solutions. Then exchanging right part of vertex 5 of both the solutions two new solutions are 1-5-6-8 and 1-2-4-5-7-8.

Note that in this process two new child solutions are generated and replaced their parent solutions in a dummy matrix of new_pop if child solutions are more fit than parent solutions. This crossover procedure is repeated $p_c \times$ pop_size number of times on new_pop and at the end dummy new_pop is copied back to original new_pop.

3.1.5 Mutation

This function systematically changes the vertices of a solution which results either a positive direction towards the optimal solution or resulted in the deviation from the optimal solution i.e. the mutated solution can be better than the previous one or not. In proposed heuristic, mutation is addressed effectively. In this process, a mutation rate or mutation probability (p_m) is fixed initially so that the frequency of application of this operator is $q = \text{pop\_size} \times p_m$. Mutation is applied on new_pop obtained after crossover for q number of times. Firstly q number of solutions are selected from new_pop by producing q different random integers between 1 and pop_size. To apply the mutation on a solution, a random number, say k, is generated between 1 and length of the path of the current solution. After that, a new path is searched using Algorithm 1 from kth vertex to the last vertex of the solution keeping vertices prior to kth vertex remain same. Now so obtained mutated solution replaced the original one if it is more fit in concerned to FSPP.

For an example, consider a randomly selected solution 1-2-4-6-8 of network shown in Fig. 4. This is called as current solution with cost = 201.67. Here path length is 5, therefore a random number (say 4) is generated between 1 and 5. Then, the first three vertices 1, 2 and 4 are remain at their places and random path is evolved starting with vertex 6 at position 4 using Algorithm 1. So the new obtained mutated solution is 1-2-4-5-7-8. The cost of this solution is 242.35. Since cost of mutated solution is more than the original one, it wouldn't be replace the parent solution (original one).

4 Tabu Search

Tabu search (TS) is a single solution based metaheuristic that guides a local search heuristic procedure to escape from local optima and to explore the solution space. This method was proposed by Grover in 1986 [12]. Tabu search is an extension of previously existing local search approaches. In fact it can be seen that basic tabu search is a combination of local search with short term memories. The process starts with initializing a random solution. A candidate list of moves is created by applying some specific steps. In each iteration, each move from candidate list generates a new solution from current solution. This set of solutions is regarded as tabu list and a solution from it is recorded as best solution if it improves the previous best and tabu list is updated. The process stops if solution remains unchanged for some specific number of iterations or the optimal solution is found.

4.1 Tabu Search for Solving Intuitionistic Fuzzy Shortest Path Problem (TSIFSP)

In this section the proposed tabu search algorithm to solve intuitionistic fuzzy shortest path problem is explained. The TSIFSP algorithm starts with generating an initial solution using Algorithm 1. Initially, this solution is called current_solution and cost of this solution is considered as current_cost and this solution is saved in a tabu list for some time period or for few number of iterations which is called tabu tenure (t) so that this solution is not selected for t number of iterations and new area of search space can be explored. The tabu list is used to prevent looping of solutions and to guide the algorithm to explore the global solution space. To improve the current_solution, N number of neighbouring solutions of current_solution are generated and best solution (solution with lowest cost) among these neighbouring solutions is determined which is referred to as best_solution and cost associated with it is called best_cost. Now this best_solution replaces the current_solution and best_cost replaces the current_cost if best_cost is less than current_cost. Once a best_solution becomes current_solution, it is inserted into the tabu list for tabu tenure t to escape from getting stuck in local optima. This process is repeated until current_solution remains unchanged for K_{max} number of iterations. After stoping the process current_solution obtained in last iteration is desired final solution and cost associated is final_cost of the problem. Some components of TSIFSP are explained below:

4.1.1 Initial Solution

Here, Algorithm 1 is used to generate an initial solution.

4.1.2 Neighbouring Solution

To generate neighboring solution of current_solution, two vertices of the current solution are selected randomly and path between these two vertices is regenerated using Algorithm 1, where these randomly selected vertices are taken as initial vertex and final vertex.

As an example, consider a path 1-2-4-5-7-8 of network shown in Fig. 4 and let two randomly selected vertices are 2 and 7. Algorithm 1 generates a new path between vertex 2 and vertex 7. Suppose this path is 2-5-7. So a neighbourhood solution of current solution 1-2-4-5-7-8 will be 1-2-5-7-8.

4.1.3 Tabu List

Once a solution is considered as current_solution, it is sent to tabu list for tabu tenure t so that the algorithm do not select same solution repeatedly as searching around some fixed solutions can prevent to explore new search space that may contain the optimal solution. Apart from this current_solution, the randomly selected vertices in an iteration which are used to find neighbouring solution, are also inserted in the tabu list for short period of time $tv = n/10$ to escape from generating similar neighboring solution repeatedly.

4.1.4 Termination Criteria

The TSIFSP algorithm stops when current_solution remains unchanged for K_{max} number of iterations i.e. current_solution is not improved for K_{max} number of iterations.

5 Experimentations and Results

All computational experiments have been carried out on an Intel i3, 2.40 GHz quad-core processor with 2 GB RAM. All algorithms in this paper are coded in C++ and compiled with Dev C++. Tuning of parameters for better performance of GAIFSP and TSIFSP is discussed as follows.

5.1 *Parameter Setting*

For parameter setting we have used randomly generated networks with vertices 20, 40, 60, 80, 100 and edges equal to 40, 78, 185, 210 and 329 respectively. Weights on each edge of graphs as trapezoidal inuitionistic fuzzy numbers are also generated randomly. For tuning the parameters used in GAIFSP, experiments have been done for setting crossover rate p_c and mutation rate p_m. Number of generation is set to 100 for all the experiments of GAIFSP. In Fig. 5a, results of tuning crossover rate p_c are shown where p_c is set to 0.2, 0.5 and 1 by fixing p_m as 0.2. From the figure it can be seen that GAIFSP performed well for $p_c = 0.5$, so this value of p_c is fixed for further experiments. Fig. 5b shows the results of tuning mutation rate p_m which is set to 0.1, 0.2 and 0.3 and it is observed that best results are obtained when p_m is 0.2 so it is fixed for further experiments. For tuning the parameters used in TSIFSP, first experiment is carried out for setting number of neighboring solutions N by fixing N equal to $n/10$, $n/5$ and $n/2$ with $K_{max} = 100$. Results of this experiment shown in

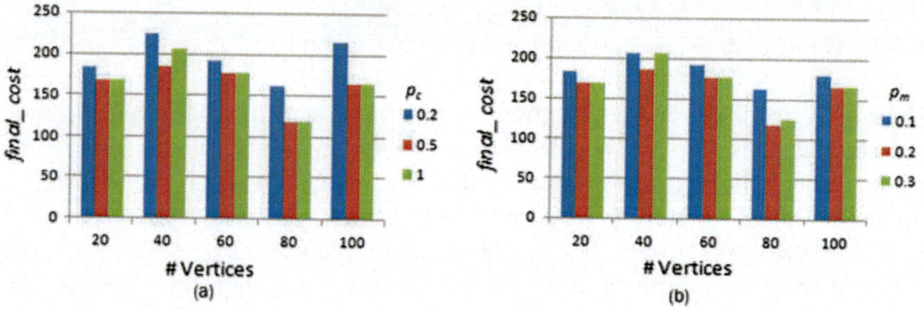

Fig. 5 Experimental results of GAIFSP with different values of **a** p_c and **b** p_m

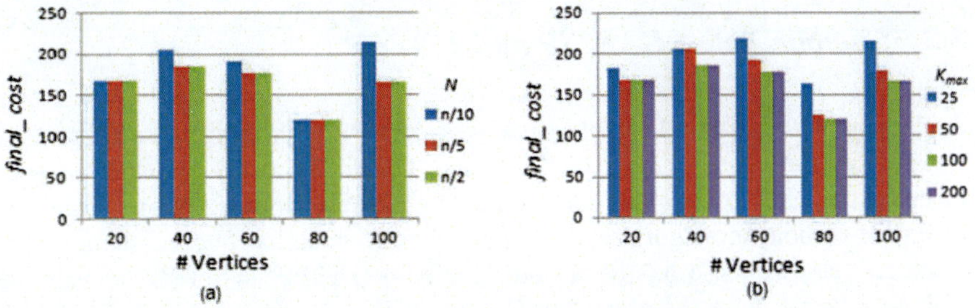

Fig. 6 Experimental results of TSIFSP with different values of **a** N and **b** $K_{\max}$

Table 2 Parameter setting

Symbols	Parameters	Value
t	Tabu tenure	$n/4$
tv	Tabu tenure for vertex	$n/10$
N	Number of neighborhoods	$n/5$
$K_{\max}$	Number of iterations	100
p_c	Crossover rate	0.5
p_m	Mutation rate	0.2
pop_size	Population size	n

Fig. 7 Elapsed time comparison of GAIFSP and TSIFSP

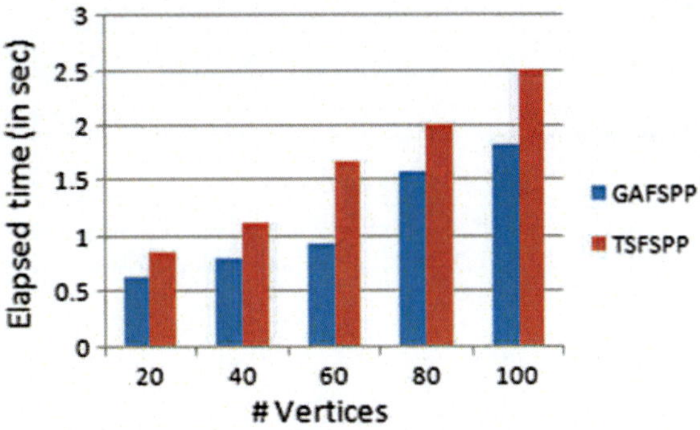

Fig. 6a clearly depict that least final_cost is obtained when number of neighboring solutions is $n/5$ or $n/2$. So N has been fixed as $n/5$. After fixing N, experiments are carried out for tuning $K_{\max}$. For this TSIFSP is executed with $K_{\max}$=25, 50, 100 and 200. From Fig. 6b it can be observed that final_cost of an instance is least for $K_{\max} = 100$. After that performance of TSIFSP is stable. So $K_{\max} = 100$ is fixed for TSIFSP. Summary of the parameters, for both the algorithms, with their fine tuned values is given in Table 2.

5.2 *Comparison of GAIFSP and TSIFSP*

As there is no benchmark graph available for FSPP with intuitionistic fuzzy arc weights, performance of both GAIFSP and TSIFSP is tested on same network instances as used to fine tune the parameters in the previous section. Since performance of both the algorithms is comparable in obtaining final_cost for all the instances, elapsed time has been noted for both GAIFSP and TSIFSP. Value of elapsed time for both the algorithms shown in Fig. 7 concluded that GAIFSP is faster than TSIFSP in terms of finding shortest path having intuitionistic arc weights in a given network.

6 Conclusion

We have considered a new shortest path problem with intuitionistic fuzzy arc weights in this paper and proposed two heuristics for solving it. One is population based and the other one is single solution based heuristic. Experiments on randomly generated graphs show that both algorithms are equally capable of finding shortest path in a given network but computation time of GAIFSP is less than that of TSIFSP. For future work, some more effective heuristics can be designed and compared for this problem.

References

1. Atanassov KT (1986) Intuitionistic fuzzy sets. Fuzzy Sets and Syst 20:87–96
2. Bharti SK (2017) Ranking method of intuitionistic fuzzy numbers. Global J Pure Appl Math 13(9):4595–4608
3. Deb K (2001) Evolutionary Algorithms. In Sheldon R, Richard W (Ed.) *Multi-objective optimization using evolutionary algorithms*. (pp. 80-128). Chichester, U. K. : John Wiley and Sons Ltd.
4. Dubois D, Prade H (1980) Fuzzy sets and systems: theory and applications. Academic Press, New York
5. Ebrahimnejad A, Karimnejad Z (2015) Alrezaamiri H. Particle swarm optimisation algorithm for solving shortest path problems with mixed fuzzy arc weights. Int J Appl Decis Sci 8(20):203–222
6. Gani AN, Jabarulla MM (2010) On searching intuitionistic fuzzy shortest path in a network. Appl Mathe Sci 4:3447–3454
7. Hassanzadeh R, Mahadevi I, Amiri NM, Tajdin A (2013) A genetic algorithm for solving fuzzy shortest path problems with mixed fuzzy arc lengths. Mathe Comput Model 57:84–99
8. Kung JY, Chuang TN, Lin CT (2006) Decision making on network problem with fuzzy arc lengths. IMACS Multi conference on computational engineering in systems Applications. CESA Beijing, China, pp 578–580
9. Lin FT (2001) A shortest-path network problem in a fuzzy environment. In: IEEE International Fuzzy System Conference, pp 1096–1100
10. Liu ST, Kao C (2001) Network flow problems with fuzzy arc lengths. IEEE Trans Syst Man and Cybern Part B: Cybern 34:765–769

11. Nehi HM (2010) A new ranking method for intuitionistic fuzzy numbers. Int J Fuzzy Syst 12(1):80–86
12. Gendreau M, Potvin JY (2010) Handbook of metaheuristics. Int Ser Oper Res Manage Sci 146:41–59 (Springer Science+Business Media)
13. Nayeem SMA, Pal M (2005) Shortest path problem on a network with imprecise edge weight. Fuzzy Optim Decis Making 4:293–312
14. Okada S, Soper T (2000) A shortest path problem on a network with fuzzy arc lengths. Fuzzy Sets and Syst 109:129–140
15. Prakash KA, Suresh M, Vengataasalam S (2016) A new approach for ranking of intuitionistic fuzzy numbers using a centroid concept. Math Sci 10:177–184
16. Sujatha L, Hyacinta JD (2017) The shortest path problem on networks with intuitionistic fuzzy edge weights. Global J Pure Appl Math 13:3285–3300
17. Tajdin A, Mahadevi I, Amiri NM, Gildeh BS (2010) Computing a fuzzy shortest path in a network with mixed fuzzy arc lengths using α-cuts. Comput Math Appl 60:989–1002
18. Wei CP, Tang X (2011) A new method for ranking intuitionistic fuzzy numbers. Int J Knowl Syst Sci 2(1):43–49
19. Wang Z, Westmant LZ (2014) New ranking method for fuzzy numbers by their expansion center. JAISCR 4:181–187
20. Yu JR, Wei TH (2007) Solving the fuzzy shortest path problem by using a linear multiple objective programming. J Chin Inst Ind Eng 24:360–365
21. Zadeh LA (1965) Fuzzy sets. Inf Control 8:338–353

Securing Genomics Data Using Blockchain Technology

Raginee Pachaury and C. Vasantha Lakshmi

1 Motivation

Genomics is the study of genes and the sequences of proteins composing them. A good alignment of such sequences helps identify common conserved regions, inherent patterns, evolutionary ancestry, functional roles, and is important for various studies related to discovering particular disease-causing protein sequences [1]. It now promises precision medicines that can be customized based on the genetic composition. This significant data is, however, held with private actors like pharmaceuticals or research organizations, thus being vulnerable to hackers. Also, an individual has negligible control over how his or her data is being shared.

When conducting genomics research, the need for data sharing to escalate its applicability for advancing scientific expedition, and the need to protect research participants' privacy requires to be balanced [2]. Data from the genomics data sets if lost or stolen may reveal the entire medical history of an individual. Interpretation of which can be misused by malicious actors for making decisions about insurance or medical treatments or to extract benefits in return for information.

Genomics is involved in Big Data. The entire genome of the human individual can now be read within a few days, generating petabytes of data, thereby posing a burden on computation and storage space. After analysis, the data is stored either in central repositories or is held only by some concerned authorities, leading to centraliza-

R. Pachaury (✉)
Department of Electrical Engineering, Dayalbagh Educational Institute,
Dayalbagh, Agra 282005, India
e-mail: ragpach2@gmail.com

C. V. Lakshmi
Department of Physics and Computer Science, Dayalbagh Educational Institute,
Dayalbagh, Agra 282005, India
e-mail: cvasanthalakshmi@gmail.com

V. H. Saran and R. K. Misra (eds.), *Advances in Systems Engineering*,
Lecture Notes in Mechanical Engineering,
https://doi.org/10.1007/978-981-15-8025-3_46

tion, thus compromising the privacy and availability of data. Various cloud services promise scalability [3] but the problem of it being a single collapse point persists.

Blockchain, which has now found applications in a variety of sectors, seems to provide a solution [4–6]. It offers decentralization and privacy [7]. In comparison with relying on a central authority, it uses various consensus protocols across a network of nodes to incorruptibly verify and record data. Thus, stealing or altering blockchain-hosted data would be almost impossible for a malicious entity. In addition, there is a concern about the sharing of genetic data with third parties by private companies with their users, that is where smart contracts come in. Instead of handing their genetic data over to a middleman, an individual can program the blockchain, to share data only with certain parties.

1.1 Organization

Section 2 presents an overview of distinct properties of blockchain technology relevant to the application being considered. Section 3 presents the implementation of a use case, and Sect. 4 discusses the future scope and challenges. Concluding remarks are present in Sect. 5.

2 Blockchain

Bitcoin, a form of an electronic coin that could be shared on a peer-to-peer network without intermediaries, introduced the concepts of this technology[8]. Blockchain, in a sense, is a distributed and immutable database shared on a network and synchronized as and when information is accessed or manipulated [9]. The participants on the blockchain network known as *nodes* or *peers*, on onboarding, receive their copy of blockhain and are able to collaborate by making *transactions*.

- A *transaction* is an event that updates the data table on every connected machine on the network. These transactions are stored on the ledger in the form of a new block after validation by all the members on the network.
- A *validator* is a node that verifies the legitimacy of the transaction and ensures that the defined protocol is being followed. Each block composed of validated transactions is broadcasted to the entire network, and the decision to permanently place it on the ledger is then made by using a consensus algorithm. This justifies the property of such data being verified and secured.

A consensus algorithm can be understood as a method to reach an agreement among the peers about the trueness of data. This establishes trust without the involvement of any third party. A transaction from the *transaction pool* is validated before storage and then cryptographically placed on the ledger in a way that it becomes read-only and is immutable.

Some pertinent features that blockchain incorporates are discussed.

1. **Decentralization**—Since all the nodes in the network have a copy of the ledger and take part in the consensus mechanisms to validate and post the transactions, the problem of there existing a single point of failure is solved.

2. **Security**—Cryptographic hashes map all the data of the block into a binary number such that even a character change in the data of the block may alter the entire binary hash. If a malicious actor comes in to alter or delete any block from the previous transactions, the hash function of that particular block changes, thereby he would also have to make changes in hash functions of the subsequent blocks, which he cannot without being noticed. Also, in order to put the block on the ledger, he needs to have enormous computational power or shall have to try random hashes, both of which are difficult problems.

3. **Privacy**—Privacy of the individual accounts is guaranteed by public key cryptography. User identities are defined by digital identifiers which are a part of the wallet (digitally owned framework). Whenever someone wants to transfer information, they prepare a receipt and sign it with their private key, which can be verified by anyone who has the public key. This handles the authenticity of transactions [10].

   Smart Contracts—Smart contracts are digitized legal contracts in tpthe form of computer programs, which on fulfillment of certain requirements execute predetermined instructions. The requirements are decided by the developer, and as the conditions are met, the transaction is verified and is submitted on to the ledger. The terms are established by the participants on the network, who also decide the representation of their data on the blockchain. In contrast to written undertakings, smart contracts provide a method of agreement which is speedy due to digitization, secure, and trustworthy due to blockchain mechanisms and is cost effective as it does not involve intermediaries.

3 Implementation

The implementation is done in the form of smart contracts in Hyperledger Fabric, a blockchain framework hosted by the Linux Foundation [11]. It is a private and permissioned framework. Smart contracts in Fabric are called chaincodes and are written in Go programming language in our implementation.

A scenario is considered with two organizations on the same blockchain network.

3.1 Research Institute

An organization where multiple sequence alignment is being performed using various techniques. We know that the aligned sequences are significant for use in medicinal, health, and genetic study purposes, which is data that need not be public. The data provided by it is in two sections, one which is public to all the participants (machine specifications, data sets being used in the problem, etc.) and the other being a private collection which consists of the alignment sequences and results.

3.2 GBlock Org

A startup company analyzes the sequencing data to provide information about what all sequence patterns are responsible for particular traits and diseases and is attempting various medicinal breakthroughs. Gblock org has access to only public data of the participants.

It has a collaboration with the institute, wherein it performs analysis on the data results provided by students, who are implementing various algorithms for multiple sequencing and keep updating their results (Fig. 1).

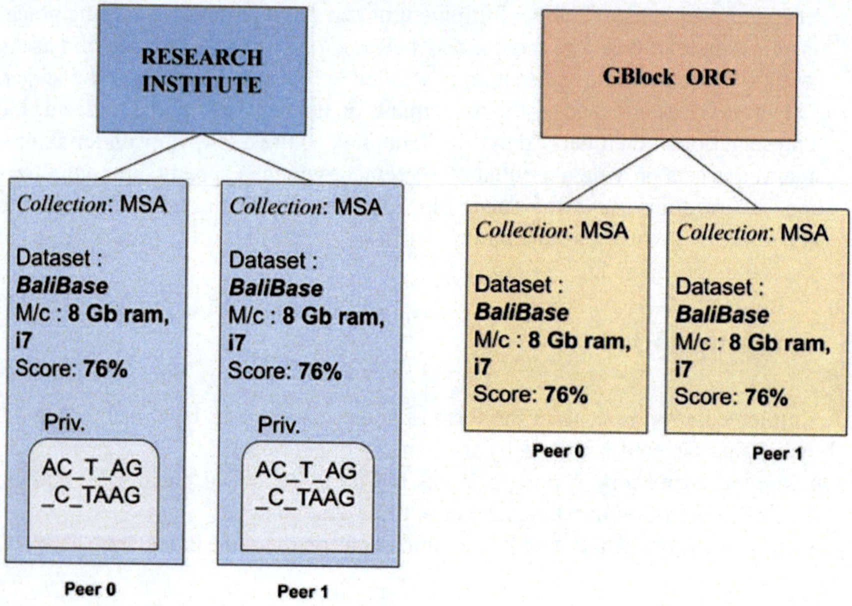

Fig. 1 Smart contract 1

Peers are nodes in the network which may represent students of the university or employees of the organization.

A network is set up on Hyperledger Fabric where *Ordering Service* (responsible for committing transactions to the ledger), *Certificate Authority* (responsible for providing validity certificates to data), and *Peers*(nodes on the network) are all on-boarded in the form of Docker containers. These containers are connected over a network. Either LevelDB or CouchDB is used as a state database to hold data as JSON. Once the network is up, chaincode is instantiated on all the peers of the organizations such that they all have a copy of the program. One can now query the database as peers. If an unauthorized peer, for instance, any peer of GBlock org, attempts to access the private sequencing data, an error is prompted. The metadata or basic information is only available to GBlock.

Both the organizations have their peers on the network with the following distinct collection definition-

- *memberOnlyRead*: The true value suggests that access to private data is allowed only to clients belonging to one of the collection member organizations.

- *blockToLive*: For sensitive information, this value reflects the length of the block life of the data in the private database. The data would remain on the private repository until the specified number of blocks and then be deleted, making it irrecoverable. One can set the blockToLive property to 0 to keep private data forever, that is, never delete private data.

Thus, the data resides securely on the blockchain network.

The solution proposed here also facilitates a layout with monetary benefits for the user as an individual or an organization. Another smart contract is employed to model transactions of incentives between GBlock org and institute wherein GBlock would be granted access to private data on a payment basis. This can evolve into a business model as well, with the involvement of cryptocurrencies like ethereum, bitcoin, stellar, etc. For instance, the demanding parties can pay a decided amount of bitcoins. If verified, the code which transfers useful information to the concerned is executed. (Fig. 2.)

Fabric provides the ability to create channels, enabling a group of participants to create a separate transaction directory. This can allow research institute to share the findings among other scholars on a separate ledger without involving any exchange of benefits. (Fig. 3.)

The system was designed over Fabric v1.4 as a framework that follows Practical Byzantine Fault Tolerance consensus algorithm and for now supports sdk in Node.js, Java, and Go [12].

4 Future Scope and Challenges

The application of blockchain technology in genomics which involves a huge amount of data has a diverse scope including *Distributed Computation*: Smart contracts working on different instances of data. This process can be parallelized *Data Storage and distribution Voting*: Vote on proposals, gene-drug delivery, and *DAPPs*: decentralized applications for maintaining independent databases or restricting ownerships. But there are technical as well as adaptation challenges that are encountered by blockchain technologies in general that need to be overcome.

- Scalability and efficiency: Most blockchains have low scalability because all the nodes on the network are required to process transactions. This also results in low transaction throughput [13].
- Rigidity of smart contracts: They cannot be changed once put onto the ledger, and real-time models might require flexibility in time.
- Storage: Since each node shares a copy, the storage requirement is ever-increasing in time.
- Consensus algorithms: Presently, incorporated algorithms require heavy resources and investments. Also, they do not offer unbiased solutions, and some particular nodes may keep winning all the time refuting all claims of decentralization [14].

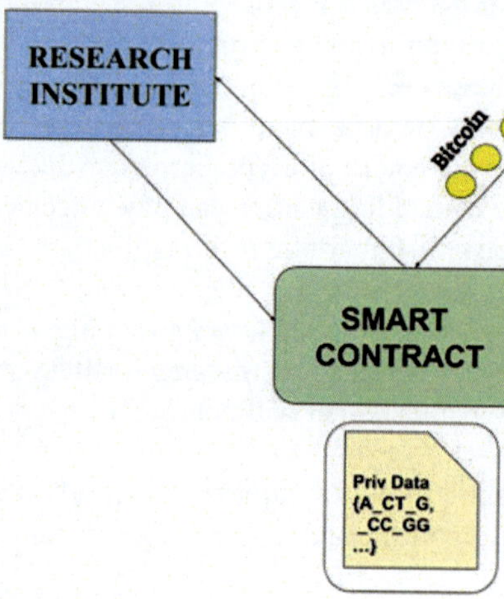

Fig. 2 Smart contract 2

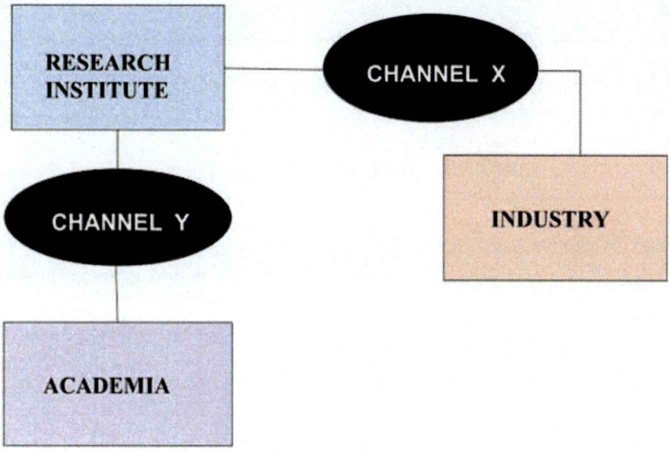

Fig. 3 Scenario with different ledger channels

5 Conclusion

Even with the challenges it poses, blockchain is a promising technology. With genomics research being inclined toward Big Data and advances in sequencing and alignment problems, integration of such a revolutionary approach seems inevitable. There are already applications that have taken a prominent place in economics and fintech. Researchers are already looking into various areas of applications like genomics, supply chain, and finance. New consensus algorithms and frameworks are emerging. Linux Foundation itself proposes six such frameworks. Other developers and industries are also pouring in investments. This has seen a recent rise in the demand for blockchain-based solutions, and if it scales well, it may well be implemented in academic research, thereby encouraging collaborative research on genomics that aims to solve the important problem of decoding the DNA and subsequent exploration.

References

1. Newberg L (2008) Significance of gapped sequence alignment. Genome Research
2. National Human Genome Research Institute. https://www.genome.gov
3. Aisling O, Jurate D, Roy D (2013) 'Big data', Hadoop and cloud computing in genomics. J Biomed Inf
4. Kulemin N, Popov S, Gorbachev A (2017) The Zenome project: blockchain-based genomic ecosystem
5. Miraz M, Ali M (2018) Applications of blockchain technology beyond cryptocurrency. arXiv:1801.03528 [cs.CR]
6. Ileri A, Ozercan H, Gundogdu A, Senol A, Ozkaya M, Alkan C (2016) Coinami: a cryptocurrency with DNA sequence alignment as proof-of-work. arXiv:1602.03031 [cs.CE]

7. Crosby M, Nachiappan, Pattanayak P, Verma S, Kalyanaraman V (2016) Blockchain technology: beyond bitcoin. Appl Innov Rev
8. Nakamoto S (2008) Bitcoin: a peer-to-peer electronic cash system
9. Tapscott D, Tapscott A (2016) Blockchain revolution: how the technology behind Bitcoin is changing money, business, and the world
10. Ozercan H, Ileri A, Ayday E, Alkan C (2018) Realizing the potential of blockchain technologies in Genomics. Genome Research
11. Buterin V (2014) A next-generation smart contract and decentralized application platform
12. Androulaki E, Barger A, Bortnikov V, Cachin C, Christidis K, Caro AD, Yellick J et al (2018) Hyperledger fabric: a distributed operating system for permissioned blockchains. In: Proceedings of EuroSys 2018 conference. arXiv:1801.10228v2 [cs.DC]
13. Scherer M (2017) Performance and scalability of blockchain networks and smart contracts. MSc thesis, Umeå University, Umeå, Sweden
14. Zyskind G, Nathan O, Pentland A (2015) Decentralizing privacy: using blockchain to protect personal data. IEEE Security and Privacy

Blockchain Technology: Architecture, Challenges, and Applications

Mahek Sangwan and A. Charan Kumari

1 Introduction

Blockchain technology is regarded as the backbone of the complete cryptocurrency structure. The most enthralling part of cryptocurrencies is the fact that it was not a directed invention. The goal of Satoshi Nakamoto, at the start, was only to build an "Electronic Peer-to-Peer Cash System" in 2008 [1]. Satoshi understood that an endeavor to create another digital centralized cash system would lead to further defeat, and thus, he shifted his focus to form an electronic cash system without a central power.

Today, blockchain technology as well as cryptocurrencies work in correspondence to undergo monetary transactions. Few hindrances in the existing banking industry are massive charges, account hijacking, counterfeit currencies generating snags of "double-spending," and frauds. E-wallets like PayPal and Paytm resolve only a minority of these concerns permitting prompt money transfers. However, they have banks in the form of a dominant power. The foremost system to cater to all these matters jointly was the concept of blockchains. A blockchain-based arrangement requires no operational fee, and all the proceedings are accomplished in minutes [2].

The remaining paper is ordered as follows: the blockchain architecture is explained in part 2; the challenges and restrictions in part 3, and lastly its applications in part 4. Part 5 concludes.

M. Sangwan (✉)
The NorthCap University, Gurugram, India
e-mail: mahek.sangwan24@gmail.com

A. C. Kumari
Dayalbagh Educational Institute, Agra, India

V. H. Saran and R. K. Misra (eds.), *Advances in Systems Engineering*,
Lecture Notes in Mechanical Engineering,
https://doi.org/10.1007/978-981-15-8025-3_47

2 Blockchain Architecture

Blockchain is a computer-generated log comprising of accounts, organized in distinct fragments holding information, termed blocks. These blocks link together by means of a cryptographic authentication, i.e., a hash function, and accumulatively establish a continuous sequence labelled as a blockchain.

2.1 *SHA-256 Hash Algorithm*

In a blockchain arrangement, transactions are primarily obtained as a basic input and are then validated with the help of a hashing procedure, resulting in a final output of a prescribed size [3]. The entire motivation for the usage of a hash function is that the ultimate result cannot be decrypted over to the former part. The fundamental hashing algorithm used in blockchains is Secure Hashing Algorithm 256 (SHA-256). It runs in the form of Message-Digest Algorithm 4 (MD4) and SHA-1. Essentially, the input is extended in a manner that the output is a multiple of 512 bits, and then, it is analyzed to 512 bit fragments $M^{(1)}, M^{(2)}, \ldots, M^{(N)}$ [4]. These segments are handled serially, beginning with a predetermined original hash $H^{(0)}$ and so forth.

It is computed as:

$$H^{(i)} = H^{(i-1)} + F_M^{(i)}(H^{(i-1)}) \quad (1)$$

Here, F signifies SHA-256 compression function and $+$ represents word-wise mod 2^{32} addition. H^n is the hash value for M [5].

2.2 *Digital Signature*

A traditional transaction is aided through cryptography and blockchain technology by generating a pair of keys. These are termed as public and private keys. The former key is accessible to everybody by means of an overtly available ledger. Contrarily, the private key is covert and is kept exclusive to its corresponding holder. The importance of using digital signatures is that it offers reliability as well as authenticates the legitimacy of a transaction. The classic algorithm used in blockchain arrangement for this is the Elliptic Curve Digital Signature Algorithm (ECDSA) [6].

2.3 Decentralized System

The first effort at constructing an electronic cash system was by David Chaum; who brought to light 'ecash' in the year 1983 as a digital currency [7]. It was executed years later by his firm Digicash. It was developed in the form of a micropayment system. Ever since, quite a lot of attempts have been made at constructing electronic currencies; however, all of these were futile. Remarkably, a common feature in all of these organizations was that they made use of a trusted third party method. Thus, the corporations at their foundation confirmed and expedited the transactions. The arrangement of blockchain, as a layout, is that of a bistributed and decentralized system. Hence, at the fundamental level, each member possesses a precise and fixed copy of the entire blockchain network that spontaneously updates itself every time a novel transaction occurs.

2.4 Distributed Ledger

Since the basic organization of a blockchain network is completely decentralized with no central authority, there is a presence of distributed ledgers. Any ledger or directory containing the data of transactions that is available to all the members of the network is termed as a distributed or a shared ledger. "A distributed ledger is a consensus of replicated, shared, and synchronized digital data geographically spread across multiple sites, countries, or institutions" [8].

2.5 Proof of Work

A prevailing problem is, if each member in the network keeps an identical copy of the ledger, who is responsible for adding subsequent blocks in this blockchain. Satoshi Nakamoto, in his narrative of Bitcoin, answers this with the help of a theory about proof of work (PoW) [1]. This idea was initially coined in the year 1999 by Jakobsson and Juels [9]. This model is analogous to resolving an intricate quantitative problem and needs an amount of computational effort. This task is performed by 'Miners.' They authenticate the transactions as well as decipher the complex problem concomitant with the formation of the block.

3 Challenges to Blockchain System

The presence of restrictions and limitations along with all the remarkable attributes in the blockchain architecture has made it extremely challenging for experts to evaluate its trustworthiness.

3.1 Hacking of Private Key

The most familiar type of attack on the blockchain architecture is hacking by theft of the private key. If a user's private key is pilfered, the entire linked currency also gets compromised. In the year of 2016, eminent blockchain networks like Shift and Krypton were subject to coercions by hackers calling themselves, "51 Crew," and were asked to pay a ransom or get their systems appropriated. The hackers took over 51% of the entire network [10]. To mitigate this, Bitcoin wallets such as blockchain.info are a better alternative as they protect the private keys.

3.2 Computing Power Takeover

In a typical proof of work, the computing power is expended in demonstration of the validity of any transaction. If a bulk of network participants approve the mining of a block, only then a transaction is interpreted as authorized. Here, anybody who can hack in the system to shift the computing power into his individual record can consequently dominate over huge parts of the blockchain network. An unidentified invader hijacked the DNS servers of the network of blockchain.info in 2016 [11]. In mitigating this, the platforms are switching to proof-of-stake unanimity. Thus, the capability to approve any transaction is dependent on the stake of members in the blockchain arrangement rather than computing power [12].

3.3 Software Vulnerability

Decentralized autonomous organization (DAO) in May 2016 was took over by an attacker who came across an ambiguity in its "smart contract" and later misused it, successively getting over $60 million [10]. Any software by individuals is susceptible to inaccuracies. For mitigation, any software should be tested proficiently and advancement must be made further vigorous.

4 Applications

In the times of today, it is somewhat blatant that a blockchain is a lot bigger than Bitcoins and is renowned in areas further than the foundational cryptocurrency.

4.1 Stablecoin

Stablecoins are basically cryptocurrencies specifically designed to minimize price volatility. Today, notable names have announced Stablecoin projects. For instance, the renowned Winklevoss brothers, who are already front runners in cryptocurrency investing, recently announced the launch of Gemini dollar, a Stablecoin developed for their trading platform Gemini [13]. Furthermore, technology tycoon IBM Corporation has also soared into the market by partnering with Stellar. IBM announced the "limited production" launch of World Wire, a global payments system connecting banks through the Stellar digital-asset network [14]. Stablecoins are unlike traditional cryptocurrencies as they do not have a restricted supply or an innate schedule. Instead, they are distributed based on market situations and finances [15].

4.2 Smart Contracts

A smart contract is a "self-enforcing piece of software that is managed by a P2P network of computers" [16]. A smart contract is essentially a bond rooted in a computer program managed by a blockchain. The program contains a set of guidelines under which all the parties of that smart contract agree to work together. If and when the previously determined rules are met, the agreement is spontaneously enforced. Smart contracts offer mechanisms for proficiently managing access rights between two or more parties [17]. One can think of it as a safety box that allows access, if and when already established conditions are true.

4.3 Smart Property

If the proprietorship of any property is managed and retrieved with the help of blockchain technology, that property is characterized as a smart property [18]. Their cataloging is done on a register or ledger along with the information on individuals with legitimate ownership. Smart keys are employed to accelerate control. As an example, one can get access to a smart-house only if they had the equivalent smart key on their device. These keys can be exchanged, and the communications could be substantiated with the aid of blockchain. As a result, the blockchain provides a procedure for recording title rights [18].

4.4 *Energy Web Foundation: Tobalaba*

EWF's goal was about building a platform in concurrence with its associates, representing an assorted group of members in the energy department. In spite of using a typical blockchain arrangement, Tobalaba, its new test network, is tailored on the inputs from EWF's shareholders. "The energy market requires a blockchain tailored to the sector itself," said Peter Bronski, EWF's marketing director [19]. In its place, the Tobalaba system makes use of blockchain to log the origin and monitor the possession that includes particulars of source, site, interval, and radiations. Present arrangements to supervise these aspects are intricate and are centrally handled by a third party. The expectation is that systems based on blockchain can expedite the procedure and deliver apparent as well as precise tracing.

4.5 *Blockchain Government*

The blockchain technology provides an arrangement for "responsive and open data." As mentioned by a 2013 report by McKinsey and Company [20], the US state of Vermont is bearing in mind the option of exhausting blockchain to track government records. Honduras, in US, is also considering the recording of property titles in a similar fashion. Till there is presence of novel shields for digital wallets and other software, governments are not capable of wholly embracing the blockchain technology.

5 Conclusion

The blockchain phenomenon has displayed its ability to create extensive modifications in countless areas from the customary financial organizations, to electronic currencies and so forth; these intriguing employments attributed to the rudimentary architecture. The paper explains the fundamental attributes of blockchain architecture establishing it a far-reaching technology. Furthermore, it mentions the limitations to this arrangement and suggests simple extenuation techniques. It was concluded by the pioneering implementations of this technology. The paper presents a comprehensive study of the blockchain network, covering all the aspects to provide an egalitarian view of the same.

References

1. Nakamoto S (2008) Bitcoin: a peer-to-peer electronic cash system. https://bitcoin.org/bitcoin.pdf
2. Appel AW (2015) Verification of a cryptographic primitive: SHA-256. ACM Trans Program Lang Syst 37(2):1–31
3. Blemus S (2017) Law and blockchain: a legal perspective on current regulatory trends worldwide. In: Revue Trimestrielle de Droit Financier (Corporate Finance and Capital Markets Law Review) RTDF N°4-2017
4. Descriptions of SHA-256, SHA-384, and SHA-512. https://www.iwar.org.uk/comsec/resources/cipher/sha256-384-512.pdf
5. Johnson D, Menezes A, Vanstone S (2001) The elliptic curve digital signature algorithm (ecdsa). Int J Inf Secur 1(1):36–63
6. Chaum D (1983) Blind signatures for untraceable payments. In: Advances in cryptology, Springer, Boston, MA
7. Walport M (2016) Distributed ledger technology: beyond block chain. UK Government Office for Science
8. Jakobsson M, Juels A (1999) Proofs of work and bread pudding protocols. Kluwer Academic Publishers 23:258–272
9. Huffington Post India. https://www.huffingtonpost.in/raja-raman/blockchain-can-transform-the-world-but-is-it-fool-proof_a_21660586/
10. NewsBTC: Blockchain.info DNS Hacking Attempt Shuts Down Service for Few Hours
11. Bitfury Group: Proof of stake versus proof of work, Version 1.0 (2015)
12. Gemini Launches in Australia, Enables all supported coins except GUSD. https://cointelegraph.com/news/gemini-launches-in-australia-enables-all-supported-coins-except-gusd
13. Things you need to know about IBM's new partnership with Stellar. https://breakermag.com/3-things-you-need-to-know-about-ibms-new-partnership-with-stellar/
14. All about Stablecoin: the new kid on the block (chain). https://economictimes.indiatimes.com/magazines/panache/all-about-stablecoin-the-new-kid-on-the-blockchain/articleshow/69367940.cms
15. Token Economy: how blockchains and smart contracts revolutionize the economy. (in press)
16. Smart Contracts. https://blockchainhub.net/smart-contracts/
17. Miraz MH, Ali M (2018) Applications of blockchain technology beyond cryptocurrency. Ann Emerg Technol Comput (AETiC) 2(1): 1–6. Print ISSN: 2516–0281, Online ISSN: 2516–029X
18. Blockchain Technology, Smart contracts and smart property. https://ieet.org/index.php/IEET2/more/Danaher20151117
19. Mechanical Engineering: blockchain enables new power markets. (in press)
20. How the Blockchain will enable self-service government. https://www.wired.co.uk/article/blockchain-is-the-new-signature

Robotic Systems

3D Printed Quadcopter

Hans Agarwal, Apar Singhal, and K. Hans Raj

1 Introduction

Advancement in the field of unmanned aerial vehicles (UAVs) have changed the way of thinking about technology in past few years and quadcopters have played an important role in it. Development of small UAVs is the interest of researchers, and they want to explore its application. Quadcopter is a drone which is an advanced form of helicopter, having better dynamic stability [1]. Quadcopter is a small type of unmanned aerial vehicle (UAV) that can be used for many purposes in present era. It is very useful for military, traffic monitoring, surveillance, weather monitoring, firefighting, research applications (reach places where humans cannot), fire sensing, search and rescue operations, and other important applications.

Quadcopters are designed using a component of electronics, mechanics, computer application, and most importantly the principles of aviation. Custom quadcopters are designed to satisfy societal needs. Quadcopter size can be decided according to the purpose for which it may be used and 3D printing allows one to manufacture high precision and miniature parts with agility.

Quadcopters are aerial robots that have four rotors that are attached to the corner ends of the frame structure, the pair of rotors in one arm, facing each other rotates in clockwise direction, whereas the other pair rotates in anti-clockwise direction [2]. Quadcopters can be controlled by varying the speed of the four rotors and no additional mechanical linkages are required to vary rotor blade pitch angles as is

H. Agarwal (✉) · A. Singhal · K. H. Raj
Faculty of Engineering, Dayalbagh Educational Institute, Agra, India
e-mail: agarwalhans.999@gmail.com

A. Singhal
e-mail: aparsinghal1998@gmail.com

K. H. Raj
e-mail: hansraj@dei.ac.in

V. H. Saran and R. K. Misra (eds.), *Advances in Systems Engineering*,
Lecture Notes in Mechanical Engineering,
https://doi.org/10.1007/978-981-15-8025-3_48

done in present-day helicopters. Quadcopters have an important feature of taking vertical take-offs, landings, and hovering over the desired areas.

3D printing is an additive manufacturing process by which three-dimensional solid objects are made by using a digital file containing the 3D design details [3]. The creation of 3D printed objects takes place in an additive manner, i.e., by laying down of material in successive layers. Each layer is a cross section of the 3D printed object. 3D printing can be used to build complex shapes using less material as compared to the traditional manufacturing.

3D printing starts with designing a 3D model, i.e., computer-aided design (CAD) model, and 3D model can be created or can be downloaded using the online 3D repositories. While creating the model, 3D scanners, apps, coding, or 3D modeling software may be used.

Nowadays, 3D printing is becoming popular in aviation industries. It is used for printing different parts of aircraft engines. Earlier, individual parts were constructed and had to be welded together but now by use of many parts 3D printing are consolidated to a single component which is lighter in weight, uses lesser material and most importantly it is about four-five times stronger. 3D printing reduces the cost of labor as well as material and improves quality and thus overall benefits the aviation industries [4]. GE Aviation is a company that has already started using 3D printers for better efficiency economic products that are lighter and stronger, and built using automated machinery with higher accuracy [5].

PLA (polylactic acid) is a green material and an alternative to petrochemical commodity plastics. Its mechanical characteristics are its superior tensile and flexural strength. PLA (polylactic acid) is very light with a density of 1.210–1.420 g cm^{-3}. PLA can be used in areas of high temperature without wear and tear as it has a high melting point of 423–433 K [6]. It is insoluble in water making it useful during rains and humid weather conditions.

3D printed quadcopter is one step forward for the aviation industries which are building quadcopters. This 3D printed quadcopter is designed using a 3D designing software SolidWorks. All the components (including propellers) are designed and then printed using the 3D printer. The material used for printing is polylactic acid (PLA).

2 Methodology

All parts are designed separately and then assembled. This design will be 3D printed after converting this solid-works file into.stl file format. Polylactic acid (PLA) material will be used to 3D print the hardware of the quadcopter. PLA is a thermoplastic derived from the renewable resources, so it is quite cheap as compared to the other plastic materials which can be used in robotics. PLA provides better strength with less weight, which makes it really important for making robots and quadcopters. PLA is semi-transparent, so it provides glossier surface which seems attractive. Some basic properties of PLA can be referred from Table 1 [7] which also gives the strength to

Table 1 Properties of PLA [7]

Properties	Value
Tensile strength	37 MPa
Density	1.3 g/cm^3
Percentage elongation	6%
Flexural modulus	4 GPa
Biodegradable	Yes, under the correct conditions
Melting point temperature [6]	423–433 K

the robots made of PLA material.

The main criteria for making this quadcopter first in separate parts and then assembling it was to make it flexible for future upgradation. If any desired change in the design of a particular part will be needed, then whole quadcopter will not be disturbed. Changes can be made at that particular part, and this upgradation will automatically transferred from the source file to the main assembly which will also save the time.

The overall process for designing the quadcopter can be summarized in Flowchart 1.

3 Design of the Quadcopter

The frame of the quadcopter was designed in SolidWorks is shown in Fig. 1, In which base, top and spider arm will be 3D printed on the Stratasys Fortus 400mc 3D printing system in the 3D printing laboratory, Dayalbagh Educational Institute, Agra, India. But for propellers, very high precision is needed because whole quadcopter balances its flight on the propellers, so Uprint SE Plus 3D printer will be used to print two clockwise and two anti-clockwise propellers. This 3D printer is situated in Advanced Manufacturing Laboratory, Engineering Department of University of Guelph, Ontario, Canada. Then, some load tests will be performed on the frame to check its durability according to the future requirements. High-pressure air will be passed through clockwise and anti-clockwise propellers at various angles for testing its accuracy and correctness. Some screenshots of the SolidWorks design of this project are given in Figs. 1, 2, and 3.

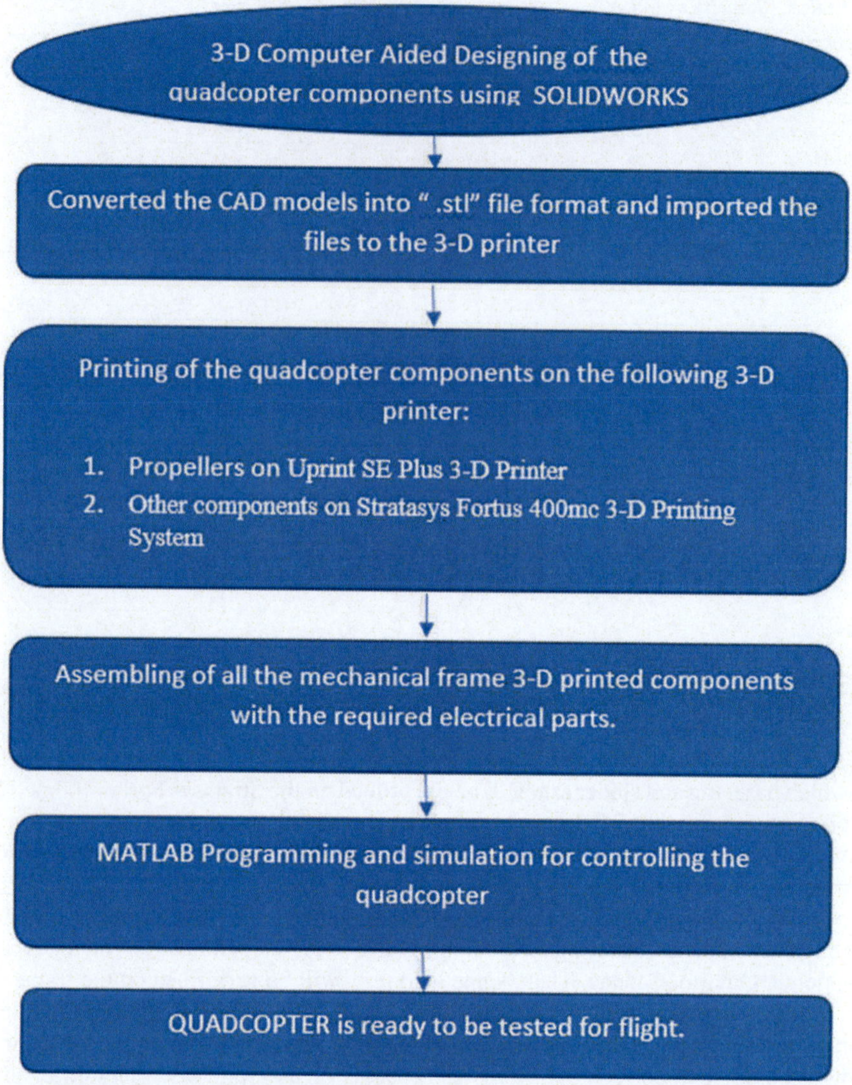

Flowchart 1 Designing process of 3D printed quadcopter

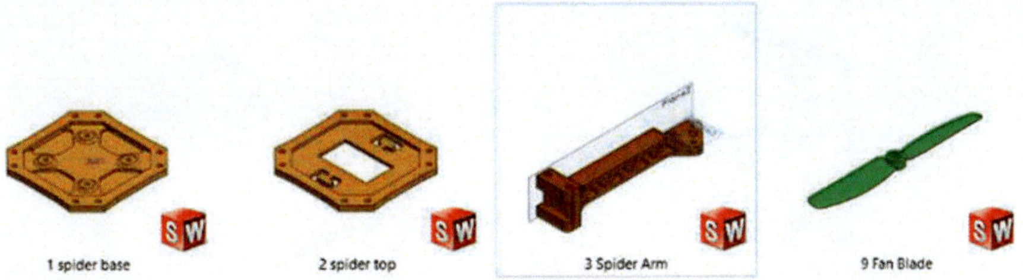

Fig. 1 SolidWorks design of the quadcopter frame

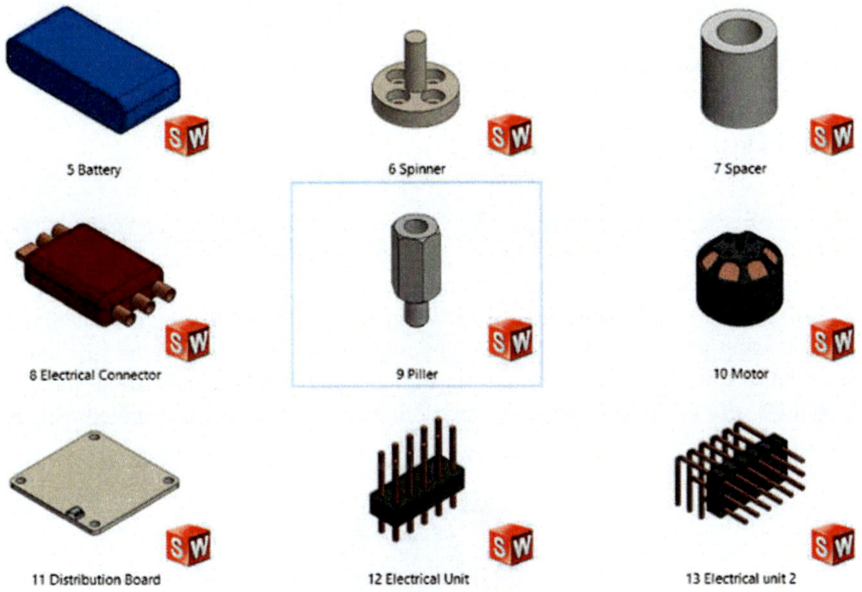

Fig. 2 SolidWorks design of the electrical and other small parts

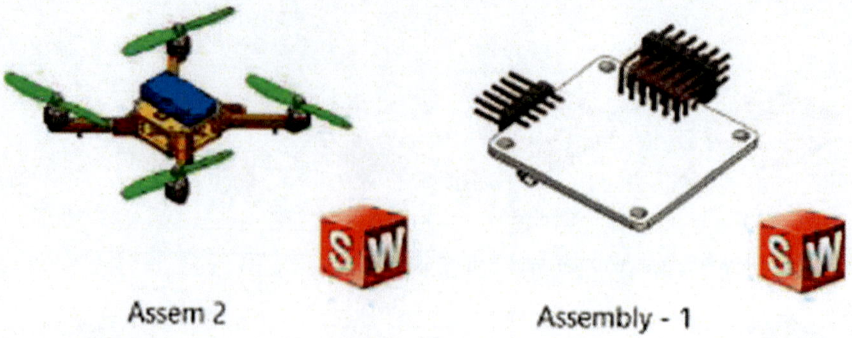

Fig. 3 SolidWorks design of the final assembly of the quadcopter

4 Working of the Quadcopter

To control the multiple motors correctly, it is mandatory to understand how contrasting speeds in the motor affect a lift from the propeller and how to lose a specific altitude and a turn in any directional axes.

Roll and pitch play important role in tilting the multirotor. This incline is achieved by increasing the speed of rotation of any two motors according to the side which we want to tilt. Through this phenomena quadcopter can be moved in all the four directions which are: forward, backward, left, and right. Camera is installed with the

Fig. 4 Representation of directions of rotation of the propellers [8]

Fig. 5 Representation of rotor speeds while quadcopter is rotating in anti-clockwise direction [8]

quadcopter with front-facing direction. So, pictures are taken by tilting it from front by increasing the speeds of the backward rotors.

Two clockwise propellers are installed in the same axis of arm in front of each other, and similarly anti-clockwise propellers are also installed in the same fashion. Quadcopter is rotated on its central axis clockwise by increasing the speeds of clockwise propellers and also rotated anti-clockwise by increasing the speeds of anti-clockwise propellers.

It can also be flipped but that operation is quite hard to performed. We can turn off two adjacent rotors for very short duration and then suddenly increase its speed after flipping, but we need skilled controller for that (Figs. 4 and 5).

A radio transmitter and receiver are used to control the quadcopter. For the good-working quadcopter, four channels (elevator, throttle, rudder, and aileron) are used. Six-channel transmitter & receiver is used to control this quadcopter. FPV screen can also be used in transmitter which can be connected to the camera mounted on the quadcopter. This provides the pilot a precise experience of flying by continuously watching in the screen [9].

Based on the momentum theory, the following formula is used for the thrust developed by the propellers of the quadcopter [10].

$$T = \pi D^2 \varrho v \Delta v / 4 \quad (1)$$

Symbols mentioned in Eq. (1) are representing the following parameters.
$T =$ thrust (N)
$D =$ propeller diameter (m)
$v =$ velocity of air at the propeller (m/s)
$\Delta v =$ velocity of air accelerated by propeller (m/s)
$\varrho =$ density of air (1.225 kg/m^3).

5 Results

The main perspective of our paper is to study the advantages of 3D printed quadcopter with the analyses of its complete designing process. Our main achievement is design of a quadrotor which is light in weight with good strength and load-carrying capacity, so that it can be used for various applications, i.e., in military, market, commercial and industrial areas, etc. [11].

Weight of this self-designed 3D printed quadcopter is much less than the conventional quadcopters of the same size. After watching simulation process of the assembly of the quadcopter in SolidWorks, it is much easier to fabricate or assemble its all the components in reasonable time. This quadcopter is easily controlled with the help of six-channel transmitter and receiver with excellent accuracy.

With the help of our supervisors in Dayalbagh Educational Institute, India, and in University of Guelph, Canada, we have got all the technical knowledge and resources to complete this project successfully. This paper describes the complete design of indigenous quadcopter.

Reference

1. Pooja Srivastava TNCP (2017) Quadcopter for rescue missions and surveillance. IOSR J Comput Eng 1:48–52
2. Hoffmann GM, Huang H, Waslander SL, Tomlin CJ, Hoffmann GM Quadrotor helicopter flight dynamics and control. In: American Institute of Aeronautics and Astronautics
3. "3D printing scales up," 3D printing, 05 09 2013. [Online]. Available: https://www.economist.com/technology-quarterly/2013/09/05/3d-printing-scales-up. Accessed 18 09 2019
4. "Tractus3D," Tractus3D, [Online]. Available: https://tractus3d.com/what-is-3d-printing/advantages-of-3d-printing/. Accessed 18 09 2019
5. "what is 3d printing," october 2018. [Online]. Available: https://3dprinting.com/what-is-3d-printing/. Accessed 19 09 2019
6. Södergård A, Stolt M (2002) Properties of lactic acid based polymers and their correlation with composition. Prog Polym Sci 6(27):1123–1163

7. Giang K (2019) "3dhubs," 3D Hubs: Local 3D printing services, [Online]. Available: https://www.3dhubs.com/knowledge-base/pla-vs-abs-whats-difference. Accessed 06 08 2019
8. "DroneBot Workshop," DroneBot Workshop, [Online]. Available: https://dronebotworkshop.com/how-does-a-quadcopter-work/. Accessed 12 09 2019
9. Kadamatt V (2019) How quadcopters work & fly: an intro to multirotors, 20 11 2017. [Online]. Available: https://www.droneybee.com/how-quadcopters-work/. Accessed 13 09 2019
10. Dickey J (2019) Air Wolf 2, quadcopterproject, [Online]. Available: https://quadcopterproject.wordpress.com/static-thrust-calculation/. Accessed 18 09 2019
11. Dutt BJ, Shah KN (2014) Quadrotor—an unmanned aerial vehicle. IJEDR 2(1):1299–1303

Control Optimization of Triple-Stage Inverted Pendulum Using PID-Based ANFIS Controllers

Kharola Ashwani and Nikam Yadnyesh

1 Introduction

Inverted pendulum on cart is widely used in many nonlinear control applications and testing several control algorithms due to its nonlinear and unstable dynamics. The system consists of a rigid pendulum mounted on a movable cart, and the aim is to stabilize all the three pendulums at vertically upright position [1]. Increasing the number of pendulums makes the dynamics of the system all the more complex and challenging [2]. The system consists of three pendulums hinged together and capable of rotating in vertical plane. Triple inverted pendulum system can be modeled as a human standing on one leg. Due to its inherent multivariability and instability, these systems are widely used for simulating performances and validation of various control techniques [3]. It has been a keen source of interest for the researchers in the past few decades. Medrano-cerda [4] designed a robust controller for control of double and triple inverted pendulum system. A DC motor was mounted on the upper and middle hinge to balance double and triple inverted pendulums, respectively. A state feedback controller was utilized, and system stability was evaluated via Nyquist plots.

A control technique which combined different control methods like stipulations, optimal control combined with knowledge-based control was proposed by Ruisen and Liang [5]. The results were obtained using nonlinear programming technique combining with experts experience for the proposed system. A swing-up control of three-link inverted pendulum was proposed by Gluck et al. [6]. The controller was designed combining a nonlinear feedforward controller with an optimal feedback

K. Ashwani (✉) · N. Yadnyesh
Tula's Institute, Dehradun, Uttarakhand, India
e-mail: ashwani.kharola@tulas.edu.in

N. Yadnyesh
e-mail: yadnyeshnikam004@gmail.com

V. H. Saran and R. K. Misra (eds.), *Advances in Systems Engineering*, Lecture Notes in Mechanical Engineering,
https://doi.org/10.1007/978-981-15-8025-3_49

controller. Real-time experiments were performed for successful implementation of the proposed system. Ananyevskiy and Anokhin [7] considered control of multi-link inverted pendulum having a fixed suspension. A control torque was applied to the first link of proposed system, and a control feedback law was developed for steering of the pendulums in equilibrium position. Prasad et al. [8] considered a Linear quadratic regulator (LQR) and PID techniques for stabilization of one link inverted pendulum system. State-space model of the system was developed, and simulations were performed using both with and without disturbance input.

Zhang et al. [9] proposed control of triple inverted pendulum via Genetic algorithm (GA)-PID Neural network (PIDNN). A nonlinear model of the system was developed which combines the merits of both GA and PIDNN control. Simulations showed that GA technique was better compared to PIDNN. Anan'evskii [10] obtained control of three-link inverted pendulum system by applying torques at the ball joints. The system has got around eight different equilibrium positions, and a control feedback module was developed to bring the system in equilibrium. In a study by Huang et al. [11], Swinging motion of three-link inverted pendulum was recorded via CCD camera and motion was further optimized with the help of Harris algorithm. The author effectively utilized the motion vision technique for stabilizing the complete system. Stabilization of triple inverted pendulum in vertical position through parametric excitation of the support has been demonstrated by Arkhipora [12]. The author applied multiple scale method and Floquet theory for controlling the system at different frequency ranges.

In this study, closed-loop feedback control strategies have been adopted for solving nonlinearity associated with the problem of balancing triple-stage inverted pendulum system at upward unstable position. This inherent unstable position gives rise to nonlinearity which needs to be linearized. The study considers a feedback linearization to compensate the nonlinearities associated with the system. A nonlinear model of the system has been derived using Newton's second law. The transient response of the system is controlled using PID controller which is further used for training of ANFIS and neural controllers. Finally, the study compares the results of the study to results of an ANFIS controller developed by Jun-wei et al. [21]. It has been clearly observed from the results that the settling time has been significantly improved using PID controller, whereas there is still a scope for improving the results of ANFIS and neural controller.

2 Mathematical Model of Triple-Stage Inverted Pendulum

The generalized equations of motion for the system were developed using a study by Eltohamy and Kuo [13]. These equations were further used for developing a Simulink model of proposed system. The equations of motion for the system were derived considering free body diagram analysis of individual sub-systems. The final nonlinear differential equations for the individual sub-systems are as follows.

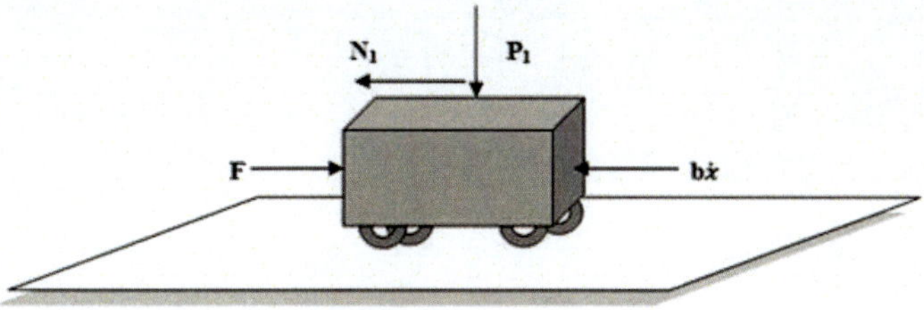

Fig. 1 Free body diagram of cart sub-system

2.1 Equations of Motion for Cart

The free body diagram of cart sub-system highlighting all the forces acting on it is illustrated in Fig. 1. In the figure, N_1 and P_1 represent the interaction forces between bottom pendulum and cart acting along horizontal and vertical directions, respectively. In the figure b, $\dot{x}$ represents frictional force acting between cart and surface during motion.

Summing forces acting on the cart in horizontal direction gives the following equation of motion,

$$\ddot{x} = \frac{1}{M}(F - N_1 - b\dot{x}) \tag{1}$$

In the above equation, $\dot{x}$ and $\ddot{x}$ denote velocity and linear acceleration of the cart, respectively.

2.2 Equations of Motion for Bottom Pendulum

The free body diagram of bottom pendulum highlighting all the forces acting on it is highlighted in Fig. 2. In the figure, N_2 and P_2 represent the interface forces amid lower and middle pendulum acting along horizontal and vertical directions, respectively. In the figure, $m_1 g$ represents the gravitational force acting on the system due to its weight.

An expression for angular acceleration $(\ddot{\theta}_1)$ of the bottom pendulum is given below,

$$\ddot{\theta}_1 = \frac{1}{I_1}\left(N_1 l_1 \cos\theta_1 + P_1 l_1 \sin\theta_1 - b_1\dot{\theta}_1 + N_2 l_1 \cos\theta_1 + P_2 l_1 \sin\theta_1\right) \tag{2}$$

where $\dot{\theta}_1$ and I_1 are angular velocity and moment of inertia for bottom pendulum.

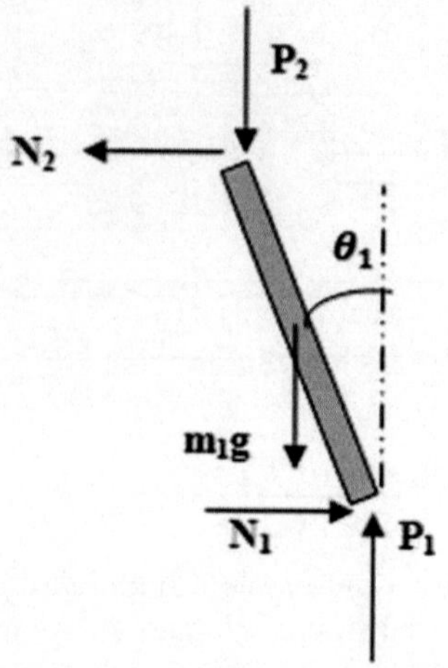

Fig. 2 Free body diagram of bottom pendulum

2.3 Equation's of Motion for Middle Pendulum

The free body diagram of middle pendulum highlighting all the forces acting on it is shown in Fig. 3. In the figure, N_3 and P_3 represent interaction forces between middle and top pendulum acting along horizontal and vertical directions, respectively. The gravitational force acting on the system due to its weight is given by m_2g, where g is acceleration due to gravity.

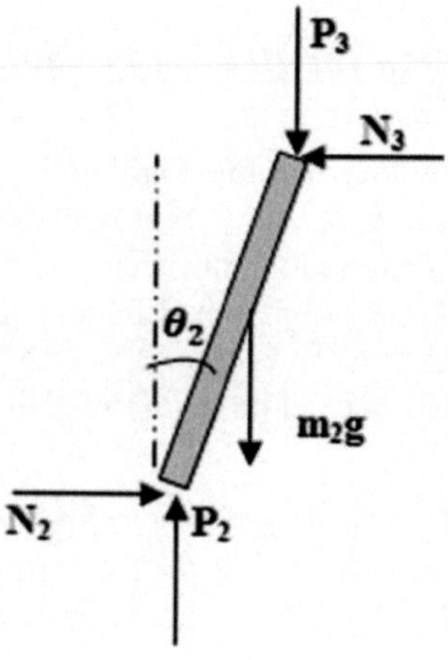

Fig. 3 Free body diagram of middle pendulum

An expression for angular acceleration ($\ddot{\theta}_2$) of the middle pendulum is given below,

$$\ddot{\theta}_2 = \frac{1}{I_2}\left(N_2 l_2 \cos\theta_2 + P_2 l_2 \sin\theta_2 - b_2\dot{\theta}_2 + N_3 l_2 \cos\theta_2 + P_3 l_2 \sin\theta_2\right) \quad (3)$$

where $\dot{\theta}_2$ and I_2 are angular velocity and moment of inertia for middle pendulum.

2.4 Equations of Motion for Top Pendulum

The free body diagram of top pendulum highlighting all the forces acting on it is illustrated in Fig. 4. In the figure, N_3 and P_3 represent interaction forces between middle and top pendulum acting along horizontal and vertical directions, respectively. The gravitational force acting on the system due to its weight is given by $m_3 g$.

An expression for angular acceleration ($\ddot{\theta}_3$) of the top pendulum is given below,

$$\ddot{\theta}_3 = \frac{1}{I_3}\left(N_3 l_3 \cos\theta_3 + P_3 l_3 \sin\theta_3 - b_3\dot{\theta}_3\right) \quad (4)$$

where $\dot{\theta}_3$ and I_3 are angular velocity and moment of inertia for top pendulum.

Equations (1)–(4) were utilized to create a Simulink of above system as given in Fig. 5.

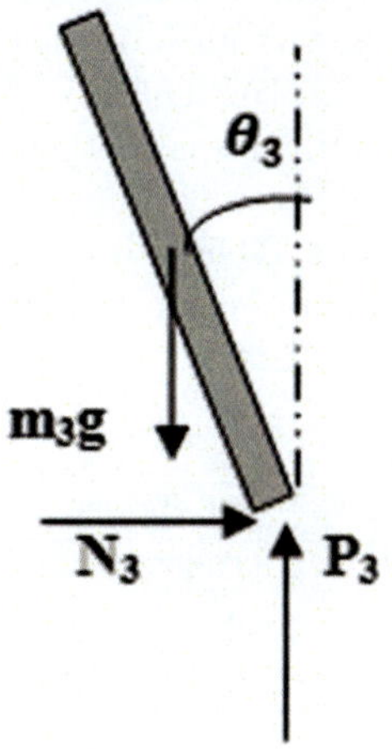

Fig. 4 Free body diagram of top pendulum

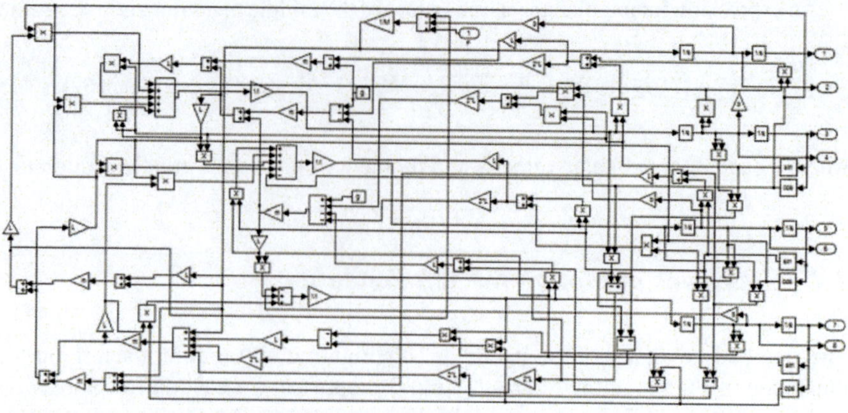

Fig. 5 Simulink of three-stage cart and pole system

3 PID Control of Triple-Stage Inverted Pendulum

A proportional–integral–derivative (PID) controller is a closed-loop feedback control mechanism widely used for various control applications [14]. The controller computes an error signal which is the difference between the desired output and measured value. The error signal is computed until the error becomes zero or minimal value is obtained. The reduction of error is performed with the help of proportional, integral, and derivative gains [15]. All the three gains were added together to calculate the output of PID controller as shown with the help of equation [16]. The gains of PID controller for triple-stage cart and pendulum have been tuned using trial and error method. The tuning procedure for gains is continued until suitable responses were obtained. The PID gains obtained after tuning diverse sub-systems of three-stage cart and pendulum system are given in Table 1. It is clear from the table that negative gains were obtained for differential gain constant, whereas proportional gain values were kept to zero for all the controllers.

Table 1 PID gains obtained after tuning for sub-systems

Sub-system	Integral gain (K_i)	Proportional gain (K_p)	Differential gain (K_d)
Top pendulum	100	0	−10
Middle pendulum	1	0	−1
Bottom pendulum	1	0	−1
Cart	10	0	−10

Table 2 Random division of data samples for training of neural controllers

Type of sample	Percentage (%)	Sample size
Training	70	103
Validation	15	22
Testing	15	22

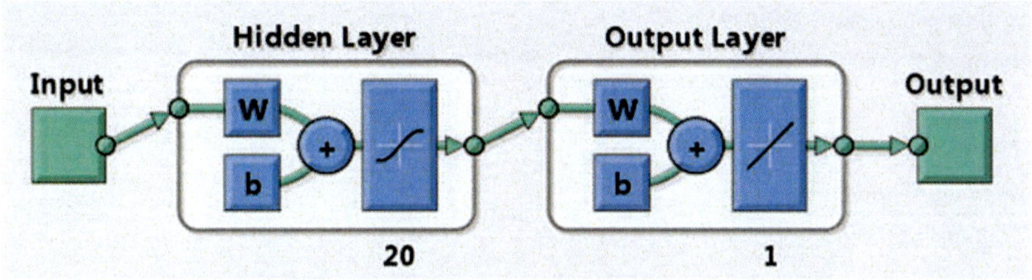

Fig. 6 Neural network architecture considered for triple-link cart and pendulum

4 Neural Control of Triple-Stage Inverted Pendulum

Neural networks are artificial learning tools based on human ability of learning. These networks are designed on the basis of learning ability of neurons of human brain. In this paper, a PID-based neural controller is designed for control of triple-link inverted pendulum. A total of 147 data samples were stored from results of PID controller and applied for training of neural controller. These data samples were further at random divided into training, testing, and validation samples in the ratio 70:15:15 in order to solve the problem of overfitting [17]. The sample size of randomly divided data samples is given in Table 2.

The neural design was constructed considering 20 neurons in the hidden layer and single neuron in output layer as shown in Fig. 6. Levenberg–Marquardt algorithm was used for training of neural controllers. The tuning was done till a regression rate of unity which was achieved for the entire samples. The actual MSE and regression values attained for each controller is given in Table 3.

5 ANFIS Control of Triple-Stage Inverted Pendulum

The term 'ANFIS' stands for adaptive neuro-fuzzy inference systems which belongs to a class of adaptive networks and is widely used for various learning and training operations [18]. These systems combine the features of both fuzzy logic controllers and artificial neural networks. ANFIS has the learning ability of artificial neural networks and logic-based reasoning of fuzzy controllers which makes them powerful learning tool [19]. The system consists of a set of nodes whose weights are adjusted

Table 3 MSE and regression values obtained for different controllers

Type of sample	MSE	Regression
Top pendulum controller		
Training	2.16e−9	1
Validation	4.09e−7	1
Testing	1.39e−5	1
Bottom pendulum controller		
Training	5.98e−9	1
Validation	1.97e−7	1
Testing	4.12e−7	1
Middle pendulum controller		
Training	5.97e−10	1
Validation	1.59e−8	1
Testing	8.01e−6	1
Cart controller		
Training	5.25e−10	1
Validation	2.66e−9	1
Testing	1.34e−9	1

during the learning process. A schematic view of ANFIS architecture is shown in Fig. 7 [21]. The results obtained from PID controller were utilized for teaching of ANFIS controller. This section further highlights the consequence of type of memberships on learning of both linear- and constant-type membership functions. Five different shape memberships, namely triangular, trapezoidal, gbell, gauss, and gauss2mf were considered and varied from 3 to 8 in number. The errors achieved incorporating these memberships for bottom pendulum controller are depicted in Table 4 which proved that the minimum training error was given by triangular memberships of constant type.

The simulation responses obtained for different sub-systems of triple inverted pendulum using above-mentioned techniques are revealed via following figures and tables. The time considered for simulation was 6 s.

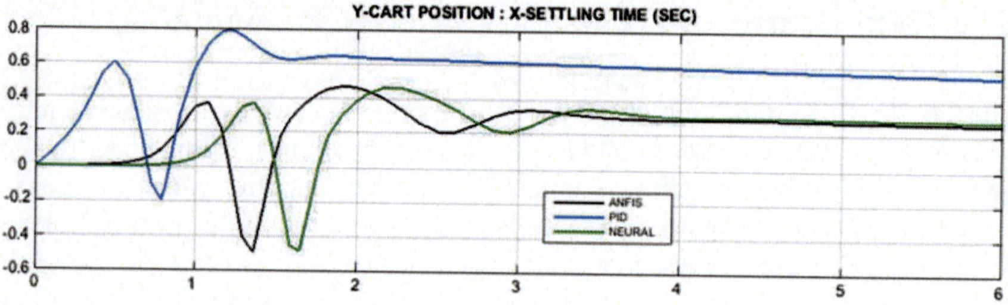

Fig. 7 Cart position response using different controllers

Table 4 Training errors obtained for bottom pendulum controller

No. of memberships	Triangular	Trapezoidal	Gbell	Gauss	Gauss2mf
Bottom pendulum controller: linear-type memberships					
3	9.52e−006	2.22e−005	3.46e−005	1.65e−005	3.27e−005
4	1.68e−005	2.98e−005	6.13e−005	2.75e−005	4.56e−005
5	1.77e−005	6.53e−005	4.15e−005	5.26e−005	6.96e−005
6	2.70e−005	2.99e−005	5.50e−005	5.98e−005	5.20e−005
7	3.51e−005	5.24e−005	4.61e−005	4.39e−005	3.52e−005
8	1.62e−005	6.57e−005	5.03e−005	5.42e−005	3.27e−005
Bottom pendulum controller: constant-type memberships					
3	4.27e−007	0.005980	0.001203	0.000321	0.001880
4	4.43e−006	0.003684	0.000950	0.000293	0.001894
5	1.64e−005	0.002785	0.000446	0.000149	0.001133
6	2.04e−005	0.001710	0.002453	0.000137	0.002234
7	0.000106	0.001926	0.000975	0.000202	0.001092
8	0.000107	0.002046	0.000664	0.000211	0.001473

6 Simulation Results

See Figs. 7, 8, 9, 10, 11, 12, and 13.

(a) Simulation response: cart position
(b) Simulation response: bottom pendulum angle
(c) Simulation response: bottom pendulum angular velocity
(d) Simulation response: middle pendulum angle
(e) Simulation response: middle pendulum angular velocity
(f) Simulation responses for top pendulum angle
(g) Simulation response: top pendulum angular velocity.

The simulation results obtained for sub-systems by Jun-wei et al. [21] are shown in Table 5.

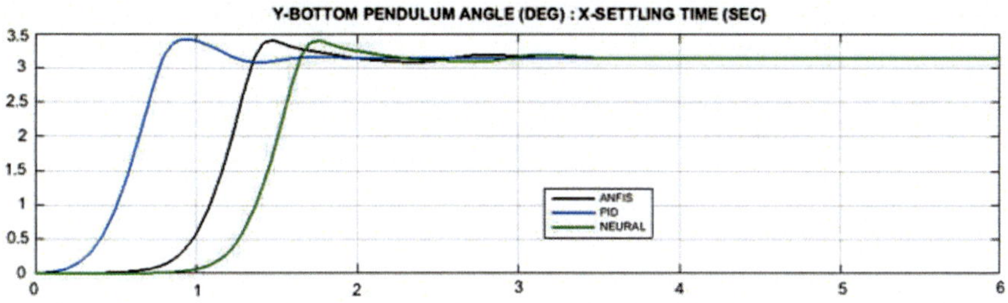

Fig. 8 Bottom pendulum angle response using different controllers

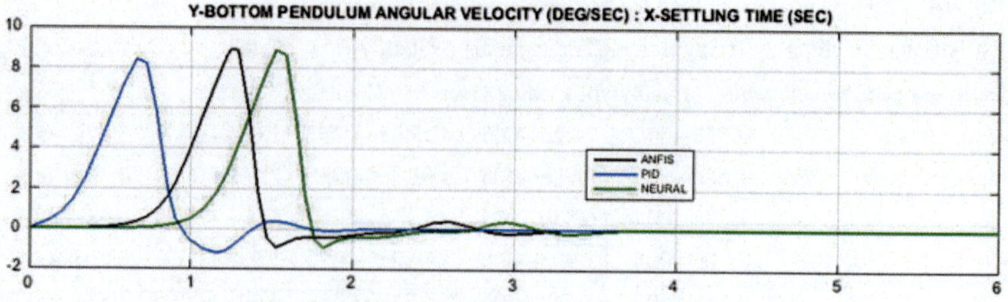

Fig. 9 Bottom pendulum angular velocity response using different controllers

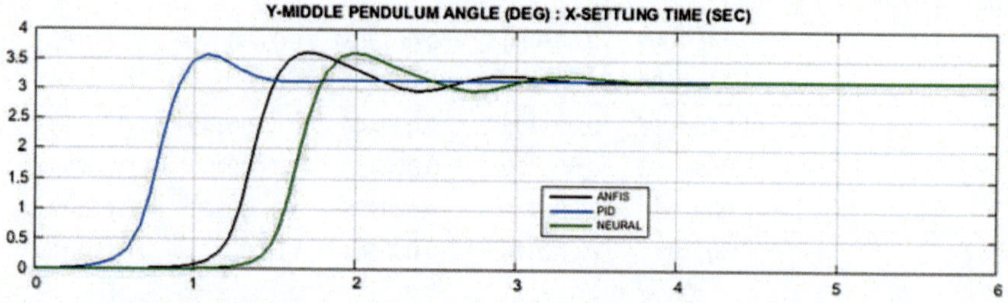

Fig. 10 Middle pendulum angle response using different controllers

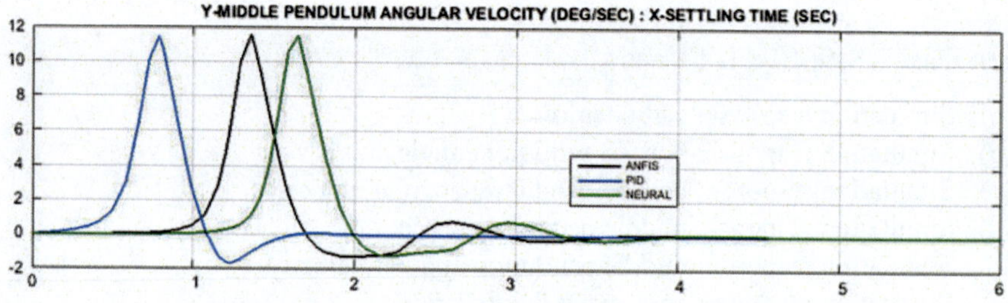

Fig. 11 Middle pendulum angular velocity response using different controllers

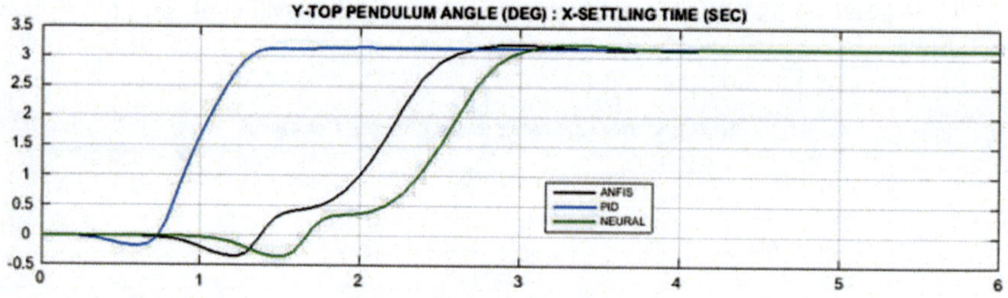

Fig. 12 Top pendulum angle response using different controllers

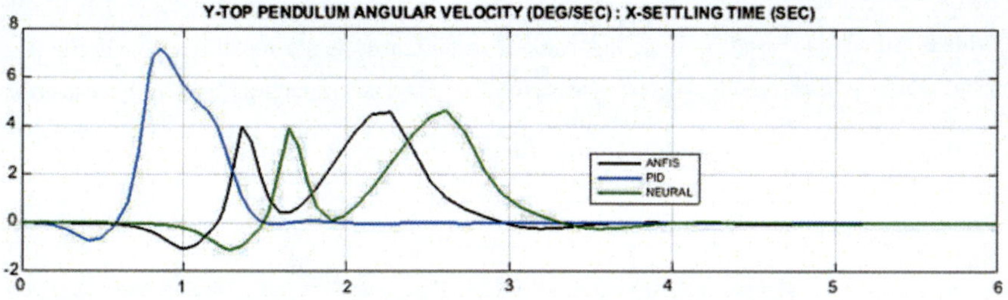

Fig. 13 Top pendulum angular velocity response using different controllers

Table 5 Simulation results obtained using ANFIS controller in study by Jun-wei et al.

Sub-system	Settling time (s)	Overshoot ranges (°)	Steady state error
Top pendulum	5.0	0.57° to −2.57°	0
Middle pendulum	4.67	0.57° to −1.38°	0
Bottom pendulum	6.0	0.57° to −1.60°	0
Cart position	5.0	NA	0

The simulation results obtained for sub-systems of three-stage cart and pole system by means of various techniques are given in Table 6.

7 Conclusion

This chapter considered different controllers based on results of fuzzy and PID control techniques for control of three-link cart and pendulum system. The outcomes of fuzzy controller were adopted to teach and train ANFIS, whereas the results of PID controller were used for training of both ANFIS and neural controllers. The results clearly proved superior response of PID-based control compared to fuzzy-based control. The fuzzy control rules can be further tuned to get better results. The simulation results for fuzzy-based ANFIS controller can be further improved by reducing the value of error tolerances either by increasing the total memberships or by changing its shape. The time required to stabilize complete system using ANFIS and neural control was 3.4 s and 4.0 s, respectively. The maximum overshoot responses for both the controllers were comparatively similar as can be inferred from the results. It is also conceived from the responses that both fuzzy and PID-based control showed excellent response toward steady-state error. The simulation results obtained for triple inverted pendulum system using PID-based ANFIS, and neural controllers were further compared to the results of ANFIS controller developed by Jun-wei et al. [20]. The comparison of results showed that the settling time taken by cart

Table 6 Simulation results of three-link cart and pendulum system using different controllers

Controller	Settling time (s)	Improvement in settling time (%) over ANFIS controller by Jun-wei et al.	Overshoot ranges (°)	Steady state error
Top pendulum angle				
PID	1.4 s	72%	3.2° to −0.2°	0
ANFIS	2.8 s	44%	3.25° to −0.3°	0
Neural	3.0 s	40%	3.25° to −0.3°	0
Top pendulum angular velocity				
PID	1.5 s	NA	7.2° to −0.8°	0
ANFIS	3.0 s	NA	4.6° to −1.1°	0
Neural	3.5 s	NA	4.5° to −1.2°	0
Bottom pendulum angle				
PID	1.4 s	72%	3.4°	0
ANFIS	2.0 s	60%	3.4°	0
Neural	2.2 s	56%	3.4°	0
Bottom pendulum angular velocity				
PID	1.8 s	NA	8.5° to −1.2°	0
ANFIS	3.0 s	NA	9.0° to −1.0°	0
Neural	3.2 s	NA	9.0° to −1.0°	0
Middle pendulum angle				
PID	1.5 s	70%	3.6°	0
ANFIS	3.4 s	32%	3.7°	0
Neural	3.7 s	26%	3.7°	0
Middle pendulum angular velocity				
PID	1.7 s	NA	11.5° to −1.8°	0
ANFIS	3.4 s	NA	11.6° to −1.4°	0
Neural	3.8 s	NA	11.6° to −1.3°	0
Cart position				
PID	1.5 s	70%	0.78° to −0.2°	0
ANFIS	3.2 s	36%	0.48° to −0.5°	0
Neural	4.0 s	20%	0.48° to −0.5°	0

and pendulum sub-systems were significantly reduced using PID-based ANFIS and neural controllers. The maximum improvement in settling time of top, middle, bottom pendulums and cart position was obtained using PID controller and was 72%, 72%, 70% and 70%, respectively.

References

1. Furut K, Ochiai T, Ono N (1984) Attitude control of a triple inverted pendulum. Int J Control 39(6):1351–1365. https://doi.org/10.1080/00207178408933251
2. Graichen K, Zeitz M (2006) A new approach to feedforward control design under output constraints applied to the side-stepping of the triple inverted pendulum. IFAC Proc 39(16):181–186. https://doi.org/10.3182/20060912-3-DE-2911.00034
3. Eltohamy KG, Kuo CY (1999) Nonlinear generalized equation of motion for multi-link inverted pendulum systems. Int J Syst Sci 30(5):505–513. https://doi.org/10.1080/002071798222811
4. Medrano-cerda GA, Eldukhri EE, Cetin M (1995) Balancing and attitude control of double and triple inverted pendulums. Trans Inst Meas Control 17(3):143–154. https://doi.org/10.1177/014233129501700306
5. Ruisen L, Liang L (1998) Intelligent control on three-stage inverted pendulums. J Shanghai Univ 2(4):284–289. https://doi.org/10.1007/s11741-998-0041-9
6. Gluck T, Eder A, Kugi A (2012) Swing-up control of a triple pendulum on a cart with experimental validation. Automatica 49(3):801–808. https://doi.org/10.1016/j.automatica.2012.12.006
7. Ananyevskiy I, Anokhin N (2012) Control of a multi-link inverted pendulum by a single torque. IFAC Proc 45(2):550–553. https://doi.org/10.3182/20120215-3-AT-3016.00096
8. Prasad LB, Tyagi B Gupta HO (2014) Optimal control of nonlinear inverted pendulum system using PID controller and LQR: Performance analysis without and with disturbance input. Int J Autom Comput 11(6):661–670. https://doi.org/10.1007/s11633-014-0818-1
9. Zhang XL, Fan HM, Zhang JY, Zhao L, Hao S (2015) Nonlinear control of triple inverted pendulum based on GA-PIDNN. Nonlinear Dyn 79(2):1185–1194. https://doi.org/10.1007/s11071-014-1735-0
10. Anan'evskii IM (2018) The control of a three link inverted pendulum near the equilibrium point. Mech Solids 53(1):516–521. https://doi.org/10.3103/S0025654418030020
11. Huang X, Wen F, Wei Z (2018) Optimization of triple inverted pendulum control process based on motion vision. EURASIP J Image Video process 73:1–8. https://doi.org/10.1186/5/3640-018-0294-6
12. Arkhipora IM (2019) On stabilization of a triple inverted pendulum via vibration of a support point with an arbitrary frequency. Vestnik St. Petersberg University, Mathematics 52(2):194–198. https://doi.org/10.1134/s1063454119020031
13. Eltohamy KG, Kuo CY (1998) Nonlinear control of a triple link inverted pendulum with single control input. Int J Control 69(2):239–256. https://doi.org/10.1080/002071798222811
14. Lee KN, Yeo YK (2010) A new predictive PID controller for the processes with time dealy. Korean J Chem Eng 26(5):1194–1200. https://doi.org/10.1007/s11814-009-0194-7
15. Koryukin AN, Voevoda AA (2015) PID controllers of some two-mass system and the double complex pairs. J Appl Ind Math 9(2):215–226. https://doi.org/10.1134/s199047915020076
16. Liang XM, Li SC, Hassan AB (2010) A novel PID controller tuning method based on optimization technique. J Cent South Univ Technol 17(5):1036–1042. https://doi.org/10.1007/s11771-010-0595-0
17. Baxter G, Srisaeng P (2018) The use of an artificial neural network to predict Australia's export air cargo demand. Int J Traffic Transp Eng 8(1):15–30. https://doi.org/10.7708/ijtte.2018.8(1).02
18. Kamil F, Hong TS, Khaksar W, Zulkifli N, Ahmad SA (2019) An ANFIS based optimized fuzzy multilayer decision approach for a mobile robotics systems in ever changing environment. Int J Control Autom Syst 17(1):253–266. https://doi.org/10.1007/s12555-017-0068-4
19. Okwu MO, Adetunji O (2018) A Comparative study of artificial neural network (ANN) and adaptive neuro fuzzy inference system (ANFIS) models in distribution system with nondeterministic inputs. Int J Eng Bus Manag 10(1):1–17. https://doi.org/10.1177/184797901187 68421

20. Al-Hmouz A, Shen J, Al-Hmouz R, Yan J (2012) Modeling and simulation of an adaptive neuro-fuzzy inference system (ANFIS) for mobile learning. IEEE Trans Learn Technol 5(3):226–237
21. Jun-wei G, Guo-qiang C, Zhi-jian J, Yong Q, Li-min J (2010) Adaptive neural fuzzy control of triple inverted pendulum. Control Theory Appl 27(2):278–282

Millimeter Wave HMSIW-Based Beam Steerable Leaky Wave Antenna for 5G Systems

Belwal Pravesh, Agrawal Rahul, Nigam Padmini, and S. C. Gupta

1 Introduction

Leaky wave antenna (LWA) is a kind of traveling wave antenna which has a use in various wireless communication system due to its appealing features of frequency dependent beam scanning capability and high radiation gain [1–8]. LWA is mainly classified into two groups: uniform LWA and periodic LWA. Uniform LWA can radiate in only one quadrant, whereas periodic LWA can radiate in backward as well as in forward direction.

Substrate integrated waveguide (SIW) has been grown as a good candidate to developing LWA in microwave and also millimeter wave due to its advantages like high efficiency and gain, low loss, high quality factor, and is possible to easily control their radiation pattern [3].

Since the introduction of half mode substrate integrated waveguide (HMSIW), it has gain a keen interest of researchers [9]. The HMSIW structure has attracted researcher in the field of antenna due to its two interesting features. First, the antenna size is reduced almost by half in comparison to the SIW, still maintaining most of the SIW's advantages, like ease in fabrication, low losses, low cost, and easy integration into/onto structures. Secondly, unlike the closed structure of SIW, the HMSIW is a open structure in one side which possesses an aperture by its own nature [10–12]. In this paper, a miniaturized millimeter wave LWA based on HMSIW is presented. The HMSIW LWA supports continuous beam steering (CBS) from −36° to +20° in the wide frequency range of 31–45 GHz including broadside radiation at 40 GHz.

B. Pravesh (✉) · A. Rahul · N. Padmini · S. C. Gupta
DIT University, Dehradun 248001, India
e-mail: pravesh.belwal@gmail.com

V. H. Saran and R. K. Misra (eds.), *Advances in Systems Engineering*,
Lecture Notes in Mechanical Engineering,
https://doi.org/10.1007/978-981-15-8025-3_50

2 Unit Cell Design

The structure of HMSIW LWA shown in Fig. 1. The substrate used here is Rogers 5880, with the relative permittivity of 2.2 and the height of 0.787 mm. The periodicity or unit length is equal to $p = 5.15$ mm. The slots are placed on top surface to leak the radiation in open space. HMSIW LWA structure has metal on both side with substrate sandwiched in between and the one-dimensional arrays of copper vias adjusted in the direction that of wave propagating at only one side of the antenna as shown in Fig. 1. To completely remove stop band, the slots are placed at a quarter distance apart in order to cancel out the reflections of each other. The parameters are marked in Fig. 1 and the values of the parameters are given in Table 1. The length of the slots is $\sim \lambda_0/8$ and width is taken $\sim \lambda_0/16$.

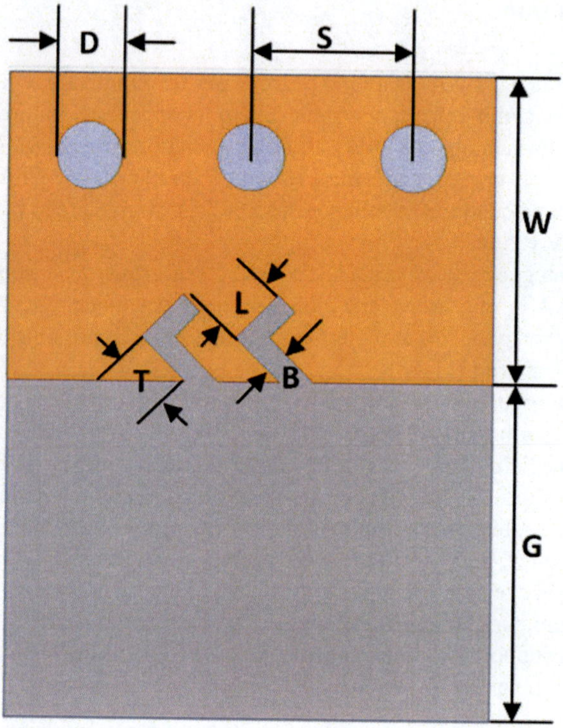

Fig. 1 HMSIW LWA unit cell

Table 1 Unit cell parameters

Parameter	Value (mm)
D	0.7
S	1.71
L	0.625
T	0.66
B	0.25
W	3.2
G	3.51

Infinite numbers of spatial harmonics are generated by the periodic LWA and the phase constant is given by [13].

$$\beta_n = \beta + 2n\pi/p \tag{1}$$

where β is fundamental phase constant of antenna structure. In presented LWA, the (spatial harmonic) $n = -1$ support fast wave. Therefore, the (phase constant) $|\beta - 1|$ contributes for radiations of LWA.

Figure 2 shows the normalized dispersion diagram of the unit cell. As seen, the (phase delay) $|\beta_{\text{eff}}.p|$ is found decreasing in the frequency range of 31–39.5 GHz and increases as the frequency increases from 39.5 to 45 GHz indicating that the antenna can beam scan in backward and forward direction. The transition of the backward direction to forward direction occurs at 39.5 GHz, having effective phase

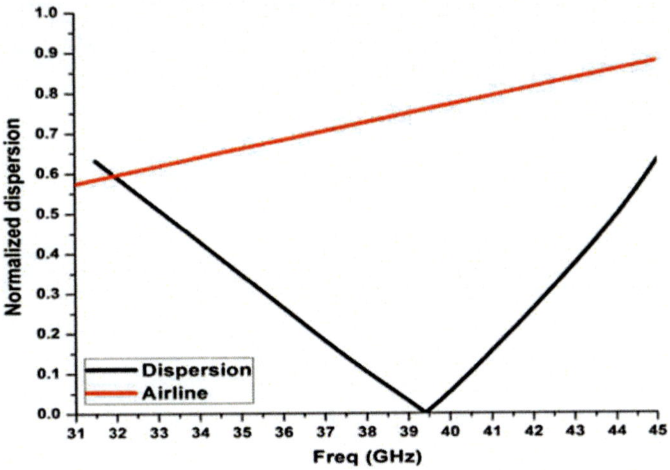

Fig. 2 Normalized dispersion diagram of the unit cell HMSIW LWA

delay $|\beta_{\mathrm{eff}}.p| = 0$ indicating the radiation in broadside direction. The main beam direction of LWA is given by.

$$\beta_n = \sin^{-1}\left(\frac{\beta_{\mathrm{eff}}}{k}\right) \tag{2}$$

where k is propagation constant of antenna structure in free space. The effective phase delay of unit cell is determined by [5].

$$\beta_{\mathrm{eff}} = \frac{1}{p}\mathrm{Re}\left[\cos^{-1}\left(\frac{1 - S_{11}S_{22} + S_{21}S_{12}}{2S_{21}}\right)\right] \tag{3}$$

3 Full Wave Analysis

The 8 unit cell is combined to form an array; total length of the array is 41.2 mm. Figure 3 shows the array of HMSIW LWA. The full wave analysis of the antenna is performed using ANSYS HFSS.

The simulated reflection coefficients $S(1, 1)$ of the HMSIW LWA are given in Fig. 4. It is seen that $S(1, 1)$ is below -10 dB in the operating frequency band and

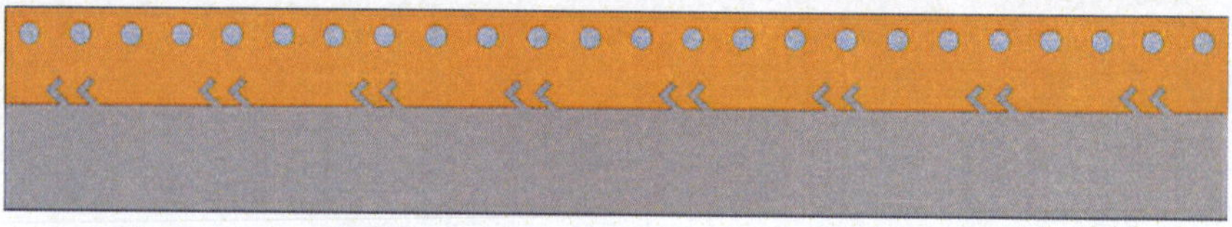

Fig. 3. 8 unit cell array of presented HMSIW LWA

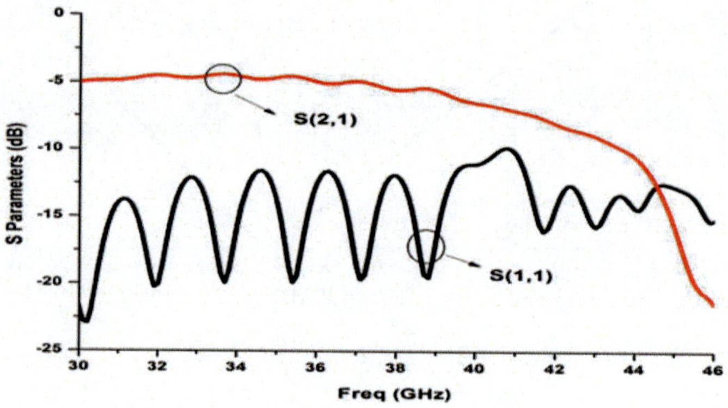

Fig. 4 *S* parameters of the presented HMSIW LWA

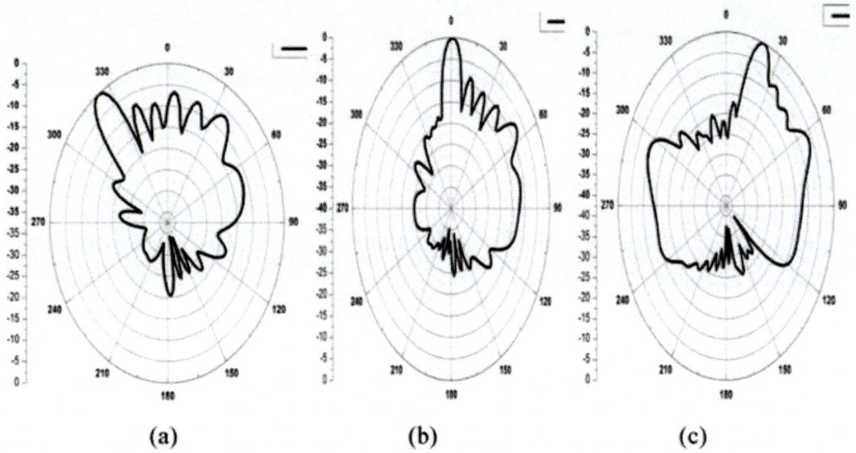

Fig. 5 Normalized radiation pattern at different frequencies **a** 31 GHz, **b** 40 GHz, **c** 45 GHz

$S(2, 1)$ is equal to −5 dB, illustrating that the good impedance matching is achieved.

The simulated normalized radiation patterns of LWA are given in Fig. 5. The main lobe in radiation pattern demonstrates the CBS from backward to forward when the frequency is increased. CBS is well achieved in the region from −36° to 23° when the operating frequency is increased from 31 to 45 GHz.

Figure 6 shows the simulated gain of the presented LWA. The simulated gain of the LWA gets its top at 14.3 dB and achieves 13.45 dB at 40 GHz (broadside direction) (Table 2).

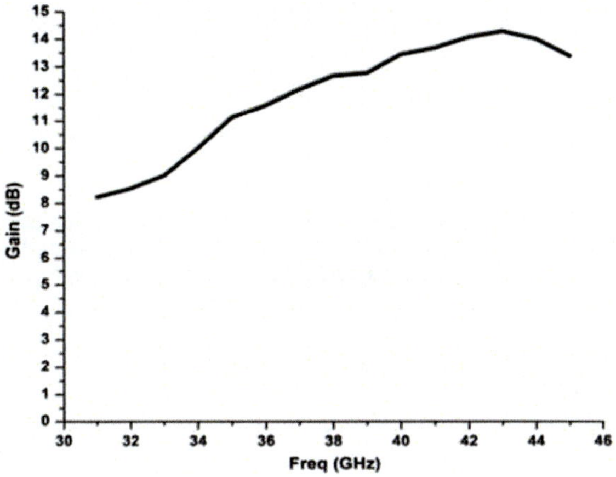

Fig. 6 Simulated gain of the HMSIW LWA

Table 2 Comparison with previously published work

References	Frequency band (GHZ)	Beam scanning (°)	Broadside	Max gain (dBi)	Antenna type
[14]	28–31	16	No	10.14	Liquid crystal
[15]	28.5	65	No	6	Varactor diode
[16]	25–28	9	No	11.7	HMSIW
[17]	26	16	No	10	Pin diode
This work	31–45	56	Yes	14.5	HMSIW

4 Conclusion

The HMSIW LWA presented here has compact in size, wide bandwidth, and CBS capability. Here, stop band has been removed by placing the slots quarter distance apart so that the reflection caused by one slot can be canceled by another slot. The dispersion diagram is studied to demonstrate and radiation patterns is also studied to verify the CBS capability from backward to forward. CBS is achieved from −36° to 20° when the operating frequency is varied from 31 to 45 GHz. The presented HMSIW LWA can be used in 5G networks of wireless mobile communication on base station tower. It will help in further improving the 5G networks over 4G networks.

Acknowledgements This work was carried at DRDO Lab at DIT University, Dehradun, India. The DRDO setup Lab was supported under grant head number ERIP/ER/1403170/M/01/1555.

References

1. Xu F, Wu K (2013) Understanding leaky-wave structures: a special form of guided-wave structure. IEEE Microw Mag 14(5):87–96. https://doi.org/10.1109/MMM.2013.2259400
2. Jackson DR, Caloz C, Itoh T (2012) Leaky-wave antennas. Proc IEEE 100(7):2194–2206. https://doi.org/10.1109/JPROC.2012.2187410
3. Xu F, Wu K (2005) Guided-wave and leakage characteristics of substrate integrated waveguide. IEEE Trans Microw Theory Tech 5(1):66–73. https://doi.org/10.1109/TMTT.2004.839303
4. Liu J, Jackson DR, Long Y (2011) Substrate integrated waveguide (SIW) leaky-wave antenna with transverse slots. IEEE Trans Antennas Propag 60(1):20–29. https://doi.org/10.1109/TAP.2011.2167910
5. Yue-Long Lyu X-XL, Wang P-Y, Erni D et al (2016) Leaky-wave antennas based on non-cutoff substrate integrated waveguide supporting beam scanning from backward to forward. IEEE Trans Antennas Propag 64(6):2155–2164. https://doi.org/10.1109/TAP.2016.2550054
6. Belwal P, Agrawal R, Gupta SC (2018) Substrate integrated waveguide leaky wave antenna with low cross polarization in X-Ku band. Int J RF Microw Comput Aided Eng (28). https://doi.org/10.1002/mmce.21590
7. Agrawal R, Belwal P, Gupta SC (2018) Asymmetric substrate integrated waveguide leaky wave antenna with open stop band suppression and radiation efficiency through broadside. Radioengineering 27:409–416. https://doi.org/10.13164/re.2018.0409

8. Agrawal R, Belwal P, Singh M, Gupta SC (2017) Continuous beam scanning in substrate integrated waveguide leaky wave antenna. Prog Electromag Res M 62:19–28. https://doi.org/10.2528/pierm17091104
9. Hong W, Liu B, Wang Y, Lai Q, Tang H, Yin XX, Dong YD, Zhang Y, Wu K (2006) Half mode substrate integrated waveguide: a new guided wave structure for microwave and millimeter wave application. In: Proceedings Joint 31st international infrared millimeter waves conference/14th international Terahertz electronic conference 2006 (IRMMW-THz 2006), p 219. https://doi.org/10.1109/ICIMW.2006.368427
10. Dong YD, Itoh T (2011) Composite right/left-handed substrate integrated waveguide and half mode substrate integrated waveguide leaky-wave structures. IEEE Trans Antennas Propag 59(3):767–775. https://doi.org/10.23919/EUMC.2009.5295967
11. Belwal P, Agrawal R, Gupta SC (2018) A half mode substrate integrated waveguide leaky wave antenna with backward forward and broadside beam scanning. In: 2018 IEEE Indian conference on antennas and propogation (InCAP), Hyderabad, India, 2018, pp 1–4. https://doi.org/10.1109/INCAP.2018.8770826
12. Liu L, Gu X, Zhu L, Rong Y, Qian H (2018) A novel half mode substrate integrated waveguide leaky-wave antenna with continuous forward-to-backward beam scanning functionality. Int J RF Microw Comput Aided Eng e21559. https://doi.org/10.1002/mmce.21559
13. Jackson DR, Oliner AA (2008) Leaky-wave antennas. In: Balanis CA (ed) Modern Antenna Handbook. Willey, New York, NY, USA. ISBN: 978-0-470-03634-1
14. Fu Z, Jiang D, Liu Y (2019) Miniaturized pattern reconfigurable HMSIW leaky wave antenna based on liquid crystal tuning technology in millimeter wave band. In: 2019 IEEE MTT-S international wireless symposium (IWS), Guangzhou, China, 2019, pp 1–3. https://doi.org/10.1109/IEEE-IWS.2019.8804055
15. Javanbakht N, Syrett B, Amaya RE, Shaker J (2019) Miniaturized reconfigurable antenna based on half-mode substrate integrated waveguide. In: IEEE International Systems Conference (SysCon). Orlando, FL, USA 2019, pp 1–4. https://doi.org/10.1109/SYSCON.2019.8836800
16. Xu J, Hong W, Tang H, Kuai Z, Wu K (2008) Half-Mode Substrate Integrated Waveguide (HMSIW) Leaky-Wave Antenna for Millimeter-Wave Applications. IEEE Antennas Wirel Propag Lett 7:85–88. https://doi.org/10.1109/LAWP.2008.919353
17. Ali SMA, Ahmed Z, Ihsan MB (2017) Digital beam scanning in HMSIW millimeterwave leaky wave antenna. In: 2017 24th international conference on mechatronics and machine vision in practice (M2VIP), Auckland, 2017, pp 1–3. https://doi.org/10.1109/M2VIP.2017.8211431

The Substrate Integrated Waveguide-Based Dumbbell-Shaped Self-diplexing Cross-Slot Antenna

Padmini Nigam, Arjuna Muduli, Sandeep Sharma, and Amrindra Pal

1 Introduction

The wireless communication now becomes the more essential part of our day to day life. For this previously, the multiple antennas were used [1]. Due to the use of multiple antennas, the cost as well as space of overall system is increased also. Because of all these drawbacks, the multiple antennas are replaced with multiband antennas, which consumed the small space, minimum frequency ratio, and compact in size also. This kind of multiband antenna is mostly required in satellite communication systems, RFID's, and narrowband Internet of things (IoT) modules. The multiband frequency antennas are usually used the single-layered patch and this turned into a compact size of transceiver systems, but a kind of the complex geometry of frequency decoupling circuit is also attached with it [2, 3].

Diplexers and triplexers are generally a type of the frequency selective or decoupling element, which is the essential part of multiband antennas to provide the higher intrinsic isolation between input ports into adjacent channels of communication [4, 5]. Nowadays, self-diplexed and self-triplexed-based antenna come under consideration due to its highly demand, compact size, and high isolation levels between input ports. In which, the diplexing and triplexing functions are performed by antenna itself.

P. Nigam (✉) · A. Pal
DIT University, Dehradun 248009, India
e-mail: nigam242@gmail.com

A. Pal
e-mail: amrindra.pal123@gmail.com

A. Muduli
K.L. Education Foundation, Guntur, Andhra Pradesh, India

S. Sharma
Centre for Reliability Sciences and Technology, Department of Electronic Engineering, Chang Gung University, Taoyuan 33302, Taiwan

V. H. Saran and R. K. Misra (eds.), *Advances in Systems Engineering*, Lecture Notes in Mechanical Engineering,
https://doi.org/10.1007/978-981-15-8025-3_51

Previously, the self-diplexing and self-triplexing antenna are based on the micro-strip technology with additional methodology like as multi-layered structures [6], electromagnetic band gap structures [7], stub-loaded resonator-based [8], and defected ground structure [9]. This kind of self-diplexed antenna has faced the problem of design complexity and higher fabrication cost. Recently, the SIW becomes the most prominent and outstanding technology for RF and microwave-based antenna for its dynamic features of planar nature, miniaturization, low losses, and lightweight used for millimeter-wave range communication systems [10, 11].

The SIW-based self-diplexing [12–18] and self-triplexing [19, 20] antennas are mostly used now because of its various merits like compact space, lightweighted, low insertion loss, and easy to fabricate with planar structures. The proposed self-diplexed dumbbell-shaped cross-slot antenna is demonstrated in this paper. The proposed antenna has the property of dual higher-order mode excitation and produces the self-isolation minimum of 21 dB at both the resonant frequencies 8.1 and 10.7 GHz.

The proposed antenna gives the uni-directional radiation pattern, minimum cross-polarization levels, and gain of 3.9 and 4.4 dBi at both working frequencies, respectively. This type of antennas also become the most interesting topic for the RF researchers and microwave antennas designers. The proposed research paper is divided into four sections. Section 2 defines antenna geometry and its analysis, Sect. 3 describes the proposed antenna results and its analysis, and at last, Sect. 4 explained about the conclusion of it.

2 Antenna Design and Operating Principle

The proposed self-diplexing antenna layout for the *X*-band applications is shown in Fig. 1. The proposed antenna comprises of SIW cavity as the four boundary walls of the conducting vias, which is used to provide the metallic connection between the upper (patch) and ground plane plate. The value of pitch (s) and diameter (d) is elected very precisely to follow the condition of the $d/s \geq 0.5$ [10, 11], by this equation, the leakage of energy and radiation losses can be minimized also.

2.1 *Antenna Working Principle*

The working phenomenon of the proposed antenna is explained by Z-parameters and visualized by the magnetic field distribution of dominating mode and perturbation of higher-order modes inside it. The |Z|-parameter is generally used to describe the two-port networks and electrically performance of linear elements. It is mostly provided the excellent quality factor and narrower bandwidth of the antenna. Where the Z_{11} and Z_{22} is self-impedance and Z_{12}/Z_{21} is mutual impedance of antenna. This is the leading requirement of the proposed self-diplexing antenna [21]. The magnetic field distribution on the upper side of antenna is used to easily understand

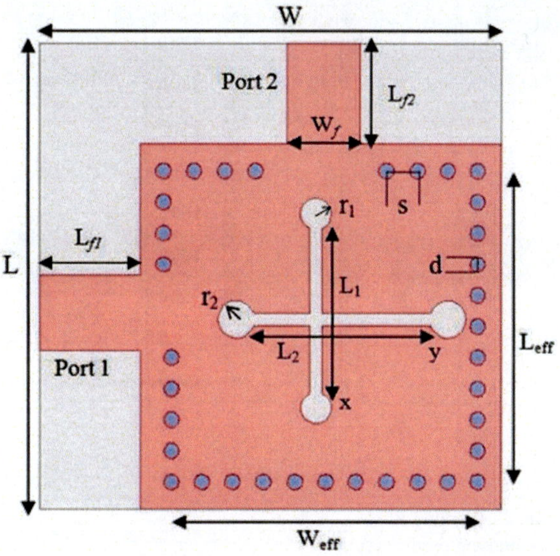

Fig. 1 Proposed self-diplexer antenna geometry. [$L = 30$, $W = 30$, $L_{\text{eff}} = 20.5$, $W_{\text{eff}} = 20.5$, $L_1 = 11.5$, $L_2 = 12.7$, $W_f = 4.8$, $L_{f1} = 6.5$, $L_{f2} = 6.2$, $s = 2$, $d = 1$, $r_1 = 1$ & $r_2 = 1.2$ (all in mm.)]

the visualization of the propagating modes. The propagating modes are achieved due to the perturbation of higher-order modes as in terms of loading effect of reactive elements.

This working methodology is divided into two parts. Firstly, the SIW cavity has no slots, and then secondly, the SIW cavity with two dumbbell-shaped cross-slots. These all are discussed in below paragraphs one by one. These two dumbbell-shaped cross-slots are of different length of L_1 and L_2, which produces the two different resonant frequencies as 8.1 and 10.7 GHz.

2.2 *The SIW Cavity working*

The working phenomenon of the SIW cavity without slots is described with the help of |Z|-parameters and vector magnetic field distribution on the top of the patch of the antenna. The dominating mode TE110 is operating at the 6.7 GHz frequency and it is justified by magnetic field distribution as shown in Figs. 2 and 3.

Here, the most important concern about numerical aspect of the proposed SIW technique-based self-diplexing antenna is given below. The effective width and absolute width of SIW-based antenna is given in Eqs. 1 and 2, while the dominant frequency is evaluated by Eq. 3 [10, 11].

$$w_{\text{eff}} = w - \frac{\text{d}^2}{0.95s} \text{or, } w_{\text{eff}} = w - 1.08 * \frac{\text{d}^2}{s} + 0.1\frac{\text{d}^2}{w} \quad (1)$$

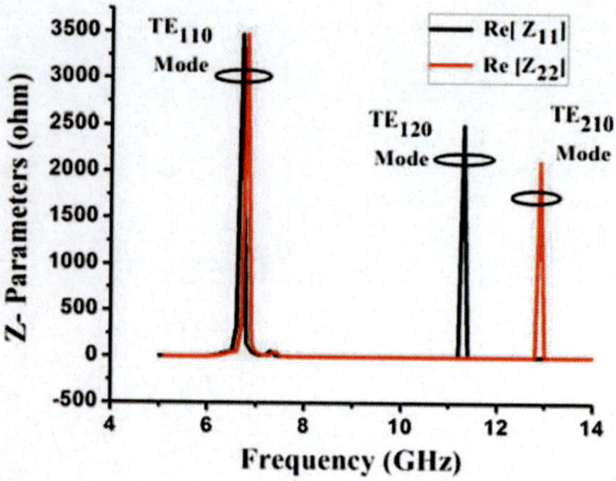

Fig. 2 *Z*-parameters of the SIW cavity

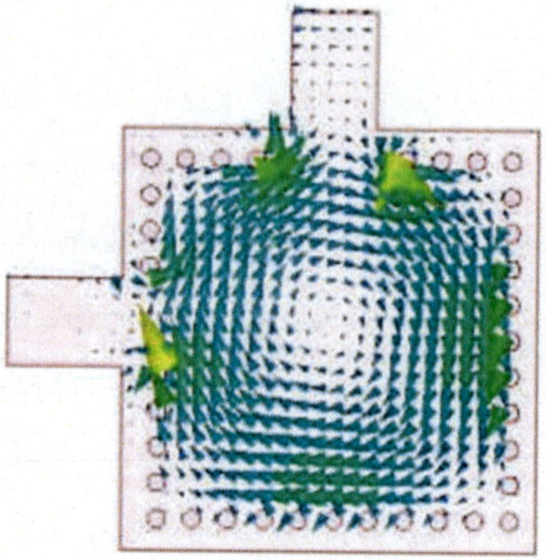

Fig. 3 Working of dominating mode TE110 inside the SIW cavity

$$w = \frac{2w_{\mathrm{eff}}}{\pi}\cot^{-1}\left[\frac{\pi s}{4w_{\mathrm{eff}}}\ln\frac{s}{2d}\right] \quad (2)$$

$$F110 = \frac{c}{2\sqrt{\mu_r\varepsilon_r}}\sqrt{\frac{\pi^2}{\mathrm{Weff}^2}+\frac{\pi^2}{\mathrm{Leff}^2}} \quad (3)$$

where w_{eff} = effective width of waveguide, d = diameter of metallic cylinder or vias, s = pitch or longitudinal gapping, as shown in Fig. 1.

The SIW cavity is also having the two higher-order modes of TE120 and TE210 at the frequency of 11.3 and 12.8 GHz. This is shown in Fig. 2 of |Z|-parameters of the proposed SIW cavity and Figs. 4 and 5 of the magnetic field distribution of TE120 mode and TE210 mode are as described below. Here, the proposed antenna

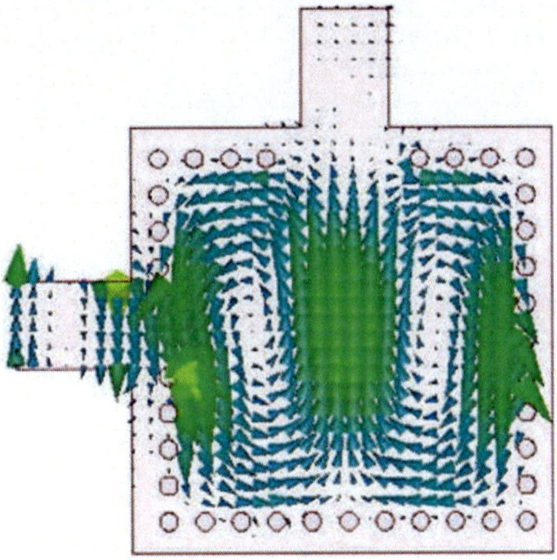

Fig. 4 Vector magnetic field distribution of higher-order mode TE120 when port 1 is ON

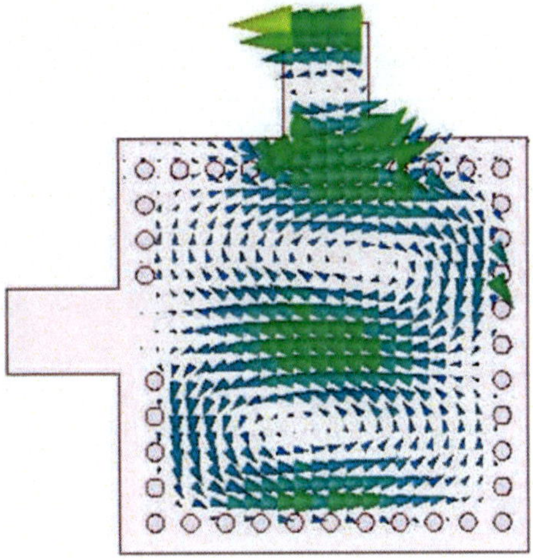

Fig. 5 Vector magnetic field distribution of higher-order mode TE210 when port 2 is ON

configuration depends on the perturbation of the higher-order mode as TE120 and TE210 mode of the SIW cavity. As when the port 1 is excited, the dominating mode at 6.7 GHz is found. Due to the perturbation of the orthogonal dumbbell slot-1 of length L_1, the higher-order mode of SIW cavity is obtained at 11.3 GHz gets to move from 11.3 to 8.1 GHz. This becomes the new radiating mode as the first resonating frequency as F_1.

When the port 2 is ON, the second higher-order mode TE210 is found at 12.8 GHz within the SIW cavity. The perturbation of transverse dumbbell slot-2 leads the shifting from higher-order mode TE210 at 12.8–10.7 GHz. This is now working as the second radiating mode at frequency F_2 of 10.7 GHz. These all two resonating modes are ready to radiate into free space. This is defining by the Z-parameters graph as shown in Fig. 6 and its magnetic field distribution diagram define in Figs. 7 and 9.

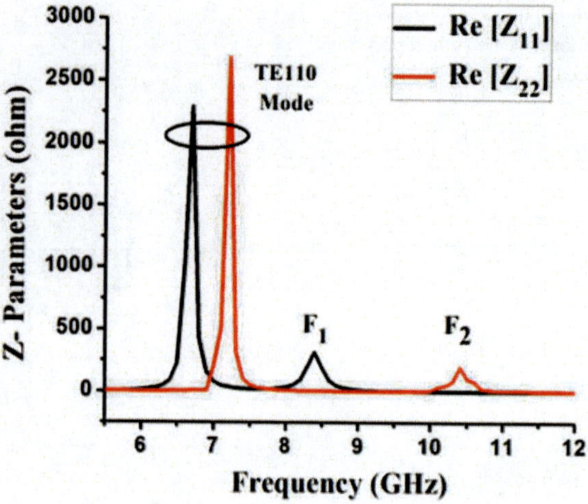

Fig. 6 *Z*-parameters of the proposed antenna

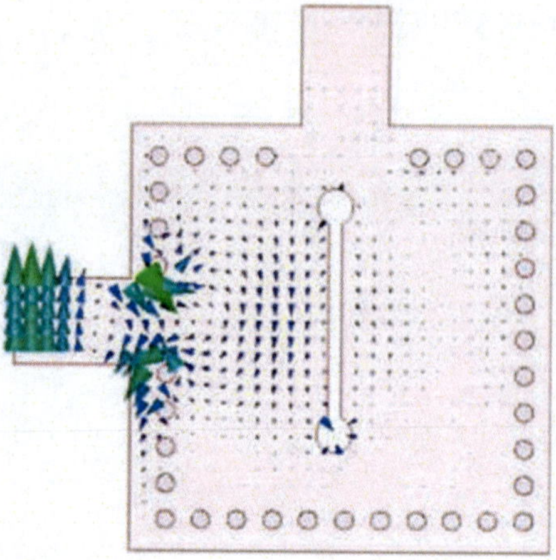

Fig. 7 Antenna radiates through slot-1 when port 1 is ON

Figure 7 defines that when port 1 is excited, the slot-1 radiates at 8.1 GHz which is the first radiating mode of the proposed antenna. Here, the only one slot as slot-1 takes the part in radiation at first resonating frequency. The field distribution of current takes part around the slot-1 only, having the no effect at the vicinity of slot-2. Here, the current distribution is about to zero or minimum as given in Fig. 8. When the port 2 is ON, slot-2 gets excited and current distribution finds out near the area of slot-2 only which is justified with the help of current distribution around slot-2 as shown in Fig. 9.

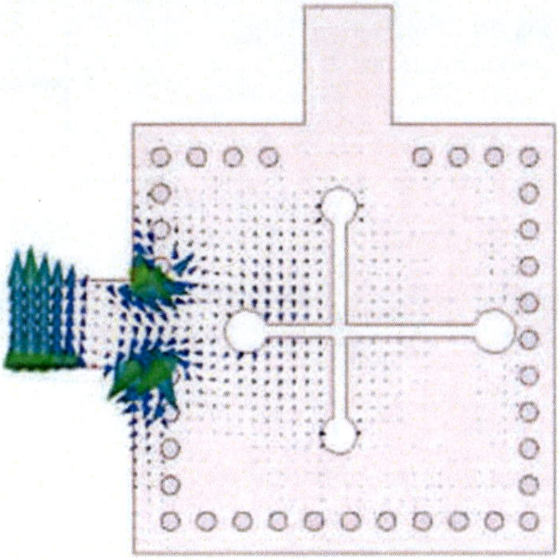

Fig. 8 When port 1 is ON only slot-1 is excited, slot-2 is unchanged

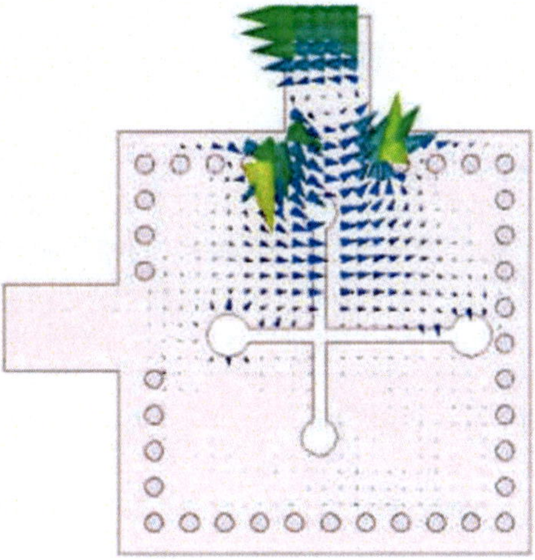

Fig. 9 Antenna radiates through slot-2 when port 2 is ON

3 Results Analysis

The proposed antenna is composed of Rogers RT/Duroid 5880 having permittivity $\varepsilon_r = 2.2$, thickness $h = 1.57$ mm, and loss tangent $\tan\delta = 0.0009$. The $|S|$-parameters of the proposed antenna is defined in Fig. 10. Where the value of the reflection coefficients ($|S_{11}|$ and $|S_{22}| \geq -20$ dB) is good and the isolation (S_{21}) is the minimum of 21 dB at port 2 and maximum of 50 dB across port 1. This is proved that the proposed antenna has the unique property of the self-diplexing to produce the intrinsic isolation between port 1 and port 2. By $|S|$-parameters, it is justified that the proposed antenna is self-diplexer. It is also having the low cross-polarization levels about ≥ -20 dB. This defines that the proposed antenna has the maximum gain in one direction without any more side lodes as leakage of energy. So, the proposed

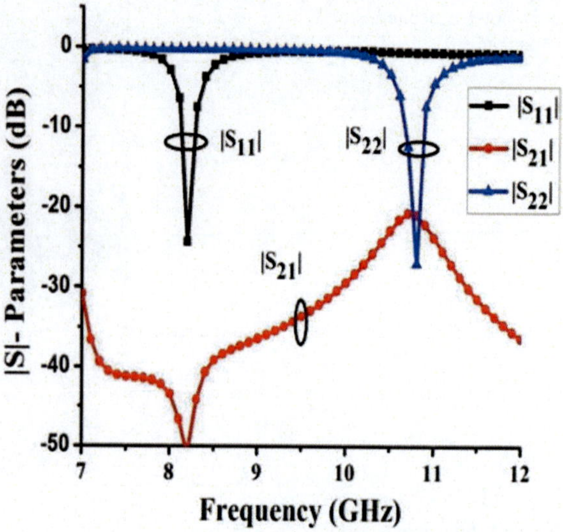

Fig. 10 |*S*|-parameters of the proposed antenna

antenna has a uni-directional radiation pattern and an average gain of 3.9 and 4.4 dBi as shown in Fig. 11. The value of front to back ratio (FTBR) of the proposed antenna is about greater than 14 dB. The proposed antenna has the normalized radiation efficiency of 67% at the both resonating frequencies as shown in Fig. 12. The comparison among self-diplexing antennas are also shown in tabular format as Table 1.

4 Conclusion

The proposed simple and compact layout of the self-diplexing antenna is designed with the help of the SIW technology is defined in this paper. Here, two dumbbell-shaped cross-slots of different lengths are etched over the cavity and fed by two orthogonal micro-strip lines of 50-Ω impedance. These two orthogonal slots are produced the two resonating frequencies at 8.1 and 10.7 GHz. The working methodology of the proposed self-diplexing antenna depends on the dominating mode and the perturbation of higher-order mode into resonating mode due to the loading effect. The proposed antenna is having the great feature of the self-diplexing with the minimum isolation of 21 dB, which eliminates the space of diplexer and triplexer circuits in RF modules. Therefore, the size and cost of RF and satellite transceiver modules are reduced, which is the extreme necessity of the transceiver systems. The proposed self-diplexing circuitry is planar in nature by which the ground plane can be used for integrated system. Here, the system in package (SiP) technology tends toward system on substrate (SoS) technology to integrate with more than one modules. It is also having the good value of front to back ratio (FTBR) as 14 dB and low cross-polarization as below of −20 dB is attained. The gain of the proposed antenna is

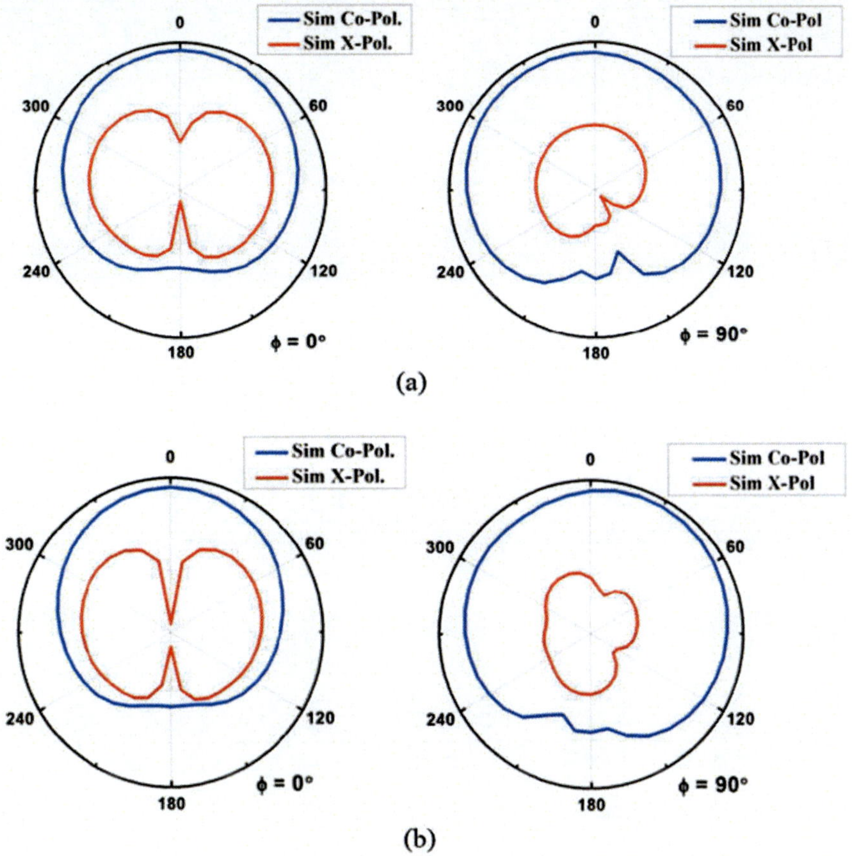

Fig. 11 2D radiation pattern of the proposed antenna, **a** when port 1 is ON, **b** when port 2 is ON

Fig. 12 Normalized radiation efficiency of the proposed self-diplexing antenna

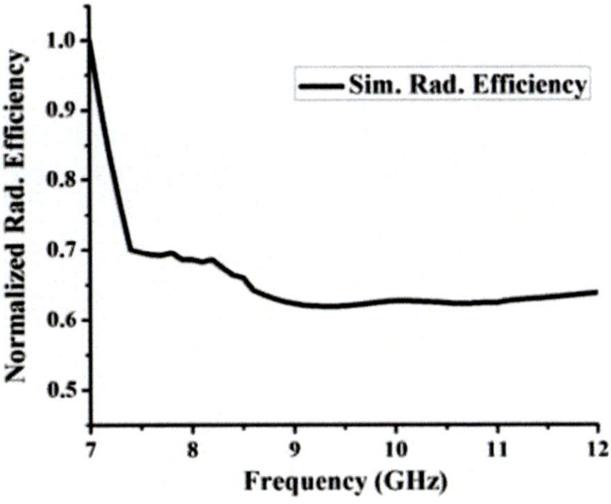

Table 1 Comparison with the proposed work and the previous works

Features		Proposed work	[15]	[17]	[14]	[16]	[12]
Operating frequency (GHz)	F_1	8.1	2.45	5.4	6.44	9.0	8.26
	F_2	10.7	5.5	5.5	7.09	11.2	10.46
Thickness (h)		1.57	1.75	1.27	1.57	1.57	0.787
Permittivity (ε_r)		2.2	4.4	10.2	2.2	2.2	2.2
Isolation $\|S_{21}\|$ (dB)	F_1	> 50	> 26	> 30	> 30	25	> 27.9
	F_2	> 21				21.5	> 30.2
Frequency ratio		1.32	2.24	1.02	1.1	1.27	1.24
FTBR (dB)	F_1	> 14	0	10	16.5	21	23
	F_2		0	10	19	19	20
Gain (dBi)	F_1	3.9	1.3	4	3.1	4.3	3.56
	F_2	4.4	4.4	4	4.8	4.2	5.24
Radiation pattern		Uni-directional	Bi-directional	Uni-directional	Uni-directional	Uni-directional	Uni-directional
Cross-polarization (dB)		< -20	N.A	< -20	< -12	< -19	N.A
					< -16		
Size		$1.23\lambda_l \times 1.2\lambda_l$	$8.3\lambda_l \times 8.3\lambda_l$	N.A	$1.46\lambda_l \times 1.46\lambda_l$	$2.6\lambda_l \times 1.0\lambda_l$	$1.3\lambda_l \times 1.1\lambda_l$
Software/model based		HFSS	HFSS	HP/Agilent Series IV (FDFD)	CST	HFSS	HFSS

λ_l—is the lowest resonant frequency
N.A.—not available

realized about 3.9 and 4.4 dBi at first and second resonating frequencies. Due to all of the positive features of the proposed antenna, it becomes well and worthy design for *X*-band satellite and wireless communications.

References

1. Deng J, Li J, Zhao L, Guo L (2017) A dual-band inverted-F mimo antenna with enhanced isolation for WLAN applications. IEEE Antennas Wirel Propag Lett 16:2270–2273. https://doi.org/10.1109/LAWP.2017.2713986
2. Wu X, Lu W, Xu J, Tong KF, Zhu H (2015) Loop-monopole composite antenna for dual-band wireless communications. IEEE Antennas Wirel Propag Lett 14:293–296. https://doi.org/10.1109/LAWP.2014.2363667
3. Yang C, Kim J, Kim H, Wee J, Kim B, Jung C (2010) Quad-band antenna with high isolation MIMO and broadband SCS for broadcasting and telecommunication services. IEEE Antennas Wirel Propag Lett 9:584–587. https://doi.org/10.1109/LAWP.2010.2053515
4. Guan X, Yang F, Liu H, Zhu L (2014) Compact and high-isolation diplexer using dual-mode stub-loaded resonators. IEEE Microw Wirel Compon Lett 24(6):385–387. https://doi.org/10.1109/LMWC.2014.2313591
5. Li W, Ye X, Ma H (2012) Design of microstrip triplexer with new miniature resonator. In: Proceedings of 2012 5th global symposium on millimeter-waves, Harbin, pp 641–643. https://doi.org/10.1109/GSMM.2012.6314415
6. Radonić V, Crnojević-Bengin V, Baskakova A, Vendik IMultilayer microwave diplexers based on dual-mode resonators for ISM/WiFi bands. In: Proceedings of 2014 Mediterranean microwave symposium (MMS2014), Marrakech, 2014, pp 1–4. https://doi.org/10.1109/MMS.2014.7088996
7. Inclan-Sanchez L, Vaquez-Roy J, Rajo-Iglesias E (2008) High isolation proximity coupled multilayer patch antenna for dual-frequency operation. IEEE Trans Antennas Propag 56(4):1180–1183. https://doi.org/10.1109/TAP.2008.919218
8. Zhang Z, Wong S, Lin J, Liu H, Zhu L, He Y (2019) Design of multistate diplexers on uniform- and stepped-impedance stub-loaded resonators. IEEE Trans Microw Theory Tech 67(4):1452–1460. https://doi.org/10.1109/TMTT.2019.2893656
9. Fallahzadeh S, Bahrami H, Akbarzadeh A, Tayarani M (2010) High-isolation dual-frequency operation patch antenna using spiral defected microstrip structure. IEEE Antennas Wirel Propag Lett 9:122–124. https://doi.org/10.1109/LAWP.2010.2043810
10. Bozzi M, Georgiadis A, Wu K (2011) Review of substrate-integrated waveguide circuits and antennas. IET Microw, Antennas Propag 5(8):909–920. https://doi.org/10.1049/iet-map.2010.0463
11. Feng Xu, Ke Wu (2005) Guided-wave and leakage characteristics of substrate integrated waveguide. IEEE Trans Microw Theory Tech 53(1):66–73. https://doi.org/10.1109/TMTT.2004.839303
12. Nandi S, Mohan A (2017) An SIW cavity-backed self-diplexing antenna. IEEE Antennas Wirel Propag Lett 16:2708–2711. https://doi.org/10.1109/LAWP.2017.2743017
13. Khan AA, Mandal MK (2019) Compact self-diplexing antenna using dual-mode SIW square cavity. IEEE Antennas Wirel Propag Lett 18(2):343–347. https://doi.org/10.1109/LAWP.2018.2890790
14. Kumar A, Chaturvedi D, Raghavan S (2018) Design of a self-diplexing antenna using SIW technique with high isolation. AEUE—Int J Electron Commun 94:386–391. https://doi.org/10.1016/j.aeue.2018.07.031
15. Lu Y, Lin Y (2012) A mode-based design method for dual-band and self-diplexing antennas using double T-stubs loaded aperture. IEEE Trans Antennas Propag 60(12):5596–5603. https://doi.org/10.1109/TAP.2012.2211852

16. Mukherjee S, Biswas A (2016) Design of self-diplexing substrate integrated waveguide cavity-backed slot antenna. IEEE Antennas Wirel Propag Lett 15:1775–1778. https://doi.org/10.1109/LAWP.2016.2535169
17. Park D, Waterhouse R, Qian Y, Itoh T (2001) Self-diplexed integrated antenna transceiver for wireless. In: IEEE antenna propagation society symposium, vol 3, pp 444–447
18. Nandi S, Mohan A (2018) SIW based cavity-backed self-diplexing antenna with plus shaped slot. Microw Opt Technol Lett 60:827–834. https://doi.org/10.1002/mop.31067
19. Chaturvedi D, Kumar A, Raghavan S (2018) An integrated SIW cavity-backed slot antenna-triplexer. IEEE Antennas Wirel Propag Lett 17(8):1557–1560. https://doi.org/10.1109/LAWP.2018.2855051
20. Kumar A, Raghavan S (2018) A self-triplexing SIW cavity-backed slot antenna. IEEE Antennas Wirel Propag Lett 17(5):772–775. https://doi.org/10.1109/LAWP.2018.2815665
21. Filinyuk NA, Ogorodnik KV, Saleh MMJ (2005)Measure of non-standard systems of *Z*- and *S*-parameters for microwave two-ports. In: 2005 15th international Crimean conference microwave & telecommunication technology, Sevastopol, Crimea, vol 2, pp 739–740. https://doi.org/10.1109/CRMICO.2005.1565116.

Implementation of Virtual Telepresence Robot Using Two Approaches Raspberry Pi and MyRIO

G. S. V. Krishna, G. Mani Babu, and V. N. Prudhvi Raj

1 Introduction

Technology can solve all the biggest problems confronting mankind, whether it is robotics, quantum or cloud computing, telemedicine, AI, etc. People faced a lot of limitations without automation technology but now people are unable to live without technology. Due to the advancement in science a technology, the concept of tele-robotics has been invented. Tele-robotics technology is the territory of apply autonomy and the control of semi-self-ruling robots from a separation utilizing remote systems like Wi-Fi, Bluetooth, and so on. Tele-robotics autonomy is a mix of teleoperation and telepresence. The examination of tele-robotics technology, which controls robots remotely using PC systems, it has been well known as of late. Telepresence refers to an innovation that permits an individual to feel as though they were available at the remote area [1, 2].

Moreover, tele-operation gives the ability to impact the remote territory. Right now, the user's position, improvements, exercises, etc., may be recognized, transmitted to the remote territory to accomplish this effect Telepresence robot can empower remote visit guides, clerical specialists, night guardians, and industrial facility assessors, among numerous different prospects. Right now, the proposed

G. S. V. Krishna · G. Mani Babu (✉) · V. N. Prudhvi Raj
Electronics and Instrumentation Engineering Department, VR Siddhartha, Engineering College, Kanuru, Vijayawada, India
e-mail: manibabu296@gmail.com

G. S. V. Krishna
e-mail: gkrish6838@gmail.com

V. N. Prudhvi Raj
e-mail: nagaprudhvi@gmail.com

V. H. Saran and R. K. Misra (eds.), *Advances in Systems Engineering*, Lecture Notes in Mechanical Engineering,
https://doi.org/10.1007/978-981-15-8025-3_52

work is to improve the working of the robot and catching the visuals from the remote area utilizing Raspberry pi/NI MyRIO [1, 3, 4]. The usage of the telepresence robot is finished utilizing the two ways to deal with decide the presentation technique of controllers.

2 Literature Survey

In this world, activity performed by a human is getting basic, in order to make it more straightforward. Right now, two terms of VIRTUAL EXPERIENCE AND TELEPRESENCE are analyzed. This investigation is as far as anyone is concerned, the first to look at the ideas of virtual experience and telepresence in a systematic manner. Virtual experience and telepresence are the exploration items to be examined. By describing them, the region of examination for the writing survey can be sketched out [5].

Virtual Experience: Virtual experience implies distinctive, including, dynamic, and full of feeling mental states happening in an individual cooperating with 3D PC reproductions. An overstated domain that is experienced through the sensors gave by a PC and in which the activities halfway figure out what occurs in the earth is known as a virtual experience. VE may likewise be viewed as a type of indirect experience.

Telepresence: Telepresence can be controlled by two mechanical determinants vividness and interactivity. Telepresence is a progressive type of robotic autonomy, remote control in which a human can have a feeling of being on that area in this way, the experience looks like virtual reality (VR). By methods for correspondence, the medium client can have the experience of being available in another condition; hence, the correspondence framework should give input promptly in light of the activities of the client [6].

2.1 Existing Tele-robots

Many attempts are made by several people on tele-robotics in early years. A review on different papers proposed to classify the tele-robotics is listed below:

The virtual model introduced in [3] Conference on Robotics and Automation. Right now, authors examined the virtual model control that utilizes recreations of envisioned mechanical segments to make powers, which are applied to genuine joint torques, accordingly making the figment. They expressed that one of the significant motivating forces of creating virtual model control is to make planning robot control calculations simpler and increasingly natural. Virtual model control has indicated that speed control and higher effectiveness is improved.

In paper [4] International Journal of networks and communications, the author proposed the structure and advancement of the correspondence and control stage for a Medical Tele-diagnosis Robot (MTR). The doctor–robot interface in MTR encourages the specialist to analyze the patients separated from the significant distance through a legitimate web association. A decentralized system called Hybrid P2P (peer-to-peer) innovation is utilized to diminish the general framework arrangement and support. MTR is a cost-effective option for telepresence robots in the clinical field for creating countries.

High-level human–system interfaces in [7]. This paper describes the beginning times and aftereffects of an undertaking which tends to the structure of regular or instinctual interfaces by consolidating telepresence and computer-generated reality advancements. It is an exploratory program of research, intended to think about the effect of new human PC interface and telepresence innovations on the presentation of the human administrator in charge of the remote automated framework.

A Remote Navigation System for a Simple Tele-Presence Robot with Virtual Reality Conference on IROS [8]. This paper proposes a remote course structure for a fundamental telepresence robot. All in all, robot components of telepresence robots are compelled, and the visual purpose of the camera on robots is limited, so they are unacceptable to use course organizations.

The framework beats the trouble by embracing framework engineering in which robots are controlled remotely through a server situated in the web, by utilizing all around structured computer-generated experience strategies for robot control and by executing practically all robot route capacities into the server. They have begun that, by conveying practically all framework capacities to a link in the Internet, correspondence traffic volume between a robot and the server diminishes, and computational preparation in the robot additionally diminishes. The power utilization of the robots can be decreased.

3 Proposed System

The project is framed as shown in Fig. 1a using Arduino Raspberry Pi [9], and Fig. 1b using MyRIO.

In Fig. 1a, Raspberry Pi is the brain of the framework. It gets input data from the cell phone by means of Wi-Fi and afterward sends controlling pulse to the servo engines to move the RPi camera. The cell phone sends the input to the Arduino Mega (Board2) with the end goal of the route or development of the virtual telepresence robot. The motor driver IC and outfitted motors are related to the completion of the route circuit. The requests to run the virtual telepresence robot can be sent through Bluetooth or RF module from the mobile phone. Right now we utilized the Bluetooth module.

In Fig. 1b, the RIO controller with GUI is used as robot controller. A virtual control framework has been created utilizing LabVIEW. RIO controller runs on continuous processor and client programmable FPGA; it can be operated using built-in Wi-Fi. The online streaming of robot vision is done using LabVIEW web server, and the camera head is controlled using sensory data (gyroscope, accelerometer, and

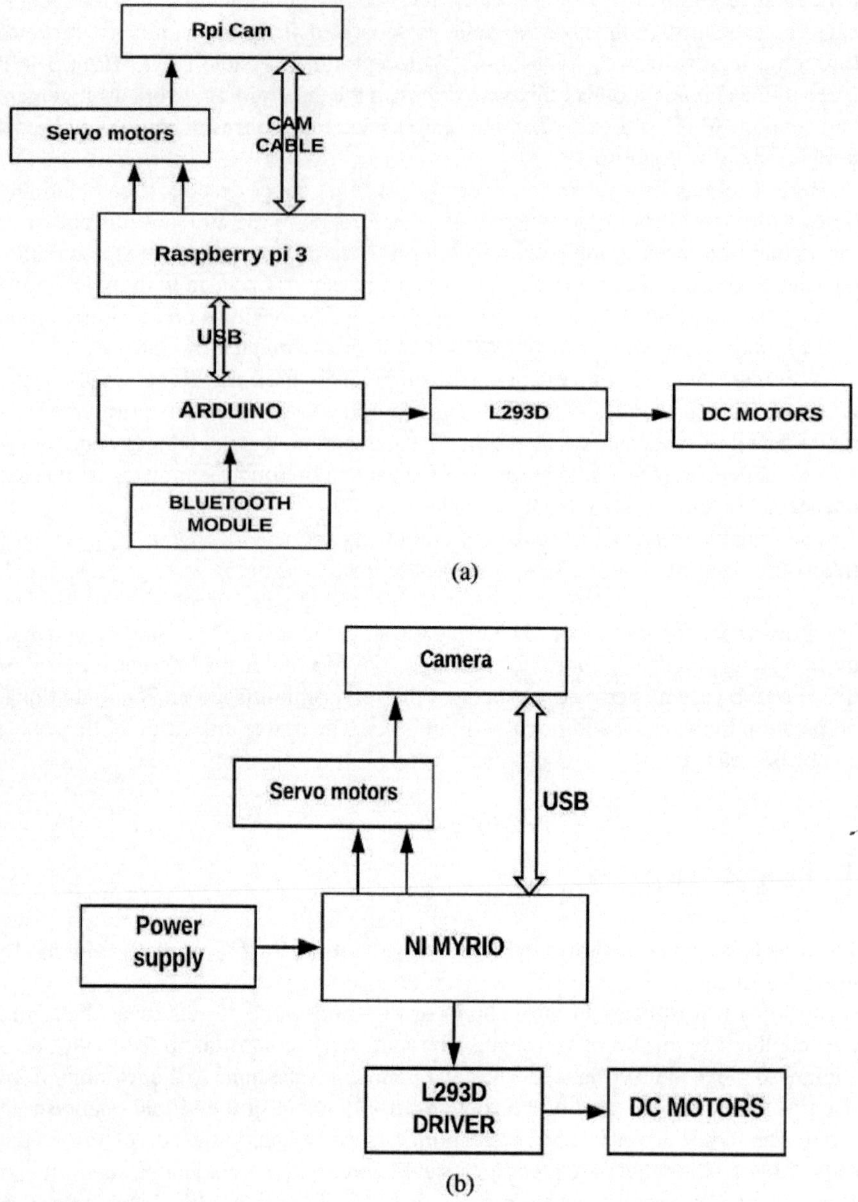

Fig. 1 **a** Proposed system with RPI, **b** proposed system with RIO

magnetometer readings) available in smart phones. It is remotely controlled using NI Data Dashboard android application for navigation purpose.

The equipment utilized in the undertaking is reasonable and simple to discover in the market. The outline of different segments, devices and software's utilized right now: 12 V charger, 8 V power supply, Arduino Mega 2560 board, servo engines, DC equipped engines, Bluetooth module, Raspberry Pi board, LabVIEW and MyRIO. Alongside these, you additionally require an Android cell phone with a reasonable application [8, 10, 11].

3.1 Navigation

The route circuit controls the development of the robot, and it contains Arduino Mega 2560 board HC05 Bluetooth module, L293D motor driver (IC1), two DC prepared motors, NI MyRIO, and NI Data dashboard. Directional information or orders from the cell phone is sent to the Arduino Mega through the HC-05 Bluetooth module/MyRIO through Wi-Fi. It is handled and afterward took care of to the engine driver IC, which drives the equipped engine in the necessary bearing. The commands to the Rpi controller are given through the Bluetooth electronics application also the Rio controller gets commands from the NI Dashboard.

1. **Arduino Mega 2560**: It is a microcontroller board based on the ATmega2560 having 54 digital input/output pins. The Arduino board acts as the directional processor of the virtual telepresence robot. Orders to drive the virtual telepresence robot in the predefined course are handled and given by the Arduino to a driver IC, which, thus, makes the engines run.
2. **HC-05 Bluetooth module**: HC-05 module is a simple to utilize Bluetooth sequential port convention module intended for straightforward remote sequential association arrangement. Its baud rate is Default and has eight information bits, one stop bit, and no parity.
3. **L239D motor driver**: L293D is a run of the typical motor driver IC that permits motor to drive; L293D is a 16-pin IC that can control a lot of two DC motors toward any path simultaneously.
4. **NI Data Dashboard**: Dash dashboard is an android application that can write values to an application in LabVIEW that contains shared variables and read values from network published shared variables.

3.2 Control

The control of servo motor and camera should be possible by utilizing RPi 3 Model B+. It is utilized, and right now it permits simple video imparting through Wi-Fi. The RPi is considered as the CPU of the virtual telepresence robot, and the RPi 3 Model B+ utilized here is the 3rd era [4].

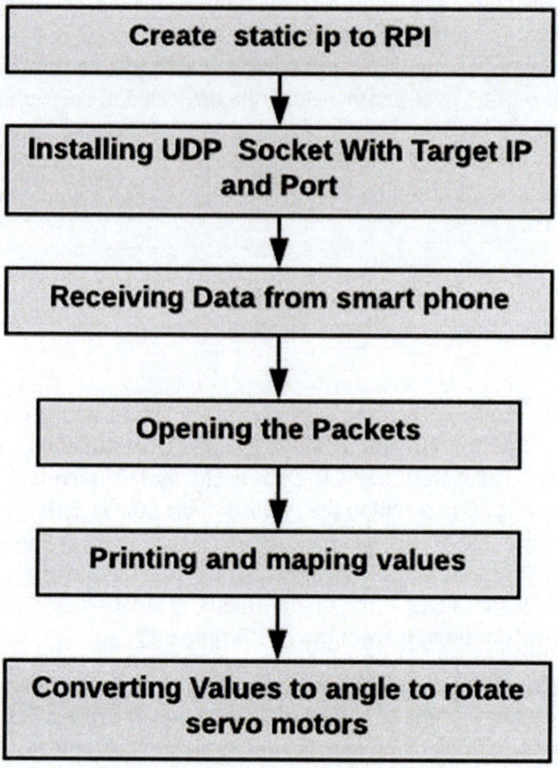

Fig. 2 Flowchart for camera control

Figure 2 shows the process of controlling servo motors with raspberry pi. We are creating Raspberry Pi static address to have unique address when connected to any router. Accelerometer data from smart phone is transmitted to Raspberry Pi through Wi-Fi using UDP protocol. The received data is manipulated and mapped with servomotors rotating angles. Camera is connected to Raspberry Pi and attached to servo motors the video acquired by Raspberry Pi is transmitted to smart phone and is displayed on dual screen application shown in Fig. 6 and is placed in VR headset for virtual reality experience.

In our second approach, NI MYRIO is used for control of servo motors and camera. The servo motors are interfaced with MyRIO port-A digital I/O pins. The sensory data from smart phone is transmitted to RPI controller through Wi-Fi **using LabVIEW webserver**.

4 Software

Raspbian OS: The Raspbian OS is flashed on to a memory card of 32 GB. Then it is inserted in the memory card slot of the Raspberry Pi board.

Apache web server: It is used in this project to configure RPi as server. Apache is a commonly used web server application that you can use on the RPi to allow it to online web pages.

Python: It is broadly used high-level programming language for general-purpose applications. These programs do not need to be compiled before running them. It is a best programming language that can be used in Raspberry Pi to interact with user.

Arduino IDE: It makes easy to navigate the robot. The C/C++ languages are used to program the Arduino.

LabVIEW: LabVIEW is used to generate a logical sequence in graphical notation.

5 Performance and Setup

In the Raspberry Pi-based design, the Arduino mega is the one which controls the movement of the robot. The directional data is sent to Arduino through Bluetooth module using Bluetooth electronics android application. The casing to hold the gear motors, servo motors, camera, and Raspberry Pi was unawed. Followed by the RPi setting, the initial configurations are performed and then connect the RPi to Wi-Fi. An IP address is assigned to the RPi and links to the setup. The captured video using a camera is transferred by the Wi-Fi. This video can be seen in your mobile screen which is connected to the matching IP address and Wi-Fi of the RPi [2].

In the MyRIO-based design, MyRIO is equipped with servo motors and geared motors are for robot movement. The USB camera is connected to the MyRIO for capturing the video at the remote location. The captured video is transmitted to the android device over Wi-Fi. The NI data dashboard application is used to design the user interface with directional keys for the movement of robot. The project has a feature of letting the user experience virtual reality by swinging the VR headset. The cell phone gives the sensor data information of the course in which client had turned. This information is sent to using Wi-Fi to the modem and to Raspberry Pi utilizing remote IMU android application which gives these qualities as a contribution to the servo motor.

6 Results and Discussions

Finally, the construction and assembling of hardware components are done. All the required software's and mobile applications are installed to control direction

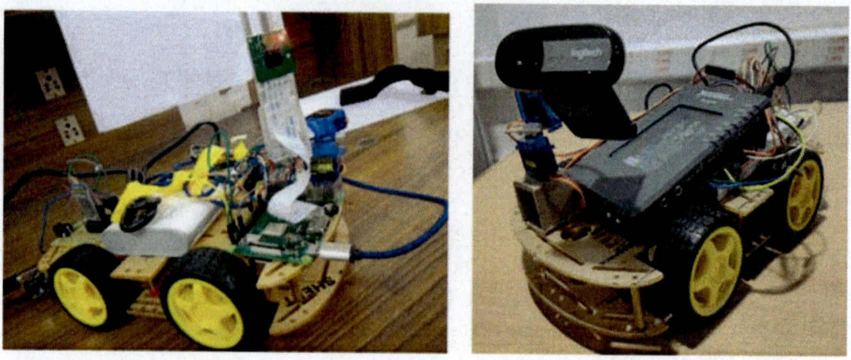

Fig. 3 Virtual telepresence robot

and navigation. We successfully demonstrated the Rpi camera can provide a clear high-quality video of 1080p and placed in VR headset for virtual experience.

Figure 3 shows the implemented virtual telepresence robot using Arduino Raspberry Pi (left) and NI MyRIO (Right).

Figure 4 shows the screenshot of wireless IMU application which is used to collect the sensor readings from IMU in Android Phone. The sensor readings were transmitted through UDP protocol in bytes to the raspberry Pi.

Figure 5 shows the screenshot of Bluetooth electronics application which is used to navigate the robot.

Figure 6 shows the video acquired by robot camera and is displayed in smart phone dual screen application. The smart phone which is to be placed in VR headset is to be installed with wireless IMU application connect it to Wi-Fi and enter the static IP Address and run application in background now open dual screen application, as we initially installed apache webserver to Raspberry Pi it creates server to pi. The video which is acquired by RPI cam can be monitored in web page by typing "192.168.0.100(static IP of raspberry pi):8000" this URL is placed on dual screen application for video monitoring and the smart phone is placed in VR box.

In Fig. 7, it shows the navigating robot with the help of Arduino Bluetooth DC motor application which is developed using MIT application inventors platform shown in Fig. 6, it sends commands to HC05 Bluetooth module which is interfaced with Arduino, based on the commands the Arduino energies DC motors for robot movement.

7 Conclusions and Future Scope

With the successful integration of the hardware and software components, the virtual telepresence robot moves almost simultaneously with the robot operator.

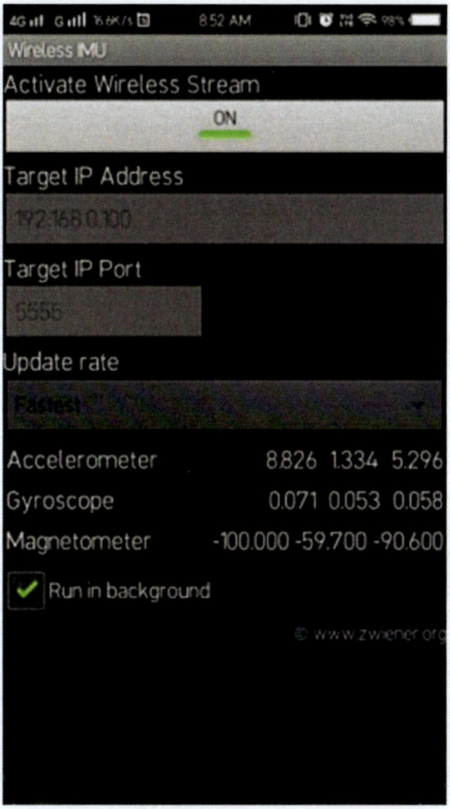

Fig. 4 Screenshot of wireless IMU in Android Mobile

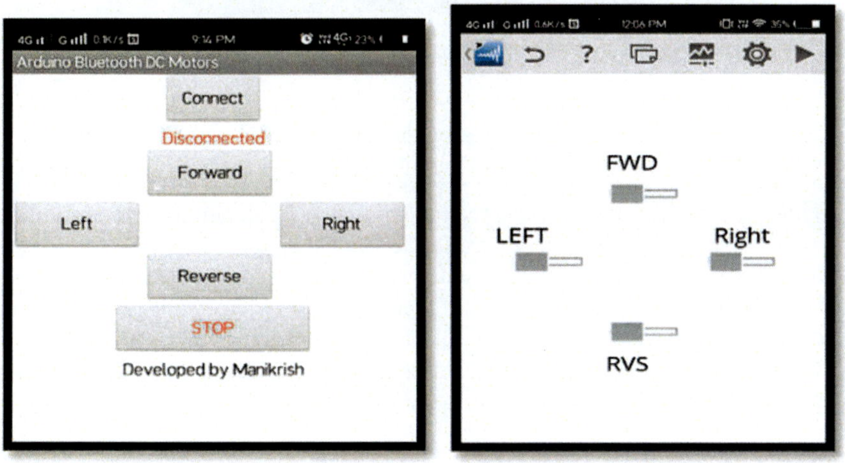

Fig. 5 Screenshot of Arduino Bluetooth DC motors application

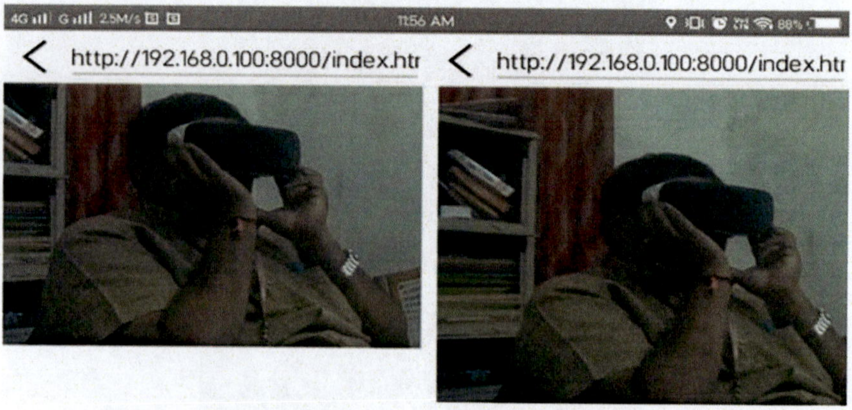

Fig. 6 Screenshot of dual screen video monitoring

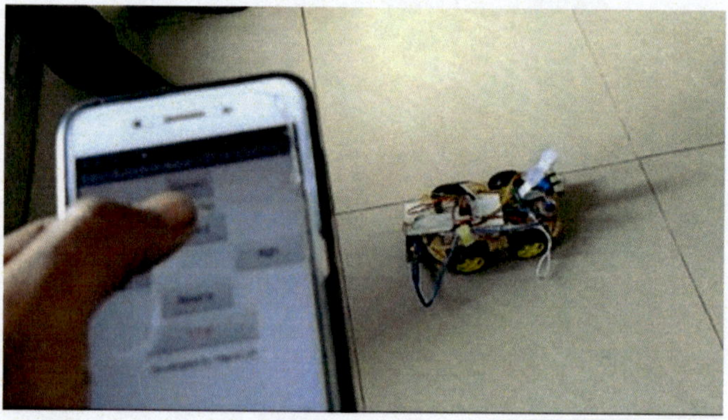

Fig. 7 Navigating robot with Bluetooth application

Positions are successfully obtained by the wireless IMU application and sent to the servo controller Raspberry Pi and the robot direction controller via Bluetooth.

- This robot can be used as a inspection/spy robot. In situations where the shop owner needs to keep an eye of what the servant is up to.
- In the case of military, instead of sending a soldier to a sight for monitoring the robot can be sent. In this way, even if there is an unexpected attack no life is lost only the robot is under damage.
- In the medical case, at times when the doctor is not able to go on rounds, the robot can be used to check the state of the patient.

On further expansion of the robot and increasing its connectivity, the robot can be used in many different applications:

- If we develop a hand-like mechanism as head movement now, we can use this robot in place of humans working in hazardous areas/environments.
- If the robot is made re proof, then the robot can be used in re and rescue operations .
- On further development and with some changes, we can make use of this robot in space research also.

Reference

1. Shaji SP (2017) Virtual telepresence robot.Int J Innov Emerg Res Eng
2. Gruter M, Myrach T (2012) Understanding virtual experience and telepresence—a review and synthesis of literature. In:ECIS 2012—proceedings of the 20th European conference on information systems, 01 2012
3. Pratt J, Pratt G, Pratt E (2000) Virtual model control of a biped walking robot.02 2000
4. Mariappan M, Khoo B (2013) Design and development of communication and control platform for medical tele-diagnosis robot (mtr).Int J Netw Commun 03 2013
5. Jitheesh P, Keeramkot F, Mohammed Ameen T, Athira PC, Madeena S, Arunvinodh C (2016) Telepresence robot doctor. In: 2016 Online international conference on green engineering and technologies (IC-GET), Nov 2016, pp 1–4
6. Kato Y (2015) A remote navigation system for a simple tele-presence robot with virtual reality. In: 2015 IEEE/RSJ international conference on intelligent robots and systems (IROS), Sep 2015, pp 4524–4529
7. Stone RJ (1991) Advanced human-system interfaces for telerobotics using virtual reality and telepresence technologies. In: Fifth international conference on advanced robotics 'robots in unstructured environments vol 1, June 1991, pp 168–173
8. Gikas V, Perakis H (2016) Rigorous performance evaluation of smartphone gnss/imu sensors for its applications. Sensors 16(8):1240
9. https://www.electronicsforu.com/electronics-projects/virtual-telepresence-robot-raspberry-pi
10. Kong W, Sessa S, Cosentino S, Zecca M, Saito K, Wang C, Imtiaz U, Lin Z, Bartolomeo L, Ishii H et al (2013) Development of a real-time imu-based motion capture system for gait rehabilitation.In 2013 IEEE international conference on robotics and biomimetics (ROBIO), IEEE, 2013, pp 2100–2105
11. Amaro JP, Patrão S, Moita F, Roseiro L (2017) Bluetooth low energy pro-le for mpu9150 imu data transfers.In: 2017 IEEE 5th Portuguese meeting on bioengineering (ENBENG), IEEE, 2017, pp 1–4

An Automated Flight Parameter Estimation Technique Using Genetic Algorithm

Nikhil Anand, A. Sabarinath, Ramesh Mettu, Chindu Mohan, S. Nisha, S. Geetha, and G. Ayyappan

1 Introduction

A launch vehicle is a complex system made of various multi-disciplinary subsystems, which work in an integrated scheme to achieve a mission objective. Most of the components of the launch vehicle are thoroughly tested and their performances are evaluated before integration into the launch vehicle [1]. However, certain vehicle subsystems such as propulsion, pyros cannot be tested completely on ground. Moreover, the interaction of the vehicle with environment such as atmosphere, wind, gravity cannot be directly evaluated. For these purposes, simulators are modeled which imitate the functioning of the physical subsystems of the vehicle and the environment in which they operate in an integrated manner [1–3].

As in any field, performance assessment of a launch vehicle after its mission in the intended environment is essential, which makes post-flight analysis (PFA) indispensable. The behavior of the vehicle in flight is in most cases marginally different from preflight prediction, due to minor perturbations in the performance of individual subsystems that cause deviations in the overall system performance [1]. One of the trivial requirements of PFA is to understand the performance of each subsystem in flight. For example, certain engine-level misalignments, center of gravity (CG) estimation error, etc. that may have occurred in flight and caused the vehicle to perform off-nominally can be brought out through PFA, and thus, the performance characteristics derived from PFA can be used for continuous improvement of the subsystems in later flights.

Given the enormous number of subsystems present in a launch vehicle, simulating the flight behavior of a vehicle by predicting the individual performance of each of

N. Anand (✉) · A. Sabarinath · R. Mettu · C. Mohan · S. Nisha · S. Geetha · G. Ayyappan
Space Transportation System, Vikram Sarabhai Space Centre, Indian Space Research Organization, Thiruvananthapuram 695022, India
e-mail: nikhil.31.anand@gmail.com

V. H. Saran and R. K. Misra (eds.), *Advances in Systems Engineering*,
Lecture Notes in Mechanical Engineering,
https://doi.org/10.1007/978-981-15-8025-3_53

these subsystems is a tedious and time consuming activity. This paper presents an automated technique for estimation of crucial parameters in each subsystem from flight telemetered data using an optimizer working in tandem with a 6DoF launch vehicle trajectory simulator. The accuracy of this estimation algorithm depends on the accuracy of optimization technique implemented. In this paper, genetic algorithm has been used as an optimizer, although techniques such as differential evolution, particle swarm, ant colony can also be used.

2 Optimization Using Genetic Algorithms (GAs)

Genetic algorithm (GA) is one such evolutionary algorithm where the process of natural selection is used to generate candidate solutions till a terminal condition (defined on the basis of optimality) is satisfied. These candidate solutions are generated through techniques inspired by natural evolution, such as inheritance, mutation, selection, and crossover [4].

A large class of problems that arise in engineering applications either cannot be solved using traditional gradient based optimization methods entirely or cannot be solved within a reasonable time [4]. GAs were first proposed by Professor John H. Holland [5] and his students in the University of Michigan in the early 1970s and have found application in a plethora of engineering problems due to their ability to attain quick solutions even for problems where classical methods cannot be implemented. GAs have been successfully applied to optimization problems like routing, scheduling, adaptive control, cognitive modeling, etc. [6].

The evolution toward an optimal solution is an iterative process and starts from a population of randomly generated individuals. The population in each iteration is called a *generation.* In each generation, the cost (usually the value of the objective function in the optimization problem being solved) of every individual in the population is evaluated. The fitter individuals (individuals with lower cost) are stochastically selected (using techniques such as Tournament Selection, Roulette Wheel Selection, etc.) from the current population, and each individual's genome is modified (recombined and possibly randomly mutated) to form a new generation [7]. Mutation ensures that the diversity of solutions is maintained in each generation which helps the algorithm to avoid getting stuck in local minima [8]. The candidate solutions from each generation are then used in the next iteration of the algorithm until either the maximum number of generations has been reached, or a satisfactory cost level is attained.

3 Formulation of Cost Function

Assume a set of time series of flight profiles y_m^T for all m belonging to $[1, 2, \ldots, N]$, where N is the number of flight profiles that are desired to be matched. The superscript

T represents the time stamps at which the corresponding data has been sampled. Let h_m^T be the time series of corresponding profiles obtained by 6DoF simulation program. Ideally, for an optimal set of input parameters, the difference between the flight and simulated profiles must be minimum. But due to the existence of noise present in the data received from telemetry, the difference between the flight and simulated profiles will be substantial even for the optimal set of input parameters. To avoid this, the noise component of data is removed from the flight profiles before computation of cost. For this purpose, data smoothers have been implemented in this study. Let φ_m^t be the smoothened flight profile obtained by,

$$\varphi_m^T = \text{Smoothing Function}\left(y_m^T\right) \tag{1}$$

The cost function can then be formulated as,

$$\text{Cost}(h_m^T) = \sum_{m=1}^{N} w_m \sum_{\substack{For \\ all t in T}} \left(\varphi_m^t - h_m^t\right)^2 (2)$$

The data smoothing technique used in this paper is highlighted in the following section.

3.1 *Smoothing Function*

Statistically, smoothing a data set involves creating an approximate function that attempts to capture important patterns in the data, while leaving out noise or other fine-scale structures/rapid phenomena. One of the most common algorithms is the 'moving average.' A moving average is commonly used with time series data to smooth out short-term fluctuations and highlight longer-term trends or cycles. The threshold between short term and long term depends on the application, and the parameters of the moving average are set accordingly [9].

4 Algorithm Formulation of the Present Problem

In this section, the formulation of the algorithm for automated parameter estimation is explained. The algorithm consists of three major modules, namely profile smoother, 6DoF dynamics simulator module, and genetic algorithm. A profile smoother smoothens the flight observered profiles before computation of the cost. This ensures a smoother cost function which aids in improving the efficiency of the parameter estimation process. The 6DoF dynamics simulator simulates the required

Table 1 Proposed format of relation matrix

	Input parameter 1	Input parameter 2	Input parameter 3	...
Flight profile 1	1	0	0	
Flight profile 2	1	1	1	
Flight profile 3	0	1	1	
⋮				

flight profiles for any set of input parameters. Since accurate modeling of dynamics is essential for post–flight analysis of a mission, a 6 degree of freedom simulator has been used to simulate the trajectory of the launch vehicle.

The first step of automated flight parameter estimation is the problem formulation. If the flight profiles received from telemetry after a mission are found to be off-nominal, the individual subsystems are identified which could have contributed to the deviation. A set of crucial parameters are then selected from these subsystems which can be perturbed to match the achieved flight profiles. Suppose n such parameters have been selected to match m flight profiles, a $n \times m$ relation matrix is drawn to understand the dependency of every flight profile on these input parameters. For example, a flight profile like lateral acceleration can be matched by suitable variations in input parameters like vehicle thrust, vehicle mass, vehicle CG offset, etc.

Table 1 shows the format of the proposed relation matrix. '1' signifies the existence of a relationship, whereas '0' signifies the absence of it. From the above matrix, it can be concluded that Flight Profile 1 depends only on Input Parameter 1, whereas Flight Profile 2 depends on Input Parameters 1, 2 and 3, and Flight Profile 3 depends only on Input Parameters 2 and 3. In such cases, the problem can be broken down into two parameter estimation problems for faster processing. The first parameter estimation problem will use the automated algorithm to estimate Input Parameter 1, the final result of which will be used to solve the second parameter estimation problem where Input Parameters 2 and 3 will be estimated simultaneously.

The algorithm begins with random initialization of the population of input parameters which marks an entry to the automation loop. Using an interfacing module, every individual of the population is converted into a format which is readable by the 6DoF simulator. The output profiles are then passed to the cost function to compute cost. This process is repeated for all the individuals in the population, which are then passed to the genetic algorithm module to create the next generation. This process is repeated till either one of the termination criteria is satisfied.

Figure 1 shows the flowchart of the automated parameter estimation process. Since the genetic algorithm and simulation software have been kept apart as individual modules, any modification can be made in either of them without affecting the remaining process.

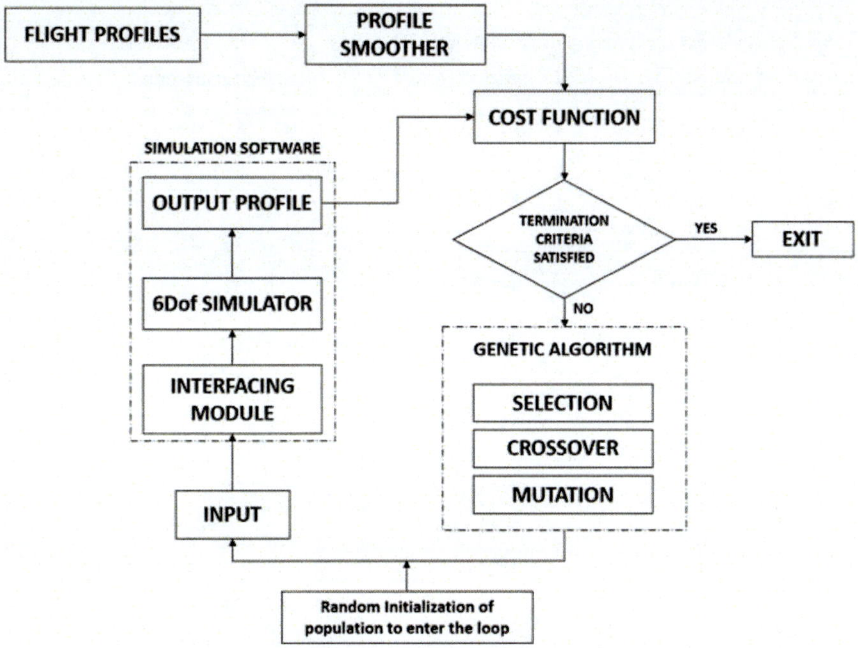

Fig. 1 Flowchart of the automated parameter estimation process

5 Algorithm Validation

This algorithm was validated by generating a flight profile by passing known perturbations to input parameters of a 6DoF simulation software. The generated profiles were then passed on to the algorithm to estimate the input parameters.

5.1 Estimating Specific Impulse Using Longitudinal Acceleration Data

A longitudinal acceleration profile was generated using a simulator by perturbing the specific impulse (Isp) of a rocket engine while keeping the mass flow rate constant. The acceleration profile was then passed on to the algorithm for estimating the specific impulse. The relation matrix in such a case is shown in Table 2.

Table 2 Corresponding relation matrix

	Specific impulse
Longitudinal acceleration	1

Table 3 Results for specific impulse estimation problem

Estimation results	Estimated value	Expected value
Thruster Isp (s)	271.00925	271.0

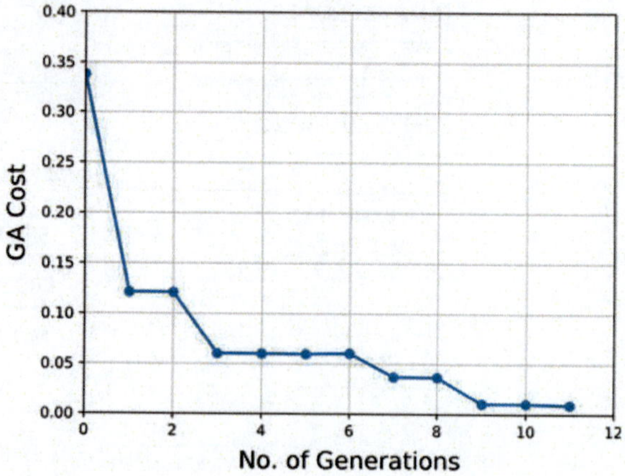

Fig. 2 Cost history for specific impulse estimation problem

It can be inferred from the above relation matrix that the above problem is a single parameter estimation problem. After 11 generations, the termination criteria of cost $<10^{-2}$ were satisfied and the results achieved are shown in Table 3. The cost history for each generation is plotted in Fig. 2.

5.2 *Estimating CG Offset and Thrust Misalignment in 2 Axis Using Digital Autopilot Commands and Lateral Acceleration*

Digital autopilot command and lateral acceleration profiles were generated using 6DoF simulator by perturbing the CG offset and thrust misalignment along pitch and yaw axis. Table 4 showcases the corresponding relation matrix in this case.

From the above relation matrix, it can be inferred that this problem can be broken down into two double parameter estimation problems. Since lateral accelerations along pitch and yaw are independent of thrust misalignments, the CG offset can be estimated first. The estimated CG offset can then be used to estimate the thrust misalignment. For the first parameter estimation problem, the termination criteria of 30 generations were satisfied and the results achieved are tabulated in Table 5.

The estimated CG offsets are passed to the second parameter estimation problem to estimate the thrust misalignments in pitch and yaw. Table 6 tabulates the results for the second parameter estimation problem, i.e., thrust misalignment estimation.

Table 4 Corresponding relation matrix

	CG offset pitch	CG offset yaw	Thrust misalignment pitch	Thrust misalignment yaw
Lateral acceleration pitch	1	0	0	0
Lateral acceleration yaw	0	1	0	0
DAP command pitch	1	0	1	0
DAP command yaw	0	1	0	1

Table 5 Results for CG offset estimation

Estimation results	Estimated value	Expected value
CG offset pitch (mm)	2.012	2
CG offset yaw (mm)	2.018	2

Table 6 Results for thrust misalignment estimation

Estimation results	Estimated value	Expected value
Thrust misalignment pitch (°)	−0.1471	−0.15
Thrust misalignment yaw (°)	0.2553	0.26

The cost history for each generation for both the parameter estimation problems is plotted in Fig. 3.

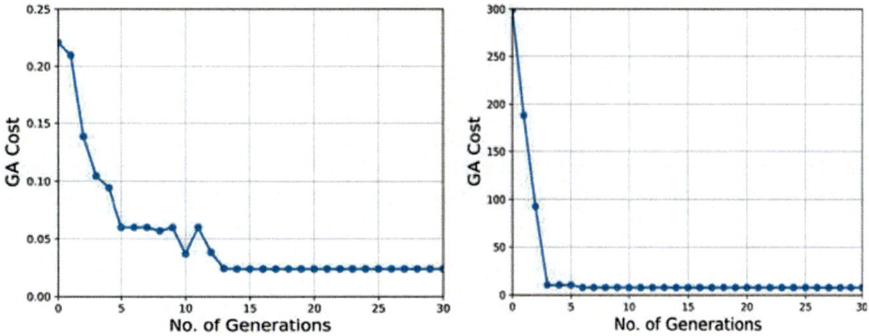

Fig. 3 (left) cost history for CG offset estimation, (right) cost history for thrust misalignment estimation using the estimated values of CG offset

6 Conclusion

This paper introduces an automated technique for parameter estimation using a simulation software working along with genetic algorithm. The flight observed performance of different subsystems can be matched using a simulation software by proper estimation of certain crucial parameters using the above technique. In order to increase the speed and accuracy of the method, certain modules such as data smoothing and relation matrix have been implemented. In this paper, a genetic algorithm-based optimization scheme has been used to demonstrate the algorithm for the estimation of.

- Thruster Isp using longitudinal acceleration profile, estimated with an error of 0.0034%.
- CG offset and thrust misalignment using lateral acceleration and digital autopilot command profiles, estimated with a maximum error of 0.9% and 1.8%, respectively.

Reference

1. Suresh BN, Sivan K (2015) Integrated design of space transportation system. Springer Publications. ISBN 978-81-322-2532-4
2. Thomson WT (1986) Introduction to space dynamics. Dover Publications. ISBN 0-486-65113-4
3. Zipfel PH (2007) Modeling and simulation of aerospace vehicle dynamics 2nd edition, AIAA Education Series. ISBN-10: 1-56347-875-7
4. Michalewicz Z (1999) Genetic algorithms+data structures=evolution programs. Springer
5. Holland JH (1975) Adaptation in natural and artificial systems. University of Michigan Press, Ann Arbor
6. Kramer O (2017) Genetic algorithm essentials. Springer. ISBN 978-3-319-52155-8
7. Yadav SL et al (2017) Study of various selection techniques in genetic algorithms. Int J Eng, Sci Math, 6(3). ISSN:2320-0294
8. Patil S et al (2014) Comparison and analysis of different mutation strategies to improve the performance of genetic algorithm. Int J Comput Sci Inf Technol 5(3):4669–4673, ISSN:0975-9646
9. James G, Witten D, Hastie T, Tibshirani R (2013) An ıntroduction to statistical learning. Springer

System Dynamics Approach for Adoption of Artificial Intelligence in Finance

Bharti Kumari, Jaspreet Kaur, and Sanjeev Swami

1 Introduction

Artificial intelligence is the simulation of human intelligence by machines. Minsky [27] defined "artificial intelligence" as "the science of making machines do things that would require intelligence if done by men". At present, artificial intelligence is used in almost every sector of our economy, and one sector which has embraced artificial intelligence most and adopted it into its processes is the financial services sector. Financial services sector includes banks, accountancy firms, credit-card companies, insurance companies, investment firms and several other related companies. Artificial intelligence has become popular in financial service sector because of its various advantages, i.e. fraud prevention, personalised financial services, customer service, better trading information, determination of credit worthiness, risk assessment and minimisation, ensure compliance to rules and regulations, automate processes, maximisation of resources, other business advice and recommendation, etc. The rise of AI in the financial industry proves how quickly it is changing the business landscape even in traditionally conservative areas.

Artificial intelligence is the single biggest revolution of our times. The growing pressure to improve performance has demanded companies to innovate further, in line with the latest Fourth Industrial Revolution or Industry 4.0, and AI is catalyst for Industry 4.0 [31]. New regulations have increased the need for AI implementation,

B. Kumari (✉) · J. Kaur · S. Swami
Department of Management, Dayalbagh Educational Institute, Agra, U.P., India
e-mail: singhbharti0608@gmail.com

J. Kaur
e-mail: jaspreet.24k@gmail.com

S. Swami
e-mail: sswami1853@gmail.com

V. H. Saran and R. K. Misra (eds.), *Advances in Systems Engineering*,
Lecture Notes in Mechanical Engineering,
https://doi.org/10.1007/978-981-15-8025-3_54

which has pushed banks to automate 24/7 and adopt new analytical tools that can include use of AI and machine learning [2, 13].

This study is significant because previous researchers have contributed towards artificial intelligence in the areas of manufacturing, health care, telecommunications, accounting, auditing, marketing and HRM [29, 31, 32, 35, 44–46], but no study has been found which has used holistic approach for AI implementation in financial services sector. The present study attempts to fill the gap by viewing the problem through bird's-eye view.

On the basis of literature review and gap analysis, researcher proposed the following research issues:

- What are the enabling and challenging factors for adoption of AI?
- While taking all the enabling, challenging, as well as positive and negative consequences of AI, should it really be adopted?
- What is the impact of adoption of artificial intelligence on financial performance and market capitalisation of a company?

Based on the above-discussed research issues, the following objectives are formed:

- To determine the feedback loops of challenges and enablers for AI adoption in financial services sector
- To examine the impact of adoption of AI on financial performance and market capitalisation of the company.

The present study is based on a nonlinear complex system, and it consists the enabling and disabling variables related with adoption of artificial intelligence in finance leading to financial performance and growth in market capitalisation of an organisation, i.e. JP Morgan Chase with a number of feedback control loops. The model developed in this study can be used in any other financial institution or company providing financial services.

The paper is further organised as follows:

Section 2 elaborates about conceptual framework, Sect. 3 includes the literature review, Sect. 4 explains the research methodology used, and Sect. 5 provides the details about results and discussion.

2 Conceptual Framework

2.1 Model of Study

The basic model for deeper understanding of the study is as follows (Fig. 1).

The basic idea behind the study is that there are some enabling factors like personal innovativeness, cheaper storage, trust and corporation of employees, competitive pressure, etc., which are pushing the adoption of this disruptive technology and

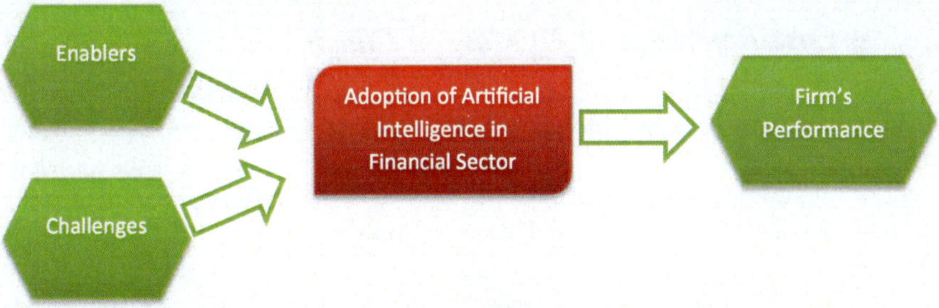

Fig. 1 Conceptual framework

global examples of its successful implementation also support this, while some challenging factors are restricting the adoption like lack of credible and quality data, lack of skilled staffs, identifying the right vendor, high cost of investment, convincing top management to deploy AI, security and privacy issues, etc. Despite of these challenges, every company is trying to adopt AI in their operations to remain in the market. This study is an attempt to know the effect of this disruptive technology on the performance of an organisation, i.e. JP Morgan Chase through financial performance and market capitalisation.

2.2 *System Dynamics Modelling and Simulation*

Based on system theory, this paper offers the dynamic approach of complex system analysis and modelling in the of system dynamics. System dynamics is a conceptual approach that enables to represent the structure and behaviour of complex systems over time, providing a method for systems description as well as a useful computational support for simulation [43]. It is concerned with using systems ideas to increase understanding of various types of phenomena and to aid in the decision-making which determines action upon the processes involved [23]. The main elements of SD are causal loop diagram, stock flow model and dynamic simulation.

The modelling and simulation process of system dynamics covers the following steps [4]:

- Description of model boundary in terms of modelling needs and priorities.
- Identification of relevant input variables and parameters.
- Model building, testing and validation.
- Scenario-sensitivity analysis.
- Policy formulation and implementation.

2.3 A Case Reference of JP Morgan Chase

America's biggest bank, JP Morgan Chase, is investing huge amount on artificial intelligence to shape its future strategies. J.P. Morgan would not disclose what it spent on this project but has said that 40% of its $10.8 billion annual technology budget is devoted to new efforts, including AI, robotic process automation and block chain [42].

The goal of J.P. Morgan AI Research is to explore and advance cutting-edge research in the fields of AI and machine learning, as well as related fields like cryptography, to develop solutions that are most impactful to J.P. Morgan's clients and businesses [21]. The bank started its virtual assistant project in 2017 after clients asked for an easier way to navigate the firm's online portal.

JP Morgan Chase introduce Contract Intelligence (COiN) platform designed to analyse legal documents and extract important data points and clauses. JPMorgan Chase has also hired a global head of machine learning from Microsoft [20].

3 Literature Review

See Table 1.

From the above literature review, enabling and challenging factors for AI adoption are identified. For any kind of adoption, there are some enabling or the driving factors that call for the need of adoption. Robust and rapid processing needs, advent of mobile technology, data availability and proliferation of open-source software offer AI a huge scope in the banking sector [33]. Competitive pressure and pressure to remain in the competition is the major factor pushing the adoption of AI [41]. AI could enhance perceived service quality, create value for customers and save employees' time [19]. One of the primary objectives using AI is to monitor all processes and data for regulatory compliance and security [40]. Several studies [16, 30, 39, 40] found that factors such as technical support, personal innovativeness, incentives, image and enjoyment with innovation have stronger influence on management of innovation. Availability of computing power owing to faster processor speeds, lower hardware costs and better access to computing power via cloud services, cheaper storage, parsing and analysis of data through the availability of targeted databases, software and algorithms, trust and corporation of employees on management's decision to implement AI tools and techniques are also enabling factors for the adoption of AI in an organisation.

It is obvious that implementation of AI will definitely give positive results to any organisation like cost reduction, increased profitability, accuracy in decision-making, accomplishment of regulatory compliances and many more, but negative impacts of AI implementation also come along with it; i.e., breach of confidential data, biasness, risk of job loss, change in job roles and many other legal and ethical issues may be faced. The main reason for non-adoption is the ineffective management,

Table 1 Synthesis matrix of literature review

Title Author(s) (Year)	Category	Focus issue				Context		Data type		Methodology		
		Banking	Financial Services (Stock Market/Portfolio Management/NBFC)	Insurance	Artificial Intelligence	Indian	Foreign	Qualitative	Quantitative	Exploratory	Descriptive	Analytical
Artificial Intelligence, Expert Systems, and Financial Planning Brandi [8]	Research Paper		✔		✔		✔	✔			✔	
Artificial Intelligence and Financial Services Pau [34]	Research Article		✔		✔		✔		✔			✔
Artificial Intelligence and Expert Systems in Accounting Databases: Survey and Extensions O'Leary [32]	Review Paper				✔		✔	✔				✔
Artificial Neural Network and the Financial Markets: A Survey Chatterjee et al. [10]	Research Artcle		✔				✔		✔			✔

(continued)

Table 1 (continued)

Title Author(s) (Year)	Category	Focus issue				Context		Data type		Methodology		
		Banking	Financial Services (Stock Market/Portfolio Management/NBFC)	Insurance	Artificial Intelligence	Indian	Foreign	Qualitative	Quantitative	Exploratory	Descriptive	Analytical
Artificial Intelligence in Portfolio Management Chan et al. [9]	Research Article		✔		✔		✔		✔			✔
Opportunities for Artificial Intelligence Development in the Accounting Domain: The Case for Auditing Baldwin et al. [5]	Review Paper				✔		✔		✔			✔

(continued)

Table 1 (continued)

Title Author(s) (Year)	Category	Focus issue				Context		Data type		Methodology		
		Banking	Financial Services (Stock Market/Portfolio Management/NBFC)	Insurance	Artificial Intelligence	Indian	Foreign	Qualitative	Quantitative	Exploratory	Descriptive	Analytical
Assessing Bank Efficiency and Performance with Operational Research and Artificial Intelligence Techniques: A Survey Fethi and Pasiouras [12]	Research Article	✔			✔		✔	✔				✔
The Role of Artificial Intelligence in the Development of Accounting Systems: A Review Moudud-Ul-Huq [28]	Review Paper				✔		✔	✔				✔

(continued)

Table 1 (continued)

Title Author(s) (Year)	Category	Focus issue				Context		Data type		Methodology		
		Banking	Financial Services (Stock Market/Portfolio Management/NBFC)	Insurance	Artificial Intelligence	Indian	Foreign	Qualitative	Quantitative	Exploratory	Descriptive	Analytical
The Effects of Artificial Intelligence and Robotics on Business and Employment: Evidence from a survey on Japanese firms Masayuki [24]	Discussion Paper				✔		✔		✔			✔
A Literature survey on Artificial Intelligence Gupta [17]	Conference Proceedings				✔	✔		✔				✔

(continued)

Table 1 (continued)

Title Author(s) (Year)	Category	Focus issue				Context		Data type		Methodology		
		Banking	Financial Services (Stock Market/Portfolio Management/NBFC)	Insurance	Artificial Intelligence	Indian	Foreign	Qualitative	Quantitative	Exploratory	Descriptive	Analytical
Artificial intelligence and machine learning in financial services: Market developments and financial stability implications Financial Stability Board [13]	Report		✔		✔		✔	✔				
Economic Policy for Artificial Intelligence Agrawal et al. [3]	Working Paper				✔			✔				✔

(continued)

Table 1 (continued)

Title Author(s) (Year)	Category	Focus issue				Context		Data type		Methodology		
		Banking	Financial Services (Stock Market/Portfolio Management/NBFC)	Insurance	Artificial Intelligence	Indian	Foreign	Qualitative	Quantitative	Exploratory	Descriptive	Analytical
Artificial Intelligence in Corporate Banking: A Closer Look at the Potential Impact on E-Business Processes Uyttendaele and Muylle [44]	Dissertation	✔			✔		✔	✔				✔
AI in Banking and Finance Goudarzi et al. [16]	Report	✔			✔	✔		✔				
The Impact of Artificial Intelligence (AI) on Financial Job Market He et al. [18]	Report				✔		✔					

(continued)

Table 1 (continued)

Title Author(s) (Year)	Category	Focus issue				Context		Data type		Methodology		
		Banking	Financial Services (Stock Market/Portfolio Management/NBFC)	Insurance	Artificial Intelligence	Indian	Foreign	Qualitative	Quantitative	Exploratory	Descriptive	Analytical
National Strategy for Artificial Intelligence #AIFORALL Roy and Kant [39]	Discussion Paper				✔	✔		✔			✔	
Artificial Intelligence for Banking, Financial Services and Insurance Sector NASSCOM & CMR [30]	Report	✔	✔	✔	✔	✔						

scarcity of resources, unskilled staff, etc. Adopting the new technology would require reengineering of the processes within the company, rewriting the service operations manuals and training staff to use the new technology, thus putting employees outside their comfort zones, so they might resist to change. They may consider the technology as a threat for their jobs [19]. The use of robots, self-service kiosks, chatbots or any other AI-enabled software or machine is not free and requires financial resources in several directions, i.e. acquisition cost, installation cost, maintenance cost, software update cost, cost for hiring specialists, cost for staff training, etc. [10].

Artificial intelligence thrives with data. The more data you have, the better your algorithms will be. However, the lack of credible and quality data is a challenging factor for AI adoption [9].

4 Research Methodology

4.1 About the Study

The study is exploratory and empirical in nature. The study used literature to identify enablers and challenges for AI adoption in Indian financial services sector. The study used secondary data for the factors identified for JP Morgan Chase. Researcher has collected data from the secondary sources comprising various reports, magazines, periodicals, articles, blogs, newspapers, research paper published in esteemed national and international journals and web references.

The study used the following data analysis tools:

- Causal loop diagram (CLD) model prepared in VENSIM software, explaining the dynamics of relationship between artificial intelligence and finance.
- The stock flow diagram (SFD) has been developed using the Stella software (V9.1.3) of system dynamics to holistically understand the system behaviours of AI in finance.

5 Results and Discussion

5.1 Building of Causal Loop Diagram for Adoption of Artificial Intelligence in Finance Sector

To map the various variables that cause the adoption of AI in finance industry and their numerous consequences, a CLD has been developed which represents causal relationships and interdependencies as well as feedback processes of the model. In this CLD (Fig. 2), there are nine reinforcing loops and two balancing loops. To remove the vagueness in the model, we have divided the entire model into two parts.

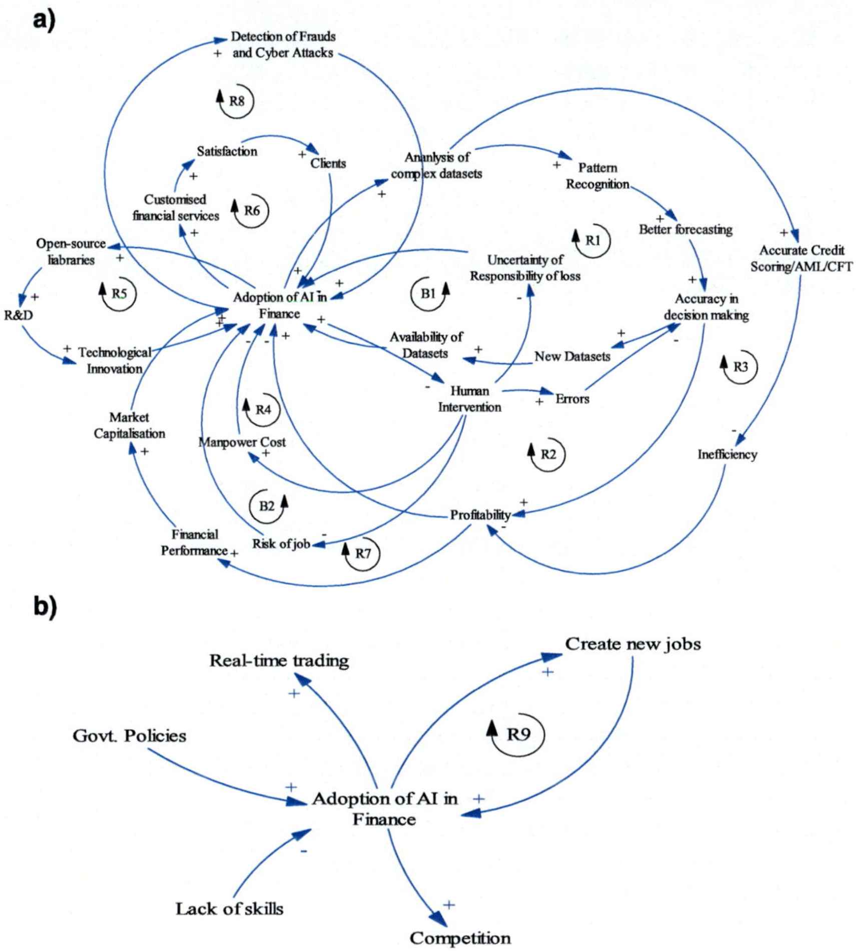

Fig. 2 **a** Causal loop diagram (Source: Author's Construction), **b** causal loop diagram (Source: Author's Construction)

R1 shows the reinforcing effect of AI adoption in financial sector as for the adoption of AI datasets play a vital role. The more data you have, the better your algorithms will be. However, just having a lot of data is not sufficient anymore. Artificial intelligence supports decision by analysing enormous information [9]. Artificial intelligence can analyse complex and huge datasets which human beings cannot do due to their limited cognitive capacity, and AI integrated with ML can predict future patterns which help in better forecasting and accurate decision-making. More accurate decision-making will lead to creation of new dataset, which will further increase available datasets for AI system. More accurate decisions lead to more profitability which further reinforces the adoption AI in financial sector as shown in R2 (Fig. 2a).

AI could allow banks and other enterprises to score creditworthiness of their potential borrowers by analysing complex data from their social media posts and Internet activity: what sites someone visits and what they purchase from ecommerce stores rather than traditional method of credit scoring taking into consideration the credit history of customers. Online behaviour can indicate whether a person is likely to pay back their loans or not. It causes lesser defaults which mean more profitability, leading to more adoption of AI (Fig. 2a, R3).

Adoption of artificial intelligence in finance sector means taking the finance sector towards automation, i.e. less human intervention and more work done by intelligent machines. By replacing manual back-office procedures with automated software solutions, banks can reduce employee errors and speed up processes. This paradigm shift can lead to smaller operational units and allow managers to concentrate on improving tasks that require human intervention. Lesser human interventions lead lesser manpower cost which also reinforces the adoption (Fig. 2a, R4).

AI-enabled open-source software is encouraging more adoption of AI by financial institutes and more research and development in the field of AI (Fig. 2a, R5).

In dynamic and aggressively competitive environment that organisations operate today, institutes or companies providing financial services need to work harder to understand their customers, to engage with them, to provide personalised financial products to them, reward them and ultimately retain them. Customers are ever-more demanding and want fast, high-quality and above all personalised service from the organisations that they deal with. Financial institutes are using AI to increase client satisfaction, improve efficiency and maintain customer loyalty in many ways [15, 20] (Fig. 2a, R6). AI-enabled tools provide financial guidance for the bank's clients through voice and text messages. This allows clients to have access to services at any time without costing more money hiring customer services personnel.

Most important, which the researcher has also depicted in the SD model (Fig. 2a), Adoption of AI will increases the financial performance, i.e. revenue from operations of the organisation that will definitely further increase the market capitalisation of the organisation, motivating the organisation to adopt more AI-enabled tools (Fig. 2a, R7).

Data security is at the top of the list whenever financial institutions are asked about their concerns. According to McAfee report [25], cybercrime and financial fraud are currently costing the global economy $600 billion. Additionally, now also have newfound ability to leverage crypto-currency exchanges to transfer and withdraw stolen money from their fraudulent activities. Such unregulated channels make it imperative for banks and financial institutions to deploy advanced techniques to fight cybercrime. One of the primary objectives using AI is to monitor all processes and data for regulatory compliance and security [40]. In an era of cybercrime and stringent regulatory requirements, a highly fraud-resistant system for protecting and authenticating almost any kind of transaction could have a revolutionary impact on the financial services industry [36] (Fig. 2a, R8).

AI will create as many new AI-based jobs as it will displace [7] which will further reinforce the adoption (Fig. 2b, R9).

The CLD also consists two balancing loops. Adoption of AI also raises legitimate complex legal questions, combined with technological concerns as to the reliability of artificial intelligence programs and software and how employees will deal with such applications in their daily tasks. For example, when computer programs are used to enter into formal contract, it is important to make sure the applicable terms and conditions are communicated to the individual at the end of the line or that a proper disclaimer is duly disclosed. The question that can be raised is who should be responsible if anything went wrong with the contract? [31]. This complex question resists the people to adopt the artificial intelligence, especially for finance-related matters (Fig. 2a, B1).

Advances in artificial intelligence and automation will transform the world. Artificial intelligence has created widespread fear of job losses, especially among the low-skilled employees [19]. According to the report from PwC [37], automated bots could take up to 38% of the jobs in the USA, 30% of the jobs in the UK, 21% in Japan and 35% in Germany. According to study from McKinsey Global Institute [26], automation will displace between 400 and 800 million jobs by 2030, requiring as many as 375 million people to switch job categories entirely. This notable concern regarding the forthcoming impact of AI on employment and the workforce is demoting the adequate adoption of AI (Fig. 2a, B2).

There are various exogenous variables those influence AI adoption in finance. These are:

i. **Lack of skills** required for AI adoption.
ii. **Govt. Policies,** i.e. incentives provided by government to encourage AI adoption by organisations in the form of exemptions and deduction in tax.
iii. **Competition,** i.e. arm race among organisations to beat each other.

5.2 *Model Parameter*

See Table 2.

5.3 *Dynamic View of Linkages Between Adoption of Artificial Intelligence, Financial Performance and Market Capitalisation of an Organisation*

As stated in dynamic model, adoption of AI is dependent on positive and negative dimensions, where positive dimensions, highlighted on the left side, include availability of data, research and development, increase in productivity, concentration on meaningful work, cost reduction and creation of new jobs. On the other hand, factors which have negative impacts are highlighted on the right side of the model, which includes risk of job loss, lack of skills, biasness and adoption cost (Fig. 3).

Table 2 Model Parameters

Parameters	Values
AI in finance	$ 43.2 million i.e. 40% of total R&D Budget of J.P. Morgan Chase $10.8 billion [22]
Financial Performance	$ 9.96 billion [22]
Market Capitalisation	$ 36.63 billion [22]
Cost Reduction	40-60% [6]
Creation of new jobs	9% [14]
Concentration on meaningful work	72% [36]
Increase in productivity	40% [1]
Competition	84% [15]
Research & Development	40% [22]
Risk of job loss	38% [37]
Lack of skills	37% [16]
Biasness	85% [49]
Adoption Cost	40% [49]

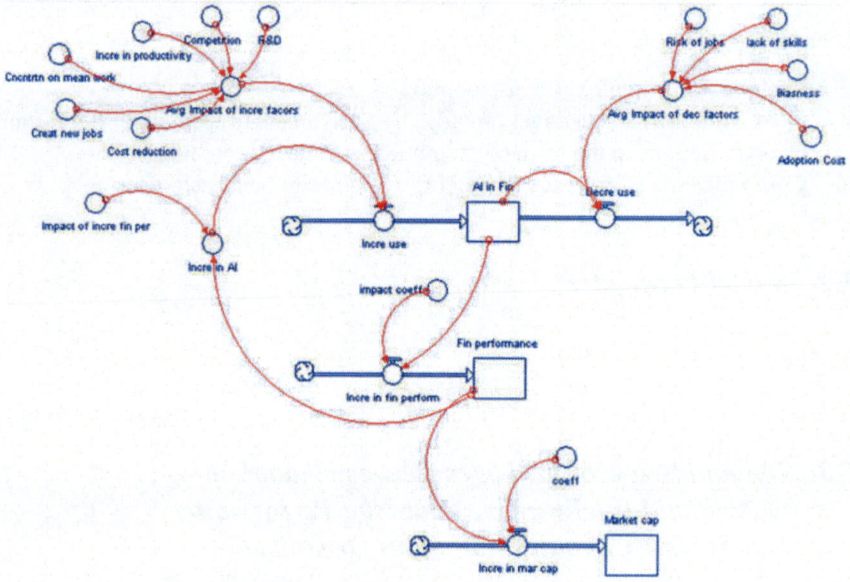

Fig. 3 Stock flow diagram (SFD) (Source: Author's Construction)

If broader focus is given, adoption of AI technology will impact financial performance of the organisation which will further impact the market capitalisation of the same. This model can be used for forecasting of AI impact for organisations other than JP Morgan Chase as well.

5.4 Discussion on Simulation Results

Artificial intelligence is the most transformative business trend in the world today. The effects of AI will be magnified in the coming decade, as virtually every industry will transform their core processes and business models to take advantage of artificial intelligence in their operations. For business leaders, it is imperative to get a plan for making AI work in the organisation. Initial AI projects may get delayed or under delivered, but the risk of businesses becoming non-competitive by ignoring AI is high. The result of simulation clearly shows that irrespective of challenges faced in adoption of AI, the use of AI in financial services sector will increase in near future because it will definitely increase the financial performance of organisations which will further increase market capitalisation of them. With the passage of time, the adoption of AI in finance industry will also increase (Fig. 4).

Summing up the simulation results, we can say that the systems dynamics-based integrative modelling approach has yielded valuable insights into the mechanics of AI adoption in finance and its impact on financial performance and market capitalisation of organisation.

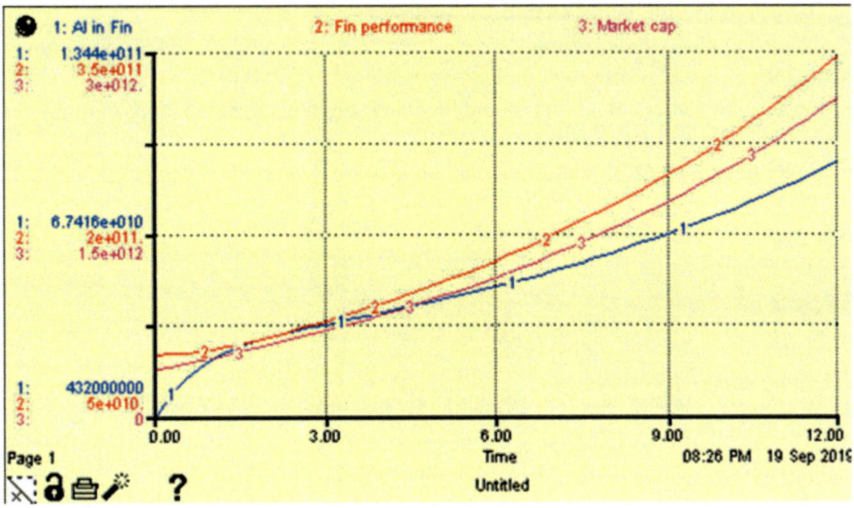

Fig. 4 Simulation of system dynamics model

6 Conclusion

Artificial intelligence has a profound impact on just about every sector of an economy, and the financial service sector is no exception. As better and more advanced artificial intelligence systems are being created, financial services sector is more adopting it to stay on top of its market and provide their customers and clients with the best service delivery possible. Almost every financial service provider has started using ANI to save time, reduce costs and add value.

This research has attempted to propose a new integrative modelling approach for adoption of artificial intelligence in finance sector in a general context. It provides a better fit for holistic demonstration of factors of adoption of artificial intelligence and consequently acquiring insight for the mutual and ever propagating interactions taking place in the system. It can be found that the proposed model, by incorporating an integrative modular framework of generating causal dynamics prevalent for AI adoption in finance sector, succeeds in acquiring the capability of short- to long-term forecasting. As can be seen here that the modelling was primarily first done based on the conceptual insight drawn through an extensive literature survey and thereafter it was put to further refinement, customisation and validation taking a case study of JP Morgan Chase. In nutshell, this study concludes that AI is playing like a game changer in the finance industry, and in the near future, it will become mandate to adopt artificial intelligence by all organisations to sustain in the market.

In the end, we can sum up the core attributes of the present study by highlighting that the undertaken research has succeeded in:

- Building a conceptual integrative model of AI adoption founded over extensive amount of literature and analytical studies.
- Capturing the "whole" along with the "parts" dynamics of AI adoption in finance sector.
- Identifying some critical incumbent feedback and feed-forward loops defining the AI adoption in finance sector.
- Developing a stock flow model based on real data sourced in from public domain.

Appendix

System Model Equation

AI_in_Fin (t) = AI_in_Fin (t − dt) + (Incre_use - Decre_use) * dt

INIT AI_in_Fin = 432000000

INFLOWS:

Incre_use = Avg_Impact_of_incre_facors + Incre_in_AI

OUTFLOWS:

Decre_use = AI_in_Fin-Avg_Impact_of_dec_factors

Fin_performance (t) = Fin_performance (t − dt) + (Incre_in_fin_perform) * dt

INIT Fin_performance = 99624000000
INFLOWS:
Incre_in_fin_perform = AI_in_Fin * impact_coeff
Market_cap(t) = Market_cap (t − dt) + (Incre_in_mar_cap) * dt
INIT Market_cap = 366301000000
INFLOWS:
Incre_in_mar_cap = Fin_performance * coeff
Adoption_Cost = 0.40
Avg_Impact_of_dec_factors = (Adoption_Cost + Biasness + lack_of_skills + Risk_of_jobs)/4
Avg_Impact_of_incre_facors = (Competition + Cncntrtn_on_mean_work + Cost_reduction + Creat_new_jobs + Incre_in_productivity + R&D)/7
Biasness = 0.85
Cncntrtn_on_mean_work = 0.72
coeff = 0.991463565
Competition = 0.84
Cost_reduction = 0.5
Creat_new_jobs = 0.09
impact_coeff = 0.42
Impact_of_incre_fin_per = 0.30
Incre_in_AI = Fin_performance * Impact_of_incre_fin_per
Incre_in_productivity = 0.54
lack_of_skills = 0.37
R&D = 0.40
Risk_of_jobs = 0.38

References

1. Accenture (2016) Newsroom. https://newsroom.accenture.com/news/artificial-intelligence-poised-to-double-annual-economic-growth-rate-in-12-developed-economies-and-boost-labor-productivity-by-up-to-40-percent-by-2035-according-to-new-research-by-accenture.htm
2. Adobe (2019) Digital Trends: Financial Services in Focus. Adobe Systems Incorporated
3. Agarwal A, Gans J, Goldfarb A (2018) Economic policy for artificial intelligence. National Bureau of Economic Research, University of Chicago Press, Toranto, pp 39–159. https://doi.org/10.1086/699935
4. Azar AT (2012) System dynamics as a useful technique for complex systems. Int J Ind Syst Eng 10(4):377–410. https://doi.org/10.1504/IJISE.2012.046298
5. Baldwin AA, Brown CE, Trinkle BS (2006) Opportunities for artificial intelligence development in the accounting domain: the case for auditing. intelligent systems in accounting, Fin Manag 14:77–86. https://doi.org/10.1002/isaf.277
6. Baumgartner T, Hatami H, Valdivieso M (2016) Why salespeople need to develop "machine intelligence". Harv Bus Rev. https://hbr.org/2016/06/why-salespeople-need-to-develop-machine-intelligence
7. BBC News (2018) AI will create as many jobs as it displaces—report. https://www.bbc.com/news/business-44849492

8. Brandi JT (1988) Artificial intelligence, expert systems, and financial planning. J Am Soc CLU & ChFC
9. Chan M-C, Wong C-C, Tse W, CheungB-S, Gordon YN (2002) Artificial intelligence in portfolio management. Springer, Berlin Heidelberg. pp 403–409. https://doi.org/10.1007/3-540-45675-9_60
10. Chatterjee A, Ayadi O, Boone B (2000) Artificial neural network and the financial markets: a survey. Manag Fin 26(12):32–45. https://doi.org/10.1108/03074350010767034
11. Davenport TH, Ronanki R (2018) Artificial intelligence for the real world. Harv Bus Rev. https://hbr.org/2018/01/artificial-intelligence-for-the-real-world
12. Fethi MD, Pasiouras F (2010) Assessing bank efficiency and performance with operational research and artificial intelligence techniques: a survey. Eur J Oper Res 204:189–198. https://doi.org/10.1016/j.ejor.2009.08.003
13. Financial Stability Board (2017) Artificial intelligence and machine learning in financial services Market developments and financial stability implications. Financial Stability Board
14. Forrester https://go.forrester.com/press-newsroom/
15. Gauvrit P (2017) How AI can deliver a personalized and predictive customer experience. Digital CX Conversations by Eptica. https://www.eptica.com/blog/how-ai-can-deliver-personalized-and-predictive-customer-experience
16. Goudarzi S, Hickok E, Sinha A (2018) AI in Banking and Finance. The Centre for Internet and Society, India
17. Gupta N (2017) A literature survey on artificial intelligence. Int J Eng Res Technol (IJERT) 5:1–5
18. He D, Guo M, Zhou J, Guo V (2018) The impact of artificial intelligence (AI) on financial job market. The Boston Consulting Group. China Development Research Foundation, China
19. Ivanov SH, Webster C, Berezina K (2017) Adoption of robots and service automation by tourism and hospitality companies. Revista Turismo and Desenvolvimento 27(28):1501–1517
20. Jackson M (2017) How artificial intelligence can deliver a personalised banking experience. Finextra. https://www.finextra.com/blogposting/14617/how-artificial-intelligence-can-deliver-a-personalised-banking-experience
21. JP Morgan https://www.jpmorgan.com/global/technology/artificial-intelligence
22. JP Morgan Chase & Co. Annual Report (2017). https://www.jpmorganchase.com/corporate/investor-relations/document/annualreport-2017.pdf
23. Keys P (1990) System dynamics as a systems-based problem-solving methodology. Syst Pract 3(5):479–493. https://doi.org/10.1007/BF01064156
24. Masayuki M (2016) The effects of artificial intelligence and robotics on business and employment: evidence from a survey on Japanese firms. RIETI Discussion Paper Series 16-E-066. Japan
25. McAfee (2018) The economic impact of cybercrime— no slowing down. https://www.mcafee.com/enterprise/en-us/assets/executive-summaries/es-economic-impact-cybercrime.pdf
26. McKinsey Global Institute (2017) Jobs lost, jobs gained: workforce transitions in a time of automation. McKinsey & Co.
27. Minsky M (ed) (1968) Semantic information processing. The MIT Press, Cambridge, MA
28. Moudud-Ul-Huq S (2014) The role of artificial intelligence in the development of accounting systems: a review. IUP J Account Res Audit Pract 13(2):7–19
29. Narain K, Swami A, Srivastava A, Swami S (2019) Evolution and control of artificial superintelligence (ASI): a management perspective. J Adv Manag Res. https://doi.org/10.1108/JAMR-01-2019-0006
30. NASSCOM (2018) Artificial Intelligence for Banking, Finanacial Services and Insurance Sector. NASSCOM-CMR
31. Omar SA, Hasbolah F, Zainudin UM (2017) The diffusion of artificial intelligence in governance of public listed companies in Malaysia. Int J Bus, Econ Law 14(2):1–9
32. O'LEARY DE (1991) Artificial intelligence and expert systems in accounting databases: survey and extensions. Expert Syst Appl 3:143–152. https://doi.org/10.1016/0957-4174(91)90095-V

33. Patel M (2018) How AI-powered mobile banking app enhances customer experience. Customer Think
34. Pau LF (1991) Artificial intelligence and financial services. IEEE Trans Knowl Data Eng 3(2):137–148. https://doi.org/10.1109/69.87994
35. Popa C (2011) Adoption of artificial intelligence in agriculture. Bulletin of the University of Agricultural Sciences & Veterinary Medicine Cluj-Napoca. Agriculture 68(1):284–293
36. PwC (2016) Financial services technology 2020 and beyond: embracing disruption. PwC
37. PwC (2017) Will robots steal our jobs? The potential impact of automation on the UK and other major economies. PwC. https://www.pwc.co.uk/economic-services/ukeo/pwcukeo-section-4-automation-march-2017-v2.pdf
38. Rijmenam Mv (2019) How artificial intelligence will disrupt the financial sector. Datafloq. https://datafloq.com/read/artificial-intelligence-disrupt-financial-sector/6417
39. Roy A, Kant A (2018) National strategy for artificial intelligence #AIFORALL. India: NITI AYOG
40. Rupali (2018) AI for banking, insurance, and financial services. DZone. https://dzone.com/articles/artificial-intelligence-for-banking-insurance-and
41. Sinclair A, Brashear J, Shacklad J (2018) AI: the momentum mindset. Accenture
42. Son H (2018) JP Morgan is unleashing artificial intelligence on a business that moves $5 trillion for corporations every day. CNBC. https://www.cnbc.com/2018/06/20/jp-morgan-is-unleashing-artificial-intelligence-on-treasury-services.html
43. Sterman J (2000) Business dynamics, system thinking and modeling for a complex world. http://lst-iiep.iiep-unesco.org/cgi-bin/wwwi32.exe/[in=epidoc1.in]/?t2000=013598/(100. 19
44. Uyttendaele J, Muylle S (2018) Artificial intelligence in corporate banking: a closer look at the potential impact on E-business processes. Gent University, Belgium
45. Vijai C (2019) Artificial intelligence in indian banking sector: challenges and opportunities. Int J Adv Res (IJAR) 7(5):1581–1587
46. Yeoh P (2019) Artificial intelligence: accelerator or panacea for financial crime? J Fin Crim 26(2):634–646. https://doi.org/10.1108/JFC-08-2018-0077
47. Yu K-H, Beam A, Kohane I (2018) Artificial intelligence in healthcare. Nat Biomed Eng 2:719–731. https://doi.org/10.1038/s41551-018-0305-z
48. Zavadskaya A (2017) artificial intelligence in finance: forecasting stock market returns using artificial neural networks. Hanken School of Economics, Helsinki
49. Rudder C (2018) The state of AI: 10 eye-opening statistics. The Enterprisers Projects. https://enterprisersproject.com/article/2018/2/state-ai-10-eye-opening-statistics

Enhanced Tracking Performances of MEMS Gyroscope with Fixed Time Convergent Controller with Unknown System States

Deepika Deepika, Sandeep Kaur, and Shiv Narayan

1 Introduction

In recent few years, the microelectromechanical (MEMS) gyroscopes have witnessed rapid advancements in diverse applications such as navigation, aerospace, platform stabilization, traffic due to their numerous merits of compact structure, low price, low power consumption and good angular velocity measuring accuracy [1, 2]. Nevertheless, their small sizes, sensitivity to temperature, and microfabrication inaccuracies induce thermomechanical noises and system parametric fluctuations which may eventually hamper their performances [3, 4]. Moreover, these uncertainties generate the oscillating frequency which does not match with two oscillating axes [5]. To this end, compensation of gyroscope's uncertainties along with accurate control of angular velocity engenders the researchers to construct a robust controller for MEMS.

In literature, the fuzzy-based controls [1–3], backstepping control [4], sliding mode control (SMC) [5–8], etc., have been widely employed for the control of gyroscope to compensate the unknown uncertainties. Owing to the intrinsic robustness and simplicity of the SMC technique, the fractional order integral terminal SMC scheme has been proposed in [5] for finite time convergence of the positions and velocities of MEMS. The authors in [6] have devised the supertwisting SMC law for tracking objective of MEMS by using PID sliding surface where PID parameters have been tuned via bat algorithm. In [7, 8], the adaptation law-based SMCs have been developed for the MEMS for faster convergence. Research contributions made in evolution of controllers for MEMS have few drawbacks. Firstly, most of the approaches aim to track or stabilize the MEMS dynamics in infinite or finite

D. Deepika (✉) · S. Kaur · S. Narayan
Electrical Engineering Department, Punjab Engineering College (Deemed to Be University), Chandigarh, India
e-mail: sharmadeepika504@gmail.com

V. H. Saran and R. K. Misra (eds.), *Advances in Systems Engineering*, Lecture Notes in Mechanical Engineering,
https://doi.org/10.1007/978-981-15-8025-3_55

time. However, in both cases, the system convergent properties are highly dependent on initial conditions [9–11]. Farther the initial position of gyroscope, more would be settling time. Secondly, in spite of excellent robustness characteristics of SMC, chattering in control signal downgrades the closed-loop performances due to discontinuous reaching law. Thirdly, the gyroscope states are assumed to be known which may not be practically feasible. Consequently, a robust strategy is required for the gyroscope which measures as well as controls the unknown system states while delivering the fixed time convergent tracking performances.

Motivated by the above discussions, a new fixed time observer is developed first in this paper for estimation of positions and velocities of the gyroscope along x and y directions. Thereafter, a new control law is constructed by merging the fixed time terminal SMC law with higher-order fixed time convergent law which would not only be freed from chattering issue, but also from the initial conditions dependency constraint. The proposed observer and controllers allow the convergence of estimation error and tracking error to zero in certain prefixed time independent of the initial conditions, respectively. Such a proposed technique has not yet been evolved in literature for the gyroscope. In addition, the acquired outcomes have been compared with the prior techniques such as PID-based SMC (PIDSMC) and fast terminal SMC (FTSMC) [3].

This paper is categorized into five sections: Sect. 2 shows the modeling for MEMS gyroscope. Section 3 renders the proposed fixed time observer-based fixed time sliding mode control theorems. Then, Sect. 4 reveals the various simulations and results for the implementation of the introduced technique. Section 5 yields the concluding remarks.

2 MEMS Modeling

Consider MEMS gyroscope composed of mass supported with sensors, electrostatic actuators to induce the oscillatory response, springs, and induction device as shown in Fig. 1. The structure on which mass is mounted moves with a constant speed and

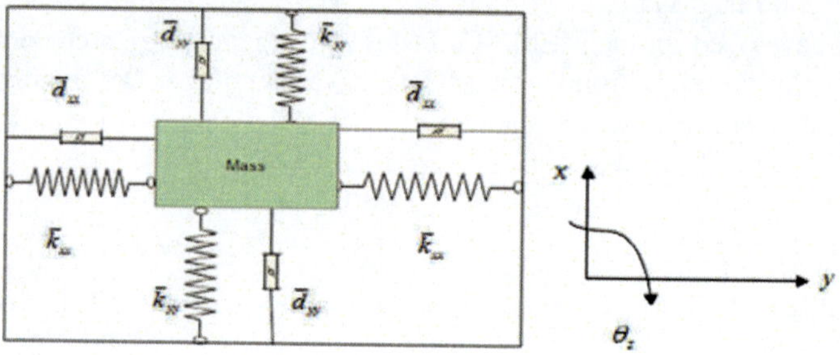

Fig.1 Assembly for a MEMS gyroscope

rotates with slowly varying angular velocity $\overline{\theta}_z$. The Coriolis force is developed in direction perpendicular to both rotational and driving axes. The dynamics for MEMS is represented by [5]:

$$\begin{aligned}\ddot{x} + d_{xx}\dot{x} + d_{xy}\dot{y} + w_{xx}^2 x + w_{xy} y &= u_x + 2\theta_z \dot{y} \\ \ddot{y} + d_{xy}\dot{x} + d_{yy}\dot{y} + w_{yy}^2 y + w_{xy} x &= u_y - 2\theta_z \dot{x}\end{aligned} \tag{1}$$

where $\overline{d}_{xx}/\overline{m}w_0 = d_{xx}, \overline{d}_{xy}/\overline{m}w_0 = d_{xy}, \overline{d}_{yy}/\overline{m}w_0 = d_{yy}, \overline{k}_{xx}/\overline{m}w_0^2 = w_{xx}^2, \overline{k}_{xy}/\overline{m}w_0^2 = w_{xy}^2, \overline{k}_{yy}/\overline{m}w_0^2 = w_{yy}^2, \theta_z = \overline{\theta}_z/\overline{m},$

$\overline{m}$ shows the mass of a proof mass, $\overline{\theta}_z$ is input angular speed, w_0^2(natural frequency square) $\langle x, y\rangle$ are positions and $\langle \dot{x}, \dot{y}\rangle$ are velocities, $\langle \overline{d}_{xx}, \overline{d}_{xy}, \overline{d}_{yy}\rangle$ are damping terms, $\langle \overline{k}_{xx}, \overline{k}_{xy}, \overline{k}_{yy}\rangle$ are spring terms and u_x, u_y are the control input forces divided by respective masses. By assuming $x_1 = x$, $x_2 = \dot{x}$, $y_1 = y$, $y_2 = \dot{y}$, (1) can be reconstructed via four differential equations introduced with certain design imperfections as system uncertainties $\zeta_1(t)$ and $\zeta_2(t)$

$$\begin{aligned}\dot{x}_1 &= x_2 \\ \dot{x}_2 &= -d_{xx}x_2 - d_{xy}y_2 - w_{xx}^2 x_1 - w_{xy}y_1 + u_x + 2\theta_z y_2 + \zeta_1(t) \\ \dot{y}_1 &= y_2 \\ \dot{y}_2 &= -d_{xy}x_2 - d_{yy}y_2 - w_{yy}^2 y_1 - w_{xy}x_1 + u_y - 2\theta_z x_2 + \zeta_2(t)\end{aligned} \tag{2}$$

Control objective: The intention of this paper is to frame an observer-based control law such that the positions and velocities along X and Y-axes of MEMS gyroscope should approach their reference values in fixed time under the influence of system uncertainties satisfying assumption. In other words, the proposed controller should be able to attain: $\lim\limits_{t\to\tau_1}\|x_1(t)\| \to x_d(t)$, $\lim\limits_{t\to\tau_2}\|\dot{x}_1(t)\| \to \dot{x}_d(t)$, $\lim\limits_{t\to\tau_3}\|y_1(t)\| \to y_d(t)$, $\lim\limits_{t\to\tau_4}\|\dot{y}_1(t)\| \to \dot{y}_d(t)$.

where $x_d(t)$ and $y_d(t)$ are the reference position signals to be tracked and their time derivatives provide the respective reference velocities. Furthermore, the velocities of the gyroscope are assumed to be unknown.

3 Fixed Time Observer-Based Control Methodology

The theorem for observer dynamics is presented below for fixed time convergence of estimated positions $\hat{x}_1(t)$, $\hat{y}_1(t)$ and velocities $\dot{\hat{x}}_1(t)$, $\dot{\hat{y}}_1(t)$ to their actual values $x_1(t)$, $y_1(t)$, $\dot{x}_1(t)$, $\dot{y}_1(t)$, respectively. Mathematically,

$$\lim_{t\to\tau_1'}\|x_1(t) - \hat{x}_1(t)\| \to 0,\ \lim_{t\to\tau_2'}\left\|\dot{x}_1(t) - \dot{\hat{x}}_1(t)\right\| \to 0,\ \lim_{t\to\tau_3'}\|y_1(t) - \hat{y}_1(t)\|$$

$$\rightarrow 0, \lim_{t \rightarrow \tau_4'} \left\| \dot{y}_1(t) - \dot{\hat{y}}_1(t) \right\| \rightarrow 0 \quad (3)$$

Theorem 1: Consider the observer dynamics given by:

$$\begin{aligned}
\dot{\hat{x}}_1(t) &= \hat{x}_2(t) - k_{11}|x_1 - \hat{x}_1|^{a_1} sign(x_1 - \hat{x}_1) - k_{12}|x_1 - \hat{x}_1|^{b_1} sign(x_1 - \hat{x}_1) \\
\dot{\hat{x}}_2(t) &= - d_{xx}\hat{x}_2(t) - d_{xy}\hat{y}_2(t) - w_{xx}^2\hat{x}_1(t) - w_{xy}\hat{y}_1(t) + u_x(t) + 2\theta_z\hat{y}_2(t) + \zeta_1(t) \\
&\quad - k_{21}|x_1 - \hat{x}_1|^{a_2} sign(x_1 - \hat{x}_1) - k_{22}|x_1 - \hat{x}_1|^{b_2} sign(x_1 - \hat{x}_1) \\
\dot{\hat{y}}_1(t) &= \hat{y}_2(t) - k_{31}|y_1 - \hat{y}_1|^{a_1} sign(y_1 - \hat{y}_1) - k_{32}|y_1 - \hat{y}_1|^{b_1} sign(y_1 - \hat{y}_1) \\
\dot{\hat{y}}_2(t) &= - d_{xy}\hat{x}_2(t) - d_{yy}\hat{y}_2(t) - w_{yy}^2\hat{y}_1(t) - w_{xy}\hat{x}_1(t) + u_y(t) - 2\theta_z\hat{x}_2(t) + \zeta_2(t) \\
&\quad - k_{41}|y_1 - \hat{y}_1|^{a_2} sign(y_1 - \hat{y}_1) - k_{42}|y_1 - \hat{y}_1|^{b_2} sign(y_1 - \hat{y}_1)
\end{aligned} \quad (4)$$

for the robust tracking of actual positions $x_1(t)$, $y_1(t)$ and velocities $\dot{x}_1(t)$,$\dot{y}_1(t)$ of gyroscope model (2) in x and y directions in fixed time independent of initial conditions. Here, the parameters $k_{ij}, i = 1, 2, 3, 4, j = 1, 2$ are the positive constants, $a_i \in (0, 1), i = 1, 2$ must satisfy $a_2 = 2a - 1, a_1 = a \in (1 - \theta_1, 1)$ for very small $\theta_1 > 0$. Similarly, the design parameter $b_i > 1, i = 1, 2$ must satisfy $b_2 = 2b - 1, b_1 = b \in (1, 1 + \theta_2)$ for very small $\theta_2 > 0$.

Proof: Due to space limitation, the proof has not been shown. The proof of convergence of the above-mentioned error dynamics can be inferred from the concept given in [9].○

In order to derive the control methodology for the control of gyroscope, the following sliding manifolds have to be considered [10]:

$$\begin{aligned}
\sigma_1(t) &= \dot{\xi}_1(t) + l_1|\xi_1(t)|^{n_1} sign(\xi_1(t)) + l_2|\xi_1(t)|^{m_1} sign(\xi_1(t)) \\
\sigma_2(t) &= \dot{\xi}_2(t) + l_3|\xi_2(t)|^{n_2} sign(\xi_2(t)) + l_4|\xi_2(t)|^{m_2} sign(\xi_2(t))
\end{aligned} \quad (5)$$

where $\xi_1(t) = \hat{x}_1(t) - x_d(t)$ and $\xi_2(t) = \hat{y}_1(t) - y_d(t)$ are the deviations from the desired position trajectories in x and y directions, respectively, and $l_1, l_2, l_3, l_4 \in \Re^+$. Moreover, $\dot{\xi}_1(t) = \hat{x}_2(t) - \dot{x}_d(t)$ and $\dot{\xi}_2(t) = \hat{y}_2(t) - \dot{y}_d(t)$.

Theorem 2: Consider the actual model of gyroscope as (2). The following robust controllers would render the robust tracking of the desired trajectories of positions $x_d(t)$,$y_d(t)$ and velocities $\dot{x}_d(t)$,$\dot{y}_d(t)$:

$$\begin{aligned}
u_x(t) &= -l_1|\xi_1(t)|^{n_1-1} n_1\dot{\xi}_1(t) sign(\xi_1(t)) - l_2|\xi_1(t)|^{m_1-1} m_1\dot{\xi}_1(t) sign(\xi_1(t)) \\
&\quad + \ddot{x}_d + d_{xx}\hat{x}_2 + d_{xy}\hat{y}_2 + w_{xx}^2\hat{x}_1 + w_{xy}\hat{y}_1 - 2\theta_z\hat{y}_2 \\
&\quad - p_1|\sigma_1(t)|^{q_1} sign(\sigma_1(t)) - p_2|\sigma_1(t)|^{q_2} sign(\sigma_1(t)) \\
&\quad - p_3|\sigma_1(t)|^{w_1} sign(\sigma_1(t)) - p_4|\sigma_1(t)|^{w_2} sign(\sigma_1(t))
\end{aligned}$$

$$u_y(t) = -l_3|\xi_2(t)|^{n_2-1} n_2\dot{\xi}_2(t) sign(\xi_2(t)) - l_4|\xi_2(t)|^{m_2-1} m_2\dot{\xi}_2(t) sign(\xi_2(t))$$

$$+\ddot{y}_d + d_{xy}\hat{x}_2 + d_{yy}\hat{y}_2 + w_{yy}^2\hat{y}_1 + w_{xy}\hat{x}_1$$
$$+ 2\theta_z\hat{x}_2 - r_1|\sigma_2(t)|^{s_1} sign(\sigma_2(t)) - r_2|\sigma_2(t)|^{s_2} sign(\sigma_2(t))$$
$$- r_3|\sigma_2(t)|^{v_1} sign(\sigma_2(t)) - r_4|\sigma_2(t)|^{v_2} sign(\sigma_2(t)) \quad (6)$$

where $q_3 = 1, q_1 = q_3q_2/(2q_3 - 1), q_2 \in (0, 1), w_3 = 1, w_1 = w_3w_2/(2w_3 - 1), w_2 > 1, s_3 = 1, s_1 = s_3s_2/(2s_3 - 1), s_2 \in (0, 1), v_3 = 1, v_1 = v_3v_2/(2v_3 - 1), v_2 > 1, p_1, p_2, p_3, p_4, r_1, r_2, r_3, r_4 \in \Re^+$.

Proof: Due to space limitation, the proof has not been shown. The proof for the control law can be inferred with the concept utilized in [10, 11]. ○

4 Simulations and Results

The implementations of the closed-loop system with different control strategies such as PID-based SMC (PIDSMC), fast terminal SMC (FTSMC) [3] and proposed control technique have been revealed in this section. In order to execute the proposed algorithm, the two steps have been taken into account. The first step is the realization of proposed theorem 1 for estimating the distinct unknown velocities. The second step involves realization of the proposed theorem 2 for the directing the positions and velocities to their respective desired values. To govern the potentiality of the proposed technique, the outcomes pertaining to four different case studies are compared with previous techniques such as PIDSMC and FTSMC [3]. The various model parameters are selected as follows [5]: $d_{xx} = 0.01, d_{xy} = 0.002, d_{yy} = 0.01, w_{xx}^2 = 355.2, w_{yy}^2 = 532.9, w_{xy} = 70.99, \theta_z = 0.1$

4.1 Observations of MEMS Positions and Velocities with Proposed Observer

The observer dynamics given in theorem 1 are implemented to obtain the positions and velocities of the gyroscope in the presence of random white noise with 0.01 W. Following values are chosen for the implementation of the observer: $a_1 = 1.8, b_1 = 0.8, k_{11} = 0.002, k_{12} = 0.003, a_2 = 2.6, b_2 = 0.6, k_{21} = 0.004, k_{22} = 0.005$ Figs. 2, 3, and 4 render the tracking of the actual gyroscope positions by an observer along x and y directions, respectively. Figures 3 and 5 show the overlapping of the actual velocities along x and y directions, respectively.

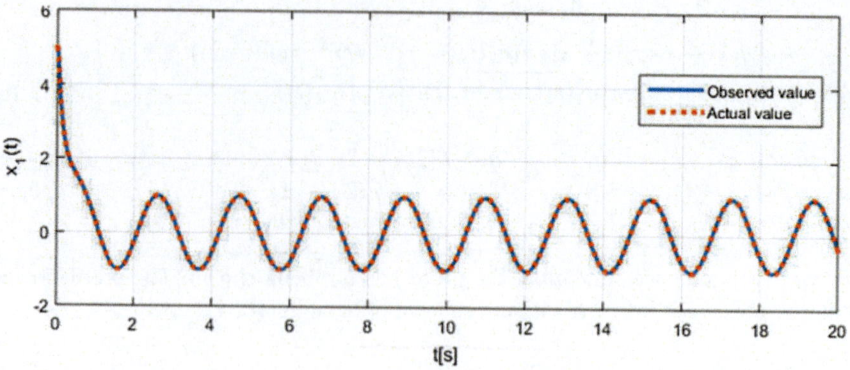

Fig. 2 x_1, $\hat{x}_1$ versus time

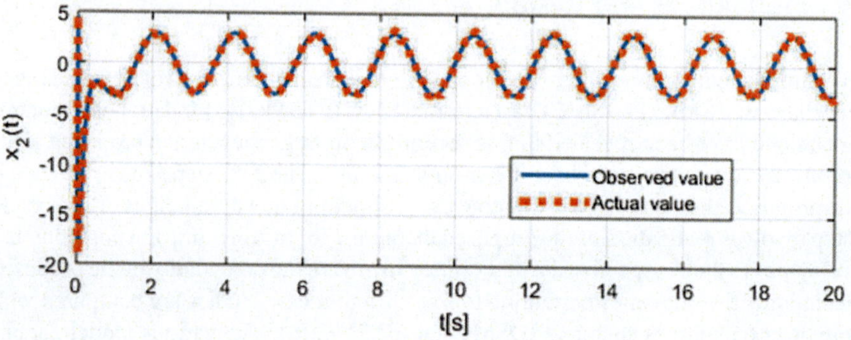

Fig. 3 x_2, $\hat{x}_2$ versus time

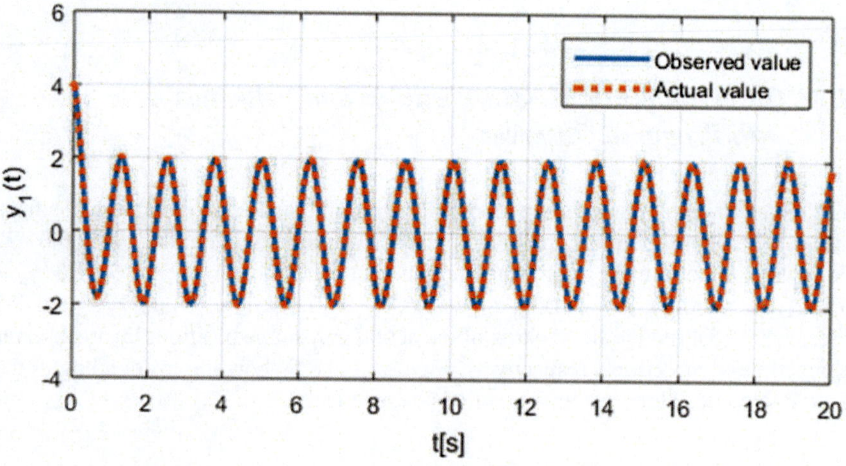

Fig. 4 y_1, $\hat{y}_1$ versus time

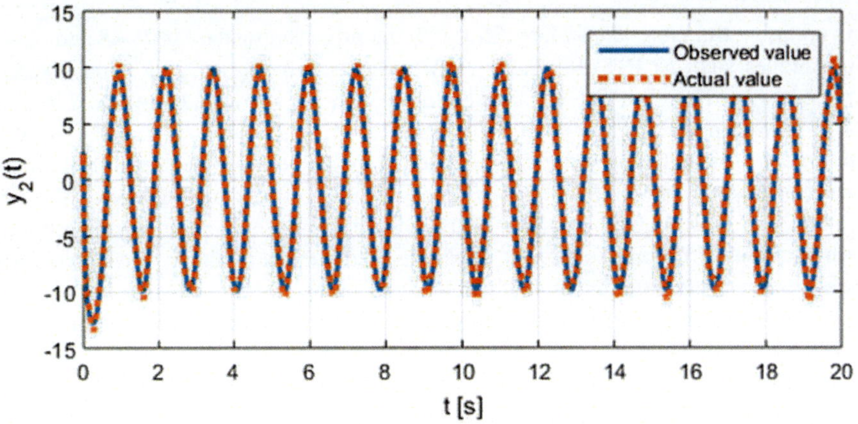

Fig. 5 y_2, $\hat{y}_2$ versus time

4.2 *Positions and Velocities Control in the Presence of Random Noises*

In this case study, the results of the proposed control strategy have been compared with the previously developed techniques such as fast terminal SMC and PID SMC. Here, the control objective is to track the desired positions and velocities of gyroscope: $x_d = \sin(3t)$, $y_d = 2\cos(5t)$ in the presence of random measurement noises. Figure 6a shows the tracking of reference position along x-direction, where the proposed technique is able to converge to reference position in 0.75 s as compared to other techniques such as FTSMC and PIDSMC in 5.1 and 10.2 s, respectively.

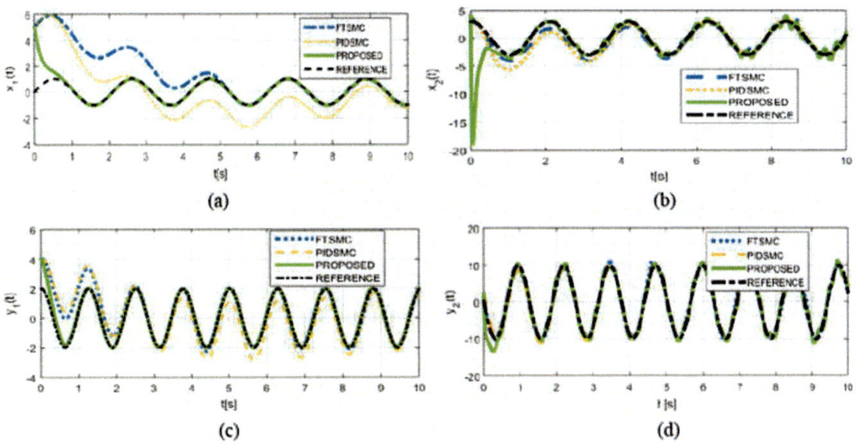

Fig. 6 **a** $x_1(t)$, **b** $x_2(t)$, **c** $x_3(t)$,**d** $x_4(t)$ versus time with FTSMC, PIDSMC, and proposed methods

In addition, the integral absolute error with the proposed method is 76.48 and other techniques such as FTSMC and PIDSMC render indices as 223.62 and 806.55, respectively. Then, in Fig. 6b, the reference velocity is tracked in 0.85, 2.9, and 4.35 s by proposed, FTSMC and PIDSMC methods, respectively. Then, Fig. 6c, the reference position along y-direction is tracked by the proposed technique with least integral absolute error (39.20) as compared to FTSMC (77.26) and PIDSMC (420.50) methods. Finally, Fig. 6d provides the tracking of the desired reference y-directional velocity with different techniques. It can be predicted from all the waveforms that the proposed technique gives the better convergence properties by shorter settling time and comparatively low error indices. This validates the efficacy of the proposed theorems.

5 Conclusions

This paper has presented a new fixed sliding mode control technique for the control of unknown gyroscope states along x and y directions. Firstly, a fixed time observer has been implemented for the gyroscope to obtain the unknown velocities in fixed time. Then, a fixed time sliding mode control law has been developed for convergence of position and velocity tracking errors to zero in fixed time independent of the position of initial conditions. In addition, the robustness against the random measurement noises has also proved the effectiveness of the proposed technique in comparison to the other existing approaches. Out of the various employed techniques, the proposed technique has given much better time domain characteristics such as settling time and integral absolute errors.

References

1. Yan W, Hou S, Fang Y, Fei J (2017) Robust adaptive nonsingular terminal sliding mode control of MEMS gyroscope using fuzzy-neural-network compensator. Int J Mach Learn Cybern 8(4):1287–1299
2. Lei D, Wang T, Cao D, Fei J (2016) Adaptive dynamic surface control of mems gyroscope sensor using fuzzy compensator. IEEE Access 4:4148–4154
3. Liang X, Li S, Fei J (2016) Adaptive fuzzy global fast terminal sliding mode control for microgyroscope system. IEEE Access 4:9681–9688
4. Xin M, Fei J (2015) Adaptive vibration control for MEMS vibratory gyroscope using backstepping sliding mode control. J Vib Control 21(4):808–817
5. Rahmani M (2018) MEMS gyroscope control using a novel compound robust control. ISA Trans 72:37–43
6. Rahmani M, Komijani H, Ghanbari A, Ettefagh MM (2018) Optimal novel super-twisting PID sliding mode control of a MEMS gyroscope based on multi-objective bat algorithm. Microsyst Technol 24(6):2835–2846
7. Rahmani M, Rahman MH (2019) A novel compound fast fractional integral sliding mode control and adaptive PI control of a MEMS gyroscope. Microsyst Technol 1–7

8. Rahmani M, Rahman MH (2019) A new adaptive fractional sliding mode control of a MEMS gyroscope. Microsyst Technol 25(9):3409–3416
9. Basin M, Avellaneda FG (2019) Continuous fixed-time controller design for dynamic systems with unmeasurable states subject to unbounded disturbances. Asian J Control 21(1):194–207
10. Zuo Z (2014) Non-singular fixed-time terminal sliding mode control of non-linear systems. IET Control Theory Appl 9(4):545–552
11. Basin M, Shtessel Y, Aldukali F (2016) Continuous finite-and fixed-time high-order regulators. J Franklin Inst 353(18):5001–5012

Investigation on ZMP Variation of 12-DoF Biped Robot in Screw Theory Framework

Navaneeth Varma, A. P. Sudheer, and M. L. Joy

1 Introduction

Area of robotics is advancing at a very fast rate around the world. Researches have been progressing in all directions of the field such as serial and parallel manipulators, wheeled, tracked and legged robots for various applications. Legged robots find their importance in various field of applications because of the discrete ground contacts, better mobility, and reliability on unpredictable terrains. The biped robot has only two legs and no torso above hip. Hence, balanced dynamic walking of biped robot is difficult and inherently nonlinear. Important methodologies for kinematic modelling of the biped robot are Denavit–Hartenberg (D–H) convention, geometric method, and Screw theory. In most of the work, dynamic model is developed using Lagrangian formulation, Newton–Euler method, Kane's method, Davies method, etc.

D–H convention is used for developing the kinematic model of 8 Degrees-of-Freedom (DoF) biped [1] and 12-DoF biped [2]. The geometric approach for kinematic modelling of five link planar biped robot is depicted in [3]. Kinematic modelling using screw theory for serial robots is considerably less in literature. The kinematic model of humanoid robot Rh-1 is developed using screws and the advantages are stated [4]. The screw-based modelling of an 18-DoF humanoid is explained [5]. The inverse kinematics is validated using differential kinematics considering each limb as a virtual chain connected to hip inertial frame.

N. Varma (✉) · A. P. Sudheer · M. L. Joy
Department of Mechanical Engineering, National Institute of Technology Calicut, Kozhikode 673601, India
e-mail: mg.navaneeth@gmail.com

A. P. Sudheer
e-mail: apsudheer@nitc.ac.in

M. L. Joy
e-mail: mlj@nitc.ac.in

V. H. Saran and R. K. Misra (eds.), *Advances in Systems Engineering*, Lecture Notes in Mechanical Engineering,
https://doi.org/10.1007/978-981-15-8025-3_56

Lagrangian formulation is used for dynamic modelling of a planar biped to study various gaits with toe-rotation [6]. The energy-based Lagrangian formulation is used to develop dynamic model of humanoid SURENA-III [7]. However, the upper hand of recursive Newton–Euler (N–E) method over Lagrangian formulation for robots with higher DoF is proved [8]. A 6-link planar biped is modelled on sagittal plane to analyze the computational complexity of both methods.

Foot rotation indicator (FRI) is a point on the foot/ground surface and the net moment about the point correlates to stability of biped robot [9]. The position of FRI indicates the postural stability of biped robot. Compared to ZMP or the center of pressure (CoP), FRI can be in the exterior of support polygon and the location can indicate the unbalanced moment acting on the biped robot. The concept of zero-moment point was introduced, and the first practical demonstration was carried out on humanoid WL-10RD, predecessor of WABOT from Japan [10]. A 12-DoF biped is kinematically modelled using D–H convention in [11]. Walking is simulated considering stability in sagittal and frontal planes based on ZMP criterion. Cycloid trajectory is used for foot motion for a given step length. The model is experimentally evaluated using a small-sized biped robot.

An algorithm for real-time generation of biped gait combining ZMP manipulation and LIP control is developed [12]. The biped center of gravity is taken as the pendulum mass and so whole-body dynamics in real scenario may differ. The mathematical model of a seven link biped robot walking on flat as well as slope changing surfaces is detailed in [13]. ZMP criteria was employed for both fixed and moving references. The simulation was limited to sagittal plane, defining the hip trajectory in terms of height variations. Similarly, a method is developed to simulate a planar 7-DoF biped robot climbing steps based on ZMP constraint [14]. The kinematic model is developed using geometric method and dynamic model using Lagrange–Euler formulation. Cubic polynomial trajectory is used for foot and a trapezoidal velocity profile is chosen for foot motion. Hip trajectory is assumed to be a straight line having slope equal to that of steps. As an extension, the dynamic walking of 7-DoF planar biped over slope surfaces is simulated using soft computing techniques such as genetic-neural (GA-NN) system and Genetic-fuzzy system (GA-FLC) [15]. ZMP criteria is employed for choosing the optimum path. The work is extended later by considering a ditch as an obstacle using the same parameters of hip and foot trajectories [16]. The aforementioned soft computing methods are employed for optimum gait based on ZMP criterion.

The influence of hip trajectory on balanced walking of biped robot is investigated considering a humanoid of 18-DoF in sagittal plane [17]. A cubic polynomial trajectory is optimized for hip using particle swarm optimization (PSO). The simulation is restricted to sagittal plane and hence ZMP variations for a three-dimensional walking of biped robot has to be studied.

ZMP tracking is also employed for developing control schemes for stable walking of biped. A combination of ZMP and LIP is used for preview control of ZMP for arbitrary foot placements [18]. Similar strategy is adopted for dynamic posture stabilization of biped robot SUBO-1 on slope surfaces [19]. In contrast to such control schemes, calculation of ZMP from resultant forces acting on the foot is presented in

[20]. A detailed study of ZMP variations is carried out using force sensors mount on foot plate. A comparison is done for ZMP trajectories between 12 DoF biped robot Mari-1 and human subjects. However, the accuracy of ZMP tracking depends on the sensor performance.

In this work, the kinematic model of a 12-DoF biped robot is developed in screw theory framework. Dynamic equations are formulated using recursive N–E method. ZMP criterion is used for analyzing the dynamic stability of biped robot. A Bezier curve is chosen as hip trajectory for the biped robot. Cycloid and Bezier curves are selected as foot trajectories for comparing the dynamic stability. The ZMP trajectory is optimized for a single support phase (SSP) based on variation of hip trajectory for given step length and step time.

2 Biped Robot Modelling

2.1 *Kinematic Model*

Kinematic model of a 12-DoF biped robot is developed using screw theory. The link lengths are selected based on human body proportions as shown in Table 1 [21]. Each leg has 6-DoF out of which 3-DoF are provided at hip, 2-DoF at ankle, and 1-DoF at knee. Screws are assigned to individual joints along the axis of rotation as shown in Fig.1. $S_1, S_2, S_3 \ldots, S_{12}$. represent the screw axes in cartesian space. The base frame {b} is attached to the right leg ankle and the end effector frame {s} is provided at the left leg ankle.

The kinematic equations are derived using product of exponential (PoE) formula and Rodrigues' equation. If θ_i is the joint variable of ith joint, the exponential mapping of screw axes can be stated as Eq. (1).

$$e^{[S]\theta} = \begin{bmatrix} e^{[\hat{\omega}]\theta} & \left(I\theta + (1-\cos\theta)[\hat{\omega}] + (\theta - \sin\theta)[\hat{\omega}]^2\right)v \\ 0 & 1 \end{bmatrix} \tag{1}$$

where $R(\hat{\omega},\theta) = e^{[\hat{\omega}]\theta} = I + \sin\theta[\hat{\omega}] + (1-\cos\theta)[\hat{\omega}]^2$ and the skew-symmetric form of normalized angular velocity as $[\hat{\omega}] = \begin{bmatrix} 0 & -\omega_3 & \omega_2 \\ \omega_3 & 0 & -\omega_1 \\ -\omega_2 & \omega_1 & 0 \end{bmatrix}$.

Table 1 Link masses and dimensions of the Biped Robot

Link	Foot	Shank	Thigh	Pelvis
Length (m)	0.021	0.375	0.373	0.206
Mass (kg)	0.725	2.769	6.079	4.597

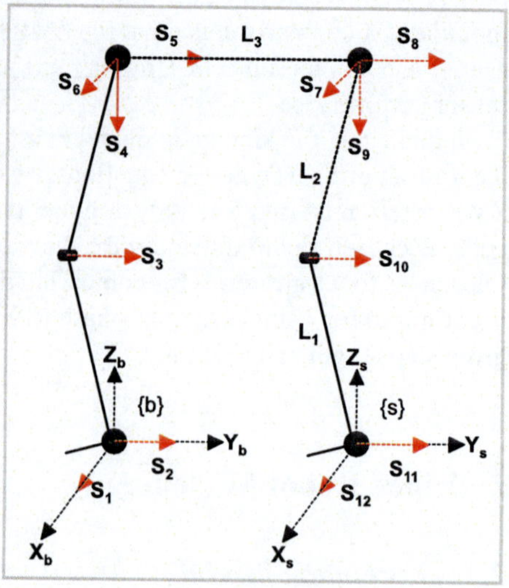

Fig. 1 Kinematic model of Biped Robot

The overall transformation matrix from left leg frame to right leg frame is given in Eq. (2).

$$T(\theta) = e^{[S_1]\theta_1}.e^{[S_2]\theta_2}.e^{[S_3]\theta_3}.e^{[S_4]\theta_4}.e^{[S_5]\theta_5}.e^{[S_6]\theta_6}.e^{[S_7]\theta_7}.e^{[S_8]\theta_8} .e^{[S_9]\theta_9}.e^{[S_{10}]\theta_{10}}.e^{[S_{11}]\theta_{11}}.e^{[S_{12}]\theta_{12}}.M \tag{2}$$

where M is the initial configuration, $= \begin{bmatrix} 1\,0\,0 & 0 \\ 0\,1\,0 & L_3 \\ 0\,0\,1 & 0 \\ 0\,0\,0 & 1 \end{bmatrix}$.

2.2 *Dynamic Model*

The dynamic modelling of 12-DoF biped robot is developed using recursive N–E method. The forward iteration provides link CoM twists and backward iteration gives the wrenches acting on corresponding joints [22]. For joints with combined DoF, a link with negligibly small mass and length is assumed among the specific joints. The frictional resistance at joints and external loads are not considered in this model. The equations for joint torque are derived in the general form shown below in Eq. (3).

$$M(\theta)\ddot{\theta} + C\big(\theta, \dot{\theta}\big)\dot{\theta} + N(\theta) = \tau \tag{3}$$

2.3 *Stability Criteria*

The ZMP is a point on the ground with respect to which the net moment of the biped robot is zero. ZMP must fall within the support polygon on the ground for the biped robot to execute dynamic stable gait. The equations to find coordinates of ZMP on the ground are shown as Eqs. (4) and (5) [23].

$$X_{\text{ZMP}} = \frac{\sum_{i=1}^{n} m_i(\ddot{z}_i + g)x_i - \sum_{i=1}^{n} m_i x_i z_i - \sum_{i=1}^{n} I_{iy}\ddot{\alpha}_{iy}}{\sum_{i=1}^{n} m_i(\ddot{z}_i + g)} \tag{4}$$

$$Y_{\text{ZMP}} = \frac{\sum_{i=1}^{n} m_i(\ddot{z}_i + g)y_i - \sum_{i=1}^{n} m_i y_i z_i - \sum_{i=1}^{n} I_{ix}\ddot{\alpha}_{ix}}{\sum_{i=1}^{n} m_i(\ddot{z}_i + g)} \tag{5}$$

3 Foot and Hip Trajectory

The foot trajectory is assumed to be cycloid and Bezier curves, respectively. Parametric equations of the curves are employed for simulating the gait. The equations of the curves are described. The parametric equation of cycloid curve is shown as Eq. (6).

$$P = [a(t - \sin t), y_{\text{foot}}, a(1 - \cos t)] \tag{6}$$

where, a is the circle radius, y_{foot} indicates the plane of left foot and parameter range is $0 \leq t \leq 2\pi$.

The parametric equation of Bezier curve is shown as Eq. (7).

$$P = (1 - t)^3 P_0 + 3t(1 - t)^2 P_1 + 3t^2(1 - t) P_2 + t^3 P_3 \tag{7}$$

where P_0, P_1, P_2, P_3 are the control points and parameter range is$0 \leq t \leq 1$.

Bezier curve is an approximation curve whereas cycloid is interpolating. The freedom of choice in the control points makes it suitable to generate complex curves. Hence, Bezier curve is chosen for the hip trajectory. The first control point is fixed at the center of pelvis and the remaining three are optimized based on ZMP criteria.

4 Results and Discussion

The biped gait is simulated in MATLAB. The SSP is executed by fixing right foot and moving the left foot through given trajectories. Foot rotation is not considered in the model. The curves are restricted to the plane of moving leg and the biped can move in

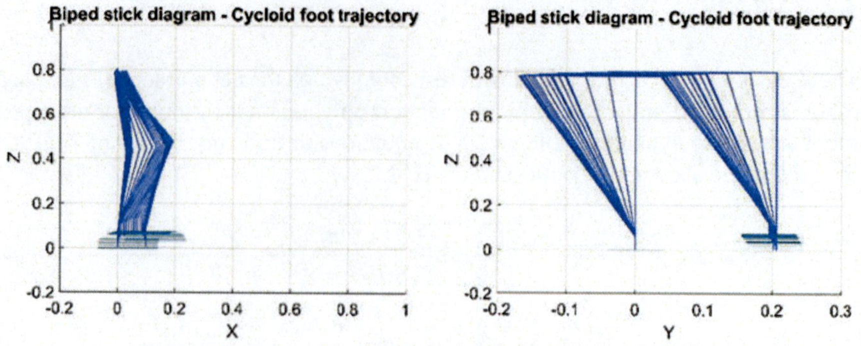

(a) (b)

Fig. 2 **a** Sagittal plane view of biped robot. **b** Frontal plane view of biped robot

three-dimensional space. The step length is chosen as 10cm. The step time is taken as 3 seconds for simulating dynamic walking. A trapezoidal velocity profile is chosen for each joint to avoid collision of the foot with ground at landing. Excluding initial and final postures, nineteen intermediate via points are used to find twists of link and ZMP coordinates. The stance foot (Right foot) represents the support polygon for the SSP model.

4.1 ZMP Variation for Cycloid Trajectory

The biped robot is simulated over a cycloid trajectory. The inverse kinematics at each intermediate point on the path is calculated iteratively using Levenberg–Marquardt (L–M) algorithm. The link CoM twists are calculated using the dynamic model and corresponding ZMP coordinates are found. ZMP trajectory is optimized by varying hip trajectory control points. Figure 2 shows the sagittal plane and frontal plane views of biped robot executing the SSP. The obtained ZMP plot for balanced gait is shown in Fig. 3.

4.2 ZMP Variation for Bezier Curve Trajectory

The biped robot is simulated over a Bezier trajectory. The inverse kinematics at each intermediate point on the path is calculated iteratively using Levenberg–Marquardt (L–M) algorithm. The location of ZMP on the ground is found as explained in previous section. Hip trajectory control points are varied for optimum ZMP trajectory. Figure 4 shows the sagittal plane and frontal plane views of biped robot executing the SSP. The obtained ZMP plot for stable gait is shown in Fig. 5.

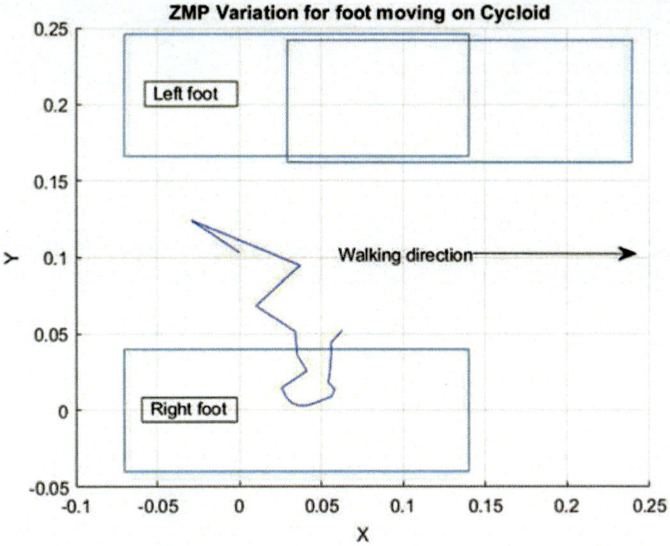

Fig. 3 ZMP plot for SSP over cycloid trajectory

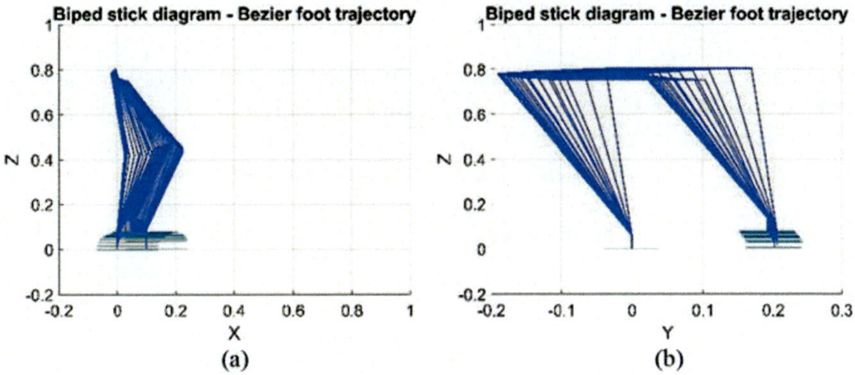

Fig. 4 **a** Sagittal plane view of biped robot. **b** Frontal plane view of biped robot

4.3 *Comparison of ZMP Trajectories*

The ZMP plots for both foot trajectories are analyzed in detail. The ZMP path obtained using Bezier curve is having more points inside the support area compared to cycloid path. Thus, the stability of dynamic gait is better in former case. Also, the control points of Bezier curve can be modified to obtain complex curves like S-curve. Hence, realistic human-like gaits can be simulated using the model.

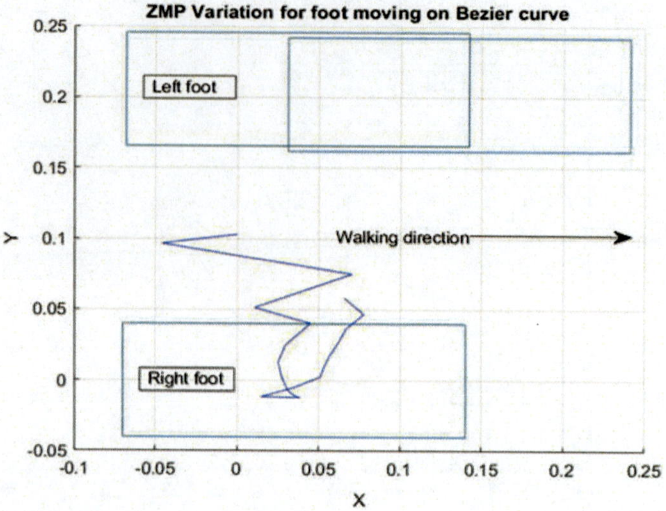

Fig. 5 ZMP plot for SSP over Bezier trajectory

5 Conclusion and Outlook

The kinematic model of biped robot is developed in screw theory framework. Dynamic model for the biped is derived using Recursive Newton–Euler method. The twist vectors of link center of mass have been found. ZMP criteria is chosen for simulating balanced dynamic walking of biped robot on flat surface. Cycloidal and Bezier foot trajectories are used for swing foot and another Bezier curve is used for hip trajectory. The hip trajectory control points are optimized based on ZMP criteria. ZMP variations for the stable SSP are plotted. The frictional loads at joints and external disturbances are not considered in this work. The work can further be extended to various gaits such as ascending and descending slopes, stepping over obstacles, jumping, running, etc. The model can be experimentally validated using a biped robot. The foot size can be optimized for various gaits. ZMP based controllers can be designed and implemented. Machine learning algorithms can be incorporated for intelligent gait generation. Human interactions and self collision issues can be addressed as well in the future.

References

1. Sarkar A, Dutta A (2015) 8-DoF biped robot with compliant-links. Robotics and Autonomous Systems 63(1):57–67. https://doi.org/10.1016/j.robot.2014.09.014
2. Shih CL, Zhu Y, Gruver (1991) Optimization of the biped robot trajectory. In: Conference proceedings 1991 IEEE international conference on systems, man, and cybernetics. IEEE, VA, USA pp 899–903. https://doi.org/10.1109/icsmc.1991.169801

3. Luxman R, Zielinska T (2017) Robot motion synthesis using ground reaction forces pattern: analysis of walking posture. Int J Adv Rob Sys 14(4). https://doi.org/10.1177/1729881417720873
4. Arbulu M, Balaguer C (2007) Real-time gait planning for Rh-1 humanoid robot, using local axis gait algorithm. In: 7th IEEE-RAS international conference on humanoid robots. IEEE, Pittsburgh, pp 563–568. https://doi.org/10.1109/ichr.2007.4813927
5. Toscano GS, Simas H, Castelan EB (2014) Screw-based modeling of a humanoid biped robot. In: Anais do XX congresso brasileiro de automática, Belo Horizonte, pp 944–951
6. Tlalolini D, Chevallereau C, Aoustin Y (2009) Comparison of different gaits with rotation of the feet for a planar biped. Robot Auton Syst 57(4):371–383. https://doi.org/10.1016/j.robot.2008.09.008
7. Sadedel M, Yousefi-Koma A, Khadiv M, Iranmanesh F (2018) Heel-strike and toe-off motions optimization for humanoid robots equipped with active toe joints. Robotica 36(6):925–944. https://doi.org/10.1017/S0263574718000140
8. Al-Shuka HFN, Corves B, Zhu WH (2014) Dynamic modeling of biped robot using Lagrangian and recursive Newton-Euler formulations. Int J Comput Appl 101(3):1–8. https://doi.org/10.5120/17664-8485
9. Goswami A (1999) Postural stability of biped robots and the foot-rotation indicator (FRI) point. Int J Rob Res 18(6):523–533. https://doi.org/10.1177/02783649922066376
10. Vukobratovic M, Borovac B (2004) Zero-moment point—thirty five years of its life. Int J Huma Rob 1(1):157–173. https://doi.org/10.1142/s0219843604000083
11. Sudheer AP, Vijayakumar R, Mohandas KP (2012) Stable gait synthesis and analysis of a 12-degree of freedom biped robot in sagittal and frontal planes. J Auto Mob Rob Intel Sys 6(4):36–44
12. Sugihara T, Nakamura Y, Inoue H (2002) Realtime humanoid motion generation through ZMP manipulation based on inverted pendulum control. In: Proceedings of the 2002 IEEE international conference on robotics and automation. IEEE, Washington, DC, USA, pp 1404–1409. https://doi.org/10.1109/robot.2002.1014740
13. Mousavi PN, Bagheri A (2007) Mathematical simulation of a seven link biped robot on various surfaces and ZMP considerations. Appl Math Mod 31(1):18–37. https://doi.org/10.1016/j.apm.2006.06.018
14. Vundavilli PR, Sahu SK, Pratihar DK (2007) Dynamically balanced ascending and descending gaits of a two-legged robot. Int J Huma Rob 4(4):717–751. https://doi.org/10.1142/S0219843607001266
15. Vundavilli PR, Pratihar DK (2009) Soft computing-based gait planners for a dynamically balanced biped robot negotiating sloping surfaces. Appl Soft Comput 9(1):191–208. https://doi.org/10.1016/j.asoc.2008.04.004
16. Vundavilli PR, Pratihar DK (2010) Dynamically balanced optimal gaits of a ditch-crossing biped robot. Robot Auton Syst 58(4):349–361. https://doi.org/10.1016/j.robot.2009.10.004
17. Mandava RK, Vundavilli PR (2017) Study on influence of hip trajectory on the balance of a biped robot. In: Attele KR, Kumar A, Sankar V, Rao N, Sarma T (eds) Emerging trends in electrical, communications and information technologies, LNEE, vol 394. Springer, Singapore, pp 265–272. https://doi.org/10.1007/978-981-10-1540-3_28
18. Kajita S, Kanehiro F, Kaneko K, Fujiwara K, Harada K, Yokoi K, Hirukawa H (2003) Biped walking pattern generation by using preview control of zero-moment point. In: Proceedings of the 2003 IEEE international conference on robotics and automation, vol 3, pp. 14-19. IEEE, Taipei, Taiwan. https://doi.org/10.1109/robot.2003.1241826
19. Cho B, Kim J (2018) Dynamic posture stabilization of a biped robot SUBO-1 on slope-changing grounds. Int J Prec Eng Man 19(7):1003–1009. https://doi.org/10.1007/s12541-018-0118-8
20. Erbatur K, Okazaki A, Obiya K, Takahashi T, Kawamura A (2002) A study on the zero moment point measurement for biped walking robots. In: Proceedings of 7th international workshop on advanced motion control. IEEE, Maribor, Slovenia, pp 431–436. https://doi.org/10.1109/amc.2002.1026959
21. Winter DA (2009) Biomechanics and motor control of human movement, 4th edn. Wiley, USA

22. Lynch KM, Frank CP (2017) Modern robotics mechanics, planning, and control, 1st edn. Cambridge University Press, USA
23. Huang Q, Yokoi K, Kajita S, Kaneko K, Arai H, Koyachi N, Tanie K (2001) Planning walking patterns for a biped robot. IEEE Trans Rob Auto 17(3):280–289. https://doi.org/10.1109/70.938385

Importance of Electronic Stability Control (ESC) System in Road Vehicles

Akshay Chittawar, Mahesh Mohan, and Sushant Kase

1 Introduction

The electronic stability control underpins the driver in about all basic driving circumstances. It involves the applications of the antilock braking system (ABS) and the traction control system (TCS) . It distinguishes vehicle sliding developments and effectively balances them. This significantly improves driving safety.

ESC mediates exactly when it recognizes a probable loss of controlling control, for instance, when the vehicle is not going where the driver is coordinating. This may happen, for example, while sliding during emergency tricky swerves, understeer or oversteer during insufficiently settled on a choice about turns on tricky lanes, or hydroplaning. ESC measures the heading of the slide and a while later applies the brakes to singular wheels unevenly in order to make torque about the vehicle's vertical center point, confining the slip and adjusting the vehicle back with the driver's told bearing. Likewise, the structure may reduce engine force or work the transmission to back the vehicle off.

All ESC makers emphasize that the system is not an introduction improvement nor a substitution for safe driving practices, yet rather a security advancement to help the driver in recovering from hazardous conditions. ESC does not extend balance, so it does not engage faster cornering (regardless of the way that it can energize better-controlled cornering). Even more, generally, ESC works inside the limits of

A. Chittawar (✉) · M. Mohan · S. Kase
Department of Mechanical Engineering, St. Vincent Pallotti College of Engineering and Technology, Nagpur, Maharashtra 441108, India
e-mail: chittawarakshay@gmail.com

M. Mohan
e-mail: mahi.ngp98@gmail.com

S. Kase
e-mail: sushantskase@gmail.com

V. H. Saran and R. K. Misra (eds.), *Advances in Systems Engineering*, Lecture Notes in Mechanical Engineering,
https://doi.org/10.1007/978-981-15-8025-3_57

the vehicle's managing and open balance between the tires and road. ESC has been demonstrated to be very effective and powerful in reducing accidents identified with loss of control, while concentrating the studies on real-life accidents demonstrates a varying impact, in general, large and consistent. The size of the effect is bigger than numerous other safety systems such as airbags and is sometimes known as the greatest advance in-car security since the induction of safety belts.

2 Situations Where ESC System Finds Its Significance

Even if the driving conditions seem safe and perfect, sometimes the driver is forced to make a sudden maneuver or negotiating a curve suddenly and that can have very serious consequences. Many things can happen on road that has the potential to turn a normal drive into a loss-of-control situation [1]. ESC system in car will not engage in the events, where accidents occur in stop and go traffic condition. However, it is more helpful when slippery conditions are the reason for loss control regardless of whether there is anyone else on the road or not. ESC would reduce the danger element while driving, in several fairly common driving situations, including some below:

- An animal appears on the roadway all of a sudden and the driver must make some evasive move to avoid a collision.
- Rough weather or poor road conditions make skidding and sliding more likely and compromises vehicle stability [1].
- When driver approaches a curve too quickly, forcing him to steer more aggressively.

3 Statistical Importance of ESC System

In specific countries, ESC has just been ordered attributable due to the incredible abatement in fatalities by its usage. For instance, since 2012, USA has ordered application of ESC in all its cars, Australia has commanded utilization of ESC in every traveler cars and SUVs from November 2011 and also implementation of same for all its vehicles from November 2013. The European Union has ordered ESC from October 31, 2014, all over Europe, New Zealand has commanded ESC in traveller vehicles, and SUVs from July 1, 2015 [2]. Numerous positive outcomes have been acquired from these nations. Various outcomes are listed below:

- In any event, 40% of deadly street mishaps are the aftereffect of skidding. Studies show that electronic stability control (ESC) would reduce skidding mishaps by up to 80%. In the European Union, it is assessed that ESC could forestall more than 4,000 casualties and 100,000 wounds every year. In the USA, these figures could ascend to 10,000 casualties and 240,000 wounds [3].

- Australia—The impact of ESC on a wide range of mishaps leading to driver injury reduced drastically to a significant 8% decrease in risk. ESC was related to a critical 8% increase in the danger of different vehicle crashes however, this impact was not obvious when limited to crashes that brought about the driver being harmed. ESC was effective at forestalling single-vehicle crashes (by 28% for all severities and 32% for mishaps that led to driver injury) and especially compelling at forestalling rollover mishaps [4]. When fitted to 4WDs, ESC decreased the danger of rollover crashes by 82%.
- USA—ESC system could forestall 40–56% of untripped rollover accidents and 14% of loss-of-control mishaps. By requiring that ESC systems be introduced on truck, tractors, and huge transports, this proposition would forestall 1807–2329 accidents, 649–858 wounds, and 49–60 fatalities at under $3 million per equivalent life saved, while producing positive net advantages [5]. NHTSA estimates that ESC will decrease single-vehicle accidents of the passenger cars by 34% and single-vehicle accidents of game utility vehicles (SUVs) by 59%, with a lot more noteworthy decrease of rollover mishaps. NHTSA estimates that ESC would save 5300–9600 lives and forestall 156,000–238,000 wounds in a wide range of accidents every year once all light vehicles are outfitted with ESC.
- Mercedes information demonstrates that the establishment of ESC as standard gear on the entirety of its vehicles has brought about a 29% decrease in single-vehicle accidents and 15% fewer mishaps overall [6]. Given these figures, the far-reaching establishment of ESC in the USA was anticipated to save approximately 5000 lives and almost $35 billion in economic losses annually. This research, which executes German government information, is particularly significant because all-new Mercedes vehicles are outfitted with ESC as a standard apparatus.

In view of all observations finished, the use of ESC in the vehicle empowers drivers to all the more likely keep up control of the vehicle. Rundown results are represented in Fig. 1

- With 100% installation of ESC, it was noticed that there was reduction in 850,000 single-vehicle crashes which has saved 8000 lives in USA. This was concluded by IIHS.
- Since 1995, overall 190,000 crashes have been avoided which involves saving 6500 lives due to ESC. This report was estimated by EU [7].
  - There is a 17% reduction in injury accidents.
  - The loss of control situations which leads to accidents are reduced by 45%.
  - There is a 70% reduction in fatal accidents in lost of control.
  - 25% fewer accidents were noted in the statistics given by the Department of Transport, in cars fitted with ESC.
- The traffic data from 2000 to 2004 was analyzed by Swedish National Road Administration and Swdish Universities and they found out that there was a 26% reduction in collison of vehicles equipped with ESC [8].

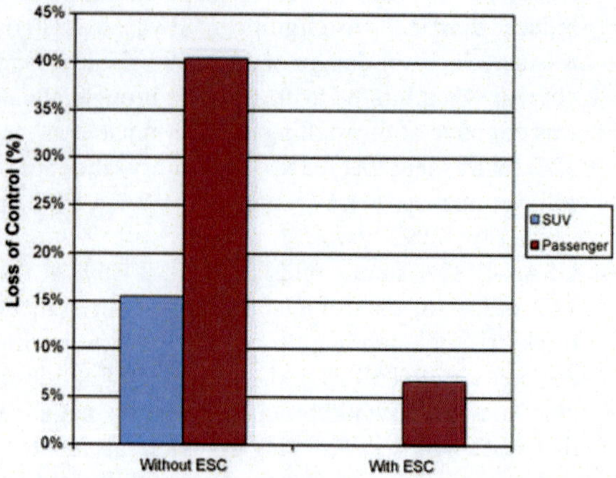

Fig. 1 Effects of vehicle and scenario events on loss of control

- There was a reduced number of impacts, injury, accidents, and rollovers in newly registered ESC-equipped Mercedes vehicles, stated by German "StatischeB undesmt."
  - Crashes of Single–vehicle is reduced by 32%.
  - Overall 16% reduction in collision.
- The frontal impacts was attained instead of off-road-side collision in vehicles which had ESC. ESC has prevented 75% of skidding accidents and 30% of fatal accidents, reported by Audi and Volkswagen [9].
- In Japan, the car manufacturing company Toyota studied that the accidents involving head-on crash and single-vehicle crashes was reduced by 33 and 42%. Mercedes confirms in their study that the ESC is in its maximum effectiveness when the vehicles are in its higher speed. This is the portion where the vehicle dynamics plays a severe role [10].
- The data which includes Toyota models equipped with ESC in Japan given by The Institute for Traffic Accident Research and Data Analysis (ITARDA) is as follows:
  - Casualties is reduced by 40%.
  - 35% reduction in head-on collision.
  - The highest effectiveness of ESC was found in the speed range of 25–65mph [11].
- A study in New Zealand states that it is expected to prevent 400 deaths and 1900 severe injuries with the help of ESC [12].
- In Sweden, 85–100 fatalities could be saved per year and 512 vehicle-related deaths could also be saved if the vehicles have ESC system in them.
- 50% fewer accidents were analyzed in cars equipped with ESC [13].

4 Result

The question that might arise now is,"Does the presence of an ESC system aid the driver in maintaining control of the vehicle in critical citations?" Previously, everything was manual. It was beyond the human ability to control or to maintain a strategic distance from mishaps. With the efficiency of the ESC framework, it has gotten conceivable to keep accidents from happening. ESC can control almost 33% of the deadly mishaps from occurring. The sensors present in the framework are customized to react post any identification of loss of control. In a couple of ESC frameworks, the engine load can likewise be reduced. The framework is proficient to the point that drivers do not have to scan for guiding redresses. This works very well for Indian streets where mishaps are frequent.

5 Conclusions

- By comprehending the above statistical data, new safety technology has been introduced massively through quite simple methods in many developing as well as developed countries. In Sweden, the ESC fitment on new cars is more than 97 %.
- Europe occupied a 32.67% production market share in 2015. The reason for stimulating the marketplace is logical proof that there is an advantage of the innovation. Both industry and society need to get engaged in making such evidence accessible.
- Important stakeholders, like the government, insurance companies, should act accordingly and only buy, rent and use vehicles with the technology that is effective. Media and industry are crucial stakeholders in the process.
- ESC was found to decrease crashes with individual injuries, particularly genuine and deadly injuries. The adequacy ran from in any event 13% for vehicle inhabitants in a wide range of accidents with a genuine or fatal result to at least 35% viability for single/approaching/overtaking genuine and deadly crashes on wet or frosty street surface.

References

1. Goodarzi A, Mehrmashhadi J, Esmailzadeh E (2009) Optimised braking force distribution strategies for straight and curved braking. Int J Heavy Veh Syet 16:78–101
2. Chen Q, Liu Y, Li X (2014) Stability control of vehicle emergency braking with tyre blowout. Int J Veh Technol 2014, Article ID 436175
3. Mishra M, Tiwari A, Maurya S (2016) Electronic stability control system and its possible contribution to Indian road safety. Int J Sci Eng Technol Res (IJSETR) 05

4. Aly AA, Zeidan E, Hamed A, Salem F (2011) An antilock braking system (ABS) control: a technical review. Intell Control Autom 186–195. https://doi.org/10.4236/ica.2011.23023
5. Mauer GF (1995) A fuzzy logic controller for an ABS braking system. IEEE Trans Fuzzy Syst 3(4)
6. Khachane D, Srivastav A (2016) Antilock braking system and its advancement. Int Res J Eng Technol (IRJET) 03 (2016)
7. Ozdalyan B (2008) Development of a slip control antilock braking system model. Int J Automot Technol 09:71–80. https://doi.org/10.1007/s12239-008-0009-6
8. Lin C-M, Hsu C-F (2003) Self learning fuzzy sliding-mode control for antilock braking system. IEEE Trans Control Syst Technol 11
9. Tseng HE, Ashrafi B (1999) The development of vehicle stability control at Ford. IEEE/ASME Trans Mechatronics, 4:223–234
10. Guntur R, Quwerkerk H (1972) Adaptive brake control sytem. In: Proceedings of the institution of mechanical engineers 186:855–880
11. Liebemann EK, Meder K, Schuh J, Nenninger G (2004) Robert bosch GmbH, the bosch electronic stability control (ESP), SAE Paper No. 2004-21-0060
12. Yu F, Feng J-Z, Li J (2002) A fuzzy logic controller design for vehicle ABS with a on-line optimized target wheel slip ratio. Int J Autom Technol 3(4), 165–170
13. Frank P (2000) Slip control at small slip value for road vehicle brake system. Periodical Ploytechnica Ser Mech Eng 44(1):23–30

Modeling and PID Control of Two-Wheel Mobile Robot with a Manipulator and Its Implementation in SIMULINK

Vasudev Gupta, Pushkar Prakash Arya, and Sohom Chakrabarty

1 Introduction

Wheel mobile robot (WMR) has a lots of practical applications for civilians as well as military purposes such as in highway maintenance, floor cleaning, agriculture, unknown environmental exploration, and hostile area surveillance [8, 10] (Fig. 1).

One of the challenge in mobile robot is to develop a simple and effective dynamic model because the mobile robot dynamic is nonholonomic; hence, the kinematic constraints are not integrable [7]. One of such model was proposed in [1, 2], the elegance of this model is that it satisfies important features usually needed in practice, first, the robot moves always in forward direction, and second, a linear model is obtained using feedback linearization, and the resulting model is decoupled in velocities. Therefore, a simple linear controller such as PID, state feedback, and LQR can be implemented. For example, in [5], the linearized model obtained in [2] was used to develop a discrete time model, a delay was introduced to depict processing timen and a pole placement based on a linear state feedback controller was designed. This paper consider the same model as in [2] to design PID controller.

In high-tech world, the demand of mobile robot has been increasing surprisingly which demands researchers to develop mobile robot for diverse applications and implement different control methodology in peruse of getting better controller and mobile robot system design. In support of this, there has been lots of contribution are

V. Gupta
SMIT, Sikkim, India
e-mail: vasugupta643@gmail.com

P. P. Arya (✉) · S. Chakrabarty
Indian Institute of Technology Roorkee, Roorkee 247667, India
e-mail: parya@ee.iitr.ac.in

S. Chakrabarty
e-mail: sohomfee@iitr.ac.in

V. H. Saran and R. K. Misra (eds.), *Advances in Systems Engineering*, Lecture Notes in Mechanical Engineering,
https://doi.org/10.1007/978-981-15-8025-3_58

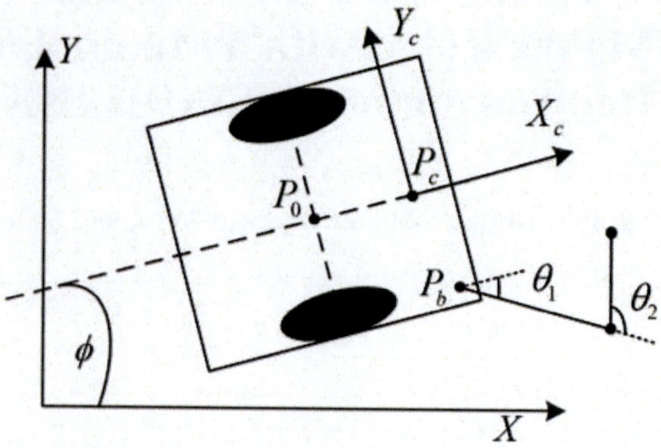

Fig. 1 Schematic diagram of TWMR

made and still researches coming out. Such as, in [3], a sliding mode control on WMR using a novel sliding mode control law and the closed loop system which is robust to bounded external disturbances is implemented, whereas in [6], a sliding mode control for a TWMR with lower center of gravity is considered. In [4], a Lyapunov and standard back stepping-based approach to achieve global regulatory property was developed. Adaptive fuzzy logic-based controller was proposed in [7], and [8] proposed a disturbance observer-based control. A kinematic analysis of WMR on cylindrical workspace is explored, controller law was proposed in [9], and a leader–follower scheme for multiple WMR in unknown environment is explored in [10].

This paper considers a modeling and PID controller design for a TWMR with a manipulator. A block diagram for the implementation is given, and the simulation is done in SIMULINK environment which gives the better insight to the different transformations used in modeling and control. Finally, a few simulations for different track is shown (Table 1).

The contents of the paper are as follows: Section 2 discusses modeling of TWMR, and Sect. 3 discusses control algorithm. In Sect. 4, a block diagram for the implementation in SIMULINK and the SIMULINK model is given, Section 5 discusses simulation results, and finally conclusions are made in Sect. 6.

2 Modeling of TWMR

The control objective is made in two part. Frist is global frame of reference, and second, local frame of reference. The local frame of reference is mounted on the center-of-mass of the robot, whereas the origin of the global frame of reference is outside the robot. The purpose of global frame of reference is to generate a path for robot, whereas the local frame of reference makes the robot move on reference path by orienting the robot with front facing.

Table 1 Nomenclature

P_o:intersection of the axis of symmetry and the axis of driving wheel
P_e : center of the mobile platform
P_b : location of the manipulator on the platform
P_r : reference point of the mobile robot with respect to the robot frame of references $X_c - Y_c$
d : distance from P_o and P_c
b : distance between the driving wheels and the axis of symmetry
r : radius of each wheel
m_c : mass of the plastform without the wheels and rotor the DC motors
m_w : mass of each driving wheel and the rotor of its motor
I_c : moment of inertia of mobile platform platform for m_c about verticle axis through P_o
I_w : moment of inertia of corresponding to m_w about the wheel axis
I_m : moment of inertia of each wheel and the motor rotor about the wheel diameter
(x_c, y_c) : coordinates of the center of mass P_c in the global frame of reference
(x_r^c, y_r^c) : coordinates of the P_r in $X_c - Y_c$ frame of reference
(x_r, y_r): coordinates of desired point in the global frame of reference
ϕ : heading angle of the mobile platform with the X-axis of the global frame of reference
θ_r, θ_l : angular displacement of the right and left wheels
τ_l, τ_r: acting torque on the left and right wheels axis generated by the left and right motors

The robot has two driving wheels with individual DC motors and have four passive wheels at corners to support the mobile robot. The following notations are used in the mathematical model of the robot.

2.1 Constraints

There are three constraints as per practical requirement from the mobile robot. First constraint ensures the robot motion must be in forward direction, the second and third constraints ensure no slip of the left and right driving wheels. These constraints are given as

$$\dot{y}_c \cos\phi - \dot{x}_c \sin\phi - d\dot{\phi} = 0;\ \dot{x}_c \cos\phi + \dot{y}_c \sin\phi + b\dot{\phi} = r\dot{\theta}_r;\ \dot{x}_c \cos\phi + \dot{y}_c \sin\phi - b\dot{\phi} = r\dot{\theta}_l \tag{1}$$

Equation (1) can be written as

$$A(q)\dot{q} = 0 \tag{2}$$

where $q = [x_c,\ y_c,\ \phi,\ \theta_r,\ \theta_l]$, and

$$A(q) = \begin{bmatrix} -\sin\phi & \cos\phi & -d & 0 & 0 \\ -\cos\phi & -\sin\phi & -b & r & 0 \\ -\cos\phi & -\sin\phi & b & 0 & r \end{bmatrix} \tag{3}$$

2.2 System Dynamics

Now, let us obtain the dynamic equations for the mobile robot in terms of Lagrange multipliers z_1, z_2, and z_3, which are given as

$$\begin{aligned} m\ddot{x}_c - m_c d(\ddot{\phi}\sin\phi + \dot{\phi}^2\cos\phi) - z_1\sin\phi - (z_2+z_3)\cos\phi = 0; \\ m\ddot{y}_c + m_c d(\ddot{\phi}\cos\phi - \dot{\phi}^2\sin\phi) + z_1\cos\phi - (z_2+z_3)\sin\phi = 0; \\ -m_c d(\ddot{x}_c\sin\phi - \ddot{y}_c\cos\phi) + I\ddot{\phi} - dz_1 + b(z_3 - z_2) = 0; \\ I_w\ddot{\theta}_r + z_2 r = \tau_r; \quad I_w\ddot{\theta}_l + z_3 r = \tau_l \\ \text{where} \quad m = m_c + 2m_w; \quad I = I_c + 2m_w(d^2+b^2) + 2I_m \end{aligned} \tag{4}$$

Writing equation of motion in (4) in a vector form, we get

$$M(q)\ddot{q} + V(q,\dot{q}) = E(q)\tau - A^T(q)\lambda \tag{5}$$

where the matrix $A(q)$ is given in (3), and matrices $M(s)$, $V(q,\dot{q})$, and $E(q)$ are given as

$$M(q) = \begin{bmatrix} m & 0 & -m_c d\sin\phi & 0 & 0 \\ 0 & m & m_c d\cos\phi & 0 & 0 \\ -m_c d\sin\phi & m_c d\cos\phi & I & 0 & 0 \\ 0 & 0 & 0 & I_w & 0 \\ 0 & 0 & 0 & 0 & I_w \end{bmatrix} \tag{6}$$

$$V(q,\dot{q}) = \begin{bmatrix} -m_c d\dot{\phi}^2\cos\phi \\ -m_c d\dot{\phi}^2\sin\phi \\ 0 \\ 0 \\ 0 \end{bmatrix}; \quad E(q) = \begin{bmatrix} 0 & 0 \\ 0 & 0 \\ 0 & 0 \\ 1 & 0 \\ 0 & 1 \end{bmatrix} \tag{7}$$

Now, represent the dynamic equation in (5) and constraints in (2) in state space with state vector q. To have $A(q)S(q) = 0$, let us define a 5×2 matrix $S(q)$ such that, then $S(q)$ is given as

$$S(q) = \begin{bmatrix} s_1(q) & s_2(q) \end{bmatrix} = \begin{bmatrix} c(b\cos\phi - d\sin\phi) & c(b\cos\phi + d\sin\phi) \\ c(b\sin\phi + d\cos\phi) & c(b\sin\phi - d\cos\phi) \\ c & -c \\ 1 & 0 \\ 0 & 1 \end{bmatrix} \tag{8}$$

where $c = r/2b$. From (8), since the columns of $S(q)$ are in null space of $A(q)$ and are linearly dependent, $\dot{q}$ is in null space of $A(q)$, $\dot{q}$ can be written as the linear combination of the two columns of $S(q)$ and given as

$$\dot{q} = S(q)\upsilon \tag{9}$$

where υ is a chosen as a set of independent velocities and chosen as $\upsilon = \begin{bmatrix} \upsilon_1 \\ \upsilon_2 \end{bmatrix} = \begin{bmatrix} \dot{\theta}_r \\ \dot{\theta}_l \end{bmatrix}$. Differentiating (9) and using (8) and pre-multiplying by S^T, we get

$$S^T(MS\dot{\upsilon}(t) + M\dot{S}\upsilon(t) + V) = \tau \tag{10}$$

The new set of state vector is $x = \begin{bmatrix} q^T & \upsilon^T \end{bmatrix}^T = [x_{c,}y_c, \phi, \theta_r, \theta_l, \dot{\theta}_r, \dot{\theta}_l]^T$. Then, the new set of robot dynamic becomes

$$\dot{x} = \begin{bmatrix} S\upsilon \\ f_2 \end{bmatrix} + \begin{bmatrix} 0 \\ (S^T MS)^{-1} \end{bmatrix} \tau \tag{11}$$

where $f_2 = (S^T MS)^{-1}(-S^T MS\upsilon - S^T V)$. By using feedback internalization, with nonlinear feedback $\tau = S^T MS(u - f_2)$, (11) can be further simplified to

$$\dot{x} = \begin{bmatrix} S\upsilon \\ 0 \end{bmatrix} + \begin{bmatrix} 0 \\ I \end{bmatrix} u \tag{12}$$

3 Control Algorithm

The relation between robot position in the world frame of reference (x_r, y_r) and the reference point in the global frame of reference (x_r^c, y_r^c) is given as

$$x_r = x_c + x_r^c cos\phi - y_r^c \sin\phi; \quad y_r = y_c + x_r^c \sin\phi + y_r^c \cos\phi \tag{13}$$

We choose output as the coordinates of the reference point in the mobile dynamics, i.e., $y = h(q) = \begin{bmatrix} x_r & y_r \end{bmatrix}^T$. The linearized system dynamics in (12) is decoupled with input–output linearization with decoupling matrix as

$$\Phi = J_h(q)S(q) = \begin{bmatrix} \Phi_{11} & \Phi_{12} \\ \Phi_{21} & \Phi_{22} \end{bmatrix} \tag{14}$$

where $\Phi_{11} = c((b - y_r^c)\cos\phi - (d + x_r^c)\sin\phi)$, $\Phi_{12} = c((b + y_r^c)\cos\phi + (d + x_r^c)\sin\phi)$, $\Phi_{21} = c((b - y_r^c)\sin\phi + (d + x_r^c)\cos\phi)$, and $\Phi_{22} = c((b + y_r^c)\sin\phi - (d + x_r^c)\cos\phi)$ and the nonlinear feedback to achieve input–output lineaization is give as

$$u = \Phi^{-1}(q)(v - \dot{\Phi}(q)\upsilon) \tag{15}$$

Then, the linearized decoupled system becomes $\ddot{y}_1 = \upsilon_1$; $\ddot{y}_2 = \upsilon_2$.

For reference tracking, consider the error, $e(t) = y_d(t) - y(t)$, where $y_d(t)$ is desired velocity, and $y(t)$ is the actual velocity of the mobile robot. The PID controller with proportional, derivative, and integral gain as K_p, K_d, and K_i, respectively, is given as

$$C = K_p e(t) + K_d \frac{d}{dt} e(t) + K_i \int e(t)dt \tag{16}$$

4 SIMULINK Model

The simulation studies is done is SIMULINK environment. The block diagram for implementation in SIMULINK is given in Fig.2(a), and Fig.2(b) is the SIMULINK model.

5 Simulation Results

The PID controller parameters are chosen as, $K_p = 30$, $K_d = 10$, and $K_i = 15$. The initial position of the robot is considered as $(x_{ro}, y_{ro}) = (0, 5)$. The simulation is done for three different tracking conditions. First, the target is to move the robot to position $(x_r, y_r) = (0, 7)$ in the 2D plane. The path followed by the robot is given in Fig.3(a). It can be observed that the robot moved from position $y = 5$ to $y = 7$ within 5 seconds.

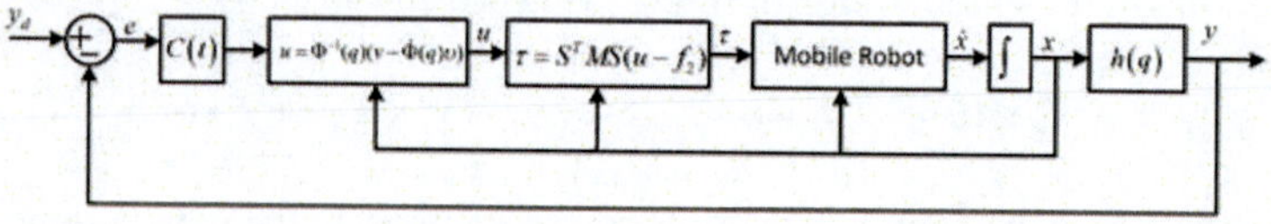

(a)

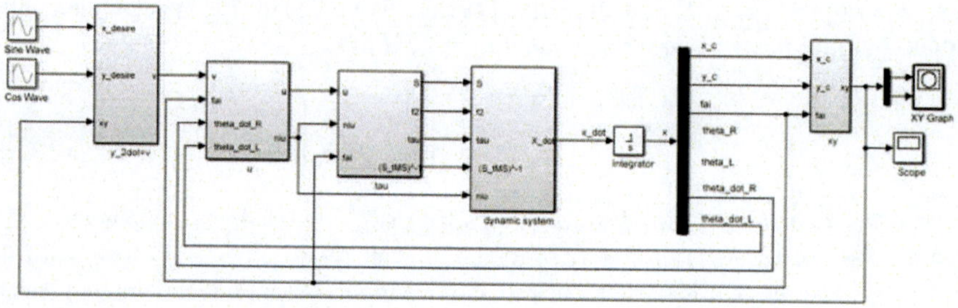

(b)

Fig. 2 Block diagram for implementation

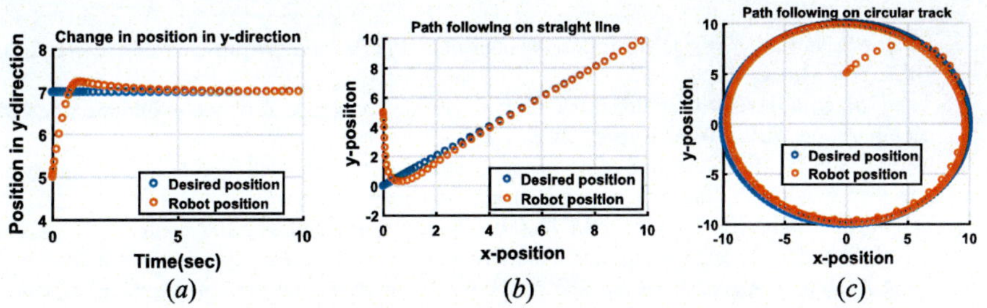

Fig. 3 Different path following results

In second case, the desired path for robot that is followed is from (0,0) to (10,10), which is a straight line. From Fig.3(b), it can be observed that the robot started from its initial position (0,5) and moved to (10,10) through desired path.

Third, robot has to move on a circular path with center at origin, i.e., (0, 0), and radius 10. The path followed by the robot is shown in Fig.3(c), and it can be seen that the robot tracked the path very close to the desired one.

6 Conclusion

In this paper, modeling and control of TWMR is discussed. A PID controller parameter is obtained by trial-and-error methodm and the simulation is done in SIMULINK environment. It is observed that PD controller cannot eliminate steady-state error, whereas PI controller gives very sluggish response, and P controller alone would not reach steady state and will be very sluggish. However, the plots are not shown as the tracking was not satisfactory. Therefore, we went for PID controller to get a satisfactory tracking. The simulation for three different tracking profile is shown in Fig. 3.

References

1. Hunag D, Zhai J, Ai W, Fei S (2016) Disturbance observer-based robust control for trajectory tracking of wheeled mobile robots. Neurocomputing 198:74–79
2. Wang Y, Wang D, Yang S, Shan M (2019) A Practical leader—follower tracking control scheme for multiple nonholonomic mobile robots in unknown obstacle environments. IEEE Trans Cont Syst Tech 27(4)
3. Das T, Kar IN (2006) Design and implementation of an adaptive fuzzy logic-based controller for wheeled mobile robots. IEEE Trans Control Syst Tech 14(3)
4. Yun , Yamamoto Y (1993) Internal dynamics of a wheeled mobile robot. In: International conference on intelligent robots and systems, Yokohama, Japan
5. Yamamoto Y, Yun X (1994) Coordinate locomotion and manipulation of a mobile manipulator. IEEE Trans Autom Control 39(6)

6. Coelho P, Nunes U (2005) Path-following control of mobile robots in presence of uncertainties. IEEE Trans Robot 21(2)
7. Yang JM, Kim JH (1999) Sliding mode control for trajectory tracking of nonholonomic wheeled mobile robots. IEEE Trans Robot Autom 15(3)
8. Yue M, Sun W, Hu P (2011) Sliding model control for two-wheeled mobile robot with lower center of gravity. Int J Innov Comput Information Control 7(2)
9. Dixon WE, Jiang ZP, Dawson DM (2000) Global exponential setpoint control of wheeled mobile robots : a Lyapunov approach. Automatica 36:1741–1746
10. Song Z, Ren H, Zhang J, Ge SS (2016) Kinematic analysis and motion control of wheeled mobile robots in cylindrical workspace. IEEE Trans Autom Sci Engg 13(2)

An Adaptive Mobile Manipulator System for Indoor Surveillance and Security Monitoring

Arindam Chakraborty, M. Naresh Naik, and Bhaskar Dasgupta

1 Introduction

CCTV camera and other static surveillance measures will soon turn out to be grossly deficient in the face of growing necessity of security monitoring in a large number of closed public places like ports and stations, halls and malls, industries and corporate houses and so on. Surveillance now requires mobility, better visibility, a degree of autonomy in operation and visual intelligence to meet the ever rising challenges of security monitoring. Last twenty years have seen a growing trend of research in mobile surveillance systems [1, 2]. Both aerial and ground vehicles have been used [3–5] from semi-autonomous to autonomous level of navigation in indoor environments. They have been deployed in stores and remote surveillance [6, 7] etc. [8–10] discusses surveillance robots from a specific purpose with fixed set of tasks but not as a self-sufficient, multi-purpose system for visual intelligence.

In this work, a sensor-loaded mobile arm platform was chosen for a security monitoring system to increase mobility and decrease the visibility related constraints in surveillance. The increased complexity due to reconfigurability is to set by scene independent operation of the system. SONAR quartet in four directions (UP, DOWN, LEFT and RIGHT) were chosen because they perform well on their own and allow range sensor-based Voronoi construction in the plane, and also, they can be easily fused with other on-board sensor modules. A PID control was employed although we leave the control aspects for a later date for detailed discussion.

A. Chakraborty (✉) · M. Naresh Naik · B. Dasgupta
Centre for Mechatronics, IIT Kanpur, Kanpur, UP 208016, India
e-mail: arindamc@iitk.ac.in

M. Naresh Naik
e-mail: moode@iitk.ac.in

B. Dasgupta
e-mail: dasgupta@iiitk.ac.in

V. H. Saran and R. K. Misra (eds.), *Advances in Systems Engineering*,
Lecture Notes in Mechanical Engineering,
https://doi.org/10.1007/978-981-15-8025-3_59

The sections in the article are organized as follows. Section 2 provides a component level description of the mobile robot system for monitoring. In 3, we list out the kinematic equations and calibration procedure. In Sect. 4, we provide an overview of SONAR quartet-based indoor navigation for KaMoMan. Section 5 presents the image acquisition and processing methods for various intelligence objectives.

2 Mechanical and Electronic Components

2.1 Mechanical Design

Each manipulator module of the mobile system is made of three parts in such a way that link length l and twist angle can be changed. Suitable acrylic sheet (that can be cast) was used for fabricating these modules. Further, spare aluminium modules were also made and kept in reserve. Manipulator base was made using mild steel on lathe machine. Two such modules were made, and servos were mounted on each one of them. It is a 'reconfigurable serial manipulator' that can operate in a cluttered environment and reach desired task space points by avoiding obstacles. The novelty and advantage of this manipulator1 is that, if the manipulator is not able to reach a particular task space point without colliding with an obstacle, the system, when reconfigured with suitable link parameters, may reach that task space point. This has been achieved by designing modules with variable DH parameters [11].

A sponge-reinforced gripper was used as the end effector of the arm to minimize the effect of arm vibration on the camera (Fig. 1).

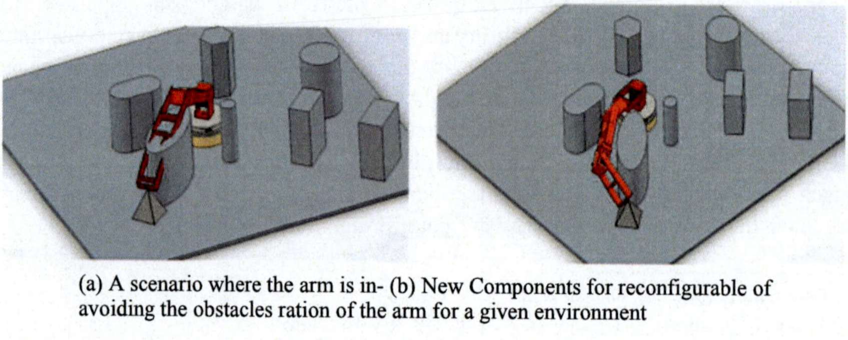

(a) A scenario where the arm is in- (b) New Components for reconfigurable of
avoiding the obstacles ration of the arm for a given environment

Fig. 1 Implementing reconfigurability of arm in the mobile system

2.2 Electronic Components and Hardware Integration

Arduino Mega 2560 was used as the microcontroller in conjunction with Hercules 6–24 V, 16 A motor driver with built-in current sensor for motor control. The driver can withstand up to 30A peak current and can be interfaced with 3.3 and 5 V logic levels. Additionally, two 10 A servo drivers were used for the robotic arm (and the gripper). The manipulator consists of three servo motors (3 DoF), with each joint (FITEC FS5515M/Futuba S3003/Blue Bird HV servo) of manipulator requires specific torque requirement to do the job. As expected lowest joint of the arm requires highest torque capability.

The mobile robot part of KaMoMan is a four-wheeled platform; each rear wheel was driven by Maxon DC motor, and front wheels (omni wheel) are passive ones. Each DC motor contains planetary gearbox GP 42 C with diameter 42 mm, 3–15 Nm torque (ceramic version). Each of them has a mounted encoder (MR, type L, 1024 pulses, 3 channel) with line driver. Arduino gives input to these servo motors through pin 2, pin 4 and pin 5. Complete system was powered by VRLA AGM 12 V lead acid battery.

HC-SR04 ultrasonic sensors have been used for avoiding obstacles and they consist of 4 pins viz. V_{cc}, trigger, echo and GND. With working current up to 15 mA, they have an effectual angle 15 and a measuring angle of 30 degrees. Input pulse has a period of 20 s, and the range of the SONAR is 3–8 m. For inertial measurement I080202 was employed (three-axis Gyro + three-axis Acceleration + three-axis Magnetic Field + Air Pressure Module) (Fig. 2).

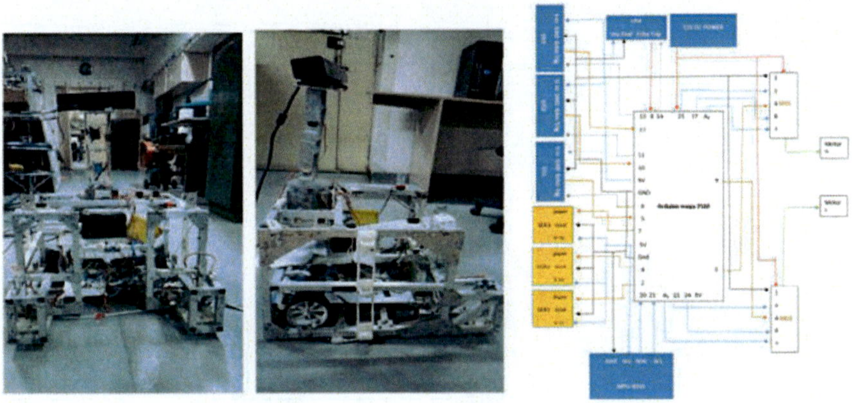

(a) (b)

Fig. 2 **a** KaMoMan holding Kinnect for image acquisition, **b** Electronic integration for the developed system

3 Kinematic Analysis and Motion Calibration of KaMoMan

3.1 Kinematics of Mobile Manipulator System

Direct Kinematics allows the end-effector position and orientation to be written as a function of the joint angles of the system, and it is a mapping from the joint angle space to the operational space of the system, where joint angle space is the space of all possible joint angles of the system, and the operational space is the space of end-effector position and orientation. We will, for the sake of brevity, only provide the results in the form of kinematic equations between mobile base and end-effector variables $X_{ee} = [x_{ee}; y_{ee}; z_{ee}]^{T}$. The vector is the vector of joint angles. Kinematic relationship is governed by the following equations (Fig. 3)

$$x_{ee} = 75c_1c_{23} + 170c_1c_2;\ y_{ee} = 75s_1c_{23} + 170s_1c_2;\ z_{ee} = 75s_{23} + 170s_2 + 65 \quad (1)$$

For the complete mobile system, the end-effector coordinate in mobile base frame M is given by

$$^{B}x_{ee} = x_B + (75c_1c_{23} + 170c_1c_2)c\phi_B - (-140 + 75s_1c_{23} + 170s_1c_2)s\phi_B \quad (2)$$

$$^{B}y_{ee} = y_B + (75c_1c_{23} + 170c_1c_2)s\phi_B - (-140 + 75s_1c_{23} + 170s_1c_2)c\phi_B \quad (3)$$

$$^{B}z_{ee} = 75s_{23} + 170s_2 + 160 \quad (4)$$

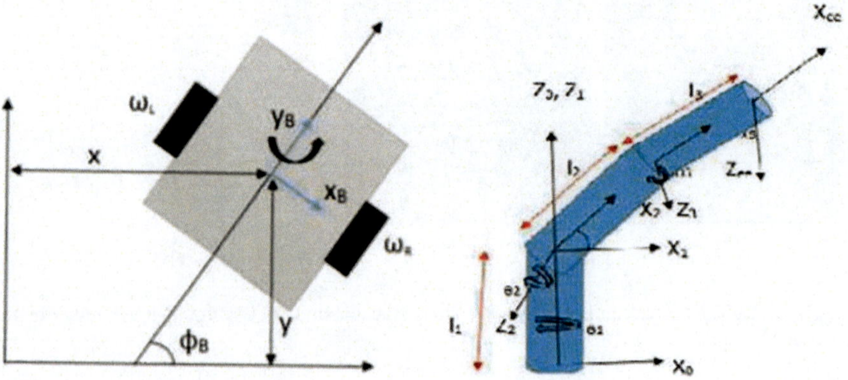

Fig. 3 Kinematic structure of KaMoMan: vehicular platform and the robotic arm in schematic (including the link parameters, length in mm)

3.2 Calibration of Motion: Finding the Drift

Right motor encoder channel A is connected to pin 44, channel B is connected to pin 45, and left motor encoder channel A is connected to pin 8, channel B is connected to pin 46 of Arduino microcontroller. The encoder has 1024 pulses/revolution; it will give 1024 pulses on channel A and 1024 pulses on channel B. For both the rising and falling edges of the channel, we will have 4096 events in one revolution (rev), giving an angular revolution !$_r$ of

$$w_r = \frac{360^\circ}{(1024.\text{pulses/Rev})(4.\text{events/pulse})} \quad (5)$$

Plugging the parameters, !$_r$ = 0:0879 and time interval between each event at 25 PWM is calculated as 17 s. Part of the calibration results performed at the laboratory test beds are shown in Fig. 4

4 Indoor Navigation

The emitter of the SONAR produces a 40 kHz sound wave, and the detector detects 40 kHz sound wave and sends electrical signal back to the microcontroller. Ultrasonic sensors give echo signal as input to the Arduino through pin 5, pin 10 and pin 12. Arduino takes this echo signal as input and controls (giving o/p to) DC motors through motor drivers. As it is visible in the block diagram below, pointing direction of arrow indicates input to that system.

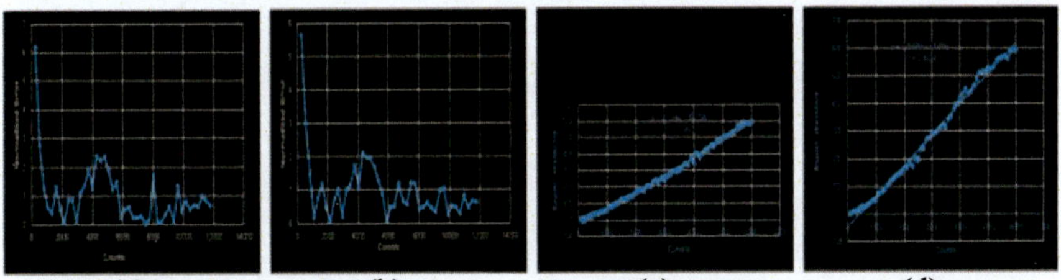

(a) (b) (c) (d)

Fig. 4 **a** and **b** Drive error calibration results for intended rotational motion in forward movement: Pulse count versus angular distance error (left and right wheel), **c** Forward Calibration (linear motion), **d** Backward calibration (linear)

4.1 Grid-Based Navigation

The grid used as a reference in the code is a 3×3 square grid having 100 cm $\times$ 100 cm. Starting node in the grid is assumed as the origin. Referring the 2D system, we have taken ARENA TOP or $+y$ axis as 1, ARENA LEFT or $-x$ axis as 2, ARENA RIGHT or $+x$ axis as 3 and ARENA BOTTOM or $-y$ axis as 4.

This algorithm takes four parameters for the navigation of KaMoMan, namely FRONT sensor distance d_1, LEFT sensor distance d_2, RIGHT sensor distance d_3, BACK sensor distance d_4. In the code, we are using 3-dimensional array. Every time while KaMoMan makes a decision, it chooses the best move possible to reach the destination. It updates the coordinates and records the direction of previous path travelled, so that it does not go to the point it came from.

4.2 Indoor Navigation

In Fig. 5, KaMoMan was moving from $(x_1; y_1)$ in the $+x$ direction by 1 m; the final coordinate is $(x_1 + 1; y_1)$. The notation we used in the flowchart (Sect. 5) for above navigation is $m[x_1 + 1][y_1]$. Where $[x_1 + 1][y_1]$ denotes final position of KaMoMan and [2] denotes previous path direction (1 for ARENA TOP, 2 for ARENA LEFT, 3 for ARENA RIGHT and 4 for ARENA BOTTOM). Every time while KaMoMan checks for an obstacle, it goes through switch case and checks each case. The respective flowcharts of each case are presented in the supplementary material.

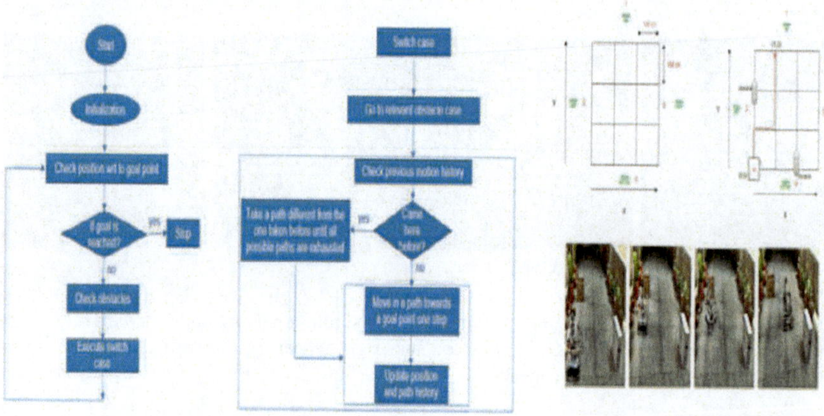

(a) Flowchart for navigation scheme in presence of obstacles

(b) Grid-based navigation and performing left and right turns

Fig. 5 Navigation flowchart and grid-based implementation for autonomous motion of KaMoMan

4.3 *Fusion-Based Navigation*

An active research idea within the working group is to provide fusion-based solution to navigation challenges. IMU sensors, SONAR and camera module data may be fused to provide a better estimate of position and other state variables of the system. Recursive Bayesian filters such as extended Kalman filter (or the information filter) can be used to generate the mean and covariance estimate of position for the vehicle. For initial estimates, a calibration data for linear and rotational motion of KaMoMan has been accrued by moving the system on chequer board patterns. A complete battle-ready version will incorporate the fusion features in controller and image processing algorithms.

5 Imaging and Analysis of Images

Images are acquired from translating wireless IP camera parallel to the scene. We fixed joint angles of third servo (end-effector servo), second servo such that IP camera can view the scene clearly, and base servo will rotate from 250 to 1200, so that, wireless IP camera attached at the end effector can view the one side of the arena clearly and take pictures of it. When base servo completes its rotation to 1200 then KaMoMan will take 900 right turn to cover the other side of arena. Base servo rotation and KaMoMan 900 right turn, this loop continues till it completes taking pictures of four sides of arena. Each image in the series acquired for panoramic image stitching partially overlaps the previous and the following images.

In Fig. 6, we have shown the low dimensional embedding [12] of the captured images and estimate the degree of freedom in motion. The knee of the curve points to the dimensionality of the total degrees of freedom involved in motion.

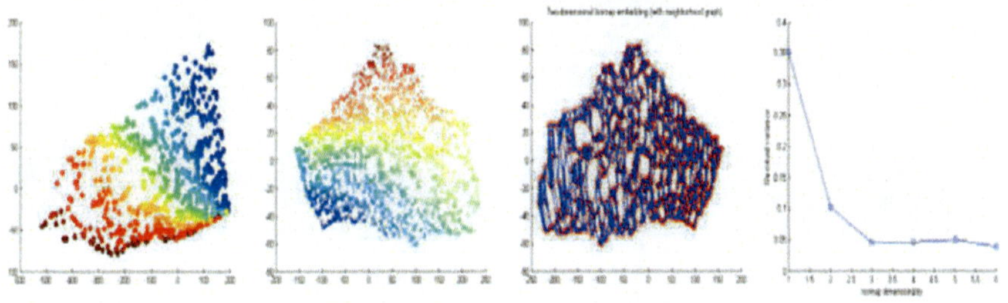

(a) (b) (c) (d)

Fig. 6 **a** Adjacency graph for two-dimensional manifold of image points, **b** Adjacency graph after adding more image points, **c** Bidirectional edges added to the graph to high-light image connectivity, **d** Residual error variance for projection across dimensions indicating dimensionality of the system

5.1 *Feature Matching for Detected Objects*

Output of SIFT descriptor for the structured environment is illustrated through gure7. Using SIFT KaMoMan is able to figure out known objects and match them in various images.

Image registration consists of extracting options, matching them and estimating transformation. Initially, we used SIFT algorithm to detect key points in a picture; then, we used RANSAC algorithm to find out best key points [13]. When SIFT descriptors [14, 15] of two images are compared, the descriptors with minimum Euclidean distance between them are said to be matched. If the ratio of first match to second match is less than 0.8, then the match is selected otherwise discarded. The ground station processor was also equipped with holography estimation of images by randomly select a sample of k key points from key points set K and represents model from this subset. The programme then determines the set of key points K_i which are within a distance threshold t of the model. Comparing the size of K_i and that of threshold largest consensus set is selected and the model is re-evaluated.

5.2 *Image Blending*

From the set of pictures, a reference image is chosen, and second image is transformed to align with the reference image. This provides the primary blending image; currently this can be treated as the reference image. Like this, all pictures are blended and a final broad image is obtained (Fig. 7 and 8).

(a) RANSAC implementation with both polygonal and smooth objects with specified features.

(b) Feature matching in structured environment; outputs from on-board process

Fig. 7 RANSAC and SIFT implementation for KaMoMan in a laboratory environment. Matched features belong to everyday objects that a manipulator system should be able to identify

Fig. 8 Blending of images captured by KaMoMan: Blended image represents various snapshots of the usual navigation environment

6 Conclusion

This work presents the design and development of a reconfigurable mobile manipulator as an artificial agent for visual intelligence and surveillance. Kinematic analysis for a such mobile system requires further calibration for slippage/drift in the expected characteristics. A template for multi-sensor electronic integration was provided for such systems, and it was emphasized that a grid-based autonomous system with range sensor-based safety margins were sufficient to generate collision-free paths. Basic image analysis algorithms, e.g. RANSAC, SIFT and low-dimensional proximity embedding etc., were implemented through ROS provides a systematic framework for image-based surveillance.

References

1. Birk A, Kenn H (2001) An industrial application of behavior-oriented robotics. In: Proceedings of IEEE international conference on robotics and automation, vol 1, p 749–754, ISBN: 0-7803-6578-1, Seoul, Korea, May 2001
2. Carnegie DA, Prakash A, Chitty C et al (2004) A human-like semi-autonomous mobile security robot. In: 2nd international conference on autonomous robots and agents, ISBN: 0-476-00994-4, Palmerston North, New Zealand, December 2004
3. Hamner B, Koterba S, Shi J et al (2010) An autonomous mobile manipulator for assembly tasks. Auton Robot 28:131 (Springer). https://doi.org/10.1007/s10514-009-9142-y
4. Capezio F, Sgorbissa A, Zaccaria R (2005) GPS-based localization for a surveillance UGV in outdoor areas. In: Proceedings of the Fifth International Work-shop on Robot Motion and Control (RoMoCo'05), pp 157–162. ISBN: 83-7143-266-6, Dymaczewo, Poland, June 2005
5. Vig L, Adams JA (2007) Coalition formation: From software agents to robots. J Intelli Robot Syst 50(1):85–118. ISSN: 0921-0296
6. Burgard W, Moors M et al (2000) Collaborative multi-robot exploration. In: Proceedings of IEEE international conference on robotics and automation, vol 1, pp 476–481. ISBN: 0-7803-5886-4, San Francisco, CA, USA, April 2000
7. Li Z, Ge SS (2013) Fundamentals in modelling and control of mobile ma-nipulators. CRC Press
8. Di Paola D, Milella A (2010) An autonomous mobile robotic system for surveillance of indoor environments. Int J Adv Robot Syst. https://doi.org/10.5772/7254

9. Milella A, Cicirelli G, Distante A (2008) RFID-assisted mobile robot system for mapping and surveillance of indoor environments. Ind Robot Int J 35(2):143–152
10. Hornung A, Philips M, Jones EG, Bennerwitz M, Likhachev M, Chitta S (2012) Navigation in three-dimensional cluttered environments for mobile manipulation. In: EEE international conference on robotics and automation, Minneapolis, Minnesota
11. Gohil MK, Chakraborty A et al (2016) Hyper-redundant robots and bioinformatics: modelling loops in RNA. In: Proceedings of the 2016 IEEE SMC 2016, vol 96, 003694–003700. IEEE SMC. https://doi.org/10.1109/smc.2016.7844808
12. Brown M, Lowe D (2007) Automatic panoramic image stitching using invariant features. Int J Comput Vis 74(1):59–73 (Springer Science + Business Media, LLC. Manufactured in the United States)
13. DL, Object recognition from local scale-invariant features. In: International conference on computer vision, Corfu, Greece, pp 1150–1157
14. Szeliski (2004) Image alignment and stitching: a tutorial. Technical Report MSRTR-2004-92, Microsoft Research
15. Hartley R, Zisserman A (2004) Multiple view geometry in computer vision, 2nd edn. Cambridge University Press, March 2004

Space Systems

System Engineering Studies on Induction of Redundant Electromechanical Engine Gimbal Control Actuation System in the Second Stage of Polar Satellite Launch Vehicle

Koppula Pramodh Reddy, C. G. Suresh Nair, and S. R. Biju

1 Introduction

Polar Satellite Launch Vehicle (PSLV) is a four-stage vehicle with solid propellant powered systems for the first and third stages and liquid propellant powered systems for the second and fourth stages. PSLV has four variants catering to different payload demands. The variants are engineered with different combinations of PSLV Strap-on Motor (PSOM). The four variants are Core Alone (CA), with Dual Strap-on (DL), with Quadruple Strap-on (QL), and six Strap-on (XL). The second stage (PS2) uses a liquid Vikas engine with propellant combination of N_2O_4 and UH25. Each stage is configured with autonomous Control power plant, catering to 3 axis control requirements during the respective flight regimes. The Pitch (P) and Yaw (Y) vehicle attitude control in the PS2 regime is by Engine Gimbal Control (EGC) system. The function of the EGC system is to control the vehicle attitudes about pitch and yaw axis during the PS2 regime of the flight. The planar gimbal is provided with a maximum vectoring capability of $\pm$ 4 degrees using actuators in $P + Y +$ and $P + Y -$ planes which are connected between the engine and the thrust frame. Electrohydraulic actuation system was the legacy system used for Engine Gimbal Control in 2nd stage of PSLV. Many single-point failure modes are present in electrohydraulic actuation system, which poses a grave concern to launch vehicle designers. Considering this, a decision was taken to replace the electrohydraulic actuation system with redundant

K. Pramodh Reddy (✉) · C. G. Suresh Nair · S. R. Biju
Indian Space Research Organization, Thiruvananthapuram 695022, India
e-mail: pramodh_koppula@vssc.gov.in

C. G. Suresh Nair
e-mail: cg_sureshnair@vssc.gov.in

S. R. Biju
e-mail: sr_biju@vssc.gov.in

V. H. Saran and R. K. Misra (eds.), *Advances in Systems Engineering*,
Lecture Notes in Mechanical Engineering,
https://doi.org/10.1007/978-981-15-8025-3_60

electromechanical actuation system, which is more fault-tolerant and fully testable on ground before integration to launch vehicle.

2 Need for the Change

In electrohydraulic actuators, oil leakage was noticed at various stages of testing. The degraded performance was noticed due to oil contamination, which is a core criterion since electrohydraulic actuators have low sliding clearances. To improve on the single-point failure modes existing in electrohydraulic actuation system, various configurations were studied and the redundant electromechanical actuation system was identified as the most suitable option. Redundant electrohydraulic actuation system which was an alternative was not chosen due to the mass and envelope constraints which became evident after preliminary studies. Electromechanical actuators (EMA) are highly reliable compared to electrohydraulic actuators (EHA). EMA does not have single-point failure modes due to their capability in incorporating redundancy through multiple electrical windings in motor end, redundancy in sensors. EMA is easy to test; the shelf life of components is higher and takes less assembly time due to less number of components compared to EHA. However, EMA Control Electronics and Checkout systems are complex compared to EHA. Another disadvantage of EMA is that it requires a constant current supply for holding in null condition, whereas EHA does not consume energy because of higher mechanical stiffness due to positive internal pressure. However, this can be circumvented by appropriate sizing of vehicle battery power.

After a trade-off study, EMA emerged as the most appropriate system for Engine Gimbal Control, and a decision was taken to induct EMA in PSLV, replacing the existing EHA system.

3 Redundancy Implementation

Redundancy and fault tolerance were given paramount consideration in the design of EMA subsystems as detailed below.

3.1 Actuator Configuration and Design

Various configurations for actuators like a screw-driven, nut-driven, and geared with multiple motors were studied before arriving at the nut-driven configuration. In this, the hollow rotor of the frameless motor is keyed to the nut of the roller screw. The rotating roller screw nut is supported at both ends using two angular contact ball bearings. The rotor is separated from the bearings through thin spacers. The stator of

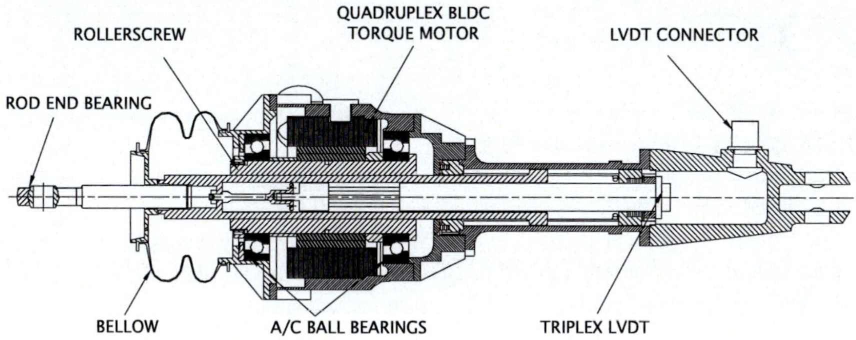

Fig. 1 Configuration of PS2 EGC electromechanical actuator

the motor is housed in a body made out of aluminum alloy and two keys positioned diametrically on the opposite sides are used to arrest the relative motion of the stator and housing. The roller screw shaft has four integral keys at its end which can travel along the key ways provided in a rigid stationary member called screw guide. This has four key ways to convert the rotary motion of the roller screw shaft to a linear one.

The roller screw shaft has been made hollow to accommodate the linear position feedback sensor, which is a linear variable differential transformer (LVDT) as shown in Fig. 1. The triple-redundant LVDT has three independent sets of windings and probes. The three probes are integrated to form a single unit using an adaptor and fastened to the moving member, which is the roller screw. The body of the LVDT is assembled to the screw guide. There are two stoppers assembled with preloaded wave springs at both ends to cushion out the impact force in case the actuator goes into an open-loop mode. Rod end bearing which is attached to the roller screw is connected to the engine. The other end of the actuator has a fork, which gets connected to the stage. Roller screw and angular contact ball bearings are preloaded to prevent backlash. The amount of preload is controlled by regulating the tightening torque of the preloading nut [1].

3.2 *Quadruplex Brush Less Direct Current (BLDC) Motor*

To avoid single-point failures, EMA is configured with quadruplex power drive. The BLDC motor has four separate three-phase windings located in the four quadrants of the stator and four sets of Hall sensors for redundancy [2]. LVDT is used as a position sensor which has triple modular redundancy. Fault detection of torque motor coils can be carried out by cross-comparison of motor coil currents. In the case of deviation beyond a threshold value, the failed coil will be isolated since it offers a drag torque due to the generator effect [3].

3.3 Control Electronics

Many single-point failure modes exist in the EHA system. On the other hand, the EMA control system provides functional redundancy for control electronics and actuators. The function of the PS2 EMA Control Electronics (CE) is to drive EMA in a closed loop. Control electronics is having a dual redundant analog servo loop with Quadruplex power drive. Servo side dual redundancy is managed by servo chain failure detection and isolation (FDI) schemes. Also, triplex LVDTs are used as position sensors where third LVDT is used for servo FDI.

3.4 Failure Detection and Isolation (FDI) Scheme

FDI scheme is designed to provide prime to redundant switch over in the event of a failure avoiding false alarms. The scheme is designed to be simple to have high reliability. The servo Failure Detection (FD) logic has three parts. The first part is the system FD logic, detecting any failure in the closed-loop system by comparing the system command and the system response at a global level. The second part is Prime and FDI feedback comparison FD Logic. The third part is Prime LVDT Va + Vb (addition of secondary coil voltages) failure detection logic. All three FD logic outputs are 'OR'ed and the 'OR'ed output is used for driving the FDI relay switchover circuits.

4 Challenges in the System Engineering

4.1 Accommodation Studies

The induction of EMA in the second stage of PSLV requires additional avionics packages and batteries compared to EHA. In EHA configuration, avionics packages are assembled in lower inter-stage between 2nd and 3rd stages (IS 2/3L) on bulkhead mounted decks and housing as shown in Fig. 2. This scheme cannot accommodate the additional packages, which called for a complete change in the package mounting scheme. Additional packages were accommodated in inter-stage subassembly on newly added honeycomb sector deck as shown in Fig. 3. Avionics packages are functionally grouped and mounted on both faces (top and bottom) of the sector deck. Existing bulkhead decks and avionic housing in inter-stage are removed. As part of optimization, all the avionics packages existing on thrust frame (TF) on the same stage were moved to the same hardware (IS 2/3L) to enable modularity and easiness of testing.

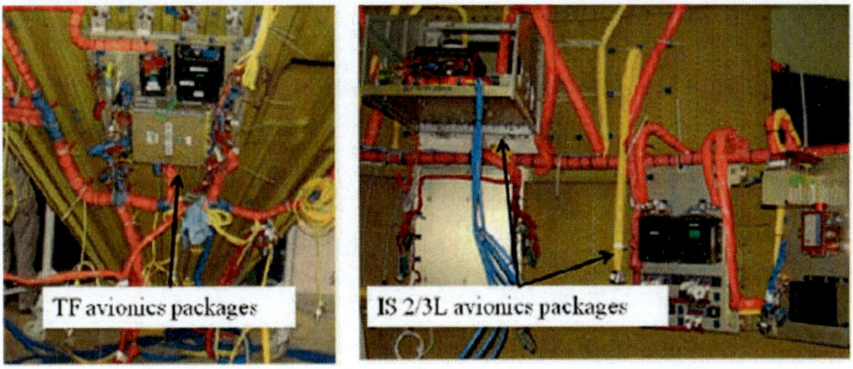

Fig. 2 Avionics packages mounting in EHA configuration

Fig. 3 Avionics packages mounting in EMA configuration

4.2 Modifications and Impact Analysis

Modifications. Due to the induction of EMA configuration, two extra cutouts (as shown in Fig. 4) are provided on the inter-stage structure to access the avionics packages at the launch pad after integration. This is essential to mount the batteries as close to launch, as well as to enable replacement of any avionics packages if

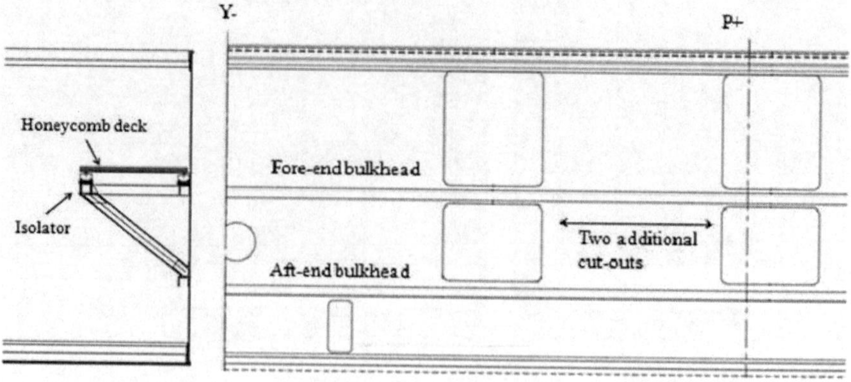

Fig. 4 Schematic of honeycomb deck and isolator

required. Cutout covers are flushed with the external surface of the vehicle, thereby not causing any change to external flow field, thereby not generating any aerodynamic loads.

The inter-stage structure is an aluminum alloy skin stringer construction stabilized by bulkheads inside. Honeycomb deck for avionics packages is supported inside inter-stage by a support ring on inner edge and fore-end bulkhead on the outer edge. The deck is mounted on isolators. The sector ring is supported by inclined members connecting sector ring and Aft-end bulkhead. To provide radial stiffness to the deck support, the inner support ring is connected to the fore-end bulkhead using radial members.

With the induction of the EMA system, the size of harness bunch from IS 2/3L to actuators through wire tunnel is increased due to an increased number of packages and redundancy implementation. Due to the increased size of harness, configuration of external cowling to the vehicle, which is used to protect harness from aerodynamic forces, is changed. The harness is held in position by ties to bond pads affixed to stage. The strength of the bond pads mainly depends on constituents and processes involved in adhesive preparation of bonding. Due to the increased mass of harness, bond pad configuration is changed to weld pad configuration to withstand higher loads. For optimizing the number of weld pads, the bridging plate is introduced between the weld pads.

Impact analysis

Vibration response analysis and Structural tests. Due to structural changes on inter-stage where avionics packages were mounted, tests were conducted to obtain responses of the modified system. The acoustic test was carried out on the inter-stage structure with all modifications to get the vibration levels on avionics package locations. Responses on avionics package mounting locations are assessed concerning vibration levels expected during flight.

Aerodynamic analysis on wire tunnel cowling. Wire tunnel cowling is used to protect harness from aerodynamic loads. Wire tunnel cowling with EMA configuration has a lower width compared to EHA cowling while radial projection has increased from 61 to 116 mm. The aerodynamic analysis was carried out at different Mach numbers to find out the Cp (pressure coefficient) distribution along the length of cowling which is used to estimate the loads. Estimating cowling loads are essential since they are transferred to structure through weld pads. Cp distribution at different Mach numbers along the length of the cowling is shown in Fig. 5. The sectional view of EHA and EMA cowling was shown in Figs. 6 and 7, respectively. The revised aerodynamic coefficients and corresponding flight loads are acceptable for the implemented design with modifications.

Propulsion system change. EHA was deriving the hydraulic power from the onboard stage itself. Pair of hot gas filters, commutator, thermal regulator, tubing's, and transducers were removed from the stage as it is not required for EMA configuration.

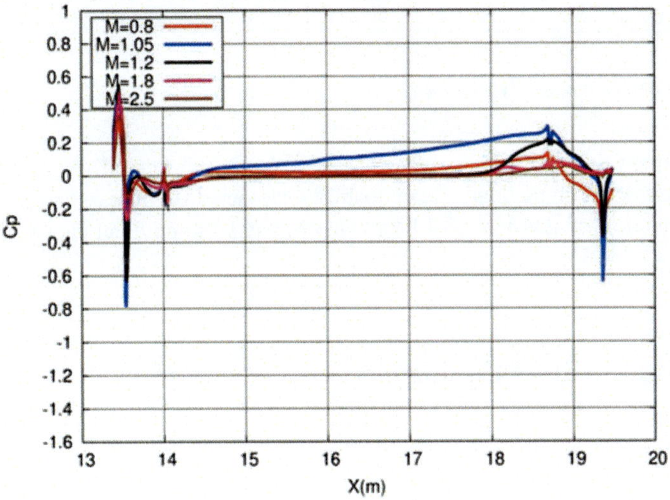

Fig. 5 Cp distribution along the length of wire tunnel cowling at different Mach numbers

Fig. 6 EHA cowling configuration

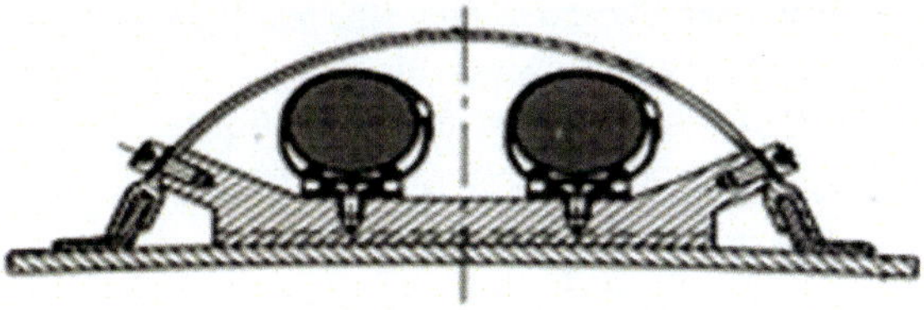

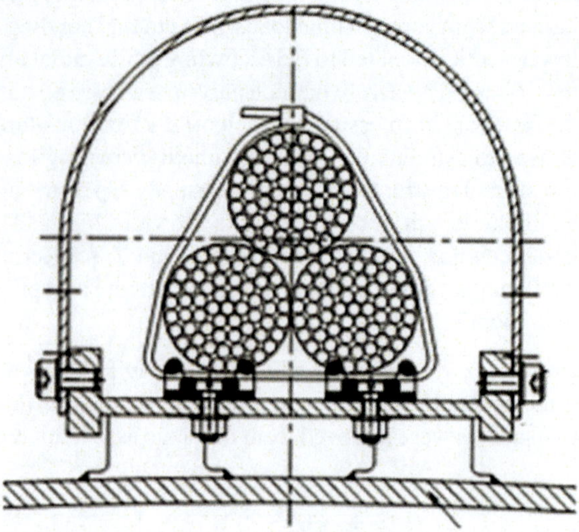

Fig. 7 EMA cowling configuration

4.3 *Overall Mass Increase*

Mass increase due to modifications is very important in launch vehicles due to its direct implication on payload mass capability. Net increase in mass in PS2 stage due to the induction of EMA is 63.17 kg. This impacts the payload capability by less than 7 kg (Table 1).

5 Validation Tests Conducted

5.1 *Acoustic Tests*

Due to modifications carried out in inter-stage structure concerning incorporation of additional cutouts, introduction of new avionic deck support structure for housing EMA-related packages, it was required to assess the dynamic characteristics of inter-stage subassembly and evaluate the response at critical locations. To assess the same, an acoustic test of subassembly was carried out in four different configurations to simulate structurally transmitted vibrations.

- Configuration-1: IS 2/3L deck mounted with isolators and without inter-stage 2/3 upper (IS 2/3U).
- Configuration-2: IS 2/3L deck mounted with isolators and with IS 2/3U.
- Configuration-3: IS 2/3L deck mounted without isolators and IS 2/3U.
- Configuration-4: IS 2/3L deck mounted without isolators and with IS 2/3U (Fig. 8).

Table 1 Additions and deletions due to the induction of EMA

Sl. No.	Additions	Mass (kg)	Sl. No.	Deletions	Mass (kg)
1.	Actuator (EMA)	45	1.	Actuator (EHA)	92
2.	IS 2/3L Avionics	89.4	2.	IS 2/3L Avionics and Housings	19.87
3.	Avionics Deck	12	3.	Decks and Brackets	12.33
4.	Supporting Structure	15	4.	IS 2/3L Harness	24
5.	Mounting Brackets	12.7	5.	TF Avionics	22.5
6.	IS 2/3L Harness	40	6.	TF Harness	36
7.	End Jackets	7.95	7.	Wire Tunnel Cover	14.6
8.	Wire Tunnel Cover	13.38	8.	End Jackets	7.95
9.	Destruct System	28.24	9.	Destruct System	31.46
10.	TF Harness	11	10.	TF Mounting Elements	18.92
11.	Low Wire Tunnel Harness	30.11	11.	IS 1/2L Avionics	2.8
12.	High Wire Tunnel Harness	64.74	12.	Low Wire Tunnel Harness	26.5
13.	Bridge Plate	19.5	13.	High Wire Tunnel Harness	8.87
14.	IS 1/2 upper (IS 1/2U) Avionics	2.5	14.	Fasteners	3.557
15.	IS 1/2L Avionics	2	15.	Propulsion Elements	11.62
16.	Fasteners	2.63		Total	332.98
	Total	396.15			

For the acoustic test, levels are derived from the external acoustic level measured in a IS 2/3U during PSLV-D3 flight. Extrapolations were carried out based on thrust at liftoff and dynamic pressure at transonic and Qmax (maximum dynamic pressure) conditions. The acoustic test was carried out with and without isolators for the deck mounting. Vibration response at all avionics package locations was verified. It was seen that the isolators are effective in reducing the vibration response on deck. Response at CE location in radial direction with and without isolators is as shown in Fig. 9. CE is an important mission-critical package that has to be tested and qualified to the level observed during the acoustic test. With this study, deck assembly with isolators was chosen for the flight

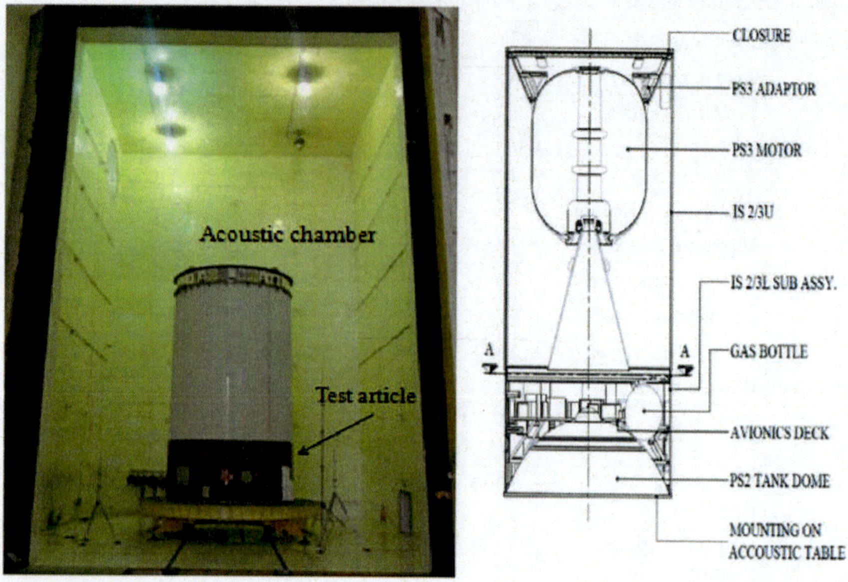

Fig. 8 Acoustic test configuration

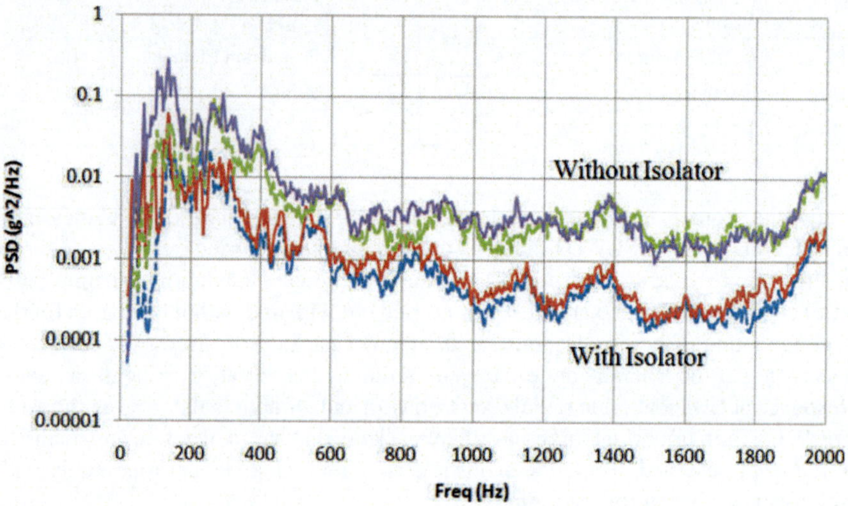

Fig. 9 Vibration response at CE location in the radial direction

5.2 *Structural Dynamic Test*

The need of a structural dynamic test for any new subassembly is to reconstruct the response seen during an acoustic test using a vibration shaker. This will avoid carrying

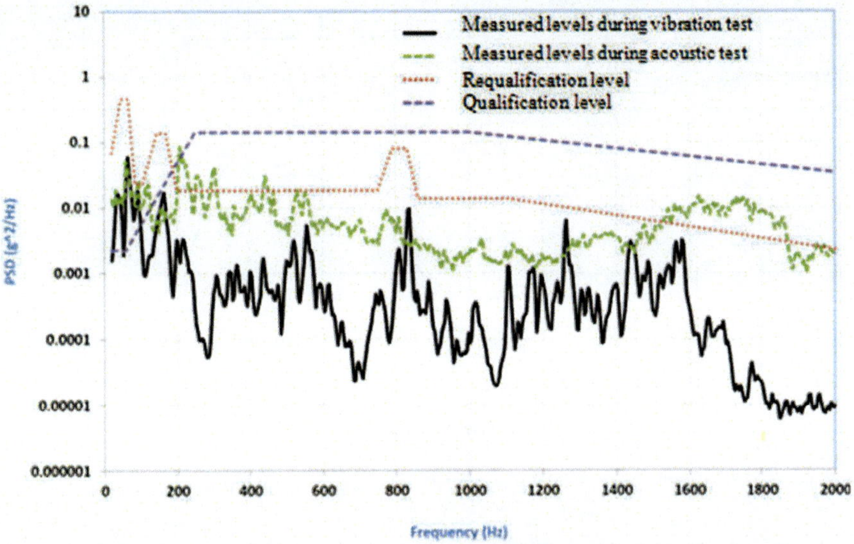

Fig. 10 CE package level re-qualification test profile based on vibration test response

out an acoustic test for every subassembly and helps in production. The structural dynamic test was carried out for IS 2/3L subassembly. Existing random vibration test profile was used to match the levels noticed during acoustic test. A good match in vibration response was observed between the acoustic test and random vibration test. Response on the deck with isolators is matched at lower frequencies up to 170 Hz. At higher frequencies, the response observed during the random vibration test (with existing test profile) was lower compared to the response observed during the acoustic test as shown in figure. Response on the propulsion modules (which are mounted on IS 2/3L without any isolators) is matched in the lower frequencies up to 350 Hz. However, the response on propulsion modules was observed to be on the higher side for higher frequencies compared to existing qualification test levels. Random vibration test levels at higher frequencies were not increased to match acoustic test levels since it affects the response at propulsion modules. Based on the response noticed during the random vibration test, CE package level qualification test levels were reworked as shown in Fig. 10. Similarly, all the avionics package level qualification test levels were reworked.

5.3 Actuator Level Tests

The signature of EMA needs to be verified in actuator level tests and system-level tests (actuator assembled to the engine) before the flight. Actuators are tested to different amplitudes of a stroke at varying frequencies. Control analysis is carried out on

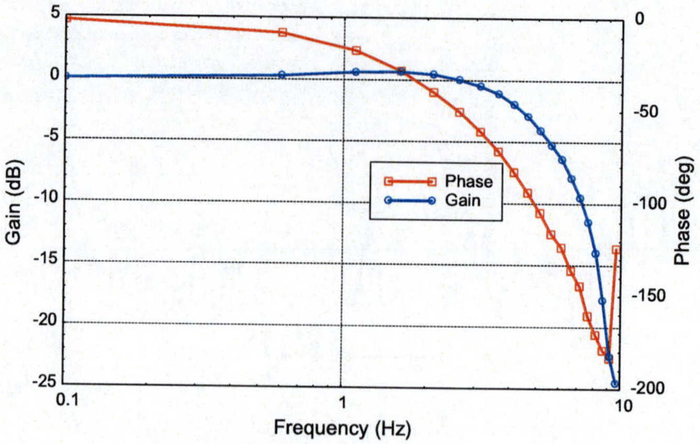

Fig. 11 PS2 EMA actuator gain and phase for 10% command amplitude

the test data to ensure stability during flight. Testability in the actual environmental conditions is one of the main advantages of EMA over EHA. The actuator is tested in the active condition during all environmental tests to prove its design margins. Qualification and acceptance test plan for EMA are similar to EHA. Frequency response tests, step response tests, ramp tests, and linearity test are carried out for various command amplitudes to assess the performance of the actuator. Gain and phase characteristics of PS2 EMA are shown in Fig. 11.

5.4 *Integrated Tests*

The integrated test is one of the most important tests in the system level where all the elements are tested together. System engineering plays a major role in finalizing the test profile during integrated tests. The test profile is designed in such a way that all the objectives of integrated tests are met with flight configuration. After integrating all the subsystems to stage with modifications, the EMA system is tested to assess the performance. The test profile designed for actuators during the integrated test is shown in Fig. 12.

6 Conclusion

All the system engineering aspects for implementation were studied in detail. Structural analysis was carried out for all the changes. Acoustic test and structural dynamic tests were carried out to assess the performance with all the modifications. The aerodynamic analysis was carried out for change in wire tunnel cowling configuration.

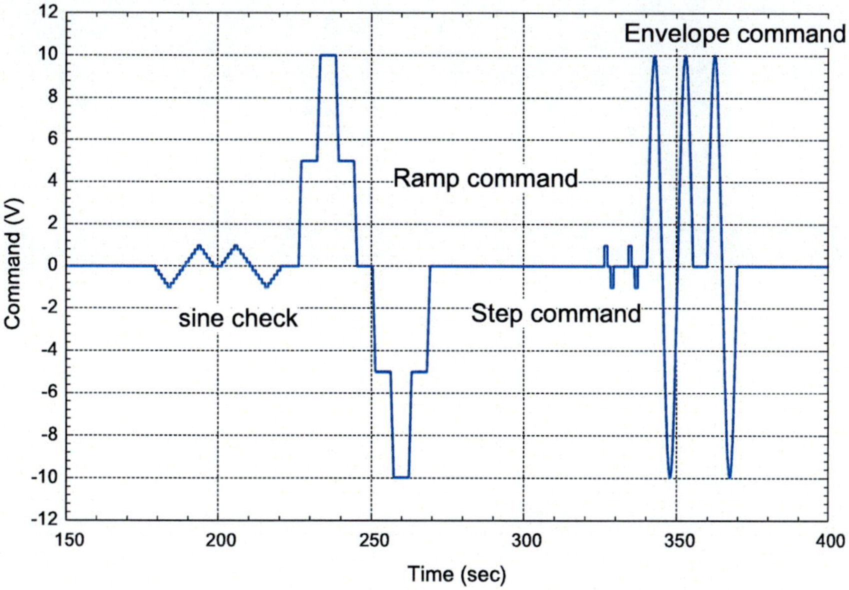

Fig. 12 Integrated test profile

Control analysis on actuator test data was carried out to ensure stability during flight. The electromechanical actuation system was successfully implemented in the second stage of PSLV, making the vehicle more reliable and tolerant of failures.

References

1. Biju Prasad B et al (2011) Optimum design of a fault tolerant linear electromechanical actuator for the lower stage thrust vector control of a satellite launch vehicle. In: 14th European Space Mechanisms & Tribology Symposium—ESMATS 2011. Constance, Germany, 28–30 Sept 2011
2. Sundstrand Energy Systems (1984) Design and test of a four channel permanent magnet brushless DC motor for electromechanical flight control actuation. Final Report S8308-R1 to NASA-Lyndon B. Johnson Space Center, December 1984
3. Cowan JR, Myers WN (1992) Design and test of a high power electromechanical actuator for thrust vector control. In: AIAA/SAE/ASME/ASEE 28th joint propulsion conference and exhibit, AIAA 92-3851

Configuration and Mission Aspects of Passive Recovery System for Payload

T. Sivamurugan, L. Sowmia Narayanan, S. Sivakumar, G. Ayyappan, and K. Kumar

Nomenclature

C_D	Coefficients of drag
H	Enthalpy, J/kg
I_{xx}, I_{yy}, I_z	Moments of inertia, kg-m^2
I_{xy}, I_{yz}, I_{zx}	Products of inertia, kg-m^2
Le	Lewis number
M	Mach number
p, q, r	Body rates, deg/s
P	Pressure, Pa
Pr	Prandtl number
q	Dynamic pressure, kPa
q_c	Convective heat transfer rate, W/cm^2
R_N	Nose radius, m
T	Temperature, K

T. Sivamurugan (✉) · L. Sowmia Narayanan · S. Sivakumar · G. Ayyappan
Space Transportation System, Vikram Sarabhai Space Centre, ISRO, Thiruvananthapuram, Kerala, India
e-mail: tsivamurugan@gmail.com

L. Sowmia Narayanan
e-mail: sowmia_narayanan@vssc.gov.in

S. Sivakumar
e-mail: s_sivakumar@vssc.gov.in

G. Ayyappan
e-mail: g_ayyappan@vssc.gov.in

K. Kumar
Human Space Flight Centre, ISRO, Bengaluru, Karnataka, India
e-mail: kkumar-hsfc@isro.gov.in

V. H. Saran and R. K. Misra (eds.), *Advances in Systems Engineering*,
Lecture Notes in Mechanical Engineering,
https://doi.org/10.1007/978-981-15-8025-3_61

V	Velocity, m/s
u, v, w	Relative velocity components in body frame (m/s)
$\sum F_{XB}, \sum F_{YB}, \sum F_{ZB}$	Sum of all the forces acting on the body in X, Y & Z-direction
$\sum M_{XB}, \sum M_{YB}, \sum M_{ZB}$	Sum of all the moments acting on the body in X, Y & Z-direction

Subscript

∞	Free stream condition
DT	Dissociation of gas mixture
T	Total condition
W	Wall condition

1 Introduction

One of the feasible options to ensure low cost access to the space is recovery of the spent stages/payload after completion of their intended role and reuse the same for multiple times with minimum turnaround time. Major recovery configuration options are winged body landing on runway, vertical landing with retro burning and splash down with minimum impact velocity based on aerodynamic decelerator. Three different kinds of aerodynamic decelerator exist, namely mechanically deployable aerodecelerator, parachute system and Inflatable Aerodynamic Decelerator system (IAD).

Sounding rockets are having inherent characteristics of lower cost, short lead time and ability to reach higher altitudes where balloons and satellites are inoperative. Sounding rockets are having passive control with spin stabilized. The present stage and payload are expendables. Studies are being carried out for recovery of payload as a first step towards the recovery of stage. This paper mainly addresses recovery of a payload from sounding rocket experiment using Inflatable Aerodynamic Decelerator. This paper briefly explains IAD configuration, mission design, structure construction and aerothermal analysis and material selection for IAD.

2 Configuration

The typical Mach number versus the drag coefficient of sphere cone, tension cone, isotensoid and parachute is as shown in Fig. 1. Figure 1 is taken from Ian G. Clark, Allison L. Hutchings, Christopher L. Tanner, Robert D. Braun.: Supersonic Inflatable

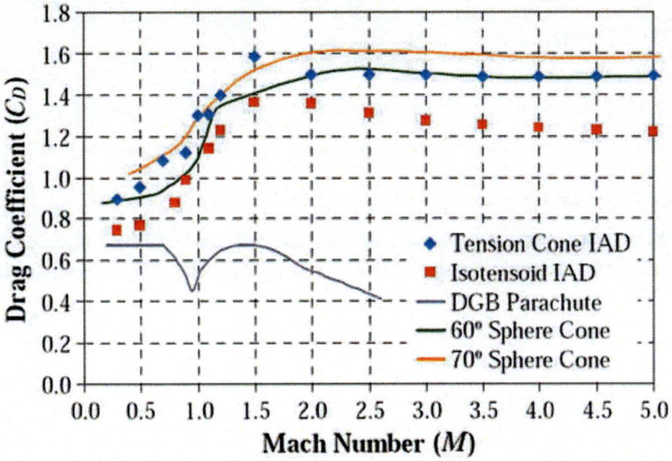

Fig. 1 Mach number versus drag coefficient for various inflatable decelerator configurations [1]

Aerodynamic Decelerators For Use on Future Robotic Missions to Mars paper [1]. Figure 1 depicts that 70° sphere cone configuration provides higher drag coefficient compared to other configurations.

The sphere cone configuration with a flare angle of 70° is chosen for considering higher drag coefficient during supersonic as well as subsonic regime. Figure 2 depicts configuration and parameters such as forward radius (Ra), base radius (Rb) and torus radius (Rt) arrived to meet the required ballistic coefficients. The deployed IAD is statically stable, which will re-orient itself and flow will be facing the forward region. Configuration parameters are varied to achieve the ballistic coefficients (β or beta) of 15, 30 and 42.5 kg/m^2 with payload mass of 50 kg.

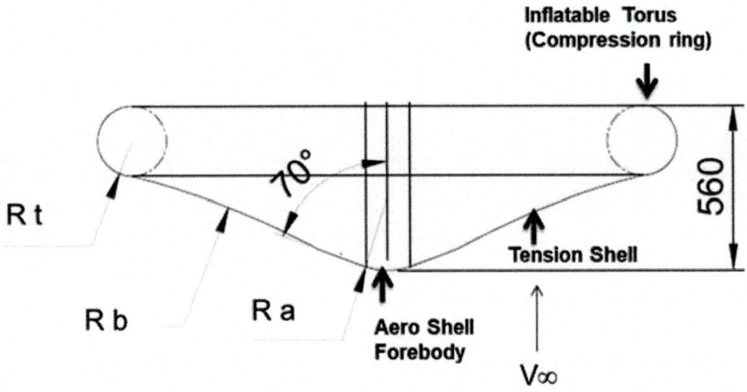

Fig. 2 Schematic IAD configuration

3 Mission Design

Mission objectives are functioning inflatable system such as aeroshell and compression ring at higher altitude, self re-orientation of IAD and touch-down velocity that should be less than 20 m/s.

Sounding rocket configurations are sized to statically stable even under performance deviations. Sounding rockets are unguided, uncontrolled and fin stabilized. Sounding rockets spinning are provided by spin rockets and then sustained due to aerodynamics of canted fin. Sounding rocket (RH-300) is a single-stage configuration, and the payload compartment is located on top of solid booster. Overall vehicle length is 5.9 m, diameter is 0.31 m, and gross mass of 500 kg with payload capability is 50 kg. Aeroshell and compression ring along with payload are in stowed condition and placed in payload compartment.

The equations of motion are solved to determine the state vectors at any instant. Equations of motion in the body frame of the vehicle are as follows.

$$m(\dot{u} - rv + qw) = \Sigma F_{XB} \tag{1}$$

$$m(\dot{v} - pw + ru) = \Sigma F_{YB} \tag{2}$$

$$m(\dot{w} - qu + pv) = \Sigma F_{ZB} \tag{3}$$

$$I_{xx}\dot{p} + (r^2 - q^2)I_{yz} + rq(I_{zz} - I_{yy}) + (rp - \dot{q})I_{xy} - (\dot{r} + pq)I_{zx} = \Sigma M_{XB} \tag{4}$$

$$I_{yy}\dot{q} + (p^2 - r^2)I_{zx} + pr(I_{xx} - I_{zz}) + (pq - \dot{r})I_{yz} - (\dot{p} + qr)I_{xy} = \Sigma M_{YB} \tag{5}$$

$$I_{zz}\dot{r} + (q^2 - p^2)I_{xy} + pq(I_{yy} - I_{xx}) + (qr - \dot{p})I_{zx} - (\dot{q} + rp)I_{yz} = \Sigma M_{ZB} \tag{6}$$

Using these equations, vehicle translation and rotational motion are determined considering all external forces and moments acting on the body.

Trajectory studies are carried out for sounding rocket (RH 300), and various trajectory parameter profiles are shown in Fig. 3. The payload reaches a peak altitude of around 90 km with help of sounding rocket and descends after accomplishing experiments, and then payload experiences higher dynamic pressure of 180 kPa at an altitude of 18 km as shown in Fig. 3. Hence, IAD deployment is scheduled at an altitude of 45 km, where sensible atmosphere exists to enhance the performance of aerodynamic decelerator and thereby dynamic pressure of the payload can be reduced significantly. Aeroshell and compression ring are deployed by injecting the inflate gas from the gas bottle. IAD will self re-orient due to monostability of the configuration and re-enters into the dense atmosphere with flow phasing towards large radius as shown in Fig. 4, which will result large aerodynamic drag, thereby

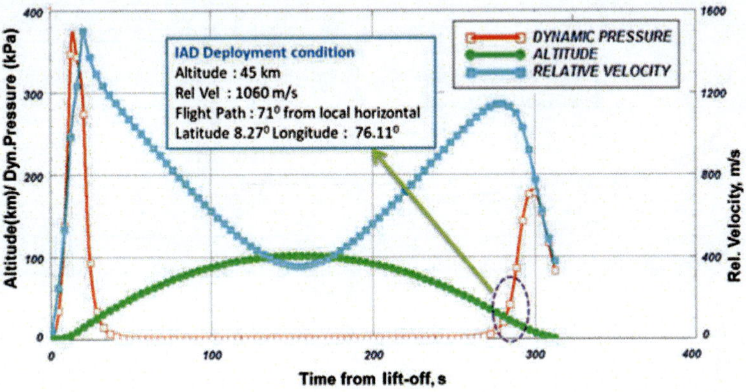

Fig. 3 Trajectory parameters for sounding rocket (RH300)

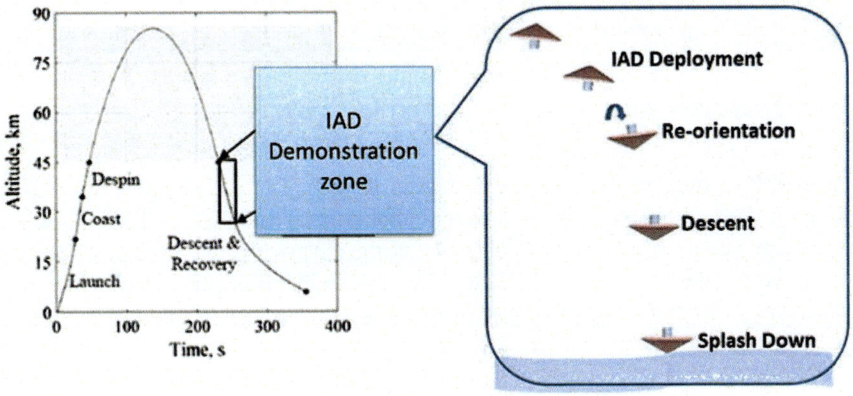

Fig. 4 Typical mission profile of sounding rocket and IAD deployment sequence

higher velocity reduces and makes soft landing without a retro brake or parachute due to ballistic coefficients.

In order to achieve the minimum touch-down speed, the various ballistic coefficients of 42.5, 30, 15 kg/m^{2}and without IAD are configured by increasing the size of IAD and total mass of 50 kg is fixed. Mission studies are carried out from peak altitude of 90 km to till touch-down phase. From Fig. 5, trajectory parameter indicates that dynamic pressure and touch-down velocity reduce by one order compared to without IAD configuration. As ballistic coefficients decrease, dynamic pressure and impact velocity also decrease, but the flight duration increases. The ballistic coefficient of 15 kg/m^2 attains the touch-down speed of 16 m/s.

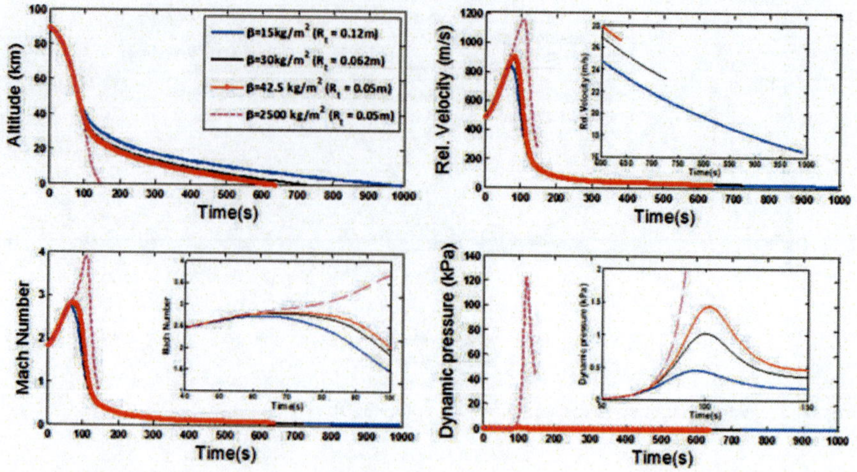

Fig. 5 Trajectory parameter profiles of payload with IAD for various ballistic coefficients

4 Structure Construction

Aeroshell is constructed as thin membrane made of ZYLON filament textile. Hemisphere is constructed as aluminium cover to hold the aeroshell. Six ribs are provided at the rear side to enhance the stability. All of the electrical device and gas injection systems are installed. Pyrotechnic cutters are used to inflation system to release pressurized gas. Internal inflation system provides gas to decelerator, and IAD system is filled with an internal pressure of 1.4 bar.

5 Aerothermal Analysis

Aerodynamic heating analysis is carried out for the IAD configuration. Heating rates at the minimum radii locations are estimated using Fay and Riddell correlation [2].

$$q_c = 0.763(\rho_T \mu_T)^{0.4}(\rho_w \mu_w)^{0.1}\left[1 + \left(L_e^{0.52} - 1\right)\frac{H_{D,T}}{H_T}\right]\left(\frac{\mathrm{d}u}{\mathrm{d}x}\right)^{0.5}\mathrm{Pr}^{-0.663} H_T - H_w$$

$$\text{Velocity gradient}\left(\frac{\mathrm{d}u}{\mathrm{d}x}\right) = \frac{1}{R_N}\sqrt{\frac{2(P_{T-}P_\infty)}{\rho_T}} \text{ for } M_\infty \geq 1$$

$$\left(\frac{\mathrm{d}u}{\mathrm{d}x}\right) = \frac{3V_\infty}{2R_N} \text{ for } M_\infty < 1$$

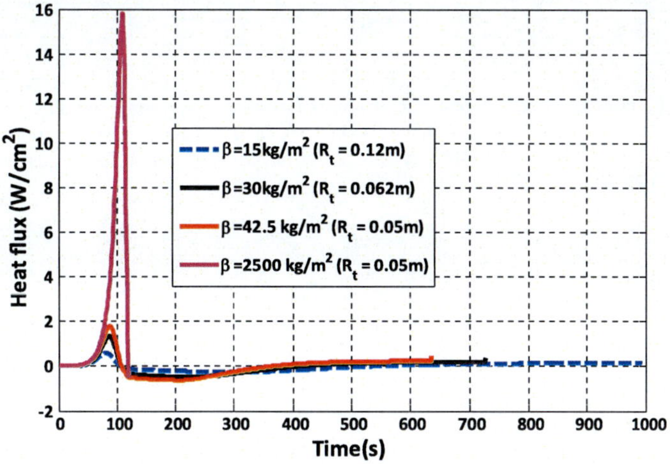

Fig. 6 Heat flux histories of payload with IAD for various ballistic coefficients

Heat flux histories for various ballistic coefficients configurations are estimated and shown in Fig. 6. The maximum heat flux of 1.8 W/cm^2 is observed, and ZYLON-based materials are selected which can withstand temperature of 100 °C. Thermal analysis is also carried out, and surface reaches the maximum temperature of 60 °C.

6 Summary

Inflatable Aerodynamic Decelerator of 70° sphere cone configuration is worked out for the total mass of 50 kg, and configuration is having merits of higher drag for all the speed regimes. Detailed mission studies are carried out by varying the ballistic coefficients and respective configurations to achieve the minimum touch-down speed of 16 m/s. Preliminary structural construction and aerothermal analysis are carried out. Aluminium is chosen for hemispherical region, and ZYLON-based composite material is chosen for aeroshell.

Acknowledgements The authors would like to express their sincere gratitude to Shri S. Somanath, Director, VSSC, for his encouragement and support towards for publication. The authors also express thanks to Dr. S. Geetha, Programme Director, STS and Shri. M. Selvaraaj, VSSC, for providing valuable suggestions and reviewing this paper.

References

1. Clark IG, Hutchings AL, Tanner CL, Braun RD (2008) Supersonic ınflatable aerodynamic decelerators for use on future robotic missions to Mars. In: IEEE Aerospace Conference 2008
2. Fay FA, Riddell FR (1958) Theory of stagnation point heat transfer in dissociated air. J Aeronautical Sci 25(2):73–85

Dual Redundant Embedded Synchroniser for 1553 Bus Synchronisation of Launch Vehicle Checkout Applications

S. Saju, M. Vysakh, C. Chandrasekharan, U. Shyamkumar, and A. Shooja

1 Introduction

The main objective of synchroniser is to ensure the minimum possible skew between communications from redundant distributed MIL 1553 bus controllers. Best way to achieve this is to have a single stable common time source triggering all bus controllers. For high reliability applications like launch vehicle static test checkout, configurations result in single point failures cannot be tolerated [1]. This demands a stable redundant clock source.

Standard PLL-based synchronisation schemes available in PXI systems were also explored. PXI chassis has a provision to synchronisation by routing clock of one chassis to another via dedicated clock sharing lines built into chassis. The synchronisation functionality was evaluated between two MIL STD 1553B cards used in bus controller configuration in two different PXI chassis and it works fine when both systems were healthy. In case of failure of master, the drift observed was 50us for every second and this system will violate the required timing requirements.

Dual FPGA master—slave trigger generation circuit pair ensures MILSTD-1553 message transmission by two bus controller units at the exact timing intervals. For human-rated launch vehicle, the requirement calls for four buses instead of two in order to accommodate for higher reliability requirements. Here, all four buses must exhibit synchronous operation. A logical extension to the dual redundant system

S. Saju (✉) · M. Vysakh · C. Chandrasekharan · U. Shyamkumar · A. Shooja
Vikram Sarabhai Space Centre (VSSC), Thiruvananthapuram, Kerala 695022, India
e-mail: s_saju@vssc.gov.in

M. Vysakh
e-mail: vysakh@vssc.gov.in

C. Chandrasekharan
e-mail: c_chandrasekaran@vssc.gov.in

V. H. Saran and R. K. Misra (eds.), *Advances in Systems Engineering*,
Lecture Notes in Mechanical Engineering,
https://doi.org/10.1007/978-981-15-8025-3_62

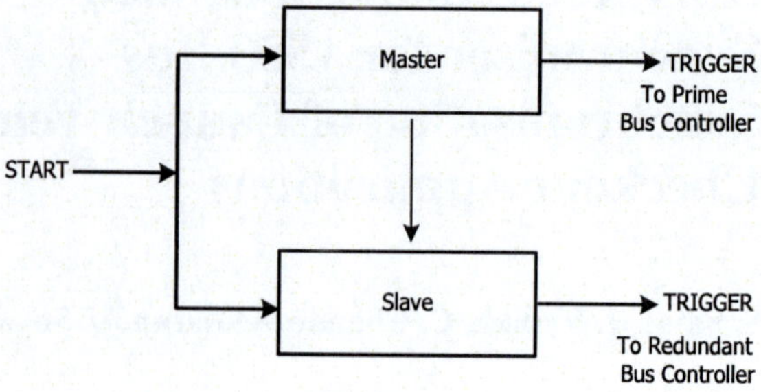

Fig. 1 Block diagram

is implemented to give a quad redundant synchroniser system to meet static test requirements for such system (Fig. 1).

The FPGA-based trigger circuits derive clock from the stable temperature compensated crystal oscillator with $\pm$ 1 ppm stability. TCXO provides temperature compensation which nullifies the temperature effect on drift [2]. With TCXO, the system drift is low, but to further improve functionality, the drift was quantified and corrected by correcting the time period of clock at certain intervals in order to get a highly stable low drift system.

2 Dual Redundant System

The FPGA-based clocking circuits were developed using ProAsic Family-based FPGAs using Libero SOC development platform with the hardware definition done by using VHDL coding [3]. Master trigger circuit has one 'START' input and two SYNC outputs. One SYNC output will be connected to bus controller and other to slave.

The slave trigger circuit has two inputs (START, SYNC_IN) and one SYNC output. After power on reset, the system will wait for START pulse. Width qualified signals are used for START signal to prevent inadvertent commanding. STOP signal is intentionally avoided to prevent accidental stop of the profile during static test.

The master and slave FPGAs are triggered simultaneously and starts counting. Master will do transition on its output on reaching the terminal count value. The slave starts sampling of the master output when its count reaches (terminal count—window duration/2) and continues till (terminal count + window duration/2) during which the master is expected to undergo state transition. If a transition is observed it will also undergo the same level transition, resets count value to zero and continue as long as master transitions in the given window. In case the window elapses without the master transitioning, the slave denotes the master as unhealthy after which it makes transitions without waiting for master (Fig. 2).

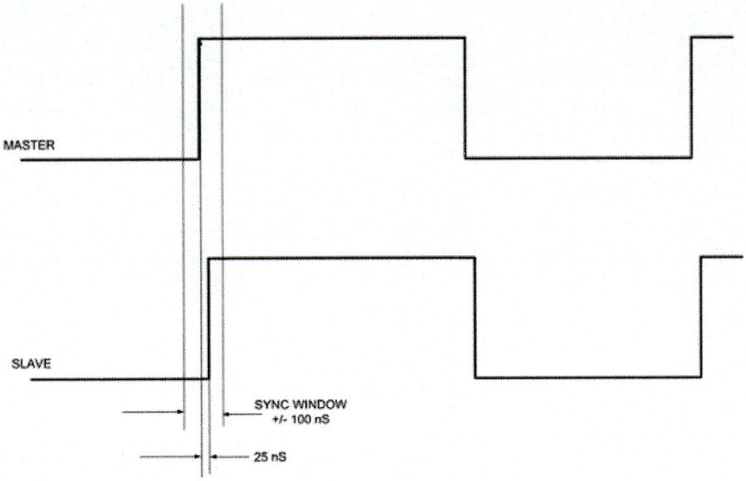

Fig. 2 Synchronised operation

3 Quad Redundant System

Quad redundant systems are used in human-rated launch vehicles. The system implementation uses quad FPGAs are used to externally trigger four synchronous MILSTD-1553 bus controllers. Each trigger circuit has connections to trigger bus and synchronisation bus. Trigger output and SYNC output of all circuits are being monitored by all other circuits. All circuits can act as either master or slave. Predefined priority settings define the initial master (Fig. 3).

In the current implementation of dual redundant synchroniser, 1 ppm clock is used to generate 20 ms trigger. In a 20 ms periodic square waveform, a transition occurs every 10 ms. The maximum amount this derived trigger can drift is 20 ns for this 10 ms interval for the given stability. Assuming the worst case where one clock is ahead by 20 ns and the other is behind by 20 ns, that is there is 40 ns delay between synchroniser (P) and synchroniser (R). This is the maximum delay possible between synchronisers and the window value should be greater than 40 ns to prevent natural clock drift from affecting system. A 200 ns window is chosen which is larger enough to ensure sync will not be lost due to natural drift present in clocks and low enough to ensure master clock drifting will not cause the slave clock to drift more than 100 ns ensuring redundant commanding chain health.

4 Logic Flow

Four FPGA clock generation circuits are implemented with one designated as master and the rest as slaves. Default priories are assigned to each trigger circuits on power

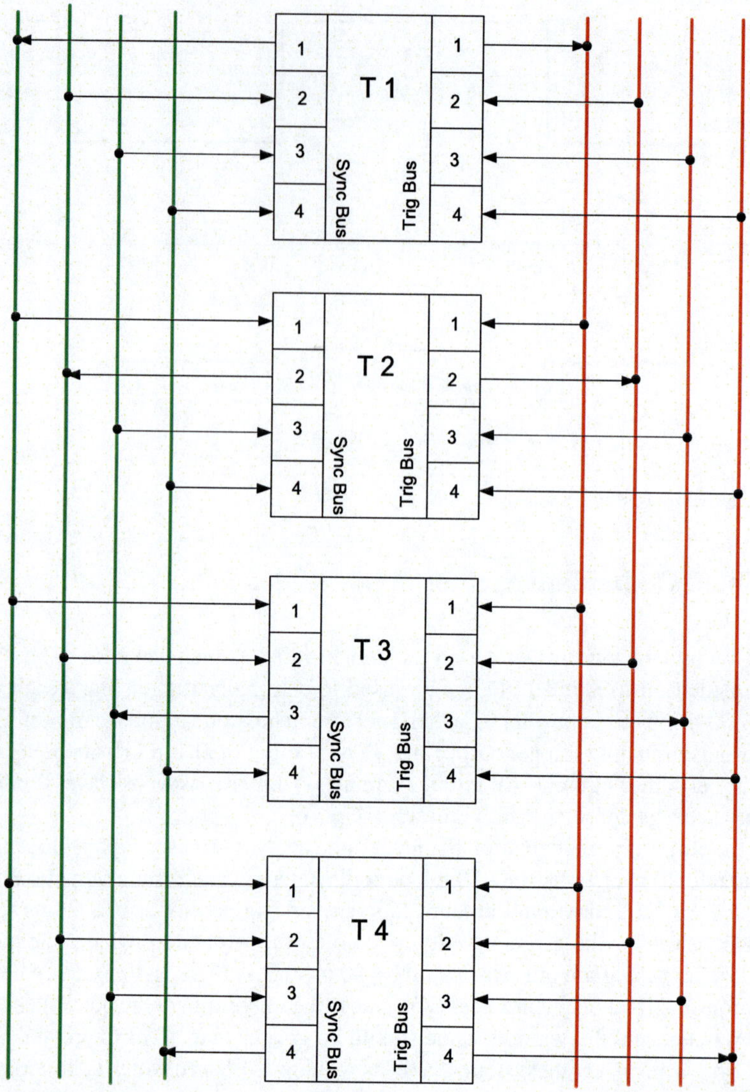

Fig. 3 Quad system

up. Highest priority circuit will act as master. Inner window and outer window zones are defined for synchronisation [4]. During normal operation, master will do transition on its output on reaching the terminal count value all slave circuits will follow master with in inner window.

Trigger signals of all circuits are connected to all other circuits. On reaching its count value with in outer window limits, the slave circuits starts sampling of all others outputs including master. If it detects transition of master during inner window, it will

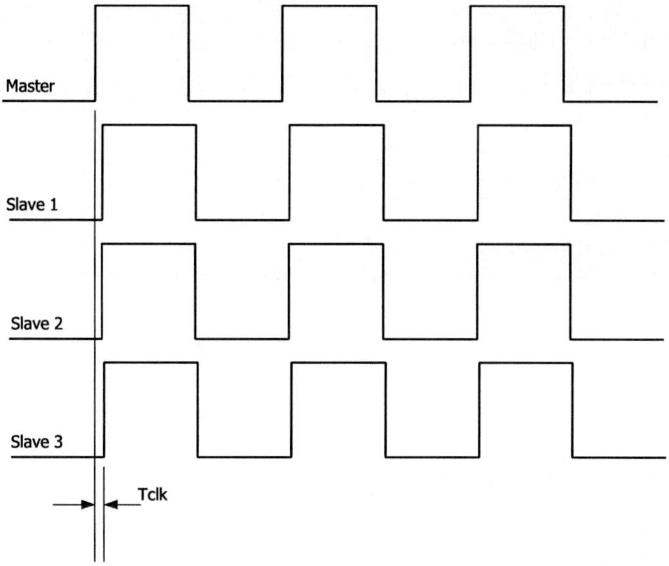

Fig. 4 Normal operation of quad systems

also do transition and resets count value and continue counting normally. If master fails, the highest priority slave is made the master on window out (Fig. 4).

5 Failure Conditions

System behaviour during failure conditions were analysed and summarised below

Case 1: Master clock is running slow/fast

If master clock is running slow/fast, all slaves will do transition on outer window. Slave circuits maintain variables to hold synchronisation status called OuterSyncWindowCount. Slave circuits count starts incrementing its OuterSyncWindowCount value. When count becomes three, the slave will output a pulse of 10 uS to indicate it is tending to sync out. Since master is having problem, other slaves will also output 10 uS syncout pulse. On confirming syncout pulse from other slaves, the highest priority slave will become master. Thus, the best clock source is ensured as master.

Case 2: Master clock is failing

If master clock fails, all slaves will not observe master's transition in their corresponding windows and all slaves will output syncout pulse. The slave with highest priority after observing all other slaves have synced out will become master. Other slaves will follow new master by referring to corresponding port.

Case 3: One Slave clock is running slow/fast

If one slave clock is running slow/fast, the slave circuit count starts incrementing its OuterSyncWindowCount value. When count becomes three, the slave will output a pulse of 10 uS to indicate it is tending to sync out. In order to prevent single sample failure from causing a proper functioning trigger circuit from being isolated, the OuterSyncWindowCount variable is defined. The OuterSyncWindowCount value is estimated to three (Fig. 5).

Since the slave is having problem, other slaves will not output 10 uS syncout pulse. By looking at the output of other slaves, it observes that others are performing optimally, and the issue lies with itself and makes syncout bit permanently high. Slaves with syncout bit high are unfit for becoming master. In case of master failing in this condition, a slave with highest priority and unset syncout bit will only become master.

Case 4: One Slave clock is failing

If one clock fails, there would not be any transitions in its output. On noticing absence of transition on any of the trigger inputs, the other slaves and master will output a pulse on its syncout bit with a pulse width equal to port number multiplied by 10 uS. On observing similar pulse width on at least one more device that corresponding trigger line is designated as unhealthy, thus, failed slave will never become master (Fig. 6).

Case 5: Two Slave clock is running slow/fast

If both the slaves are having different clock skews, it is taken care in case 3. It is highly improbable that two clock sources having equal skew in the same direction, under such conditions there is means of isolating the failure and all will go to free running mode.

6 Test Results

The dual redundant system was tested by recording 10 s data in an oscilloscope at nanosecond precision. The measurement samples were taken periodically every 30 min for three days under different temperature conditions. The time between 499 pulses which could be precisely measured was taken, and the expectant time for it without drift would have been 9.9800000 s and using this as baseline, and the drift was quantified. The results are tabulated as follows:

Time measured (s)	Drift in micro-seconds
9.9800000	0
9.9800004	0.4
9.9800000	0

(continued)

(continued)

Time measured (s)	Drift in micro-seconds
9.9800000	0
9.9799996	−0.4
9.9799994	−0.6
9.9799994	−0.6
9.9799994	−0.6
9.9799994	−0.6
9.9799991	−0.9

The synchroniser was extensively tested before static test using a test matrix which covers all failure modes and the performance was found to be optimal. Long duration stabilisation tests were also carried out; the synchronisers were always in sync for the entire test duration.

7 Conclusion

The dual redundant synchroniser system was implemented and was tested with control electronics and it held in sync condition for the entire eight-hours of testing. The slave and master were made to run in free running mode also for eight hour duration. Even in free running mode, the trigger generation circuits catered to the timing requirements of control electronics and commands were always properly received. The timing drift of trigger circuits was analysed and the system is found to be superior to all other available alternatives.

Static test of the solid motor was successfully completed using the synchroniser.

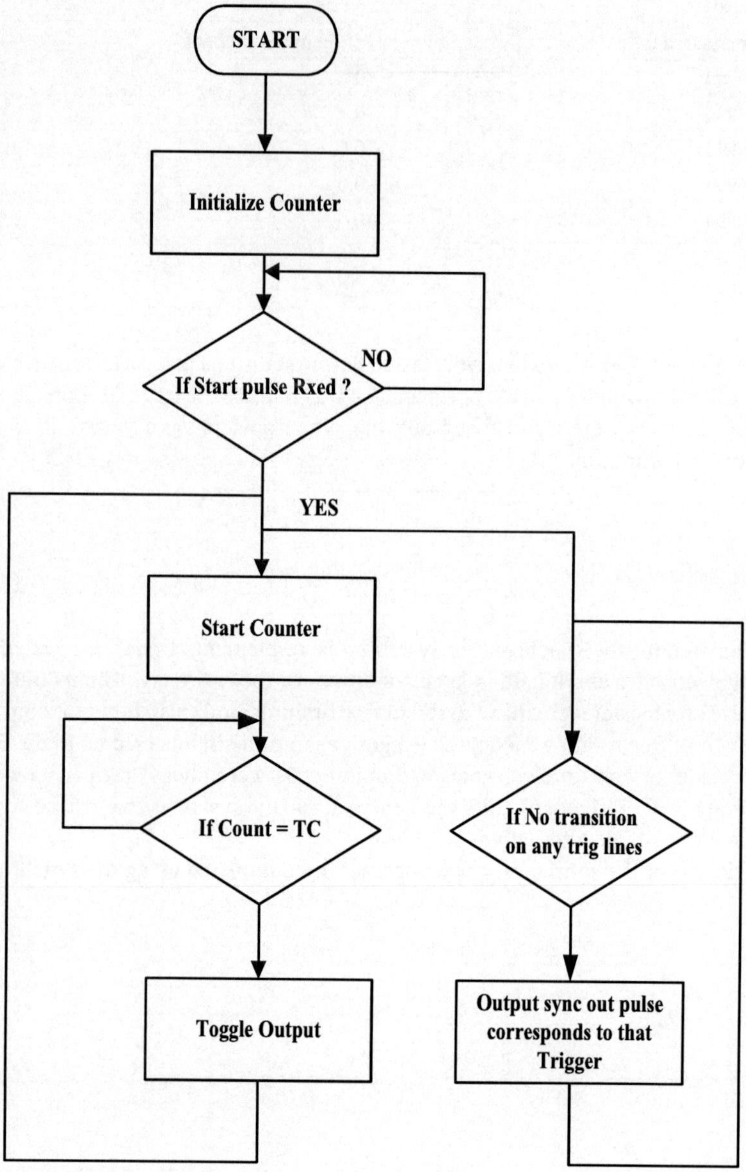

Fig. 5 Logic flow of master

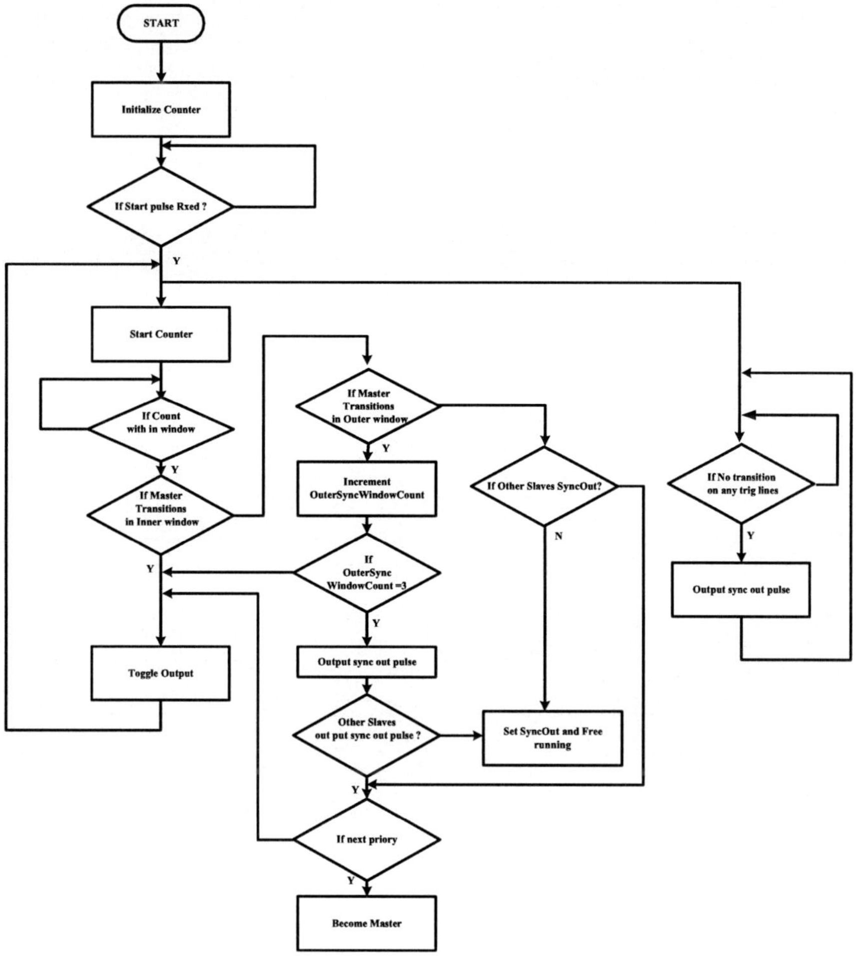

Fig. 6 Logic flow of slave

References

1. User Manual of Guidance Systems
2. Datasheet TCXO IT5300DC
3. Datasheet "ProAsic 3"
4. User Manual of Control Electronics

Frequency Scaling of Limit-Cycle Oscillation in a Launch Vehicle Electrohydraulic Actuation System

Baby Sebastian, Priya C. Kurian, Sunanda Lona, J. Venkateswaran, U. P. Abhijith, B. Jaya, and M. N. Namboodiripad

1 Introduction

The complex mission requirements of satellite launch vehicles call for highly reliable onboard systems with accurate performance. The launch vehicle under consideration for a specified flight zone is steered in pitch, yaw, and roll directions by means of thrust vector control using electrohydraulic servo actuators. These actuators make use of a combination of linear force motor (LFM)-driven direct drive valve (DDV) and hydraulic amplifier valve (HAV) for flow control.

The integrated test of the actuation system in the flight-equivalent simulated test rig as part of actuator in loop simulation for a mission exhibited an extra movement in the feedback during 0.5% slow sine excitation [1]. Subsequent analysis revealed that the frequency with which DDV oscillates varies as a function of different nonlinearities residing in the actuator. Limit cycling frequency varies during commanded time and null held duration, and this limit-cycle frequency at certain operating condition interacts with the resonance mode of the total flex nozzle control (FNC) system. Hence, there is a need to determine the cause and limit the consequences of this effect on the overall performance of the system. This paper focuses on identifying the cause and mitigating the effects of 'varying frequency limit-cycle oscillations' observed during various operating zones of the FNC system.

The system modeling and associated nonlinearities are briefed in Sect. 2. The treatment of the nonlinearities using describing function approach and estimation of limit cycle is explained in the next section. Design and validation of various compensator schemes to effectively scale the limit-cycle frequencies beyond a specified minimum bound is done in the next section. Further, the effect of introduction of

B. Sebastian (✉) · P. C. Kurian · S. Lona · J. Venkateswaran · U. P. Abhijith · B. Jaya · M. N. Namboodiripad
VSSC, Indian Space Research Organisation, Thiruvananthapuram 695022, India
e-mail: baby_sebastian@vssc.gov.in

V. H. Saran and R. K. Misra (eds.), *Advances in Systems Engineering*, Lecture Notes in Mechanical Engineering,
https://doi.org/10.1007/978-981-15-8025-3_63

this compensator in the overall closed loop performance was studied, and its stability over the entire operating range is also discussed in the last section.

2 Actuation System Model

The functional model of the hydraulic actuation system (Fig. 1) consists four layers of analog loops. The high-gain current loop around the LFM is the innermost loop. Position loop of DDV which is the second loop provides excellent tracking performance as well as high bandwidth for the DDV position w.r.t. DDV command. The third control loop is the HAV position feedback loop. Even under the influence of large flow force disturbance, HAV loop will ensure excellent tracking performance of the final stage spool valve and sufficient bandwidth. The last control loop is the outer position loop. The nozzle deflection is measured by two sets of LVDTs in the FNC system, one LVDT is mounted in the mirror image position using suitable bracket, and the main sensor is mounted along with actuator. The average of the mirror and main LVDTs will provide the actual deflection of the nozzle discarding the axial nozzle movement.

2.1 *Nonlinearities in Actuation System*

The hydraulic actuator consists of multiple non-linearities such as hysteresis in LFM, valve overlaps, and radial clearance in DDV and HAV, saturation in LFM power, valves and actuator stroke limit, friction, saturation of signals in servo electronics, etc. These nonlinearities are existing at multiple subsystems and combinedly contribute toward selfsustained oscillations in the system response.

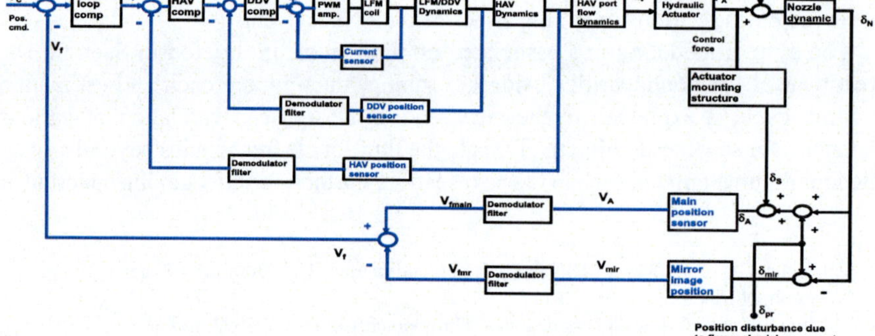

Fig. 1 Functional block diagram of an electrohydraulic actuation system

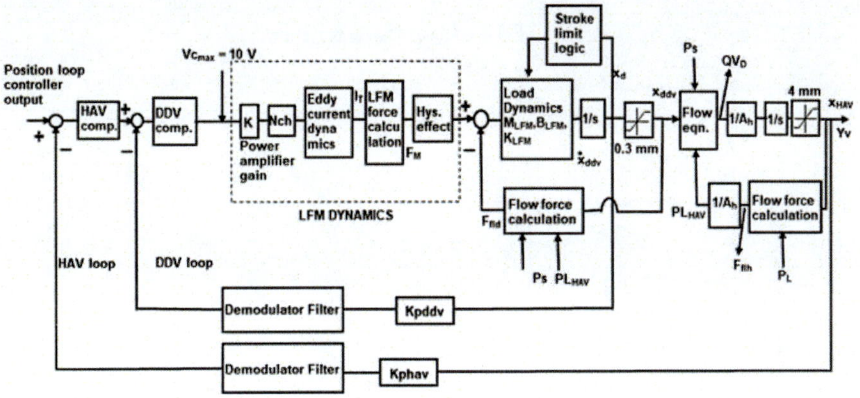

Fig. 2 Block schematic representation of valve

The cylindrical samarium–cobalt magnets used in LFM show hysteresis characteristics like any other magnets. This leads to a familiar hysteresis characteristic of magnetic flux density lagging behind magnetic field intensity and gets exhibited in the developed force–current relationship. The block schematic representation of valve is shown in Fig. 2. The hysteresis characteristics is modeled as backlash as shown in Fig. 3. From hereon, we consider this nonlinearity as a backlash acting on a servo loop. In a servo system, backlash may cause sustained oscillations or chattering phenomena, and the system may even turn unstable for large backlash [2, 3].

The controlled flow through DDV to HAV is

$$Q_{\text{ddv}} = K_{\text{Nddv}} x_{\text{d}} \text{sign}(P_{\text{s}} - \lambda P_{\text{LHAV}}) \sqrt{|P_{\text{s}} - \lambda P_{\text{LHAV}}|} \quad (1)$$

where K_{Nddv} is the flow gain constant of DDV, x_{d} is the valve movement, and P_{s} and P_{LHAV} are supply and load pressures. λ is the sign of HAV movement from null. The flow gain of DDV around null is dependent on its port shape, underlap/overlap conditions, radial clearance between spool and sleeve, etc., and will have an impact over the low amplitude response of the valve. It varies from valve to valve. This nonlinearity is captured by modeling the effective port opening as a nonlinear function of spool movement. The relationship between spool movement and effective port opening is modeled in a piecewise linear form and is shown in Fig. 4. For ideal valve, the gain of port opening to spool movement is unity.

3 Model Reduction and Analysis

The describing function analysis of the hysteresis nonlinearity and the Nyquist plot of the linearized system are used to establish the existence of limit-cycle oscillations, and the frequency of the limit-cycle oscillations is also computed.

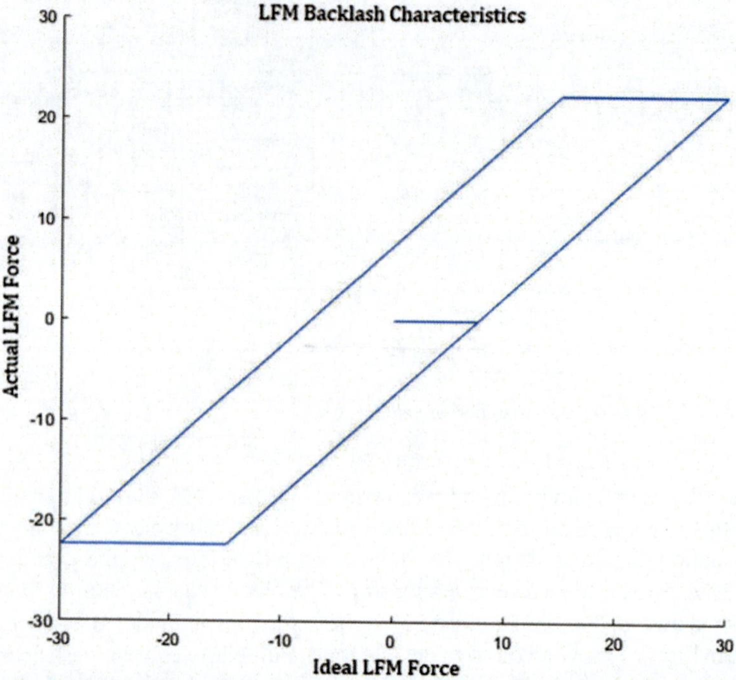

Fig. 3 Hysteresis valve nonlinearity

In order to carry out the describing function analysis, the nonlinearity under consideration is separated from other components which are then lumped together to arrive at a simplified linear equivalent. Since the current analysis is concerned with the limit-cycle oscillations triggered by the combined effects of nonlinearities including hysteresis in the LFM loop and the PI controller in the HAV loop, it is sufficient to limit the analysis within the HAV-DDV loop.

Hysteresis being a nondifferentiable nonlinearity does not allow linear approximation. However, since it is a separable nonlinearity, its dynamics is considered separately from the rest of the model which is now a linear one. Based on the linearization performed for various elements, the simplified linear equivalent model can be represented as in Fig. 5.

On superimposing the polar plot with the critical locus $(-1/K_N(X))$, an intersection point is obtained, thus indicating the presence of a limit cycle (Fig. 6) [1, 4, 5].

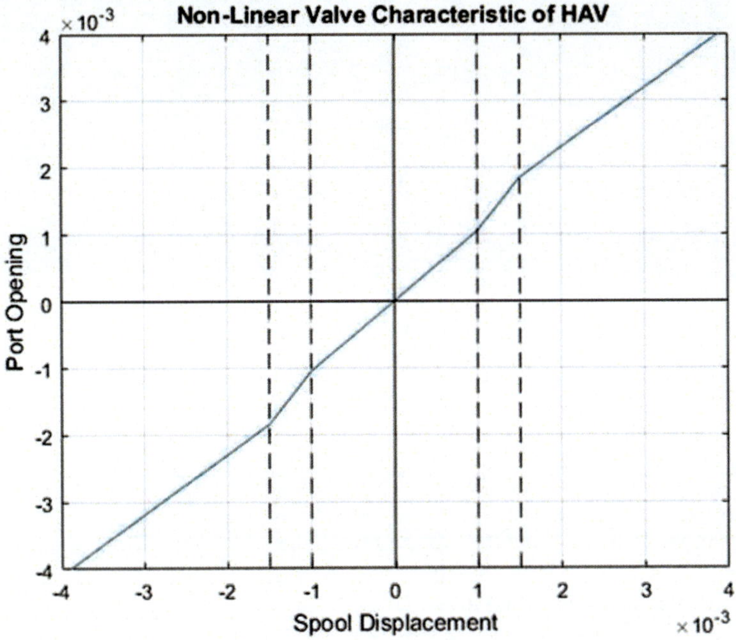

Fig. 4 Nonlinear valve gain

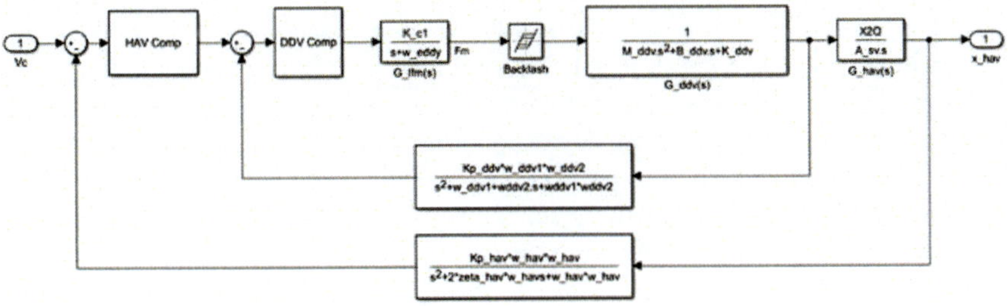

Fig. 5 Simplified equivalent model of valve

3.1 Effect of Valve Nonlinear Characteristics on Limit Cycle

During the hardware testing phase, the system was found to oscillate with varying frequency depending upon the region of operation of DDV. During the null (0–2%) and near-null (2–10%) operating conditions, the frequency of oscillations was observed to be around 12 Hz and 16 Hz, respectively, and the frequency remained at around 22 Hz at all other operating conditions. As is evident from previous analysis, the 22 Hz oscillation can be attributed to the effect of hysteresis present in linear force motor. However, in order to incorporate the observation of frequency variation

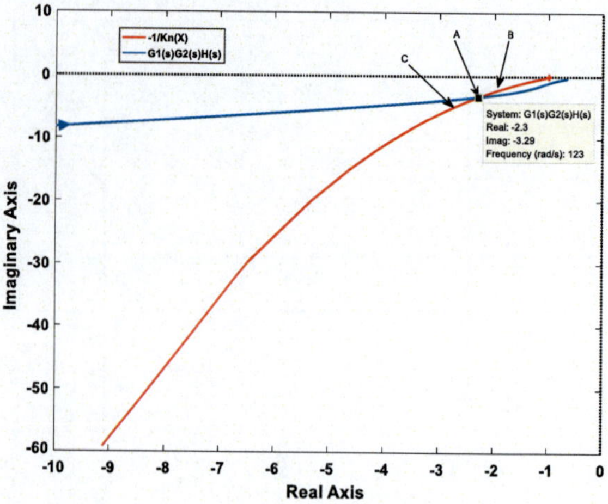

Fig. 6 Describing function analysis

into the model, the DDV overlap effects need to be addressed. With the introduction of DDV overlap nonlinearity, the limit-cycle frequency variations were replicated in the mathematical model simulation.

4 Frequency Scaling of Limit Cycle

Due to the presence of hysteresis in LFM combined with DDV overlap, limit cycling at various frequencies was observed with variations in operating zones of DDV. It is imperative to isolate the frequency of the limit cycles from entering the position loop of the actuation system to prevent possible resonant excitations. This section discusses various approaches in design of compensators which scales the limit cycling frequencies over the entire DDV operating zone beyond a certain minimum bound. This in turn would ensure suppression of limit-cycle oscillations impact in position loop. Since the overlap introduces a varying loop gain at various operating zones, the frequency of oscillation was also found to be varying. Thus, a straight forward approach to maintain the frequency constant would be to maintain the loop gain constant. At the same time in order to ensure that the limit cycling frequency remains well above the specified minimum bound, the loop gain needs to be increased by a fixed value as well.

4.1 Valve Operating Zone-Based Gain Scheduling

The simplest and direct way of ensuring a constant frequency for limit-cycle oscillations is to maintain the loop gain of the HAV-DDV loop constant. Since the DDV

Fig. 7 Variable valve gain

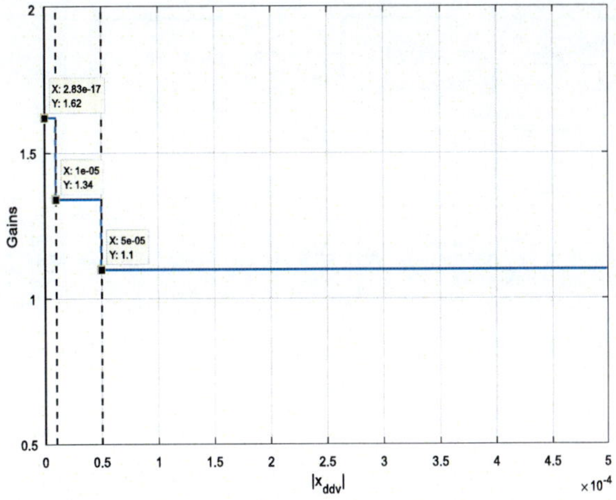

overlap model introduces a varying loop gain at different x_{ddv} limits, the loop gain can be maintained constant if the forward path gain of the HAV loop is appropriately scheduled. This would ensure that the slope induced by DDV overlap gets compensated appropriately, so that a constant relative loop gain is maintained. In the absence of DDV overlap (ideal DDV), the limit-cycle oscillations were found to be at a frequency of 22 Hz. Through simulation studies, it was observed that if the limit cycling frequencies with ideal DDV is maintained above 25 Hz, its effects in position loop get minimized. During ideal operation, a 10% increase in loop gain guarantees that the limit-cycle frequency remains at 26 Hz. Based on the above discussion, an x_{ddv}-based gain scheduling was implemented. The gains for different DDV operating zones are Fig. 7.

4.2 *Slope Estimation-Based Gain Boost Technique*

Valve operating zone-based gain boosting was focusing on scaling the limit cycling oscillation frequency accurately to a specified minimum bound by demanding the DDV overlap model as a prerequisite. However, these methodologies being model dependent does not qualify as a generalized solution. In order to overcome the shortcomings of the previous methodology, in this approach, an attempt is made to estimate the overlap characteristics of the DDV indirectly. This is achieved by calculating the ratio between DDV displacement as measured by the LVDT (x_{ddv}) before DDV overlap and $(x_{\mathrm{ddv}})'$ measured after DDV overlap. As is evident from Figs. 8 and 9, this ratio is a direct measurement of the DDV overlap. Since a direct measurement of x_{ddv} after DDV overlap block is unavailable, this needs to be estimated from HAV displacement. Based on the estimator, a gain boost compensator is designed which is represented in Fig. 10.

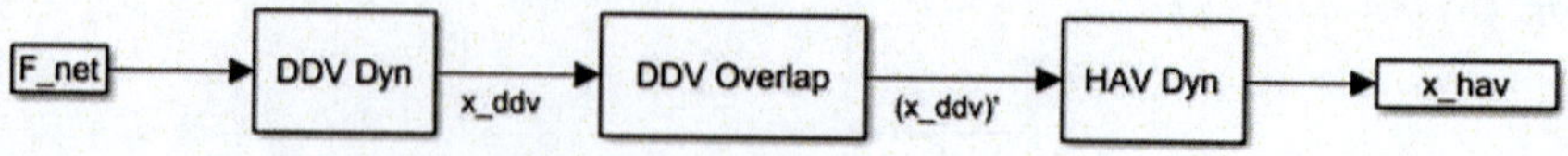

Fig. 8 Effect of DDV overlap on x_{ddv}

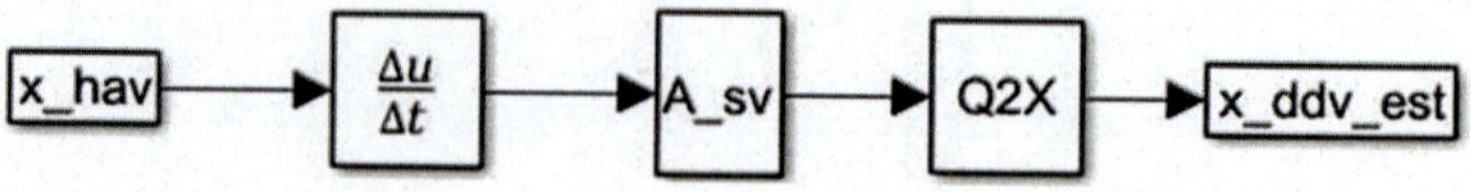

Fig. 9 Valve movement estimation

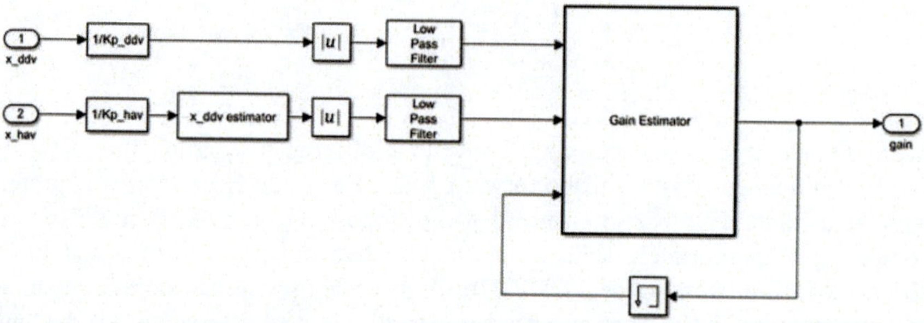

Fig. 10 Variable gain computation from measured and estimated valve movement

The ratio between estimated DDV displacement and actual DDV displacement is calculated using the above block diagram. This signal is then used to generate the required gain boost so as to maintain a net relative gain of 1.1. Compared to the previous frequency scaling techniques, this method does not require an accurate DDV overlap model. At the same time, it ensures that the limit cycling frequency remains at 26 Hz. This compensator thus ensures efficient frequency scaling without any additional hardware requirements. The effect of addition of this compensator in the HAV loop is as shown in Fig. 11. The gain values provided by the slope estimation-based gain boost compensator is as shown in Fig. 12. It can be observed that during steady state, (the DDV operates around its null region) a DC gain of 1.62 is maintained so as to ensure the net relative gain is 1.1, and the frequency of oscillation is above the specified minimum bound of 26 Hz.

The effect of addition of the slope estimation-based gain boost compensator on the outer position loop can be visualized from the position loop response shown in Fig. 13. It can be also observed that the frequency scaling of limit cycles in the inner loops led to an effective suppression of limit-cycle oscillation in the outer position loop. To emphasize the effect of slope estimation-based gain boost compensator in the position loop, a phase portrait of the states actuator velocity ($\dot{x}_{\mathrm{a}}$) w.r.t actuator displacement (x_{a}) is plotted as Fig. 14. A 75% suppression of excursions in actuator displacement is observed.

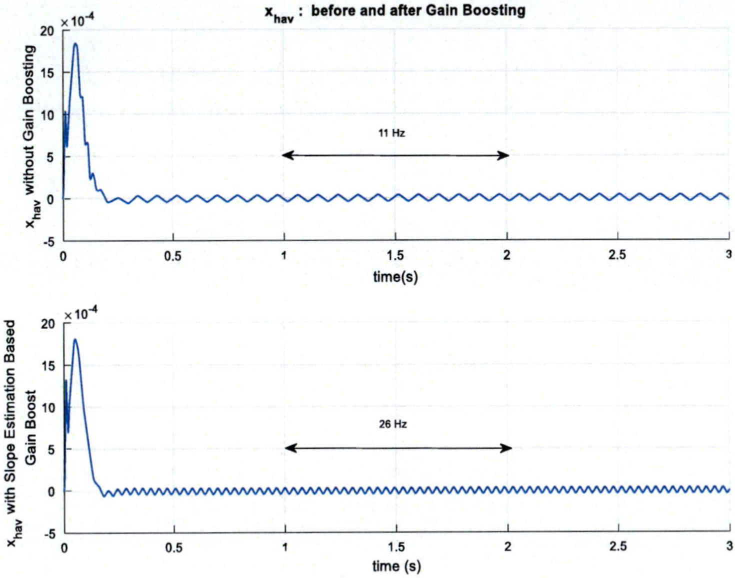

Fig. 11 Performance with gain boosting

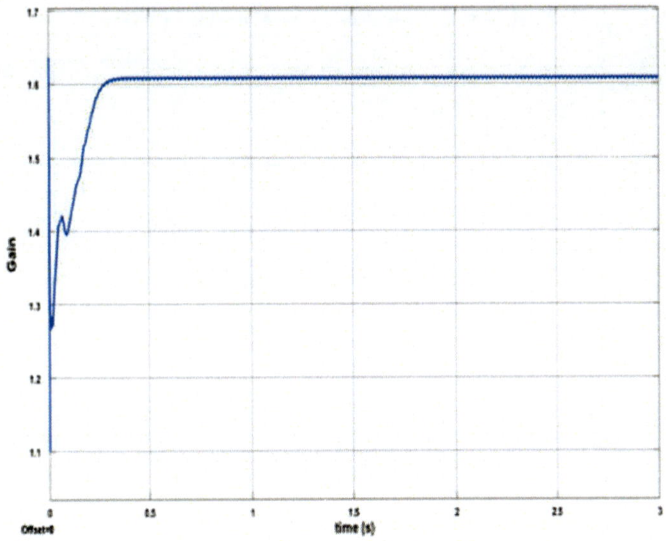

Fig. 12 Gain output

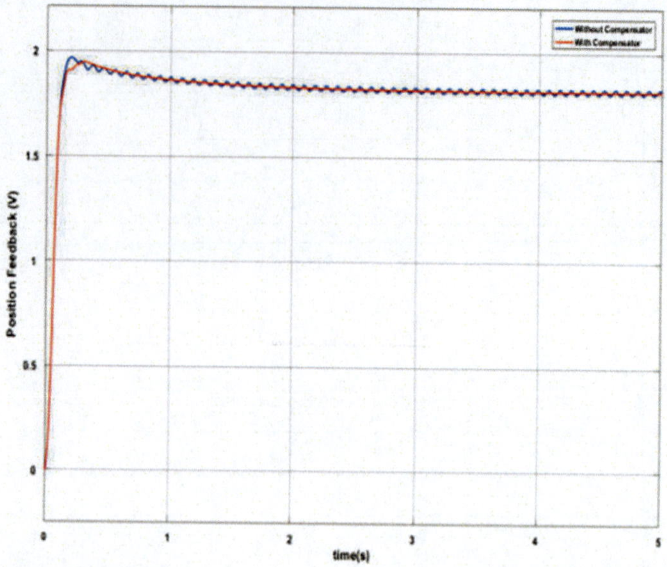

Fig. 13 Actuator position output

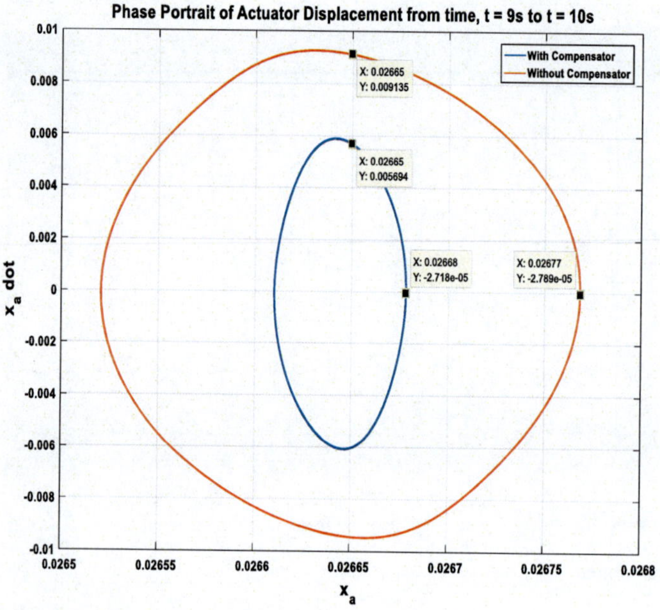

Fig. 14 Phase portrait of actuator

5 Conclusion

A thorough analysis of the varying frequency limit-cycle oscillations induced in the DDV-HAV loop due to the presence of LFM hysteresis with DDV overlap of an electrohydraulic actuator was carried out. The DDV overlap nonlinearity was brought in as a piecewise linear model to simulate the effect of varying frequency limit cycles observed during actual hardware tests. An effective way of frequency scaling of limit cycle is established. The performance developed method has been evaluated through simulation.

This study proposes a generalized compensator for frequency scaling of limit cycle oscillations.

References

1. Impram ST, Munro N (1996) Describing function analysis of nonlinear systems with parametric uncertainties. In: UKACC international conference on control, pp 112–116
2. Slotine JJE, Li W (1991) Applied nonlinear control. Prentice Hall Inc., Upper Saddle River
3. Nagrath IJ, Gopal M (1994) Control systems engineering. Wiley Eastern Ltd., Hoboken
4. Choudhury SK, Atherton DP (1974) Limit cycle in higher order nonlinear systems. Proc Ind Electr Eng 121:717–724
5. Wang C, Yang M, Zheng W, Kun Hu, Dianguo Xu (2017) Analysis and suppression of limit cycle oscillation for transmission system with backlash nonlinearity. IEEE Trans Ind Electr 64(12):9261–9270

Development of IoT Based Underwater Drone for Maritime Security and Surveillance

Narri Yadaiah, Nagireddy Ravi, and Garnapalli Shreya

1 Introduction

The security and surveillance system is essential and necessary in maritime borders especially for India due to peninsular country. The sea border is a vulnarable area where any trespassing can take place. Maritime surveillance is most important in ensuring the safe use of the sea and in securing India's maritime borders to meet the challenges and threats relating to the safety of navigation, marine pollution, law obligation, and overall security [1].

Drone has become a very interesting research topic and has brought innovation and new market [2]. Many researchers have been focused on unmanned autonomous vehicles (UAVs) and developed UAVs for various application in aerial, land, and underwater applications such as non-military, agriculture, and surveillance [3–5].

The motivation for developing a low-cost underwater drone is from development of underwater robot for environmental monitoring [6]. In the past decade, many techniques have been developed for the maritime security and surveillance. A novel

N. Yadaiah
Department of Electrical and Electronics Engineering, Jawaharlal Nehru Technological University, Hyderabad, Telangana 500085, India
e-mail: svpnarri@yahoo.com

N. Ravi (✉)
Pulichintala Hydro Electric Scheme, Telangana State Power Generation Corporation Limited, Suryapet, Telangana 508246, India
e-mail: nagireddyravi@rediffmail.com

G. Shreya
G. Narayanamma Institute of Technology and Science (for Women), Jawaharlal Nehru Technological University, Hyderabad, Telangana 500104, India
e-mail: shreyagarnapalli@gmail.com

V. H. Saran and R. K. Misra (eds.), *Advances in Systems Engineering*,
Lecture Notes in Mechanical Engineering,
https://doi.org/10.1007/978-981-15-8025-3_64

method is proposed to detect foreign body in the marine waters using laser beams, and the system is a combination of laser and PIR sensor that measures infrared light radiating from objects [7]. In [8], the advanced RISC machine microcontroller with global system for mobile and radio frequency identification is implemented for monitoring intruder border crossing.

The objective of this paper is development of Internet of Things (IoT)-based underwater drone with light detection and ranging (LiDAR) sensor to serve the future requirements of coastal security and surveillance. IoT is a technology in telecommunications that interrelates the computer devices, mechanical, digital devices provided with the unique identifiers and ability to exchange data. The LiDAR sensor makes the data collection fast and most accurate compared to SONAR and RADAR as it does not have any geometric distortions. The LiDAR sensor is cheaper and has high measurement accuracy. It gives the precise distance of the target and sent to the security personnel with an alert notification, whereas the satellites may not be responded precisely due to bad weather.

2 Internet of Things

The term IoT was first coined by Kevin Ashton in 1999. The IoT is a system in which Internet extends into the real-world things such as computing devices, mechanical and digital machines, or people that have distinct identities and capacity to transfer data over a network without need of human-to-human or human-to-computer interaction [9].

The global market for IoT is expected to grow to $581 billion by 2020 at an annual growth rate of 15–23%. The pictorial representation of IoT is as shown in Fig. 1. The characteristics of IoT are (i) dynamic and self-adapting, (ii) self-configuration, (iii) interoperable communication protocols, (iv) unique identity, and (v) integrated

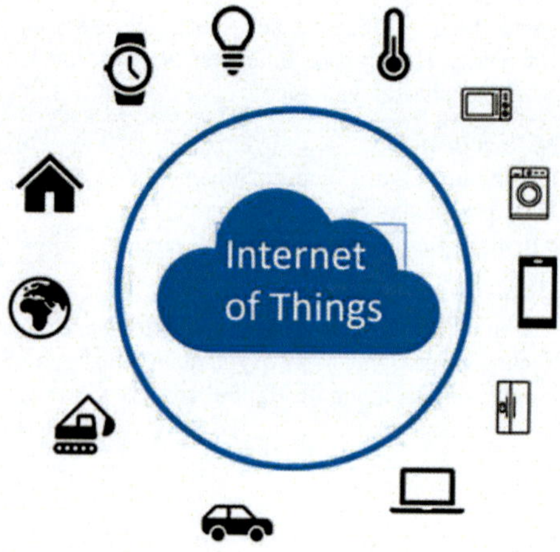

Fig. 1 Representation of Internet of Things

information network. In the past decade, the application of IoT has been increasing covering wide range of applications like healthcare, utilities, transport, etc. It has found wide range of applications in developing Smart Home, Wearables, Smart City, Connected Health, Smart Grids, Industrial Internet, Connected Car, and Smart Retail.

3 Development of IoT Based Underwater Drone

In this section, the development of underwater drone is presented and described its operation and main components. The underwater drone is waterproof and will operate below the water. It is designed and developed for remote opertation. The navigation system is in manual mode and programmed to do the job efficiently. The hardware components utilzed for development of the system are shown in Fig. 2a–f. These components are integrated with each other providing the underwater drone a solid, convenient, and transportable. The IoT system has been proposed for online monitoring of objects for sea border security. The IoT system is ideal for connecting various things for interacting and communication and makes the system easy and safe. It is adaptable as most of the corporations, industries having the Internet, mobiles, so that transferring of data will become much easier and less cost through Internet.

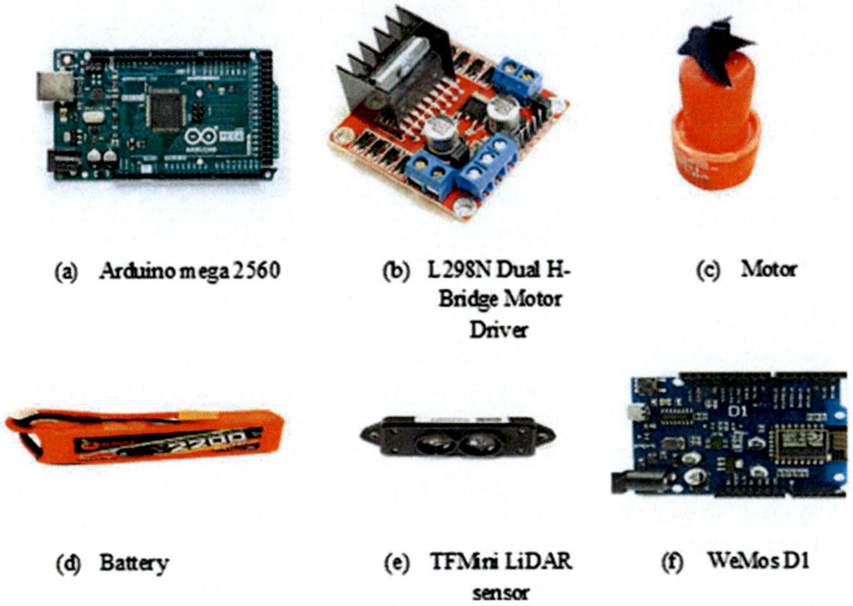

(a) Arduino mega 2560 (b) L298N Dual H-Bridge Motor Driver (c) Motor

(d) Battery (e) TFMini LiDAR sensor (f) WeMos D1

Fig. 2 Popular and low-cost components used in development of IoT underwater drone

Fig. 3 Structure of underwater drone

3.1 Design of Hull

The hull of the underwater drone is constructed using polyvinyl chloride (PVC) a synthetic thermoplastic material made by polymerizing vinyl chloride. The material was considered due to the very low cost and ease of availability with all types of fittings in various diameters. In order to balance, sink, and to get orientation for underwater drone, some weights such as metal rods and bars are added to main structure. The assembled underwater drone is shown in Fig. 3.

3.2 System Development

Figure 4 shows the schematic diagram for operation of underwater drone. The joysticks are used as a remote controller and sends signal to Arduino mega 2560 microcontroller for the movement of drone. The Arduino mega 2560 is brain of the underwater drone which acts as central processing unit, and two H-bridge drivers are connected to drive the motors.

The L298N dual H-bridge motor driver which is a bidirectional motor driver is used to drive the motors of the underwater drone, and each H-bridge circuit is able to control two motors in both directions. In this application, two drivers are employed one to move underwater drone forward and backward directions and other to move underwater drone ascent and descent. The control of the underwater drone made using the PWM signals which receives from microcontroller. The Arduino mega 2560 is programmed with the Arduino software, and software serial library allows for serial communication [10].

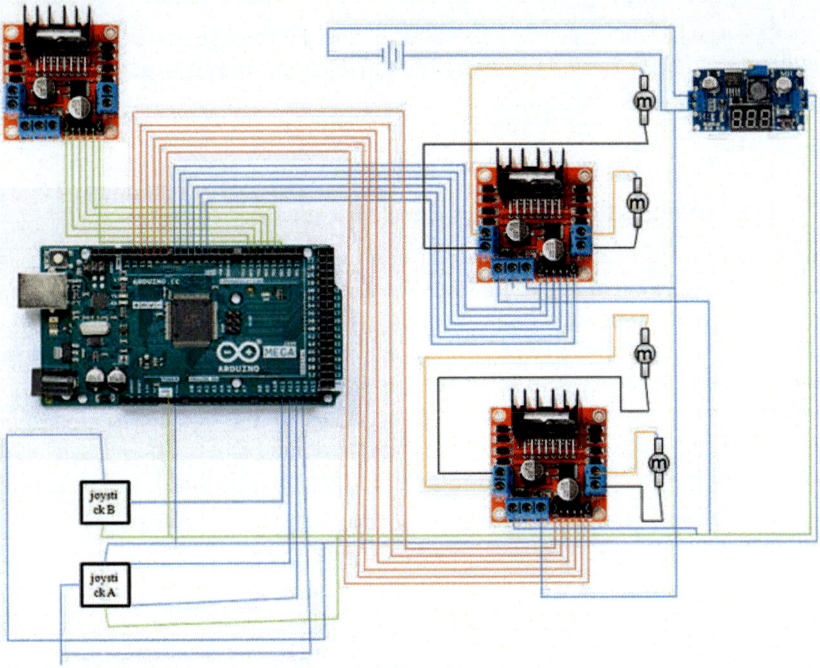

Fig. 4 Schematic diagram of underwater drone

3.3 Sensors and Data Communication

LiDAR is a remote sensing technique that uses light in the form of a pulsed laser to determine ranges. It will work based on time of flight (TOF) principle. The TFMini micro LiDAR distance sensor is equipped on underwater drone which meaures the distance of the target. Figure 5 shows the proposed IoT based monotoring system.

In order to enable IoT, a WeMos D1 R2 ESP8266 Wi-Fi based development board is used. The board is controlled by the ESP8266 chip and has a 32-bit processor. The data is transmitted through a Wi-Fi which is incorporated in the WeMos D1 R2

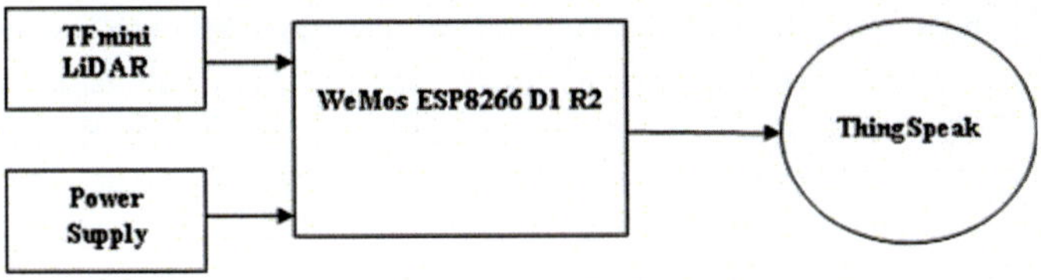

Fig. 5 Block diagram of proposed IoT system

board. The data from sensor is transmitted to board is through cables, data can access online using ThinkSpeak Web site, and available tools can be used to communicate for actions.

4 Implementation, Results, and Discussion

The IoT based system has been proposed for online monitoring of the objects for maritime security and surveillance. The IoT is ideal for building blocks and consists of sensors, processor, gateways, and applications. The IoT enabled underwater drone is equipped with LiDAR sensor which is an eye of the drone and is used for detecting objects. The IoT enabled underwater drone continuously monitors the detection of objects and provides the distance of the object to remote access. The steps involved in detecting the objects and monitoring may be summarized in the following algorithm:

4.1 Algorithm

Step 1: The movement of the underwater drone is set by remote station.
Step 2: The sensor collects the distance of target and serially send it to microcontroller.
Step 3: Based on the distance of the target, the system decides whether the target is within permissible area.
Step 4: If the target crosses the limit, an alarm is displayed on the monitoring system, and a notification is sent to an authorized user.

4.2 Results and Discussions

In order to illustrate the IoT based underwater drone for maritime security and surveillance, a scheme shown in Fig. 6 is considered. The system consists of a sensor, processing module microcontroller, and an open-source Web portal.

The LiDAR sensor captures the distance of the object in the form of analog signals. The signals are sent to WeMos D1 R2 microcontroller, and the data collected is processed through WeMos D1 R2 microcontroller. The ESP8266 Wi-Fi which is inbuilt in the microcontroller transmits the data to central server. The inbuilt Tx pin is used to send data through Wi-Fi and can upload the data to the central server.

To monitor the data on the Internet, ThingSpeak an open-source application which is vital tool for IoT based projects is used [11]. By using ThingSpeak site, the data can be observed over the Internet, using the channels and Web pages provided by ThingSpeak. ThingSpeak collects the data from the sensors that can be analyzed and visualizes the data and acts by activating a reaction.

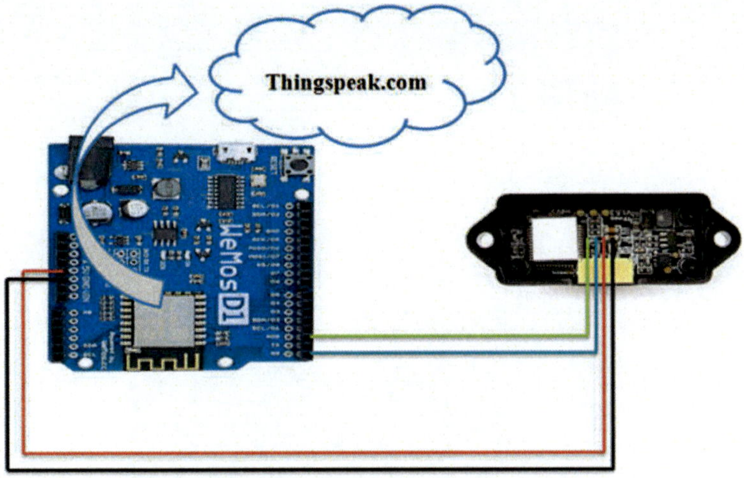

Fig. 6 Scheme of IoT based alarm system

The authorized users can monitor and access the distance of the object by logging on ThingSpeak Web site as shown in Fig. 7. It illustrates the distance that can be monitored continuously through ThingSpeak Web site, and an alarm alert is activated when the object is in between 200 and 210 cm from the underwater drone.

The *If This Then That also known as IFTTT* is a free Web-based service to create statement with conditions, called applets. An applet is activated by changes that occur within other Web services. IFTTT is a platform used to connect ThingSpeak for message service to mobile/email, so that alert message can be sent whenever the distance of the object is critical and mobile alert will be delivered to the concerned authorized persons which is shown in Fig. 8.

In order to validate the performance of the proposed underwater drone, the toy submarine is moved toward underwater drone with different speeds and ranges that

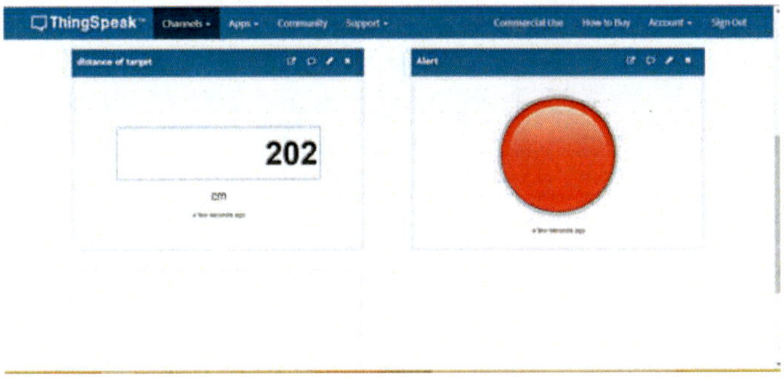

Fig. 7 Data and alarm on ThingSpeak channel

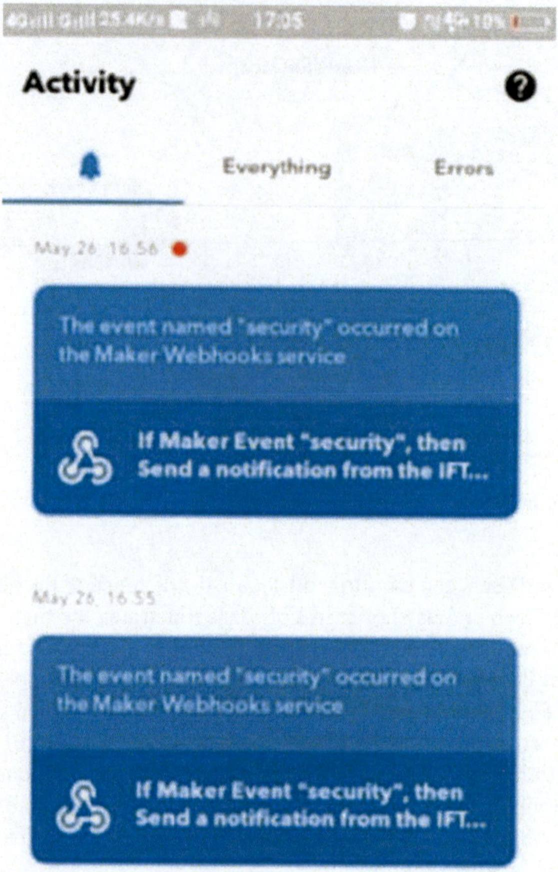

Fig. 8 Alert on mobile of authorized user

are measured, some instantaneously moving objects are also tested, and results are given in Table 1.

Table 1 Demonstartion of results

Speed of object	Distance of object and underwater drone (cm)	Alert message
Instantaeous	200	No
3 cm/s	200	Yes
5 cm/s	200	Yes

5 Conclusions

This paper presented a low-cost, efficient, and real-time IoT enabled underwater drone for maritime security and surveillance has been implemented and tested. The experimental results showed that it is able to recognize simulated objects and measures its distance from drone and able to send messages to remote user correctly. The developed product may be used in real-time application with high accuracy.

References

1. Brusch S, Lehner S, Fritz T, Soccorsi M, Soloviev A, van Schie B (2011) Ship surveillance with TerraSAR-X. IEEE Trans Geosci Remote Sens 49(3):1092–1103
2. Sarris Z (2001) Survey of UAV application in civil market. In: IEEE Mediterranean conference on control and automation
3. Dumaik M (2017) Lice hunting underwater drone protects salmon. IEEE Spectr 54(4):9–10
4. Mian AA, Daobo W (2008) Nonlinear flight control strategy for an underactuated quadrotor aerial robot. In: Proceeding of the IEEE international conference on networking, sensing and control (ICNC), Sanya, 6–8 Apr 2008, pp 938–942
5. Olivares-Mendez MA, Mejias L, Campoy P, Mellado-Bataller I (2013) Cross-entropy optimization for scaling factors of a fuzzy controller: a see-and-avoid approach for unmanned aerial systems. J Intell Robot Syst 69(1):189–205
6. Ming L, Hirayama T, Oyangi S (2016) Underwater-drone with panoramic camera for automatic fish recognition based on deep learning. IEEE Access 4:1–7
7. Mitra N, Maji A, Agarwalla P (2017) Integrated maritime security & surveillance. In: 2nd International conference on devices for integrated circuit (DevIC). IEEE Xplore Digital Library, pp 766–768
8. Kamalakannan B, Naresh K, Sakthivel P (2016) Protecting fishermen's by detecting and warning them while crossing sea borders using GSM AND RFID technologies. In: Online international conference on green engineering and technologies (IC-GET)
9. Zanella A, Bui N, Castellani A, Vangelista L, Zorzi M (2014) Internet of Things for smart cities. IEEE Internet Things J 1(1):22–32
10. www.arduino.cc/Guide/documentation
11. www.thingspeak.com. ThingSpeak—an open IoT platform

Green Energy Systems

Fabrication of Three-Electrode Lithium Cell Using Solid Polymer Electrolyte

P. Tamilselvi and M. Hema

1 Introduction

Implementation of green technology by the manufacturing of environmentally friendly materials is an alternative way to reduce the pollution problem [1]. Lithium-ion polymer batteries have received the recent interest among the most promising energy storage devices. It plays the crucial role in electric vehicles, electrochemical storage devices, cell phones, tablets, digital cameras, and laptops [2–4]. Due to enhancing demand of lightweight, flexible, and thin lithium batteries, it motivates the researchers to focus the research on portable electrodes and flexible polymer electrolyte. Usage of polymer electrolyte in lithium battery offers many advantages than the liquid electrolyte such as non-leakage, flexible, less flammability, high energy density, and wide electrochemical stability window [5, 6]. Numerous researches have been carried out in the solid polymer electrolytes during the last few decades. However, maximum ionic conductivity, higher transference number, and suitable electrochemical stability at ambient temperature are the challenging properties for the development in lithium-ion batteries.

Preparation of cathode materials in lithium batteries includes oxides of transition metals. During the removal of lithium, it may undergo oxidation to higher valences. One of the most common cathode material used in lithium batteries is $LiCoO_2$. It forms the α-$NaFeO_2$ structure, (distorted rock-salt structure). This material cation has trigonal structure in alternating planes. It is helpful for lithium ions for lithiation

P. Tamilselvi (✉)
Department of Physics, Bannari Amman Institute of Technology, Sathyamangalam, Tamil Nadu 638401, India
e-mail: tamilselvinew@gmail.com

M. Hema
Department of Physics, Kamaraj College of Engineering and Technology, Virudhunagar, Tamil Nadu 626001, India

V. H. Saran and R. K. Misra (eds.), *Advances in Systems Engineering*,
Lecture Notes in Mechanical Engineering,
https://doi.org/10.1007/978-981-15-8025-3_65

and delithiation [7]. Stoichiometric surface content of $LiCoO_2$ with reduced particle size is essential to improve the performance of the cycling.

In the present study, optimized polymer electrolyte of 80PVA:20PVdF:15$LiCF_3SO_3$ has been taken for electrochemical investigation which was reported in our earlier studies as highest ionic conducting polymer electrolyte [8]. AC impedance spectroscopy was used to analyze the cell impedance and electrochemical stability of the polymer electrolyte. The coin cell was constructed by sandwiching the polymer electrolyte between $LiCoO_2$ as cathode and Li metal as anode to investigate the electrochemical impedance and electrochemical stability of prepared polymer electrolyte.

2 Experimental

2.1 *Materials Used*

PVA (125,000), PVdF (530,000) have purchased from S.d.fine, lithium trifluromethane sulfonate, $LiCF_3SO_3$, purchased from Alfa Aesar, N, N-Dimethyl formamide, (DMF) from Merck, and Lithium cobalt oxide, $LiCoO_2$ from Sigma Aldrich.

2.2 *Preparation*

Solution-casting technique was used for the preparation of optimized 80PVA:20PVdF:15$LiCF_3SO_3$ solid polymer electrolyte. At first, the mentioned quantities of polymers and salt were dissolved in distilled DMF solvent at 60 °C. Both polymer solution and salt solutions were mixed together at the same temperature. The obtained solution was stirred continuously until homogeneous solution has been obtained. The homogenous solution was casted onto the glass petri dishes and dried at 60 °C. Dried electrolyte film was transparent and free-standing nature. Thickness of the films has been calculated in the range of 100 μm.

2.3 *Measurements*

Transference number was calculated by sandwiched the polymer electrolyte between two stainless steel electrodes. In this method, DC voltage across the sample and the corresponding DC current versus time has been calculated. Impedance spectroscopy was studied using Biologic, SP-300 from the frequency 42 Hz to 1 MHz at 303 K by sandwiching the optimized polymer film between aluminum-blocking electrodes.

Electrochemical stability of polymer electrolyte was evaluated by linear sweep voltammetry (LSV) using electrochemical analyzer Biologic, SP-300 with the scan rate of 5 mV s^{-1} using Li/Polymer electrolyte/Stainless steel. Cyclic voltammetry was measured in the same instrument at room temperature over a voltage range of 3.2–4.3 V with the scan rate of 0.1 mV s^{-1} for the cell Li/Polymer electrolyte/$LiCoO_2$.

2.4 Cell Assembly

Coin cell was constructed to investigate the performance of the prepared polymer electrolyte. $LiCoO_2$ was used as cathode, lithium metal was as counter and reference electrode, and polymer electrolyte was as both electrolyte and separator in the coin cell.

Fine particles of 80 wt% $LiCoO_2$ material and 10 wt% of PVDF (binder) is mixed with 10 wt% acetylene black in N-methylpyrrolidone (NMP) solvent for the preparation of cathode electrode. The homogenous slurry has been prepared by adding the solvent slowly, and it was coated onto well-cleaned aluminum foil. It was vacuum dried for 24 h at 120 °C, and after that the dried electrode was cut into disks. The obtained circular disks are compressed by a roller at room temperature to make a smooth and compact electrode structure. Then, it has been stored in argon-filled glove box to avoid the absorption of moisture.

3 Results and Discussion

3.1 Ionic Transference Number Measurement

The transference number of mobile species in the prepared polymer electrolyte was calculated using DC polarization technique [9]. The transference number of an ion is calculated by the fraction of the total current carried by the respective ion across the material. Each ion has different mobility which carries extremely different portions of the total current [10].

Figure 1 shows the polarization curve of 80PVA:20PVdF:15$LiCF_3SO_3$ solid polymer electrolyte. Transference number (t_{ion}) has been calculated using the following relation, and it has been found to be 0.97.

$$t_{ion} = (I_i - I_f)/I_i$$

where t_{ion}, I_i, and I_f are the transference number, initial current after switching on the DC voltage, and residual current taken for the final evaluation, respectively. The value approaching close to unity indicates that the charge transport in the prepared polymer electrolytes is predominantly due to the ions not electrons.

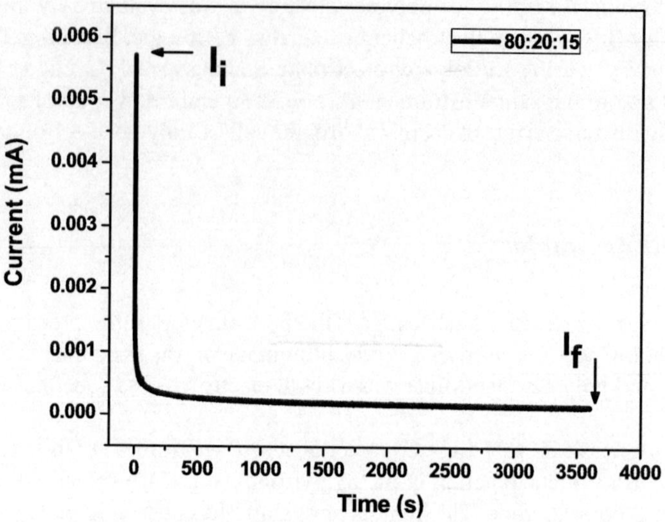

Fig. 1 Current versus time plot of 80PVA:20PVdF:15LiCF$_3$SO$_3$ polymer electrolyte at 303 K

3.2 *Impedance Analysis*

Figure 2 shows the room-temperature complex impedance spectra of

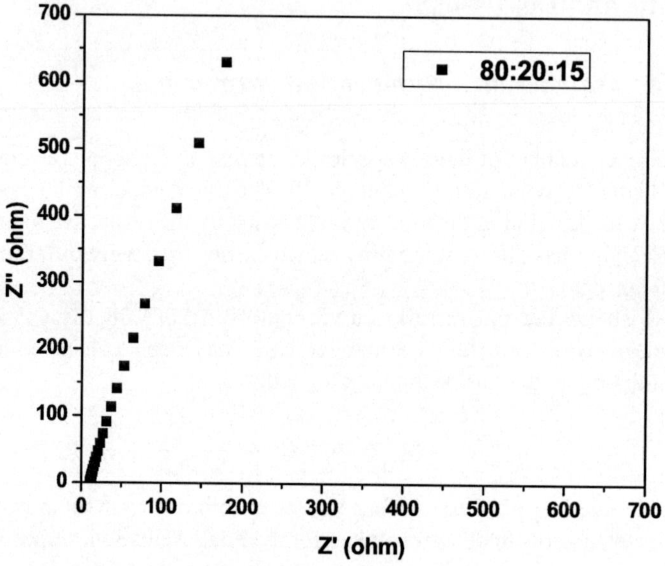

Fig. 2 Complex impedance plot of polymer electrolyte 80PVA:20PVdF:15LiCF$_3$SO$_3$ at 303 K

80PVA:20PVdF:15$LiCF_3SO_3$ polymer electrolyte. It exhibits the spike at lower frequency region corresponding to the bulk effect of the blocking electrodes [11, 12]. The intercept of the semicircle or spike with the real axis implies bulk resistance (R_b) of the electrolyte. In our earlier report, conductivity studies have been discussed [8] for different concentration of $LiCF_3SO_3$ doped in 80PVA:20PVdF blend polymer electrolytes. Highest conductivity of 2.7×10^{-3} Scm^{-1} at 303 K has been obtained for 80PVA:20PVdF:15$LiCF_3SO_3$ polymer electrolyte (optimized system) and was taken for the present investigation.

Using the optimized polymer electrolyte, coin-type cell has been constructed with the configuration of Li/Polymer electrolyte/$LiCoO_2$. As soon as constructed the cell, more number of free ions is drifted from one electrode to another electrode thereby absorbed in the lithium electrode forming charge accumulation at the electrode–electrolyte region.

3.3 Linear Sweep Voltammetry Studies (LSV)

Figure 3 depicts the current voltage response of the prepared polymer electrolyte at ambient temperature. Current in the anodic scan is related to the decomposition of polymer electrolyte which implies the electrochemical stability. In the present study, electrochemical stability has been found to be greater than 4.7 V (vs. Li^+/Li), which is well suitable for high-voltage lithium polymer/lithium-ion polymer battery. The

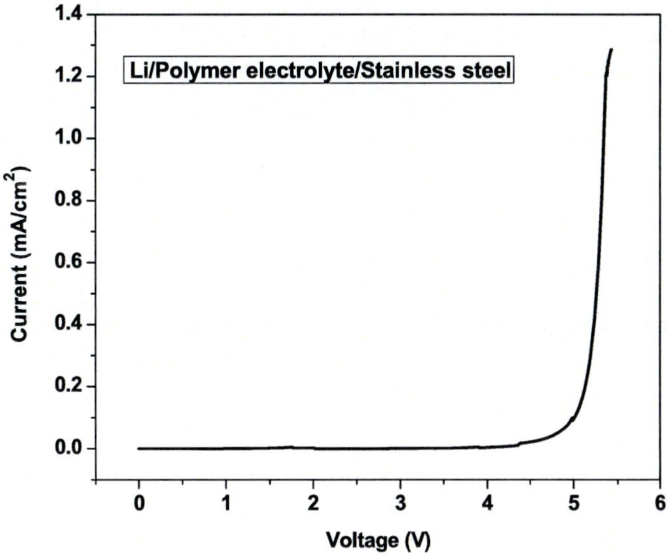

Fig. 3 LSV curve of 80PVA:20PVdF:15$LiCF_3SO_3$ polymer electrolyte versus L/Li^+

high stability of the film revealed that the prepared sample has no impurities. The PVA-PVdF blend-based polymer electrolyte stability obtained in the present study is higher than the other blends reported by Zhai et al. [13, 14] and Ghosh et al. [10]. Thus, the prepared polymer electrolyte stability (above 4.7 V) can be best suited for practical lithium-ion polymer battery.

LSV does not provide information regarding an oxidation reaction between a polymer electrolyte and cathode material. So, to have a further in-depth knowledge about electrochemical phenomenon of the prepared polymer electrolyte against the oxidation of cathode material, cyclic voltammetry has been studied and discussed.

3.4 *Cyclic Voltammetry (CV)*

Cycling performance of Li/Polymer electrolyte/$LiCoO_2$ coin cell has been depicted using cyclic voltammetry as shown in Fig. 4.

Cyclic voltammetry shows the redox peaks corresponding to the intercalation and de-intercalation of lithium ion in the anode electrode. Also, it is evident that the polymer electrolyte under present investigation has the ability to transport more mobile ions with enhanced mobility at ambient temperature as confirmed from high conductivity value.

The anodic oxidation appeared between 4–4.1 V corresponding to the removal of Li^+ ions from $LiCoO_2$ material, (host) during anodic scan. The observed indistinct oxidation peak appeared in the CV curve may be due to the side reactions between

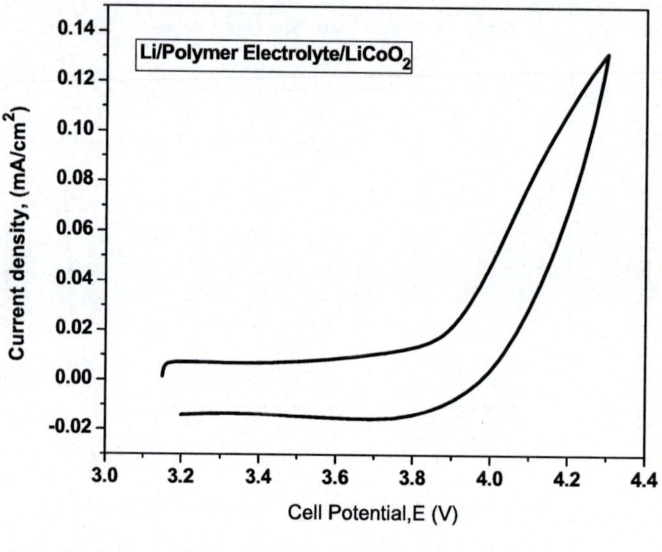

Fig. 4 Cyclic voltammograms of $LiCoO_2$/Li cell in prepared 80PVA:20PVdF:15$LiCF_3SO_3$ polymer electrolyte

polymer electrolyte and fresh lithium metal, thereby forms the interface layer in the first cycle which leads to increase of impedance. Purity of the prepared polymer electrolyte has been supported to flow very low current up to the anodic breakdown voltage. Small and broad reduction peak appeared near 3.7 V due to the slow reduction state of anions of lithium triflate [15]. The oxidation and reduction peak corresponding to $LiCoO_2$ in the present investigation is good in accordance with the result reported by Park et al. [16] and Ma et al. [17]. In cathode materials, Li ions can be reversibly intercalated and de-intercalated [18]. In the present investigation, very less lithium ions have been extracted out from the cathode material. This is mainly due to unwanted interfacial reaction between the charged $LiCoO_2$ and prepared polymer electrolyte [19] which leads to the formation of resistive layer. The resistive layer may obstruct charge transport at the electrode active particles and electrolyte interface; as a result, it weakens the cyclic performance of the cell [20–22]. However, from the electrolyte point of view, the prepared polymer electrolyte has been stable above 4.4 V versus Li/Li^+ which is well in accordance with LSV curve. These results show that the prepared polymer electrolyte have high anodic stability and satisfactory compatibility with cathode material [23]. It makes the prepared polymer electrolyte to be highly stable for high-voltage lithium-ion batteries.

4 Conclusion

Electrical and electrochemical properties of thin and flexible polymer electrolyte: $80PVA{:}20PVdF{:}15LiCF_3SO_3$ (optimized system) has been studied. The calculated ionic conductivity is about 2.7×10^{-3} Scm^{-1} at 303 K. The coin type cell with the configuration of $LiCoO_2/80PVA{:}20PVdF{:}15LiCF_3SO_3/Li$ has been fabricated using the prepared polymer electrolyte. Transference number studies revealed the influence of ions that are predominant in the polymer electrolyte. Electrochemical stability of the polymer electrolyte has been found to be higher than 4.7 V versus Li/Li^+ as confirmed from LSV. Excellent cyclic and reversible performance has been obtained for prepared polymer electrolyte from the CV studies that makes it well suitable for lithium-ion battery application.

Acknowledgements All authors in the paper acknowledge Prof. G. Hirankumar, Ramakrishna Mission Vivekanandha College, Chennai, Tamil Nadu for impedance studies.

References

1. Siracusa V, Rocculi P, Romani S, Rosa MD (2008) Biodegradable polymers for food packaging: a review. Trends Food Sci Technol 19:634–643

2. Wang Q, Zhang B, Zhang J, Yu Y, Hu P, Zhang C, Ding G, Liu Z, Zong C, Cui G (2015) Heat-resistant and rigid-flexible coupling glass-fiber nonwoven supported polymer electrolyte for high-performance lithium ion batteries. Electrochim Acta 157:191–198
3. Dunn B, Kamath H, Tarascon JM (2011) Electrical energy storage for the grid: a battery of choices. Science 334:928–935
4. Tarascon JM, Armand M (2001) Issues and challenges facing rechargeable lithium batteries. Nature 414:359–367
5. Kurc B, Jesionowski T (2015) Modified TiO_2-SiO_2 ceramic filler for a composite gel polymer electrolytes working with $LiMn_2O_4$. J Solid State Electrochem 19:1427–1435
6. Kim SH, Choi KH, Cho SJ, Kil EH, Lee SY (2013) Mechanically compliant and lithium dendrite growth-suppressing composite polymer electrolytes for flexible lithium-ion batteries. J Mater Chem 1:4949–4955
7. Fergus JW (2010) Recent developments in cathode materials for lithium ion batteries. J Power Sour 195:939–954
8. Tamilselvi P, Hema M (2014) Conductivity studies of $LiCF_3SO_3$ doped PVA: PVdF blend polymer electrolyte. Phys B 437:53–57
9. Hema M, Selvasekarapandian S, Arunkumar D, Sakunthala A, Nithya H (2009) FTIR, XRD and ac impedance spectroscopic study on PVA based polymer electrolyte doped with NH4X (X=Cl, Br, I). J Non-Cryst Solids 355:84–90
10. Ghosh A, Wang C, Kofinas P (2010) Block copolymer solid battery electrolyte with high Li-ion transference number. J Electrochem Soc 157:A846–A849
11. Su'ait MS, Ahmad A, Hamzah H, Rahman MYA (2009) Preparation and characterization of PMMA–MG49–$LiClO_4$ solid polymeric electrolyte. J Phys D 42:055410
12. Tamilselvi P, Hema M, Asath Bahadur S (2018) Investigation of nanocomposite polymer electrolytes for lithium ion batteries. Polym Sci Ser A 60:102–109
13. Park J-H, Kim J-S, Shim E-G, Park K-W, Hong YT, Lee Y-S, Lee S-Y (2010) Polyimide gel polymer electrolyte-nanoencapsulated $LiCoO_2$ cathode materials for high-voltage Li-ion batteries. Electrochem Commun 12:1099–1102
14. Zhai W, Zhu H, Wang L, Liu X, Yang H (2014) Study of PVDF-HFP/PMMA blended microporous gel polymer electrolyte incorporating ionic liquid [BMIM]BF4 for Lithium ion batteries. Electrochim Acta 133:623–630
15. Yang L, Furczon MM, Xiao A, Lucht BL, Zhang Z, Abraham DP (2010) Effect of impurities and moisture on lithium bisoxalatoborate (LiBOB) electrolyte performance in lithium-ion cells. J Power Sour 195:1698–1705
16. Park Y, Shin SH, Hwang H, Lee SM, Kim SM, Choi HC, Jung YM (2014) J Mol Struct 122:157–163
17. Ma X, Huang X, Gao J, Zhang S, Peng Z, Deng Z, Suo J (2014) Anionic polymer electrolyte with enhanced electrochemical performance based on surface-charged latex nanoparticles for flexible lithium-ion batteries. J Power Sour 272:259–266
18. Wang Y, Liao H, Wang J, Qian X, Zhu Y, Cheng S (2013) Effects of current collectors on electrochemical performance of FeS_2 for Li-ion battery. Int J Electrochem Sci 8:4002–4009
19. Parka J-H, Choa J-H, Kimb J-S, Shimb E-G, Lee S-Y (2012) High-voltage cell performance and thermal stability of nanoarchitectured polyimide gel polymer electrolyte-coated $LiCoO_2$ cathode materials. Electrochim Acta 86:346–351
20. Lee KS, Sun YK, Noh J, Song KS, Kim DW (2009) Improvement of high voltage cycling performance and thermal stability of lithium–ion cells by use of a thiophene additive. Electrochem Commun 11:1900–1903
21. Sun YK, Cho SW, Myung ST, Amine K, Prakash J (2007) Effect of AlF_3 coating amount on high voltage cycling performance of $LiCoO_2$. Electrochim Acta 53:1013–1019
22. Chang W, Choi JW, Im JC, Lee JK (2010) Effects of ZnO coating on electrochemical performance and thermal stability of $LiCoO_2$ as cathode material for lithium-ion batteries. J Power Sour 195:320–326
23. Li B, Liu Y, Zhang X, He P, Zhou H (2019) Hybrid polymer electrolyte for Li–O_2 batteries. Green Energy Environ 4:3–19

HVAC Design and Operation for Green Buildings

S. Sriram, K. Hema, and N. Karthiga Shenbagam

1 Introduction

As herbal resources swiftly dissipate around the arena, the want for green or environmentally pleasant building practices is becoming a necessity. The so-called sustainable homes are built or renovated beneath sustainable development, a layout method that reduces the harmful effect on herbal sources and looks on the lifestyles-cycle costs of the power. The advantages of sustainable homes are not tough to understand. By following green design practices, building proprietors and builders can do it economically and eco-friendly. Green buildings have decrease running prices than the latter. Second, sustainable layout enables minimize vast environmental impacts, consisting of water usage, ozone layer depletion and raw materials utilization.

This paper describes the basic principles of green building and discusses the role of heating, ventilating and air-conditioning (HVAC) for ensuring high overall performance green buildings in layout and operation. The design techniques for effective and sustainable HVAC structures are explained, and the brand new rising HVAC technologies for sustainable homes are described. It is hoped that HVAC designers and different building specialists may want to develop a better model of green buildings and observe effective strategies.

S. Sriram · K. Hema · N. Karthiga Shenbagam (✉)
Bannari Amman Institute of Technology, Sathyamangalam, India
e-mail: karthis47@gmail.com

S. Sriram
e-mail: sriram.ce18@bitsathy.ac.in

K. Hema
e-mail: hema.ce18@bitsathy.ac.in

V. H. Saran and R. K. Misra (eds.), *Advances in Systems Engineering*,
Lecture Notes in Mechanical Engineering,
https://doi.org/10.1007/978-981-15-8025-3_66

2 What is Green Building?

Despite all the discussions and advances within the global expertise, precisely what constitutes a green constructing is still not that properly understood—even among many experts. In simple terms, sustainable homes are strength and useful resource efficient, non-wasteful and non-polluting, especially adaptable for long-term functionality; they may be smooth to operate and keep and are supportive of the productiveness and well being of the occupants. The layout of a sustainable building utilizes a whole building, lifecycle technique. It must be stated that a green constructing is a continually evolving concept, a purpose. What makes any particular construction "sustainable" is a unique answer that responds to the particular purposeful necessities and the climatic situations of the construction place. In other words, "green" is distinctive for every construction. And, not every construction is completely green in each aspect of its design. The crucial thing is to preserve shifting toward the purpose.

From the point of view of HVAC engineer, excessive performance sustainable layout is a common feel method to constructing design. High overall performance design also takes into account the power and environmental performance of the building in the course of its entire lifestyles cycle, which includes site selection, construction, operations and preservation, renovations, demolition and substitute (Fig. 1).

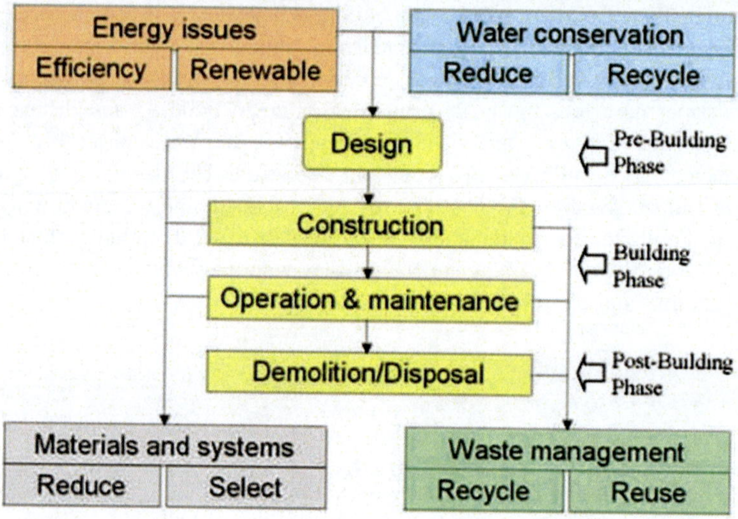

Fig. 1 Building life cycle and sustainable construction

3 The Role of HVAC

HVAC structures can also harm the environment with the aid of needless use of energy which results in depletion of non-renewable resources, mainly fossil fuels, either via the generation of power or thermal strength, both of which make contributions to environmental pollutants.

Environmental harm by means of HVAC systems will also be because of noise and by means of the release of infected water and air containing chemical compounds, lubricating oils, refrigerants, warmth switch fluids, particulate or gaseous rely, or microbiological organisms. In most conditions, HVAC structures will drastically effect how "sustainable" a building is. Therefore, the mission group must not overlook the capacity and have an impact on while developing the design.

HVAC engineers can play a crucial role in environmental design via being the technical/analytical resource for the crew and with the aid of encouraging the development of more rigorous assessment tools which are suitable, realistic and pleasant and defensible. To make a distinction for the green layout purpose, they have to push architects to layout higher envelopes, inspire HVAC selections be primarily based on lifestyles-cycle costing, insert their impact earlier in the technique and aid the right commissioning of buildings. They must also try to train building proprietors or builders the cost of a green/sustainable layout in order that the sustainable system may be achieved efficiently.

4 Design Strategies

Although HVAC systems offer many opportunities for recovery and re-use of thermal energy, the preferred solution is to use less energy in the first place. This is achievable by more energy-efficient buildings, systems and equipment and through improved operating and maintenance procedures. More attention should be paid to the thermal characteristics of building and strategies for minimizing internal loads, examining in detail the opportunities for natural ventilation and day lighting and exploring ways to reduce the energy requirements of HVAC [1]. For example, the use of high-performance glazing, insulation and effective sun control can substantially reduce cooling and heating loads and the size of HVAC systems and components required to meet these loads. The ultimate objective is to minimize refrigeration requirements, or do away with refrigeration altogether and to make use of renewable energy resources.

In order to design a better building, a green project team must rethink the design and construction process in terms of improving building performance. In addition to improving energy efficiency and environmentally sound operations, high-performance sustainable design also seeks to provide an indoor environment that enhances the productivity of the occupants. Current research information suggests that good indoor environmental quality increases alertness and improves performance, while reducing illnesses and absenteeism. Sustainable design may thus provide significant lifecycle cost savings and productivity gains in addition to energy and resource savings.

It is essential for the design team to reexamine basic engineering principles in order to develop appropriate system design in support of a green building. Techniques such as natural ventilation, thermal mass storage, radiant cooling and passive solar control are applications of basic thermodynamic principles. The designers must also help drive innovation and be an expert in understanding the technical characteristics and performance of architectural components, construction practices and maintenance/operational practices in terms of energy use, environmental impact and indoor air quality.

Financial analysis techniques are used inorder to determine the optimal configurations of architectural and engineering components, while meeting the typical budget, schedule and aesthetic requirements. For example, good indoor air quality is achieved through a collaborative process between the architect and engineer. It begins with the effort to avoid potential sources of contaminants within the space. The engineer should assist the architect, the owner and the construction team in understanding the impact on indoor air quality of things like finish materials selections, contaminant-producing processes in the space (such as printing), HVAC system design and configuration, construction methods and operations and maintenance practices. The HVAC engineer must then design a system to achieve ventilation effectiveness for a given space layout.

The analysis of green building design can be time-consuming due to the many factors involved. A clear definition and holistic understanding of the design problem is important since the environmental equation is complex. For example, newly developed refrigerants, while more environmentally friendly, may be hazardous (e.g., flammable), be less energy efficient, require extensive re-engineering of existing systems or depend on the development of new refrigerating machines. When the energy requirements of new refrigerants are considered over the lifecycle of the equipment, the potential environmental damage resulting from the higher energy usage may outweigh that from possible leakage/disposal of the original refrigerant.

5 New Technologies

Apart from the concept of sustainable, whole-constructing layout as the "right element to do," advances in technology are attracting the imaginations of building designers in recent times [2]. Under floor air structures, radiant cooling and heating, building-incorporated photovoltaic's, gasoline cells, sulfur and magnetic-induction-based lights systems, excessive performance (low emissivity, spectrally selective) glazing, and increasingly more state-of-the-art virtual controls and sensors are being considered on many modern projects. New structures and delivery methodologies are evolving, and they may provide huge technical advantages to a assignment. A few crucial technologies which have the ability to enhance the environmental overall performance of HVAC systems are described underneath.

5.1 *Under Floor Air-Supply Systems*

This is a subject that has acquired a lot of coverage currently. Figure 2 indicates a easy schematic of an underneath ground air-supply device. Like maximum structures, there are each tremendously appropriate and less than perfect programs. Some tempo-setting and sustainable designers discover it thrilling on the grounds that all the out of doors air is delivered to the building volume in near proximity to the occupants. Perhaps the maximum compelling packages are massive, high areas like conference centers and airport terminals, wherein probably drafty air flows might be of teen concern to what is a usually brief occupancy. This idea of turning in cooling and ventilation air down low is becoming known as displacement ventilation.

Another beneath floor air-supply approach makes use of a plenum area that contains cabling structures for delivering conditioned air and avoids the need for air ducting. It is exceptionally flexible and has the potential to provide better thermal consolation, indoor air nice and lifecycle price financial savings for today's modern-day workplaces. In recent years, with the growing needs of facts technology system and community, the use of raised-floor production has attracted an awful lot attention in many cities. Combined with an under floor air gadget, the raised-floor design can

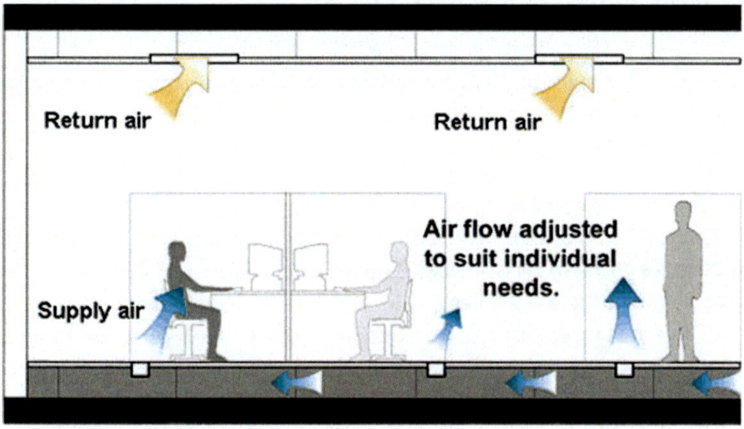

Fig. 2 Underfloor air-supply system. *Credit* Center for the Built Environment, University of California, Berkeley

offer a powerful answer for workplaces and industrial areas. However, issues about the hassle of dust stirring from the ground and the possible soreness caused by draft and vertical temperature differences nonetheless want to be investigated.

5.2 Demand-Controlled Ventilation

Improved detection of pollutants in the building air stream will improve indoor air quality and energy efficiency of the HVAC system, particularly by using advanced sensors and controls technology. Demand-controlled strategies adjust the ventilation rate so that when there is less demand for ventilation (i.e., fewer occupants), less fresh air is supplied to the building in order to save energy. The most common of these strategies provides ventilation air in proportion to the indoor CO_2 concentration [3]. The use of CO_2 sensors that allow for demand ventilation of spaces based on actual occupancy can reduce unnecessary overventilation. However, since these sensors will be used to modulate the amount of outdoor air supplied to occupants, their long-term reliability will be critical. You may encounter resistance to their use by local code officials and others. In general, the greatest savings in electrical usage for cooling with the addition of demand-controlled ventilation occur in situations where the opportunities for economizer cooling are less. This is true for warm and humid climates and for buildings that have relatively low internal gains (i.e., low occupant densities).

5.3 Solar Cooling and Refrigeration

It is feasible to use solar thermal energy or solar energy to energy a cooling appliance or a refrigerator. Active sun power systems use a mechanical or electric device to switch sun power absorbed in a solar collector to some other aspect inside the gadget. There are a few types of the systems now getting used or advanced, which include solar absorption air-conditioning, solar-assisted desiccant cooling, photovoltaic-powered evaporative cooling, photovoltaic-powered warmness pumps, air-conditioners and refrigerators and Rankine-cycle warmness engine.

The cooling call for and availability of sun strength tend to coincide in maximum countries. For instance, throughout summers at the same time as cooling load increases, the sun depth also will increase. Hence, a herbal synchrony between the weight and the input would be performed, and this can offer an awesome possibility for call for-side control. Aside from being a easy renewable energy supply, using sun electricity reduces our reliance on polluting fossil fuels.

5.4 Desiccant-Based Cooling Systems

Most mechanical cooling system designs attempt to address both the sensible and latent cooling loads with the same pieces of equipment. This does not have to be the case. Many applications can be addressed through the use of desiccants and other methods to control latent loads. Desiccant-based system is designed to provide cooling without refrigeration [4]. In these systems, a desiccant removes moisture from the air, which releases heat and increases the air temperature. The dry air is cooled using either evaporative cooling or the cooling coils of a conventional air conditioner. The adsorbed moisture in the desiccant is then removed (the desiccant is regenerated, or brought back to its original dry state) using thermal energy supplied by natural gas, electricity, waste heat, or the sun. Commercially available desiccants include silica gel, activated alumina, natural and synthetic zeolites, titanium silicate, lithium chloride and synthetic polymers.

6 Conclusion

Green constructing method allows building proprietors and managers to reduce electricity intake, improve the paintings surroundings and reduce the environmental impacts of constructing operations. If the building can reduce operating costs, growth occupant productiveness and decrease fitness proceedings, in addition to be environmentally accountable, it is a sustainable constructing. To ensure this, HVAC systems have an important function to play since most of the sustainable constructing factors are without delay or not directly suffering from the overall performance of the HVAC structures. An included and holistic design method beginning at a task's inception is needed to optimize the HVAC layout and operation for green buildings.

Currently, those technology are being used efficaciously in a few niche packages, and performance enhancements will maintain to force down gadget expenses and accelerate integration of the new systems within traditional HVAC systems. HVAC and constructing designers are accountable for bringing this into reality and contributing to the sustainable revolution.

References

1. Bobenhausen W, Lahiri D (1999) HVAC design for green buildings. Heating Pip Air Conditioning 71(2):43–50
2. Centre for Construction Ecology (1998) Environmentally friendly cooling systems. Technical Memorandum TM 16/98, Building Services Research and Information Association, Bracknell, Berkshire, UK
3. Hartman T (2000) Designing greener buildings. Automated Buildings, Nov 2000. Available at https://www.automatedbuildings.com/

4. NREL (2001) Advanced desiccant cooling & dehumidification program. National Renewable Energy Laboratory (NREL). https://www.nrel.gov/desiccantcool/
5. Sullivan CC (2001) Zero waste and green buildings. Consult Specifying Eng 29(3)

Design of Green Buildings

A. Mohamed Suhail, V. M. Shebi Rithika, and N. Karthiga Shenbagam

1 Introduction

This chapter offers an outline of green homes and is divided into three sections. The first phase discusses what makes buildings green and gives several definitions of inexperienced constructing. The 2nd phase discusses the environmental impacts of traditional homes and explains not unusual green building practices with recognize to siting, power efficiency, water performance, building substances, occupant fitness and nicely being, and production and demolition waste, and the function of legal professionals in the green constructing discipline. This chapter presents a short evaluation of troubles that are discussed in greater elements. The negative impacts of buildings and construction activities are also well recognized. These include the noise, dust, traffic congestion, water pollution, and waste disposal during the construction stage. A large quantity of natural and human resources will be consumed. Once completed, buildings continue their impacts on the environment.

A. Mohamed Suhail · V. M. Shebi Rithika · N. Karthiga Shenbagam (✉)
Bannari Amman Institute of Technology, Sathyamangalam, India
e-mail: karthigashenbagam@bitsathy.ac.in

A. Mohamed Suhail
e-mail: mohamedsuhail.ce18@bitsathy.ac.in

V. M. Shebi Rithika
e-mail: shebirithika.ce18@bitsathy.ac.in

V. H. Saran and R. K. Misra (eds.), *Advances in Systems Engineering*,
Lecture Notes in Mechanical Engineering,
https://doi.org/10.1007/978-981-15-8025-3_67

2 What Makes Buildings Green?

The beginning of the twenty-first century has ushered inside the technology of inexperienced buildings. Green building, also known as a sustainable building, is a structure that is designed, built, renovated, operated, or reused. In an ecological and resource-efficient manner. According to a few estimates, there are about 81 million buildings in the USA. Most of those homes use energy inefficiently, generate massive amounts of waste in their production and operation, and emit huge quantities of pollution and greenhouse gases. In assessment to traditional homes, inexperienced homes searching for land and power effectively, conserve water and different assets, improve indoor and outdoor. While green buildings nevertheless constitute a tiny subset of current buildings, their numbers are increasing rapidly. In November 2006, the U.S. Green Building Council, the nonprofit organization responsible for the advent of the Leadership in Energy and Environmental Design (LEED) green constructing rating device, introduced that 623 buildings had completed a few stage of LEED certification. As of December 2009, this variety had grown to more than 2400, and over 35,000 buildings were in the method of achieving some stage of LEED certification. The U.S. Dept. of Energy Buildings Technology Program [1] gives the details of various funding opportunities and methodologies for validating the buildings, the general guidelines, and provisions which a building should have to convert it as green building or to build a building as green building are given in detail [2, 3], the planning and architectural aspects of the green building in the middle east are clearly explained [4, 5], it identifies the general procedures and how to construct building that are constructed with sustainable materials and also energy efficient structures and its methodologies for effective construction [6–8].

3 Definitions of "Green Building"

While the definition of what constitutes a green building is continuously evolving, the Office of the Federal Environmental Executive offers a useful operating definition. This company defines this time period as: the practice of (1) increasing the efficiency with which buildings and their websites use energy, water, and substances and (2) decreasing building impacts on human health and the environment, through better siting, design, construction, operation, renovation, and elimination—the entire building life cycle.

Similarly, the Environmental Protection Agency (EPA) defines green constructing as follows: the practice of creating structures and the use of procedures which can be environmentally responsible and aid-green at some point of a building's life cycle from siting to design, creation, operation, preservation, upkeep, and deconstruction. This exercise expands and enhances the classical constructing design concerns of economy, utility, sturdiness, and luxury. Green constructing is also referred to as a sustainable or "high overall performance" constructing. Both of these definitions mention life cycle evaluation (LCA).

4 The Most Important Element: The Efficient Use of Energy

The most important element of green building is energy efficiency. Higher levels of energy efficiency reduce carbon emissions—both from power plants and the home's own energy systems. Energy efficiency is the key to achieve sustainability in green building. Lowering the energy consumption in construction is starting to become a significant improvement chance for many organizations.

In truth, for the reason that the term "inexperienced building" can be relatively unclear, a few humans choose to use the time period Overview of Green Buildings so-called excessive overall performance building. However, inexperienced buildings, on an average, use much less strength than traditional homes. In reality, there is a growing debate whether homes that gain a few degree of LEED certification are more green of their use of strength than everyday buildings [8]. Fortunately, there are numerous methods to enhance a construction's power performance, from insulating walls to putting in computerized shutoff switches for lighting fixtures. Energy performance may be often mandated through local and nation energy codes, which require that new and appreciably renovated buildings comply with more and more stringent power efficiency necessities.

5 The Reality of the Built Environment: The Problem of Existing

The truth is that the huge majority of homes are not green, and these buildings will continue to be used for many years to return. Improving the energy efficiency of existing buildings generally includes a procedure referred to as retrofitting, which can mean whatever from putting in more strength-green furnishings to growing the amount of insulation in a constructing. The U.S. Green Building Council has a score well-known particularly focused on existing homes, called LEED-EBOM (EBOM stands for "existing buildings operation and protection").

6 Impacts of Conventional Buildings that Green Buildings Seek to Rectify

The environmental affects of buildings are giant. Conventional homes use large amounts of electricity, land, water, and uncooked substances for his or her creation and operation. They are chargeable for large greenhouse fuel (GHG) emissions as well as emissions of other harmful air pollutants. They also generate big quantities of creation and demolition waste and have serious influences on vegetation and wildlife. An analysis of those problems demonstrates the scope of the problem.

7 Energy Use in Buildings

Worldwide, buildings devour big quantities of strength. The United Nations Environment Programmer has said that 30–40% of all primary strength produced global is utilized in buildings. America uses twice as a whole lot power in step with unit of monetary output as Germany, and nearly 3 times as a great deal as Japan. Fortunately, there are numerous approaches to enhance a building's electricity performance. New innovations in insulation can lessen the strength used in production insulation and permit insulation to be recycled or biodegradable. Mineral, fibrous, and cellulose-derived substances are now to be had for insulation purposes. Another huge consumer of energy is a construction's heating, air flow, and air-conditioning (HVAC) machine. Properly designed and mounted HVAC systems can reduce the quantity of strength used for heating and cooling a building. Electric lights consumes about one-quarter to 1-3 of the electricity in a standard commercial building. Building facts modeling (BIM) permits constructing layout and construction groups to draw and take a look at the building's running structures, along with electricity or hot water, in a single laptop version.

8 Greenhouse Gas Emissions and Indoor Air Pollution

Greenhouse gases are present in the atmosphere naturally, released by natural sources, or formed from secondary reactions taking place in the atmosphere. They include carbon dioxide, methane, nitrous oxide, and ozone. These extra emissions are increasing greenhouse gas concentrations in the atmosphere, enhancing the natural greenhouse effect, which is believed to be causing global warming. While man-made greenhouse gases include carbon dioxide, methane, and nitrous oxide, some like the CFCs are completely new to the atmosphere. Given that homes use massive amounts of strength and given that most of this strength comes from the burning of fossil fuels, it is not always unexpected that buildings accountable for many millions of heaps of GHG emissions annually. On a percent basis, buildings in the USA are accountable for approximately 40% of the country's overall GHG emissions. Unfortunately, greenhouse gases are not the only harmful pollutants that homes emit. Indoor stages of air pollutants might also greatly exceed outdoor tiers. Indoor air pollutants are particularly crucial given that we spend maximum of our time interior. The EPA has estimated that indoor stages of pollution may be two to five instances better, and sometimes extra than one hundred instances higher, than outdoor air pollutants stages. This pollutants can come from a wide sort of sources. One way to lessen the presence of these pollution is to ensure that indoor air is regularly changed by way of out of doors air and to make certain that this outdoor air is well filtered. Unfortunately, homes are frequently poorly ventilated and do no longer sufficiently filter the air this is recirculates, main to air that is doubtlessly harmful to building occupants' health. A number one attention of green buildings is the fitness and well being of their occupants. Many older buildings suffer from what's commonly known as "sick constructing syndrome."

8.1 Building Water Use

As is the case with energy, buildings use staggering amounts of water during their operation. Buildings also usually displace vegetation that can capture and absorb precipitation. The net result is that municipal sewer systems are often overburdened. During rainfall events, billions of gallons of water flow into these sewer systems as runoff, carrying contaminants with them. Many older municipalities have combined sewer systems that capture both storm water runoff and wastewater from buildings. These combined sewer systems are not designed to treat the massive amounts of water that flow into them during heavy rainfall events. Thus, they are equipped with combined sewer overflows (CSOs), which act as safety valves and deposit much of this water, which contains raw sewage and other contaminants, into waterways. One of the primary uses of water in a building is for toilets, sinks, showers, and similar uses. The byproduct of these uses is wastewater. Bathrooms can be installed or retrofitted with low flow or waterless urinals and toilets that use considerably less water for flushing. Dual-flush toilets that use less water for liquid than solid waste are also available. Another water-related problem in buildings is storm water runoff. Buildings and landscapes can be designed to maximize the amount of catchment area, and water can be collected in cisterns, barrels, or swales. The collected water can be detained, retained, and routed for use in building evaporative coolers and toilets, and for irrigation purposes. "Gray water" can also be used in building operations.

8.2 *Sustainable Building Materials*

In building, environmentally friendly materials (also known as green building materials) are those in which, for their production, placing, and maintenance, actions of low environmental impact have been performed. They have to be durable, reusable or recyclable, include recyclable materials in their composition and have to be from resources of the area where the building activity will take place—they have to be local materials. These materials also have to be natural (soil, adobe, wood, cork, bamboo, straw, sawdust, etc.) and must not be spoilt by cold, heat or humidity. They have to be highly durable and can incorporate different technologies, such as capturing energy, capturing CO_2 while removing pollution. They are used when, in the long term, they have a lower environmental cost than the natural materials. Timber is the material with the lowest environmental impact on its production and life cycle. There is a reuse of wastes that are generated in other sectors for the production of building materials, such as quarry wastes (marble, slate, etc.). Also, we have wastes generated by industrial processes like ashes or mud that are used, or the urban solid ones. There are other original techniques, such manufacturing concrete with recycled tire rubber; using the mud from sewage plants for making bricks, or wood and cork remains (pruning, sawmills, sanding dust, etc.), and especially vegetal fibers (bamboo, coconut, etc.) that, once mixed with cement, also work as insulators. In order to achieve a sustainable construction, we have to put an end to those bad long-standing habits of squandering natural resources.

9 Overview of Green Buildings Construction, Operation, and Demolition Waste

Building C&D waste within the USA totals approximately 136 million heaps yearly, accounting for nearly 60% of overall non-business waste era. By manner of assessment, the whole amount of municipal waste generated inside the USA each year totals 209.7 million lots. According to some estimates, 4 heaps of waste are generally deposited into a landfill throughout the construction of a brand new 2000-square-foot home. Construction waste is composed in general of lumber and manufactured wooden merchandise (35%), drywall (15%), masonry materials (12%), and cardboard (10%). The rest is a mixture of roofing materials, metals, plaster, plastics, foam, insulation, textiles, glass, and packaging. Although an awful lot of this cloth is recyclable, maximum of it is miles deposited into landfills. Green homes commonly are looking for to minimize the quantity of C&D waste they generate. Building deconstruction is the orderly dismantling of building additives for reuse or recycling. In assessment to building demolition, deconstruction includes taking aside quantities of buildings or doing away with their contents with the number one goal being reused.

10 Conclusion

Green constructing is a financially, health, and most importantly environmentally responsible idea that more people want to undertake. Many constructing substances and renewable energy resources exist to reduce one's effect upon the environment. Green buildings brings together a vast array of practices and techniques to reduce and ultimately eliminate the impact of building and the environment and human health. It often emphasizes taking advantages of renewable resource and photovoltaic techniques and using plants and trees. Many other techniques, such as packed gravel or permeable concrete instead of conventional concrete or replenishment of ground water, are used as well.

References

1. U.S. Dept. of Energy Buildings Technology Program, Obama Administration Launches New Energy Efficiency Efforts (29 June 2009). Available at https://www1.eere.energy.gov/buildings/news_detail.html?news_id=12607
2. Yudelson J (2007) Where are all the LEED projects? Table 2 (12 July 2007). Available at https://www.sustainablefacility.com/Articles/Leed/BNP_GUID_9-5-2006_A_10000000000000134921. LEED is discussed in more detail in Chapter 2
3. See U.S. Green Building Council. About USGBC. Available at https://www.usgbc.org/DisplayPage.aspx?CMSPageID=124
4. Office of the Federal Environmental Executive, The Federal Commitment to Green Building: Experiences and Expectations. Available at https://www.ofee.gov/sb/fgb_report.asp
5. EPA, Green Building. Available at https://www.epa.gov/greenbuilding/pubs/about.htm
6. In May 2008, The International Organization for Standardization (ISO) released a set of principles for evaluating building products throughout their life cycle. The standard, referred to as Sustainability in Building Construction—General Principles (ISO 15392:2008), is available for purchase at https://www.iso.org/iso/iso_catalogue/catalogue_tc/catalogue_detail.htm?csnumber=40432
7. See U.S. Department of Energy, Office of Energy Efficiency and Renewable Energy, About High Performance Buildings. Available at https://www.eere.energy.gov/buildings/highperformance/design_approach.html
8. Gifford H (2008) A better way to rate green buildings: LEED sets the standard for green building, but do green buildings actually save any energy?

Modern Green Building Concept and Sensors

A. Madhupreetha, H. S. Jenith, and N. Karthiga Shenbagam

1 Introduction

Construction industry has its own significant environmental, social and economic impact on the society. Positive impacts of construction activities includes building facilities to satisfy the human needs, giving out employment opportunities directly or indirectly and contributing toward the economy. The construction industry in Australia contributes 7.5% to the GDP and provides 1 million jobs. Negative impacts of construction are water pollution, waste disposal, noise. According to the World Business Sustainable Development, building BLOC accounts for 40% of total energy consumption. Buildings produce greenhouse gas emission which results in global warming. The carbon emission of buildings across the world will reach 42.4 billion tones in 2035, adding 43% on the level of 2007. The increasing demand for landfill presents a new challenge to all the areas. Robichaud and Anantatmula pointed out that they are four pillars of green buildings (i.e.) minimization of impacts on the environment, enhancing the health conditions of occupants, the return on investment to developers and local community, and the life cycle consideration during the planning and development process, mainly life cycle perspective, environmental sustainability, health issues, and impacts on the community.

A. Madhupreetha · H. S. Jenith · N. Karthiga Shenbagam (✉)
Bannari Amman Institute of Technology, Sathyamangalam, India
e-mail: karthigashenbagamn@bitsathy.ac.in

A. Madhupreetha
e-mail: madhupreetha.ce18@bitsathy.ac.in

H. S. Jenith
e-mail: jenith.ce18@bitsathy.ac.in

V. H. Saran and R. K. Misra (eds.), *Advances in Systems Engineering*,
Lecture Notes in Mechanical Engineering,
https://doi.org/10.1007/978-981-15-8025-3_68

2 What is Green Building?

Green building or green construction or sustainable building refers to both a structure and the application of processes that is resource efficient throughout a building's lifetime. This requires close cooperation with the site engineers, the architects, the client, and everyone involved in the construction of the building. And, it is the practice of increasing the efficiency of the energy, water, and materials which are used in the construction site. And, also it reduces the impact on health of the humans and the surrounding for the entire life.

> We cannot solve our problems with the same thinking used when we created them
> —Albert Einstein, Physicist.

3 What is Green Concrete?

A concrete which uses less energy and less emission of carbon dioxide than other concrete is green concrete. Highly eco-friendly [1].

3.1 Materials for Green Concrete

1. Locally available: Materials found locally and transportation facility.
2. Salvaged, re-furnished, or remanufactured: Includes saving materials from disposal and renovating, repairing, restoring, improving the quality or value of the product.
3. Reusable or recyclable: Select materials that can be easily dismantled and reused and recycled at the end of their useful life.
4. Recycled demolition waste aggregate.
5. Recycled concrete aggregate.
6. Blast furnace slag.
7. Manufactured sand.
8. Glass aggregate.
9. Fly ash.

3.2 Mix Design

Packaging of materials is based on the aggregate size which helps in the binding. Aggregates replace excess cement paste to give improved stability and increase strength and durability. Less cement generates less heat of hydration. Optimum proportions of aggregates are known to have the best and dense concrete in both fresh and hardened stage of concrete.

3.3 Advantages

User-friendly, easier handling better consistency and easier finishing, reduction in creep and shrinkage, uses local products and recycled products, heat of hydration of green concrete are significantly lower than traditional concrete, lower temperature rise in large concrete and increases durability and lower permeability, high strength per kilogram of cement.

4 Why Bamboo is More Stronger than Steel Reinforcement

Bamboo has high tensile strength so it is comparable with steel. Also, it has a impermeable protective layer on the outerside which protects it from rotting due to water.

4.1 Use of Bamboo and Its Strength

Bamboos tensile strength is roughly 28,000 per square inch versus steels 23,000. Bamboo is more stronger that is 6 times that of steel. Also, the energy needed to produce steel is almost 50 times that of its natural product.

5 Hybrid Fiber Concrete with GGBS

Ordinary Portland cement (OPC) being a significant material in the production of concrete forms an indigenous substance to bind aggregate. The manufacturing of OPC necessities firing a great quantity of fuel for the decomposition of limestone, resulting in the discharge of carbon dioxide [2]. It causes pollution to the environment and depletion of raw materials (limestone) [3].

Ground Granulated Blast Furnace Slag (GGBS) is a byproduct from the blast furnaces used to produce iron. It is highly eco-friendly. Hence, this work explores the possibility of using the steel fiber and glass fiber with GGBS in concrete which is in the grade of M25. To dilute the chemical onslaught and to increase thermal insulation in concrete, glass fiber (GF) was added [4]. Various experimental studies on compressive strength, split tensile strength, flexural strength, density of hybrid concrete were done. The maximum density of hybrid fiber concrete is 2556 kg/m^3 occurred at 2% volume fraction of hybrid fibers, and the density of the hybrid concrete is 65% higher than the control concrete. At 28 days, 65.9 and 15.35% increase in flexural strength was occurred than the control concrete and steel fiber concrete. The overall strength of concrete is enhanced by the inclusion of hybrid fiber than the control concrete. Inclusion of hybrid fiber does not affect the workability of the concrete. Hence, hybrid fibers can be used with GGBS in concrete to enhance the strength of the concrete without affecting the workability.

6 Chilled Beam Technology

Let us start with "chilled beam" system. A passive chilled beam consists of a fin-and-tube heat exchanger, contained in a housing that is suspended from the ceiling. Chilled water passes through the tubes. Warm air from the space rises toward the ceiling, and the air surrounding the chilled beam is cooled causing it to descend back toward the floor, creating convective air motion to cool the space. This allows a passive chilled beam to provide cooling without the use of a fan. An active chilled beam also consists a spin-and-tube heat exchanger contained in a housing that is suspended from, or recessed in the ceiling. The primary difference is that the active chilled beam requires a primary air supply. The primary air passes through nozzles, which induce from the space up through the cooling coil. The primary air also provides the code required outdoor ventilation air to the space. For this reason, active chilled beams are suggested.

Both the active and passive chilled beams are designed to provide sensible cooling only, so the dehumidification must be provided by a separate dedicated outdoor air unit.

6.1 Benefits of Chilled Beams

- Chilled beams have no terminal unit fan, so the overall systems fan energy is lower when compared to other systems requiring terminal equipment fans.

- Chilled beam have no condensate fan or filter, so the overall systems maintenance requirements are reduced compared to the systems with terminal equipment filters are condensate fans.
- The system requires minimal air-side infrastructure because central air systems are only required for outdoor air ventilation.
- The result is reduced floor and space requirement for the central air handling equipment and vertical duct rises (Fig. 1).

Various methods of design exist, each possessing their own set of risk and reward. Chilled beam systems can provide several benefits, including

- Potential reductions in initial costs of equipment and construction material.
- Increase in occupant comfort beyond that which is achievable through traditional systems, space adaptability, energy efficiency, and simple operation and maintenance.

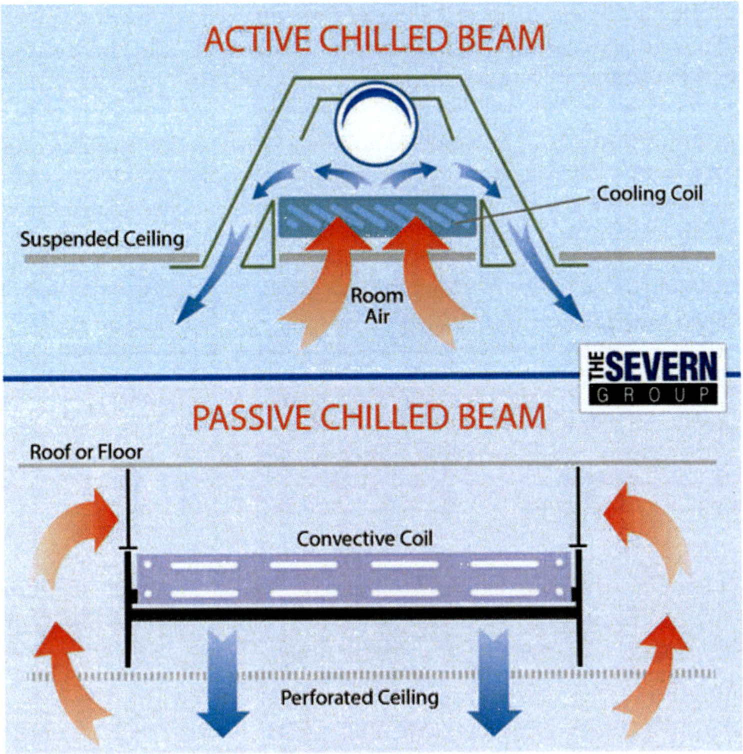

Fig. 1 Active and passive chilled beam

7 Incinerated Toilets and Optibo

Incinerating toilets can also be waterless. Instead of breaking down waste biologically, these toilets torch it. They send the waste to an incinerator, where it is burned to sterile ash [5]. The toilet sits in your bathroom and has an electric exhaust pipe that exits through your roof. To run, it needs batteries or can be plugged into a wall outlet. You use the toilet normally, toilet paper and all. But before you flush, you must close the lid, for reasons that will soon be clear. Next, you decide whether to press the "urine" or "[solid] waste" button on the control panel. Then, the toilet fires up. Flushing is handled by some type of dry method, like an auger (essentially a large screw) that turns to push the waste into the incinerator. A propane, diesel, or natural gas tank feeds into the incinerator. The incinerator injects fuel and ignites your waste, burning it in one example, at 800 F (427 °C). Urine cooks for up to 10 min; solid waste takes about half an hour (source: Ecojohn). Optibo is the product of a collaboration between Swedish firms to solve the housing industry's problem of availability of space caused by high land prices. Optibo's main architect was inspired when he saw the Disney cartoon, Mickey's Trailer, on TV.

The project has led to the construction of an apartment which is only 25 m^2 (270 square feet) in area. The single-room living space has the furniture built into the floor, and the room can be changed from a living room to a bedroom, to a dining room, and back. The kitchen area is affixed to the wall and does not change.

8 Sensors in Green Buildings

See Fig. 2.

9 Paralink Features

- Ultra high strength materials are used to have gretaer durability and life in the structure. MD strength up to 1350 kN, high modulus, tough durable polyethylene sheath, low creep characteristics, easy to install, highly resistant to chemicals, micro-organisms, UV radiation and mechanical damage.

9.1 Paralink

It is an ultra-strong reinforcement geogrid for civil engineering applications which is manufactured from **paraweb**. Paraweb consists of discrete bundles of parallel high-strength synthetic fibers encased in a tough and durable polymeric sheath. Paraweb is evenly spaced, using bonded cross-machine direction webs, to create 2D structures

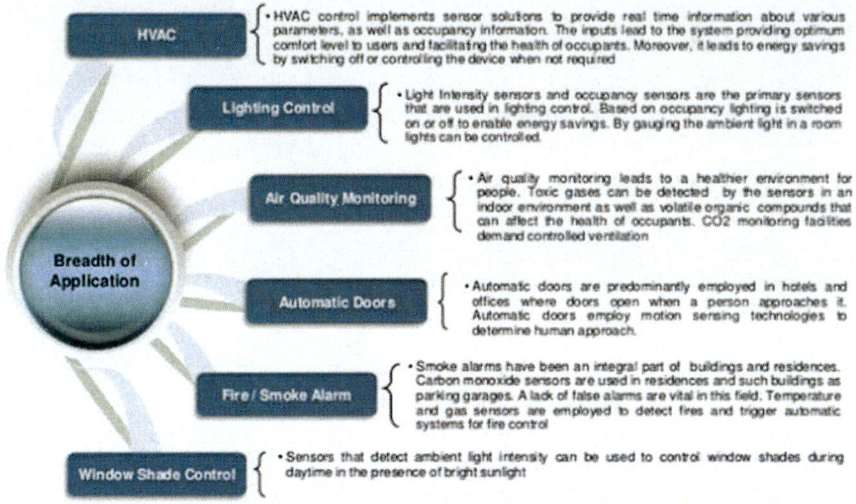

Fig. 2 Application of sensors

up to 4.5 m wide. The standard maximum ultimate tensile strength of paralink is 1350 kN/m wide (Fig. 3).

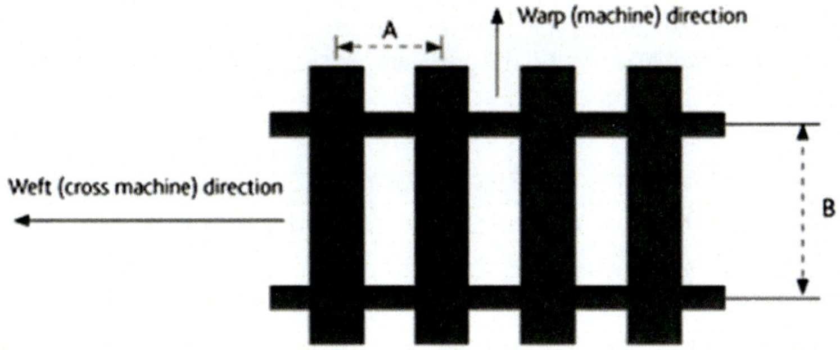

Fig. 3 Paraweb

10 ASHRAE

The American Society of Heating, Refrigerating and Air Conditioning is an American professional association seeking to advance heating, ventilation, air conditioning, and refrigerator systems design and construction [6].

11 ETP

- Effluent treatment plant (ETP) is a process design for treating the industrial waste water for its reuse or safe disposal to the environment.
- Influent: Untreated industrial waste water.
- Effluent: Treated industrial waste water.
- Sludge: Solid part separated from waste water by ETP.

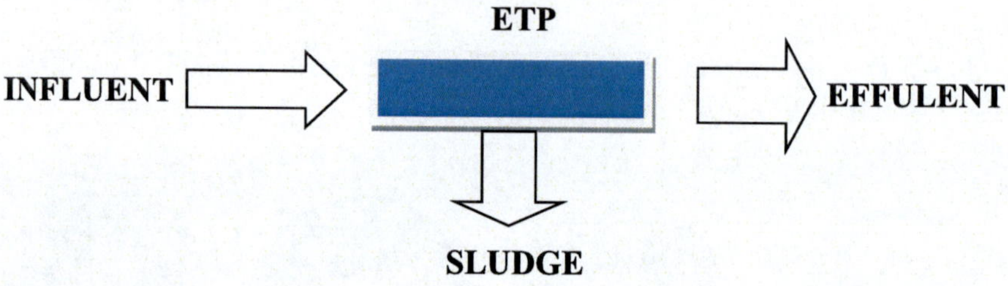

11.1 Need of ETP

- Effluent treatment plant is used to clean industry effluent and recycle it for further use, reduce the usage of fresh/potable water in industries, to cut expenditure on water procurement, to meet the standards for emission of environmental pollutants from various industries set by the government, and avoid hefty penalties [7].

11.2 Toward Safety from Toxic Gases in Underground Mines Using Wireless Sensor Networks and Ambient Intelligence

In the real-time gas explotion structutres, applications of remote sensing is widely used in the currect developing period. Modeling of theory in WSNs, Ohm's law, mobile robot sensing, concepts of static, curvilinear gas calibration equation.

Ambient intelligence based on temporal statistical method. Knowledge in improving safety with an assurance of assistance whenever required. The common mine gases are carbon dioxide, hydrogen, carbon monoxide, methane, nitrogendioxide. Building Information Modeling is used to monitor the Indoor Environmental Quality (IEQ) by monitoring the indoor air quality and Particulate Matter (PM).

12 Auto-claved Aerated Concrete

By using AAC, we can get reduction in the dead load of the construction materials which could result in a decrease in cross section of concrete structural, and subsequently, reinforcement can be reduced due to the lightweight of the concrete [8, 9]. Zero pollution manufacturing process is the one that reduces the amount of pollution and helps in producing zero emmision of gases. It is lighter than normal clay bricks, therefore, simpler and economical to transport. Lighter weight makes construction easier and faster. ACC materials can be sanded, routed, or cut into any desirable size on site using some powered tools with carbon steel cutters [10, 11].

12.1 Top Ten Emerging Trends in 2018

These things help in molding our twenty-first century,

1. Net zero/net positive energy
2. Biomimicry
3. Circular economy and the influence of indigenous world views
4. High-tech/high touch
5. Passive design/active design
6. Urban density
7. Human health and well-being
8. Reimagine
9. Reimagine the supply chain and the blockchain
10. Renewed emphasis on design and design thinking.

13 Conclusion

In around 40% annual energy conspumtion nealy 1/3 of the CO_2 can be reduced which help in saving our environment. Mostly, carbon emission comes from buildings world. As we heard in our provincial budgets and our municipal budgets, there is a limited resource for building new, and in fact, if we just keep on building new, it is not enough. So we have to reimagine the supply chain, so we are actually procuring the design and construction of buildings differently so that buildings can be designed differently, buildings that harness gifts of nature, building that need less energy, buildings that breathe. This was the ultimate vision that created GRIHA or Green Rating for Integrated Habitat Assessment. Therefore, architects and civil engineers should have the utmost responsibility to rebuilt our traditional buildings by introducing new technologies without causing any traditional, cultural aspects. It should be socially economical and affordable. And being a civil engineers, we need to enhance the safety of the workers in the site so sensors can be installed in monitoring the actions. These activities help in environmental benefits, economic benefits, social benefits. This research identified the emerging development taking place in our country. It ultimately helps in reducing the waste, pollution, and environmental degradation.

References

1. www.civilengineeringdiscoveries.com
2. Kong DLY, Sanjayan JG (2008) Damage behavior of geo-polymer composite exposed to elevated temperature. Cement Concr Compos 30(10):986–991
3. Joshua Daniel A, Sivakamasundari S, Abhilash D (2017) Comparitive study on the behavior of geopolymer concrete with hybrid fibers under static cyclic loading. Procedia Eng 173:417–423 (Proceedings of the 11th international symposium on plasticity, and impact mechanics. Symp. Implast 2016)
4. Bhalachandra SA, Bhosle AY (2013) Properties of glass fiber reinforced geo-polymer concrete. Int J Mod Eng Res 2007–2010
5. Cindrella Forbrenningstoalett—Forsiden. www.cindrellaeco.no. Retrieved 28 Nov 2016
6. About ASHRAE. Retrieved 18 June 2018
7. Walters JK, Wint A (1981) Industrial effluent treatment, vol 1. Applied Science Publishers Ltd. London
8. Autoclaved aerated concrete. www.cement.org. Retrieved 28 Nov 2018
9. Product specifications—AIRCRETE. Aircrete—Europe.com. Archived from the original on 02 Apr 2015. Retrieved 16 July 2014
10. Indian Cement Review Magazine/AAC can bring down construction cost by up to 20 percentage. Retrieved 28 Nov 2018. Cs1 maint: extra punctuation
11. Using autoclaved aerated concrete correctly. Masonry Magazine, 01 June 2008. Retrieved 28 Nov 2018

Carbon Capture Technology a Sustainable Energy Production Technique: A Review

Rajat Kawalkar, Harrsh Kumar Dubey, and Satish Lokhande

1 Introduction

CO_2 is one of the predominant gases in our atmosphere which does not allow any sort of radiation of the sun's heat back into space which indirectly contributes to the greenhouse effect and causes global warming. However, as per United Nations Environment Programme (UNEP), if it was not for this effect to occur naturally, the temperature of earth would be insignificant to sustain life on this planet, i.e., 30 °C [1]. This fact invariantly signifies the impact with which this natural ecosystem balance of gases possess in climatic change of our planet.

Capturing CO_2 in modern industrial world has become one of the most eminent and eye catching areas to be addressed by international unions and scientific community. However, with proper applied methodologies and support of organizations investing in large capital, there is a substantial amount of reduction in emissions by approximately up to 80% in carbon content [2]. However, there are varied factors that are taken into consideration by authorities before application of such vital technology which prominently features the cost, energy accessibility and storage combined with knowledge development and sharing of the process [3].

However, various industrial processes that use fossil fuel as a source of energy are mainly involved in different sectors such as power generation, iron steel and cement industries and have henceforth resulted in a large amount of greenhouse gases emissions which primarily features a higher percentage of carbon dioxide content [4]. Hence, to reduce this emission rate at large point sources, various technologies have been established in many of the concerned nations which involves trapping CO_2

R. Kawalkar (✉)
TCS, Nagpur, India
e-mail: rajatkawalkar123@gmail.com

H. K. Dubey · S. Lokhande
Mechanical Engineering, Priyadarshini College of Engineering, Nagpur, India

V. H. Saran and R. K. Misra (eds.), *Advances in Systems Engineering*,
Lecture Notes in Mechanical Engineering,
https://doi.org/10.1007/978-981-15-8025-3_69

at its initial emission source (which usually turns to be combustion area of biomass or fossil fuels], transporting it through pipelines to storage location by compressing it so that it changes its state of matter and isolating it from contaminating the environment [5].

2 Technologies Involved in Capturing CO_2

The combustion of natural gases acts as a major contributor to production of CO_2. This captured carbon dioxide has varied applications in manufacturing cement and food industries [6]. The cost involved in installing a full-fledged CCS plant incorporates around 3/4th of total cost including capturing, storage and transportation [7].

Processes in which coals are turned into powder, usually in major plants across the continent, are in which carbon dioxide contents are separated through proper concentration from the balance of flue gases (Fig. 1).

There are three basic types of carbon capture processes: pre-combustion, post-combustion and oxy-fuel with post-combustion which we will discuss in detail.

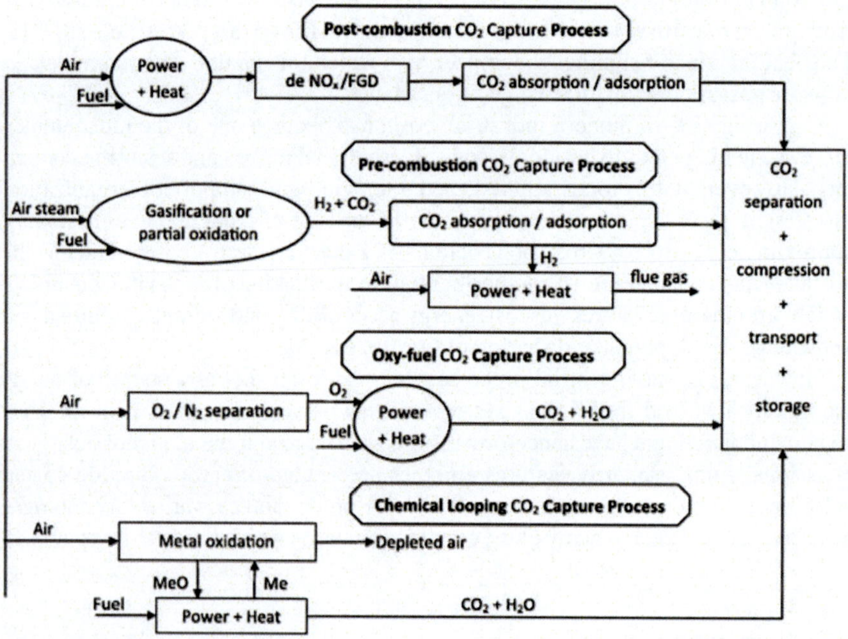

Fig. 1 CO_2 capture technologies [2]

Fig. 2 Chemical processes involved in pre-combustion carbon capture process [2]

$$\text{Coal} \xrightarrow{\text{Gasification}} CO + H_2 \quad (1)$$

$$CO + H_2 \xrightarrow{\text{Gas shift reaction}} H_2 + CO_2 \quad (2)$$

$$CH_4 + H_2O \xrightarrow{\text{Reformation}} CO + H_2 \quad (3)$$

2.1 *Pre-combustion*

As the terminology suggests, this process is carried out for the removal of CO_2 before the combustion of fossil fuels. Firstly, the fuel (usually coal) is converted into a state of matter which can be easily captured. Gasification process involving coal as source fuel is partially oxidized in steam and oxygen under high pressure and temperature to form synthesis gas also referred to as syngas (Eq. 1) [8]. This syngas is further made to undergo the combustion process and hence is used to produce power in steel plants. The power generated is usually applied in process of integrated gasification combined cycle (IGCC) [5]. The syngas generated primarily constituting of carbon monoxide, hydrogen, CO_2 and other gaseous components in relatively smaller quantities undergoes water–gas shift reaction with steam forming more quantity of H_2while CO gas will be converted into CO_2 (Eq. 2) (Fig. 2).

Natural gas, as it mainly contains CH_4, can be reformed to syngas containing H_2 and CO (Eq. 3). Reforming is carried out by the conversion of gaseous fuel into syngas in the presence of oxygen [9]. Due to the rich content of CO_2 at high pressure (around 1/5th of entire content), easier removal of gas is facilitated before H_2 combustion. Glycol solvent such as Selexol, which is used to trap CO_2 through chemical processes, and various technologies involved in the installation of IGCC plants make this process preferred and valued in the scientific community. Once the sorbent pressure is dropped, CO_2 is released which involves physical absorption during the combustion process. Loss in energy due to a shift reactor has its limitations in IGCC plants [5] (Fig. 3; Tables 1 and 2).

2.2 *Post-combustion*

Post-combustion carbon capture refers to capturing of carbon dioxide from complete combustion of fossil fuels generated gases. Modern appliances and applications of combustion in coal-sourced power plants are usually carried out in a furnace [12]. The major gases arising out from the coal based furnace consists of nitrogen and water vapour. Furthermore, the by-products of incomplete combustion for the furnace based operation mainly consists of sulphur dioxide, nitrogen oxide and fly ash (consisting of major constituents as silicon Oxide (SiO_2), aluminium oxide

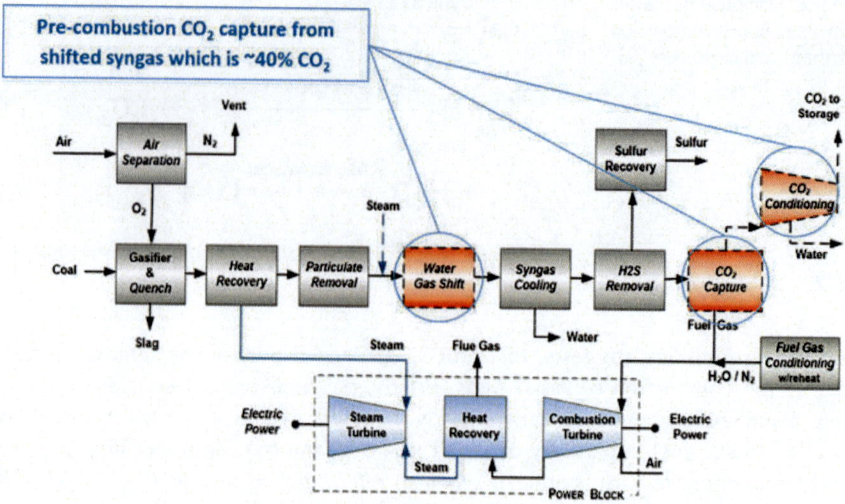

Fig. 3 Process diagram carbon capture and compression for an IGCC power plant [10]

Table 1 Power plant thermal efficiencies [11]

Type of fuel used	PGT used	Capturing technology used	Efficiency achieved (in %) (in LHV)
Coal	Pulverized coal	None	44
		Post-combustion	35.3
		Oxy-combustion	35.4
	Dry feed coal in IGCC	None	43.1
		Pre-combustion	34.5
	Slurry feed coal in IGCC	None	38.0
		Pre-combustion	31.5
Gas	Cycle combined with gas turbine application	None	55.6
		Post-combustion	49.6
		Oxy-combustion	44.7

Note High heating value (HHV) efficiencies of the coal-fired plants are 0.956 times the lower heating value (LHV) efficiencies of the gas-fired plants are 0.904 times the LHV efficiencies

(Al_2O_3), and calcium oxide (CaO). These products are major constituents of global warming and degrade the ecological balance [13]. Solvents, such as amine solution, for instance, monoethanolamine (MEA), are chemically reacted with the flue gases which indeed facilitate the reaction [14]. Absorber, a chamber in which flue gases are scrubbed, is used to separate nearly 90% of CO_2 captured. Thereafter, the soluble CO_2 is transferred into another equipment named as regenerator utilizing steam. Finally, the gas is compressed under high pressure and is transported and stored while the solvent used in the combustion process is recycled back to absorber [5].

Table 2 CO_2 emissions data [11]

Fuel used	PGT	Capturing technology used	CO_2 emissions generated (g/kWh)	CO_2 actually captured (g/kWh)	CO_2 avoided using same tech baseline (g/kWh)	CO_2 avoided using PF baseline (g/kWh)	CO_2 avoided using NGCC baseline (g/kWh)
Coal	PF	None	743	–	–	–	–
		Post-combustion	92	832	651	651	287
		Oxy-combustion	84	831	659	659	295
	Dry IGCC	None	763	–	–	–	–
		Pre-combustion	142	809	621	601	237
	Slurry IGCC	None	833	–	–	–	–
		Pre-combustion	152	851	681	591	227
Gas	NGCC	None	379	–	–	–	–
		Post-combustion	63	362	316	680	316
		Oxy-combustion	12	403	367	731	367

Major applications of such kind of technology are continuously been observed at highly retrofitted plants which generally use pulverized coal as fuel in power plants [15]. However, challenges arise in this kind of technological usage wherein the solvent used is comparatively small in quantity and requires upscaling which indirectly impacts on overall power output by nearly 30%. Moreover, the high performance requirement by sorbent technology has immense impact on a larger scale [16] (Figs. 4, 5 and 6).

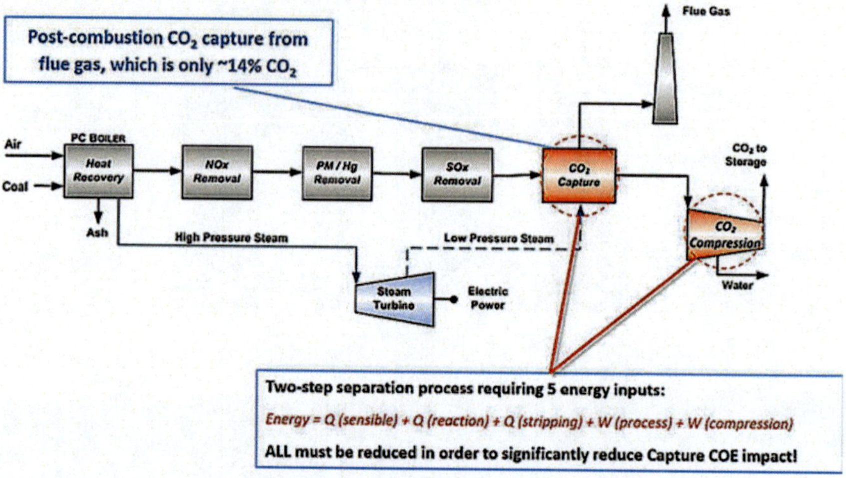

Fig. 4 Process diagram of carbon capture and compression in a coal-fired power plant [17]

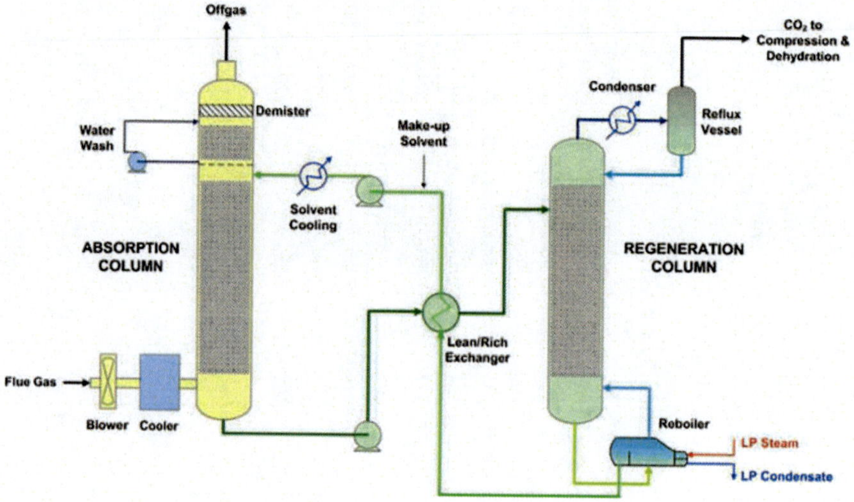

Fig. 5 Post-combustion scrubbing technology process [18]

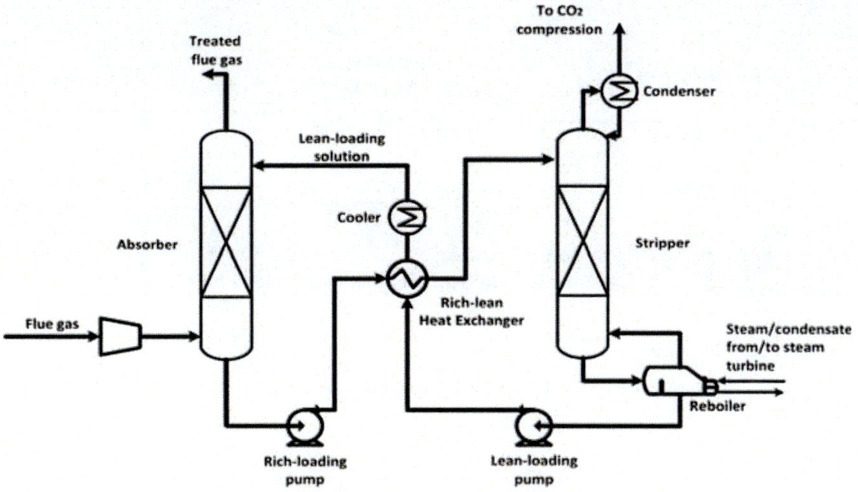

Fig. 6 Post-combustion CO_2 capture process [19]

2.3 *Oxy-fuel Combustion*

Oxy-fuel combustion is the process burning of fuel in the presence of nearly entire oxygen instead of air [20]. This combustion causes a high concentration of CO_2 in flue gas which results in easier purification of CO_2 than other methodologies [4, 21–24]. Reduced NO_x emission is one of the prominent advantages of this technology due to O_2 [6]. Usually, high content of oxygen (more than 90%) is supplied employing cryogenic air separation unit. Combustor is used to recover thermal heat caused due to recycled high temperature flue gases. Flue gases which are made up of CO_2, water vapour and sulphur dioxide are then passed through a unit of compressed and reduced temperature as well as pressure changes which eliminates water vapour [25]. This leaves the flue gas with CO_2 content which is transported and stored. High carbon capture efficiency due to a higher composition of concentration in carbon dioxide makes it one of the most convenient technologies. However, high cost of air separation unit along with possible uncertainties in corrosion of equipment may have an impact on its application [16] (Fig. 7).

3 CO_2 Separation Technologies

With regard to CO_2 to be separated from flue gases, in this context, we will discuss the prominent methods which are currently used in industrialized processes in separating CO_2 before transportation from plants to facilities.

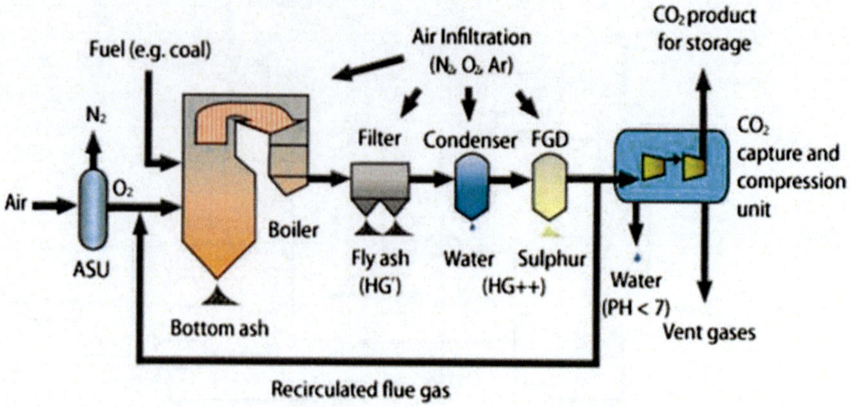

Fig. 7 Oxy-fuel combustion CO_2 capture process [19]

3.1 *Physical Absorption*

Physical absorption is the process of absorption of CO_2 from flue gases without undergoing chemical reaction with particular solvents having suitable solubility under high pressure and low temperature. The CO_2 is made to pass through an absorbing container whose primary function is to act as a chamber for the mixture between CO_2 from the absorbent and to separate CO_2 from the gases. Thereafter, this CO_2-rich streamline flow is made to pass through a series of flash tanks also known as regenerator wherein with degrading pressure CO_2 gets desorbed and regeneration of solvent takes place which is again recycled back to absorber for further usage and regenerated CO_2 is compressed and moved to storage. Low operation costs in this technology give it a comprehensive edge compared to other methods. However, this method can only be used if the absorbent is used under low thermal temperatures for better physical absorption. The major absorbents (solvents) used for this process are Estasolvan (Tributyl phosphate or TBP), Fluor Solvent (propylene carbonate or PC), Purisol (normal methyl pyrrolidone or NMP), Rectisol and Ifpexol (methanol), Selexol (dimethyl ether of polyethylene glycol or DEPG) and Sepasolv-MPE (mixture of polyethylene glycol dialkyl ethers). This theory of mixing is governed by Henry's law [26, 27] (Fig. 8).

3.2 *Physical Adsorption*

This technique involves trapping of gas molecules that come in contact with the surface due to accordance with different state of matters particulate. In this technique, a stream of gas is made to pass through a column coated with suitable adsorbent. Due to the affinity, CO_2 is adsorbed and separated from the flue gases by processes

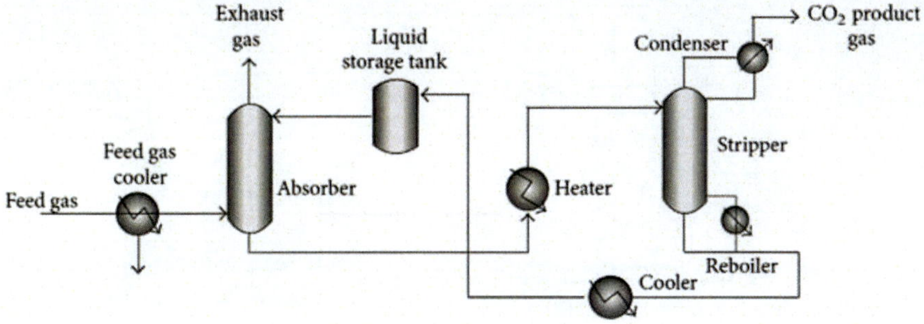

Fig. 8 A CO_2 absorption plant [28]

of pressure swing adsorption (PSA), temperature swing adsorption (TSA), vacuum pressure swing adsorption (VSPA) and electric swing adsorption (ESA) [28–30]. This corresponds to the regeneration of adsorbent as well. The methods mentioned above have major contributions in cost which is relatively low and depends vastly on the gradient of changes in temperature and pressure. Various materials and methods involved are zeolite 5 A, zirconium-benzenedicarboxylate and activated carbon. The productivity of the CO_2 adsorption rate is maximum observed in ESA followed by TSA, PSA, VSA and VSPA for single bed configurations which depend highly on the regeneration of adsorbent [28] (Fig. 9).

3.3 *Cryogenic Separation*

In this process, CO_2 is separated from flue gases while undergoing substantial changes in compression, cooling and expansion. The major by-product of this process is the formation of liquid CO_2 [30]. The major advantage of using this process is that CO_2 in the liquid state can be easily stored and transported without any further usage of any solvents/absorbents. Additionally, corrosion issues are invariantly avoided through this technology. However, the cost involved in the application of this technology is comparatively high since the requirements of this process reside on operating at lower temperature ranges. This also adds upon the freezing of water causing blocking problems in pipelines [31]. It has to be carefully noted that the flue gases, after having nitro-oxides, sulphur-oxides and water removed, only enter the compressor (Fig. 10).

3.4 *Membrane Separation Technology*

This technology involves the usage of membranes to separate CO_2 from flue gases depending on sizes and permeability of sizes of particles. In modern industrialized

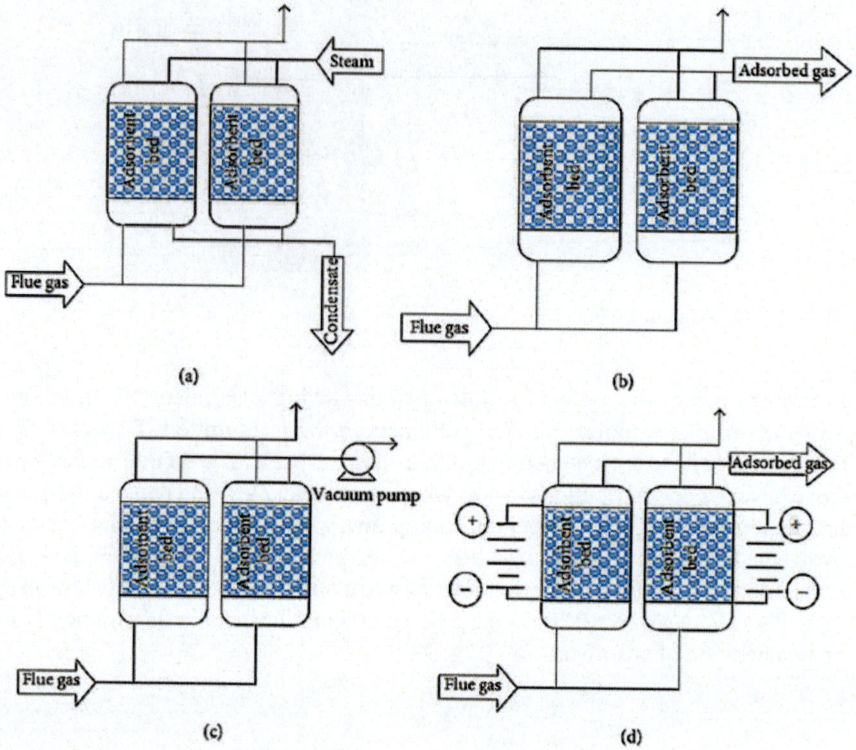

Fig. 9 Various adsorption cycles **a** TSA, **b** PSA, **c** VSPA, **d** ESA [28]

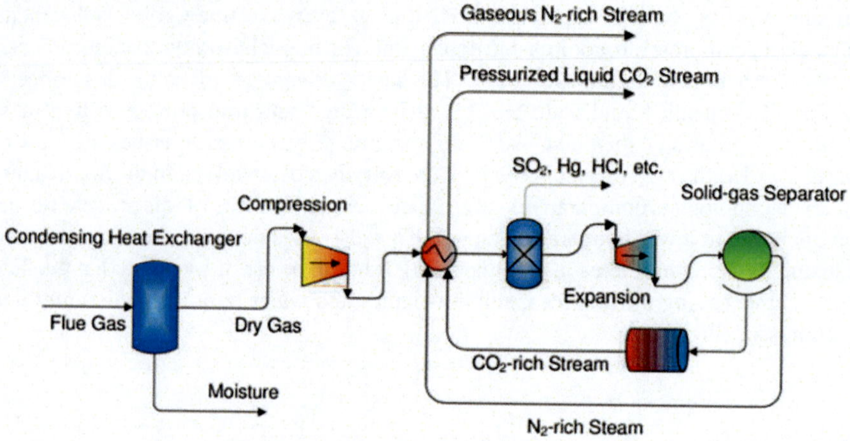

Fig. 10 Schematic diagram for cryogenic separation in CCS

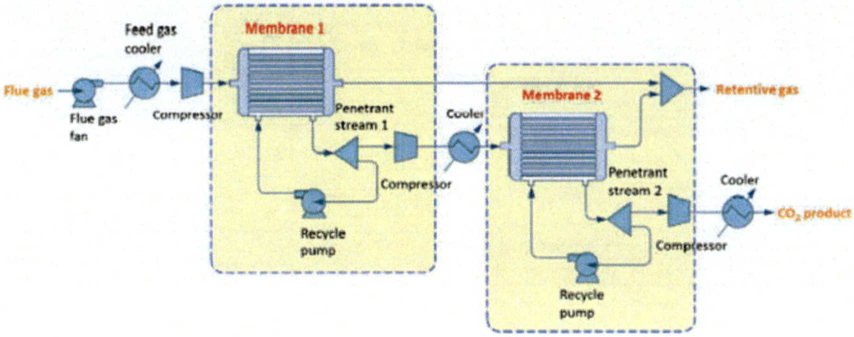

Fig. 11 Flow diagram of flash gas by membranes in CCS [4]

processes, inorganic, metallic, polymeric and solid, etc., types of membranes are used widely [32, 33]. The process involves a repeated stream of gases that are passed through a series of membranes through which CO_2 is separated and flue gases are recycled over again [4]. The movement of particles is based on the theory of solution–diffusion or absorption–diffusion process. Additionally, the movement of particles inside the chamber between from one membrane to another is due to partial differences in pressures [2, 33]. The advantage one gets using this technology is the avoidance of contamination of any other particles with a tolerance rate of 1/10 of original design. However, the limitations such as temperature range for membranes, corrosion threat and non-resilience to industrial practices have had a major impact on its development and applications [34]. Recent breakthrough in the selection of membrane technology such as transport membrane separation technique and mixed matrix membrane technology facilitates the movement of CO_2 in bicarbonate form [35, 36]. The most suitable materials used for the manufacturing of membranes are ceramics, metals, gas absorption and polymers such as zeolite, mesoporous silica and zeoliticimidazolate (Figs. 11 and 12; Tables 3 and 4).

4 CO_2 Transportation and Storage

Once captured, another major aspect of the entire methodology is the process of transporting the CO_2 and its storage. The major challenges faced during its planning are to give prior importance to three prominent factors: (1) compression of the gas which is scheduled over 8 MPa and accounts for the majority of total costs involved, (2) corrosion impacts to pipelines and (3) power consumption based on fluid [5]. Therefore, to achieve the above objectives, a distance of around 200 km is used for recompression of the pipeline. Pipelines necessarily act as the equipment for transporting the captured CO_2 over 40 Mtpa in 2500 miles [39]. Road tankers (operating at −20 °C and 2 Mpa pressure) and ships (operating at around 0.7 Mpa pressure) are also a feasible mode of transportation. However, the pipeline technology makes

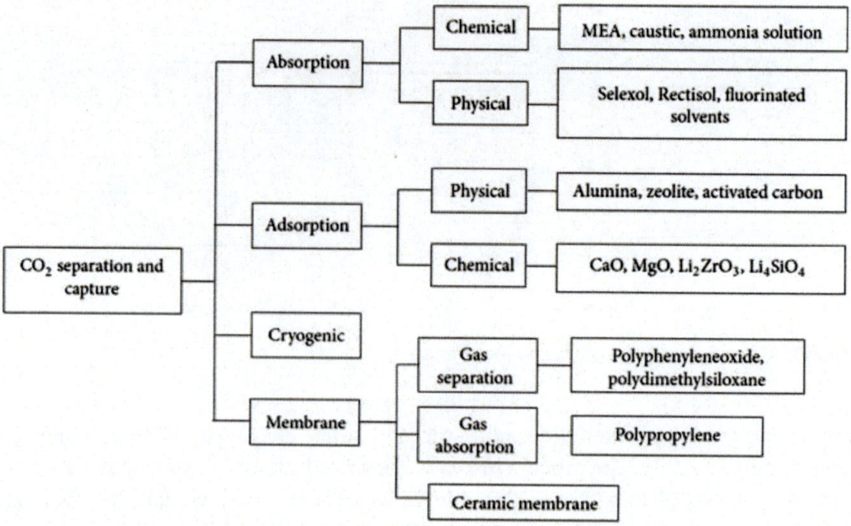

Fig. 12 Different CO_2 separation technologies [28]

it a pretty obvious choice for stakeholders which substantially has an overall effect in costs involved. The pipelines are made to operate in conditions at high pressure and low temperature. Usually, compressors are used to drive the flow. In spite of all, route selection, over-pressure protection, leak detection and other design factors of the pipeline have a vital role to play [13].

Sequestration as it is referred to, the method of permanently storing the injected carbon dioxide beneath the ground has been one of many keen areas of activities where researchers are exploring all possibilities. Saline formations and gas fields (for gaseous forms), synthesis of gas to produce carbonates (for liquid forms), geological storage such as coal seams, depleted basins and oil recovery methods are widely used [40, 41]. However, safety measures of these sites are also been given the appropriate attention for being robust over time. Porous and permeability of these geological sites involving measures such as usage of cap rock for avoiding leakage during trapping and containment of $CO_2$are also taken into consideration during selection [42, 43]. Finally, trapping mechanisms such as structural and stratigraphic trapping (happening beneath permeable rocks), residual trapping (involving porous rocks), solubility trapping (gas trapping) and mineral trapping (pore changing) have helped in avoiding leakage that may occur due to mobility of CO_2 particles [40] (Table 5).

5 List of Currently Operational Large Scale and Pilot Demonstrations CCS Facilities Across the World [45]

See Table 6.

Table 3 Comparison between different carbon dioxide separation technologies

Parameters	Physical absorption	Physical adsorption	Membrane technology	Cryogenic separation	References
Advantages	Most suitable for pre-combustion capture; solvent regeneration possible	Reversible methodology comprising with redeemable adsorbent; good efficiency	Easily commercially available; good efficiency	Highly applied industrial technology; easy recovery of CO_2in liquid form	[2]
Disadvantages	High thermal heat requirement for regeneration; disturbance in ecology due to solvent degradation	Scarcity in finding high temperature adsorbent; CO_2desorption not easily achievable	Non-performance of membranes in higher temperatures; issues in operation	Only suitable for highly concentrated CO_2 mixture; high requirement of energy for low temperature surroundings required by technology	
Carbon dioxide separation yield (in %)	92–96	83–92	80–90	More than 98	
Energy requirement (in MJ/kg CO_2)	4.2–6	2.1–2.9	0.3–0.7	7–12	[27]

6 List of Policies Undertaken in Various Unions/Countries [46]

6.1 European Commission

As per emissions trading schemes (ETS), 450 million allowances have been set apart from innovation fund which will primarily focus on developmental activities pertaining to renewable and CCUS energy demonstration projects, energy storage and low-carbon innovation in energy-intensive industry. Out of this 450 million, 50 million are allocated and available as per the 2013–2020 New Entrant Reserve Fund (NER 300). If modern day estimates are to be considered, this funding is worth over €9 billion.

6.2 Norway

The government of Norway has allocated over 280 million NOK for advancement in CCS deployment, including funds for support to Front End Engineering Design (FEED) studies for two full-chain CCS facilities. Each of these facilities is forecasted to capture around 400,000 tons per annum of CO_2 for storage below the North Sea seabed. Additionally, the government has also established Gassnova, a state-owned research organization focused on CCS, and the Technology Centre Mongstad, a CO_2 capture technology testing facility.

Table 4 Comparison between different carbon capture technologies [37, 38]

Aspect	Post-combustion capture	Pre-combustion capture	Oxy-fuel combustion capture
Technology	Full-scale operational plants with application of this technology	Applications in industries with different processes; full operational loaded plants still under constructional phases in the majority of world	Still upcoming technology with not full-scaled operational plants
Advantages	Majorly applied in widely used pulverized coal plants for power generation. Still scope for efficient plants	Low energy requirements involved in the separation process; low water usage and major availability of wide ranging process; other useful by-products generation	No emissions due to non-presence of chemical pollutants; robust and compatible with all kinds of fuels; highest capture efficiency than other technologies; easy installation; compatible with high pressured steam and highly technological advanced equipment used widely
Disadvantages	Incompatible with low pressured gas; sometimes upscaling required for amine technology causing sufficient involvement of energy in compression and scrubbing process; high water consumption	Energy losses occurring frequently due to regeneration; operations still not completed in IGCC cycle-operated power plants; thermal issues due to temperature variations and impacting efficiency due to turbine application	Discontinuous developmental activities observed till date; reduced outputs due to the presence of ASU units; air leakage acts as a major concern due to high pressurized operations; corrosion issues may occur on a rare basis

(continued)

Table 4 (continued)

Aspect	Post-combustion capture	Pre-combustion capture	Oxy-fuel combustion capture
Economy involved	Installation activities causing major expenditure *Capital costs involved:* Gas fired: USD 830 per kW Coal fired: USD 1950 per kW *Electricity costs involved:* Gas fired: USD 0.075 per kW Coal fired: USD 0.07 per kW	Sorbent technology consumes majority cost in IGCC plants *Capital Costs Involved:* Gas fired: USD 1120 per kW Coal fired: USD 1870 per kW *Electricity costs involved:* Gas fired: USD 0.092 per kW Coal fired: USD 0.063 per kW	Air separation apparatus involves a majority stake in overall costs involved *Capital costs involved:* Gas fired: USD 1500 per kW Coal fired: USD 2250 per kW *Electricity costs involved:* Gas fired: USD 0.11 per kW Coal fired: USD 0.07 per kW

Table 5 Costs involved in CO_2 storage [44]

Possible alternatives	Cost estimates (US$/ton CO_2 stored)	Cost estimates (US$/ton C stored)
Geological storage[a]	0.5–0.8	2–29
Geological monitoring	0.1–0.3	0.4–1.1
Ocean[b]		
Pipeline	6–31	22–114
Ship	12–16	44–59
Mineral carbonation[c]	50–100	180–370

[a]Does not include monitoring costs
[b]Includes offshore transportation costs; range represents 100–500 km distance offshore and 3000 m depth
[c]Unlike geological and ocean storage, mineral carbonation requires significant energy inputs equivalent to approximately 40% of the power plant output

6.3 United Kingdom

By 2030, the deployment of CCUS was undertaken by the UK Government's Clean Growth Strategy to bring down the costs sufficiently. CCUS Cost Challenge Taskforce a report delivered in 2018 acknowledged that CCS/CCUS is vitally important to decarbonizing major growing industries such as steel, cement, fertilizer, petrochemicals and flexible natural gas and also identified the need for stable long-term supportive policy. It further lays the foundation for the UK to move to a new energy economy with decarbonized heavy industry and hydrogen fuels that complement renewable deployment.

Table 6 Large scale and pilot demonstrations CCS facilities across the world

Plant name	Country	Operation year	Plant industry	Capture capacity (in Mtpa)
Abu Dhabi CCS (Phase 1 being emirates steel industries)	UAE	2016	Iron and steel production	0.8
Air products steam methane reformer	USA	2013	Hydrogen production	1
Boundary dam carbon capture and storage	Canada	2014	Power generation	1
Century plant	USA	2010	Natural gas processing	8.4
CNPC Jilin oil field CO_2 EOR	China	2018	Natural gas processing	0.6
Coffeyville gasification plant	USA	2013	Fertilizer production	1
Enid fertilizer	USA	1982	Fertilizer production	0.7
Gorgon carbon dioxide injection	Australia	2019	Natural gas processing	3.4–4
Great plains synfuels plant and Weyburn-Midale	USA	2000	Synthetic natural gas	3
Illinois industrial carbon capture and storage	USA	2017	Ethanol production	1
Lost cabin gas plant	USA	2013	Natural gas processing	0.9
Petro nova carbon capture	USA	2017	Power generation	1.4
Petrobras Santos basin pre-salt oil field CCS	Brazil	2013	Natural gas processing	1–2.5
Quest	Canada	2015	Hydrogen production	1
Shute creek gas processing plant	USA	1986	Natural gas processing	7
Sleipner CO_2 storage	Norway	1996	Natural gas processing	1
Snohvit CO_2 storage	Norway	2008	Natural gas processing	0.7
Terrell natural gas processing plant	USA	1972	Natural gas processing	0.4–0.5

(continued)

Table 6 (continued)

Plant name	Country	Operation year	Plant industry	Capture capacity (in Mtpa)
Uthmaniyah CO_2-EOR demonstration	Saudi Arabia	2015	Natural gas processing	0.8
Arkalon CO_2 compression facility	USA	2009	Ethanol production	0.17–0.29
Bell Creek-incidental CO_2 storage associated with a commercial EOR project	USA	2010	Natural gas processing	4
Bonanza bio-energy CCUS EOR	USA	2012	Ethanol production	0.1
CarbFix project	Iceland	2012	Power generation	
CIUDEN: CO_2 storage technology development plant	Spain	2015		0.01
CO_2 CRC Otway	Australia	2008	Natural gas processing	0.01
Core Energy CO_2– EOR	USA	2003	Natural gas processing	0.3–0.35
Course 50-CO_2 ultimate reduction in steelmaking process by innovative technology for cool earth 50	Japan		Iron and steel production	0.01
Cranfield project	USA	2009		4.7
Daqing oil field EOR demonstration project	China	2003	Industrial applications	0.2
Drax bioenergy carbon capture pilot plant	UK	2019	Power generation	
Farnsworth unit EOR field project development phase	USA	2013	Ethanol and fertilizer production	

(continued)

Table 6 (continued)

Plant name	Country	Operation year	Plant industry	Capture capacity (in Mtpa)
Geothermal plant with CO_2 Reinjection	Croatia	2018	Power generation	0.05
Husky energy Hashburn and Tangleflags CO_2 Injection in heavy oil reservoirs project	Canada	2012	Ethanol production	0.08
ITRI calcium looping pilot	China	2013	Cement production	
Karamay Dunhua oil technology CCUS EOR project	China	2015	Methanol production	0.1
Michigan basin large-scale injection test	USA	2013	Natural gas processing	0.1–0.24
NET power clean energy large-scale pilot plant	USA	2018	Power generation	
PCS nitrogen	USA	2013	Fertilizer production	0.2–0.3
Petro China Changqing Oil Field EOR CCUS	China	2017	CTL	0.05–0.1
Sinopec Shengli oil field carbon capture utilization and storage pilot project	China	2010	Power generation	0.03–0.04
Sinopec Zhongyuan oil field carbon capture utilization and storage pilot project	China	2006	Chemical production	0.12
STEPWISE pilot of SEWGS technology at Swerea/Mefos	Sweden	2007	Iron and steel production	
Tomakomai CCS demonstration project	Japan	2016	Hydrogen production	0.1

6.4 United States of America

In February 2018, due to the passed Bipartisan Budget Act, an increment is increased over the tax credits for the geological storage of CO_2established by section 45Q of the Internal Revenue Code. Originally created in 2008, CCUS tax credit was worth US$10 per tonne of CO_2used for EOR and US$20 per tonne of CO_2 stored in saline formations. The 2018 amendments will 15.

Increase the current tax credit for CO_2that is captured and used for EOR or natural gas recovery to US$35 per tonne in 2026;

Increase the current tax credit for CO_2that is captured and stored in saline formations to US$50 per ton in 2026;

Ramp up the tax credits from the original values to the new values in 2026, after which they will be indexed by inflation and remove the 75 Mt cap on the programme.

The 45Q amendments create a good business case for investment in CCS and provide the policy confidence that investors require. A value of up to US$50 per ton of $CO_2$is likely to incentivize a new wave of new CCS facilities in the US over the next five years.

6.5 China

China restructured its ministry named "The ministry of ecology and environment (MEE)" which focuses on mitigating climate change, has long recognized the importance of CCS and has recently implemented a suite of measures designed to accelerate CCS deployment. The tasks include:

Creation of national carbon market in the electricity sector;
Promoting low-carbon technologies, with an emphasis on CCUS;
Supporting CCUS pilots and near-zero carbon emissions pilots
Provision of funding for CCS research projects promoted by the Ministry of Science and Technology;
Bringing amendations in the Environmental Impact Assessment Guidelines to better address CCUS projects and
Establishing a CCUS capacity building project for government officials and researchers directly involved in CCUS.

6.6 Canada

Canada's clean growth and climate action plan ensures addressing to plan carbon. The price must start at CAN$10 per ton of CO_2 (or greater) and rise to CAN$50 per ton by 2022. A federal carbon pricing system came into effect on 1 January 2019, as a backstop for any jurisdiction that does not put in place a carbon pricing system that meets the federal standard. All direct revenue from carbon pricing will go back to the jurisdiction. Funding for CCS projects is available and regulations addressing the geological storage of CO_2 have been promulgated in Alberta.

6.7 Japan

Led by the Ministry of the Environment and the Ministry of Economy, Trade and Industry, the Japanese Government supports a wide range of studies on the investigation of potential CO_2 storage sites, CCS feasibility studies, the assessment of legal and regulatory structures necessary for the management of long-term liability for stored CO_2 and the environmental, economic and social impacts of CCS. Fundings for pilot CCS demonstration facilities, including the Hydrogen Energy Supply Chain, which was announced in 2018 is also available. The Japanese Government is promptly committed through policy to supporting activity designed to systematically identify and defeat barriers to CCS deployment.

7 Challenges

Coal, which is frequently termed as the most realistic source of highly efficient power generation, is also one of the reasons for major emissions. Hence, implementation of CCS technology in these plants will successfully help to mitigate this concern. The significance of this can be understood by the fact that in the modern world in the power sector if more than 50% of CCS technology is implied it can simply account for capture of a billion tons of CO_2. However, these statistics can alone not subsidize the limitations this technology can account. CCS installations, as discussed earlier has major implications on overall efficiency and subsequently has a higher consumption of energy in compression of gases [12] (Fig. 13).

The above figure depicts the difference between current, emerging and advanced technologies. The dotted line predicts the electricity prices. It is quite evident from the above figure that the prices in current technologies are the most expensive when compared to the costs. However, this lone approach is not feasible in subsidizing the overall reduction in CO_2 levels while energy generation [47]. Hence, due to the installation of CCS technology whose cost accounts for 75% of total cost, it has direct implications on overall increment I range of about 30–90%. Moreover, another reason is also the energy loss incurred due to the compression process [48].

The cost involved in capturing CO_2 from a new gasification infrastructure is usually cheaper than a coal plant with post combustion capture [49]. However, the major challenges involved are (1) the operation time period for a gasifier and production facility must be same; (2) IGCC is comparatively more expensive than coal plants with application of CCS technology in either of them [50]; (3) finance involved is also directly proportional to the height and source coal fuel.

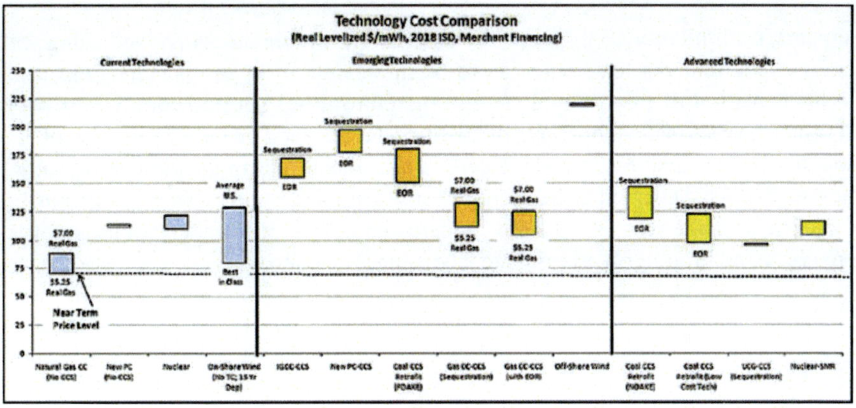

Fig. 13 Comparison laid between various technologies in electricity generation [12] (*Note* Incremental growth for real levelized cost scheduled at 2.5% annually from 2018)

Coal power plants generated power is more costly than post combustion capture. Hence, the price of electricity is likely to hike 80% of the costs. (1) Without CCS remodelling, cost of present plants will be more than the power plant; (2) energy required to heat solvents and then compress CO_2 from stack involves in reducing 30% of efficiency; (3) little developments from researchers using commercial pilot plants have resulted in significant increase in efficiency which furthermore increases the pulverized coal combustion processes finances [51]. Sequestration is one of the safest methods when used as per the governance of science and technology such as oil recovery in which CO_2 is injected into brine formations [52]. But multiple injections in one site requires knowledgeable application combined with development. Hence, programmes for the conservation and development of these sequestration sites and avoiding ecological concerns become one of the most vital challenges which are very well supported by employing institutional practices [53–55].

8 Conclusion

In this paper, we have successfully discussed the different technologies used widely in CCS technology. Additionally, various techniques for carbon separation is very well versed discussed in the above literature. Methodologies such as pre-, post- and oxy-fuel combustion are discussed and evaluation results reveal that each method has its value and recognition as per the requirements. Since pre- and post-combustion carbon capture are more preferable technologies from the gas purification point of view, they can also be used for small-scale pilot plant or demonstration plants which have yet not been materialized. Oxy-fuel technology is one of the promising and upcoming methods of carbon separation which has already acquired appraisals from the research and development communities. Furthermore, discussions on their advantages, limitations are hereby discussed in detail. Finances involved along with energy requirements have emerged to be the main factors of consideration in the implementation of this technology and therefore it has evolved into one of many significant areas of developments to facilitate the market demands. In spite of all, it is found to be conclusive from the review that retrofitted CCS applications in prevailing power plants have found to be more expensive than the new power plants. Finally, the current challenges and policies undertaken by various organizations/countries are discussed thoroughly to mitigate environmental concerns.

References

1. Ronca D (2008) How carbon capture works. How Stuff Works
2. Leung DYC, Caramanna G, Maroto-Valer MM (2014) An overview of current status of carbon dioxide capture and storage technologies. Renew Sustain Energy Rev 39:426–443
3. Global CCS Institute (2015) Global Status of CCS: 2015. Melbourne: Global CCS Institute, pp. 1–18
4. Song C, Liu Q, Deng S, Li H, Kitamura Y (2019) Cryogenic-based CO_2 capture technologies: state-of-the-art developments and current challenges. Renew Sustain Energy Rev 101:265–278
5. Wilberforce T, Baroutaji A, Soudan B, Al-alami AH, Olabi AG (2019) Outlook of carbon capture technology and challenges. Sci Total Environ 657:56–72
6. Rubin ES, Mantripragada H, Marks A, Versteeg P, Kitchin J (2012) The outlook for improved carbon capture technology. Prog Energy Combust Sci 1–42
7. Blomen E, Hendriks C, Neele F (2009) Capture technologies: improvements and promising developments. Energy Procedia 1:1505–1512
8. U.S Department of Energy (2020) Pre-combustion carbon capture research. Department of Energy. https://www.energy.gov/fe/science-innovation/carbon-capture-andstorage-research/carbon-capture-rd/pre-combustion-carbon
9. Michailos S, Emenike O, Ingham D, Hughes KJ, Pourkashanian M (2019) Methane production via syngas fermentation within the bio-CCS concept: a techno-economic assessment. Biochem Eng J 150:107290
10. U.S Department of Energy. Pre-Combustion CO_2 Capture. netl.doe.gov. https://www.netl.doe.gov/coal/carbon-capture/pre-combustion
11. Lewis C, Adams D, Cover JD (2007) Capturing CO_2 in IEA Greenhouse Gas R & D Programme. In: Performance of the power plants, pp 1–23
12. Herzog H, Smekens K, Dadhich P, Dooley J, Fujii Y, Hohmeyer O, Riahi K (2006) The IPCC special report on carbon dioxide capture and storage, in ECOS 2006, Sources of CO_2. In: Carbon Dioxide Capture and Storage, Metz B, Davidson O, de Coninck H, Loos M, Meyer L (eds) Proceedings of the 19th International Conference on Efficiency, Cost, Optimization, Simulation and Environmental Impact of Energy Systems, pp 65–85
13. Balbus JM, Wilson ML (2000) Human health & global climate change: a review of potential impacts in the United States
14. Agbonghae EO, Hughes KJ, Ingham DB, Ma L, Pourkashanian M (2014) Optimal process design of commercial-scale amine-based CO_2 capture plants. Ind Eng Chem Res 53:14815–14829
15. Ali U, Font-Palma C, Akram M, Agbonghae EO, Ingham DB, Pourkashanian M (2017) Comparative potential of natural gas, coal and biomass fired power plant with post-combustion CO_2 capture and compression. Int J Greenh Gas Control 63:184–193
16. Theo WL, Lim JS, Hashim H, Mustaffa AA, Ho WS (2016) Review of pre-combustion capture and ionic liquid in carbon capture and storage. Appl Energy 183:1633–1663

17. Post-combustion CO_2 Capture (2020) netl.doe.gov. https://www.netl.doe.gov/coal/carbon-capture/post-combustion
18. Chisholm G, Gardiner R, Bryant M, Hunt M, Black S (2011) Development of post combustion capture technology. In: Doosan Power Systems, pp 1–26
19. Wang Y, Zhao L, Otto A, Robinius M, Stolten D (2017) A review of post-combustion CO_2 capture technologies from coal-fired power plants. Energy Procedia 114:650–665
20. Yang X et al (2018) Prediction of the radiative heat transfer in small and large scale oxy-coal furnaces. Appl Energy 211:523–537
21. Stanger R et al (2015) Oxyfuel combustion for CO_2 capture in power plants. Int J Greenh, Gas Control
22. Mathekga HI, Oboirien BO, North BC (2016) A review of oxy-fuel combustion in fluidized bed reactors. Int J Energy Res 40:875–902
23. Wappel D, Gronald G, Kalb R, Draxler J (2010) Ionic liquids for post-combustion CO_2 absorption. Int J Greenh Gas Control 4:486–494
24. Savile CK, Lalonde JJ (2011) Biotechnology for the acceleration of carbon dioxide capture and sequestration. Curr Opin Biotechnol 22:818–823
25. Fauth DJ, Gray ML, Pennline HW, Krutka HM, Sjostrom S, Ault AM (2012) Investigation of porous silica supported mixed-amine sorbents for post-combustion CO_2 capture. Energy Fuels 26:2483–2496
26. Ban Z, Lau K, Azmi M (2014) Physical absorption of CO_2 capture: a review. Adv Mater Res 917:134–143
27. Mondal MK, Balsora HK, Varshney P (2012) Progress and trends in CO_2 capture/separation technologies: a review. Energy 46:431–441
28. Songolzadeh M, Soleimani M, Ravanchi MT, Songolzadeh R (2014) Carbon dioxide separation from flue gases: a technological review emphasizing reduction in greenhouse gas emissions. Sci World J 1–34
29. Abdullah A, Idris I, Shamsudin IK, Kim J, Othman MR (2019) Carbon dioxide separation from carbon dioxide-methane gas mixture using PSA utilizing inorganic and organic adsorbents. AIP Conf Proc 2124:020058–1–020058–13
30. Dai Z, Deng L (2016) Membrane absorption using ionic liquid for pre-combustion CO_2 capture at elevated pressure and temperature. Int J Greenh Gas Control 54:59–69
31. Zheng J, Zhang P, Linga P (2016) Semiclathrate hydrate process for pre-combustion capture of CO_2 at near ambient temperatures. Appl Energy
32. Satyapal S, Filburn T, Trela J, Strange J (2001) Performance and properties of a solid amine sorbent for carbon dioxide removal in space life support applications. Energy Fuels 15:250–255
33. Aaron D, Tsouris C (2005) Separation of CO_2 from flue gas: a review. Sep Sci Technol 40:321–348
34. Zhang Y, Sunarso J, Liu S, Wang R (2013) Current status and development of membranes for CO_2/CH_4 separation: a review. Int J Greenh Gas Control 12:84–107
35. Kang Z et al (2015) Mixed matrix membranes composed of two-dimensional metal-organic framework nanosheets for pre-combustion CO_2 capture: a relationship study of filler morphology versus membrane performance. J Mater Chem A 1–10
36. Park SH, Lee SJ, Lee JW, Chun SN, Bin Lee J (2014) The quantitative evaluation of two-stage pre-combustion CO_2 capture processes using the physical solvents with various design parameters. Energy pp 1–9
37. Mills S (2012) Coal-fired CCS demonstration plants, 2012. Available: https://www.usea.org/sites/default/files/102012_Coal-firedCCSdemonstrationplants%2C2012_ccc207.pdf
38. MacDonald M (2006) CO_2 capture as a factor in power station investment decisions technical study: IEA Greenhouse Gas R&D Programme, pp 1–144
39. Borhani TNG, Akbari V, Afkhamipour M, Hamid MKA, Manan ZA (2015) Comparison of equilibrium and non-equilibrium models of a tray column for post-combustion CO_2 capture using DEA-promoted potassium carbonate solution Chem Eng Sci 122:1–25
40. Borhani TNG, Akbari V, Hamid MKA, Manan ZA (2014) Rate-based simulation and comparison of various promoters for CO_2 capture in industrial DEA-promoted potassium carbonate absorption unit J Ind Eng Chem

41. Parliament office of science and technology (2009) CO_2 Capture, Transport and Storage. Postnote Number 335, Parliamentary Office of Science and Technology, London (June 2009), https://www.parliament.uk/documents/post/postpn335.pdf
42. McCoy S (2009) Carbon capture and sequestration: Framing the issues for regulation - An interim report from the CCSReg Project, pp 1–162
43. Kay D, Faaij A, Turkenburg W (2008) Health, safety and environmental risks of underground CO 2 storage–overview of mechanisms and current knowledge. Clim Change 74(1–3):289–318
44. Herzog H, Smekens K, Dadhich P, Dooley J, Fujii Y, Hohmeyer O, Riahi K (2006) The IPCC special report on carbon dioxide capture and storage, in ECOS 2006, Cost and economic potential. In: Carbon Dioxide Capture and Storage, Metz B, Davidson O, de Coninck H, Loos M, Meyer L (eds), Proceedings of the 19th International Conference on Efficiency, Cost, Optimization, Simulation and Environmental Impact of Energy Systems, pp 339–362
45. Global CCS Institute (2018) Facilities—global CCS Institute. [Online]. Available: https://co2re.co/FacilityData. Accessed: 02 Sep 2019
46. Global CCS Institute (2018) The global status of CCS, pp 1–84
47. Programme IG (2007) Environmental assessment for CO_2 Capture and storage
48. Dooley JJ et al (2006) Carbon dioxide capture and geologic storage—a core element of a global energy technology strategy to address climate change. Joint Global Change Res Inst (JGCRI) 1–37
49. Ansolabehere S (2007) The Future of Coal: OPTIONS FOR A CARBON-CONSTRAINED WORLD. In: Massachusetts Institue of Technology, pp 1–192
50. Socolow R, Pacala S, Greenblatt J (2004) International conference on greenhouse gas control technology, Vancouver, Canada, September 5–9, 2004. In: GHGT-7, 7th International conference greenhouse gas control Technologies, Vancouver, Canada, Sept. 5-9, 2004, pp. 1–4
51. Document 52006DC0843. [Online]. Available: https://eur-lex.europa.eu/legal-content/EN/TXT/?uri=celex:52006DC0843. Accessed: 02 Sep 2019
52. Carbon Capture and Storage (CCS) and low carbon fossil fuels. [Online]. Available: http://www.fossiltransition.org/. Accessed: 02 Sep 2019
53. Zhang Z, Li Y, Zhang W, Wang J, Soltanian MR, Olabi AG (2018) Effectiveness of amino acid salt solutions in capturing CO_2: a review. Renew Sustain Energy Rev 98:179–188
54. Zhang Z, Cai J, Chen F, Li H, Zhang W, Qi W (2018) Progress in enhancement of CO_2 absorption by nanofluids: a mini review of mechanisms and current status. Renew Energy 118:527–535
55. Elliot TR, Celia MA (2012) Potential restrictions for CO_2 sequestration sites due to shale and tight gas production. Environ Sci Technol 46(7):4223–4227

The Effect of Friction Stir Welding on the Mechanical Properties of Hybrid $Mg/(SiC_p + Al_2O_{3p} + Gr_p)$-MMC

Jaspreet Hira, Alakesh Manna, S. K. Mangal, and Pushpinder Kumar

1 Introduction

Friction stir welding also known as solid-state welding technique to join/weld metals has gained tremendous responsiveness in research area as well as manufacturing industries. Latest technology applications such as automotive, defence, manufacturing and microwelding use this friction stir welding technique. The most attractive feature of this welding is the solid-state welding process in which the work material is not melted to allow low temperature and lower heat output welding compared to what is needed to melt the work material. This attractive feature overcomes the conventional fusion welding where excessive heat is required to melt the work material. As it requires less heat friction, stir welding is safer, economical and easy operated welding technique. The friction stir welding combines lightweight materials such as aluminium alloys, magnesium alloys, copper and titanium alloys, which by traditional welding are very hard to weld. One of the major attractive features of this welding technique is it requires lesser number of equipment, eliminates the use of filler material and improves weld ability. Ugende et al. [1] concluded that the stainless steel proved to be better tool material for FSW, which provided fine-grained

J. Hira (✉)
Mechanical Engineering Department, Amity University Noida, Noida, Uttar Pradesh, India
e-mail: jhira@amity.edu

A. Manna · S. K. Mangal
Mechanical Engineering Department, PEC University of Technology, Chandigarh, India
e-mail: kgpmanna@rediffmail.com

S. K. Mangal
e-mail: skmangal_pec@rediffmail.com

P. Kumar
Thapar University Patiala, Patiala, India
e-mail: pushpindergera@gmail.com

V. H. Saran and R. K. Misra (eds.), *Advances in Systems Engineering*, Lecture Notes in Mechanical Engineering,
https://doi.org/10.1007/978-981-15-8025-3_70

microstructures and better mechanical properties, and SS tool material with rotational speed of 1120 rpm exhibited superior tensile strength properties. Gungor et al. [2] with the low friction speed both similar as well as dissimilar alloys of aluminium such as 5083-H111 and 6082-T651 result in higher fatigue limits and good tensile strength. Gibson et al. [3] discussed various methods such as self-reacting (SR-FSW), stationary shoulder, friction stir processing (FSP), friction stir spot welding (FSSW), assisted FSW and pulsed FSW and their impact on aerospace, car and ship building applications. Devanathana et al. [4] concluded that the axial speed contributes 35 %, followed by traverse speed and spindle speed which are 25% and 12%, respectively. Zhao et al. [5] found out that with the welding speed of 80 mm/min and rotation speed of 1200 pm, sound welding with good mechanical properties was produced between 6013 Al alloy and AZ31 Mg. Fu et al. [6] concluded that with the combinations of intermediate rotation rate ($\omega = 600$–800 rpm) and low traverse speed ($v = 30$–60 mm/min) by placing Mg on advancing side, tool offsetting to Mg 0.3 mm good mechanical properties is obtained using dissimilar metals of 6061-T6 aluminium alloy to AZ31B magnesium alloys. Rao et al. [7] concluded that the welds with larger upper sheet thickness with small interfacial hook height generated the highest lap shear strength. They also found out that at rotational speed of 1500 rpm and 0.2 mm tool shoulder plunger depth exhibits the maximum lap shear strength. Carlone et al. [8] concluded that strong joint can be achieved at 1000 rpm and 120 mm/min by using FSW, also grain refinement was observed in the fusion zone of TIG joint and nuggets zone of FSW of ZE41A magnesium alloy. Gok et al. [9] found that the best results are obtained from both experimental and simulation studies, under a rotation speed of 1964 rpm and a transverse speed of 20 mm/min. Rao et al. [10] found out that it is feasible to FSSW cast AM60B Mg alloy to AA6022-T4 Al alloy and the lap shear failure load increased with increase in tool shoulder plunge depth at 1000 rpm tool rotation rate partially due to increased weld bond width, while discontinuous intermetallic compounds (IMCs) resulted in improved weld strength.

From the above literature review, a lot of research is been carried through on friction stir welding. Therefore, our focus is to investigate the effect of FSW on prepared hybrid Mg/($SiC_p + Al_2O_{3p} + Gr_p$)-MMC and find out the optimal parameter for welding MMC with sound joint with higher values of tensile strength and microhardness.

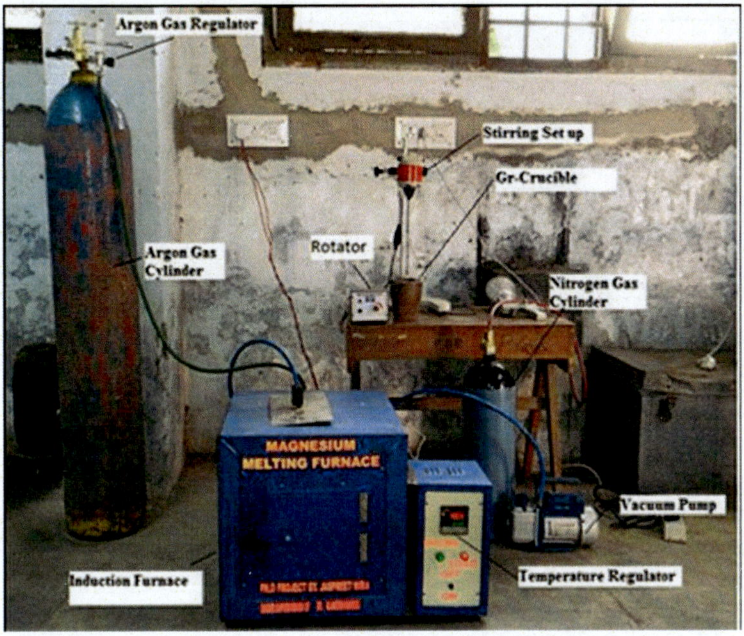

Fig. 1 Fabricated gas injection liquid stirring Mg–MMC casting set-up

2 Preparation of Hybrid Magnesium Metal Matrix Composite with Hard Reinforced Particles

The fabricated gas injection casting set-up has been shown in Fig. 1 [11].

3 Design and Development of Friction Stir Welding Set-Up

Friction stir welding consists of rotational cylindrical tool, work material to be welded, backing plate and clamp holder as shown in Fig. 2. A vertical milling machine is used for joining of Mg/($SiC_p + Al_2O_{3p} + Gr_p$) plates in butt joint using the process of friction stir welding. The machine range is 400, 800 and 1400 R.P.M with 2 HP power capacity and these three levels of rotational speeds are designed for current planning and experimentation, which are adjusted with the help of belt and pulley arrangement. The vertical movement of the machines manually powered and lock nut restrains the vertical motion. The Mg–MMC workpiece is placed on the sliding table and both the plates are held together with the clamp and holder and closed together with the help of nut screw arrangement. The arbour fitted in the vertical chamber can vary the shoulder diameter. The rotational cylindrical tool is mounted in the arbour with fixed shoulder diameter. The table has a sliding arrangement fitted

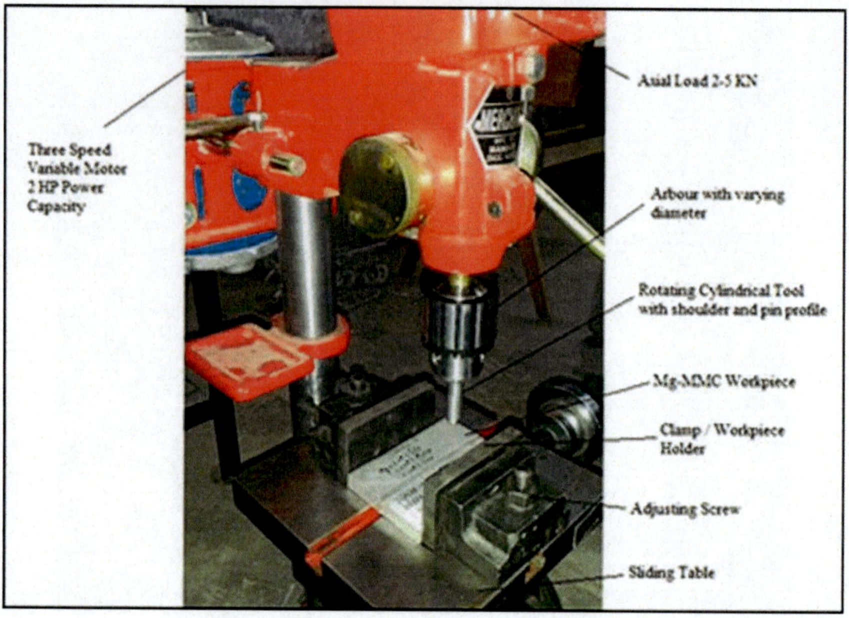

Fig. 2 Friction stir welding set-up

with a motor and sliding mechanism, and different feed rate is achieved with varying the voltage regulator fitted with the set-up.

4 Tool Design

The design consists of the combination of two cylinders of a specific radius ratio known as shoulder and smaller radius pin or probe as shown in Fig. 3. The height of the pin or probe is more than the height of the work material thickness but it is not equal to the overall thickness. Proper pin penetration is very important in joining the metal, and back plate supports the pin geometry and the force with which the tool rotates and penetrates into the welded region should be optimal for welding. Improper clamping will lead to gaps which lead to the formation of cavities and worm holes in the weldment.

In this research work, we have used different tool designs for friction stir welding. Figure 4 shows the four types of tool designed for welding of hybrid (Mg/SiC + Al_2O_{3p} + Grp) metal matrix composite. High carbon and high chromium steel tools having hardness of 270 HV were manufactured based on the experimental design. The tools were fabricated on heavy-duty lathe machine using solid carbide inserts. The tool material selection was based on literature review. The attractive feature of this tool is that it is easy to machine into the desired shape and its hardness value

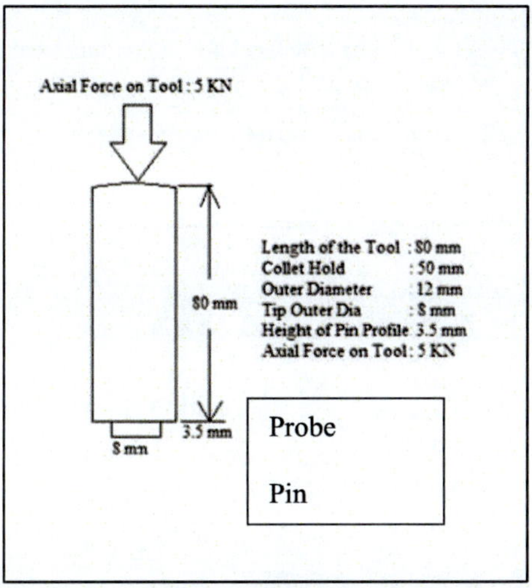

Fig. 3 Design of the rotational cylindrical tool used in friction stir welding

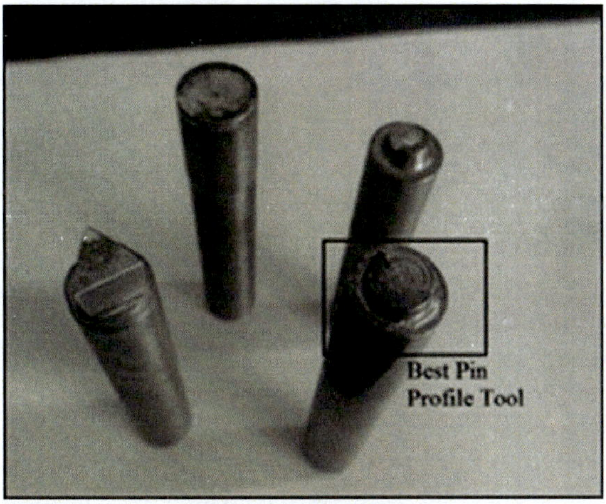

Fig. 4 Best pin profile tool with excellent welding capability

is four times of that of magnesium MMC. The important parameters considered for designing the tool are the tool shoulder diameter, tool pin profile and tool tip profile. Figure 4 shows the best pin profile tool after performing a number of experiments. The chemical composition of the high carbon and high chromium steel is presented in Table 1.

Table 1 Chemical composition of high carbon and high chromium steel tool

Component	C	Si	Mn	P	S	Cr	W	V	Ni	Cu	Fe
Wt % Max	2.18	0.60	0.60	0.03	0.03	12.25	1.00	1.00	0.30	0.25	Bal

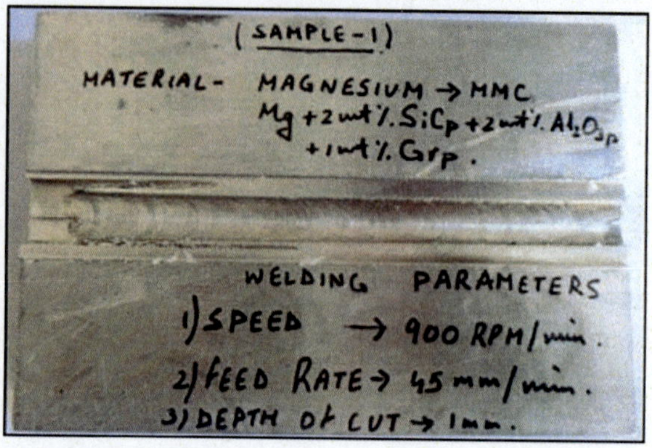

Fig. 5 Welding of Mg/(2 wt%SiC + 2 wt%Al_2O_3 + 1 wt%Gr)-MMC at welding speed 900 rpm, feed rate 45 mm/min and depth of cut 1 mm

5 Results and Discussion

5.1 *Welding of Mg/ (2 wt% SiC + 2 wt%Al_2O_3 + 1 wt%Gr)-MMC*

Figure 5 shows the welding of Mg/(2 wt%SiC + 2 wt%Al_2O_3 + 1 wt%Gr) by friction stir process with the tool profile circular, circular pin 8 mm and shoulder diameter 12 mm profile prepared on the fabricated vertical milling machine, the welding speed was set at 900 rpm/min and feed rate was set at 45 mm/min with the depth of cut 1 mm. It was observed that the average microhardness of the welded segment is 45 HV which is similar to the microhardness value of Mg/(2 wt % SiC + 2 wt%Al_2O_3 + 1 wt%Gr) MMC and the tensile strength value of the 160 MPa.

5.2 *Welding of Mg/(4 wt%SiC + 2 wt%Al_2O_3 + 1 wt%Gr)-MMC*

Figure 6 shows the welding of Mg/(4 wt%SiC + 2wt%Al_2O_3 + 1wt%Gr) by friction stir process with the tool profile circular, circular pin 8 mm and shoulder diameter 12 mm profile prepared on the vertical milling machine, the welding speed was set at 1400 rpm/min and feed rate was set at 45 mm/min with the depth of cut 1 mm.

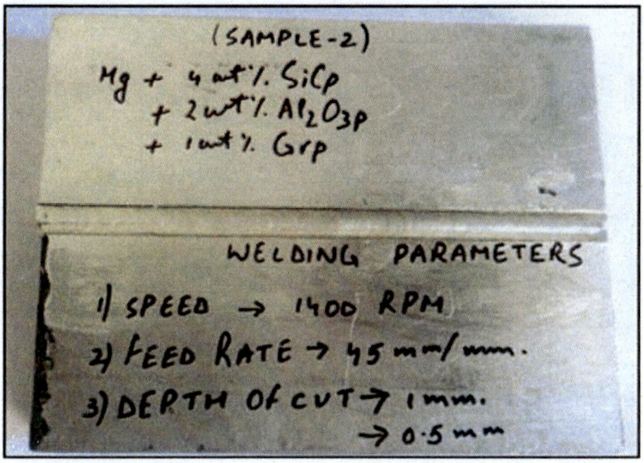

Fig. 6 Welding of Mg/(4 wt%SiC + 2 wt%Al_2O_3 + 1 wt%Gr)-MMC at welding speed 1400 rpm, feed rate 45 mm/min and depth of cut 0.5 mm and 1 mm

It was observed the average microhardness of the welded segment is 50 HV and its tensile strength value is 200 MPa which is greater as compared to 2 wt % of SiC particulates.

5.3 *Welding of Mg/(6 wt%SiC + 2 wt%Al_2O_3 + 1 wt%Gr)-MMC*

Figure 7 shows the welding of Mg/(4 wt%SiC + 2 wt%Al_2O_3 + 1 wt%Gr) by friction stir process with the tool profile circular, circular pin 8 mm and shoulder diameter

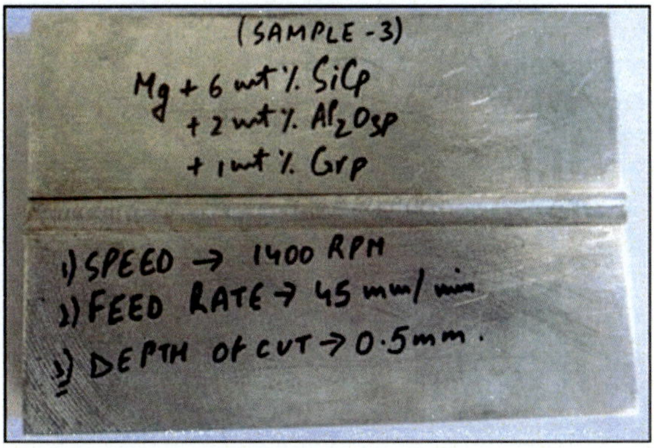

Fig. 7 Welding of Mg/(4 wt%SiC + 2 wt%Al_2O_3 + 1 wt%Gr)-MMC at welding speed 1400 rpm, feed rate 45 mm/min and depth of cut 0.5 mm

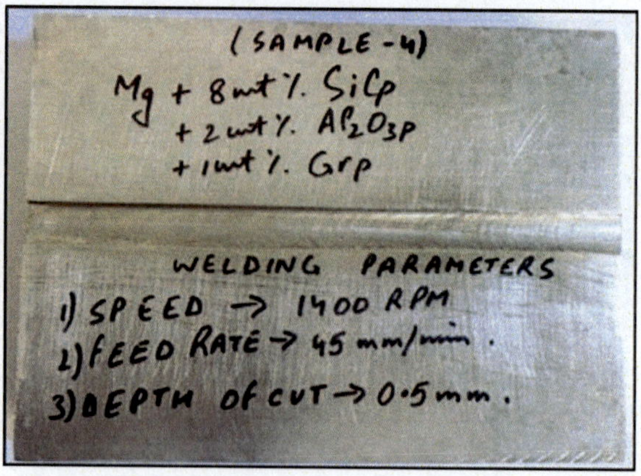

Fig. 8 Welding of Mg/(8 wt%SiC + 2 wt%Al_2O_3 + 1 wt%Gr)-MMC at welding speed 1400 rpm, feed rate 45 mm/min and depth of cut 0.5 mm

12 mm profile prepared on the vertical milling machine, the welding speed was set at 1400 rpm/min and feed rate was set at 45 mm/min with the depth of cut 0.5 mm. It was observed that the average microhardness of the welded segment is 49 HV and its tensile strength value is 210 MPa.

5.4 *Welding of Mg/(8 wt%SiC + 2 wt%Al_2O_3 + 1 wt%Gr)-MMC*

Figure 8 shows the welding of Mg/(8 wt% SiC + 2 wt%Al_2O_3 + 1 wt%Gr) by friction stir process with the tool profile circular, circular pin 8 mm and shoulder diameter 12 mm profile prepared on the vertical milling machine, the welding speed was set at 1400 rpm/min and feed rate was set at 45 mm/min with the depth of cut 0.5 mm. It has been observed the average microhardness of the welded segment is 46 HV and its tensile strength value is 220 MPa.

5.5 *Indentation Mark on Non-welded and Welded Hybrid Mg/(8 wt% SiC + 2 wt%Al_2O_3 + 1 wt%Gr) Metal Matrix Composite*

Figure 9a shows the indentation on the non-welded and (b) shows the indentation on welded zone. As it is clear from Fig. 9 that the microlevel images that indentation diameters are different for the non-welded and welded zone. The parallel lines

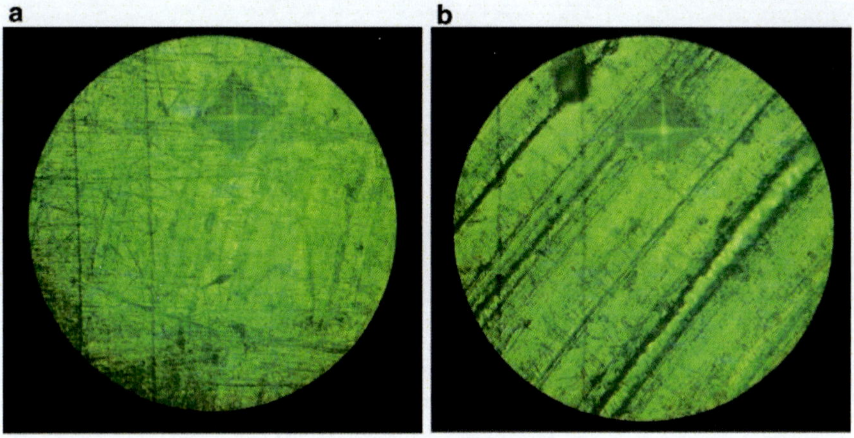

Fig. 9 **a** Indent on non-welded zone. **b** Indent on friction stir welded zone

in Fig 9b show the effect of friction stir welding. The presence of hard reinforcement particulates shows the small size of the indent on non-welded zone of hybrid Mg/(8 wt% SiC + 2 wt%Al_2O_3 + 1 wt%Gr) metal matrix composite. In addition, after friction stir, welding the indenter size is larger and more clear which means that the after effect of joining is more uniform and even distribution at microscopic level.

6 Conclusions

1. Tool profile circular with circular pin 8 mm and shoulder diameter 12 mm is the best pin profile for achieving maximum hardness and maximum tensile strength for joining process.
2. The highest value of the microhardness (HV) is obtained at Mg/(4 wt%SiC + 2 wt%Al_2O_3 + 1 wt%Gr) sample 3 by friction stir welding. The 4wt % contributes to higher value of hardness as compared to all the data. And these values are obtained at 1400 rpm, feed rate 45 mm/min and depth of cut 0.5 mm with circular pin profile with pin diameter 8 mm and shoulder diameter 12 mm.
3. The maximum tensile strength (MPa) of the welded MMC is for Mg/(8 wt%SiCp + 2wt%Al_2O_3p + 1 wt%Grp) which is 220 MPa highest amongst the all wt% of particulates. In addition, the tool profile for this highest tensile strength value is 1400 rpm, feed rate is 45 mm/min and depth of cut is 0.5 mm with circular pin profile with pin diameter 8 mm and shoulder diameter 12 mm.

References

1. Ugender S, Kumar A, Reddy AS (2014) Microstructure and mechanical properties of AZ31B magnesium alloy by friction stir welding. Procedia Mater Sci 6:1600–9. https://doi.org/10.1016/j.mspro.2014.07.143
2. Gungor B, Kaluc E, Taban E, Sik A (2014) Mechanical, fatigue and microstructural properties of friction stir welded 5083–H111 and 6082–T651 aluminum alloys. Mater Des 56:84–90. https://doi.org/10.1016/j.matdes.2013.10.090
3. Gibson BT, Lammlein DH, Prater TJ, Longhurst WR, Cox CD, Ballun MC, Dharmaraj KJ, Cook GE, Strauss AM (2014) Friction stir welding: Process, automation, and control. J Manufact Proc 16(1):56–73. https://doi.org/10.1016/j.jmapro.2013.04.002
4. Devanathan C, Babu AS (2014) Friction stir welding of metal matrix composite using coated tool. Procedia Mater Sci 6:1470–5. https://doi.org/10.1016/j.mspro.2014.07.126
5. Zhao Y, Lu Z, Yan K, Huang L (2015) Microstructural characterizations and mechanical properties in underwater friction stir welding of aluminum and magnesium dissimilar alloys. Mater Des 65:675–81. https://doi.org/10.1016/j.matdes.2014.09.046
6. Fu B, Qin G, Li F, Meng X, Zhang J, Wu C (2015) Friction stir welding process of dissimilar metals of 6061–T6 aluminum alloy to AZ31B magnesium alloy. J Mater Process Technol 218:38–47. https://doi.org/10.1016/j.jmatprotec.2014.11.039
7. Rao HM, Rodriguez RI, Jordon JB, Barkey ME, Guo YB, Badarinarayan H, Yuan W (2014) Friction stir spot welding of rare-earth containing ZEK100 magnesium alloy sheets. Mater Des 56:750–4. https://doi.org/10.1016/j.matdes.2013.12.034
8. Carlone P, Palazzo GS (2015) Characterization of TIG and FSW weldings in cast ZE41A magnesium alloy. J Mater Process Technol 215:87–94. https://doi.org/10.1016/j.jmatprotec.2014.07.026
9. Gök K, Aydin M (2013) Investigations of friction stir welding process using finite element method. Int J Adv Manufact Technol 68(1–4):775–80. https://doi.org/10.1007/s00170-013-4798
10. Rao HM, Yuan W, Badarinarayan H (2015) Effect of process parameters on mechanical properties of friction stir spot welded magnesium to aluminum alloys. Mater Des 66:235–45. https://doi.org/10.1016/j.matdes.2014.10.065
11. Hira J, Mangal SK, Manna A (2015) Fabrication of Hybrid Mg/(Al_2 O_{3p} + SiC_p + Gr_p) metal matrix composite on developed gas injection liquid stir casting setup. Arab J Sci Eng 40(9):2729–38. https://doi.org/10.1007/s13369-015-1755-1

Thermal Systems

Modeling of Gardon Gage Heat Flux Sensor Under Aerothermal Environment

H. M. Ankita, K. S. Lakshmi, P. Anoop, B. Sundar, K. K. Raveendra Babu, L. Sowmianarayanan, and G. Ayyappan

1 Introduction

Launch vehicles face aerodynamic heating during the ascent phase of their trajectory. Payloads are enclosed within the payload fairing (PLF) of a launch vehicle, through the sensible atmosphere. The payload fairing separates when the heat flux on the PLF falls to less than or equal to 1135 W/m^2. The aerodynamic heating on and the thermal response of the PLF has to be quantified for the optimal design of a thermal protection system that satisfies the thermal, structural andaerodynamic constraints. Accurate flight measurements of the same are essential to validate the TPS design.

In flight, aerodynamic heating is measured using Gardon gage type sensors which are circular foil copper-constantan thermal transducers, and are placed flush with the

H. M. Ankita (✉) · K. K. Raveendra Babu · L. Sowmianarayanan · G. Ayyappan
Space Transportation System, Vikram Sarabhai Space Centre, Thiruvananthapuram, India
e-mail: hudedagaddi_ankita@vssc.gov.in

K. K. Raveendra Babu
e-mail: kk_raveendrababu@vssc.gov.in

L. Sowmianarayanan
e-mail: sowmia_narayanan@vssc.gov.in

G. Ayyappan
e-mail: g_ayyappan@vssc.gov.in

K. S. Lakshmi · P. Anoop · B. Sundar
AERO, Vikram Sarabhai Space Centre, Thiruvananthapuram, India
e-mail: lakshmi_kota@vssc.gov.in

P. Anoop
e-mail: p_anoop@vssc.gov.in

B. Sundar
e-mail: b_sundar@vssc.gov.in

V. H. Saran and R. K. Misra (eds.), *Advances in Systems Engineering*,
Lecture Notes in Mechanical Engineering,
https://doi.org/10.1007/978-981-15-8025-3_71

external surface of the PLF. Aerodynamic heating in the ascent phase of the launch vehicle is convective in nature. Gardon gage sensors are calibrated for a radiative source and are known to be associated with a degree of uncertainity in results when used to measure heat flux inconvective environments. This is due to the dependence of the incoming convective heat flux on the sensor surface temperature and an increased nonuniformity in heat flux across the sensor surface due to a parallel bulk flow [1]. Across flights, the flight measured heat flux is observed to be lesser than the expected incident heat flux and a trend change in the heat flux history is recorded for certain flights. The present work focuses on developing a methodology to estimate sensor measured heat flux from the thermal response of the sensor model and the expected flight environment. The results have been validated with flight measurements.

2 Methodology of Heat Flux Estimation

Theoretical methods are used for estimation of aerodynamic heating rates along the flight path of the vehicle. Flow properties are computed as time-dependent variables for vehicle's altitude, relative velocity, and pressure distribution data for various Mach numbers with standard atmospheric model and vehicle's geometry. These flow field parameters are used for the calculation of heat transfer parameters using engineering methods.

Aerodynamic heat flux is estimated at various locations of the payload fairing. On the cylindrical region of the payload fairing, method of Van Driest [2, 3] is used for the estimation of aerodynamic heat flux. This method is based on boundary layer flow over a flat plate.

2.1 Modeling

A Gardon gage consists of a circular constantan foil (radius, R) mounted on an annular copper cylinder as shown in Fig. 1. The sensor works on the priciple of Seebeck effect where the temperature gradient across the center and edge of the constantan foil results in an electromotive force (EMF) generation. This EMF signal is converted to the incident heat flux (q), using a calibration constant.

This Gardon gage when mounted on the external surface of the PLF measures aerodynamic heat fluxexperienced during the ascent phase of the flight.

To capture the flight-measured aerodynamic heat flux, a circular foil copper-constantan heat flux transducer has been modeled in ANSYS [4]. The construction of a Gardon gage sensor is axisymmetric in nature. An axisymmetric model of the sensor is created in ANSYS as shown in Fig. 2. It is subject to flight thermal environment, and the thermal response of the sensor is used to derive the expected heat flux measurements.

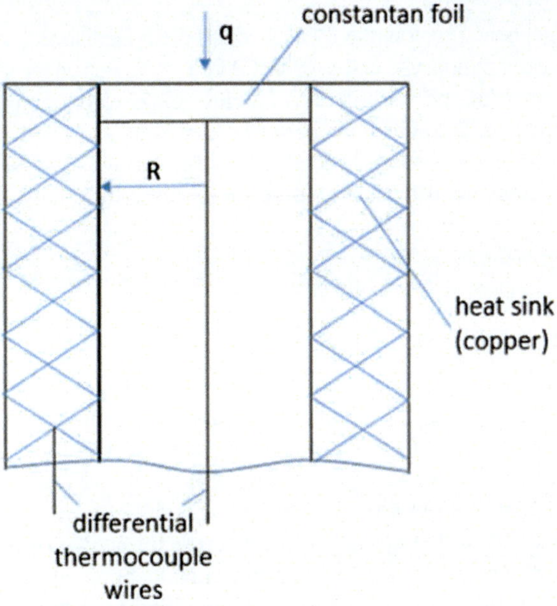

Fig. 1 Cross-sectional view of a Gardon gage sensor

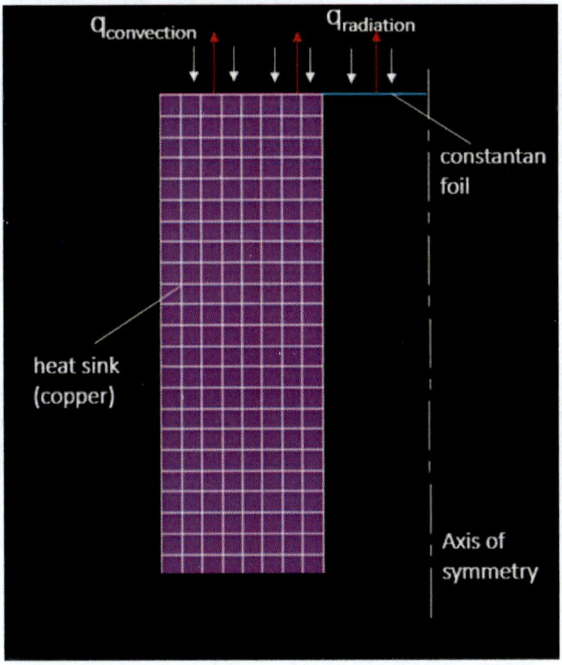

Fig. 2 Axisymmetric finite element model of a Gardon gage sensor

Aerodynamic heat flux on the PLF is computed for launch vehicle trajectories of typical geostationary transfer orbit (GTO) and sun-synchronous polar orbit (SSPO) missions using the method of Van Driest. Cold-wall aerodynamic heat flux is computed using a constant wall temperature. Computed cold-wall heat flux history is applied as a convective boundary condition on the surface of the sensor. Radiation losses to the ambient are also accounted. A transient thermal analysis is then carried out.

2.2 *Derivation of Heat Flux*

The cold-wall heat flux, q_{coldwall}, computed by assuming a constant wall temperature, is applied uniformly across the surface of the sensor model (Eq. 1). Convective heat transfer coefficient, h and the temperature of the gas, T_g, is derived using the method of Van Driest. In reality, the sensor experiences a non-uniform heat flux across its surface due to the thermal gradient across the sensor surface. In addition, the temperature of the sensor surface at any particular location influences the effective incoming aerodynamic heat flux (Eq. 2). Hence, total heat flux measured by the sensor, q_{measured}, is a function of the thermal response of the sensor. Substituting the constant wall temperature in the cold-wall heat flux with the sensor surface temperature, T_{sensor}, provides the heat flux expected to be measured by the sensor based on its thermal response to the mixed environment.

$$q_{\text{coldwall}} = h\left(T_{\text{g}} - T_{\text{coldwall}}\right) \tag{1}$$

$$q_{\text{measured}} = h\left(T_{\text{g}} - T_{\text{sensor}}\right) \tag{2}$$

3 Results

The methodology of heat flux estimation is validated with three different flight measurements of geostationary transfer orbit mission (GTO) (Flight 1) and medium (Flight 2) and low-altitude (Flight 3) sun-synchronous polar orbit (SSPO) missions. Figure 3 compares the trajectories of these flights.

3.1 *Flight 1*

A comparison of the computed cold-wall heat flux on the PLF, the expected measurement of heat flux as computed from the sensor thermal response, and actual flight

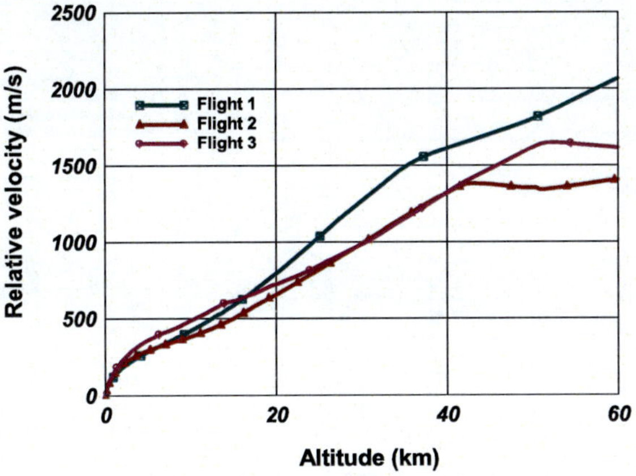

Fig. 3 Comparison of flight trajectories for GTO and medium and low earth SSPO missions

measured heat flux history for a typical GTO mission trajectory is as shown in Fig. 4b. The total gas temperature and the sensor surface temperature from the analysis are as shown in Fig. 4a. It is observed that the heat flux obtained from the sensor thermal response has a good match with the flight-measured heat flux history. This validates the methodology for estimation of aerodynamic heat flux.

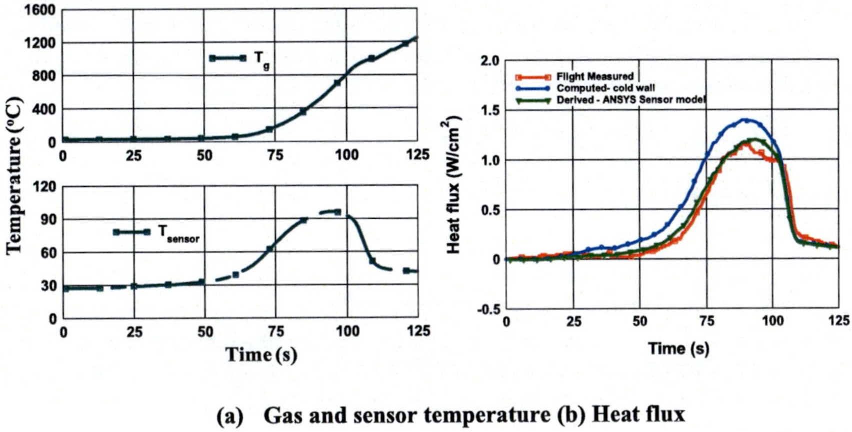

Fig. 4 Sensor temperature and heat flux histories for Flight 1

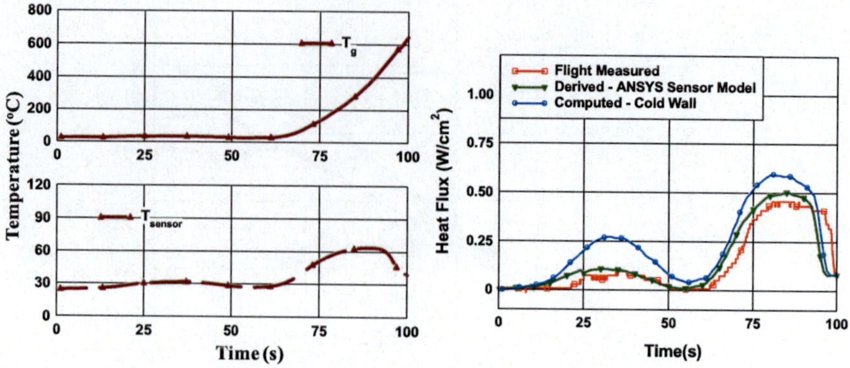

(a) Gas and sensor temperature (b) Heat flux

Fig. 5 Sensor temperature and heat flux histories for Flight 2

3.2 Flight 2

A comparison of the computed incident cold-wall heat flux on the PLF, the expected measurement of heat flux as computed from the sensor thermal response, and actual flight observed readings for a typical medium altitude sun-synchronous polar orbit (SSPO) mission is as shown in Fig. 5b. The total gas temperature and the sensor surface temperature from the analysis are as shown in Fig. 5a. It is observed that the heat flux obtained from the sensor thermal response and the expected flight environment has a good match with the flight measured readings.

3.3 Flight 3

A comparison of the computed incident cold-wall heat flux on the PLF, the expected measurement of heat flux as computed from the sensor thermal response, and actual flight observed readings for a typical low-altitude sun-synchronous polar orbit (SSPO) mission is as shown in Fig. 6b. The total gas temperature and the sensor surface temperature from the analysis are as shown in Fig. 6a. It is observed that the heat flux computed from the sensor thermal response and expected flight environment has a fairly good comparison with the flight observed values. However, the measured peak heat flux occurs at a later time than that expected from the cold–wall computation. There exists a difference in trend between the computed cold-wall heat flux and the flight measurements. This is a result of the significant increase in sensor temperature at the time of occurrence of peak cold-wall heat flux. The effect of the sensor temperature becomes less apparent when the gas temperatures is considerably higher at the corresponding time. This explains the inability of the sensor to capture

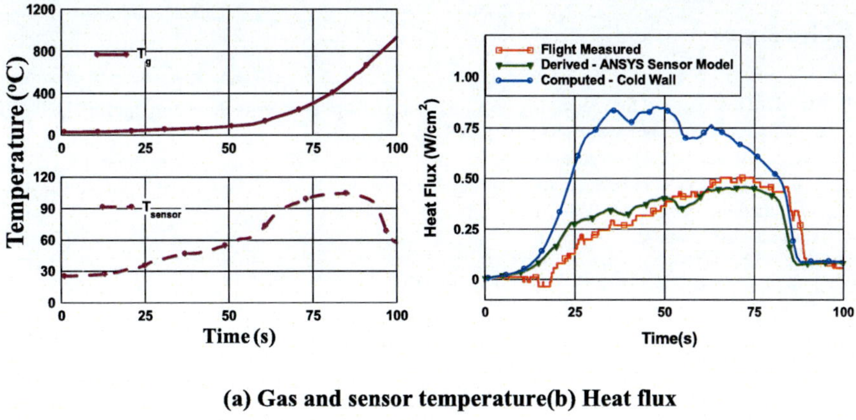

Fig. 6 Temperature and heat flux histories for Flight 3.

the peak heat flux at 35–40 s while the heat flux captured at a later time has a closer match with the cold-wall heat flux.

4 Conclusions

A circular foil copper-constantan transducer mounted flush with the surface of the payload fairing of a launch vehicle is used to measure the aerothermal heat flux experienced during flight. The thermal response of a finite element model of a circular foil copper-constantan heat flux transducer and the flight expected thermal environment has been used to compute the heat flux expected to be measured by the sensor during flight. The results have been validated with flight measurements.

- For a convective environment, the sensor surface temperature participates significantly in the measured heat flux. The measured values of heat flux are lower than the computed cold-wall heat flux due to the participation of the sensor thermal response. The thermal response of the remainder of the PLF may be different from that of the sensor, and thus, the sensor measurement may misrepresent the actual thermal environment experienced.
- The effect of the sensor temperature on the measured heat flux is considerable when the gas temperature is low. This may result in the incorrect capture of the trend in heat flux history in the initial phases of the trajectory.

Thus, the thermal response of the Gardon gage heat flux sensor is to be considered during the validation of heat flux estimation in an aerothermal environment.

References

1. Standard test method for measuring heat flux using a Copper-Constanatan circular foil, heat-flux transducer, C.: Designation E 511 – 01, ASTM International.
2. Van Driest ER (1951) Turbulent boundary layer in compressible fluids. J Aeronaut Sci 18
3. Van Driest ER (1957) Investigation of laminar boundary layer in compressible fluids using the cross method. NACA TN 2597, 1957. Curve fits from NASA-TMX-2058.
4. ANSYS User manual.

Varying Area Vibrating Structure in a Fluid for Energy Gain

Igors Tipans, Janis Viba, Shravankoundinya Vutukuru, and Martins Irbe

1 Introduction

Major advancements in the field of mechatronics and swift proliferation of robots gave birth to autonomous underwater vehicles (AUVs). AUVs are the vehicles intended to perform underwater tasks without on-board pilot and are mainly equipped with on-board sensors and instruments and mostly used for under water surveillance and data recording (Fig. 1). Although AUVs stand as an excellent alternative for the giant traditional vessels, their mission capability is handicapped by power capacity of the vehicles battery effecting both the range and duration. Placing multiple batteries on-board resulted in additional space that accounted for overweight and other important design considerations for the vehicle. All in all, before completing the commanded task, the AUVs are required to undergo many recovery stages for recharge. It is also realized that the recovery process/stages are time consuming and require highly skilled labors at the support/docking site (Figs. 1 and 2).

The present article is devoted only to suggest an efficient alternative for recharging the on-board power pack of a robotic fish with the help of a varying working area vibrating structure (tail or wing) (Fig. 2) that has the ability to scavenge energy from the surrounding medium (fluid).

I. Tipans (✉) · J. Viba · S. Vutukuru (✉) · M. Irbe
Riga Technical University, Riga, Latvia
e-mail: igors.tipans@rtu.lv

S. Vutukuru
e-mail: vshravankoundinya1989@gmail.com

J. Viba
e-mail: janis.viba@rtu.lv

M. Irbe
e-mail: martins.irbe@rtu.lv

V. H. Saran and R. K. Misra (eds.), *Advances in Systems Engineering*,
Lecture Notes in Mechanical Engineering,
https://doi.org/10.1007/978-981-15-8025-3_72

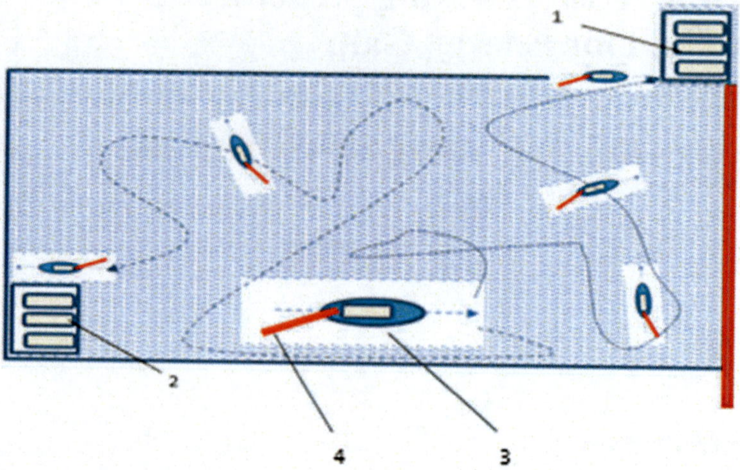

Fig. 1 Actions and motion trajectory of autonomous underwater vehicle (AUVs): 1—battery recharge or exchange station (dry dock); 2—battery recharge or exchange station (underwater); 3—fish hull; 4—perforated tail fin

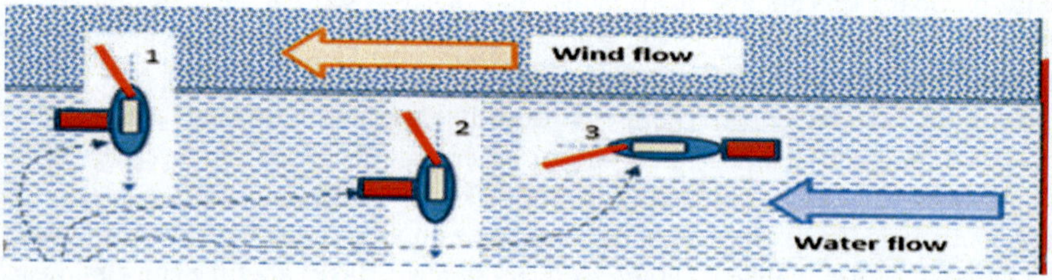

Fig. 2 Power pack charging system of AUVs without docks: 1—charging system using flapping tail from the wind flow; 2, 3—charging system using flapping tail perpendicular and parallel to water flow (a future concept)

Thanks to the recent advancements in the field of energy gain/extraction that made possible to realize that a flapping/vibrating structure can be used from a routine propulsive concept to energy extraction process. It is important to emphasize on the fact that these flapping/vibrating structures (flat plate/perforated plates) are easy to be manufactured/maintainable as they are in most cases untwisted in structure and, additionally, can be operated in shallow waters. A complete and detailed focus on the state of the art research findings in the direction of energy gain/energy extraction by making use of flapping or oscillating structures (foils) from the surrounding medium is discussed at length in the following paragraph.

Possibility for energy extraction from fluids through non-traditional technique using variations in additional area of a vibrating object by Viba et al. [1], it was suggested in the study that by control of the additional area an efficient system for energy gain/extraction could be achieved. Further, taking into account the flow type effects, two-dimensional flow over semi-actuated flapping foil in gusty flow was performed by Zhan et al. [2], the conclusion was that the energy extraction ability for the overall system increased to that of the uniform flow. Investigation on the shape effect of the foil, improvement in performance for energy was achieved by combining the two different aerofoils that are symmetric in nature, thick leading edge of one foil with the thin trailing edge of another foil in an unsteady and incompressible flow and it was performed by Boudis et al. [3]. The point of connection of the two foils was crucial to play a significant role to achieve good efficiency. Studies on deforming foils have also been performed, solutions for two-dimensional Navier–Stokes equation for a time varying deforming camber flapping foil using Fluent was performed to obtain two different efficiencies for power extraction and propulsion by Hoke et al. [4], and the results showed that with right deformation of foil camber the efficiency of the overall system increased. Foils and their power extraction effect in tandem arrangement were studied by Karbasian et al. [5], the inference was that there should be limited number of flapping foils to extract desired energy/power and the overall efficiency just after the second foil decreased. Xie et al. [6] using an oscillating elliptical foil in 2D with thickness of 10% that of the chord is investigated for energy extraction process through a modified flapping motion. The conclusion was that for high reduction in frequency and for low pitching amplitude there resulted in good overall performance. Passive actuation of a flat plate tail, hinged to the symmetric foil with the help of a spring is studied by using lattice Boltzmann method (LBM) and immersed boundary method by Liu et al. [7], two cases for the tail—rigid and flexible were investigated and the inference was that certain appropriate frequency must be achieved to obtain increase in average power efficiency. Commendable research on flexible wing (foil) was performed by Liu et al. [8], two different cases, one with a flexible and the other with dual flexible wings were performed. Both cases showed at typical flapping frequency, an increase of power efficiency at a near low angle of attack. In a similar way, the effect of flexibility and active control for flapping flat plate power extraction was studied by Tian et al. [9], and four different cases were considered for the plate and concluded that for the flat rigid plate with good active control showed higher efficiency as compared to flexible flat plate. Appreciable work by Usoh et al. [10] mentioned about advantages of flat plate over aerofoil.

From the above-mentioned scientific works, it is clear that there is a possibility of energy gain/extraction from a vibrating/flapping flat plate, but this technique needs to be investigated extensively for a better overall power efficiency. It is anticipated that a flat perforated plate working area that would switch between clean plate and perforated plate with help of a mechatronic system would result in good power efficiency, and the present article is completely devoted to this innovative method of energy extraction. A similar perforated plate as a new propulsive fin type device was studied by Viba et al. [11]. In the present work, two approaches for energy gain being investigated, initial technique that deals with the fin in curvilinear translation motion and the other with the fin in rotational motion with a nonlinear spring. The working medium was only air; as a future concept, working medium as water was also shown in Fig. 2. The fish was so designed that when there was a need to recharge, the tail would project out of water and either flapping or rotating tail fin is used to scavenge energy from the surrounding fluid (air).

In order to carry out the synthesis of the proposed new technique introduced in the present work, it is necessary to note that it is solely a problem in modeling involving the movement or oscillation of objects (tail fin) in air or water. Fluid–body interaction phenomenon is partly solved in earlier engineering works [12–14]. The main idea here involves an analytical method in which the space-time modeling problem is roughly replaced by dividing the solid–fluid interaction into two zones: one is the frontal pressure zone; the second is the rear suction zone (or vacuum side). Method pertaining to zones is used in this work. This article does not cover or takes into account all phases of robotic fish activity which includes swimming dynamics and also the techniques that help in stopping and anchoring at the time of charging in the fluid stream (air/water). Therefore, only the interaction of anchored structures with fluid flow will be considered. In addition, it is realized that there are many possible techniques for charging systems, the simplest systems only are considered in the article: Tail fin plane translation motion system with liner spring; system of rotation of the tail fin around a fixed axis with the deployment of nonlinear spring.

2 Tail Fin Translation Motion System with Linear Spring

Model of a perforated plate or the tail fin model for the robotic fish is shown in Fig. 3. Plate or a tail fin with width B and radial length L, the system is assumed to be of a single degree of freedom where the perforated plate with respect to the wind (fluid flow) is at an angle α; and the rods 1 and 2 at angle φ aligned to wind direction. A linear spring with stiffness c is used to ensure effective rotation along the z-axis. An electric generator serves to generate energy is placed parallel to z-axis.

Wind flow interaction forces Q can be calculated according to the approximate calculations [14] as per the theory of mechanics as given by Eq. (1):

$$Q = B \cdot L \cdot MCP \cdot \rho \cdot (1 + c) \cdot [V \cdot \sin(\alpha) - \omega \cdot r \cdot \cos(\alpha - \varphi)]^2 \cdot \mathrm{Sign}[V \cdot \sin(\alpha) - \omega \cdot r \cdot \cos(\alpha - \varphi)], \quad (1)$$

where B is width of plate; L is length of plate; MCP is function of area mechatronics control of perforated plate; ρ is density of fluid (air); V is wind velocity; α is plate angle with respect to the wind (fluid flow); r is length of rods 1 and 2; φ is rods 1 and 2 angle; ω is angular velocity of rods 1 and 2; C is constant, taking into account suction (downstream of plate interactions).

Equation (1) can be understood by considering the tail fin system in x-y plane. The projection of the relative velocity of the tail fin and fluid flow to normal, n is calculated (Fig. 4). The projection of the relative velocity depends on the direction and magnitude of the flow velocity vector V as well as on the wing translation velocity $V0 = \omega \cdot r$ (Fig. 4). According to theory of mechanics, the interaction components

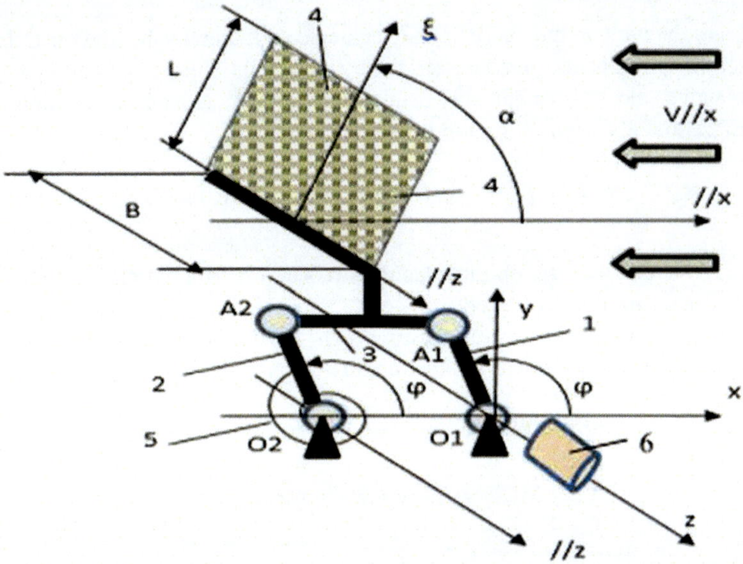

Fig. 3 Vibrating perforated tail fin with linear spring model: 1, 2—two rods of equal length; 3—three-dimensional traverse in XY and YZ planes; 4—perforated plate with width *B* and length *L*; 5—torsion spring with rotational stiffness *c*; 6—electric generator

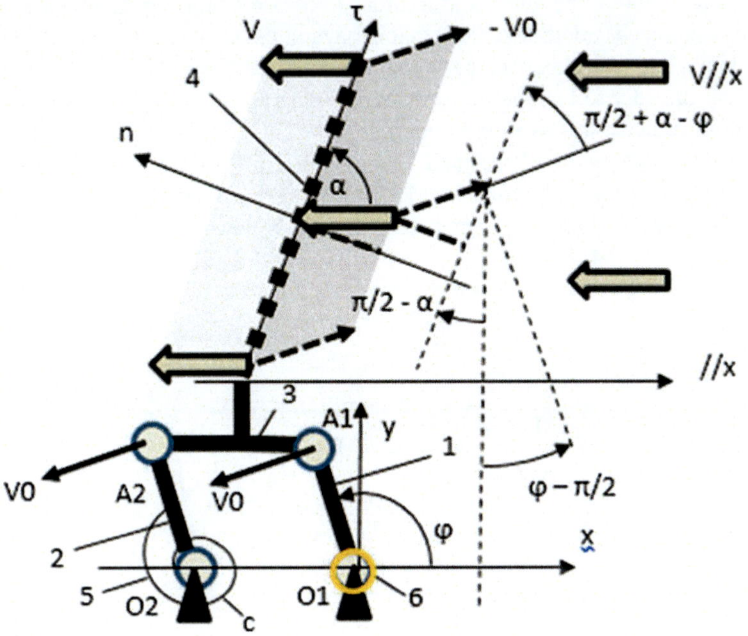

Fig. 4 Side view in plane x-y plane for the perforated tail fin model

parallel to the wing in the axis of τ may be neglected; thereby, behind the tail fin the interaction components are characterized by a constant C Eq. (1), [14].

For a given electro-mechanical (mechatronic) system with one degree of freedom, the differential equation of motion is as follows (Fig. 4).

$$Jz \cdot \ddot{\varphi} = Q \cdot r \cdot \cos(\varphi - \alpha) - c \cdot (\varphi - \varphi 0) - b \cdot \omega, \tag{2}$$

where Jz—a constant which includes the mass and inertia moments of the system elements in terms of kinetic energy. $\ddot{\varphi}$—angular acceleration of rods 1, 2 or generator 6; c—stiffness of linear torsional spring, $\varphi 0$—spring neutral position angle, b—coefficient of generator linear interaction, ω—angular velocity of generator or rods 1&2.

In this work, the MCP function was selected as follows for Eq. (1):

$$\text{MCP} = C1 + C2 \cdot \text{sign}(\omega - \Delta), \tag{3}$$

where $C1$, $C2$—positive constants, always less than 1 and $C1 > C1$, Δ—small parameters of angular velocity varying from $+1$ or -1.

Equation (2) is nonlinear and its solution must be sought in numerical form. In addition, it should be noted that there are many parameters in the equation whose values should be optimized in order to obtain better energy efficiency. This optimization needs additional work that is beyond the scope of the current article.

A small plate with dimensions $L = 0.25$ m and $B = 0.4$ m was chosen for numerical modeling in order to test the performance of the system under consideration. The wind flow velocity was assumed to be medium at $V = 10\ \text{m s}^{-1}$. Accordingly, the mechatronic parameters of the area change were: $C1 = 0.\ 55$, $C2 = 0.45$, $\Delta = 0.1$ rad. The lengths of the rods 1 and 2 and the spring stiffness were: $r = 0.25$ m, $c = 40\,\text{N m rad}^{-1}$. The other parameters were as follows: $\varphi 0 = \pi \cdot 2^{-1}$ rad, $\alpha = \pi \cdot 2^{-1}$ rad, $b = 0.4\,\text{N m s rad}^{-1}$, $C = 0.5$, $\rho = 1.25\,\text{kg m}^{-3}$.

The following graphs show the characteristic function of the system motion with linear spring (Figs. 5, 6, 7, 8, 9 and 10).

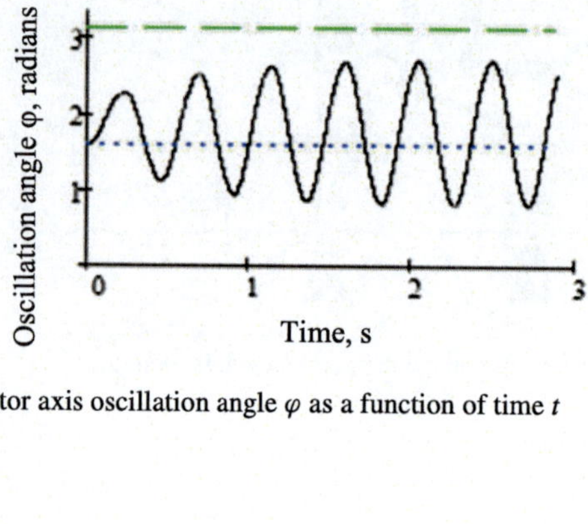

Fig. 5 Electric generator axis oscillation angle φ as a function of time t

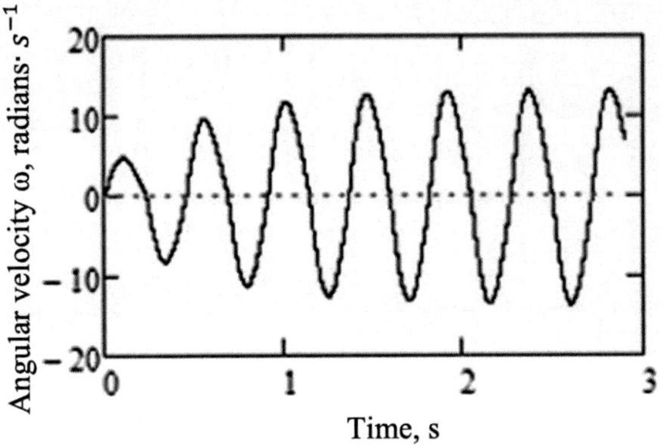

Fig. 6 Electric generator angular velocity ω as a function of time t

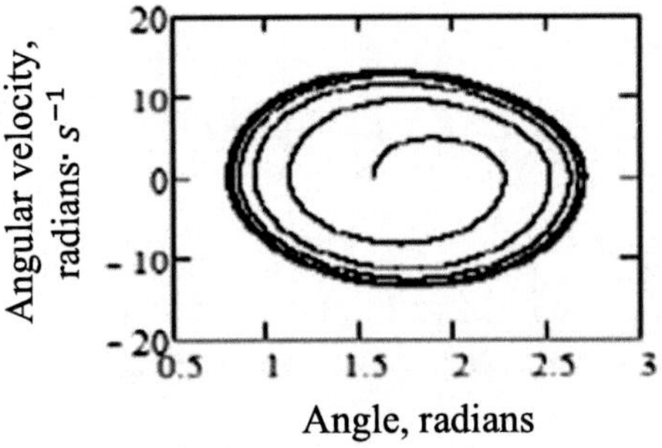

Fig. 7 Motion in phase plane

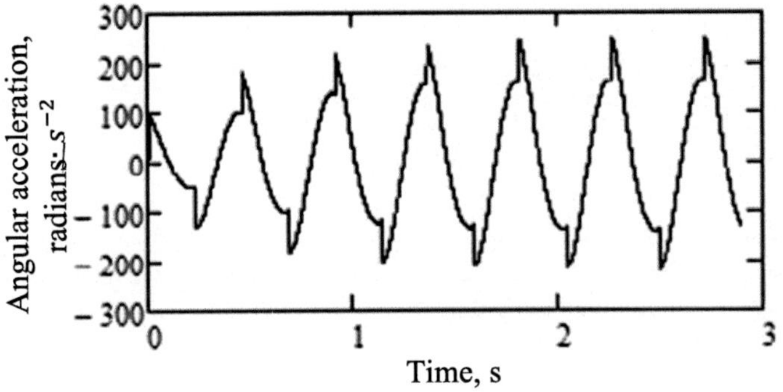

Fig. 8 Generator angular acceleration ε as a function of time t

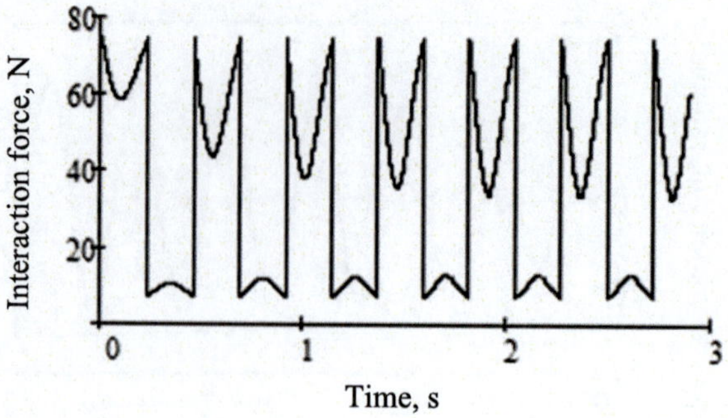

Fig. 9 Wind flow interaction forces Q as a function of time t

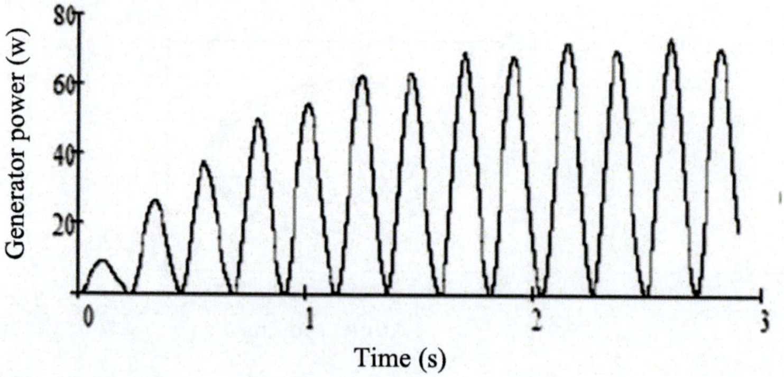

Fig. 10 Instantaneous power of the generator by the formula $b \cdot \omega^2$ (watts)

3 Tail Fin Rotation Motion System with Nonlinear Spring

The model consists of the similar perforated plate (tail fin) with width B and radial length L. The plate has single degree of freedom with angle φ. It can rotate around a fixed z-axis. A nonlinear spring is used in this model as shown in Fig. 11.

According to classical mechanics, the wind interactions force moment Mz against rotation z-axis can be given by Eq. (4):

$$Mz = B \cdot \mathrm{MCP} \cdot \rho \cdot (1 + C) \cdot \int_0^L \left[(V \cdot \sin(\varphi) - \xi \cdot \omega)^2 \cdot \xi \cdot \mathrm{d}\xi \cdot \mathrm{sign}(V \cdot \sin(\varphi) - \xi \cdot \omega)\right] \tag{4}$$

where ξ is the radial coordinate of the small square of the plate, $\mathrm{d}\xi$ is differential of coordinate ξ (Fig. 11). The other parameters in Eq. (4) are the same as in earlier Eqs. (1) and (2). It should be noted here that for low velocities V, the relative velocity

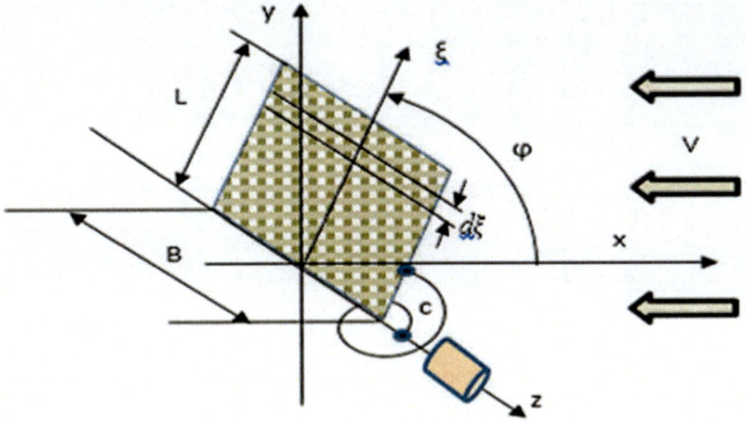

Fig. 11 Vibrating tail fin with nonlinear spring model

direction in Eq. (4) can change and vary from +1 to −1. Taking the positive value, we obtain the Eq. (5).

$$Mz = B \cdot \mathrm{MCP} \cdot \rho \cdot (1 + C) \cdot \left[\frac{L^4 \cdot \omega^2}{4} - \frac{2 \cdot L^3 \cdot V \cdot \sin(\varphi) \cdot \omega}{3} + \frac{L^2 \cdot V^2 \cdot [\sin(\varphi)]^2}{2} \right] \tag{5}$$

Then, in this case, the differential equation of motion of the perforated plate is:

$$Jz \cdot \ddot{\varphi} = Mz - c(\varphi) - b \cdot \omega \tag{6}$$

where $c(\phi)$—is nonlinear torque of the torsion spring. The other values in Eq. (6) are same as mentioned earlier. In this study, a piecewise linear torsion spring was used, which has the following characteristic:

$$-c(\varphi) = -c \cdot (\varphi - \varphi 0) - c \cdot K \cdot (\varphi - \varphi 0) \cdot \left[0.5 + 0.5 \cdot \mathrm{sign}(\varphi - \varphi 0)\right], \tag{7}$$

where K is constant for nonlinearity.

According to Eqs. (6) and (7), the rotational motion of the perforated plate in the phase plane with $K = 10$ is shown in Fig. 12. Similarly, the torque moment curve is given in Fig. 13. The projection value of the relative velocity Vr of the plate edge to normal is:

$$Vr = V \cdot \sin(\varphi) - L \cdot \omega \tag{8}$$

According to the value from Eq. (5) found in the integral in Eq. (4), the projection in Eq. (8) must be positive:

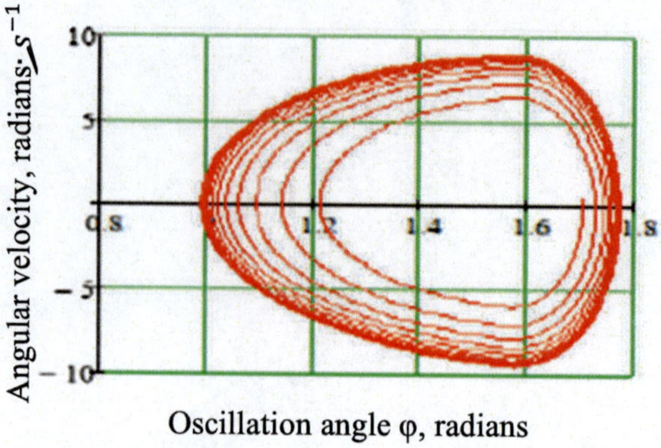

Fig. 12 Rotational motion of the perforated plate in the phase plane with $K = 10$

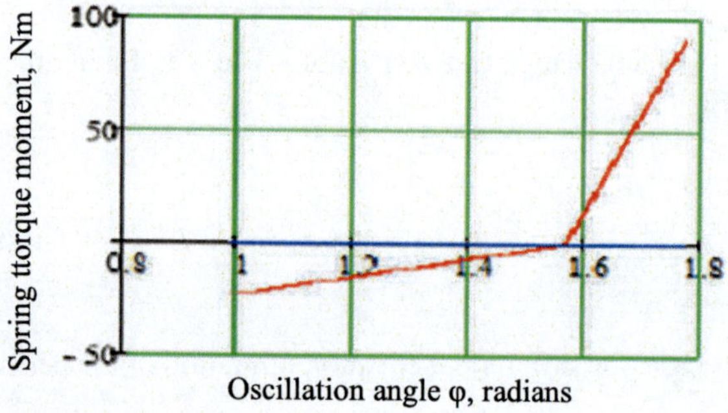

Fig. 13 Nonlinear torque of the torsion spring

$$Vr = V \cdot \sin(\varphi) - L \cdot \omega \geq 0 \tag{9}$$

In this simulation example as in Eq. (9), the inequality rule is satisfied, as is clear from Fig. 14. At the end of the rotational model analysis, a graph showing the instantaneous power of the electric generator is analyzed from (Fig. 15).

4 Robotic Fish Prototype

The current section provides illustrative information about a robotic fish prototype manufactured and tested by the RTU under the FILOSE project. The prototype was created with radio remote control (Fig. 16). Inside the robot is a power unit, a signal capture unit and three actuators: an actuator for tail movement or float speed change; tail steering actuator; a side fin actuator that allows the robot to float or dive.

Initial testing of the prototype took place in a special RTU water tank (Fig. 17).

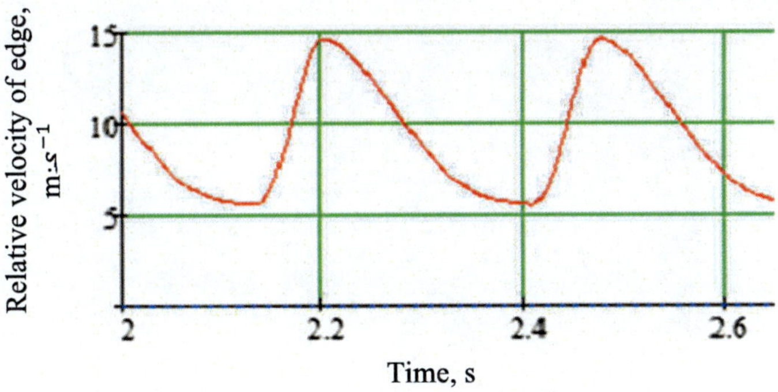

Fig. 14 Relative velocity *Vr* of plate edge as a function of time t

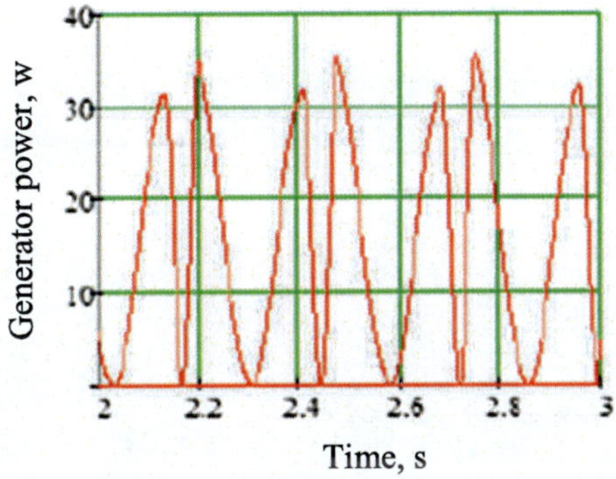

Fig. 15 Electric generator power as a function of time t

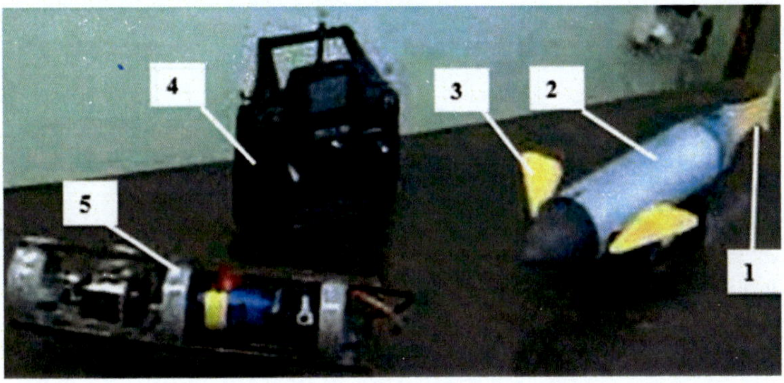

Fig. 16 Robot fish prototype: 1—tail, 2—hull, 3—side fins, 4—remote control, 5—three actuators inside (taken with permission from Tsyfansky et al. [15])

Fig. 17 Water tank at RTU test facility (taken from Tsyfansky et al. [15])

Fig. 18 Robot fish test in Kisezers (Kish lake) (taken from Tsyfansky et al. [15])

Radio remote control options were tested. Maneuvering options such as diving, floating, turning and changing speed could be performed effectively by the robotic fish.

In real environment conditions, the experiments proved the effectiveness of the robot dive and swimming control over long distances, returning the object to its starting position as shown in Fig. 18.

5 Results and Conclusion

Only preliminary experiments were performed without the perforated plate model for the tail fin, and satisfactory results were obtained. It was found out that the robot maneuvers could be performed effectively in shallow waters. The only drawback of theoretical results is that the viscous effects of the fluid medium are not taken into consideration. The following understandable conclusions can be drawn from the above theoretical results: Stable oscillations of the structures like tail fin or wing from a constant wind flow can be induced if the action area of the perforated plate is changed by mechatronics. Stationary oscillations occur very quickly, within a few cycles. The obtained generator's instantaneous power is not large the obvious reason being that the area of the perforated plate is small. To increase the efficiency, the parametric optimization is a must.

From the analysis of the theoretical results, it is possible to realize a practical model of the robotic fish with good power efficiency, and thereby open new doors for design of efficient autonomous underwater vehicles.

Acknowledgements The authors express their gratitude to RTU leading researcher Semyon Cifansky for the constructive solution of the robot fish as well as to the engineer Vladimir Yakushevich for providing experimental work.

References

1. Viba J, Shtals L, Eiduks M (2009) Energy extraction from air or water by vibrations. Latvian J Phys Tech Sci Phys Tech Probl Energetics 46(2):3–12. https://doi.org/10.2478/v10047-009-0005-2
2. Zhan J, Xu B, Wu J, Wu J (2017) Power extraction performance of a semi-activated flapping foil in gusty flow. J Bionic Eng 4:99–110. https://doi.org/10.1016/S1672-6529(16)60381-5
3. Boudis A, Benzaoui A, Ouall H, Guerri O, Bayeul-Laine AC, Coutier Delgosha O (2018) Energy extraction performance improvement of a flapping foil by the use of combined Foil. J Appl Fluid Mech 11(6):1651–1663. https://doi.org/10.29252/jafm.11.06.29099
4. Hoke CM, Young J, Lai JCS (2015) Effects of time-varying camber deformation on Flapping foil propulsion and power extraction. J Fluids Struct 56:152–176. https://doi.org/10.1016/j.jfluidstructs.2015.05.001
5. Karbasian HR, Esfahani JA, Barati E (2015) Simulation of power extraction from tidal currentsby flapping foil hydrokinetic turbines in tandem formation. Renew Energy 81:816–824. https://doi.org/10.1016/j.renene.2015.04.007
6. Xie Y, Lu K, Zhang D (2014) Investigation on energy extraction performance of an oscillating foil with modified flapping motion. Renew Energy 63(C):550—557. https://doi.org/10.1016/j.renene.2013.10.029
7. Liu Z, Lai JCS, Young J, Tian FB (2016) Numerical study on the performance of a flapping Foil power generator with a passively flapping flat plate. In: 20th Australasian fluid mechanics Conference, Perth, Australia
8. Liu W, Xiao Q, Cheng F (2013) A bio-inspired study on tidal energy extraction with flexible flapping wings. Bioinsp Biom 8. https://doi.org/10.1088/1748-3182/8/3/036011
9. Tian FB, Young J, Lai JCS (2013) Improving power-extraction efficiency of a flapping plate From passive deformation to active control. J Fluids Struct 51:384–392. https://doi.org/10.1016/j.jfluidstructs.2014.07.013
10. Usoh CO, Young J., Lai JCS, Ashraf MA (2012) Numerical analysis of a non-profiled plate for flapping wing turbines. In: 18th Australasian fluid mechanics conference, Launceston, Australia, 3–7 December 2012
11. Viba J, Beresnevich V, Tsyfansky S, Kruusmaa M, Fontaine JG, Megill W (2011) New designs of fin type propulsive devices of robotic fish. In: 15th international conference on advanced robotics, Tallinn University of Technology, pp 625–630. https://doi.org/10.1109/ICAR.2011.6088600

12. Tipans I, Viba J, Irbe M, Vutukuru SK (2019) Analysis of non-stationary flow interaction with simple form objects. Agron Res J 17:1227–1234. https://doi.org/10.15159/AR.19.083
13. Vutukuru SK, Viba J, Tipans I, Viksne I, Irbe M (2019) Analysis of flat plate vibrations by varying frontal area to the flow. Eng Rural Dev, 1408–1414. https://doi.org/10.22616/ERDev2019.18.N147
14. Tipans I, Viba J, Vutukuru SK, Irbe M (2019) Vibration analysis of perforated plate in non-stationary motion. Vibroengineering Procedia 25:48–53. https://doi.org/10.21595/vp.2019.20801
15. Tsyfansky S, Viba J, Jakushevich V (2015) Radio controlled fish robot RR-9. J Transp Eng Mech, 98–103. ISSN 1407-8015

Identification, Analysis and Innovative Solution to the Performance Degradation of Thermal System at Advanced Thermovacuum Test Facility

Sandeep Yadav, J. Prijith, Mathew Varghese, V. Murugesan, and S. Syamlal

1 Introduction

Advanced thermovacuum test facility (ATVF) is one of the largest thermovacuum facility (chamber Dia. 5 m) for testing launch vehicle and satellite components at Vikram Sarabhai Space Center (VSSC). Being such a large chamber, the thermal system of ATVF is divided into seven different zones. Each zone is controlled by an independent thermal conditioning unit (TCU) which has the capability of simulating 90–400 K inside the thermovacuum chamber with a spatial temperature difference of 5 K on such large shrouds. To achieve this stringent temperature requirements on the thermal shroud with controlled rate of change of temperature, a large number of control elements, viz., cryo valves, blower, heater, etc. have to be operated within a closed loop control system [1]. At ATVF, the thermal system control is tuned in such a manner that achieved temperature on the shroud follows the computed temperature with the required ramp rate [2].

The main objective of thermovacuum testing is to subject test article to specified temperatures in vacuum, without any violations of upper and lower temperature

S. Yadav (✉) · J. Prijith · M. Varghese · V. Murugesan
QRMIG/SR, Vikram Sarabhai Space Centre, Thiruvananthapuram, India
e-mail: Sandeep_yadav@vssc.gov.in

J. Prijith
e-mail: J_Prijith@vssc.gov.in

M. Varghese
e-mail: Mathew_Varghese@vssc.gov.in

V. Murugesan
e-mail: V_Murugesan@vssc.gov.in

S. Syamlal
Arya Electronics, Thiruvananthapuram, India
e-mail: syamlal.kjr@gmail.com

V. H. Saran and R. K. Misra (eds.), *Advances in Systems Engineering*,
Lecture Notes in Mechanical Engineering,
https://doi.org/10.1007/978-981-15-8025-3_73

limits [3]. Any temperature violation from the specified limits on test article may lead to the rejection of test hardware in post-thermovacuum cycling (TVC) checks and may have further impact on the mission schedule. Therefore, it is very important to carefully control the test article temperature inside the thermovacuum chamber. The temperature on test article in thermovacuum chamber is achieved using thermal shroud, whose temperature is controlled by thermal conditioning unit (TCU). Therefore, any deviation from the expected trends of temperature on thermal shroud is not acceptable in the interest of test article being tested inside thermovacuum chamber. After many years of operation and successful testing of large number of test articles, the thermal system of ATVF showed signs of temperature deviations from the expected trends on shroud. This deviated and degraded thermal system performance is not acceptable as it could lead to temperature limits violations on test article. Therefore, a detailed analysis of thermal system data was carried out to identify the root cause for this problem and the same is being discussed in this paper.

2 System Description

Advanced thermovacuum facility (ATVF) systems are divided into vacuum, thermal and auxiliary systems, which are controlled with programmable logic controller (PLC) in a coordinated manner with necessary safety interlocks. Inside vacuum chamber, thermal shrouds are used as the heat sink/source for controlling the test article temperature. TCUs work as a closed loop control system with gasous nitrogen medium, cryo valves, cold gas generator (CGG), blower and heater as per the schematic as shown in Fig. 1 [2].

The liquid nitrogen is converted into cold gas in CGG which is been further heated using heater to raise the shroud temperature or allowed to go as cold gas to bring the shroud temperature down. The two regulation valves as shown in Fig. 1 are operated by the PLC to maintain the mixed gas temperature going to shroud in accordance with the ramp rate selected for the shroud. ATVF thermal system has seven shrouds but eight TCUs. The eighth TCU is used as spare to take care of any operational TCU failure in the middle of ongoing test. The implementation scheme of spare TCU is shown in Fig. 2.

Here, it can be seen that the shroud to be controlled remains the same but the TCU controlling its temperature is replaced by the spare TCU by operating certain valves. The different TCUs of ATVF are identified with the numbers starting from 310 to 380 for first TCU to last (spare) TCU, respectively. All eight TCUs are independent of each other and are isolated except for the vent line which is common to all. The parameters of any specific TCU will be referenced with the corresponding TCU number. TCU control flow diagram at ATVF is shown in Fig. 3. Here, T_{set} is the set temperature required on shroud and is a user selectable parameter. Based on current shroud temperature (T_{shroud}) and set temperature (T_{set}), the controller will calculate the theoretical temperature (T_{th}), which the TCU will try to attain at T_{GN2} (mixed

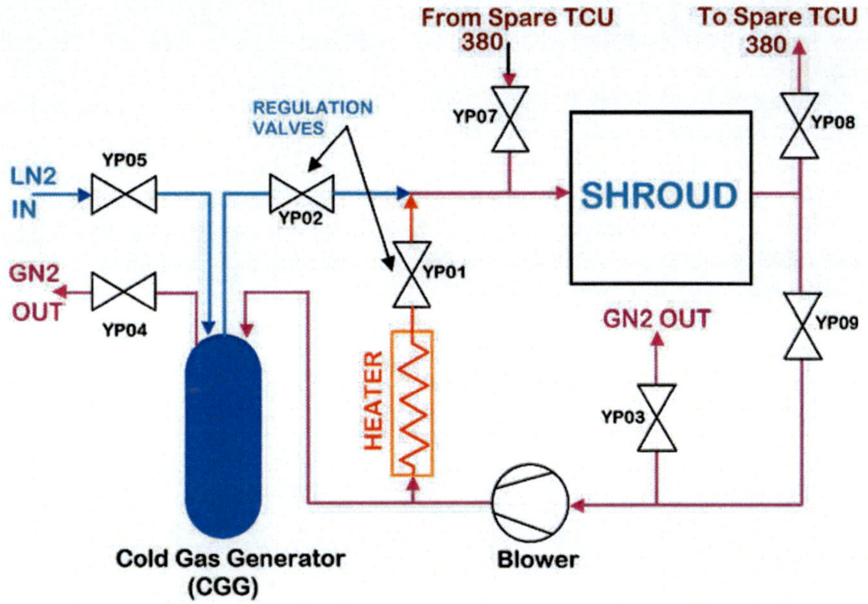

Fig. 1 TCU schematic

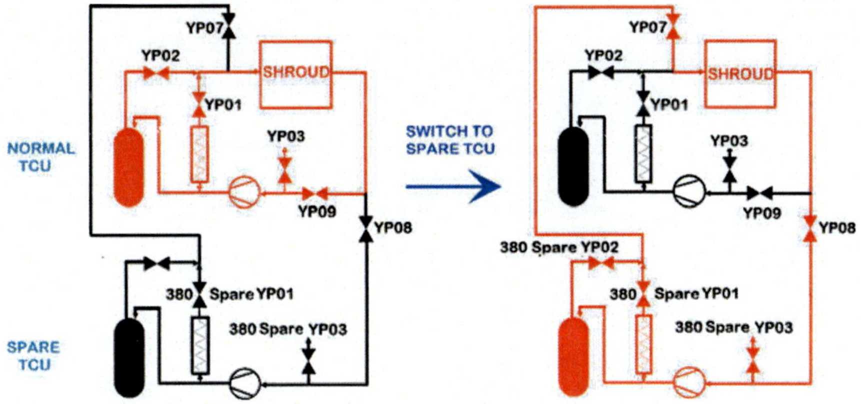

Fig. 2 Spare TCU configuration

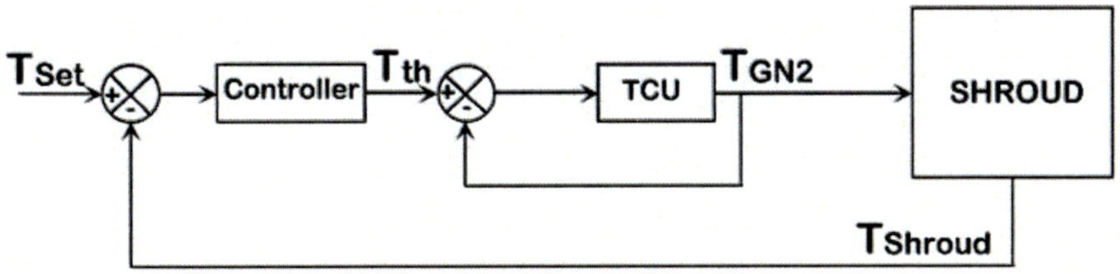

Fig. 3 Control flow diagram for TCUs

gas temperature) using the CGG cold gas, heater, blower and valves as depicted in Fig. 1.

During round the clock thermovacuum cycling tests, the thermal system performance is evaluated by monitoring various parameters. Some of the critical parameters being monitored are given in Table 1 with their expected values.

The expected trends of TCU parameters when its performance is normal is shown in Fig. 4. All these values were tuned by the contractor (M/s AMOS and M/s CSL) of the facility during its installation for the optimum consumption of liquid nitrogen and for required system performance.

Table 1 TCU parameters and their expected values

Parameter	Legend	Expected values
Shroud set temperature	T_{set}	User control
Shroud avg. temperature	T_{avg}	Based on T_{set}
Controller theoretical temperature	T_{th}	As per selected ramp rate
CGG level	L_{LN2}	1300 mm (below 150 K) 100 mm (above 150 k)
CGG pressure	P	7500 mbar(above 150 K), 4000 mbar(below 150 K)
Blower speed	S	20 Hz (min) 160 Hz (max)
Shroud spatial temperature variation	ΔT	5 K (min)

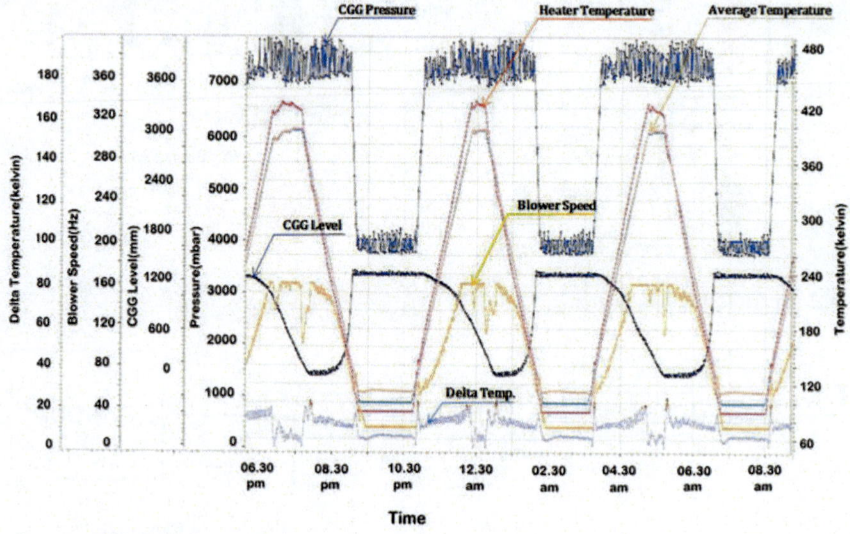

Fig. 4 Nominal trends for TCU 330

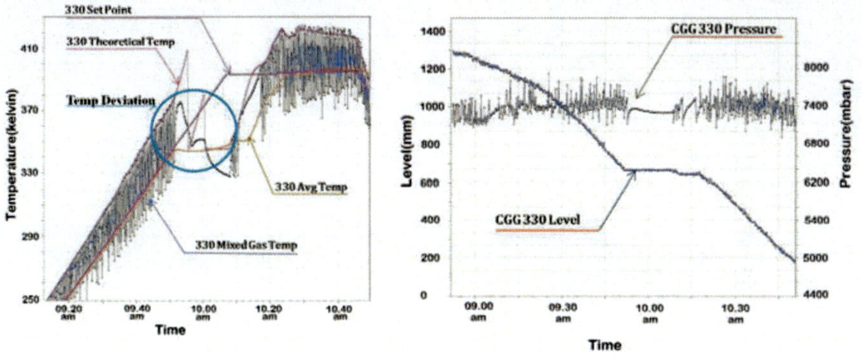

Fig. 5 TCU 330 shroud deviated average temperature, pressure and CGG level

3 Problem Statement

After few years of operation, during one of the test, it was observed that the shroud average temperature (T_{avg}) is deviating from the expected trend and not following user-specified (T_{set}) temperature for TCU 330. At the same time, the expected level and pressure value of TCU also were found to be deviating from their usual values. This appeared to be a control runaway condition with T_{th} going very high and putting heater to a full time use with hot regulation valve in open condition. This could have led to a situation of uncontrolled temperature rise at shroud and might result in a condition which could be dangerous to the test article. The observed trends of such behavior for TCU 330 are shown in Fig. 5. Due to its abnormal performance, TCU 330 was turned OFF and spare TCU (380) was started for continuing with the test. Interestingly, the spare TCU also could not achieve the desired set point and follow the set temperature trend. It was also observed that the pressure trend in the spare TCU is also not as expected. Spare TCU performance trends are shown in Fig. 6.

With the failure of TCU 330 and TCU 380, the entire thermal system of ATVF was put OFF and testing of the test article stopped to analyze and troubleshoot the existing problem.

4 System Performance Analysis

Thermal system of ATVF by design had redundancy in terms of spare TCU against the failure of any other TCU. Failure of both working and spare TCU to meet the functional requirements was cause for concern. To troubleshoot and find the root cause of the problem, the role of each element in the TCU was accessed. From the available plots and data, prima facie the doubtful elements were heater, regulation valves and vent valve. All these elements were tested individually and their performance was found satisfactory. Further, on a detailed analysis of data from other

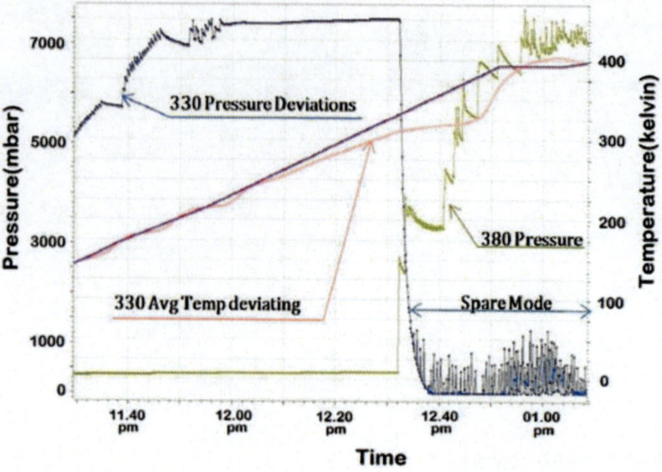

Fig. 6 TCU 380 erroneous pressure and level trends

TCUs, it was found that the required pressure in thermal conditioning units (TCU) is not maintained as expected at the time when thermal system is working. Moreover, the other TCUs which were not in use were also showing a pressure rise to a level more than expected (5 bar) as shown in Fig. 7.

Since the TCUs are isolated except for the vent, the observation led to a possibility of leak between different TCUs. The valve YP08 alone connects the spare TCU to other TCUs as shown in Fig. 2. The status of these valves in SCADA was found to be normal and in closed condition during the test. As a confirmatory test to rule out leak through these valves, GN2 flushing at 7.5 bar for only one TCU was performed. During flushing, pressure buildup signature was observed in other TCUs also which were not selected for flushing. This confirmed the leak path among different TCUs. The thermal system of ATVF is tuned to operate at 7.5 bar to ensure sufficient gas mass flow takes place through the shroud [4]. Any leak from the TCU may lead to a lesser gas mass flow than required, stressing blower, heater and regulation valve to their maximum limits to attain controller set point. Despite maximum efforts put up

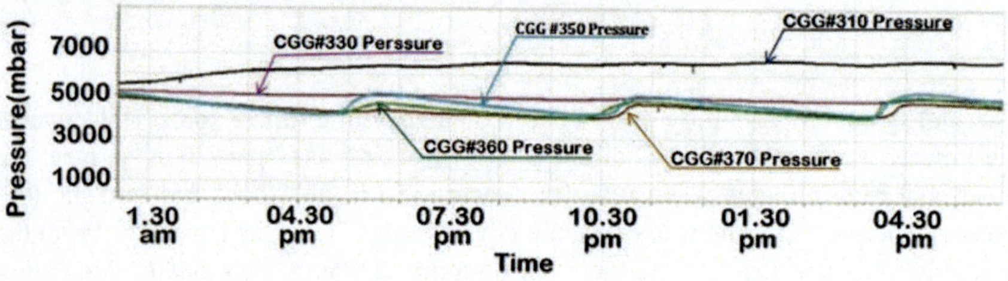

Fig. 7 High pressure levels in unused TCUs

by TCU elements, the controller set point was not getting achieved and a deviation in temperature trend was observed. This leak among TCUs due to valve YP08 was probably the root cause for the observed problem.

5 Solution-Modified System Configuration

Once the probable cause of the problem was identified, the valve in question was analyzed. The valve (YP08) in question is a pneumatically operated cryo valve with flow from under its seat as shown in Fig. 8. The valve as per the manufacturer specification is suitable for working up to a pressure of 10 bar medium [5]. Compressed spring force in valve actuator is designed to keep the valve in NC condition against the 8-bar pressure. The actuator of the valve has two chambers, the pilot pneumatic valve is connected to the bottom chamber so that pressure acts against the spring force to open the valve when commanded. Cryo valve operational conditions are specified in Eq. 1 and 2, respectively, for normally closed (NC) and actuated condition.

$$\left(F_{\text{spring}}\right) > F_{\text{medium at 8 bar}} \tag{1}$$

$$\left(F_{\text{spring}}\right) < \left(F_{\text{compressed air at 7 bar}}\right)_{\text{bottom chamber}} + F_{\text{Flow at 8 bar}} \tag{2}$$

The valve is kept in NC condition with the spring force in the actuator which appears to be insufficient in long run due to aging and continuous usage. This could have resulted in the underperformance of this valve actuator YP08 which is unable to

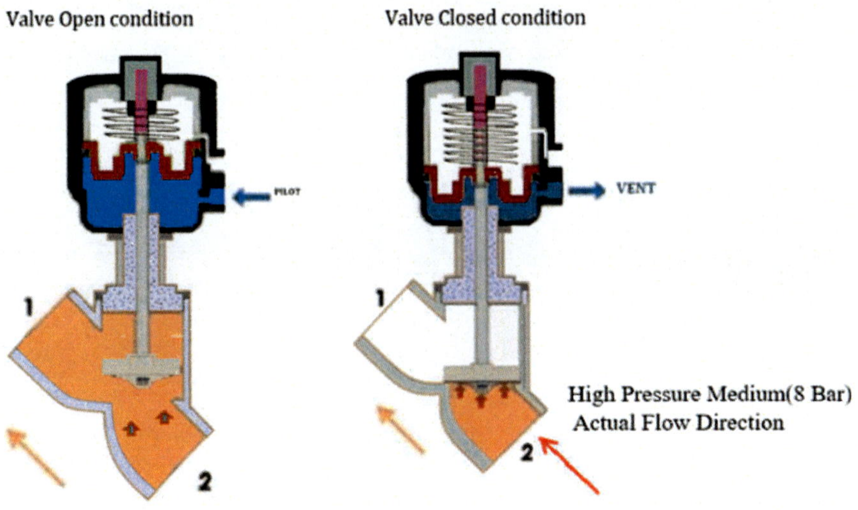

Fig. 8 Actuation scheme of YP08

maintain the valve in a perfectly closed condition, rather when the medium under 8-bar working pressure is applied, and the valve seat gets lifted from its position leading to a leak of GN_2 medium to the other TCUs. Examination of the valve seats revealed that this issue is not attributed to wear and tear. These valves are very essential for the functioning of the TCU and hence the problem associated with them led to a complete shutdown of the thermal system.

By looking at the scheme of pneumatic actuation system of this valve, an innovative solution was evolved for rectifying the existing problem. As per original 3/2 port configuration of the pilot solenoid valve (Fig. 9), port 2 is connected to port 3 in non-actuated condition and to port 1 in actuated condition. Compressed air was used to open the valve based on a solenoid valve command. It was felt that if compressed air is used to assist the actuator spring, then it will keep the valve in closed condition, which otherwise is leaking due to TCU operating pressure. Since, the actuator is with double actuating piston, the solenoid valve ports to actuator chamber are re-assigned so that, a fractional pressure of compressed air through a pressure regulator was fed to the actuator to assist the actuator spring (Fig. 9). In this modified 5/2 scheme, in NC condition, port 1 is connected to port 2, and at the same time, port 4 is vented through port 5. In the actuated condition, port 1 is connected to port 4 and port 2 is vented through port 3. The air pressure in this 5/2 scheme is adjusted to a value so that normal opening of valve is not affected. Modified operational conditions for the cryo valve are specified in Eqs. 3 and 4, respectively, for normally closed (NC) and actuated condition.

$$\left(F_{\text{spring}}\right) + \left(F_{\text{compressed air at 4 bar}}\right)_{\text{Top chamber}} > F_{\text{medium at 8 bar}} \quad (3)$$

$$\left(F_{\text{spring}}\right) < F_{\text{Flow at 8 bar}} + \left(F_{\text{compressed air at 4 bar}}\right)_{\text{Bottom chamber}} \quad (4)$$

The scheme adopted worked well and all leaks between the TCUs were arrested by making these changes in all the TCUs.

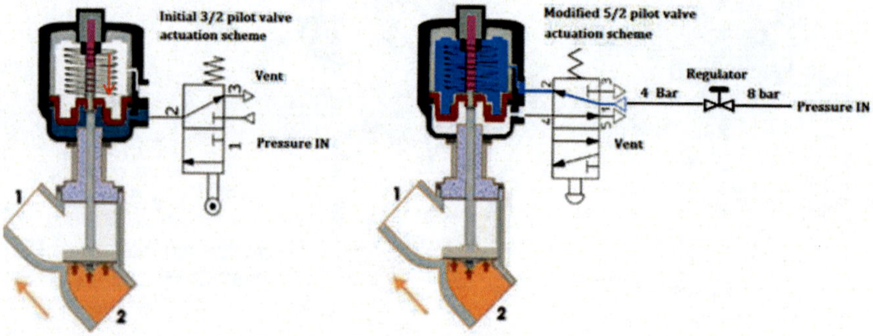

Fig. 9 Modified actuation scheme of YP08

6 Results

After changing the actuation scheme for cryo valve YP08 for all TCUs, once again GN2 flushing of TCU 330 was performed at 7.5 bar. This time no unwanted pressure rise is seen in other TCUs except for TCU 330, as shown in Fig. 10.

The performance trial test of TCU 330 for one cycle from 100 to 358 K was found perfectly alright after this modification. Pressure, CGG level and average temperature on shroud were all as expected, observed trends are shown in Fig. 11.

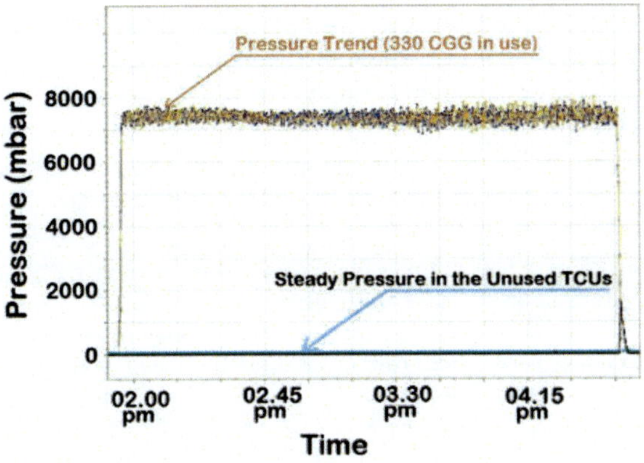

Fig. 10 Steady pressure in unused TCUs

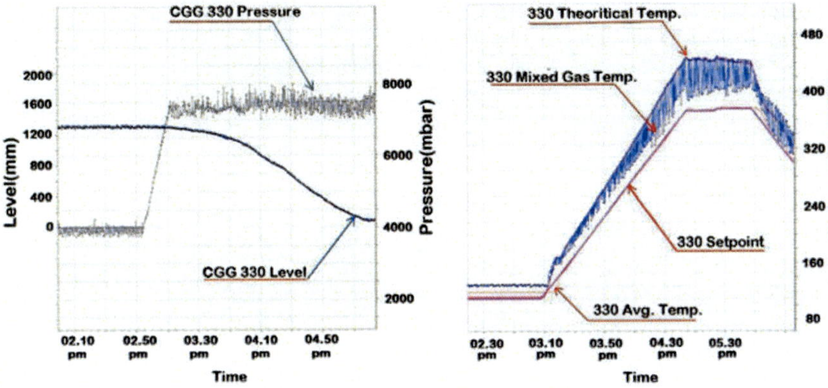

Fig. 11 Normal performance by TCU 330 after YP07 actuation modification

7 Conclusion

Since ATVF is a large and complex facility, where all operations are automated through PLCs and commands are issued through consoles, a knowledge and expertise of different engineering disciplines are required for its operation and maintenance. The problem associated with thermal system as discussed in this paper could have been solved by replacing all the YP08 valves once they are identified as problematic. But changing these valves is a very difficult activity, if not impossible, as the entire multi-layered thermal insulation has to be removed from all the TCU and relaid after valve replacement. Moreover, the stringent test schedule requirement cannot be met. Hence, it was pertinent to address the design deficiency in the thermal system with regard to the choice of valve YP08. Nevertheless, the systematic approach and multidisciplinary approach led to a good engineering solution within the shortest time frame and with no extra cost to resolve the problem observed in the thermal system.

Acknowledgements The authors are sincerely grateful to the support and guidance extended by DD, VSSC (SR) in troubleshooting of the ATVF thermal system. Authors also want to acknowledge the contractors who realized the facility, viz., M/s AMOS, Belgium and CSL, Belgium for their documentation which they have handed over to VSSC at the time of its commissioning.

References

1. Pollara RA (2014) Improved thermal vacuum chamber temperature performance via gaseous nitrogen thermal conditioning units. SSMET 2014
2. ATVF thermal system design document: Unpublished manuscript.
3. Soler Chisabas RS et al (2017) Development of a thermal vacuum chamber for tesing in small satellites. In: 47th intenational conference on environmental systems (ICES-2017–228)
4. Thirumaleshwar M ()1978 Design of a thermal system for thermovacuum chambers. Cryogenıcs, November 1978
5. Operating instructions: Burkert Type 2000, 2002

Determining the Working Behaviour of Hydraulic System Using Support Vector Machine

Jatin Prakash and P. K. Kankar

1 Introduction

The hydraulic systems are one of the most critical systems in the industries because of its better capabilities. The use of hydraulic systems varies from weight lifting using a single-stage hydraulic cylinder to the applications in heavy earth moving equipment [1]. A simple hydraulic system consists of the actuators and the hydraulic fluid flow circuit, whereas a complex system may consist of various working circuits like actuating circuit, cooling circuit, etc. Proper functioning of the cooling circuit is as desired as the functioning of the actuating circuit. Very few research progress has been made in the field of condition monitoring of the cooling circuit of the hydraulic system. Till date, attempts have been made to identify the leakage detection from hydraulic cylinders [2]. Besides, a neural network has been developed for the fault diagnosis of complex fluid power systems and pumping machinery [3, 4]. Observations of various literature suggest the implementation of different machine learning models for the fault categorization and diagnosis [5]. The process of gathering the information contained by the signals acquired in data-driven technique is called the feature extraction. To extract the useful information from the captured signal, it is a crucial step in the condition monitoring of mechanical component or machinery. Feature ranking techniques are espoused in order to indicate the importance of data. Information gain, Fisher score, Wilcoxon ranking, etc., have been used for feature selection for the condition monitoring of rotary bearing [6, 7]. These extracted features are fed to the system so as to monitor the working behaviour of

J. Prakash · P. K. Kankar (✉)
System Dynamics Lab, Discipline of Mechanical Engineering, Indian Institute of Technology Indore, Indore 453552, India
e-mail: pkankar@iiti.ac.in

J. Prakash
e-mail: phd1901103007@iiti.ac.in

V. H. Saran and R. K. Misra (eds.), *Advances in Systems Engineering*, Lecture Notes in Mechanical Engineering,
https://doi.org/10.1007/978-981-15-8025-3_74

mechanical systems. In a recent study, a deep neural network model has been adopted to inspect the condition of the cooling circuit of a hydraulic system [8]. Model-based fault recognition technique and data-based fault detection technique are two of the most widely used fault detection techniques in mechanical machinery. Because of the complexity of the mathematical modelling, model-based fault detection cannot be frequently used, thus, later one is more preferred.

In this work, the support vector machine (SVM) has been employed to supervise the health condition of the hydraulic system. The cooling circuit of the hydraulic system has been kept under investigation using the pressure signals. Four different statistical features have been extracted from the raw pressure signal acquired. All the features have been fed to the classifier as input for the behaviour classification. The correlation and their dependency plot have been used to visualize the data features and relationship among them. To achieve better accuracy, two different methods, namely tenfold cross-validation and grid search cross-validation methods have been employed. The results show better accuracy is achieved in the grid search cross-validation method.

2 Methodology

2.1 Datasets

The predictive maintenance of the mechanical component or machinery can be performed using a historical dataset collected over time. Helwig et al. developed a test rig for the automated condition monitoring of the complex hydraulic system [9]. The whole circuit consists of various components. The data gathered for the pressure variation in the cooling circuit of the developed test rig has been employed in this study. The schematic view of the considered hydraulic system is shown in Fig. 1. 2205 instances of pressure variation data are gathered for 60 s at the sampling frequency of 100 Hz.

The flowing oil is entered in the cooling circuit after passing through a contamination filter and exits to the sink flowing out through a relief valve. The signals obtained from the pressure sensor belong to three dissimilar classes as mentioned

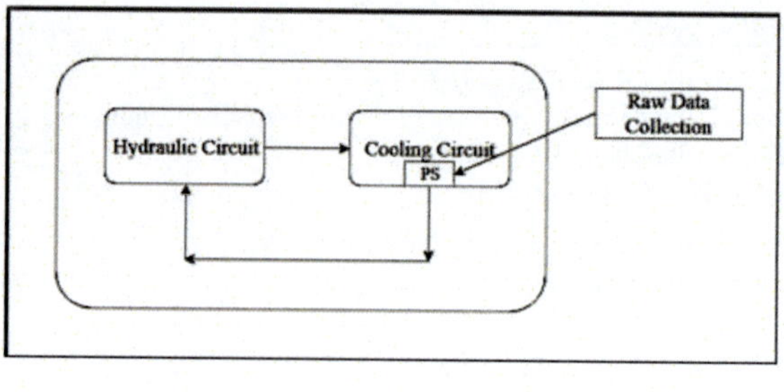

Fig. 1 Block diagram showing the hydraulic and cooling circuit

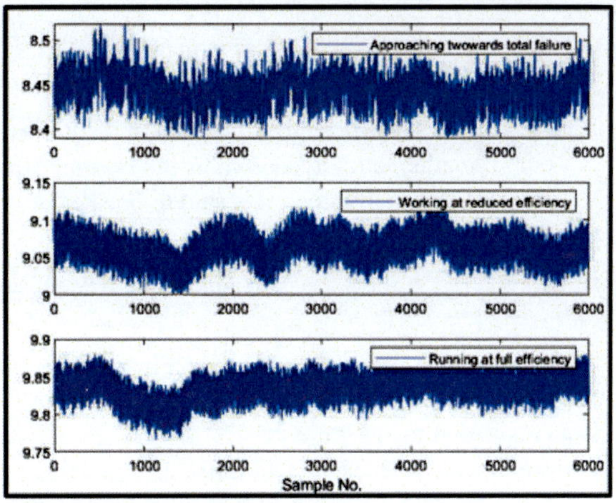

Fig. 2 Sampled signature of pressure variation [8]

by Prakash and Kankar [8]. Figure 2 shows the captured pressure signal with three diverse working conditions.

2.2 *Feature Extraction*

The very initial process to abstract out the useful information from the raw signals involves feature extraction from the signal. As such, there is no mathematical formulation to determine the optimal number of features. Thus, the selection of features to be extracted is of much significance. The features carry useful data about the raw data, but not all features are relevant. In [8], four optimal features have fed into the deep neural network and very good accuracy has been reported. So, in this study, four relevant features explained in brief below have been extracted.

i. Root Mean Square (RMS): It is the square root of the arithmetic mean value of the squares of the signal readings. It is also known as the quadratic mean.

$$\text{RMS} = \sqrt{\frac{1}{n}\left(x_1^2 + x_2^2 + \cdots + x_n^2\right)} \tag{1}$$

ii. Standard Deviation: It is static that measures the distribution of dataset with respect to its mean. It is the square root of the variance.

$$\text{Standard Deviation} = \sqrt{\frac{\sum_{i=1}^{n}(x_i - \overline{x})^2}{n-1}} \tag{2}$$

iii. Kurtosis: Kurtosis is derived from the Greek word *"Kurtos"* and it infers how heavily the tails of distribution differ from the tails of a normal distribution.

$$\text{Kurtosis} = \frac{\sum_{i=1}^{n} \frac{(x_i - \overline{x})}{n}}{\sigma^4} \tag{3}$$

iv. Skewness: Skewness is the distortion or asymmetry in a symmetrical bell-shaped curve, mathematically known as the normal distribution. If the curve is shifted towards either end, then it is said to be skewed.

$$\text{Skewness} = \frac{\sum_{i=1}^{n} (x_i - \overline{x})^3}{n} \tag{4}$$

These four extracted features constitute the feature matrix. The sample feature values used to train the SVM have been mentioned in Table 1. The feature matrix is segregated into the ratio of 4:1 for the training and the testing purpose. Once the training of the classifier is finished, the testing data is fed to it. *k*-fold cross-validation method is employed to eliminate the biaseness accompanied by the grid search method. The grid search method has been explained in Sect. 4.

In this work, tenfold cross-validation (for $k = 10$) is used to visualize how the model performs when a new dataset is fed to it, whereas grid search is a way to choose the best among the family of models, parameterized using a grid of parameters.

Table 1 Sample input feature values for SVM

	RMS	Standard deviation	Kurtosis	Skewness
Feature amplitude	9.185596	0.576802	1.541731	0.144612
	9.185666	0.577158	1.541128	0.142409
	9.185822	0.57733	1.541655	0.141214
	9.186052	0.57726	1.542304	0.140638
	9.186017	0.576846	1.5426	0.142243
	9.185874	0.576423	1.54277	0.143977
	9.185496	0.576332	1.542723	0.145696
	9.174669	0.573275	1.540543	0.151498
	9.174173	0.573551	1.540006	0.153071
	9.173866	0.573793	1.539785	0.153478
	9.173932	0.573795	1.540218	0.153755
	9.175028	0.572664	1.541951	0.15128
	9.174916	0.572707	1.541823	0.150424
	9.187213	0.574811	1.541986	0.148613
	9.187459	0.5747	1.541223	0.148226
	9.187597	0.574699	1.541951	0.147909
	9.187061	0.575101	1.541723	0.148787
	9.187267	0.575085	1.541632	0.14829
	9.187387	0.574991	1.542063	0.148439

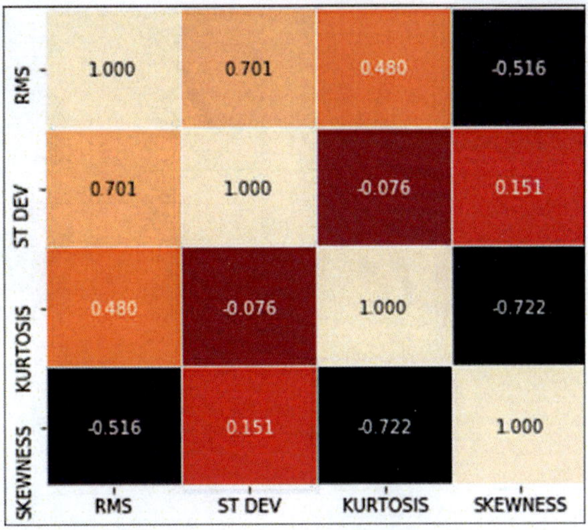

Fig. 3 Correlation matrix for extracted features

2.3 *Correlation Coefficient*

The correlation matrix is a table that shows the correlation coefficient between two variables. It is used to summarize a large amount of data to visualize the patterns and their mutual dependency. In the proposed manuscript, the correlation factor (r) has been calculated for the four different extracted features with respect to each other and themselves also. It is obvious that the data depends on itself fully, so 'r' for itself is equal to unity and represented in diagonal elements. For two given variables, suppose x and y, the correlation coefficient can be calculated using Eq. 1. The correlation matrix of the extracted features is shown in Fig. 3. Higher values in the block show that these particular two properties are highly correlated.

$$r = \frac{n\left(\sum xy\right) - \left(\sum x\right)\left(\sum y\right)}{\sqrt{\left[n\sum x^2 - \left(\sum x\right)^2\right]\left[n\sum y^2 - \left(\sum y\right)^2\right]}} \quad (5)$$

2.4 *Dependency Plot*

Figure 4 shows the plots among the four different features extracted from pressure signals. It can be treated as the data visualization technique. It shows the scattering of feature matrix data with respect to itself as well as with respect to each other feature. The histogram is obtained when distribution of feature vector is obtained with respect to itself. The histograms as shown in Fig. 4 can be generally classified

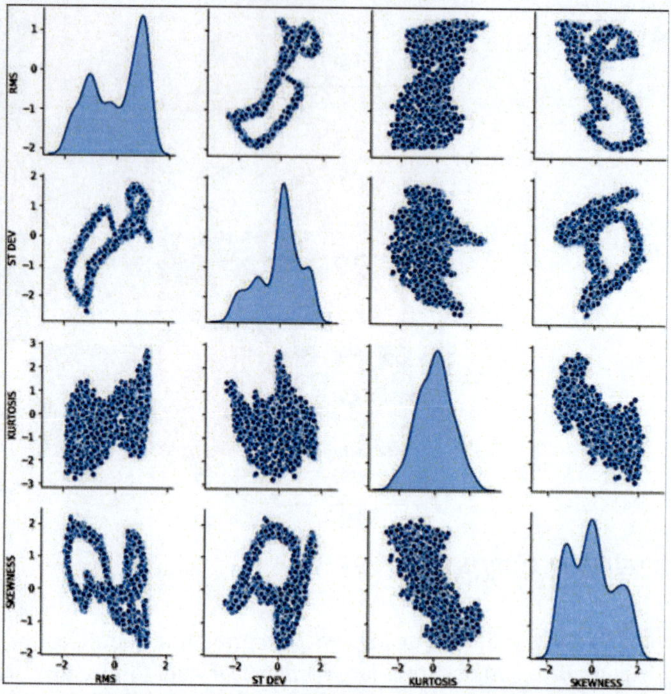

Fig. 4 Dependency plot for four extracted features

into two classes, i.e. unimodal and bimodal histogram. Unimodal histogram infers the spread of data about a single value, whereas the bimodal histogram concludes the spread of data about two peak values. Most of the histograms (except Kurtosis) can be inferred as a bimodal histogram. Rest other corresponding plots show the distribution of the data with respect to another one. The trend observed in dependency plot also well satisfies the interpretations drawn from the correlation matrix in Sect. 2.3.

3 Support Vector Machine (SVM)

SVM is categorized as a supervised machine learning process proved efficient for the classification or regression problems. It exercises an extremely complex data transformation to define an optimal hyperplane between the possible categories of outputs using various kernels. SVM attempts to define a boundary between the two different categories in such a way that the marginal distance between two different classes is maximum, so that the generalization error is minimum. The adjacent data points used to describe the margin are known as support vectors. For example, in a given N number of sample set $\{(x_i, y_i)\}$; $i = 1$ to N. It is required to determine a separation plane among different possible planes that distinct input samples into their

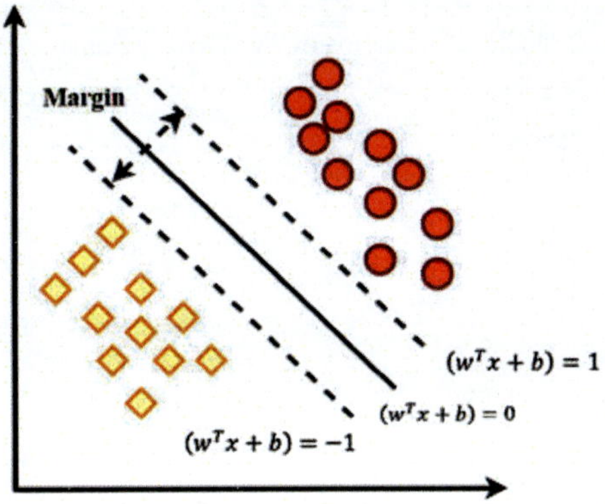

Fig. 5 Representation of hyperplane classifying two-class problem by SVM

classes by least generalization error. For two different classes of data, the data can be labelled as $y_i = 1$ and $y_i = +1$ associated with the respective class. Slack variables (ξ) are considered non-negative for a non-separable type of data. The hyperplane defined by $f(x) = 0$ separating the given data can be attained as a solution using mentioned optimization problem [5].

Minimize $\frac{1}{2}\|w\|^2 + c\sum_{i=1}^{N} \xi_i$.

subjected to $\begin{cases} y_i\left(w^T x_i + b\right) \geq 1 - \xi_i \\ \xi_i \geq 0, i = 1, 2, 3, \ldots N \end{cases}$ where c is the error penalty constant.

Figure 5 shows the two classes of data separated by a hyperplane.

4 Grid Search Method

The grid search method is an approach or process of scanning the data to configure optimal parameters for an artificially intelligent model. It tunes the methodically build model for various parameters and evaluate it with each combination of algorithm parameters stated in a grid. In this study, four different values of error penalty constant (c) in combination with two different kernels have been employed. Nine discrete values of 'gamma' have been also used in the grids. Table 2 mentions the combination of different SVM hyperparameters used in the grid search method.

The higher value of the penalty parameter (C) tries to minimize the misclassification of the training data, whereas the lower value of penalty parameter attempts to maintain a smooth classification. At the same time, if the value of the penalty parameter is set to be very high to correctly classify all the data, there is always a higher chance of overfitting in the model. So, in this study, the maximum value of C has been kept at 1000.

Gamma (γ) is the inverse of the standard deviation of the Gaussian function that controls the trade-off between error due to bias and variance in the SVM model. For large values of γ, support vector does not have much impact on the classification of data.

Table 2 Combination of SVM hyperparameters for grid search

Kernel	*C* (Penalty parameter constant)
Linear	1 10 100 1000
Rbf	1 10 100 1000
Gamma	0.1, 0.2, 0.3, 0.4, 0.5, 0.6, 0.7, 0.8, 0.9

5 Results and Discussion

5.1 Confusion Matrix

A confusion matrix often called as an error matrix is generally used to designate the attainment of a classification model on a particular set of data for which the actual values are known. In the ongoing study, support vector classifier parameters are set as penalty parameter, $C = 1.0$, with '*rbf*' kernel and the value of Υ to be auto-deprecated. This classifier is used for the test data (20% of the total data, 441 in number) and the accuracy achieved is 95.91%. The corresponding confusion matrix is publicized in Fig. 6.

Further, to enhance the accuracy of the SVM classifier, tuning of the hyperparameters is required. In this work, tenfold cross-validation followed by grid search method has been performed. The cross-validation employing grid search finds out

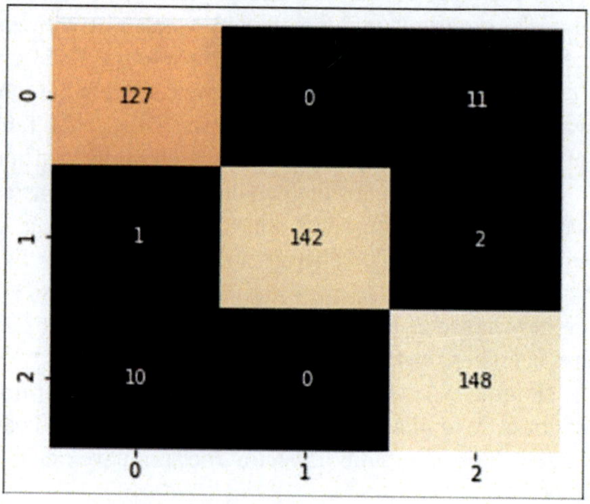

Fig. 6 Confusion matrix for support vector classifier

Table 3 Combination of the best hyperparameters according to grid search method

Hyperparameter	Value
Penalty parameter (C)	1000
Gamma	0.6
Kernel	Rbf

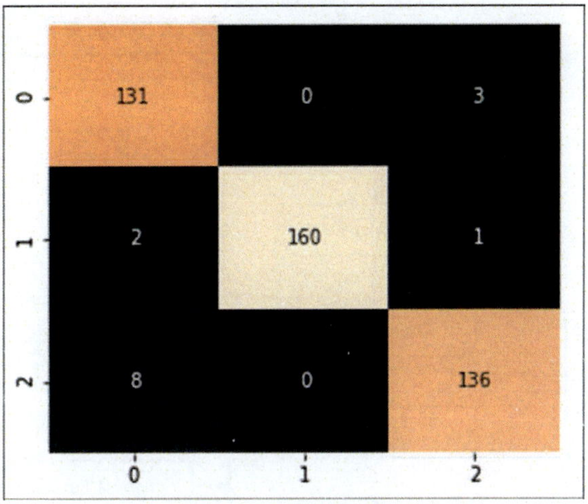

Fig. 7 Confusion matrix with the best parameters

the best parameters for the better accuracy of the classifier by applying the various parameters to the classifier in one grid. The best parameters for this data in the support vector machine as determined by using grid search method are mentioned in Table 3. The combination of these parameters yields the accuracy of 98.80%. The corresponding confusion matrix is shown in Fig. 7. The flowchart of the methodology adapted is shown in Fig. 8.

6 Conclusions

In this study, the following conclusions and observations can be drawn:

- Working behaviour of the hydraulic cooling circuit has been determined using pressure signature employing support vector machine (SVM).
- The correlation matrix and the dependency plots have been shown in a good argument with each other.
- The grid search method can successfully be employed to determine the best hyperparameters combination.
- The accuracy has been boosted to 98.80% from 95.91% employing the best parameters as determined by grid search method.

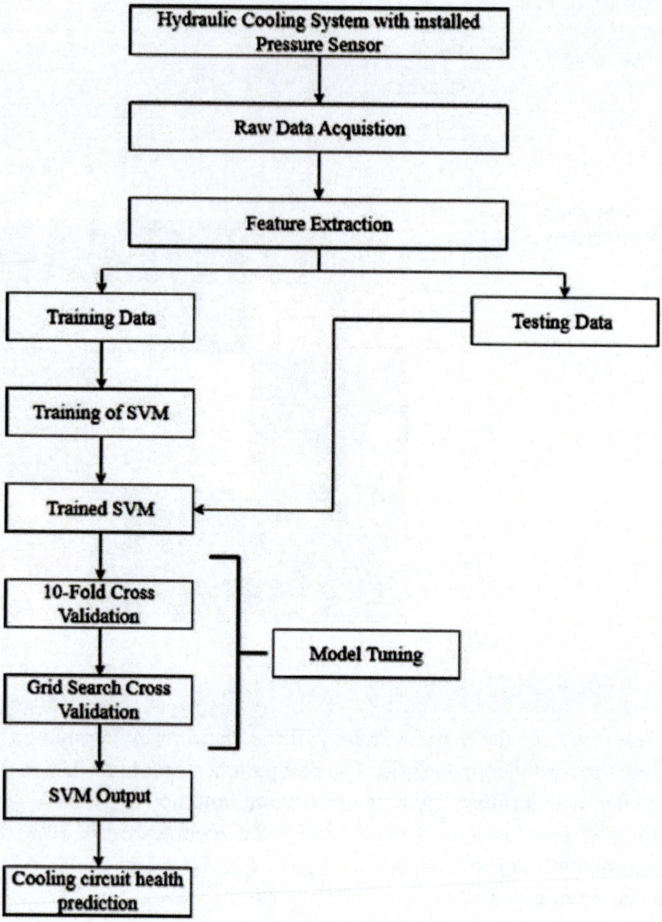

Fig. 8 Flowchart of cooling circuit condition monitoring

- The best accuracy of the developed model by Prakash and Kankar [8] using feature ranking technique is 99.54% which is not much less than that achieved in this study. Thus, it can also be concluded that feature ranking plays an important role in boosting the accuracy of the machine learning models.

References

1. Prakash J, Gupta SK, Kankar PK (2020) An analytical approach to evaluate the maximum load carrying capacity for pin-mounted telescopic hydraulic cylinder. Proc Inst Mech Eng Part C J Mech Eng Sci, 0954406220916524

2. Yao J, Yang G, Ma D (2014) Internal leakage fault detection and tolerant control of single-rod hydraulic actuators. Math Probl Eng
3. Ramdén T, Krus P, Palmberg JO (1995) Fault diagnosis of complex fluid power systems using neural networks. In: Proceedings of the fourth Scandinavian international conference on fluid power, Tampere, Finland, pp 26–29
4. Ilott PW, Griffiths AJ (1997) Fault diagnosis of pumping machinery using artificial neural networks. Proc Inst Mech Eng Part E J Process Mech Eng 211(3):185–194
5. Vakharia V, Gupta VK, Kankar PK (2015) Ball bearing fault diagnosis using supervised and unsupervised machine learning methods. Int J Acoust Vibr 20(4):244–250
6. Upadhyay R, Manglick A, Reddy DK, Padhy PK, Kankar PK (2015) Channel optimization and nonlinear feature extraction for Electroencephalogram signals classification. Comput Electr Eng 45:222–234
7. Vakharia V, Gupta VK, Kankar PK (2016) Bearing fault diagnosis using feature ranking methods and fault identification algorithms. Procedia Eng 144:343–350
8. Prakash J, Kankar PK (2019) Health prediction of hydraulic cooling circuit using deep neural network with ensemble feature ranking technique. Measurement, 107225
9. Helwig N, Pignanelli E, Schütze A (2015) Condition monitoring of a complex hydraulic system using multivariate statistics. In: 2015 IEEE international instrumentation and measurement technology conference (I2MTC) proceedings. IEEE, pp 210–215

Design of Automatic Bolt Washer Assembly

Irwin Osmond Toppo, Rishu Sharma, and Bhajneet Singh

1 Introduction

Study is done for the automotive industry where thousands of bolts of different sizes are being used. In automobile manufacturing industry, it is being observed that to manufacture a single truck, nearly 1500 bolt of different sizes are being used with washer in it. The assembly of bolt and washer is done by the worker manually which creates hindrance in the work flow. Bolt and washer are being brought to the respective shop floor separately where the worker working in the shop floor assembles the bolt and washer. On an average, to assemble bolt with the correct number of washers in, it takes nearly 6 s to the worker. Extra number of bolts to the washer causes manufacturing defect in production line.

Following are the design consideration for designing the prototype

1. The correct number of the washer in every bolt.
2. Proper use of space near the workplace.
3. To avoid wastage of washer.
4. To put the washer in specified bolt only.

I. O. Toppo (✉) · R. Sharma · B. Singh
GB Pant Government Engineering College, New Delhi, India
e-mail: irwinosmond@gbpec.edu.in

R. Sharma
e-mail: rishusharma@gbpec.edu.in

B. Singh
e-mail: bsingh@gbpec.edu.in

V. H. Saran and R. K. Misra (eds.), *Advances in Systems Engineering*,
Lecture Notes in Mechanical Engineering,
https://doi.org/10.1007/978-981-15-8025-3_75

By taking the example of automobile industry, it was observed that considerable man hours will be saved by using this automatic assembly prototype.

2 Methodology

The prototype arranges the specified number of washers to each nut with the use of electric circuit and pneumatic system. The various components of the prototype are two parallel conveyors, four (4) non-pneumatic cylinders, electric motor, one insulating disc with conducting wires attached along the radius in it at a specified distance to set the timing of actuation. The assembly of washer and bolt is done in four stages (Fig. 1).

1. Aligning the washer one over another in a stack.
2. Allowing the correct number of the washer to fall in the bolt.
3. The pneumatic system mechanism works to maintain correct order and correct time.
4. Aligning the bolt in order that washer fall directly on it.

The schematic of the arrangement shows that the movement of the four plates is controlled by pneumatic system. The distance between two plates is equal to total number of washer required to be put on the bolt provided thickness of the single washer is known. The movement of plates is governed by the plunger of the pneumatic system. The sequence and timing of operation of particular plunger are controlled by specially designed electric circuit (Fig. 2).

2.1 *Time Calculation for a Cycle*

2.1.1 Assumption

No air resistance, considering small distance covered by washer to bolt with initial speed = 0 m/s^2
g (gravity) = 9.8 m/s^2
Number of washer assumed for calculation = 4.
Let the thickness of single washer = T (m)
Therefore, total thickness = $4T$ (m)

For time calculation (2 steps)
Step 1: Leaving of washer from plate 2 (bottom most)
Step 2: Dropping down of washer from stack, from plate 1 (top most) to plate 2

Distance between plate 1 and plate 2 = $4T$
Step1: Leaving of washer from plate 2 to fall in bolt.

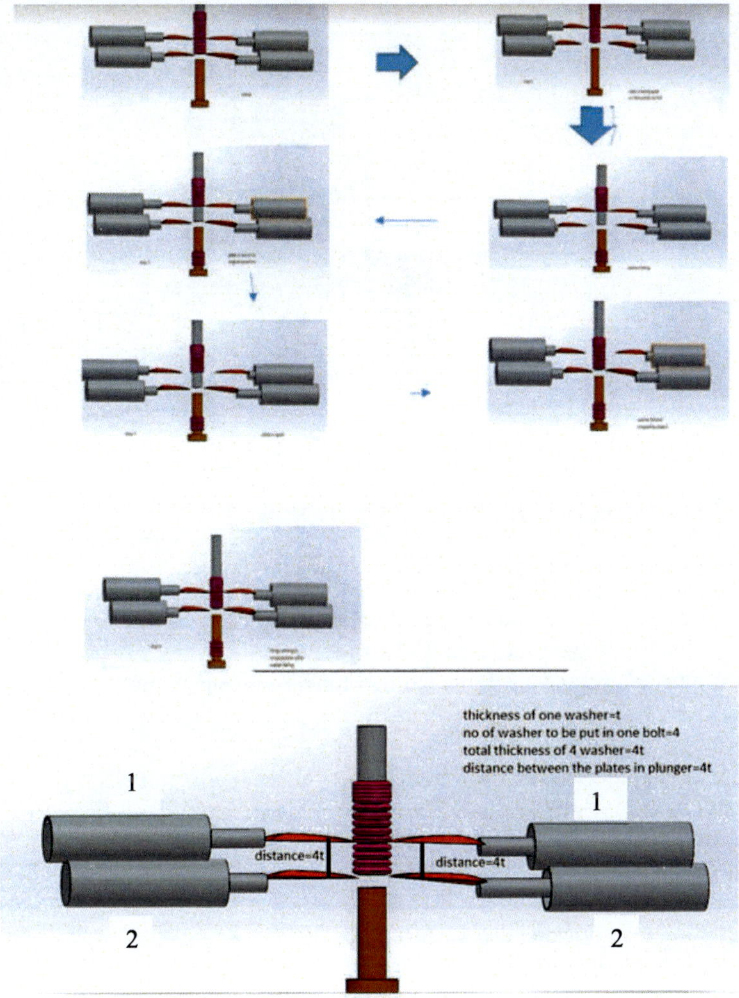

Fig. 1 Schematic of pneumatic system

Distance (S) travelled by washer to fall on bolt $= 4T$

$S = (u * t) + (1/2) * (a * t^2)$, t is time (s), a is acceleration due to gravity, $u = 0$ (m/s^2)

$4T = (1/2)\,(a * t^2)$

$t = \sqrt{8T/a}$

$t = \sqrt{8T/9.8}$

$t = (0.9035 * \sqrt{T})$ s.

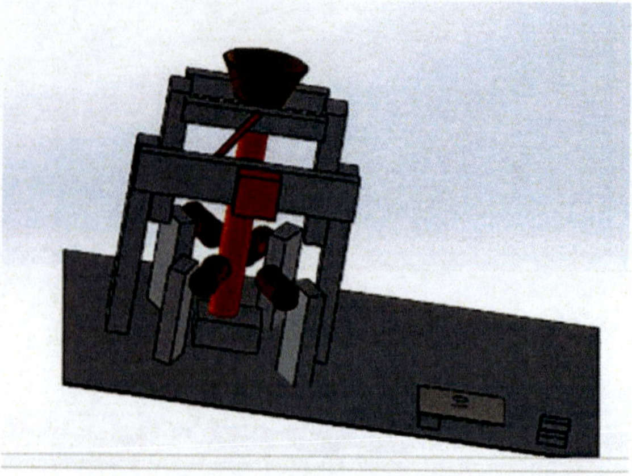

Fig. 2 Schematic of prototype

2.2 *Time Calculation Plunger Operation*

Let the time taken = 0.25 s (experimental value referred from manual)
For 1 cycle, number of plunger movement = 4
Total time for plunger movement = 1 s (4 * 0.25)
Total time taken for 1 cycle including plunger movement and falling of washer time
$t = (1 + 1.87\sqrt{T})$ seconds.
t = minimum time taken for 1 complete cycle, i.e., putting of washer in one bolt.
Output rate
(Maximum and ideal considering all kinds of friction) = 3600/ $(1 + 1.87\sqrt{T})$ bolts/h.
Let the thickness of washer = T (m)
Therefore, total thickness = $4T$ (m)
The time to compete the cycle is approx. $t = \sqrt{8T/a}$ (s)
Based on the calculated time, the plunger operation time is approximately 3600/$(1 + 1.87\sqrt{T})$ Bolts/h (Fig. 3).

2.3 *Movement of Pneumatic System in Correct Time and Correct Order*

In order to control the movement of plunger, special mechanism is designed. The mechanism consists of insulating disk, electric motor, and copper wire. The copper wire is attached to the insulating disk along the radius. The insulating disk is rotated by the motor at specified RPM; the four wires which are attached to the disk along the radius act as switches to pneumatic system. Movement of pneumatic system:

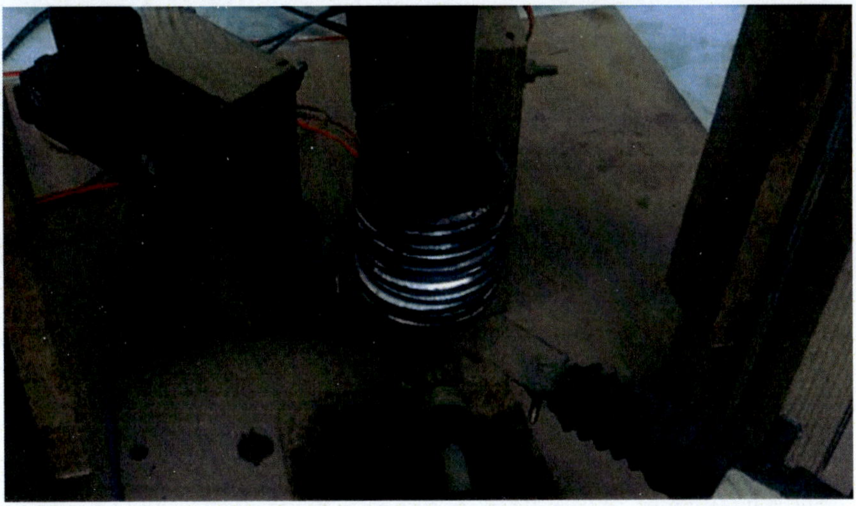

Fig. 3 Arrangement of washer and bolt

Plate-2 move apart
Plate-2 move in original location
Plate-1 move apart
Plate-1 move in original location

Reference: Plate-2 moving at a particular (Θ).
Step 1: Disengagement of plate-2 and falling of washer from plate-2 to bolt.
Total time taken = ($0.25\ \text{s} + 0.9035\sqrt{T}$) seconds before circuit 2 is activated.

In ($1 + 1.87\sqrt{T}$) seconds plate moves to 360° (Fig. 4).
In 1 s, plate moves to $360/1 + 1.87°\sqrt{T}$

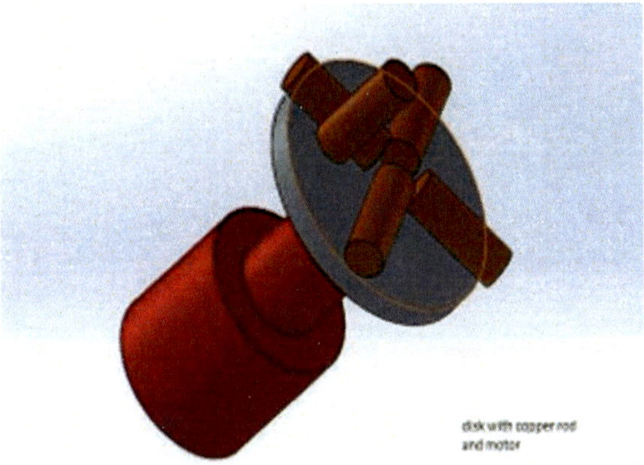

Fig. 4 Schematic for electric circuit

In $(0.25 + 0.9035\sqrt{T})$ second plate moves to: $360 * (0.25 + 0.9035\sqrt{T}) \Big/ (1 + 1.87\sqrt{T})°$

Θ_2 for wire $2 = 360 * (0.25 + 0.9035\sqrt{T}) \Big/ (1 + 1.87\sqrt{T})°$

Θ_3 for wire 3: Total time taken for plate-2 to come to initial condition $= 0.25$ s

$\Theta_3 = (\Theta_2 + 360 * 0.25/1 + 1.87\sqrt{T})$ (from reference point)

Θ_4 for wire 4: Total time taken $=$ movement plate 1 $+$ falling of washer from plate-1 to plate-2

$\Theta_4 = \Theta_3 + (360 * 0.25 + 0.9035\sqrt{T}) = 1/1.87\sqrt{T}$.

2.4 RPM of Motor (for Plunger Movement)

For completion of 1 cycle, time taken $= (1 + 1.87\sqrt{T})$ s.

For completion of 1 cycle, the insulating plate has to do 1 rotation; Total time taken for 1 rotation $= (1 + 1.87\sqrt{T})$ s. One complete rotation of plate takes place in $(1 + 1.87\sqrt{T})$ s. In 1 s $(1/\,(1 + 1.87\sqrt{T})$ rotation. In 60 s $= 60/\,(1 + 1.84\sqrt{T})$ rotation. RPM of insulating plate motor $= (60/\,(1 + 1.87\sqrt{T}))$ RPM.

3 Discussion

The time studies done on the prototype predict considerable reduction in cycle time-saving men hours along with reduction in wastage of washers and avoiding the manufacturing defects. The prototype put correct desired number of washers to each bolt to the washers. The total space occupied by the whole system is less compared to required bolt and washer assembly area. Analytical equation derived for calculating various timing is validated with the prototype.

4 Conclusion

The prototype designed is suitable for all the industries were bolts and washers are used. The prototype can be scaled for actual industrial application. The prototype reduces time and cost of cycle efficiency. Instead of solenoid piston, pneumatic system can be utilized to decrease the cycle time. The material of the plate attached to the piston could be changed to a flexible material with high stiffness to improve the efficiency of the prototype.

Solar Energy Systems

Sparsely Connected DenseNet for Malaria Parasite Detection

Soniya, Sandeep Paul, and Lotika Singh

1 Introduction

Malaria is a perniciousness diseases. According to World Health Organization (WHO), 219 million malaria cases and 435,000 malaria deaths were reported worldwide in 2017 [12]. It is caused by transfer of plasmodium parasite organism of an infected female Anopheles mosquito. Identification of the malaria disease can be done by one of the tests like thick and thin blood smears, rapid diagnostic test, molecular test, antibody test, drug resistance test, and blood test. For each test, a manual process is carried out by clinicians which required a huge effort and expertise of the domain. Machine learning techniques can be used to automate this process and to assist the clinicians which may fasten the manual process of identification and can be used for societal benefits specifically in outreach areas.

The work reported in [7] had showed that as the depth of network increases the performance of network improves. But with large number of layers, the problem of degradation occurs with no further significant improvements in performance. Highway nets [9], residual nets [1], and densely connected convolutional network [3] had efficiently solved this problem with the help of skip connections which allow a smooth flow of gradients in backward direction and provide an efficient training of deeper network. Densely connected convolutional neural network (DenseNet) [3] is considered as an efficient deep network due to a unique connectivity pattern and high performance. The network has the significant feature of the concatenation of previous

Soniya · S. Paul (✉) · L. Singh (✉)
Dayalbagh Educational Institute, Agra, India
e-mail: spaul@dei.ac.in

L. Singh
e-mail: lotikasingh@dei.ac.in

Soniya
e-mail: dei.soniya@gmail.com

V. H. Saran and R. K. Misra (eds.), *Advances in Systems Engineering*,
Lecture Notes in Mechanical Engineering,
https://doi.org/10.1007/978-981-15-8025-3_76

layer information and passes into the subsequent layers in contrast to the summation of information done in other popular models like residual networks (ResNets) [1]. Specifically, for DenseNet, it is observed that not all nodes are equally significant and the feature maps generated by each node may not contribute equally to the desired task such as classification. This gives the motivation of developing the strategy which provides the reduced architecture and hence a simplified topology for otherwise extremely complex DenseNet. Recent reported studies [2, 10, 13] had explored the design of DenseNet in order to improve performance of the network. Work done in [2, 13] explored different aggregation topologies to design sparse network. In contrast to [2, 13], the report in [10] studied the effect of channel-wise dropout, layer-wise dropout, and unit-wise dropout on DenseNet in order to reduce overfitting. This paper presents a novel approach to design simpler topologies of DenseNet by introducing sparsity in the structure with the help of DropConnect method. The rest of the paper is organized as follows: Sect. 2 gives a detailed description of the proposed approach; Sect. 3 discusses the implementation platform details, the experiments and results are discussed in Sect. 4; and Sect. 5 concludes the paper.

2 The Proposed Approach

DenseNet [3] is considered as one of the efficient high end convolutional neural networks which consists of multiple blocks and within each block, each node has collective information of previous nodes. With the help of skip and sequential connections, a node combines information of all previous nodes via concatenation operation. Since each node in the block performs convolutional operation without reducing the size of incoming feature maps, the block is also known as normal cell. Each of the blocks is followed by a transition block. The transition block (also called as reduction cell) is responsible for reducing the spatial dimension of feature maps with the help of convolutional and pooling operations.

DropConnect technique [11] has been integrated with dense network for generating sparse connections. In dropout [8], to prevent co-adaption between the neurons, few neurons of a fully connected network are dropped randomly with some threshold value. In DropConnect, instead of dropping a neuron, few incoming signals of the neuron are dropped randomly. In this work, the DropConnect technique has been integrated with the DenseNet at node level in order to combine the feature maps from all preceding nodes. This helps in generating sparsity in weight connections of a block.

Let f_k denote the feature map of the kth node, node k can combine the feature maps of all previous $(k-1)$ nodes as described in Algorithm 1. In the algorithm, rand(0,1) denotes a random number generated from uniform distribution and pth represents a predefined threshold value. The term CFM denotes the combined feature maps and $S(\cdot)$ represents activation function. In order to receive the feature maps at node k from any of the $(k-1)$th nodes, a random number is generated and if this number is greater than equal to predefined threshold value pth, then only feature maps of the $(k-1)$th node will be considered and play a pivotal role towards the collective information of node *k*. The procedure repeats for each node of all blocks of a network. In the end, the network will consist of selective feature maps of each node only. For the extreme value of *p*th as 0, the proposed approach gives a complete DenseNet without any sparsity in skip connections of a block. However, for another extreme value of *p*th as 1, the network would consist of sequential connections only. For intermediate value of *p*th, variety of network structures can be designed with different levels of sparsity. The proposed study was validated on malaria parasite detection dataset as discussed in the next section.

Algorithm 1: Pseudocode of the proposed approach

```
Define p_th : predefined threshold value
for each node k in block b do
    Define C_FM = {}
    for i = 1:{k-1} do
        if rand(0,1) ≥ pth
            C_FM.concatenate(f_i)
        endif
    end for
    f_k = S(C_FM)
end for
```

3 Experiments

The development of the proposed approach was done using PyTorch 1.0.0 [5] on HP Z8 G4 workstation. The workstation has 128 GB RAM and NVIDIA Quadro GP100 GPU with CUDA toolkit 9.0.

An extensive set of experiments was performed to understand the behavior of dense network with different values of *p*th as shown in Table 1. In this study, the malaria parasite detection dataset [6] was used. The dataset has a total of 27,558 colored images of cells of size 100 × 100 which were divided into five-folds. The aim was to identify the normal cells from infected parasite cells. For a particular value of *p*th, an individual network was trained for each fold. Due to nature of the dataset, performance of network was evaluated in terms of accuracy (total number of correctly classified parasitized and normal cells), sensitivity (percentage of parasitized cells that are classified as parasitized), specificity (percentage of normal cells that are identified as normal), F1-Score (measure test's accuracy), and MCC (correlation coefficients between the desired and observed binary classifications). With the help of true positive (Tp), true negative (Tn), false positive (Fp), and false negative (Fn), the above-mentioned parameters can be defined as follows.

Table 1 Effect of *pth* on malaria parasite detection dataset

pth	Fold	Parameters (K)	Accuracy	Sensitivity	Specificity	F1-Score	MCC
0.1	F1	474	0.951949	0.9375	0.9595	0.9485	0.8969
	F2	476	0.963735	0.9410	0.9857	0.9623	0.9283
	F3	472	0.964609	0.9500	0.9786	0.9634	0.9295
	F4	476	0.942750	0.9477	0.9455	0.9450	0.8930
	F5	476	0.959202	0.9483	0.9690	0.9565	0.9182
Average			0.955775	0.9449	0.96766	0.95514	0.91318
0.9	F1	398	0.953762	0.9179	0.9742	0.9449	0.8928
	F2	399	0.961922	0.9483	0.9715	0.9590	0.9204
	F3	402	0.963702	0.9519	0.9751	0.9625	0.9276
	F4	396	0.948369	0.9383	0.9578	0.9463	0.8967
	F5	396	0.951042	0.9291	0.9707	0.9473	0.9022
Average			0.955759	0.9371	0.96986	0.9520	0.90794

$$\text{Sensitivity} = \text{Tp}/(\text{Tp} + \text{Fn});\ \ \text{Sensitivity} = \text{Tn}/(\text{Tn} + \text{Fp});$$

$$\text{Accuracy} = (\text{Tp} + \text{Tn})/(\text{Tp} + \text{Tn} + \text{Fp} + \text{Fn});$$
$$\text{F1 - Score} = 2\text{Tp}/(2\text{Tp} + \text{Fp} + \text{Fn})$$

$$\text{MCC} = ((\text{Tp} \cdot \text{Tn}) - (\text{Fp} \cdot \text{Fn}))/\text{sqrt}((\text{Tp} + \text{Fp})(\text{Tp} + \text{Fn})(\text{Tn} + \text{Fp})(\text{Tn} + \text{Fn}))$$

Tp shows the number of parasitized cells identified as parasitized whereas Tn is the number of normal or uninfected cells identified as normal. Similarly, Fp denotes the number of normal cells which are identified as parasitized cells and Fn is the number of parasitized cells identified as normal. The used dense network has two blocks and each block has ten nodes. All nodes in a block are connected via sequential as well as skip connections. In a block, each node performs convolutional operation with receptive field of size 3 × 3, stride of 1, and padding of 1. The transition block consists of one convolutional layer followed by a pooling layer. The convolution layer used receptive field of size 3 × 3 with stride of 2 and pooling layer performed max pooling operation with receptive field of size 2 × 2 and stride of 2. The training of network was done by the Adam learning law [4] with batch size of 50 with learning rate of 0.0001 for 50 epochs.

Initially, experiments were conducted without introducing *pth* in the dense network. For each fold, a network was trained. On the average, the test accuracy of 0.949701 was achieved with 485 K weight parameters.

Next, experiments were carried out with different values of *pth* as shown in Table 1. In order to clearly show the effect of *pth*, experiments were conducted with extreme values as 0.1 and 0.9. With respect to fold 1, for *pth* = 0.1, the obtained network shown an accuracy of 0.951949 with 474 K weights parameters. On the other hand, with *pth*

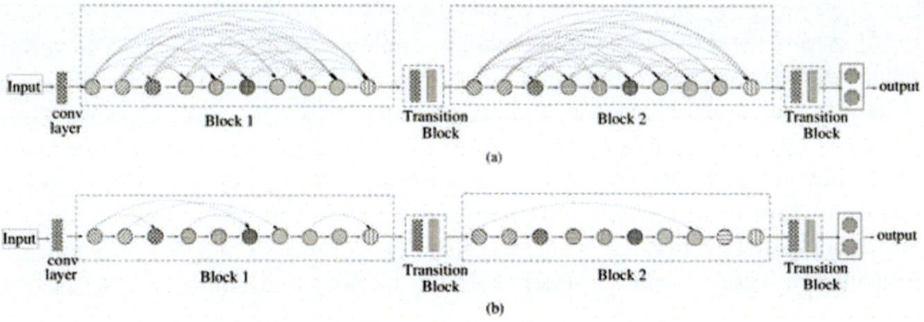

Fig. 1 Designed network structure for fold 1 with **a** $pth = 0.1$ and **b** $pth = 0.9$

= 0.9, almost same amount of accuracy of 0.953762 was achieved by 398 K weights parameters only. As shown in Fig. 1, the obtained network for high value of *p*th introduces more sparsity in the skip connections of block and hence the network has simpler structure with less parameters and without compromising in performance. From Table 1, the same effect can be seen for other folds also. On the average, the test accuracy of 0.9557 has been achieved with much simpler network structures.

4 Results and Discussion

A comparison between the performance of the obtained networks for different values of *p*th and fully dense network was performed. It was found that on the average with $pth = 0.9$, accuracy of 0.955759 was achieved with very less complex structures. The obtained networks for high value of *p*th follow the unique connectivity pattern as of DenseNet and with the help of sparse connectivity within the normal blocks, the overall network structure becomes much simpler. A comparison of the proposed approach with other existing approaches has been done in Table 2. The work done

Table 2 Malaria parasite detection: Comparison with other models as reported in [6]

Model	Parameters (M)	Accuracy	Sensitivity	Specificity	F1-Score	MCC
AlexNet	62	0.944	0.947	0.941	0.944	0.886
VGG-16	138	0.959	0.949	0.969	0.959	0.916
ResNet-50	25.6	0.959	0.947	0.972	0.959	0.917
DenseNet-121	0.9	0.952	0.960	0.944	0.953	0.902
Customized Net	0.4	0.927	0.905	0.951	0.928	0.884
Proposed work $pth = 0.1$	0.47	0.955	0.944	0.967	0.955	0.913
Proposed work $pth = 0.9$	0.39	0.955	0.936	0.969	0.952	0.907

in [6] used transfer learning method with multiple pretrained networks. The performance of designed networks using our approach is better than the performance of AlexNet, DenseNet-121, and Customized net as reported in [6]. With respect to VGG-16 and ResNet-50, the obtained results are comparable. Moreover, these comparable results are obtained at the cost of number of parameters. Our approach yields simpler architecture with only 0.39 million parameters which is 2.30, 65.64, and 353.84 times less as compared to parameters of DenseNet-121, ResNet-50, and VGG-16 network, respectively. Thus, it can be concluded that the proposed approach can generate economical simpler network structures without compromising much in accuracy.

5 Conclusion and Future Scope

This paper proposed a unique method for generating sparse dense networks with the help of DropConnect method. At a particular node, the concatenation of feature maps was performed on the basis of DropConnect technique with predefined threshold value *p*th. The high value of *p*th yields sparse connections in blocks and generates very simpler network topologies.

In this work, different network topologies were designed with the help of predefined threshold value *p*th. Also, only arrangements of skip connections were studied in this work. In the future, more flexible network topologies can be designed by considering the number of nodes of a block along with skip connections. Also, for a given dataset, the optimal topology of both normal and reduction blocks along with the above-mentioned parameters can be considered for exploring more flexible and lightweight network topologies.

Reference

1. He K, Zhang X, Ren S, Sun J (2016) Deep residual learning for image recognition. In: 2016 IEEE conference on CVPR, Las Vegas, NV, USA, June 27–30, 2016, pp 770–778
2. Hu H, Dey D, Giorno AD, Hebert M, Bagnell JA (2017) Log-densenet: how to sparsify a densenet. Computing Research Repository (CoRR) (2017)
3. Huang G, Liu Z, van der Maaten L, Weinberger KQ (2017) Densely connected convolutional networks. In: 2017 IEEE conference on CVPR, pp 2261–2269
4. Kingma DP, Ba J (2014) Adam: a method for stochastic optimization. Computing Research Repository (CoRR) arXiv abs/1412.6980
5. Paszke A, Gross S, Chintala S, Chanan G, Yang E, DeVito Z, Lin Z, Desmaison A, Antiga L, Lerer A (2017) Automatic differentiation in PyTorch. In: NIPS-W
6. Rajaraman S, Antani SK, Poostchi M, Silamut K, Hossain MA, Maude RJ, Jaeger S, Thoma GR (2018) Pretrained networks as feature extractors toward improved malaria parasite detection in thin blood smear images. PeerJ
7. Simonyan K, Zisserman A (2014) Very deep convolutional networks for large-scale image recognition. Computing Research Repository (CoRR) arXiv/abs/1409.1556
8. Srivastava N, Hinton GE, Krizhevsky A, Sutskever I, Salakhutdinov R (2014) Droupout: a simple way to prevent neural networks from overfitting. JMLR 15(1):1929–1958

9. Srivastava RK, Greff K, Schmidhuber J (2015) Highway networks. Computing Research Repository (CoRR) abs/1505.00387
10. Wan K, Feng B, Xie L, Ding Y (2018) Reconciling feature-reuse and overfitting in densenet with specialized dropout. Computing Research Repository (CoRR)
11. Wan L, Zeiler M, Zhang S, LeCun Y, Fergus R (2013) Regularization of neural networks using dropconnect. In: Proceedings of the 30th ICML, vol 28, pp III-1058–III-1066
12. World malaria report 2018. World Health Organization (2018). https://apps.who.int/iris/handle/10665/275867
13. Zhu L, Deng R, Maire M, Deng Z, Mori G, Tan P (2018) Sparsely aggregated convolutional networks. In: 15th Europran conference, Munich, Germany, Sept 8–14, Proceedings, Part XII, pp 192–208

Template Security Using Fuzzy Extractor: A Review

Naveen Kumar Gupta, Sunita Meena, Pulin Kumar, Vipin C. Dobhal, and Manvjeet Kaur

1 Introduction

In recent years, authentication using biometric system has become an active area of research for verifying the identity of a person. Passwords are traditional approach of authentication. However, they are not a secure solution because they suffer from various flaws such as the need to be memorized by human being or can easily be guessed by an attacker. Another technique, cryptography, is used to provide the authentication and secrecy of data. But drawbacks still persist in this technique like key management and key bit length problem. Due to these drawbacks, authentication using biometric data came into existence. The use of biometrics is necessary for the control of cryptographic keys. Biometrics uses behavioral and physical attributes for recognition of a person instead of ID cards and passwords. Biometric characteristics such as uniqueness, universality, permanence, measurability, and circumvention have led to widespread deployment of biometric authentication systems in many applications [1]. To make biometric recognition system publicly acceptable, security issues are needed to be reviewed. Breach in the security has always been a concern in

N. K. Gupta (✉) · S. Meena · P. Kumar · V. C. Dobhal
Central Research Laboratory, Bharat Electronics Ltd, Ghaziabad, Uttar Pradesh, India
e-mail: naveenkumargupta@bel.co.in

S. Meena
e-mail: sunitameena@bel.co.in

P. Kumar
e-mail: pulinkumar@bel.co.in

V. C. Dobhal
e-mail: vipinchandradobhal@bel.co.in

M. Kaur
PEC University of Technology, Chandigarh, India
e-mail: manvjeetkaur@pec.ac.in

V. H. Saran and R. K. Misra (eds.), *Advances in Systems Engineering*,
Lecture Notes in Mechanical Engineering,
https://doi.org/10.1007/978-981-15-8025-3_77

authentication system and same is the case with biometric authentication systems. At any stage in the system, security attacks can occur, but attack on the stored templates can be potentially most damaging for any biometric system. So, biometric-based authentication can be used to facilitate cryptography for confidentiality, encryption, and decryption of messages with the help of biometric key templates [2]. The fusion of two technologies biometrics and cryptography generates a system known as bio-cryptosystem (BCS). The idea behind working of BCS is either generation of cryptographic key from biometric template or providing security to crypto-graphic key through biometric template [2, 3]. These systems need to generate auxiliary information known as helper data from the biometric template in order to validate the user. Helper data does not provide any information about the template.

The rest of this paper is organized as in Sect. 2, and template protection schemes are classified into different categories. The comprehensive explanation of BCS technique and fuzzy extractor is explained in Sect. 3. Section 4 represents the review of fuzzy extractor on different modalities of biometric. Finally, the conclusion is presented in Sect. 5.

2 Template Protection Scheme

A template is a compressed form of the enrolled biometric that contains distinct characteristics which have been extracted from a biometric sample. Biometric data in the form of a template reduces the storage requirement to a large extent. Template protection is required to preserve the privacy and to enhance security of biometric data. Unlike passwords, if template is compromised it must be revocable and cancelable.

The following vulnerabilities can occur due to attacks on the template:

1. Replacement of an original template by an impostor's template,
2. Unauthorized access to the system if physical spoof is created from stolen template. Classification of various biometric template protection techniques are shown in Fig. 1.

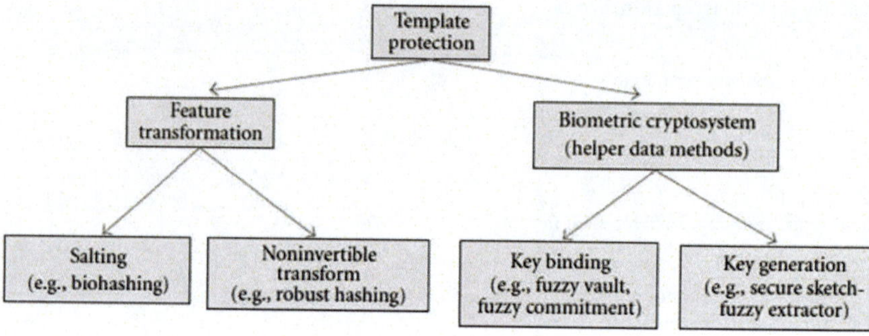

Fig. 1 Various biometric template protection techniques

The following four properties should be satisfied by an ideal biometric template protection scheme [4].

1. *Diversity:* For ensuring the users privacy, the same protected template should not allow cross matching across different applications.
2. *Revocability:* If the template is compromised, it *should* be revoked, and then a new protected template must be reissued based on the same biometric data.
3. *Security:* It must be *difficult* to obtain the original biometric template from the stored template and helper data.
4. *Performance:* The *recognition* performance matrices false acceptance rate (FAR) and false rejection rate (FRR) of the biometric system should not degrade while matching.

The template protection schemes are categorized into: feature transformation scheme or cancelable biometrics (CB) and BCS scheme.

2.1 Feature Transformation

Feature transformation is a technique in which a transformation function scheme is used to transform the biometric template and then the transformed template is stored in the database. Based on the characteristics of the transformation function, feature transformation is of two types.

1. *Invertible (salting)*: An invertible transformation function is used to transform the template with the help of a key. If the transformed template and the key are compromised, it is possible to regenerate the original template. Therefore, this scheme is secure until the key is compromised [4, 5].
2. *Non-invertible transforms*: On the contrary, a non-invertible transformation function is used to transform the template which is hard to invert. Regeneration of original template is not possible even if the transformed template and the key are compromised. Therefore, this scheme provides better security [6].

2.2 Biometric Cryptosystem

Biometric cryptosystems either bind a random key to a biometric or generate a key from a biometric [7]. To verify the user, some information from the template called helper data is publicly stored [4]. The helper data is chosen in such a way that it does not disclose the information of the actual template but contains enough information to perform a match [8]. There are two types of biometric cryptosystems.

1. *Key binding*: In this approach, a random key that does not depend on bio-metric features is bound with the biometric template from which helper data is obtained and stored in database. By applying an appropriate key retrieval algorithm, keys

Table 1 Advantages and disadvantages of feature transformation scheme

Salting	Non-invertible
Advantage 1. Low FAR occur, due to the introduction of key 2. It allows revocability and diversity	*Advantage* 1. An intruder cannot retrieve an original template with the help of compromised key 2. User-specific and application-specific transformation functions are used to attain revocability and diversity
Disadvantage 1. If the key is compromised, the original biometric template can be recovered by an attacker using the compromised key and invertible transformation function 2. Recognition performance gets degraded, due to large intra-user variations	*Disadvantage* 1. The transformation function should preserve the discriminability of the feature set 2. It is hard to design a transformation function that fulfills both the non-invertibility and discriminability conditions simultaneously

are obtained from the helper data during authentication [9]. An authentication is said to be successful if the correct key is extracted from helper data during matching. Fuzzy commitment and fuzzy vault are two approaches for key binding.

2. *Key generation*: In this approach, the key is derived directly from the biometric input rather than using a random key [4]. Helper data-based key generation schemes are referred as FE [10]. A fuzzy extractor reliably extracts a uniform random string (key) from user's biometric input, and helper data is stored in database to assist the reconstruction.

The various template protection techniques with their advantages and disadvantages are given in Tables 1 and 2.

3 Fuzzy Extractor

Fuzzy extractor is one of the best approaches to secure biometric features. It extracts the key from biometric template itself. Biometric features are noisy and non-uniform. Even two biometric samples belonging to the same individual are not completely identical, but are close. Thus, in order to preserve security, some error tolerating scheme is needed. Fuzzy extractor resolves both the problems of nonuniformity of inputs and error tolerance. It allows reliable extraction of some randomness R (can be used as a key) from its biometric input data w. The key R is highly uniform and can be reproduced same as the original one even though the biometric input changes from w to w' [11, 12]. The fuzzy extractor has two phases.

1. *Generation (Gen) procedure*: It consists of two sub-phases, namely secure sketch (SS) and strong extractor (Ext). SS is an essential condition for practical implementation of fuzzy extractor. It generates some public information about its input (w) that does not reveal any information about input (w). Extractor outputs

Table 2 Advantages and disadvantages of bio-cryptosystem scheme

Fuzzy commitment (key binding)	Fuzzy vault (key binding)	Fuzzy extractor (key generation)
Advantage 1. Ordered set of input; key is generated from the input data using extractor	*Advantage* 1. Accuracy is on high, which only depends on the radius of matching box	*Advantage* 1. Use ordered set of inputs 2. Key is generated from biometric template 3. Requires less storage space (no key storage)
Disadvantage 1. It does not allow modification of keys	*Disadvantage* 1. Enrollment and authentication of the template in this scheme is pre-aligned 2. Memory space gets increased due to chaff points 3. Matching accuracy is degraded due to alignment issues and nonlinear distortions	*Disadvantage* 1. Key should be highly stable. Key entropy and key stability are inversely proportional. So, it is more difficult to maintain it 2. It cannot be reused multiple times for same biometric signal due to low key entropy

a highly uniform randomized string which can be used as a secure key and is independent of weak random input string with low entropy.

2. *Reproduction (Rep) procedure*: It has two sub-phases, namely recovery (Rec) phase and extractor (Ext) phase. This procedure allows to recover same w from w' only if $d\ (w, w') \leq$ threshold. The recovered w along with already stored x is given as an input to Ext to output a string R.

Construction of fuzzy extractor from secure sketch is shown in Fig. 2.

Basic fuzzy extractor algorithm [10]:

1. *Gen procedure:*

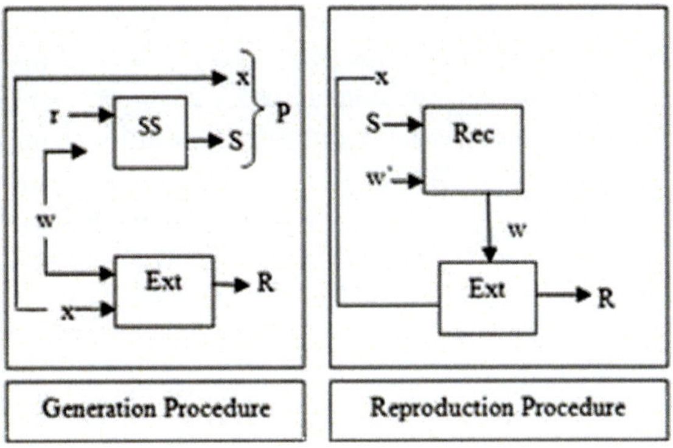

Fig. 2 Construction of fuzzy extractor from secure sketch

a. *Secure Sketch SS (w, r)*: Pass the input w and r to the secure sketch function (SS) to get output S, where w denotes template, r is random number, and S is sketch.
b. *Extractor Ext (w, x)*: It is used to generate a random string R by using w and x as input, where x is a common reference string.
c. S *and r* are stored in database as helper data *P*(*S*, *r*).

2. Rep procedure:

a. *Rec (w′, S):* This phase will return the w using w′ if $d(w, w') \leq$ (threshold).
b. Extractor Ext is used to generate a random string R by using w and x as input, where x is a common reference string.

4 Analysis of Fuzzy Extractor Technique Based on Different Traits and Approaches

In this section, a review of fuzzy extractor template protection technique, for various unimodal traits and multimodal traits is presented. Various other combinational approaches (like fuzzy extractor with biohashing, fuzzy extractor with random projection, fuzzy extractor with smartcards) are also been explored.

4.1 Unimodal Traits

In unimodal biometric approach, fuzzy extractor is implemented on single biometric modality. This section contains a review of fuzzy extractor based on single biometric modality that signifies the importance of fuzzy extractor in biometric template protection as summarized in Table 3.

Tong et al. [13] have proposed a fingerprint-based method. When fingerprints are used, extracted features are unordered. Fuzzy extractor cannot use these features as it requires ordered set of inputs. The “localization of the morphological center of fingertip” technique is used to convert unordered features into ordered form known as “finger codes.” These finger codes are used to retrieve a random value, which can be used as a key in fingerprint-based security applications. This method combines finger code and error-correcting codes to avoid the alignment issues in fingerprint.

Arakala et al. [14] have proposed minutiae-based fingerprint authentication to perform pre-alignment free matching in the encrypted domain with the help of fuzzy extractor. In this scheme, each minutia extracted from fingerprint is defined by descriptor. Quantization and digitalization representation of minutiae measurements is inferred in a new way so that a construct, pin sketch, can be applied to the minutiae. On comparison of two fingerprints, the global position of corresponding minutia is compared first. If there is a match in global position, it then qualifies for match in

Table 3 Review of fuzzy extractor scheme based on unimodal trait

Name of Author	Year	Modality	Technique	Database	Remarks
Tong et al. [13]	2007	Fingerprint	Fuzzy extractor	IBM-GTDB	• Converts unordered fingerprint minutiae points into ordered fingercode to remove minutiae related drawbacks • Euclidean distance is used as a matching technique
Arakala et al. [14]	2007	Fingerprint	Fuzzy extractor	FVC2000	• The minutiae measurements are digitally represented by a construct called PinSketch • Matching is done using local and global features
Khalil et al. [15]	2008	Fingerprint	Fuzzy extractor	FVC2000	• Cross-matching algorithm is used as feature extraction technique
Yang et al. [16]	2010	Fingerprint	Fuzzy extractor	UPEK of CASIA and DB2 of FVC2000	• Robust against stored biometric template attacks Modification in stored helper data is detected using common reference string model
Alvarez et al. [17]	2012	Iris	Fuzzy extractor and fuzzy vault	CASIA	Ordered iris template is converted into unordered template called iriscode, which is given as an input to fuzzy vault
Meng et al. [18]	2016	Finger vein	Fuzzy extractor and fuzzy vault	Standard fingervein database	• Shows the comparison between two bio-cryptosystem technique, fuzzy vault and fuzzy extractor, based on various performance metrices FRR, FAR, complexity of force attack

(continued)

Table 3 (continued)

Name of Author	Year	Modality	Technique	Database	Remarks
Sutcu et al. [19]	2009	Face	Fuzzy extractor	ORL face	• Investigates the effects of randomization on the security and performance

the shape of local structure. A successful match is considered when the number of minutia matches is sufficient enough in both global and local structures.

Khalil et al. [15] have proposed a technique to extract minutiae features of fingerprint by normalization, thinning, and feature filtering. To get the ordered set of minutiae features, cross-numbering technique is used. Fingerprint ordered features in the form of a vector are used as input to fuzzy extractor for generating a key.

It may be possible that the stored helper data at server side can be modified by adversary. To know about these modification, a nontrivial fingerprint key extraction is proposed by Yang et al. [16]. This scheme is robust towards stored biometric template attacks. It uses fuzzy extractor to generate a key and message authentication code (MAC) with key manipulation to know about modification in stored helper data. In order to provide protection to a random number used in MAC, fuzzy vault is also used. This paper shows that both security and performance can be improved by message authentication code with key manipulation and fuzzy vault.

Alvarez et al. [17] have proposed a crypto-biometric system in order to retrieve either a previously saved key or a secret key generated by fuzzy extractor, using user's iris biometric template. Fuzzy vault and fuzzy extractor schemes are combined in this approach. Firstly, the features are extracted from iris which is an ordered set of data. This ordered set of data is converted into an unordered set to use it as an input to fuzzy vault. A different approach, rather than clustering, is proposed by authors to generate the iris code (unordered) from a set of iris template(ordered).

Meng et al. [18] have proposed a two-finger vein encryption scheme using fuzzy extractor and vault. This paper enhances the scheme by removing redundancy point at the bifurcation of finger vein. The comparison has been done on two bases, first being the fuzzy vault and second being the combination of fuzzy extractor and fuzzy vault with parameters specified as FAR, FRR, re-release performance, and irreversibility. Design and analysis of fuzzy extractor for faces are given by Sutcu et al. [19]. They pointed out limitations of existing fuzzy extractor technique which are not able to handle entropy estimations, feature selection, and continuous feature spaces. This scheme is based on principal component analysis (PCA) features technique. It has two steps: feature vector extraction and quantization of the feature vector to compute sketches from faces to extract constraint keys.

4.2 Multimodal Traits

In multimodal approach, fuzzy extractor is implemented with the combination of more than one modality. Use of multiple modalities in fuzzy extractor as an input improves the security. The review of papers is summarized in Table 4.

An efficient approach for generating cryptographic keys from iris and face biometrics has been proposed by the authors Abuguba Saad et al. [20]. The proposed system employs three modules: feature extraction module, face feature extraction module, and key generation module. In feature extraction module, the preprocessing steps are applied on iris and face images using the Gabor filter and PCA, respectively.

Table 4 Review of fuzzy extractor scheme based on multimodal traits

Name of author	Year	Modality	Technique	Database	Remarks
Abuguba saad et al. [20]	2015	Face and Iris	Fuzzy extractor	CASIA-IrisV4 and ORLfor face	• Key is generated using feature level fusion of Gabor filter and PCA of iris and face
Jagadeesan et al. [21]	2010	Face and Iris	Fuzzy extractor	CASIA and Fingerprint images from publicly available source	• Key generated from multimodal biometric traits increases security
Marimuthu et al. [22]	2010	Dual Fingerprint	Fuzzy extractor	FVC2000	• Key is generated by the fusion of unique minutiae points of same user for encryption and decryption
Chen [24]	2014	Iris and Fingerprint	Fuzzy extractor	IITD Iris Database and PolyUHRF for fingerprint	• Fuzzy extractor focuses on correction of intra-user variety by extracting stable codeword • Secret share scheme is used for key binding and recovery Virtualmultimodal database
Gupta [25]	2017	Iris and Finger vein	Fuzzy extractor	CASIAv4 And Live Iris Database and PKU for finger vein	• A commonreference string model is used to extract the secure cryptography key • Robustsystemas it ensures about changes in stored secure sketch using MAC-KMS scheme Live database use for iris

(continued)

Table 4 (continued)

Name of author	Year	Modality	Technique	Database	Remarks
Kaur Taranpreet [26]	2017	Iris and Dual finger print	Fuzzy extractor	CASIAv4 and Live Iris Database and Live for fingerprint	• An algorithm on conversion of unordered fingerprint minutiae points into ordered minutiae points • Hash algorithm is used for protection of key Live database use for iris and fingerprint

Then the extracted features are fused to form a multimodal template. This multimodal template can be used as an input to generate strong 256-bit cryptographic key. The future research will be focused on reducing the FRR and implementing more advanced face feature extraction algorithms based on normalized principal component analysis and Gabor filters.

Jagadeesan et al. [21] used multimodal biometrics (iris and fingerprint) traits as an input for generation of secure key. In order to provide better security, key is generated from biometric features. Initially, the preprocessing of fingerprint and iris images is performed to extract minutiae points and texture properties, respectively, and then values are stored in x- and y-coordinates as two vectors for fingerprint and similarly two vectors for iris. Then fusion of extracted features at feature level is done by concatenation or merging of vectors in order to construct the multi-biometric template which is used as an input to fuzzy extractors. Finally, a 256-bit cryptographic key is generated for encryption and decryption of messages.

Marimuthu et al. [22] have used dual fingerprints of human being as a biometric input, so as to enhance security. Fingerprints are acquired, processed, and finally true minutiae points are extracted from both fingerprints in the form of vector. Then, the distance between minutiae points of a fingerprint is calculated using Euclidean distance method suggested by Nandakumar et al. [23]. The unique distance values are stored in two different vectors for two different fingerprints. Then, these arrays are concatenated to form dual fingerprint template which is used as an input to fuzzy extractors to generate 256-bit cryptographic key. In [24], Chen et al. have proposed a scheme by combining two technologies, namely fuzzy extractors and secret share. Using each enrolled biometric, fuzzy vector is generated and then this vector is used to get a stable codeword using fuzzy extractor. Secret sharing technology is used for key binding and recovery of key using the above-generated codewords. This scheme is better in terms of matching accuracy and key entropy.

In [25], Gupta has proposed a scheme, using iris and finger vein, which is robust against adversary attack on stored secure scheme. In this approach, SS is secured using MAC-KMS with a random value r and r is secured using fuzzy commitment. Fuzzy extractor uses a common reference string model in random extractor function for extraction of the secure cryptographic key. The security analysis proves that the system is robust and secure.

In [26], Kaur has proposed a scheme, using iris and dual fingerprint. In this paper, an algorithm is designed to convert dual fingerprint unordered dataset into ordered dataset as fuzzy extractor works on ordered dataset. In order to store the key, a hash function is used.

4.3 Combinational Approach

In combinational approach, fuzzy extractor is implemented along with some other approach as summarized in Table 5.

Table 5 Review of fuzzy extractor scheme with other techniques

Name of author	Year	Modality	Technique	Database	Remarks
Karthi et al. [27]	2013	Fingerprint and Iris	Fuzzy extractor And Random projection	CASIA and FVC2004	• By using fuzzy extractor, key is generated from bio-metric traits Key as a token along with biometric data is used by random projection method to generate same biometric template
Imamverdiev [28]	2012	Fingerprint	Fuzzy extractor and turbo codes	DB1, DB2 and DB3 of FVC2004	• Different keys are generated from same biometric using pseudorandom permutation of the biometric template blocks
Nguyen et al. [29]	2015	-	Fuzzy extractor and authentication protocol	–	• System is immune from man-in-middle attack and reply attack New biometric key generated by altering/changing the random parameter x if required
Karthi et al. [30]	2016	Fingerprint	Fuzzy extractor and bio hashing	FVC2002 and FVC2004	• Fuzzy extractor used to provide security and privacy to biometric template by not storing key in database directly • Biohashing provide revocability and non-invertible property to system
Zhang et al. [31]	2014	-	Fuzzy extractor And smart card and password	–	• Lower computation cost at server side and higher computation cost at client side as compare to Lee scheme [32] • Provide higher security • Help to achieve key agreement

(continued)

Table 5 (continued)

Name of author	Year	Modality	Technique	Database	Remarks
Duan et al. [33]	2011	-	Fuzzy extractor and smart card	–	• Opposes the guessing attack and the parallel attack Unable to withstand the masquerade attack [34]

Karthi et al. [27] have proposed a hybrid technique which uses combination of fuzzy extractor and random projection (RP) method. For authentication process, iris and fingerprint traits are used as multimodal biometric. First, the features are extracted from the iris image to form the iris code for key generation. Fuzzy extractor generates the key which is used as a token by RP. Fingerprint features and extracted key are used as an input to RP to get a secure template. The drawbacks of feature transformation method are overcome by this approach as the privacy of the template stored in database cannot be hindered.

Imamverdiev et al. [28] have proposed the key generation method from local and global features of fingerprint using fuzzy extractor, pseudo-zernike moments, and turbo codes. Pseudo-random generator is used in order to increase the entropy of generated key by fuzzy extractors. Pseudo-zernike movements are used for recognition of various biometric traits and to represent the image by a relatively small number of moments. Fuzzy extractor generates only a single key from biometric data. But by using pseudo-random permutation of the biometric template blocks, different keys from same biometric data can be generated if it required.

In [29], Nguyen et al. have proposed a biometric authentication protocol based on the Kerberos scheme. This protocol is resistant to replay attack and man-in-middle attack. The key extracted with the help of fuzzy extractor is used to authenticate the user. If biometric key is unveiled, attacker will not be able to infer biometric due to high security of fuzzy extractor scheme.

Karthi et al. [30] addressed the security issues and privacy of the stored template in the database. Two separate approaches namely fuzzy extractors and biohashing are used. Fuzzy extractor provides security and privacy by not storing key in database and biohashing provides non-invertible and revocable properties to the templates. During enrollment phase, a uniform binary string key is produced by fuzzy extractors and converted into a biocode with the help of biohashing technique. In authentication phase, uniform key and biocode extract features from query template Q and match them with the stored one. Remote three-factor authentication scheme combining biometric with password and smartcard based on fuzzy extractors has been proposed by Zhang et al. [31]. In this paper, authors point out design flaws of Lee scheme [32]. This scheme has four phases: initialization, registration, login authentication, and password change. This scheme emphasizes on achieving key agreement. For accessing different privileges by different users, an access control method has been introduced.

Sultan et al. [35] have proposed an online voting scheme based on fuzzy extractors and passwords. Fuzzy extractor provides authentication based on biometric, and secret password provides protection to the voter. This scheme has three phases: initialization phase, voting phase, and vote counting phase. Authors have also compared the functionality of pair voting and related work using different parameters like key length, auditability and verifiability, authentication voting length. Pair voting scheme neither needs any special software nor special hardware, and therefore, scheme is cost-effective.

5 Conclusion

In this paper, a review on the BCS technique called fuzzy extractor using different modalities has been presented. The technique emphasis on reduction of the risk of biometric template theft, as no biometric data and key value are stored at the time of enrollment. This technique uses the query biometric data to decommit the secret values at the time of verification. The review of various implementation of fuzzy extractor on the basis of unimodal traits, multimodal traits, and combinational approaches has been presented in this paper. This technique opens a way towards the ample use of biometric for security beyond recognition.

References

1. Jain AK, Ross R, Prabhakar S An introduction to biometric recognition. IEEE Trans Circ Syst Video Technol 14(1):4–20
2. Hao F, Anderson R, Daugman J (2006) Combining crypto with biometrics effectively. IEEE Trans Comput 55(9):1081–1088
3. Kanade SG, Petrovska-Delacretaz D, Dorizzi B (2012) Enhancing information security and privacy by combining biometrics with cryptography. Synth Lect Inf Secur Privacy Trust 3(1):1–140
4. Jain AK, Nandakumar K, Nagar A (2008) Biometric template security. EURASIP J Adv Signal Process 2008:113
5. Rathgeb C, Uhl A (2011) A survey on biometric cryptosystems and cancelable biometrics. EURASIP J Inf Secur 2011(1):3
6. Ratha NK, Connell JH, Bolle RuudM (2001) Enhancing security and privacy in biometrics-based authentication systems. IBM Syst J 40(3):614–634
7. Cavoukian A, Stoianov A (2011) Biometric encryption. In: Encyclopedia of cryptography and security. Springer, pp 90–98
8. Jain AK, Ross A, Uludag U (2005) Biometric template security: challenges and solutions. In: 2005 13th European signal processing conference. IEEE, pp 1–4
9. Uludag U, Pankanti S, Prabhakar S, Jain AK (2004) Biometric cryptosystems: issues and challenges. Proc IEEE 92(6):948–960
10. Dodis Y, Reyzin L, Smith A (2004) Fuzzy extractors: how to generate strong keys from biometrics and other noisy data. In: International conference on the theory and applications of cryptographic techniques. Springer, pp 523–540
11. Dodis Y, Ostrovsky R, Reyzin L, Smith A (2008) Fuzzy extractors: how to generate strong keys from biometrics and other noisy data. SIAM J Comput 38(1):97–139
12. Verbitskiy EA, Tuyls P, Obi C, Schoenmakers B, Skoric B (2010) Key extraction from general non-discrete signals. IEEE Trans Inf Forensics Secur 5(2):269–279
13. Tong VVT, Sibert S, Lecoeur J, Girault M (2007) Bio-metric fuzzy extractors made practical: a proposal based on fingercodes. In: International conference on biometrics. Springer, pp 604–613
14. Arakala A, Jeffers J, Horadam KJ (2007) Fuzzy extractors for minutiae-based fingerprint authentication. In: International conference on biometrics, pp 760–769. Springer, 2007.
15. Khalil MS, Muhammad D, Masroor Ahmed M (2008) Converting fingerprint local features to public key using fuzzy extractor. In: The international Arab conference on information technology. Citeseer
16. Li X, Yang Bo, Zhang M, Yang D (2010) Construction and performance of robust fingerprint key extractor. J Inf Comput Sci 7(6):1287–1299

17. A´lvarez Marin˜o A, Hern´andez A´lvarez F, Hern´andez Encinas L (2012) A cryptobiometric scheme based on iris-templates with fuzzy extractors. Inf Sci 195:91–102
18. Meng PFG (2016) Two schemes of finger vein cryptosystem
19. Sutcu Y, Li Q, Memon N (2009) Design and analysis of fuzzy extractors for faces. In: SPIE defense, security, and sensing. International Society for Optics and Photonics, pp 73061X–73061X
20. Abuguba S, Milosavljevic MM, Macek N (2015) An efficient approach to generating cryptographic keys from face and iris biometrics fused at the feature level. Int J Comput Sci Netw Secur (IJCSNS) 15(6):6
21. Jagadeesan A, Duraiswamy K (2010) Secured cryptographic key generation from multimodal biometrics: feature level fusion of fingerprint and iris. *arXiv preprint* arXiv:1003.1458
22. Marimuthu M, Kannammal A (2015) Dual fingerprints fusion for cryptographic key generation. Int J Comput Appl 122(23)
23. Nandakumar K, Jain AK, Pankanti S (2007) Fingerprint-based fuzzy vault: implementation and performance. IEEE Trans Inf Forensics Secur 2(4):744–757
24. Chen C, Wang C, Yang T, Lin D, Wang S, Hu J (2014) Optional multi-biometric cryptosystem based on fuzzy extractor. In: 2014 11th international conference on fuzzy systems and knowledge discovery (FSKD). IEEE, pp 989–994
25. Gupta NK, Kaur M (2017) A robust and secure multitrait based fuzzy extractor. In: 2017 8th international conference on computing, communication and networking technologies (ICCCNT). IEEE, pp 1–6
26. Kaur T, Kaur M (2017) Cryptographic key generation from multi-modal template using fuzzy extractor. In: 2017 tenth international conference on contemporary computing (IC3). IEEE, pp 1–6
27. Karthi G, Azhilarasan M (2013) Hybrid multimodal template protection technique using fuzzy extractor and random projection. IJRCCT 2(7):381–386
28. Imamverdiev YN, Sukhostat LV (2012) A method for cryptographic key generation from fingerprints. Autom Control Comput Sci 46(2):66–75
29. Nguyen TAT, Dang TK (2015) Combining fuzzy extractor in biometric-kerberos based authentication protocol. In: 2015 International conference on advanced computing and applications (ACOMP). IEEE, pp 1–6
30. Govindharaju K, Ezhilarasan M (2016) Securing biometric template using a hybrid scheme. In: Proceedings of the international conference on informatics and analytics. ACM, pp 105
31. Zhang M, Zhang J, Zhang Y (2015) Remote three-factor authentication scheme based on fuzzy extractors. Secur Commun Netw 8(4):682–693
32. Lee C-C, Hsu C-W (2013) A secure biometric-based remote user authentication with key agreement scheme using extended chaotic maps. Nonlinear Dyn 71(1–2):201–211
33. Yang D, Yang B (2009) A new password authentication scheme using fuzzy extractor with smart card. In: International conference on computational intelligence and security, 2009 (CIS'09), vol 2. IEEE, pp 278–282
34. Duan X, Li X (2011) Security of a new password authentication scheme using fuzzy extractor with smart card. In: 2011 IEEE 3rd international conference on communication software and networks (ICCSN). IEEE, pp 282–284
35. Sultan NH, Barbhuiya FA, Sarma N (2015) Pair-voting: a secure online voting scheme using pairing-based cryptography and fuzzy extractor. In: 2015 IEEE international conference on advanced networks and telecommuncations systems (ANTS). IEEE, pp 1–6

A Framework for Assessing the Viability of Solar PV in Province-2 of Nepal Using System Dynamics Approach

Swatantra Kumar Karna and Rhythm Singh

1 Introduction

Nepal has high hydropower potential but only 2% of the estimated hydropower potential has been tapped till date [1]. It is a major source of generation accounting to more than 90% [2, 3] of the total power generation in Nepal. This generation however is not able to meet the domestic demand and the forced load-shedding due to the demand–supply gap has been managed by power import from India [3]. Power-sector policies in Nepal have traditionally been more hydropower friendly. In case of solar PV, [4] has led to the formulation of guidelines for grid-connected PV systems [5] in year the 2018 A.D. Although Nepal has 2100 MW of urban solar PV and 1829 MW of concentrated solar power (CSP) potential [6], no utility-scale solar PV plant has been installed so far [3]. On the other hand, Province-2, not having any generation of its own, has focused on solar power generation for power self-sufficiency [7].

Solar PV installation depends on multiple factors, viz. technical, geographical, socio-economic and political factors. Insolation variability, grid availability and quality, temperature, shading, voltage drop and power loss are the technical factors which are very much dependent on the geographical location [8–13]. Economic factors constitute good return on equity due to investment friendly policies. Financial barriers and risks associated with solar PV development are unique in nature, and they escalates in case of low-income countries like Nepal where experience in the solar PV sector is low [14, 15]. Proper financial instruments such as venture capital equity,

S. K. Karna · R. Singh (✉)
Department of Hydro and Renewable Energy, Indian Institute of Technology Roorkee, Roorkee 247667, Uttarakhand, India
e-mail: rhythm@hre.iitr.ac.in

S. K. Karna
Department of Electricity Development, Sano Gaucharan, Kathmandu, Nepal

V. H. Saran and R. K. Misra (eds.), *Advances in Systems Engineering*,
Lecture Notes in Mechanical Engineering,
https://doi.org/10.1007/978-981-15-8025-3_78

capital grants, subsidy, result-based financing are often used to promote investment [15].

System dynamics approach captures the interactions between the different variables and helps to understand the behaviour of the system by simulating stock-flow models. Causality in system dynamics refers to both events and behaviours. Causal-loop diagram only talks about events in general, whereas behaviours of the system can be analysed after quantifying the stock and flow diagram [16]. Systems approach has widely been used for energy security studies and renewable energy studies [9, 13, 14, 17–22].

In context of Nepal, Ghimire and Kim [23] had identified twenty-two barriers categorized into six types, prominent being (i) economic and (ii) policy and political barrier, that hinder renewable energy development in Nepal. Nepal [24], on the other hand, concludes that demand–supply constraints should be addressed for the adoption of renewable energy technology at a large scale. In an attempt to assess the feasibility of rooftop solar PV in urban areas, Gautam et al. [25] estimated a generation of 3338 MWh of energy from 492 MWp capacity of solar PV and recommends policy changes. The subsidy being provided by the government in rural areas of Nepal has been effective in improving the affordability for renewable energy technology [26, 27]. Bhandari and Stadler [28] conclude that solar PV systems are not economic solutions for the grid-connected urban areas of Nepal, whereas considering social aspects, solar PV system is a convincing alternative for rural village electrification in Nepal.

Province-2 has no generation of its own and so it is aiming for self-sufficiency in energy through solar power. Studied literature confines solar PV to rural sector only in context of Nepal. Study for utility-scale solar PV generation has not been carried out even on national level. Apart from that, no study on provincial level has been done so far. This paper presents a conceptual framework using systems approach for studying the viability of utility-scale solar PV plant in Province-2, details of which are mentioned in next sections. Section 2 deals with the energy scenario in Nepal and Province-2 of Nepal. Section 3 explains the conceptual framework which is in the form of causal-loop diagram, whereas Section 4 concludes the paper with scope of future work.

2 Energy Scenario in Nepal and Province-2 of Nepal

Nepal is endowed with sufficient amount of hydropower resources with an estimated generating potential of 83 GW [1]. As given in Table 1, fuelwood contributes the highest (70.47%) to total primary energy consumption, whereas the contribution of electricity to total primary energy consumption is only 3.39% [29]. This indicates a major challenge for the policy makers to shift traditional energy consumption pattern to electricity.

Hydropower is a major source of power in the country. As shown in Table 2, generation from hydropower sums up to 1,128,705 kW which is 95.47% of the total

Table 1 Distribution of energy consumption by fuel types in FY 2014-15

S. No.	Fuel types	Amount (000 GJ)	Percentage share
1	Fuelwood	352,229.10	70.47
2	Agriculture residue	17,408.43	3.48
3	Animal dung	18,401.96	3.68
4	Coal	19,819.09	3.97
5	Petroleum	62,618.27	12.53
6	Electricity	16,932.75	3.39
7	Renewable	12,430.26	2.49

Source Electricity Demand Forecast Report 2014–40 by WECS, Nepal

Table 2 Power scenario of Nepal in year 2018

Peak Demand (MW)	1,508.16
Generation from NEA owned Hydropower Plant (GWh)	2,308.24
Generation from NEA owned Thermal Power Plant (GWh)	0.13
Total Generation from NEA owned plants (GWh)	2,308.37
Power Purchase from India (GWh)	2,581.80
Generation from IPP (GWh)	2,167.76
Available Energy (GWh)	7,057.93
Hydropower Installed Capacity (kW)	1,128,705
Thermal power Installed Capacity (kW)	53,410
Solar Power Installed Capacity(kW)	100
Total Installed Capacity (kW)	1,182,215
Transmission and Distribution Losses (%)	20.45
Under construction Hydropower Projects by NEA (kW) (FY 2018/19)	975,100
Under construction Hydropower Projects by IPP (kW) (FY 2018/19)	2,613,895

Source Nepal Electricity Authority: A year in Review-Fiscal Year 2018/2019

installed capacity [3]. However, this is only 2% of the estimated hydropotential [1]. Most of the hydropower plants installed are of run-of-river type and the generation in dry season drops sharply when the load is considerably high [2]. In the fiscal year 2018–19, peak load recorded in the month of Poush (Dec.–Jan.) was 1243.57 MW when the internal power generation was only 590.1 MW leading to an import of 653.47 MW from India. This was the maximum power import from India during that fiscal year [3]. There are four transmission linkages between Nepal and India for power exchange with a major Dhalkebar–Muzaffarpur 400 kV transmission line. Rest three are 132 kV lines from Tanakpur, Balmikinagar and Kataiya [30]. During the year 2018, 36.58% of the available electricity has been imported from India through these links. Also during fiscal year 2018–19, an export of 42,497.96 MWh energy was done to India [3]. As shown in Table 3, three major sectors of energy

Table 3 Sector-wise electricity consumption in FY 2016–17

Sector	Electricity consumption (GWh)	Percentage
Household	2147.8	45.08
Business	350.6	7.36
Industrial	1719.3	36.08
Other	547.0	11.48
Total	4,764.7	

Source Economic survey 2017/18, Ministry of finance Nepal

consumption in Nepal have been identified, out of which household sector is the leading energy consumer followed by industrial sector [31].

Regarding Province-2, demand data has not been published yet by the authority. Province-2 of Nepal has 79.77% electrification status compared to the national grid connectivity status of 77.49% [1]. It has currently zero power generation of its own and all of its power need has to be fulfilled by the national grid. With no prominent hydropower source, unlike other provinces, solar PV is seen as a contributing factor for the provincial energy demand [32]. Provincial government has declared a target of 300 MW of solar installation [33]. Total 206 MW of survey licence and 10 MW of construction licence for solar PV plant have been issued by Department of Electricity Development till date for different places in Province-2 [34].

To conclude, in terms of national demand and supply, it is seen that almost one-third of the total electricity required is imported from India with occasional exporting. So sufficient generation planning is necessary and more plants need to come up to reduce import and meet peak load. Also, household sector is found to be a larger consumer of electricity than the industrial sector; policies to encourage industrial demand seem necessary. Similarly, power demand shall also increase with the extension of grid both at the national and provincial levels.

3 Conceptual Framework

On the basis of literature survey and consultation with experts for the viability of solar PV in Province-2, four sub-systems were identified with 45 directly or indirectly influencing parameters as mentioned in Table 4. Their interactions together form an integrated causal-loop diagram as shown in Fig. 1. The four sub-systems are further described in the subsections below.

Table 4 Parameters influencing different sub-systems

S. No.	Sub-Systems	Influencing parameters
1	Demand	Residential demand, commercial demand, industrial demand, agricultural demand, transport sector demand, purchasing power, GDP, commercial growth rate, population growth rate, industrial growth rate, industrialization policy, agricultural growth rate, emigration rate, transport policy, agricultural policy
2	National Power Generation	Demand, demand–supply gap, power exchange with India, energy exchange price, generation from hydropower, generation from other intermittent sources
3	Solar PV Installed Capacity in Province-2	National power generation, generation from other intermittent sources, available usable land, land-use policy, investment, solar insolation availability, demand–supply gap, grid capacity, generation from SPV, jobs due to SPV
4	Investment	Land cost, transmission facility, equipment cost, energy storage, subsidy, renewable energy policy, PPA rate, energy exchange price, purchasing power, cost of generation, local tariff, financial risks and barriers (comprised of equity financing, selection of appropriate financial instrument, adoption to new technology, capital markets and long-term financing, project scale), return on equity

3.1 Demand

Power demand of Province-2 comprises of five sectoral demands, namely residential, commercial, industrial, agricultural and transport sector demand. All demands are influenced by population growth rate as well as corresponding sectoral growth rates which are obviously impacted by corresponding policies. Exemption of taxes to be levied by province on industries for 20 years [35] is more likely to increase industrial demand in Province-2. Residential and commercial demands, on the other hand, are also impacted by purchasing power of the people along with the above-mentioned factors. Sectoral growth rate has a positive impact on the gross domestic product of the country which feedbacks with significant impact on purchasing power of people. With the growth of agriculture and industry in the province, people going out of the province for maintaining their livelihood is likely to reduce which would further tend to increase the population in the region. This has been indicated by the auxiliary variable named "Emigration Rate".

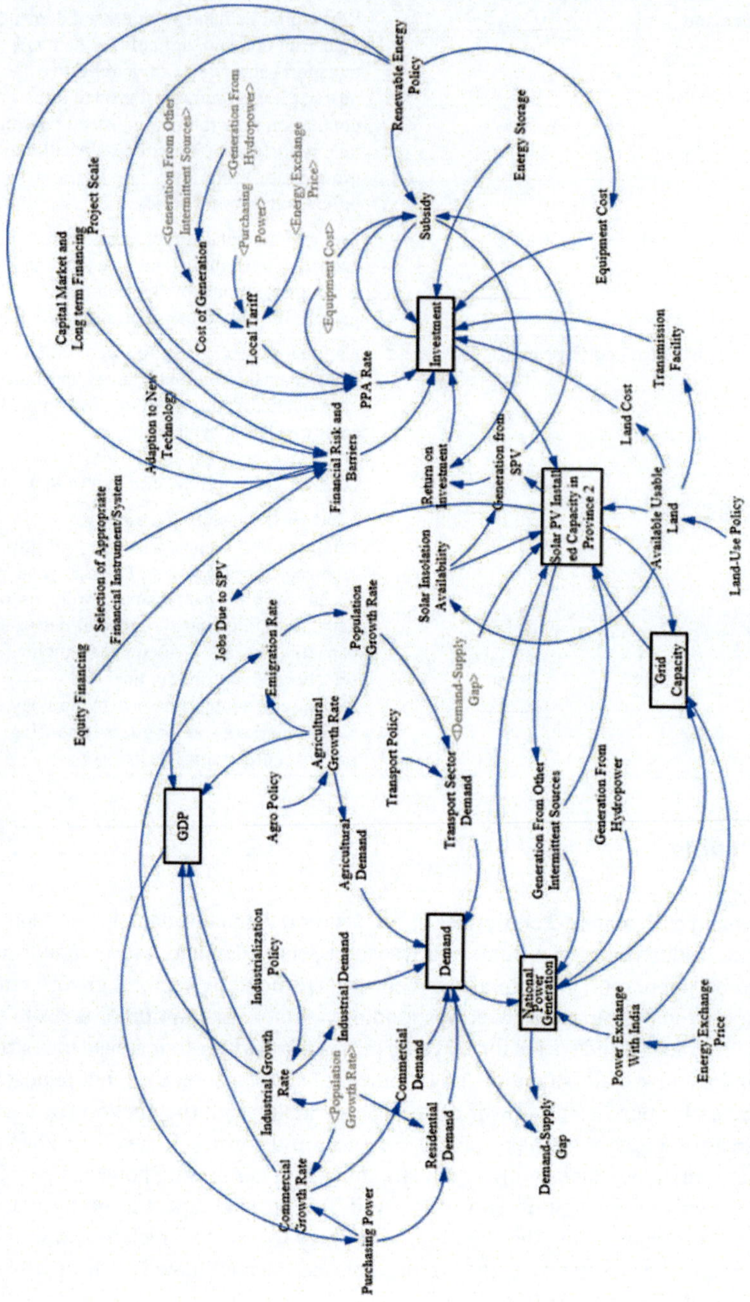

Fig. 1 Causal-loop diagram indicating interlinkages between different factors influencing solar PV installation in Province-2

3.2 National Power Generation

There are seven identified factors which are directly associated with national power generation. Those are demand, demand–supply gap, power exchange with India, grid capacity, generation from hydropower, generation from other intermittent sources and solar PV installed capacity of Province-2. Increase in demand of Province-2 compels the national power generation to increase. Generation from hydropower, generation from other intermittent sources and power exchange with India contribute to national power generation. Power exchange with India, however, will depend on the energy exchange price mutually agreed between two countries. National power generation and energy exchange with India will use the national grid which, in turn, might constrain the quantity of power being transmitted from solar PV capacity of Province-2. Generation from solar PV plants in Province-2 shall contribute to generation from other intermittent sources and thus increase national power generation.

3.3 Solar PV Installed Capacity in Province-2

Solar PV installation in Province-2 depends on many factors. In this paper, only utility-scale solar PV plants have been considered. Gap in demand and supply will dictate the required capacity of solar PV to be installed. As per [5], the cumulative power purchase agreements for grid-connected alternative energy sources (solar, wind and bio-gas) should not exceed 15% of the total generation capacity. Thus, generation from other intermittent sources, together with national power generation, shall limit the solar PV installation in Province-2. Land-use policy shall dictate the usable land for power generation which again shall impact the total installed capacity of solar PV. Solar power generated in Province-2 shall add load to the grid. This, along with national power generation and cross-border power exchange, might result in grid congestion, limiting further installation of solar PV in Province-2. This has been indicated by a feedback arrow from grid capacity stock to "solar PV Installed capacity in Province-2" stock. Generation from the solar PV plant depends on installed capacity and available solar insolation, which also depends on the location of the solar PV plants. Higher will be the generation, greater will be return on investment which will further boost investment in the solar sector. This has been clearly indicated by the dependency arrows in causal-loop diagram shown. Also, more solar sector flourishes it would create more jobs in solar PV sector which in turn shall contribute to the GDP.

3.4 Investment

Development of solar power plant is dependent on the investment in the sector which is influenced by multiple factors such as different costs incurred, return on investment, financial risks and barriers, renewable energy policy, power purchase rate as depicted in the causal-loop diagram. Land cost and cost of transmission infrastructure for connecting to the nearest point of grid availability have to be borne by the developers themselves [5]; whereas if energy storage system is adopted, it would add extra burden to the investment. Higher investment is encouraged by subsidy, at the same time, subsidy would lead to higher return on investment promoting further investment.. Similarly, higher return on investment as a result of investing in solar PV technology would further encourage investing in this sector as a learning effect.

Power purchase agreement is said to occur in "Take or Pay" mode which is encouraging for investment. However, the current PPA rate of Nepali Rupees 7.30 is valid only till February 2021, and the policy talks of no escalation in prices thereafter [5].

Financial risk and barriers pose a hindrance for investment in renewable energy sector [15]. Nepal's financial sector has a great deal of expertise in hydropower sector but solar sector is still in nascent stage. Thus adaptation to this new technology is a barrier for financing. Scale of solar PV project is highly dependent on available land, and so very high scale projects is rare which again hinders investment. Similarly in low-income countries like Nepal, there are limited numbers of large-scale project sponsors with the ability and willingness to fund renewable energy projects on corporate finance basis so lack of equity financing is also a barrier. Likewise financial institutions are more directed towards those investments offering the highest returns at the lowest risk also in the shortest period of time which does not encourages capital market and long-term financing in solar sector. These risks and barriers can be mitigated by the selection of proper financial instruments such as capital grants, venture capital, appropriate guarantees and insurance and result-based financing, and investment can be boosted up subject to the renewable energy policy .

4 Conclusion and Scope of Future Work

Factors influencing the installation of solar PV in Province-2 of Nepal have been identified and presented in the causal-loop diagram. Four sub-systems, namely demand, power generation in Nepal, solar PV installed capacity in Province-2 and investment, have been identified, and their interconnections have been established. This gives a clear picture of the contributing factors and their influence on the solar PV installed capacity, although this information is not sufficient to synthesize the behaviour of the system. For concluding the viability of solar PV in Province-2, the above causal-loop diagram needs to be converted into stock and flow diagram; the parameters need to be quantified; and the resulting model needs to be simulated. Furthermore, sensitivity analysis needs to be performed by varying the crucial parameters to identify the policy recommendations . This paper provides a conceptual framework based on which further proceedings can be done.

References

1. Nepal Electricity Authority (2019) Nepal electrification statistics: 2019
2. Nepal Electricity Authority (2019) Year book of generation directorate fiscal year 2018/19
3. Nepal Electricity Authority (2018) Nepal Electricity Authority A year in review-Fiscal year 2018/19
4. Ministry of Energy, W.R. and I.N.: National Energy crisis mitigation and electricity development decade concept paper, 2015.
5. Nepal, M. of energy water resources and irrigation: grid connected alternative energy development guideline, 2018. Nepal (2018)
6. Resource WE (2008) United Nations Environment Programme Global Environment Facility Solar and Wind Energy Resource Assessment in Nepal (SWERA). Final Report (GIS PART)
7. Province-2 Nepal: Budget of province 2 FY 2018/19 (2018)
8. Rocher L, Verdeil E (2019) Dynamics, tensions, resistance in solar energy development in Tunisia. Energy Res Soc Sci 54:236–244. https://doi.org/10.1016/j.erss.2019.04.010
9. Guo X, Guo X (2015) China's photovoltaic power development under policy incentives: a system dynamics analysis. Energy 93:589–598. https://doi.org/10.1016/j.energy.2015.09.049
10. Sindhu S, Nehra V, Luthra S (2017) Investigation of feasibility study of solar farms deployment using hybrid AHP-TOPSIS analysis: case study of India. Renew Sustain Energy Rev 73:496–511. https://doi.org/10.1016/j.rser.2017.01.135
11. Zhang F, Deng H, Margolis R, Su J (2015) Analysis of distributed-generation photovoltaic deployment, installation time and cost, market barriers, and policies in China. Energy Policy 81:43–55. https://doi.org/10.1016/j.enpol.2015.02.010
12. Fthenakis V, Mason JE, Zweibel K (2009) The technical, geographical, and economic feasibility for solar energy to supply the energy needs of the US. Energy Policy 37:387–399. https://doi.org/10.1016/j.enpol.2008.08.011
13. Lyu H, Li H, Wallin F, Xv B (2017) Research on Chinese solar photovoltaic development based on green-trading mechanisms in power system by using a system dynamics model. Energy Procedia 105:3960–3965. https://doi.org/10.1016/j.egypro.2017.03.823
14. Liu X, Zeng M (2017) Renewable energy investment risk evaluation model based on system dynamics. Renew Sustain Energy Rev 73:782–788. https://doi.org/10.1016/j.rser.2017.02.019
15. The World Bank (2013) Climate investment funds: financing renewable energy: options for developing financing instruments using public funds. Energy Dev, 1–60
16. Schaffernicht M (2007) Causality and diagrams for system dynamics. In: Proceedings of the 2007 international conference on systems and dynamics society, pp 1–24
17. Tukulis A, Pakere I, Gravelsins A, Blumberga D (2018) Methodology of system dynamic approach for solar energy integration in district heating. Energy Procedia 147:130–136. https://doi.org/10.1016/j.egypro.2018.07.042
18. Walters J, Kaminsky J, Gottschamer L (2018) A systems analysis of factors influencing household solar PV adoption in Santiago, Chile. Sustainabilty 10:1–17. https://doi.org/10.3390/su10041257
19. Saavedra MR, Cristiano CH, Francisco FG (2018) Sustainable and renewable energy supply chain: a system dynamics overview. Renew Sustain Energy Rev 82:247–259. https://doi.org/10.1016/j.rser.2017.09.033
20. Zaman R, Brudermann T (2018) Energy governance in the context of energy service security: a qualitative assessment of the electricity system in Bangladesh. Appl Energy 223:443–456. https://doi.org/10.1016/j.apenergy.2018.04.081

21. Hsu CW (2012) Using a system dynamics model to assess the effects of capital subsidies and feed-in tariffs on solar PV installations. Appl Energy 100:205–217. https://doi.org/10.1016/j.apenergy.2012.02.039
22. Mutingi M, Mbohwa C, Kommula VP (2017) System dynamics approaches to energy policy modelling and simulation. Energy Procedia 141:532–539. https://doi.org/10.1016/j.egypro.2017.11.071
23. Ghimire LP, Kim Y (2018) An analysis on barriers to renewable energy development in the context of Nepal using AHP. Renew Energy 129:446–456. https://doi.org/10.1016/j.renene.2018.06.011
24. Nepal R (2012) Roles and potentials of renewable energy in less-developed economies: the case of Nepal. Renew Sustain Energy Rev 16:2200–2206. https://doi.org/10.1016/j.rser.2012.01.047
25. Gautam BR, Li F, Ru G (2015) Assessment of urban roof top solar photovoltaic potential to solve power shortage problem in Nepal. Energy Build 86:735–744. https://doi.org/10.1016/j.enbuild.2014.10.038
26. Gurung A, Ghimeray AK, Hassan SHA (2012) The prospects of renewable energy technologies for rural electrification: A review from Nepal. Energy Policy 40:374–380. https://doi.org/10.1016/j.enpol.2011.10.022
27. Mainali B, Silveira S (2012) Renewable energy markets in rural electrification: country case Nepal. Energy Sustain Dev 16:168–178. https://doi.org/10.1016/j.esd.2012.03.001
28. Bhandari R, Stadler I (2011) Electrification using solar photovoltaic systems in Nepal. Appl Energy 88:458–465. https://doi.org/10.1016/j.apenergy.2009.11.029
29. Water and Energy Commission Secretariat (2017) Government of Nepal Water and Energy Commission Secretariat Electricity Demand Forecast Report
30. Nepal Electricity Authority (2012) Year book-fiscal year 2018/2019 of transmission project management directorate
31. Ministry of Finance (2018) Economic survey 2017/2018
32. National Planning Commission, G. of N (2018) Final Report on "Study and analysis of optimal distributed generation for access to grid electricity for all in five years with participation from local-level government"
33. Service H.N. (2019) Provincial Target to produce 300 MW in fiscal year 2019–20.pdf, https://thehimalayantimes.com/nepal/province-2-govt-presents-policies-and-programmes-for-upcoming-fiscal/
34. Department of Electricity Development, Issued License Status of Solar PV, www.doed.gov.np
35. Shrestha PM (2019) Two provinces make the most of Investment Summit. https://kathmandupost.com/national/2019/04/01/two-provinces-make-the-most-of-investment-summit (2019)

Review of Solar Powered Organic Rankine Cycle for Indian Conditions

Trushil Patel, Nishant Mandal, and Parth Prajapati

Nomenclature

ORC	Organic Rankine Cycle
FPC	Flat Plate Collector
LFR	Linear Fresnel Reflector
PTC	Parabolic Trough Collector
ODP	Ozone Depletion Potential
GWP	Global Warming Potential
CFC	Chlorofluorocarbon
HCFC	Hydro chlorofluorocarbon
TES	Thermal Energy Storage

1 Introduction

India is emerging as the developing nation in every sector. Be it technology, education, and economical; today India is competing with every developed nations in the world. With development the demand of the energy is increasing day by day. According to global energy consumption India lies at 5th position consuming 3.9% of world's commercial energy [1]. Also the industries, power plants, combustion engines emit numerous amount of exhaust gases in the form of waste heat. Waste heat is regarded as low grade of energy which can be converted to high grade in the form of electric

T. Patel (✉) · N. Mandal · P. Prajapati
Department of Mechanical Engineering, Pandit Deendayal Petroleum University, Gandhinagar 382007, India
e-mail: trushil.pmc18@sot.pdpu.ac.in

V. H. Saran and R. K. Misra (eds.), *Advances in Systems Engineering*, Lecture Notes in Mechanical Engineering,
https://doi.org/10.1007/978-981-15-8025-3_79

energy thus making Organic Rankine Cycle (ORC) a suitable technology for the job of energy generation [2].

Organic Rankine Cycle is a thermodynamic process that uses the working fluids in a closed loop to convert heat energy into mechanical energy using an evaporator and expander as shown in Fig. 1. The obtained mechanical energy is then converted into electrical output when coupled with generator. The working fluids used here are generally carbon containing elements having lower boiling point and higher molecular mass. The working temperature of the ORC operates at low temperature as compared to conventional steam Rankine cycle which requires higher magnitude of heat input, thus making ORC more suitable in operations at low temperature (Fig. 2).

The heat energy required to operate ORC can be obtained from different sources like biomass, solar pond, geothermal, solar thermal technologies, and waste heat from diesel engine, industries, etc. Among many possible heat sources, renewable sources of energy are sustainable source than fossil fuels for generation of electric energy. Some common forms of renewable energy are tidal energy, geothermal, wind and solar energy. Sun is regarded as perpetual source of energy and considering the technological advancement in the field of solar energy, organic Rankine cycle can be operated with it. Since most part of the country lies in tropical and sub-tropical region, the solar energy is in abundance from geographical perspective of India.

In solar powered ORC, the thermal fluid is heated by using solar collectors which is then passed in heat exchanger where the working fluid gains energy turning it into superheated vapor. The vapor is passed to the expander to generate the output in form of electrical energy. The vapor is then condensed in the condenser and is then pumped back in heat exchanger and the cycle repeats.

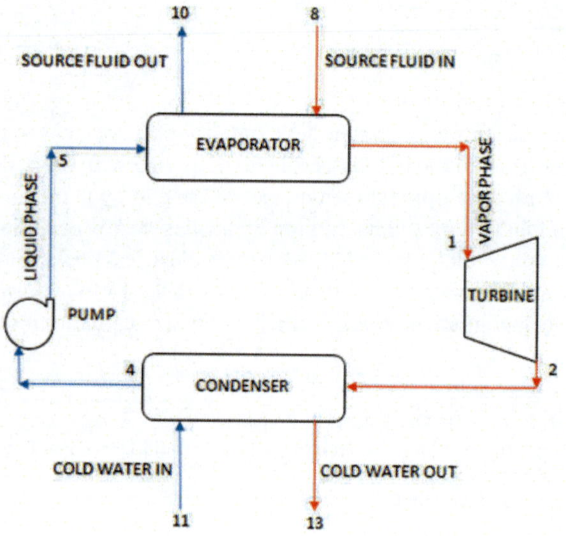

Fig. 1 Conventional organic Rankine cycle

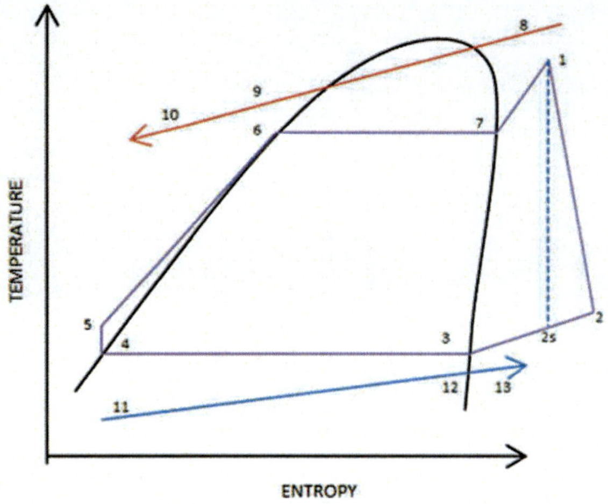

Fig. 2 T s diagram

Table 1 Solar irradiance in India [3]

State	Average direct normal irradiance (kWh/m²/day)	Average global horizontal irradiance (kWh/m²/day)
Gujarat	5.95	5.83
Rajasthan	5.68	5.79
Himachal Pradesh	5.64	5.15
Madhya Pradesh	5.57	5.62
Karnataka	5.52	5.92

Magnitude of solar energy is quantified in terms of solar irradiance. Solar irradiance is the amount of solar energy received per unit area per day. Table 1 shows annual solar irradiance for different states of India and it can be observed that Gujarat receives maximum amount of solar irradiance followed by Rajasthan making these places suitable for solar powered ORCs.

Further studies show that during January to May, states considered in the Table 1 receive maximum radiation and lowest during June to September. On the other hand, during winter season duration of the day is shorter as compared to summer season. Furthermore it was observed that according to Indian geography, some places receive good amount of sunlight throughout the year but has no sufficient irradiance (Fig. 3).

2 Solar Thermal Technologies

Solar thermal technologies are used to trap the solar radiation to heat the working fluids which increases the thermal energy of the fluid making it superheated vapor.

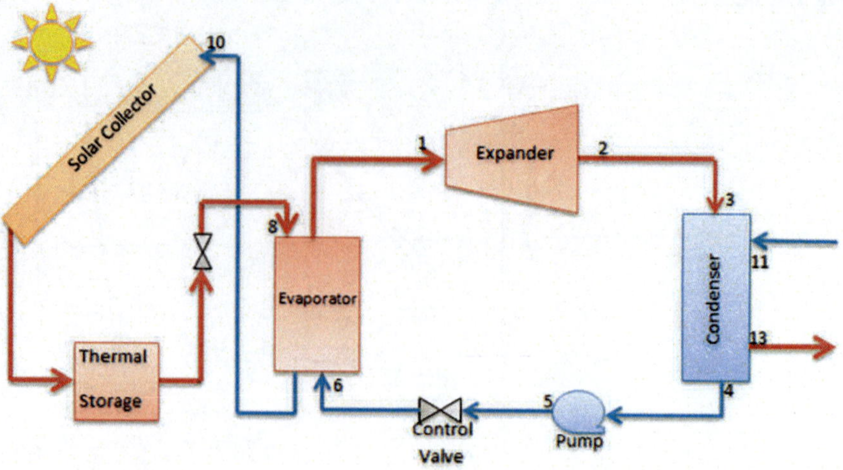

Fig. 3 Solar powered ORC

The fluid in superheated form drives the turbine to convert thermal energy to mechanical energy eventually converted to electrical energy using a generator. The solar radiations are collected using different solar collectors. In the following section different types of solar collectors used in the conventional technologies are described.

2.1 *Flat Plate Collector*

Flat plate collectors are solar thermal collectors which are mostly in the form of box like structure as shown in Fig. 4. In this the solar radiation is passed through one of

Fig. 4 Flat plate collector [5]

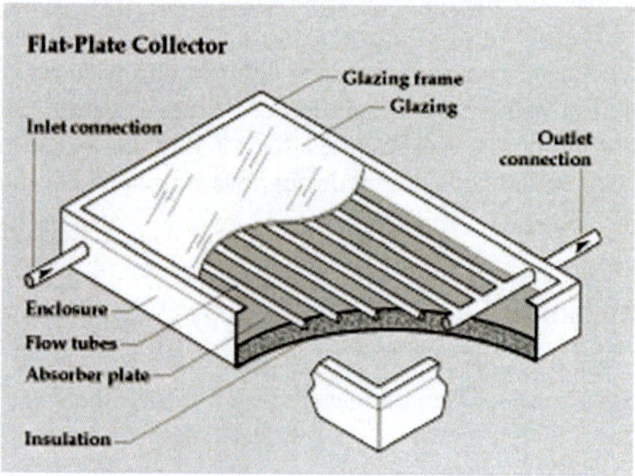

the face of the box which is transparent and mostly of glass. The rays on entering the FPC is absorbed by the tube which have high absorptivity, which help in increasing the temperature of the fluid inside the FTC and this fluid is then transferred to the storage area. The fluid inside FTC can reach temperature up 10 to 80 °C. Wang et al. [4] found that using flat plate collector R24fa and R123 could prove to be most suitable as working fluid for the system due to low operational pressure and high system performance in which the system had total of 300 solar collector of 1.76 * 1.6 * 1.1 m^3 (Figs. 5 and 6).

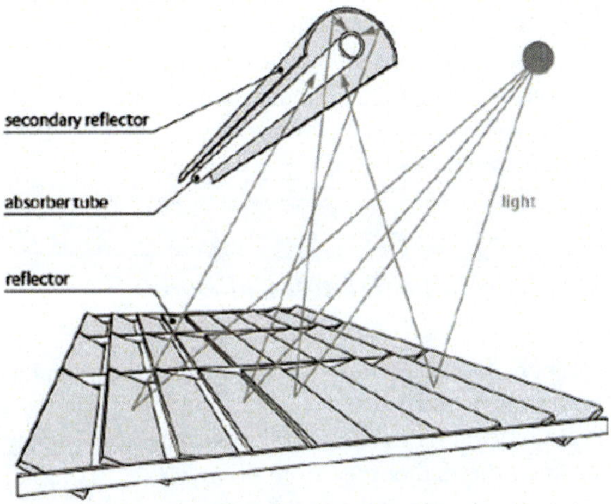

Fig. 5 Linear Fresnel reflector [6]

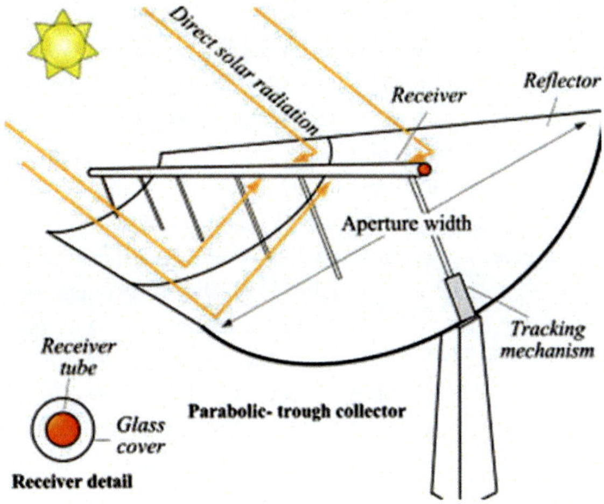

Fig. 6 Parabolic trough collector [18]

2.2 *Linear Fresnel-Reflector (LFR)*

Linear Fresnel reflector is one of the aspect of solar technology which helps in reflecting solar energy to a pre-defined position of tube through which thermal fluid flows. In this system the solar rays are concentrated to a specific point mostly a tube by reflecting type mirror which are adjustable in nature. Temperature range of 60–250 °C can be attained with LFR depending on the solar radiation and the type of material used. The energy conversion of solar to electrical energy varies from 10 to 200 MW with an average efficiency of 8–10%. The largest operational LFR concentrating solar power plant was installed in 2014 in India, with capacity of 125 MW which can produce electricity of 280,000 MWh/year [7]. Similarly, solar thermal power plant in Kimberlina US, Italy, Australia - The Liddell and the Puerto Errado 2 were installed with capacity of 5 MW, 1 MW, 9 MW and 30 MW respectively for commercial uses [8].

2.3 *Parabolic Trough Collector*

Parabolic trough collector is a type of solar thermal technology in which a parabolic primary reflector is used to concentrate the solar to a linear tubular receiver which is at the focal length of the primary reflector. The temperature of the working fluid can gain thermal energy from 60 to 300 °C using PTC with different material and concentration of parabolic reflector. The concentration can be analyzed by the ratio of area of parabolic reflector to the area of the receiver. As per the direction of sun the PTC can be placed with axis in North south during the summer and in winter the axis in east west direction which tracks the sun in north south direction to obtain maximum output [9]. Zhy et al. [10] have found that trancal oil when used as working fluid, it will produce maximum efficiency of 66% under movable receiver. Wang et al. [11] found that methanol can produce 16% efficiency increase in methanol reforming process with variable concentration ratio along PTC length with initial concentration 37. PTC has higher electrical potential of 16.12% then LFR that is 10.92% [12]. In India yearly potential 1584 kwh/m^2 is available with 23.16% of electrical potential in PTC and 12.17% in LFR [13].The difference between LFR and PTC is that the reflective mirror used in LFR is rectangular whereas in PTC it is a parabolic type. Sagar et al. [14] developed solar based ORC of 200 kW$_t$ that can be operated during night time with the help of thermal storage system (TES). The concept is helpful in remote places of India where sufficient sources of energy are not viable and hence solar energy can be stored to be used after sunset. Bellos et al. [15] used PTC for operating solar driven trigeneration system. Solar energy collected is used to drive the ORC and heat rejected from the condenser is used to operate absorption heat pump. Developed system could generate optimum power of 177.6 kW$_{el}$ along with heating and cooling outputs of 972 kW and 398.8 kW respectively. Wang et al. [16] has used dimethyl ether as a fuel with PTC with the efficiency of 9.87% which is higher compare to ORC system. Antonio et al. [17] modeled ORC system which runs from the exhaust gases coming out of gas turbine. PTC is used as solar collector which uses molten salt as heat transfer fluid.

3 Working Fluids

The fluids used in the ORCs are organic fluid of higher molecular masses containing carbon or their derivatives. These fluids are selected based on factors like critical temperature, safety factors for environment, toxicity, flammability and state of fluid (dry, wet or isentropic). Most of these fluids have low boiling point compared to water, thus making them preferable as working fluids of ORCs. Organic fluids are usually classified as wet, dry and isentropic fluid which represents magnitude of dT/dS on T-s curve. dT/dS corresponds to the condition of organic fluid at the end of the expansion process and it is recommended that organic fluid remains in superheated state to prevent partial condensation of fluid which might affects turbine blade and eventually reduces turbine life cycle. For the same reason, dry fluids are always preferred to be used in system with $dT/dS > 1$.

Table 2 shows the working fluid with different efficiency for an ORC plant of 1 kW in which the highest efficiency was found 8.55% for methanol which was attained by the high thermal efficiency of 12.22% with the volume flow rate at 0.95 L/s.There are many other fluids like R11, CFC, HCFC, R12, etc. which gives better output but due to their toxicity and poor environmental performance, they were not considered.

Table 2 Cycle efficiency for different fluids

Fluids	Critical Temperature (°C)	Boiling point (°C)	Critical pressure (MPa)	ODP	GWP (100 years)	Efficiency for 1 kW (%)
R113 [19]	214.1	47.6	3.439	0.8	5820	7
R245fa [20]	154.2	15.14	3.65	0	820	8.23
Methanol [20]	240.2	64.7	8.1	–	–	8.55
R141b [20]	204.2	32	4.25	0.12	725	8.23
R600a [20]	134.7	−11.7	3.64	0	–	5.28

Table 3 Consideration of organic fluid for different solar collectors

Fluid	Critical temperature (°C)	Boiling temperature (°C)	Linear	Flat	Parabolic
R134a	101	−26.3	✗	✓	✗
R11	198	23.77	✗	✓	✗
R141b	204.2	32	✓	✓	✗
Benzene	289	80.1	✗	✗	✓
Toluene	320	111	✗	✗	✓

4 System Capacity and Economic Consideration

The capacity of the ORC can be determined based on the geography, solar irradiance, fluid type and requirement of the output power. So with respect to requirement of the output power the capacity of the orc is divided into three scales.

- Small scale (1–20 kW)
- Medium scale (20–80 kW)
- Large scale (80 kW above).

ORC power plant that generates power of 1–20 kW is most feasible in today's market due to its relatively low cost, low maintenance and efficient performance. The thermodynamic model was studied and data were simulated on the working fluids keeping the mass flow rate at 1 kg/s, leading to production of 17.5 kW and 22.5 kW power output by R134a and R245fa at 70 °C. Further increasing the temperature at 99 °C, the output power was found to be 25 kW and 30 kW for R134a and R245fa respectively. In both the cases the efficiency was found to be 10% [21]. The cost for R134a and R245fa is $0.17/KWh and $0.14/KWh, with payback period of 10.5 years and 7.5 years respectively. Electrical potential of ORC is 4.4 GW in India [22]. On very small scale this system can be incorporated with vehicles to generate power from exhaust gases emitted by vehicles which can be used to heat the working fluid. The power generated can be used in batteries.

Table 3, shows the working fluids that can give better performance with particular solar collector, based on the critical temperature and boiling point as of fluid with low critical temperature cannot be used as the working fluid for e.g. R34a if used in PTC then it would reach temperature above critical temperature and hence cannot be used.

5 Conclusion

In the present work, review of the scope of solar powered organic Rankine cycle is reported. Depending on the solar irradiance of the place, different types of solar collectors can be installed as per need. Maximum solar irradiance is observed for

Gujarat and Rajasthan and eventually opens the scope of technological advancement in the field of generation of energy using solar powered ORC. For small scale production flat plate collector can be used with R134a due to its low temperature range of FPC and for large scale production Parabolic trough collector can be used with toluene as working fluid. R134a and R245fa are more preferable in small scale production because of their better performance and cost efficiency. Majority of the solar based ORC systems are operated using parabolic trough collectors because of their versatility in operation and enhanced energy conversion efficiency. India is developing day by day and conventional sources of energy alone cannot fulfill the need making solar powered ORCs suitable for present and future scenario as alternative source.

References

1. Popat Y, Shrivastava S, Saxena H (2018) Current scenario and future scope of solar energy in India. SAR J 1(2):57–66, ISSN 2619-9955. https://doi.org/10.18421/SAR12-05
2. Prajapati PP, Patel VK (2019) Thermo-economic optimization of a nanofluid based organic Rankine cycle: a multi-objective study and analysis. Thermal Sci Eng Progr. https://doi.org/10.1016/j.tsep.2019.100381
3. https://www.synergyenviron.com/tools/solar-irradiance. Last accessed on 15 Sept 2019
4. Wang M, Wang J, Zhao Y, Zhao P, Dai Y (2013) Thermodynamic analysis and optimization of a solar-driven regenerative organic Rankine cycle (ORC) based on flat-plate solar collectors. Appl Thermal Eng 50(1):816–825. https://doi.org/10.1016/j.applthermaleng.2012.08.013
5. https://www.daviddarling.info/encyclopedia/F/AE_flat_plate_solar_thermal_collector.html. Last accessed on 15 Sept 2019
6. Poullikkas A, Hadjipaschalis I, Kourtis G (2013) A comparative overview of wet and dry cooling system for Rankine cycle based CSP plant. Trends in Heat and Mass Transfer 13
7. Müller-Steinhagen H, Trieb F, Trieb F (2004) Concentrating solar power: a review of the technology. Available from https://www.dlr.de/tt/Portaldata/41/Resources/dokumente/institut/system/publications/Concentrating_Solar_Power_Part_1.pdf. Last accessed on 15 Sept 2019
8. SolarPACES (2016) CSP projects around the world. Available from https://www.solarpaces.org/csp-technology/csp-projects-around-the-world. Last accessed on 15 Sept 2019
9. Duffie JA, Beckman WA (2006) Solar engineering of thermal processe, 3rd edn. Wiley, Hoboken, NJ, USA
10. Zhu Y, Shi J, Li Y, Wang L, Huang Q, Xu G (2016) Design and experimental investigation of a stretched parabolic linear Fresnel reflector collecting system. Energy Convers Manag 126:89–98
11. Wang Y, Liu Q, Sun J, Lei J, Ju Y, Jin H (2017) A new solar receiver/reactor structure for hydrogen production. Energy Convers Manage 133:118–126
12. Bellos E, Tzivanidis C (2018) Assessment of linear solar concentrating technologies for Greek climate. Energy Convers Manage 171:1502–1513
13. Kumar KR, Reddy KS (2012) 4-E (energy–exergy–environmental–economic) "57 analyses of line-focusing stand-alone concentrating solar power plants." Int J Low-Carbon Technol 7(1):82–96
14. Lakhani S, Raul A, Saha SK (2017) Dynamic modelling of ORC-based solar thermal power plant integrated with multitube shell and tube latent heat thermal storage system. Appl Therm Eng 123:458–470. https://doi.org/10.1016/j.applthermaleng.2017.05.115
15. Bellos E, Tzivanidis C (2017) Parametric analysis and optimization of a solar driven trigeneration system based on ORC and absorption heat pump. J Clean Prod 161:493–509. https://doi.org/10.1016/j.jclepro.2017.05.159

16. Wang S, Fu Z (2019) Thermodynamic and economic analysis of solar assisted CCHP-ORC system with DME as fuel. Energy Convers Manage 186:535–545. https://doi.org/10.1016/j.enconman.2019.02.076
17. Pantaleo AM, Camporeale SM, Sorrentino A, Miliozzi A, Shah N, Markides CN (2017) Solar/biomass hybrid cycles with thermal storage and bottoming ORC: system integration and economic analysis. Energy Procedia 129:724–731. https://doi.org/10.1016/j.egypro.2017.09.105
18. Cabrera FJ, Fernández-García A, Silva RMP, Pérez-García M (2013) Use of parabolic trough solar collectors for solar refrigeration and air-conditioning applications. Renew Sustain Energy Rev 20:103–118. https://doi.org/10.1016/j.rser.2012.11.081
19. Saitoh T, Yamada N, Wakashima SI (2007) Solar Rankine cycle system using scroll expander. J Environ Eng 2:708–719
20. Baral S, Kim KC (2014) Thermodynamic modeling of the solar organic Rankine Cycle with selected organic working fluids for cogeneration. Distrib Gener Alternative Energy J 29(3):7–34. https://doi.org/10.1080/21563306.2014.10879015
21. Baral S (2019) Experimental and techno-economic analysis of solar-geothermal organic Rankine cycle technology for power generation in Nepal. Int J Photoenergy, Article ID 5814265
22. Sarkar J, Bhattacharyya S (2015) Potential of organic Rankine cycle technology in India: working fluid selection and feasibility study. Energy 90:1618–1625. https://doi.org/10.1016/j.energy.2015.07.001

Miscellaneous Systems

Systematic Diagnostics for C4I Systems: A Conceptual Framework

Aanchal Punia, Saurabh Bansal, and Pradeep Kr. Malik

1 Introduction

C4I systems are the state of art military software intensive systems for providing decision making and situation awareness (information about the location and status of enemy and friendly forces) for operations. The operational behavior of the system is inherently complex. Defense forces (army, navy, air force) and Homeland Security use C4I technologies for military strategy and operations. C4I system generally comprises of different sub-systems communicating over a wired or wireless network, where each sub-system is developed independently, and the unit testing of respective sub-system is the responsibility of the development and testing team. Post-Software Quality Testing (SQT) and prior to deployment, integration of modules is done in a known environment. During integration in simulated environment, these sub-systems communicate with each other and are tested for their specific requirements. Due to the inter-dependencies of the sub-systems, various issues such as data mismatch, invalid/wrong data, network address mismatch may arise during integration. Identification and diagnostics of any fault/issue occurring during integration and onsite operations is time-consuming and difficult. Normally, in absence of adequate data and under time pressure, developer & integration team uses ad-hoc methods to identify the problems in the system. In this paper, the problems which arise during lab-integration and onsite operations are identified. A framework is then proposed to provide efficient and time-saving solutions for these problems.

A. Punia (✉) · S. Bansal · P. Kr. Malik
Central Research Lab, Bharat Electronics Ltd, Ghaziabad, India
e-mail: aanchal.nith@gmail.com

S. Bansal
e-mail: saurabhbansal@bel.co.in

P. Kr. Malik
e-mail: pradeepkrmalik@bel.co.in

V. H. Saran and R. K. Misra (eds.), *Advances in Systems Engineering*,
Lecture Notes in Mechanical Engineering,
https://doi.org/10.1007/978-981-15-8025-3_80

Bosco et al. [1] proposed integrated diagnostic and repair information system (IDRIS) which automated the maintenance procedure and increased operational system readiness. Billy Kidwell [2] presented a decision support system which classified code faults and analyzed the source code.

In this paper, a diagnostic framework is introduced for identifying the system faults. The framework aims at solving the system problems in less time. This framework should automate the manual task of issue identification and recovery. The framework should work external to the existing C4I system. The existing C4I system software shall not be modified. The proposed diagnostic framework (DF) should troubleshoot C4I systems, a distributed mission critical system, in a systematic way. Diagnostic framework shall help in discovering problems in the system related data, exploring set-up details and giving information to support team in the desired format. The collected diagnostic data helps in problems identification, isolation and diagnosis of the fault when they occur. The framework should enhance the system performance by diagnosing a problem in minimum time, and hence achieving lesser down-time and a stable system.

Section 2 addresses the various C4I problems identified. Section 3 explains the diagnostic framework in detail. Section 4 discusses the application of this framework in a C4I system through a case study. Section 5 shows mapping of C4I system problems with the framework features and concludes the work and Section 6 briefly describes the future work possible.

2 Challenges Addressed

The identified existing problems in a C4I system are mentioned below.

A. Replication of issue at developers' side: Most of the identified problems require the reproducibility of the issue on the system for resolving the issue, i.e., developer has to see how the problem occurred. And it is not always possible to reproduce the problem in the main deployed system; also, the main system may not be available for debugging purpose. Developer cannot replicate any issue reported from the site at his side and developer cannot debug the deployed system. This forms a deadlock.

B. System logs: Every sub-system stores the logs on its own, but the logs are stored on the system itself. Centralized access of these logs is not available. These logs are to be transferred manually, when required. Also, each sub-system has its logs in its own format, which may differ from other sub-systems.

C. Faults during integration: Integration teams follow an exhaustive integration test procedure to verify integrated system functionalities. Whenever a system failure occurs, identification and diagnosis of the fault are a difficult task. A fault may arise because of invalid/wrong data, data-structure mismatch, wrong network protocols, message drop, etc. may be called interface error. Each sub-system has to be checked manually to identify the sub-system responsible for the fault.

D. Sub-system unavailability: Testing of the system during integration requires all sub-systems to be active. There are cases when any sub-system is not developed completely or is under maintenance; hence, testing of the system as a whole gets delayed.
E. Database monitoring: All data of the C4I systems is maintained in a database. Each sub-system can perform read and write operations on the database. Multiple writers can be writing data for the same attribute to the database. In current scenario, only the last change is known, but the writer is unknown. Also, DBA cannot know which other writers have also changed the data corresponding to this attribute.

Along with the abovementioned problems, time taken in solving these problems is also of concern. Identification and recovery of error in a C4I system is a mundane task. It requires deployment site visits by developers to see how the issue arises, debugging at site may not be possible, after resolution, it has to be again tested at site. Therefore, a proactive approach is required to isolate and resolve the issues in the system as they occur in minimum time and effort. The proactive approach should suggest a way of reproducibility of the problem on the developer's side, without having a need of reproducibility on the main deployed system. To ease and automate the issue/fault diagnosis, a diagnostic framework is proposed in this work.

3 C4I Diagnostics: Proposed Diagnostic Framework

Diagnostic framework is a set of tools and classes which provides adaptive diagnostics and troubleshooting. The aim of the diagnostic framework is to find the source of problem sooner and hence reducing the downtime of the system. It gives detailed information about the problem, so that the support team/developers can fix the problem in first attempt. If issues are detected early in the product lifecycle, they cause less harm and are easy to resolve; therefore, this framework is most useful during the integration phase. It identifies and isolates the problem. Diagnostic framework should be centralized to the C4I system, so that it can communicate with every sub-system. The diagnostic framework tool should capture and monitor logs from different sub-systems. This framework aims at minimizing the efforts done by the developers and integration teams during discovery and resolution of a system issue.

Figure 1 depicts the diagnostic framework components. The framework is divided into various components, so that each problem can be handled separately by an individual component. The different components combined together are capable of solving most of the C4I systems' problems. Therefore, the issues can be identified, diagnosed and resolved in comparatively less time than the traditional approach. This approach saves time, thus ensuring fast delivery of a reliable and stable system.

Based on the understanding of the problem, diagnostic framework shall consist of the following features:

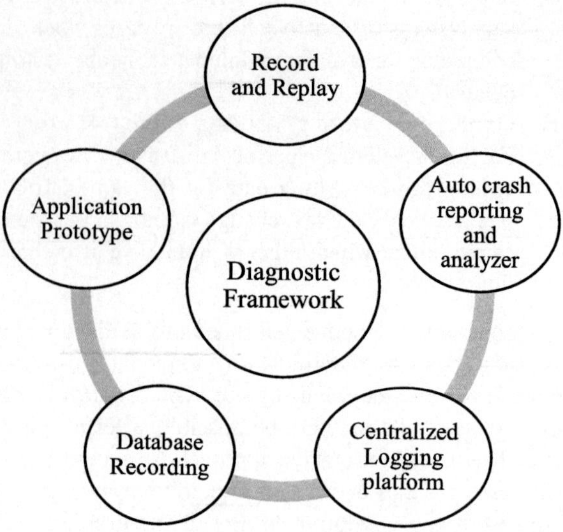

Fig. 1 Diagnostic framework: components

3.1 Record and Replay

The framework should provide record and replay feature. System events on-line recordings can be replayed off-line for inspection of system [3]. Recording feature provides recording of the ongoing communication and scenarios. Replay feature provides simulation of the actual scenario at some later point of time for issue reproducibility or analysis. Data communication and display screen should be recorded. This feature provides simulation of the system behavior at some specific time. This simulation shall help in double checking the activities happened at a particular time and analyze how the system behaves for different scenarios and in different environment.

3.1.1 System Screen Record and Replay

System screen is recorded in this feature. Ongoing display screen activities like mouse click, mouse hover, system status, etc. will be captured. Screen recording records the sequence of user actions. Replay of these recordings tell developers the steps performed when the problem occurred, hence recreating the exact scenario.

3.1.2 System Data Record and Replay

Real-time data is recorded. Recording should be done based on a configurable recording file, which specifies the data to be recorded, the sub-systems involved, etc. Data replay will initiate the recorded communication and system will behave as if the communication is going on in real.

This feature shall facilitate recording of system data in real time and generation of the same system behavior later/off-site via replay. The replay feature allows the developer team to replicate any issue occurring on-site at their end, thus can resolve the same without any need of actual site visit, just to observe when and how the issue arises. Data records keep a track of in and out messages. Data recordings are database files, which can be analyzed and checked, in case wrong message is suspected. Record and replay can be used for training purpose as well [4].

3.2 Autocrash Reporting and Analyzer

Whenever a system/application crashes, the core dump is saved in the system's memory. To get that core dump, the system has to be accessed manually. This framework suggests that a centralized server for logging should exist in the C4I system. Centralized server shall collect the relevant data such as core files and other data of forensic utility when application is crashed. Logging to the server shall happen automatically whenever an application is crashed. Back trace from the core dump will be automatically generated. Function names from the back trace on which function is crashed will be saved in a file.

Automatic Bug Reporting Tool (ABRT) is a daemon and collection of tools which can handle crashes and monitor logs for errors. ABRT stores and processes the crashes occurred [5].

This feature shall facilitate autodetection of a system crash or failure. It should generate a crash report, which shall be used by the developers and integration team for finding the cause as soon as possible.

3.3 Centralized Logging Platform

Logs from each sub-system should be maintained on a centralized logging server. These logs can be filtered on the basis of different log levels such as debug, info, fatal, error, warn, critical. Sub-systems application should use pre-defined logging architecture for this feature to work better. The collected logs can be analyzed to find any bug in the system application, to find any error during the integration process or onsite.

Logging mechanisms like rocket-fast system for log processing (RSYSLOG), log4j, log4net, etc. can be used for collecting the logs. Rsyslog is an open-source logging program [6]. Rsyslog daemon on a client machine sends logs over to a remote log server. It can use both UDP and TCP protocols. This log server shall keep logs from all the sub-systems. P.K. Sahoo et al. [7] discussed log management in detail and proposed a solution for centralized log storage using syslog.

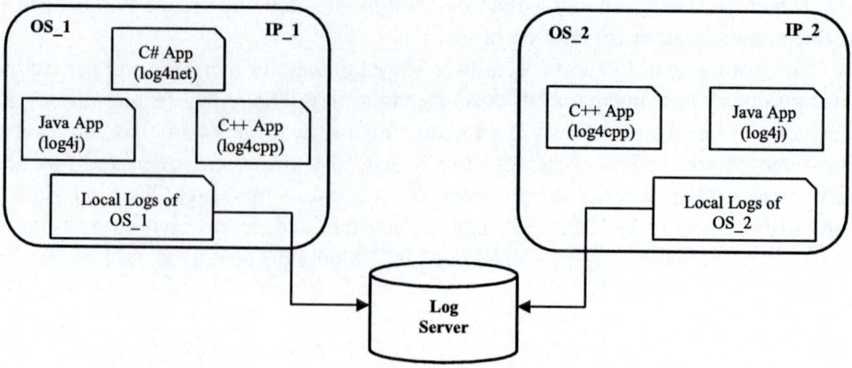

Fig. 2 Centralized logging

Figure 2 shows how logging shall be done in a system consisting of two subsystems, each having unique IP address and works on different operating systems (OS). The modules responsible for logging within an application will send logs to the log server. Log server shall maintain logs from different machines depending on the IP address of the machine. The next directory level shall consider the application name (appName) and then message.log file will be created for each running application.

Over time, the size of the log file will keep on increasing because of new data entries. Log files should be compressed periodically for better memory organization. Operating systems use "logrotate" to rotate, compress and maintain logs automatically [8]. Cron is a Linux-based scheduler. The cron services run on commands mentioned in Crontab [9]. Cron daemon calls logrotate after specific intervals.

This feature shall facilitate centralized logging. The logs provided by different sub-systems should be in a standard format as decided by the architecture team. These logs will help in finding any bug or exception error, i.e., troubleshooting any system issue.

3.4 Database Recording

Multiple users read and write simultaneously from the same database, and it is difficult to know who has done what changes in the database. DBA can see only the changes done but is unaware of who has made the change. DBA should know who has made what change to the database. The purpose of this feature is to use the server and database-level DDL triggers for recording the changes as well as the details about the changes made [10]. The article "Monitoring and recording DDL changes on SQL 2005 (NARC)" [10] suggests that database recording can be done in four steps. The first step includes creation of two tables; one for the changes at the server level and other for changes at the database level. The proposed framework suggests duplication of the main database table with some additional fields (hostname, username, action). Once the table is created, then database triggers have to be created which shall record the DDL events (create, update, delete).

This feature shall maintain the details (who, when, what) of the changes done to database. It shall detect any change made by unauthorized personnel.

3.5 Application Prototype

This facility shall send messages on behalf of a non-active sub-system during testing and integration. During integration testing, it substitutes for a sub-system which was supposed to be active but is not functional currently. This facility can send and receive messages from a sub-system. It also checks for valid message fields. It should generate alerts when a message attribute is not in the defined range. It keeps an eye on the live message sequences moving through the system. This facility should provide testers a method to check the messages over the network and verify the structure of the message.

This feature shall facilitate testing of interfaces between two sub-systems. Also, it provides data in and out of a sub-system, which can be used during development as well.

4 Case Study

BEL provides C4I systems to defense forces. These systems are deployed at various locations. The required system comprises of various sub-systems/modules. Each subsystem has its respective team of system architects and designers, developers and testers. This case study involves three cases. Case 1 discusses the use of simulator during development phase. Case 2 discusses the steps taken for resolving a system failure during system deployment phase. Case 3 talks about optimization issues which occurs during both development and deployment phases.

Case 1: System development phase: In general, every module develops and maintains its own simulator for unit testing. A requirement change affecting a particular module leads to updation of its respective simulator. Application prototyping uses a single simulator for different modules. Any new change will be updated once in simulator. This saves time and effort.

Case 2: System deployment phase: Generally, a system crash is reported along with log files from the customer site. In a multi-module system, isolating the fault on responsible module is a complex task. If the faulty module is not identified with the help of logs, then developer team visits the site for identification and isolation of the faulty module.

Site team records the system failure scenario using the record feature of DF Tool and sends the recordings along with log files. The support team can debug the issue and identify the faulty module by replaying the same scenario recording, without any need of site visit, and release the updated version. This leads to reduction in issue resolution time.

Case 3: Some unexpected scenarios like maximum CPU utilization, memory over-use, network bandwidth mismatch occur during development and deployment phase. These scenarios indicate that system has to be optimized. System recordings, database recordings and logs captured from site are then analyzed to find a solution.

5 Results and Conclusion

The proposed framework, explained in Sect. 3, resolves the C4I system problems mentioned in Sect. 2. This framework minimizes the developer's efforts of resolving an issue. It accelerates the fault diagnosis process. It helps in delivering a robust and reliable C4I system which is capable of autofault detection. This framework supports the debugging process. Each component of the diagnostic framework is designed to solve one or more problems.

Table 1 shows the mapping between the problems identified for a C4I system in Sect. 2 and the diagnostic framework components mentioned in Sect. 3.

The diagnostic framework described in this paper is designed for a C4I system. This framework will collect data, record site scenarios, capture crash reports and logs. Along with it, it also provides replaying of the recorded scenarios, reproducing issues for developers' to debug. Also, during integration phase, pin-pointing the faulty sub-system requires ad-hoc methods. This framework collects the logs from different sub-systems, which can then be used to identify the faulty sub-system. This framework can be used during debugging and testing a system. To summarize, we can say that this framework should diagnose any issue during integration and system set-up phase and also when the system is deployed at site.

Table 1 C4I problems—Framework components mapping

S. No.	C4I problems	Framework components
1	Replication of issue at developers' side	Record and replay
2	System logs	Centralized logging
3	Faults during integration	Autocrash reporting and analyzer, Application prototype
4	Sub-system unavailability	Application prototype
5	Database monitoring	Database recording

6 Future Work

In this paper, we have discussed how the framework should be used for a C4I system. Logs and communicated data are collected. The collection of logs and data over time may provide a rich dataset from which a knowledge base can be built. This knowledge base may then provide solutions to the complex and unseen problems. Data analysis tools can be used to get inference from the recorded data. This feature shall alert for abnormal system behavior based on analysis of the recorded data. Framework may also provide the option of viewing the recorded data for further analysis.

References

1. Bosco CD, Su LP, Darty MA, Kusek SM, Griffin BA, Lee TC (1995) Integrated diagnostic and repair information system (IDRIS)—an open system approach. In: Conference Record AUTOTESTCON '95. 'Systems readiness: test technology for the 21st century', 8–10 Aug 1995
2. Kidwell B (2011) A decision support system for the classification of software coding faults: a research abstract. 21–28 May 2011
3. Thane H, Hansson H (2000) Using deterministic replay for debugging of distributed real-time systems. In: Proceedings 12th Euromicro conference on real-time systems, Euromicro RTS 2000, 19–21 June 2000
4. Aggarwal V, Sarita (2017) Testing and debugging of C4I Systems using Record and replay. Computing for Sustainable Global Development, 1st–3rd March 2017
5. How ABRT works, https://abrt.readthedocs.io/en/latest/howitworks.html
6. Cezar M, How to setup Rsyslog client to send logs to Rsyslog server in CentOs 7. https://www.tecmint.com/setup-rsyslog-client-to-send-logs-to-rsyslog-server-in-centos-7/
7. Sahoo PK, Chottray RK, Jena G, Pattnaik S (2012) Syslog a promising solution to log management. Int J Adv Res Comput Sci, May-June, 2012, 584–588
8. Logrotate(8)—Linux man page, https://linux.die.net/man/8/logrotate
9. How to use Cron in Linux, https://opensource.com/article/17/11/how-use-cron-linux
10. Greaves J, Monitoring and recording DDL changes on SQL 2005 (NARC). https://www.sqlservercentral.com/articles/monitoring-and-recording-ddl-changes-on-sql-2005-narcJamesGreaves, 29 Feb 2008

Bio-Info-Sensor Image Processing Approach: Disaster Pre-alarm for Earthquake

Mahima Yadav, Shubham Chaudhary, and Anshul Agarwal

1 Introduction

The most frightening and hazardous phenomena of nature is earthquake. Earthquake is a hazardous disaster on the earth. The word earthquake means shaking of earth. An earthquake occurs naturally or due to some human activities. Most of the natural occurring earthquakes caused due to geographical faults, some due to volcanic activities, landslides, etc. Nuclear tests and mine blasts are those human activities which sometimes results in occurrence of earthquake. It is the sudden, rapid shaking of the earth, caused by the breaking and shifting of rock as it releases strain that has accumulated under the earth's crust for over a long time. The forces of tectonic plates have shaped the earth, as the huge plates that form the earth's surface slowly move over, under and past each other. Sometimes, the movement is gradual. At different occasions, the plates are bolted together, unfit to discharge collected vitality. When the accumulated energy grows strong enough, the plates break free and causes an earthquake [1]. It is measured using observations from seismometers. The minute size is the most well-known scale on which quakes bigger than around 5 are accounted for the whole globe. The more various quakes littler than extent 5 revealed by national seismological observatories are estimated for the most part on the neighborhood

M. Yadav (✉)
Department of Electrical Engineering, Dayalbagh Educational Institute, Agra, India
e-mail: mahimayadav611998@gmail.com

S. Chaudhary
G. N. N. College Agra, Agra, India
e-mail: shubham27bt@gmail.com

A. Agarwal
Department of Applied Science (Chemistry), Faculty of Engineering and Technology (FET), Agra College, Agra, India
e-mail: anshul_chem@yahoo.com

V. H. Saran and R. K. Misra (eds.), *Advances in Systems Engineering*,
Lecture Notes in Mechanical Engineering,
https://doi.org/10.1007/978-981-15-8025-3_81

greatness scale, additionally alluded to as the Richter scale. These two scales are numerically comparative over their scope of legitimacy. Extent 3 or lower quakes are generally practically intangible or feeble and size 7 and over conceivably causes genuine harm over bigger zones, contingent upon their profundity [2].

When the epicenter of a large earthquake is located offshore, the seabed may be moved to cause a tsunami. Earthquakes may also cause landslides, and occasionally volcanic activities. Earthquake prediction efforts have failed over past years. Efforts to make longer-term deterministic forecasts have not given valid results [3]. There are five stages of earthquake prediction which depend on duration. It can be predicted at different time intervals such as long-term (tens of years), intermediate-term (years), short-term (months to weeks), and immediate (days and less) [4]. Many efforts are being made to detect it previously. Neural network approaches are also used to determine time lag between the seismic electric signals (SES) and the seismic event itself [5] and to liquefaction-induced lateral spreading during strong earthquake [6]. Earlier nonlinear finite element method (FEM) and seismic vulnerability curves were used to determine the damage during earthquake [7]. Based on research among feed-forward Levenberg–Marquardt backpropagation (LMBP) neural network, a recurrent neural network, and a radial basis function (RBF) neural network RNN gives high accuracy [8]. Earthquake are not continuous, its collection of segments that are separated by discontinuities and is steps in map view. Hence, one of the methods is to detect the earthquake area exactly and controlling the length of earthquake, and this can be done by stopping rupture propagation which will limit fault step [9].

2 Animals as a Biosensors

Based on various researches regarding earthquake detection, we use word "precursors." Precursors are the sensors which can detect earthquake signals prior to its occurrence. Based on various researches made over the biosensors, here, biosensors are referred as the animals and many researches prove that animals show abnormal behavior in seismically active region and in our system, we are making use of these bio-precursors as sensors for earthquake detection. Animals show unusual and abnormal behavior from some days before occurring the earthquake. Rats, weasels, snake, and centipedes apparently left their homes and set out toward well-being a few days before a dangerous tremor. Episodic proof proliferates of creatures, fish, winged animals, reptiles, and creepy crawlies displaying weird conduct anyplace from weeks to seconds before a quake. Be that as it may, predictable and dependable conduct before seismic occasions, and a system clarifying how it could function, still escapes us [10]. For the most part, seismic tremor is gone before by certain sign like geophysical antecedents or other like anomalous creature conduct [11]. The recognition of unusual animal behavior in a systematic way can lead us to predict large and potentially destructive earthquakes [12].

The tremors transmit different sorts of seismic waves. The "p" (essential) waves travel through the world's hull about twice as quick as the "s" (auxiliary) waves, so they show up first. The more noteworthy the separation, the more noteworthy the postponement between them. The P waves are likewise flimsier, and frequently unnoticed by individuals. Along these lines, the indication of caution revealed in the creatures was without a doubt incited by the p waves [13]. When rocks were under very high levels of stress just before an earthquake, they release charged particles. These charged particles can flow out into the surrounding rocks and when they arrive at the Earth's surface, they react with the air, converting air molecules into charged particles known as ions. These positive airborne ions increase the level of serotonin (A Phenolic Amine $C_{10}H_{12}N_2O$ that is a powerful vasoconstrictor and is found specially in the blood serum and gastric mucosa of mammals [14]), a stress hormone, in the blood of animals, which causes the abnormal behavior in animals [15].

3 Behavior Recognition System

This system is technically based on image recognition. Image recognition, a domain, based on computer vision technology that includes methods for acquiring, processing, analyzing, and understanding images and, in general, high-dimensional data from the real world in order to understand information from image data using models constructed with the aid of geometry, physics, statistics, and learning theory. Here, the main emphasize is given on real, scientific, anomalous animal behavior, e.g., in the forms of decisions realizing from their change in the effect of stimuli, respectively [16–18]. Earthquake prediction though the living organism is great enhancement over existing technique [19]. Even neural network approach can be used where nonlinear changes of the eight parameters of earthquake are measured. A three-layer Levenberg–Marquardt feedforward learning algorithm can also work to model the earthquake prediction [20] (Fig. 1 and Table 1).

4 Datasets

Datasets play a major role in machine learning. The efficiency of the project depends on quality and quantity of dataset. In this section, details of the datasets are discussed. Three labeled datasets are used in this paper. First dataset contains the images of animal's versus empty images (does not contain animals). Second dataset contain images of different species of animals, for example, Cows, cat, dog, elephant, fox, wolf, rat etc., and the last contains different event and states of animals like moving, standing, eating, running, etc.

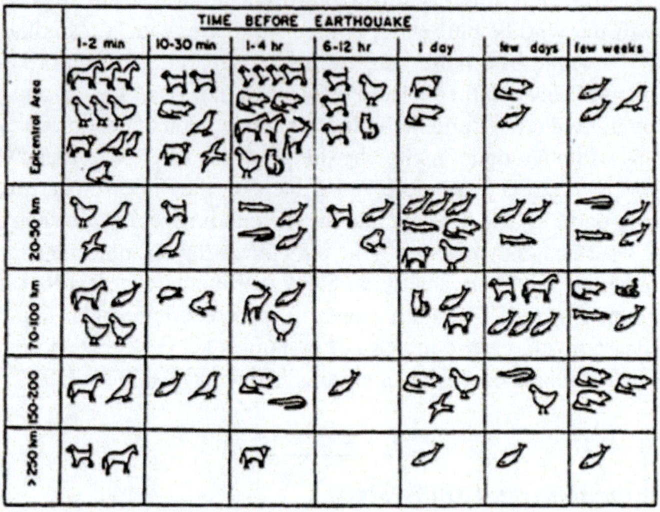

Fig. 1 Distribution of animal behavior according to distance from the epicenter and the time before the main shock of 36 different earthquakes in Europe, Asia, North America, and South America. Symbols indicate reports on the following animals: catfish, eels, other fish, frogs, snakes, turtles, sea birds, chickens, other birds, dogs, cats, deer, horses, cows, rats and mice. Data are non-systematic and collected from many sources [21]

5 Algorithm

Real-time images are captured through camera. And image frames are extracted from videos. Each extracted image pass through preprocessing which is to remove noise from the image. Then Animal detection and recognition of its category based on feature take place. Deep neural network is used for animal behavior recognition and classification.

The problem is tacked by merging following techniques.

5.1 Detection of an Animal

This is two-class classification problem which used dataset of first category animal vs empty. VGG model is used for this problem. Visual geometry group (VGG) is deep convolutional neural network model whose convolutional filter is of size 3 * 3 [23]. VGG 16 is pretrained CNN which has 16 layers. It uses ImageNet dataset to train model. Following steps are implemented in MATLAB to identify animal.

Load images and retrain the network with labeled images. In our case, two labeled are empty and animal.

Table 1 Abnormal animal behavior prior to earthquakes and other contexts in which similar behavior has been observed [21]

Animal	Behavior reported before earthquake	Behavior reported in another context	Age group	Days before EQ. OCC
Cats	Constant hiding, refusal to go outside	Psychogenic shock	Adult	Few days before [10]
Chickens	Flying to high perches, crowding together, hysteria	Sudden darkness, loud explosion	Adult	–
Dogs	Barking, following owner constantly from room to room	territorial response, response to strangers, overdependent pet	Adult	–
Fish	Jumping out of water	Quick turns during swimming, twilight hunting	Adult	Few weeks before [10]
Mice	Behave as if drunken, convulsions	Audiogenic seizure caused by noise of 4–80 kHz, 90–130 dB	Adult	2 Weeks [22]
Mussels	Move to higher attachment site on seashore	Rising water before hurricane	Adult	–
Pigs	Biting each other's tails	Overcrowded conditions	Adult	–
Rats	Vigilance, jumpiness, vertical leaping, crouch-like gesture, muscle contraction	Alarm response to ground predators, acoustic startle response	Adult	2 weeks [22]

An image frame is captured from video and passed in this as an input image. Then image is read and resized using imread and imresize, respectively.

Classify the image in one of the two classes using classify. Then, display the output image with predicted label [24].

5.2 *Species Identification*

After detection of animal, we identify its species which is multiclass classification problem. For this issue, we use following approach.

Get second dataset and split into training and testing. In our case, training data is 70% and testing data is 30%.

Train SVM classifier using attributes and features. Support vector machine is efficient method for multiclass classification.

Use the output image and pass as input in above classifier to predict label. Measure efficiency and accuracy.

5.3 Animal Behavior Recognition

In this last step, we will recognize the state of the animal or the action which animal is performing. This is also multiclass classification problem which will use third category of dataset [25]. The algorithm is same as in species identification only dataset is changed and the input image in this case is the output of previous step. Result will predict whether earthquake will occur or not based on animal's behavior [26].

6 System Development

6.1 Learning Element/System

Learning system can be defined as the training element which defines or covers the part of system's way of exhibiting physical functionalities and exceptional behaviors. For example, in the implementation of system development, we will design an artificial lake consisting of all aquatic animals and also the terrestrial animals covering a few portions of the land. A video-tracking system is made in use to keep track of all the animal's usual and exceptional activities being categorized on their behavior basis, taking input from the atmosphere and analyzing the animal's behavior (Fig. 2).

Learning system also contains all the detailed information about earthquake conditions revealing the adverse impact of disaster to the environment. This system mainly emphasizes on the concept of automatic animal behavior filtration which can be done with the help of behavioral analysis algorithm filtered systematically based on the detailed, elaborated disastrous conditions. This training element is connected directly to the central database server which in turn, plays another significant role in system's development phase.

6.2 Tracking System

Tracking system would be the actual system/setup responsible for the detection of an earthquake prior to its occurrence. This system includes a central database/knowledge base server, an alarm, an automatic behavior recognition system and planted cameras at various suspected sites color (Fig. 3).

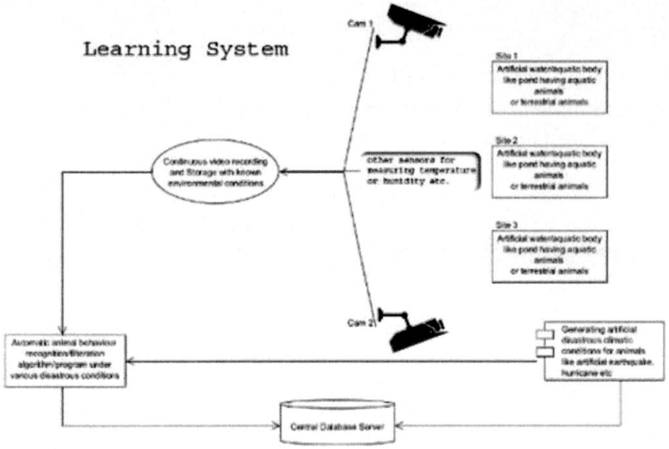

Fig. 2 Overview of learning system

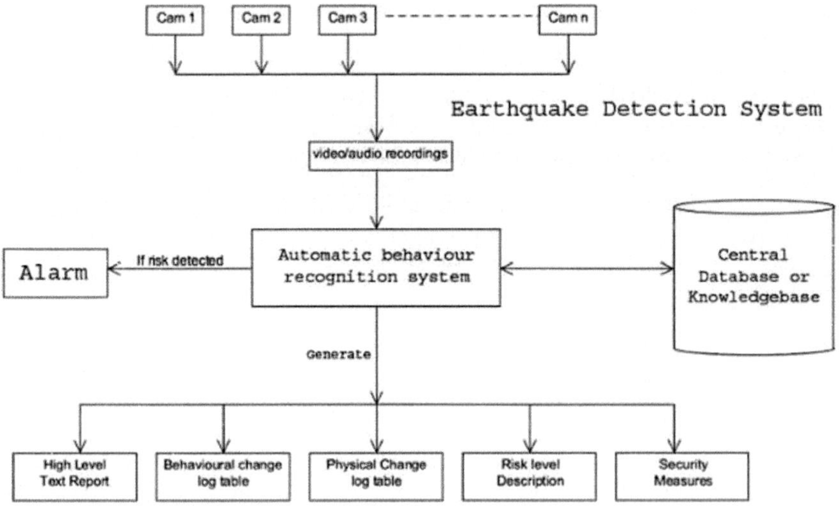

Fig. 3 Flowchart of automatic behavior recognition system

6.3 Central Database/Knowledgebase Server

It is an RDBMS Server which keeps all the information/results analyzed by the learning system and all other previous research results regarding the same criteria. Earthquake Detection System (EDS) will use this database for the mapping of recognized behavior with the respective event or physical change.

6.4 Alarm

An indicating system which will raise an alert when there is any dangerous activity recognized by automatic detection system.

6.5 Supplementary Result

Supplementary Result consists of five detailed reports are as follows:

High-level text report consists of all the information regarding each activity time to time in a txt or doc file in a layman readable language and is available openly.

Behavioral change log table consists of all the information about the changes in the behavior of animals under observations in a tabular format under different circumstances.

Physical change log table consists of all the information regarding physical change, i.e., humidity, temperature, pressure, etc., about the earth and its atmosphere.

Risk-level description gives you the detailed analysis like severity, intensity, expected damage of all the upcoming risks or disasters based on behavioral changes in animals and previous history stored in central database.

Security measure is a report that provides us the various ways or methods or activities to be performed under the emergency conditions.

7 Conclusion

Exact prediction of earthquake is not possible till this date. But various flying birds, domestic birds, domestic animals, reptiles, water animals (including eels and other contents), etc., show abnormal behavior acting as precursors or sensors can be used to a good deal. At the same time, modern computer technology (supervision + application) is also very useful to collect the relevant date in order to seek a fact-finding mission ultimately.

Acknowledgements We gratefully acknowledge Harshita Agarwal, Akash Gupta, Smriti Mathur and our research group members (360°prg) for their critical evaluation of the Manuscript, useful comments and suggestions.

References

1. Campaign R (2018) READY: prepare, plan, stay informed. DHS Seal- FEMA. [Online]. Available: https://www.njwtpl.org/community-guides-directories/us-government/ready-prepare-plan-stay-informed
2. I. f. C. Studies (2011) Earthquake FAQ, 6832 Ellison Hall, University of California, Santa Barbara, 24 July 2011. [Online]. Available: https://crustal.eri.ucsb.edu/outreach/faq.php
3. Robert JG (1997) Earthquake prediction: a critical review. Geophys J Int 131(3):425–450
4. Vladimir K-B (2002) Earthquake prediction: state-of-the-art and emerging Possibilities. Ann Rev Earth Planet Sci 30:1–33
5. Moustra M, Avraamides M, Christodoulou C (2011) Artificial neural networks for earthquake prediction using time series magnitude data or Seismic Electric Signals. Expert Syst Appl 38(12):15032–15039
6. Baziar MH, Ghorbani A (2005) Evaluation of lateral spreading using artificial neural networks. Soil Dyn Earthq Eng 25(1):1–9
7. de Lautoura OR, Omenzetterb P (2009) Prediction of seismic-induced structural damage using artificial neural networks. Eng Struct 31(2):600–606
8. Panakkat A, Adeli H (2007) Neural network models for earthquake magnitude prediction using multiple seismicity indicators. Int J Neural Syst 17(01):13–33
9. Wesnousky SG (2006) Predicting the endpoints of earthquake ruptures. Nature 444:358–360
10. E. H. Program, Animals & earthquake prediction. USGS- science for a changing world, [Online]. Available: https://earthquake.usgs.gov/learn/topics/animal_eqs.php.
11. Bhargava N, Katiyar VK, Sharma ML, Pradhan ML (2009) Prediction through animal behavior: a review. Indian J Biomech, 159–165
12. George P, The use of animals in earthquake prediction. Rense.com, [Online]. Available: https://rense.com/general61/use.htm
13. Imster E (2011) How animals at National Zoo reacted to August 23 D.C. earthquake. EarthSky, 24 August 2011. [Online]. Available: https://earthsky.org/earth/how-animals-at-the-national-zoo-reacted-to-earthquake
14. Merriam, Webster's New Colligate Dictionary. Merriam-Webster, Incorporated, 2019. [Online]. Available: https://www.merriam-webster.com/
15. Gill V (2011) How animals predict earthquakes (BBC). Science reporter, BBC Nature
16. Shapiro LG, Stockman GC (2001) Computer vision. Prentice Hall, Washington
17. Morris T (2004) Computer vision and image processing. Palgrave Macmillan
18. Jähne B, Haußecker H (2000) Computer vision and applications, A guide for students and practitioners. Academic Press
19. Pirmagomedov R, Blinnikov M, Amelyanovich A (2018) IoT based earthquake prediction technology. In: Internet of things, smart spaces, and next generation networks and systems, Russia
20. Külahcı F, İnceöz M, Doğru M, Aksoy E, Baykara O (2009) Artificial neural network model for earthquake prediction with radon monitoring. Appl Radiat Isot 67(1):212–219
21. Tong WK (1988) Abnormal animal behavior and the prediction of earthquake. ESCI-402: Current problems in Earth Science, Chicago
22. Polyakov YS, Ryabinin GV, Solovyeva AB, Timashev SF (2015) Is it possible to predict strong earthquakes? Pure Appl Geophys 172(7):1945–1957
23. Simonyan K, Zisserman A (2015) Very deep convolutional networks for large-scale image recognition. In: ICLR
24. "vgg16," Mathworks, [Online]. Available: https://www.mathworks.com/help/deeplearning/ref/vgg16.html;jsessionid=432681aa4ed60f1e9ee9bf594add
25. Norouzzadeha MS, Nguyenb A, Kosmalac M, Swansond A, Palmere MS (2018) Automatically identifying, counting, and describing wild animals in camera-trap images with deep learning. PNAS 115(25):E5716–E5725
26. Trnovszky T, Kamencay P, Orjesek R, Benco M, Sykora P (2017) Animal recognition system based on convolutional neural network. Adv Electr Electron Eng 15:517–525

Effect of Variant Seat Suspension Systems on Human Subject in Agricultural Conditions

Arvinder Singh, Harkirat Singh, Ishbir Singh, Sachin Kalsi, and V. H. Saran

1 Introduction

Indian economy depends mainly on agricultural and its products that contribute large percentage in gross domestic products (GDP). Also, occupation of maximum population depends mainly upon agriculture, and for this, they work for 6–8 h in variant field operations. With change in technology, farm tractor has become more useful and effective in cultivating large amount of agricultural land in a short span of time. Due to off road conditions, it generates lot of vibrations while doing agricultural operations like harrowing, plowing, and harvesting and these vibrations get directly transferred to operator that causes many harmful effects on his health generally at low frequencies. According to International Organization of Standardization (ISO-2631-1) vibrations within frequency range of 0–20 Hz, whole-body vibrations cause many ill-effects in human body like lower back pain, motion sickness, and digestion problem [1].

Many researchers have contributed in reducing the effect of vibrations on human body at variant environmental conditions using different methods. Jalil and Griffin [2] studied using experimental study an effect of random fore-and-aft vibrations

A. Singh · H. Singh · I. Singh (✉) · S. Kalsi
Chandigarh University, Mohali 140413, India
e-mail: ishbir@rediffmail.com

A. Singh
e-mail: arvindernagra2@gmail.com

H. Singh
e-mail: harkiratsinghdhadly@gmail.com

V. H. Saran
Indian Institute of Technology Roorke, Roorkee 247667, India
e-mail: saranfme@iitr.ernet.in

V. H. Saran and R. K. Misra (eds.), *Advances in Systems Engineering*,
Lecture Notes in Mechanical Engineering,
https://doi.org/10.1007/978-981-15-8025-3_82

when exposed to twelve subjects under frequency range of 0–20 Hz. Three resonance frequencies were found in which the first was observed 1–2 Hz at lower and middle back, second and third were found to be in the range of 4–5 and 5–8 Hz at upper back, lower and middle back, respectively. Bhoi et al. [3] designed and fabricated a hydraulic collapsible tractor operator enclosure to provide more comfort to driver during different weather conditions. This apparatus has been observed with the comfort level of 8.75 out of in 10 (cold conditions 18–220 °C), 7.75 out of 10 in hot conditions (32–350 °C) and 7.5 out of 10 in dusty conditions of comfort scale.

Catania et al. [4] performed experiment to find out the frequency weighted acceleration by using different equipment with slope of 0% and 30%. It has been found that rototiller and vibro farmer produced more vibration than chisel plow and flail mover. Cutini et al. [5] performed experiment on three farm tractors and six different terrains. The frequency and acceleration were measured in x-, y-, and z-directions. It was found that tractor forward speed and track roughness were the most dominant factor while the maximum amplitude was in z-axis. Dewangan et al. [6] conducted experimental study on thirty male and twenty-seven female subjects to find out the effect of anthropometric and gender effect on vibration transmissibility in terms of apparent mass. It was found that primary resonance frequencies were higher in male subject due to more body mass. Goglia et al. [7] calculated the vibration transmissibility to the human body through the tractor steering in terms of frequency weighted acceleration. Gupta and Gupta [8] developed a thirteen degree of freedom (DOF) model in standing posture by using 50th percentile US anthropometric data and found experimental response near to the theoretical transmissibility ratio. Langer et al. [9] conducted experiments using farm tractor with large square baler. The level of exposed vibration was found more when the test run was in downhill mode rather than uphill moving and also, when the four-wheel mode was engaged, it increases the vibration dose value on driver. Nawayesh and Griffin [10] experiments were conducted on twelve male subjects, and it was found that at backrest angle of 150 negligible effect of apparent mass was observed. The angles of inclination were 00, 50, 100, and 150 between frequency range of 0–20 Hz with acceleration magnitude of 0.125, 0.25, 0.625, 1.25 m/s^2. Naser Nawayseh [11] developed a two degree of rotation and translational lumped mass model to compare the apparent mass calculated by experimental study. Primary peak was observed due to the rotation and second peak due to the translational motion of developed model.

Park et al. [12] evaluated the health risk issues of vehicle drivers. It was found that people who work in agricultural field has higher chance of health problem rather than the driver of work in recreational vehicle in the exposure level of more than 8 h stated by ISO-2631. Rahmatalla and DeShaw [13] conducted experiments on ten subjects when exposed to fore-and-aft and vertical vibrations using a simulator. The transmissibility found to have higher value when positions are considered to be in armrest with supports rather than steering wheel holding posture. Rakheja et al. [14] performed experiment with twelve male and female subjects and exposed to acceleration of 0.25, 0.5, and 1.5 m/s^2 rms in vertical direction. It was observed that body mass and hand position during seated conditions has higher influence in apparent mass calculations. Scarlett et al. [15] conducted experiments to predict exposure of WBV during different field operations; it was found that emission of vibration from tractor was mostly dependent upon operations and less dependent upon the suspension system.

Servadio et al. [16] calculated the vibration transmissibility from ground to seat of a tractor during forward speed operations. Thuong and Griffin [17] conducted experiments on sixteen subjects in octave frequency band of 0.5–16 Hz while the subjects were considered to be in standing posture. Below the frequency of 3.15 Hz subjects was feeling uncomfortable and found to be unstable to stand. Mehta and Tiwari [18] conducted experiments on seat cushion material by changing thickness and density of polyurethane (PU) foam and synthetic rubber foam. It was found that at higher frequencies (more than 5 Hz) thin PU foam has higher transmissibility than thicker foam. They had recommended that synthetic rubber foam should have thickness around 100 mm and density of 69.72 kg/m^3 for seat pan to lower the transmissibility of vibration.

From the literature review, it has been found that most of the work has been performed on European and US anthropometric data experimentally on seat cushion and suspension system. The objective of current study is to find out the effect of WBV on human subject with variant suspension system of tractor seat.

2 Methodology

Large amount of the Indian population is dependent on agricultural occupation and also, spend maximum time in agricultural activities. Taking this into consideration and health problems due to vibration generated while operating agricultural activities, current study has been performed to find out the apparent mass and vibration transmissibility in Indian human subject when exposed to low-frequency vibration in variant Indian agricultural conditions. A driver of 76 kg that corresponds to 95th percentile of the Indian male population has been considered for current study to find out the effect of vibrations on human subjects. A CAD model of tractor seat and human subject with realistic properties along with variant joints has been modeled.

3D CAD model of farm tractor seat has been modeled using reverse engineering technique; for this original seat has been dismantled as shown in. Figure 1a shows original tractor seat and Fig. 1b and each dimensions measured with the help of various available measuring tools. Figure 1c shows a CAD model of tractor seat modeled in SOLIDWORKS 2016 using measured physical dimensions of each part of it.

The physical dimensions have been taken from 95th percentile anthropometric for Indian male subject for 76 kg of mass available in existing literature [19]. Human subject with mechanical properties of bone, muscle, and skin has been modeled into CAD model using 3-layer approach as shown in Fig. 2b. In this approach, average cross section of bone, skin, and muscles have been considered along with average biomechanical properties [20] as per existing literature shown in Table 1.

As shown in Table 2, seat cushion has been given different properties of

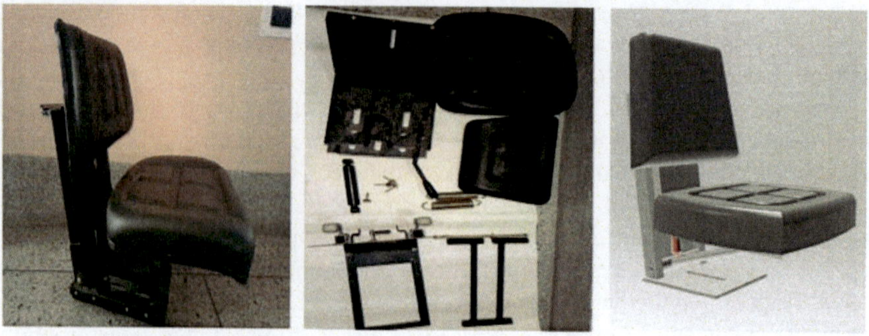

(a) (b) (c)

Fig. 1 Tractor seat **a** original; **b** dismantled parts **c**) CAD model

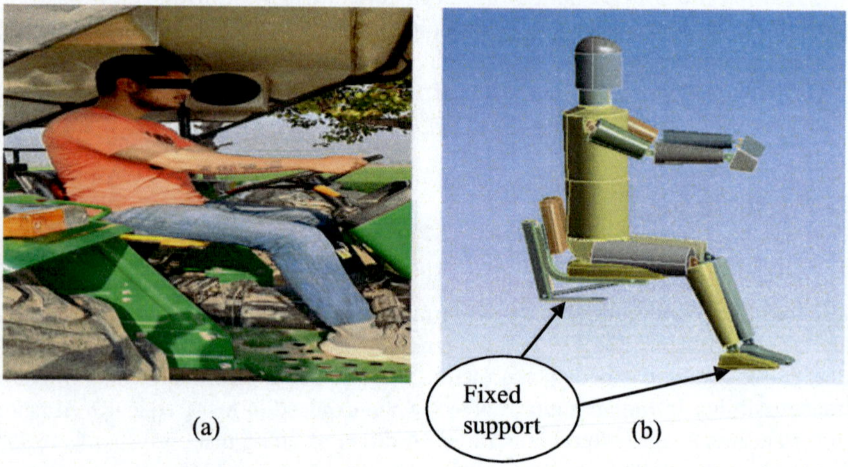

Fig. 2 **a** Actual representation of human subject on tractor seat. **b** CAD model of human subject on tractor seat

Table 1 Biomechanical properties of human body [20]

S. No.	Biomechanical material	Young's modulus (MPa)	Density (kg/m^3)	Poisson's ratio
1	Bone	17,000	1850	0.3
2	Muscle	13	1062	0.3
3	Skin	0.15	1000	0.46
4	Posture bone	3500	1400	0.25

Table 2 Material properties of cushion material [18]

Seat parts	Thickness (mm)	Density (kg/m^3)	Young' modulus (MPa)
Medium-density PU foam (P_1) for seat base pan	88	12.66	2–3
Synthetic rubber foam (R_1) for seat base pan	101	69.72	0.07–2
Frame	–	7850	210
Sub-frame	–	7850	210
Spring	–	7860	21,000

polyurethane foam material and synthetic rubber foam with various thickness and density taken from the existing literature. The variety of foam has been taken to find the transmissibility of vibration and to predict the behavior of apparent mass when subjected to different accelerations. The vibration behavior for different spring stiffness has been analyzed using ANSYS 18.2 software.

The condition of the Indian farm tractor driver has been considered as shown in Fig. 2a, b, and the feet of subject and seat base has been considered to be fixed. As, in real conditions the seat base is fixed with the chassis of tractor and feet are placed on the tractor floor area while driving. Hands have also been considered to be in position of holding the steering wheel. The rms acceleration values 0.5–1.5 m/s^2 have been given from the seat base as sinusoidal input using ANSYS 18.2.

Figure 2a shows actual representation of human subject siting on a tractor seat with the feet are touching the tractor body which has been considered to be fixed while in Fig. 2b represents CAD model of stated actual conditions.

3 Meshing

Meshing has been done on the geometry using ANSYS 18.2. Meshing is the basic terminology of creating small element on any solid CAD model to refine the results when the model has inner geometry and intricate shapes. In this study, hexahedral and tetrahedral types of mesh element have been used to evaluate the results. Hexahedral elements have been used on the head section and seat pan area where the maximum results have been evaluated. In most of the parts that have intricate and contouring shapes, tetrahedral elements have been used, because these elements are found to be stiffer and cover most areas when geometry is found to be complex. Figure 3 represents the complete meshed model of human subject and tractor seat model. In inner shape, hexahedral elements have been used to reduce the count of nodes and elements, also to reduce time for calculation in software during simulation.

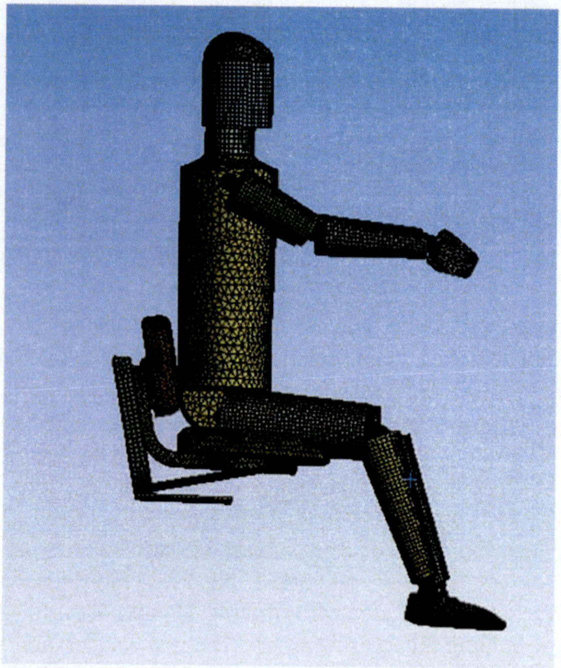

Fig. 3 Meshed model human subject on tractor seat

4 Results and Discussion

In this present work, the seat transmissibility at different spring stiffness value of k_1 0.3, k_2 0.55, and k_3 0.7 kg/mm has been evaluated with human subject of 76 kg mass at acceleration of 0.5 m/s^2. The second evaluation of transmissibility in terms of apparent mass has been evaluated at different excitation of 0.5, 0.75, and 1.0 m/s^2. Seat transmissibility using different cushion material of polyurethane and synthetic rubber foam has been predicted by exciting the seat at different magnitude of vibration.

4.1 Seat Transmissibility at Different Spring Stiffness

In any farm tractor, seat is the only component which is provided with damper and spring to counteract the effect of vibration on the human body. In present study, tractor seat has been excited with magnitude of 0.5 m/s^2 from the base of seat and transmissibility of vibration has been evaluated at the seat base pan where the human body comes in contact with the seat. Three different spring stiffness values of k_1 (0.3), k_2 (0.55), and k_3 (0.7 kg/mm) have been considered to predict the behavior.

It has been be observed from Fig. 4, transmissibility is increasing for the low-

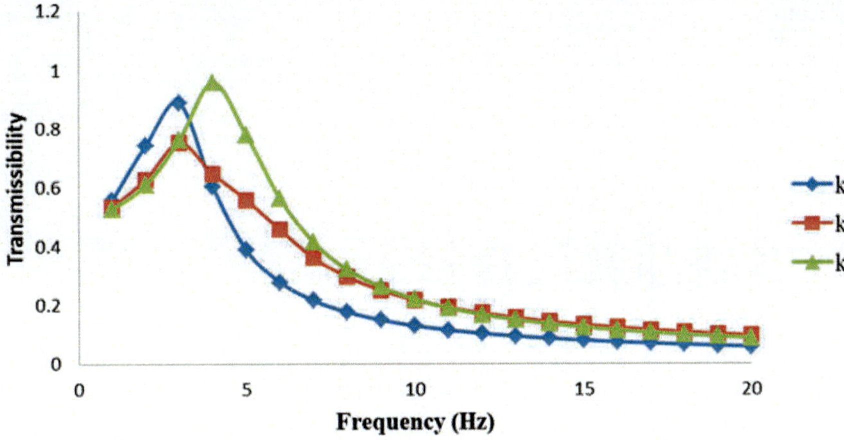

Fig. 4 Seat base pan transmissibility at different spring stiffness under 0.5 m/s^{-2}

frequency range and maximum effect for all the stiffness values found to be in the range of 2–6 Hz. In all the results, the maximum effect has been seen in value of k_3 when excited with the acceleration of 0.5 m/s^2. It has been found that stiffness value of k_2 has least effect of vibration transmissibility from the base to the seat pan. Lower value of transmissibility to seat base pan corresponds to lesser amount of vibrations transmitted to human body.

4.2 *Effect on Apparent Mass*

Evaluation of apparent mass at acceleration of 0.5 and 0.75 m/s^2 has been done, and the results are validated with the study conducted by Dewangan et al. [6]. It has been found from Fig. 5 that apparent mass tends to increase for the frequency range between 0 and 6 Hz. As per Fig. 5, when seat has been excited at magnitude of 0.5 m/s^2 the maximum transmissibility has been found to be in the range of 100–120 kg between frequency ranges of 4–6 Hz, while the same has been found to be 90–100 in the study conducted by Dewangan et al. [6]. At magnitude of 0.75 m/s^2, maximum transmissibility has been observed in the range of 120–150 kg. There is one secondary peak in frequency range of 10–12 Hz in Dewangan et al. [6], but in current study there is only one primary peak in the range of 4–6 Hz. The reason behind this may be due to the unavailability of all physical and environmental parameters which is not considered during the FEM analysis.

In the study conducted by Dewangan et al. [6], the mass of human subject was considered as 81 kg and the response spectrum has two peaks with primary and small secondary peak where the frequency range found to be secondary peak between 4–6 Hz and 8–12 Hz, respectively, probably the trend has been observed to be same in

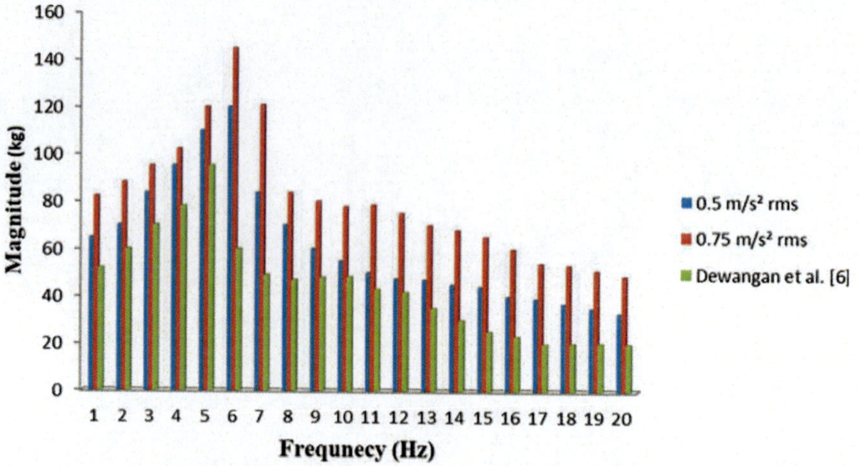

Fig. 5 Comparison of current study and Dewangan et al. [6] for magnitude of apparent mass transmissibility in human subject with 76 kg mass at excitation of 0.5 and 0.75 m/s^2

the present study with consideration of 76 kg mass human subject of 95th percentile of Indian anthropometric data.

4.3 *Effect of Seat Cushion Material on Transmissibility*

Seat cushion material transmissibility has been evaluated and validated with the experimental results calculated by Mehta and Tiwari [18]. Mehta and Tiwari [18] experimentally measured the transmissibility of vibration exposure on seat cushion material by taken different cushioning material of variable thickness and density.

The frequency was taken in the range of 0–8 Hz with sinusoidal input of exposure that has been provided using lab testing equipment [4]. In the present simulated study, the evaluation has been done by taking the same value of thickness and density for both PU foam and synthetic rubber foam. As shown Fig. 6, the seat has been excited with acceleration value of 0.5 m/s^2 in the frequency range 0–8 Hz. The maximum

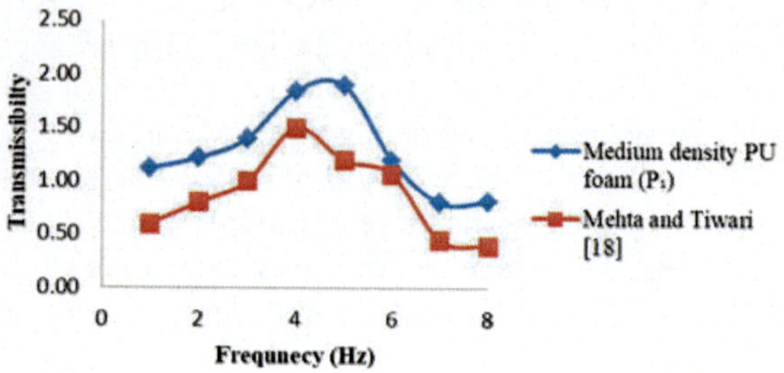

Fig. 6 Validation of results of medium-density PU foam (P_1)

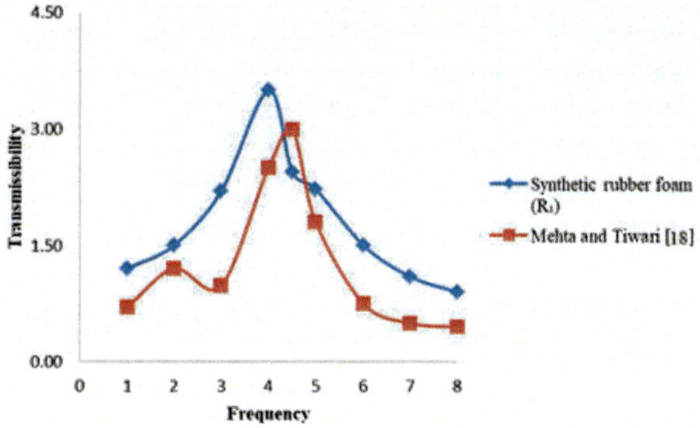

Fig. 7 Comparison of present study with experimental results of Mehta and Tiwari [18] for transmissibility of synthetic rubber foam (R_1)

transmissibility has been around 4–5 Hz while in study conducted by Mehta and Tiwari [19] dominated frequencies were found to be 4–6 Hz. The evaluation of results has been done for medium-density polyurethane foam (P_1). The trend in both studies has been found to be almost same but the simulated study has higher values than experimental measurements. In the present study, the peak is observed at 5 Hz as against 4 Hz in Mehta and Tiwari [19]. Figure 7 shows the transmissibility of synthetic rubber foam (R_1) with thickness 101 mm at same acceleration of 0.5 m/s^2, and the frequency has been maintained in the range of 0–8 Hz. The resonance for the current study observed at 4 Hz while the same peak has been observed at 4.5 Hz in measurement taken by Mehta and Tiwari [18].

It has been observed that for both the evaluation, the primary frequency has been found to be low in synthetic rubber foam R_1 and it decreases after the resonance frequency. So, the synthetic rubber foam attenuates more low frequencies rather than medium-density PU foam while the PU foam has slightly higher frequency range for attenuation of low-frequency vibration. So when the human body works under low-frequency vibration condition most likely said to be in agricultural conditions.

5 Conclusion

In this current study, 3D CAD model of tractor seat and Indian human subject of 76 kg of mass have been developed using solid modeling. The seat has been modeled with reverse engineering technique and for developing the human CAD model data has been taken from the existing literature of 95th percentile Indian anthropometric data. Resonance frequency evaluation of seat with different foam material has been performed, and the frequencies have been found to be 5 Hz and 4 Hz for medium-density PU foam P_1 and for synthetic rubber foam R_1. Apparent mass of human subject of 76 kg has been evaluated in terms of transmissibility, and primary frequency has been observed in the range of 4–6 Hz. Seat suspension transmissibility has been evaluated using different types of spring stiffness, and it has been found that spring with stiffness k_2 is effective in low-frequency attenuation. The results for apparent mass evaluation have been validated with Dewangan et al. [6] and found to be in the range of existing literature. For seat cushion material, comparison of results has been done with study conducted by Mehta and Tiwari [18] in which the synthetic rubber has been found to be effective in attenuation of vibration and for providing more comfort for the driver. For the future purpose, study can be beneficial to select material while designing the seat for vehicle.

References

1. ISO 2631-1 (1997) Mechanical International Organization for standardization, Geneva, Switzerland
2. Jalil NAA, Griffin MJ (2007) Fore-and-aft transmissibility of backrests: variation with height above the seat surface and non-linearity. J Sound Vibr 299:109–122
3. Bhoi PK, Tiwari VK, Dhar R, Dewangan KN (2007) The effect of all season collapsible tractor operation enclosure on driver comfort in three adverse environmental conditions. Int J Ind Ergon 37:479–487
4. Catania P, Vallone M, Alleri M, Morello G, Sparta G, Febo P (2013) Whole body vibration during field operation in vineyard. J Agric Eng 143:710–714
5. Cutini M, Costa C, Bisaglia C (2016) Development of a simplified method for evaluating agricultural tractor's operation whole body vibration. J Terrramech 63:23–32
6. Dewangan KN, Shahmir A, Rakheja S, Marcotte P (2013) Seated body apparent mass response to vertical whole body vibration: gender and anthropometric effect. Int J Ind Eng 43:375–391
7. Goglia V, Gospodaric Z, Kosutic S, Filipovic D (2003) Hand-transmitted vibration from the steering wheel to driver of a small four-wheel drive tractor. Appl Ergon 34:43–49
8. Gupta M, Gupta TC (2017) Modal damping ratio and optimal elastic moduli of human body segment for anthropometric vibratory model of standing subjects. ASME J Biomech Eng 139:1–13 (2017)
9. Langer TH, Ebbesen MK, Kordestani A (2015) Experimental analysis of occupational whole-body vibration exposure of agricultural tractor with large square baler. Int J Ind Ergon 47:79–83
10. Nawayseh N, Griffin MJ (2003) Non-linear dual-axis biodynamic response to vertical whole-body vibration. J Sound Vibr 268:503–523
11. Nawayseh N (2016) A mathematical model of the apparent mass of the human body under fore-and-aft whole-body vibration. Int J Autom Mech Eng 13:3613–3627
12. Park M, Fukuda T, Kim T, Maeda S (2013) Health risk evaluation of whole-body vibration by ISO 2631–5 and ISO 2631–1 for operator of agricultural tractor and recreational vehicles. Ind Health 51:64–70
13. Rahmatalla S, DeShaw J (2011) Effective seat to head transmissibility in whole-body vibration: effects of posture and arm position. J Sound Vib 330:6277–6286
14. Rakheja S, Stiharu I, Boileau PE (2002) Characteristics under automotive postures and vertical vibration. J Sound Vibr 253(1):57–75
15. Scarlett AJ, Price JS, Stayner RM (2007) Whole-body vibration: evaluation of emission and exposure level arising from agricultural tractor. J Terramech 44:65–73 (2007).
16. Servadio P, Marsili A, Belfiore NP (2007) Analysis of driving seat vibration in high forward speed tractor. Biosyst Eng 97:171–180
17. Thuong O, Griffin MJ (2011) Vibration discomfort of standing person: 0.5–16 Hz fore-and-aft, lateral and vertical vibration. J Sound Vibr 330:816–826
18. Mehta CR, Tiwari VK (2002) Vibrational characteristics of tractor seat cushion material and ride comfort. J Low Freq Noise Vibr Active Control, 77–86
19. Chakrabarti D (1997) Indian anthropometric dimensions: for ergonomic design practice. National Institute of Design, Ahmedabad

20. Dong RG, He L, Du W, Cao ZK, Huang ZL (2019) Effect of sitting posture and seat on biodynamic response of internal human body simulated by finite element modelling of body-seat system. J Sound Vibr 438:543–554

Vibration Response of Fir Tree Root Blades with the Variation in Fixing Condition on Blade Root Interfaces

Rajesh K. Bhamu, Akash Shukla, S. P. Harsha, and Satish C. Sharma

1 Introduction

Since the inspiration of using power of steam to rotate the prime mover, the steam turbine has played a foremost role as a key source of power production globally. Power produced by the steam-based thermal power plants in India is about to 65% (CEA, 2019) [1]. Crucial components of steam power plant are blades which are responsible to convey the high pressure and high temperature steam's linear motion into a rotary motion of the turbine shaft. Due to uninterrupted work of energy conversion, blades experience the major brutal dynamic surroundings throughout their existence duration. The last stage LP steam turbine blades, having huge size rotating airfoils, experience considerable centrifugal forces throughout operating situation which designed to extract the final remnants energy from steam. To keep away from any shattering, failure of blade dynamics analysis is compulsory. From literature, it is found that low pressure steam turbine blade is most vulnerable as proved by various researchers [2–4]. In 1952, Lo and Renbarger [5] developed the equation of motion for the vibration of the rotating cantilever beam whereas Rao [6] found dynamic response of single free position blade assuming it like a pre-twisted cantilever beam. Although it is good for simply supported beam/cantilever but unable to provide better response for intricate tapered and twisted blade of fir tree root.

First time in 1967, blade analysis using FEM had done by Dokumaci [7] to find the first two modes of a free standing gas turbine blade using 20 node shell elements. Rieger [8] modelled the natural frequencies and steady stresses of a steam turbine

R. K. Bhamu (✉) · S. P. Harsha · S. C. Sharma
Department of Mechanical & Industrial Engineering, Indian Institute of Technology Roorkee, Roorkee 247667, India
e-mail: rbhamu@me.iitr.ac.in

A. Shukla
BHEL, Haridwar, India

V. H. Saran and R. K. Misra (eds.), *Advances in Systems Engineering*, Lecture Notes in Mechanical Engineering,
https://doi.org/10.1007/978-981-15-8025-3_83

6-blade I.P. group on a segment of the disk by using the 8 node 3D solid element in ANSYS. Ma et al. [9] discussed the natural frequency of rotating blade considering the eccentric force resulted from rotation at high speed. Shukla et al. [10, 11] carried out relative study of steam turbine blade with help of model analysis through analytically and mathematically along with finding the changes in the crack size with variation in vibration response of fir tree root blades. Heidari et al. [12] analyzed a 3D steam turbine blade model varying in length and thickness and concluded that maximum Von Mises and strain were experienced by longer blade than shorter one.

To obtain the blade's dynamic response characteristics, modal analysis is a cost efficient process. Natural frequency and mode shape of the blade were obtained by this at room temperature. Since FE-based modal analysis be able to forecast the blade's dynamic response under operational situation, hence, it should be adopted for such blades.

Fir tree root of the object blade has fitted with rotor groove flank in a manner that three faces (Top, Middle, and Lower) of blades at both side, i.e., suction side and pressure side will remain in contact. However, there is very small clearance of 0.3 mm available between all faces of rotor groove flanks and fir tree root faces. Sometimes, there is an error in blade root fixity occurring due to which blade will not come into proper contact with rotor flank that leads to catastrophic failure in blade.

2 Modal Analysis of FIR Tree Root Blade

Shape of final stage blade of low pressure turbine is tapered and twisted due to which frequency calculation is complicated. 250 MW LP steam turbine last stage blade with fir tree root geometry has been modeled in solid works and further analyzed through FEM-based ANSYS software (Fig. 1).

2.1 Modal Analysis of LP Steam Turbine Blade Based on FE Technique

Blade vibration motion equation analyzed with the help of FEM, numerical model's free vibration equation is as follows [10]:

$$M\ddot{u} + Ku = 0 \tag{1}$$

where

M—Mass matrix, K—Stiffness matrix, and u—Node displacement vector.

The structure can be imagined to vibrate in free harmonic form with absence of externally applied force

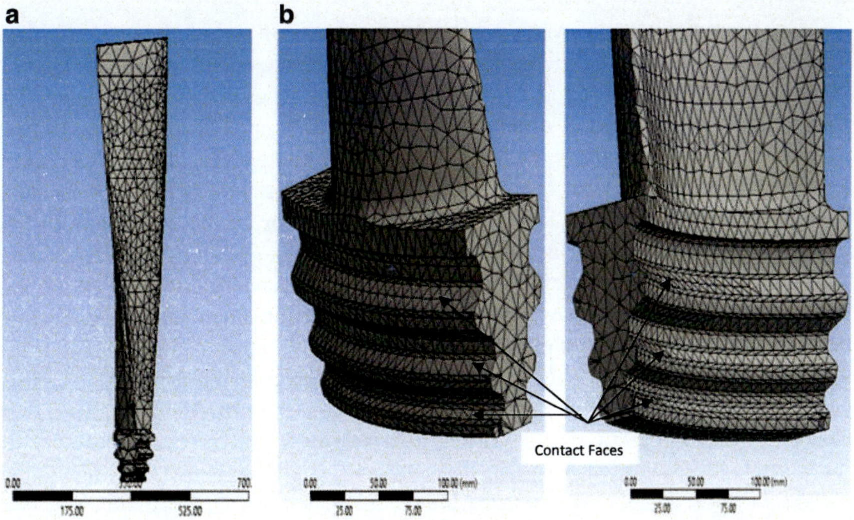

Fig. 1 **a** LP last stage blade meshed model. **b** Contact faces of fir tree root blade

$$u(t) = \varphi \sin(\omega t + \theta) \tag{2}$$

which leads to the Eigen value problem:

$$\left[[K] - \omega^2[M]\right]\{\varphi\} = \{0\} \tag{3}$$

where ω is the natural frequency and φ is structure's mode shape correspondingly. The mode shape and natural frequency will be achieved by solving the Eigen value problem.

Last stage blade of L P turbine of 250 MW power plant modeled in Solid Works. Modal analysis of free positioning sole blade with fir tree root has done with the help of FE-based solver of ANSYS.

Material properties of low pressure turbine blade assumed as a isotropic elastic and other properties are considered as below:

Blade material	X10CrNiMoV1222
Young's modulus	2.16×10^5 MPa
Density	7850 Kg/m^3
Poisson ratio	0.3

For complete blade meshing and maximum relevance at all the sharp corners and profile radius, a continuous 3D 10 nodded tetrahedral elements SOLID 187 methods is used. Complete blade model is separated in 258,129 elements and computed the outcome at 151,990 nodes.

Meshed model of a pre-warped free positioning steam turbine fir tree root blade assigned its boundary condition as its fitting in rotor's groove, known as fixed support boundary condition. Out of all the contact faces, two faces of blade are assigned as frictionless sliding supports which act as a collar for zone of blade profile. Fir tree root of the object blade has three faces (Top, Middle and Lower) of different contact areas at both side of blades, i.e., suction side and pressure side. There is very small clearance of 0.3 mm available between all faces of rotor groove flanks and fir tree root faces. The available fitting clearance is bridged during running condition due to centrifugal action. Blade is axially locked by the fitting piece followed by the locking plate. Now, the blade can move only in the radial direction due to the inherent design flexibility. Hence, in running condition, it is moderately convenient to imagine that all faces will stay parallel to the matching faces. In view of that following, 12 contact configurations as mentioned in Table 1 are considered and free vibration response has been studied in all these cases of blade root fixity.

Table 1 Different blade root fixity condition (Con = Contact, Non-Con = Non contact)

Contact cases	Suction side faces of blade root (S)			Pressure side faces of blade root (P)			Remarks
	Top face (T)	Middle face (M)	Lower face (L)	Top face (T)	Middle face (M)	Lower face (L)	
I	Con	Con	Con	Con	Con	Con	The entire face in contact
II	Con	Con	Con	Non-Con	Non-Con	Non-Con	Suction side all face in contact
III	Non-Con	Non-Con	Non-Con	Con	Con	Con	Pressure side all face in contact
IV	Con	Non-Con	Non-Con	Con	Non-Con	Non-Con	Two corresponding faces in both side contact
V	Non-Con	Con	Non-Con	Non-Con	Con	Non-Con	
VI	Non-Con	Non-Con	Con	Non-Con	Non-Con	Con	
VII	Con	Non-Con	Non-Con	Non-Con	Non-Con	Non-Con	Only one face of root is in contact
VIII	Non-Con	Con	Non-Con	Non-Con	Non-Con	Non-Con	
IX	Non-Con	Non-Con	Con	Non-Con	Non-Con	Non-Con	
X	Non-Con	Non-Con	Non-Con	Con	Non-Con	Non-Con	
XI	Non-Con	Non-Con	Non-Con	Non-Con	Con	Non-Con	
XII	Non-Con	Non-Con	Non-Con	Non-Con	Non-Con	Con	
XIII	Non-Con	Non-Con	Non-Con	Non-Con	Non-Con	Non-Con	No face in contact

3 Result and Discussion

After carrying out the modal analysis based on FE of normal fir tree root blade, natural frequencies and mode shapes are achieved as follow. The natural frequencies and mode shape for their first 6 modes have shown in Fig. 2 out of which primary three modes are flap-wise bending, chord-wise bending, and torsion modes, respectively, and remaining are higher modes known as overlapping modes.

Blade dynamic analysis for variable fixing condition has been done and natural frequencies and mode shapes are obtained as shown in Fig. 3a. Frequency variation of different blade fixity conditions in comparison with normal blade as given in Fig. 3b.

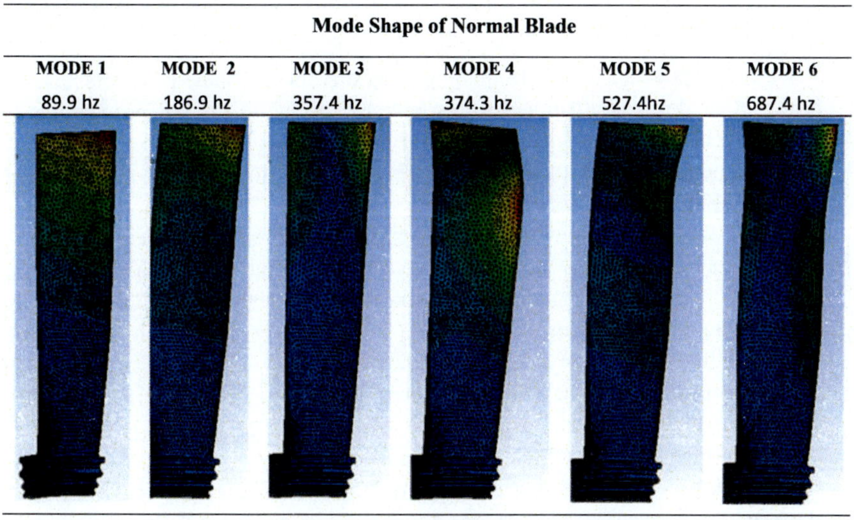

Fig. 2 Obtained natural frequencies and mode shapes for fir tree root blade (when all faces are in contact)

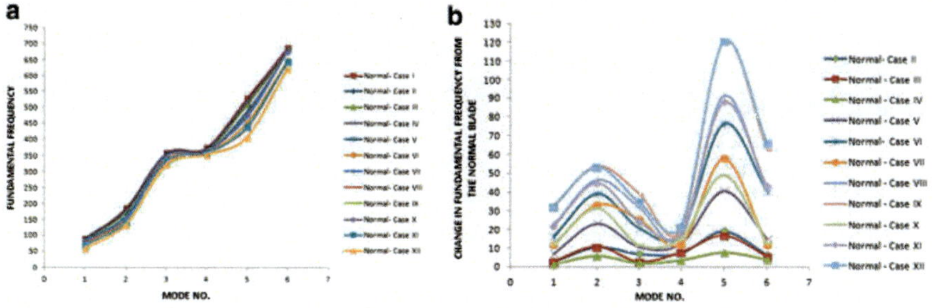

Fig. 3 **a** Frequency and mode shape at different fixity cases. **b** Variation of frequency in comparison with normal blade

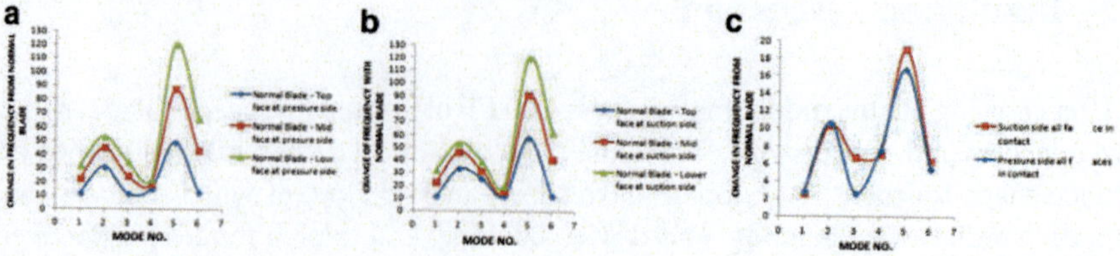

Fig. 4 **a** Variation of frequency for suction- /pressure-side contact. **b** Variation of frequency with the normal blade at suction side contact. **c** Variation of frequency for suction- /pressure-side contact

After observing the frequency and relative frequency plot obtained through modal analysis data of low pressure last stage steam turbine blade having fir tree root (Refer Figs. 3 and 4) can draw following points:

- There is an incessant fall in frequency from the normal blade in all the cases of fixing of blade; it is attributed to the effect of reduction in stiffness (imparted slackness) due to non- constrained surfaces on the root faces.
- The obtained natural frequency in case IV (only both top face of root in contact) is closer to the normal blade fixing condition of the blade. It can be concluded by this result that the top face contributes maximum in the root stiffness.
- In case II and III, where one side (suction and pressure side) of all the faces is in contact the natural frequency variation depends on the directional flexibility imparted against the corresponding mode shape. However, it is also evident from the graph that variation in frequency is more in case II during 2nd and 4th mode.
- It is clear from Fig. 4 that the maximum drop in frequency observed in cases where only lower flank of faces are in contact. Lower flank of the root is the farthest contact surface of blade which imparts more flexibility.
- Hence, whenever the clearance are higher in top and middle faces and if the blade stays only in contact with lower face, then the blade will face the maximum drop in frequency, which may reach within the operating frequency of turbine which may be catastrophic in failure. However, in case of top and middle faces, the blade will run in comparatively safer side.
- The completely loose blade behaves simple as unstable structure.

4 Conclusion

The above study of normal blade fixity conditions with rotor based on FE provides an outlook that distinct change in dynamic behavior of normal blade in all the cases of fixity of blade specially in the case of contact with lower face in which maximum drops of frequency occurs which may lead to catastrophic failure. This study can be further extended by clubbing experimental model of blade laxity and irregular hold up for coupled dynamic and static analysis on blade root of complex structure of blades throughout in service situation.

References

1. Government of India (2019) All India installed capacity (in Mw) of Power stations installed capacity (in Mw) of power utilities in the states/Uts located in
2. Kubiak Sz. J, Urquiza B. G, García C. J, Sierra E. F (2007) Failure analysis of steam turbine last stage blade tenon and shroud. Eng Fail Anal 14(8 SPEC. ISS.):1476–1487
3. Segura JA, Castro L, Rosales I, Rodriguez J A, Urquiza G (2017) Dignostic and failure analysis in blades of a 300 MW steam turbine. Eng Fail Anal 82:631–641
4. Mazur Z, García-Illescas R, Porcayo-Calderón J (2009) Last stage blades failure analysis of a 28 MW geothermal turbine. Eng Fail Anal 16(4):1020–1032
5. Lo H, Renbarger JL (1952) Bending vibrations of a rotating beam. In: Proceedings of the first U. S. National Congress of Applied mechanics, A.S.M.E., pp 75–79.
6. Rao DK, Rao JS (1974) Equation of motion of rotating pre-twisted blades in ending-bending-torsion with effect of warping, shear, rotary inertia etc. In: Proceeding of the Silver jubilee conference, Aeronautical Society India, Bangalore, p 4
7. Dokumaci E, Thomas J, Carnegie W (1967) Matrix displacement analysis of coupled bending–bending vibrations of pretwisted blading. Arch J Mech Eng Sci 1959–1982 (vols 1–23) 9:247–254
8. Rieger NF, Nowak WJ (1977) Analysis of fatigue stresses in steam turbine blade groups. In EPRI workshop on steam turbine availability
9. Ma YX (2005) Vibration analysis, testing and diagnosis of turbocharger blade. Test Diagnosis 25(2)
10. Shukla A, Harsha SP (2016) Vibration response analysis of last stage LP turbine blades for variable size of crack in root. Procedia Technol 23:232–239
11. Shukla A, Harsha SP (2015) An experimental and FEM modal analysis of cracked and Normal steam turbine blade. Mater Today Proc 2(4–5):2056–2063
12. Heidari M, Amini K (2017) Structural modification of a steam turbine blade. IOP Conf Ser Mater Sci Eng 203:12007

Effect of Foundation on Free Vibration of Tapered Functionally Graded Material Plate

V. Kumar, S. J. Singh, V. H. Saran, and S. P. Harsha

1 Introduction

Due to high strength and high rigidity, advanced multilayered composite materials are commonly used in many mechanical, civil, and aerospace engineering systems. For example, a homogeneous elastic layer of ceramic material can be attached to a metallic structure's surface and act as a thermal barrier in a high-temperature environment. Stress concentration and delamination between two interfaces arise due to sudden changes in material properties. One way to resolve this shortage is by using functionally graded materials with continually variable material properties that have noticeable benefits over homogeneous and layered materials to maintain the structure's integrity [1]. The ceramic grading method along with metal introduced in Japan marked the start of considering the option of using functionally graded material (FGM) for different structural applications [2]. A number of research papers are available on a functionally graded structure with static and dynamic behavior. A third-order shear deformation theory has been developed by Reddy [3] to investigate the static analysis of functionally graded (FG) plates. Buckling and vibration analysis of

V. Kumar (✉) · V. H. Saran · S. P. Harsha
Mechanical and Industrial Engineering Department, Indian Institute of Technology Roorkee, Roorkee, Uttarakhand 247667, India
e-mail: vineetrana81@gmail.com

V. H. Saran
e-mail: saranfme@iitr.ac.in

S. P. Harsha
e-mail: surajfme@iitr.ac.in

S. J. Singh
Mechanical Engineering Department, Pandit Deendayal Petroleum University, Gandhinagar 382009, Gujarat, India
e-mail: simran.singh@sot.pdpu.ac.in

V. H. Saran and R. K. Misra (eds.), *Advances in Systems Engineering*,
Lecture Notes in Mechanical Engineering,
https://doi.org/10.1007/978-981-15-8025-3_84

FG polygon plates are presented by Cheng and Batra [4]. Hosseini-Hashemi et al. [5] have obtained the results for free vibration analysis using first-order shear deformation theory, and FG plate was resting on an elastic foundation. Ferrrira et al. [6] have presented a first and third order shear deformation theory model for free vibration analysis of the rectangular FGM plate using Mori-Tanaka homogenization technique. Zenkour [7] has investigated the free vibration of simply supported FG sandwich plates using the sinusoidal shear deformation theory.

Furthermore, rectangular plates supported on the elastic foundation has been widely adopted by many researchers to model real engineering applications during the last few years. Various models have been developed to show the interaction between plate and elastic foundation. Winkler suggested the first elastic foundation model for continuum. But it had some discontinuity involving spring displacement adjacent to it. Later, it was improved by Pasternak to explore the interaction between adjacent spring displacement in longitudinal as well as in lateral direction by introducing a new parameter. Closed-form solution of free vibration of Mindlin plate resting on Pasternak foundation was investigated by Xiang et al. [8]. Free vibration of clamped thick plate resting on a two-parameter foundation was presented by Zhou et al. [9]. Yang and Shen [10] presented a model for vibration analysis of initially stressed FG plate resting on elastic foundation. A FG plate accorded with simple power law distribution was clamped at boundaries during the study. Lee et al. [11] have been presented the bending effect using refined higher-order shear and normal deformation theory for power law (P-), exponential (E-), and sigmoid (S-) FGM plates. Chebyshev polynomials and Ritz's method were used to investigate the free vibration of FG plate resting on Winkler foundation [12]. Singh and Harsha [13] studied the free vibration and buckling analysis of sandwich plate resting on an elastic foundation. Different boundary conditions were applied with new sigmodal function. Lü et al. [14] had obtained free vibration analysis of a simply supported FG plate resting on the Pasternak foundation using state-space method. The free vibration analysis of thin to thick annular plate supported on Winkler–Pasternak foundation [15] was investigated. Baferani et al. [16] presented the vibration analysis of FGM plate resting on a two-parameter elastic foundation using third-order shear deformation plate theory. Singh and Harsha [17] investigated the nonlinear behavior of S-FGM sandwich plate resting on Pasternak foundation. Airy's stress function with Galerkin method was employed to investigate the frequency parameter of a plate under thermal environment. Thai and Choi [18] presented a refined shear deformation theory for vibration analysis of FGM plate resting on Pasternak foundation. Vibration analysis of functionally graded plate was examined using new higher-order shear deformation theory, and equation of motion for FGM plate resting on elastic foundation was derived using displacement-based theory [19].

A lot of research has been done on rectangular functionally graded plates using classical plate theory, first-order shear deformation theory, higher-order shear deformation theory, and refined plate theories. Some of the work has been reported on one or two-parameter elastic foundation incorporated with rectangular FGMs plates. As per the author's knowledge, work is not published on rectangular FGMs plate with variable thickness resting on an elastic foundation. So, the motivation of work is to

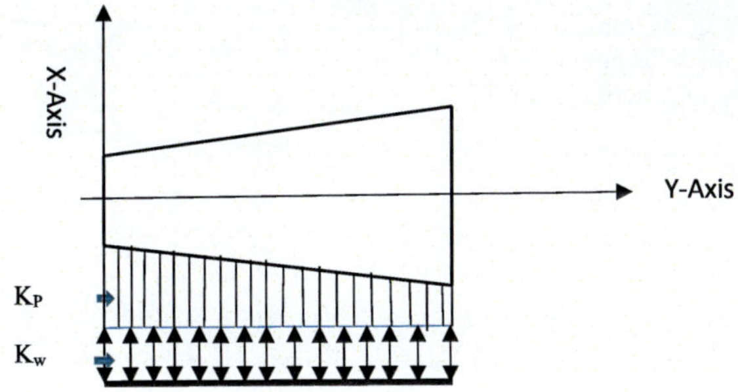

Fig. 1 Tapered FGM plate resting on two-parameter elastic foundation

find out the free vibration analysis of P-FGM plates with variable thickness in one direction. Simply supported plate is resting on Winkler–Pasternak foundation.

2 Mathematical Formulation

A functionally graded material plate with variable thickness in *Y*-direction resting on the Winkler–Pasternak foundation is depicted in Fig. 1. The plate with length *a*, width *b* and height *h* has edges simply supported.

3 Material Properties

FGM is a type of advanced composite material wherein mechanical properties vary in the thickness direction as shown in Fig. 2. Here, Poisson's ratio is assumed constant throughout the study, while Young modulus and density are varying according to simple power law [20–23] as follows.

$$E(x, y, z) = E_m + (E_c - E_m)\left(\frac{z}{h(x, y)} + \frac{1}{2}\right)^p \quad (1)$$

$$\rho(x, y, z) = \rho_m + (\rho_c - \rho_m)\left(\frac{z}{h(x, y)} + \frac{1}{2}\right)^p \quad (2)$$

Aluminum and alumina are taken as the constituent materials during the investigation of FGM plate that has the mechanical properties as $E_m = 70$ GPa, $\rho_m = 2702$ kg/m^3 and $E_c = 380$ GPa, $\rho_c = 3800$ kg/m^3 for metal and ceramic, respectively. Here, *p* is the volume fraction exponent and *h*(*y*) is the taper function for the variable thickness of the plate.

$$h(y) = h_0\left(1 + \eta\left(\frac{y}{b}\right)\right) \quad (3)$$

Fig. 2 Variation of Young modulus in the thickness direction of P-FGM

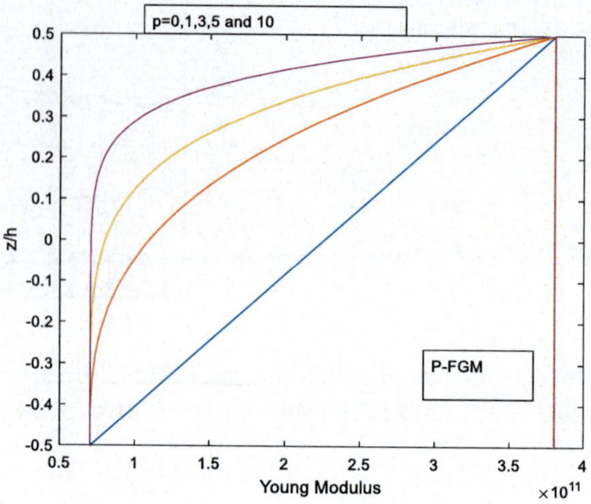

And, η is a *taper ratio* which scaled up or scaled down the thickness along the y-direction.

$$\eta = \frac{h_b}{h_0} - 1 \tag{4}$$

where h_0 and h_b are the thickness of the FGM plate at $y = 0$ and $y = b$.

4 Constitutive Relations

There are five relations to solve the stress–strain relationship for a plate problem.

$$\begin{bmatrix} \sigma_{xx} \\ \sigma_{yy} \\ \tau_{yz} \\ \tau_{xz} \\ \tau_{xy} \end{bmatrix} = \begin{bmatrix} C_{11} & C_{12} & 0 & 0 & 0 \\ C_{12} & C_{22} & 0 & 0 & 0 \\ 0 & 0 & C_{44} & 0 & 0 \\ 0 & 0 & 0 & C_{55} & 0 \\ 0 & 0 & 0 & 0 & C_{66} \end{bmatrix} \begin{Bmatrix} \varepsilon_{xx} \\ \varepsilon_{yy} \\ \gamma_{yz} \\ \gamma_{xz} \\ \gamma_{xy} \end{Bmatrix} \tag{5}$$

$$C_{11} = C_{22} = \frac{E(x, y, z)}{1 - \nu^2}, C_{12} = \frac{\nu E(x, y, z)}{1 - \nu^2} = \frac{\nu E(x, y, z)}{1 - \nu^2},$$

$$C_{44} = C_{55} = C_{66} = \frac{E(x, y, z)}{2(1 + \nu)}$$

5 Formulation

The present formulation is based on the variational principle. Non-polynomial higher-order shear deformation-based displacement function is chosen for free vibration analysis of tapered FGM plate resting on an elastic foundation where u and v are the displacements in in-plane direction and w is displacement in transverse directions, respectively, while Φ_x and Φ_y are the rotations expressed as:

$$u(x, y, z, t) = \overline{u}(x, y, t) - z\left(\frac{\partial \overline{w}}{\partial x}\right) + \left(\tanh^{-1}\left(\frac{rz}{h}\right) - z\frac{r/h}{1 - r^2/4}\right)\Phi_x(x, y, t)$$
$$v(x, y, z, t) = \overline{v}(x, y, t) - z\left(\frac{\partial \overline{w}}{\partial y}\right) + \left(\tanh^{-1}\left(\frac{rz}{h}\right) - z\frac{r/h}{1 - r^2/4}\right)\Phi_y(x, y, t)$$
$$w(x, y, z, t) = \overline{w}(x, y, t) \quad (6)$$

The shape function [24] in present formulation is given as$\overline{\lambda}(z) + z\vartheta$ where$\overline{\lambda}(z) = \tanh^{-1}(rz/h)$ and $\vartheta = -(r/h)/(1 - r^2/4)$, $r = 0.088$.

$$\begin{Bmatrix} \varepsilon_{xx} \\ \varepsilon_{yy} \\ \gamma_{yz} \\ \gamma_{xz} \\ \gamma_{xy} \end{Bmatrix} = \begin{Bmatrix} \varepsilon^{(0)}_{xx} \\ \varepsilon^{(0)}_{yy} \\ 0 \\ 0 \\ \gamma^{(0)}_{xy} \end{Bmatrix} + z\begin{Bmatrix} \varepsilon^{(1)}_{\text{b}\ xx} \\ \varepsilon^{(1)}_{\text{b}\ yy} \\ 0 \\ 0 \\ \gamma^{(1)}_{\text{b}\ xy} \end{Bmatrix} + f(z)\begin{Bmatrix} \varepsilon^{(1)_{xx}}_{\text{s}\ xy} \\ \varepsilon^{(1)}_{\text{s}\ yy} \\ 0 \\ 0 \\ \gamma^{(1)}_{\text{s}\ xy} \end{Bmatrix} + f'(z)\begin{Bmatrix} 0 \\ 0 \\ \gamma^{(0)}_{yz} \\ \gamma^{(0)}_{xz} \\ 0 \end{Bmatrix} \quad (7)$$

5.1 Equation of Motion

Hamilton's principle was used to obtain the equations of motion for tapered FGM plate resting on the Winkler–Pasternak foundation.

$$\int_{t_1}^{t_2} (\delta U + (\delta W_{\text{b}} + \delta W_{\text{e}}) - \delta K)\text{d}t = 0 \quad (8)$$

where δU, δW_{b}, δW_{e}, and δK are the virtual strain energy, work done due to normal load, potential energy due to elastic foundation and kinetic energy, respectively.

Now, the virtual kinetic energy can be written as

$$\delta U = \iiint \left[\sigma_{xx}\delta\varepsilon_{xx} + \sigma_{yy}\delta\varepsilon_{yy} + 2(\sigma_{xy}\delta\varepsilon_{xy} + \sigma_{yz}\delta\varepsilon_{yz} + \sigma_{zx}\delta\varepsilon_{zx})\right]\text{d}v \quad (9)$$

Substituting Eqs. 7 and 5 in Eq. 9 yields the following:

$$\delta U = \int_A \begin{bmatrix} \left(N_{xx}\delta\varepsilon_{xx}^{(0)} + N_{yy}\delta\varepsilon_{yy}^{(0)} + N_{xy}\delta\gamma_{xy}^{(0)}\right) + \\ \left(M_{xx}^{\mathrm{b}}\delta{\varepsilon_{\mathrm{b}}^{(1)}}_{xx} + M_{yy}^{\mathrm{b}}\delta{\varepsilon_{\mathrm{b}}^{(1)}}_{yy} + M_{xy}^{\mathrm{b}}\delta{\gamma_{\mathrm{b}}^{(1)}}_{xy}\right) \\ \left(M_{xx}^{\mathrm{s}}\delta{\varepsilon_{\mathrm{s}}^{(1)}}_{xx} + M_{yy}^{\mathrm{s}}\delta{\varepsilon_{\mathrm{s}}^{(1)}}_{yy} + M_{xy}^{\mathrm{s}}\delta{\gamma_{\mathrm{s}}^{(1)}}_{xy}\right) + \\ \left(M_{xx}^{\mathrm{q}}\delta\Phi_x + M_{yy}^{\mathrm{q}}\delta\Phi_y\right) + \vartheta\left(Q_x\delta\Phi_x + Q_y\delta\Phi_y\right) \end{bmatrix} \mathrm{d}x\mathrm{d}y \tag{10}$$

where

$$\begin{Bmatrix} N_{xx} & M_{xx}^{\mathrm{b}} & M_{xx}^{\mathrm{s}} \\ N_{yy} & M_{yy}^{\mathrm{b}} & M_{yy}^{\mathrm{s}} \\ N_{xy} & M_{xy}^{\mathrm{b}} & M_{xy}^{\mathrm{s}} \end{Bmatrix} = \int_{-h/2}^{h/2} (1, z, \lambda(z)) \begin{Bmatrix} \sigma_{xx} \\ \sigma_{yy} \\ \tau_{xy} \end{Bmatrix} \mathrm{d}z \tag{11.a}$$

$$\begin{Bmatrix} Q_x & M_{xx}^{\mathrm{q}} \\ Q_y & M_{yy}^{\mathrm{q}} \end{Bmatrix} = \int_{-h/2}^{h/2} \left(1, \lambda'(z)\right) \begin{Bmatrix} \tau_{xz} \\ \tau_{yz} \end{Bmatrix} \mathrm{d}z \tag{11.b}$$

The above relations can be further simplified and written in concise form as follows:

$$\begin{Bmatrix} \mathbf{N} \\ \mathbf{M^b} \\ \mathbf{M^s} \end{Bmatrix} = \begin{bmatrix} \mathbf{A} & \mathbf{B} & \mathbf{F} \\ \mathbf{B} & \mathbf{D} & \mathbf{H} \\ \mathbf{F} & \mathbf{H} & \mathbf{J} \end{bmatrix} \begin{Bmatrix} \boldsymbol{\varepsilon} \\ \boldsymbol{\varepsilon_b} \\ \boldsymbol{\varepsilon_s} \end{Bmatrix} \tag{12.a}$$

$$\mathbf{Q} = (\vartheta\mathbf{A} + \mathbf{K})\boldsymbol{\gamma}_{\mathbf{s}} \tag{12.b}$$

$$\mathbf{Q^q} = (\vartheta\mathbf{K} + \mathbf{L})\boldsymbol{\gamma}_{\mathbf{s}} \tag{12.c}$$

where $\{A_{ij}, B_{ij}, D_{ij}, F_{ij}, H_{ij}, J_{ij}, K_{ij}, L_{ij}\} = \int_{-h/2}^{h/2} C_{ij}(x, y, z) \cdot \{1, z, z^2, \lambda(z), z\lambda(z), \lambda(z)^2, \lambda'(z), \lambda'(z)^2\}\mathrm{d}z$

Virtual potential energy of the plate due to the elastic foundation

$$\delta W_{\mathrm{e}} = \iint (K_{\mathrm{w}} - K_{\mathrm{p}}\nabla^2)\overline{w}\delta\overline{w}\mathrm{d}x\mathrm{d}y \tag{13}$$

K_{w} is the spring stiffness of the Winkler-type foundation, and K_{P} is the shear modulus of Pasternak-type foundation.

The variation of kinetic energy is expressed as

$$\delta K = \int_A \int_{-h/2}^{h/2} \rho(\dot{u}\,\delta\dot{u} + \dot{v}\,\delta\dot{v} + \dot{w}\,\delta\dot{w})\mathrm{d}z\mathrm{d}A \tag{14}$$

The virtual kinetic energy in terms of the inertia is obtained after substitutinging Eq. 6 into Eq. 14.

$$\delta K = \int\limits_A \left[\begin{array}{l} I_0\left(\dot{\overline{u}}\delta\dot{\overline{u}} + \dot{\overline{v}}\delta\dot{\overline{v}} + \dot{\overline{w}}\delta\dot{\overline{w}}\right) - I_1\left(\dot{\overline{u}}\frac{\partial\delta\dot{\overline{w}}}{\partial x} + \frac{\partial\dot{\overline{w}}}{\partial x}\delta\dot{\overline{u}} + \dot{\overline{v}}\frac{\partial\delta\dot{\overline{w}}}{\partial y} + \frac{\partial\dot{\overline{w}}}{\partial y}\delta\dot{\overline{v}}\right) \\ +I_2\left(\frac{\partial\dot{\overline{w}}}{\partial x}\frac{\partial\delta\dot{\overline{w}}}{\partial x} + \frac{\partial\dot{\overline{w}}}{\partial y}\frac{\partial\delta\dot{\overline{w}}}{\partial y}\right) + J_1\left(\dot{\overline{u}}\delta\dot{\Phi}_x + \dot{\Phi}_x\delta\dot{\overline{u}} + \dot{\overline{v}}\delta\dot{\Phi}_y + \dot{\Phi}_y\delta\dot{\overline{v}}\right) \\ -J_2\left(\frac{\partial\dot{\overline{w}}}{\partial x}\delta\dot{\Phi}_x + \dot{\Phi}_x\frac{\partial\delta\dot{\overline{w}}}{\partial x} + \frac{\partial\dot{\overline{w}}}{\partial y}\delta\dot{\Phi}_y + \dot{\Phi}_y\frac{\partial\delta\dot{\overline{w}}}{\partial y}\right) + J_3\left(\dot{\Phi}_x\delta\dot{\Phi}_x + \dot{\Phi}_y\delta\dot{\Phi}_y\right) \end{array} \right] \mathrm{d}x\mathrm{d}y$$

where $\{I_0, I_1, I_2, J_1, J_2, J_3\} = \int_{-h/2}^{h/2} \rho\{1, z, z^2, f(z), zf(z), f(z)^2\}\mathrm{d}z$.

5.2 Governing Equations

The governing differential equation is obtained for free vibration analysis of tapered FGM plate.

$$\delta\overline{u} : \quad \frac{\partial N_{xx}}{\partial x} + \frac{\partial N_{xy}}{\partial y} = I_0\ddot{\overline{u}} - I_1\frac{\partial\ddot{\overline{w}}}{\partial x} + J_1\ddot{\Phi}_x, \tag{15.a}$$

$$\delta\overline{v} : \quad \frac{\partial N_{yy}}{\partial y} + \frac{\partial N_{xy}}{\partial x} = I_0\ddot{\overline{v}} - I_1\frac{\partial\ddot{\overline{w}}}{\partial y} + J_1\ddot{\Phi}_y \tag{15.b}$$

$$\begin{aligned} \delta\overline{w} : &\left(\frac{\partial^2 M^{\mathrm{b}}_{xx}}{\partial x^2} + \frac{\partial^2 M^{\mathrm{b}}_{yy}}{\partial y^2} + 2\frac{\partial^2 M^{b}_{xy}}{\partial x\partial y}\right) = \hat{N}_{\mathrm{e}} + I_0\ddot{\overline{w}} + I_1\left(\frac{\partial\ddot{\overline{u}}}{\partial x} + \frac{\partial\ddot{\overline{v}}}{\partial y}\right) \\ &- I_2\nabla^2\ddot{\overline{w}} + J_2\left(\frac{\partial\ddot{\Phi}_x}{\partial x} + \frac{\partial\ddot{\Phi}_y}{\partial y}\right) \end{aligned} \tag{15.c}$$

$$\delta\Phi_x : \frac{\partial M^{\mathrm{s}}_{xx}}{\partial x} + \vartheta\frac{\partial M^{\mathrm{b}}_{xx}}{\partial x} + \frac{\partial M^{\mathrm{s}}_{xy}}{\partial y} + \vartheta\frac{\partial M^{\mathrm{b}}_{xy}}{\partial y} - M^{\mathrm{q}}_{xx} - \vartheta Q_x = J_1\ddot{\overline{u}} - J_2\frac{\partial\ddot{\overline{w}}}{\partial x} + J_3\ddot{\Phi}_x \tag{15.d}$$

$$\delta\Phi_y : \frac{\partial M^{\mathrm{s}}_{yy}}{\partial y} + \vartheta\frac{\partial M^{\mathrm{b}}_{yy}}{\partial y} + \frac{\partial M^{\mathrm{s}}_{yx}}{\partial x} + \vartheta\frac{\partial M^{\mathrm{b}}_{yx}}{\partial x} - M^{\mathrm{q}}_{yy} - \vartheta Q_y = J_1\ddot{\overline{v}} - J_2\frac{\partial\ddot{\overline{w}}}{\partial y} + J_3\ddot{\Phi}_y \tag{15.e}$$

Furthermore, L_{ij}'s represents the linear differential operator as shown below.

$$\tilde{L}_{11}\overline{u} + \tilde{L}_{12}\overline{v} - \tilde{L}_{13}\overline{w} + \tilde{L}_{14}\Phi_x + \tilde{L}_{15}\Phi_y = I_0\ddot{\overline{u}} - I_1\frac{\partial\ddot{\overline{w}}}{\partial x} + J_1\ddot{\Phi}_x \tag{16.a}$$

$$\tilde{L}_{21}\overline{u} + \tilde{L}_{22}\overline{v} - \tilde{L}_{23}\overline{w} + \tilde{L}_{24}\Phi_x + \tilde{L}_{25}\Phi_y = I_0\ddot{\overline{v}} - I_1\frac{\partial\ddot{\overline{w}}}{\partial y} + J_1\ddot{\Phi}_y \tag{16.b}$$

$$\tilde{L}_{31}\bar{u}+\tilde{L}_{32}\bar{v}-\tilde{L}_{33}\bar{w}+\tilde{L}_{34}\Phi_x+\tilde{L}_{35}\Phi_y-\hat{\mathrm{N}}_{\mathrm{e}}=I_0\ddot{\bar{w}}+I_1\left(\frac{\partial\ddot{\bar{u}}}{\partial x}+\frac{\partial\ddot{\bar{v}}}{\partial y}\right)$$
$$-I_2\nabla^2\ddot{\bar{w}}+J_2\left(\frac{\partial\ddot{\Phi}_x}{\partial x}+\frac{\partial\ddot{\Phi}_y}{\partial y}\right) \quad (16.c)$$

$$\tilde{L}_{41}\bar{u}+\tilde{L}_{42}\bar{v}-\tilde{L}_{43}\overline{w}+\tilde{L}_{44}\Phi_x+\tilde{L}_{45}\Phi_y=J_1\ddot{\bar{u}}-J_2\frac{\partial\ddot{\overline{w}}}{\partial x}+J_3\ddot{\Phi}_x \quad (16.d)$$

$$\tilde{L}_{51}\bar{u}+\tilde{L}_{52}\bar{v}-\tilde{L}_{53}\overline{w}+\tilde{L}_{54}\Phi_x+\tilde{L}_{55}\Phi_y=J_1\ddot{\bar{v}}-J_2\frac{\partial\ddot{\overline{w}}}{\partial y}+J_3\ddot{\Phi}_y \quad (16.e)$$

5.3 Analytical Solution

Consider a simply supported tapered rectangular plate on elastic foundations. The boundary conditions in accordance with the two-parameter elastic foundation problem can be given as:

$$\left.\begin{array}{l}\bar{u}(x,0)=\bar{u}(x,b),\ M_{yy}(x,0)=M_{yy}(x,b)\\ \bar{v}(0,y)=\bar{v}(a,y),\ M_{xx}(0,y)=M_{xx}(a,y)\\ \Phi_x(x,0)=\Phi_x(x,b),\ N_{yy}(x,0)=N_{yy}(x,b)\\ \Phi_y(0,y)=\Phi_y(a,y),\ N_{xx}(0,y)=N_{xx}(a,y)\end{array}\right\}=0$$

The displacement fields satisfying the boundary conditions in terms of undetermined coefficients and trigonometric functions are given as:

$$\{\overline{u}(x,y),\phi_x(x,y)\}=\sum_{m=1}^{\infty}\sum_{n=1}^{\infty}\{U^{mn},\theta_x^{mn}\}\frac{\partial\xi^m(x)}{\partial x}\xi^n(y)\mathrm{e}^{i\omega t} \quad (17.a)$$

$$\{\overline{v}(x,y),\phi_y(x,y)\}=\sum_{m=1}^{\infty}\sum_{n=1}^{\infty}\{V^{mn},\theta_y^{mn}\}\xi^m(x)\frac{\partial\xi^n(y)}{\partial y}\mathrm{e}^{i\omega t} \quad (17.b)$$

$$\overline{w}(x,y)=\sum_{m=1}^{\infty}\sum_{n=1}^{\infty}W^{mn}\xi^m(x)\xi^n(x)\mathrm{e}^{i\omega t} \quad (17.c)$$

$\xi^m(x)$ and $\xi^n(y)$ are the Eigen function which satisfied the boundary condition and defined for simply supported boundary condition when the boundaries are parallel to y-axis $\xi(0)=\xi''(0)=0,\ \xi(a)=\xi''(a)=0$ and parallel to x-axis $\xi(0)=\xi''(0)=0,\ \xi(b)=\xi''(b)=0$, respectively, as:

$$\xi^m(x)=\sin(\alpha x),\ \xi^n(y)=\sin(\beta y).$$

where $\alpha = \frac{m\pi}{a}$, $\beta = \frac{n\pi}{b}$.

Free vibration equation is represented in matrix form as:

$$\left(\kappa_{ij} - \omega^2 M_{ij}\right)\chi_j = 0 \quad (i, j = 1, 2, 3, 4, 5)$$

And, finally, we get

$$\left|\kappa_{ij} - \omega^2 M_{ij}\right| = 0 \tag{18}$$

where k and M are representing the stiffness and mass matrix, respectively.

$$\kappa_{ij} = \begin{bmatrix} \kappa_{11} & \kappa_{12} & \kappa_{13} & \kappa_{14} & \kappa_{15} \\ \kappa_{21} & \kappa_{22} & \kappa_{23} & \kappa_{24} & \kappa_{25} \\ \kappa_{31} & \kappa_{32} & \kappa_{33} & \kappa_{34} & \kappa_{35} \\ \kappa_{41} & \kappa_{42} & \kappa_{43} & \kappa_{44} & \kappa_{45} \\ \kappa_{51} & \kappa_{52} & \kappa_{53} & \kappa_{54} & \kappa_{55} \end{bmatrix},$$

$$M_{ij} = \begin{bmatrix} -I_0\mu_6 & 0 & I_1\mu_6 & -J_1\mu_6 & 0 \\ 0 & -I_0\mu_2 & I_1\mu_2 & 0 & -J_1\mu_2 \\ -I_1\mu_9 & -I_1\mu_3 & -I_0\mu_1 + I_2(\mu_3 + \mu_9) & -J_2\mu_9 & -J_2\mu_3 \\ -J_1\mu_6 & 0 & J_2\mu_6 & -J_3\mu_6 & 0 \\ 0 & -J_1\mu_2 & J_2\mu_2 & 0 & -J_3\mu_2 \end{bmatrix}$$

$$\kappa_{11} = \int_0^a \int_0^b (A_{11}\mu_{12} + A_{66}\mu_8)\mathrm{d}x\mathrm{d}y$$

$$\kappa_{12} = \int_0^a \int_0^b ((A_{12} + A_{66})\mu_8)\mathrm{d}x\mathrm{d}y$$

$$\kappa_{13} = -\int_0^a \int_0^b (B_{11}\mu_{12} + (B_{12} + 2B_{66})\mu_8)\mathrm{d}x\mathrm{d}y$$

$$\kappa_{14} = \int_0^a \int_0^b ((\vartheta B_{11} + F_{11})\mu_{12} + (\vartheta B_{66} + F_{66})\mu_8)\mathrm{d}x\mathrm{d}y$$

$$\kappa_{15} = \int_0^a \int_0^b \mu_8(\vartheta(B_{12} + B_{66}) + (F_{12} + F_{66}))\mathrm{d}x\mathrm{d}y$$

$$\kappa_{21} = \int_0^a \int_0^b (A_{12} + A_{66})\mu_{10}\mathrm{d}x\mathrm{d}y$$

$$\kappa_{22} = \int_0^a \int_0^b (A_{66}\mu_{10} + A_{22}\mu_4)\mathrm{d}x\mathrm{d}y$$

$$\kappa_{23} = -\int_0^a \int_0^b (B_{22}\mu_4 + (B_{12} + 2B_{66})\mu_{10})\mathrm{d}x\mathrm{d}y$$

$$\kappa_{24} = \int_0^a \int_0^b \mu_{10}(\vartheta(B_{12} + B_{66}) + (F_{12} + F_{66}))\mathrm{d}x\mathrm{d}y$$

$$\kappa_{25} = \int_0^a \int_0^b ((\vartheta B_{66} + F_{66})\mu_{10} + (\vartheta B_{22} + F_{22})\mu_4)\mathrm{d}x\mathrm{d}y$$

$$\kappa_{31} = \int_0^a \int_0^b (B_{11}\mu_{13} + (B_{12} + 2B_{66})\mu_{11})\mathrm{d}x\mathrm{d}y$$

$$\kappa_{32} = \int_0^a \int_0^b (B_{22}\mu_5 + (B_{12} + 2B_{66})\mu_{11})\mathrm{d}x\mathrm{d}y$$

$$\kappa_{33} = -\int_0^a \int_0^b (D_{11}\mu_{13} + (2D_{12} + 4D_{66})\mu_{11} + D_{22}\mu_5 + K_w\mu_1 - K_P(\mu_3 + \mu_9))\mathrm{d}x\mathrm{d}y$$

$$\kappa_{34} = \int_0^a \int_0^b (\mu_{13}(H_{11} + \vartheta D_{11}) + \mu_{11}((H_{12} + 2H_{66}) + \vartheta(D_{12} + 2D_{66})))\mathrm{d}x\mathrm{d}y$$

$$\kappa_{35} = \int_0^a \int_0^b (\mu_5(H_{22} + \vartheta D_{22}) + \mu_{11}((H_{12} + 2H_{66}) + \vartheta(D_{12} + 2D_{66})))\mathrm{d}x\mathrm{d}y$$

$$\kappa_{51} = \int_0^a \int_0^b \mu_{10}(\vartheta(B_{12} + B_{66}) + (F_{12} + F_{66}))\mathrm{d}x\mathrm{d}y$$

$$\kappa_{52} = \int_0^a \int_0^b ((\vartheta B_{66} + F_{66})\mu_{10} + (\vartheta B_{22} + F_{22})\mu_4)\mathrm{d}x\mathrm{d}y$$

$$\kappa_{53} = -\int_0^a \int_0^b (\mu_4(H_{22} + \vartheta D_{22}) + \mu_{10}((H_{12} + 2H_{66}) + \vartheta(D_{12} + 2D_{66})))\mathrm{d}x\mathrm{d}y$$

$$\kappa_{54} = \int_0^a \int_0^b \mu_{10}(\vartheta(\vartheta D_{12} + H_{12}) + (\vartheta H_{12} + J_{12}) + \vartheta(\vartheta D_{66} + H_{66}) + (\vartheta H_{66} + J_{66}))\mathrm{d}x\mathrm{d}y$$

$$\kappa_{55} = \int_0^a \int_0^b ((\vartheta^2 D_{22} + 2\vartheta H_{22} + J_{22})\mu_4 + (\vartheta^2 D_{66} + 2\vartheta H_{66} + J_{66})\mu_{10} - (\vartheta^2 A_{44} + 2\vartheta K_{44} + L_{44})\mu_2)\mathrm{d}x\mathrm{d}y$$

$$\{\mu_1, \mu_3, \mu_5, \mu_7, \mu_9, \mu_{11}, \mu_{13}\} = \left\{\xi_x\xi_y, \xi_x\xi_y'', \xi_x\xi_y^{iv}, \xi_x'\xi_y', \xi_x''\xi_y, \xi_x''\xi_y'', \xi_x^{iv}\xi_y\right\}\xi_x\xi_y$$

$$\{\mu_2, \mu_4, \mu_{10}\} = \left\{\xi_x\xi_y', \xi_x\xi_y''', \xi_x''\xi_y'\right\}\xi_x\xi_y', \{\mu_6, \mu_8, \mu_{12}\} = \left\{\xi_x'\xi_y, \xi_x'\xi_y'', \xi_x'''\xi_y\right\}\xi_x'\xi_y$$

where

$$\tilde{L}_{11} = A_{11}\frac{\partial^2}{\partial x^2} + A_{66}\frac{\partial^2}{\partial y^2};\ \tilde{L}_{12} = (A_{12} + A_{66})\frac{\partial^2}{\partial x \partial y};$$

$$\tilde{L}_{13} = B_{11}\frac{\partial^3}{\partial x^3} + (B_{12} + 2B_{66})\frac{\partial^3}{\partial x \partial y^2}$$

$$\tilde{L}_{14} = (\vartheta B_{11} + F_{11})\frac{\partial^2}{\partial x^2} + (\vartheta B_{66} + F_{66})\frac{\partial^2}{\partial y^2};$$

$$\tilde{L}_{15} = (\vartheta(B_{12} + B_{66}) + (F_{12} + F_{66}))\frac{\partial^2}{\partial x \partial y}$$

$$\tilde{L}_{22} = A_{22}\frac{\partial^2}{\partial y^2} + A_{66}\frac{\partial^2}{\partial x^2};\ \tilde{L}_{23} = B_{22}\frac{\partial^3}{\partial y^3} + (B_{12} + 2B_{66})\frac{\partial^3}{\partial x^2 \partial y};$$

$$\tilde{L}_{24} = (\vartheta(B_{12} + B_{66}) + F_{12} + F_{66})\frac{\partial^2}{\partial x \partial y}$$

$$\tilde{L}_{25} = (\vartheta B_{66} + F_{66})\frac{\partial^2}{\partial x^2} + (\vartheta B_{22} + F_{22})\frac{\partial^2}{\partial y^2};$$

$$\tilde{L}_{33} = D_{11}\frac{\partial^4}{\partial x^4} + D_{22}\frac{\partial^4}{\partial y^4} + 2(D_{12} + 2D_{66})\frac{\partial^4}{\partial x^2 \partial y^2}$$

$$\tilde{L}_{34} = (\vartheta D_{11} + H_{11})\frac{\partial^3}{\partial x^3} + (H_{12} + \vartheta(D_{12} + 2D_{66}) + 2H_{66})\frac{\partial^3}{\partial x \partial y^2}$$

$$\tilde{L}_{35} = (\vartheta D_{22} + H_{22})\frac{\partial^3}{\partial y^3} + (H_{12} + \vartheta(D_{12} + 2D_{66}) + 2H_{66})\frac{\partial^3}{\partial x^2 \partial y}$$

$$\tilde{L}_{44} = (\vartheta(H_{11} + \vartheta D_{11}) + (J_{11} + \vartheta H_{11}))\frac{\partial^2}{\partial x^2} + (\vartheta(H_{66} + \vartheta D_{66}) + (J_{66} + \vartheta H_{66}))\frac{\partial^2}{\partial y^2} - \left(L_{55} + \vartheta^2 A_{55} + 2\vartheta K_{55}\right)$$

$$\tilde{L}_{45} = (\vartheta(H_{12} + \vartheta D_{12}) + (J_{12} + \vartheta H_{12}) + \vartheta(H_{66} + \vartheta D_{66}) + (J_{66} + \vartheta H_{66}))\frac{\partial^2}{\partial x \partial y}$$

$$\tilde{L}_{55} = (\vartheta(H_{66} + \vartheta D_{66}) + (J_{66} + \vartheta H_{66}))\frac{\partial^2}{\partial x^2} + (\vartheta(H_{22} + \vartheta D_{22}) + (J_{22} + \vartheta H_{22}))\frac{\partial^2}{\partial y^2} - \left(L_{44} + \vartheta^2 A_{44} + 2\vartheta K_{44}\right)\Phi_y$$

6 Results and Validation

A lot of research has been done on rectangular FGM plate with elastic foundation and without foundation. But results are not available for tapered FGM plate with elastic foundations as per author's knowledge. So the formulation is validated with rectangular FGM plates on elastic foundation. A numerical example is taken for validation of rectangular P-FGM plates with ceramic/metal constituent material. Aluminum/alumina FGM plate is resting on a two-parameter elastic foundation. A refined plate theory was developed with the material properties vary in the thickness direction using simple power law. Table 1 shows the results for the non-dimensional frequency parameter ($\widehat{\omega} = \omega a^2 \sqrt{\rho_m / E_m}/h$) of P-FGM plate have good agreement with [18]. The material properties are as follows:

$E_c = 380$ GPa and $\rho_c = 3800$ kg/m^3 for ceramic material (alumina), $E_m = 70$ GPa and $\rho_m = 2702$ kg/m^3 for metal (aluminium), while the Poisson ratio (0.3) is taken constant for the study. Non-dimensional elastic foundation is assumed as:

Table 1 Comparative study of P-FGM square plates with simply supported boundary condition resting on a two-parameter elastic foundation, span ratio *a*/*h* = 10

K_w	K_P	Source	Power law (*p*)				
			0	1	2	5	10
0	0	Present	11.3351	8.6824	7.8763	7.4034	7.1453
		Ref.[a]	11.3351	8.6824	7.8763	7.4034	7.1453
0	100	Present	15.9731	14.9401	14.7799	14.8692	14.9162
		Ref.[a]	15.9732	14.9401	14.7799	14.8692	14.9162
100	0	Present	11.6147	9.1035	8.3643	7.9521	7.7293
		Ref.[a]	11.6147	9.1035	8.3643	7.9521	7.7293
100	100	Present	16.1728	15.1887	15.0455	15.1497	15.2045
		Ref.[a]	16.1728	15.1887	15.0455	15.1497	15.2045

[a]Reference [18] used for validation

$$K_{\mathrm{W}} = \frac{\widehat{K}_{\mathrm{W}} a^4}{D_{\mathrm{n}}},\ K_{\mathrm{P}} = \frac{\widehat{K}_{\mathrm{P}} a^2}{D_{\mathrm{n}}},\ D_{\mathrm{n}} = \frac{E_{\mathrm{m}} h^3}{12(1-\nu^2)}$$

Results are obtained for tapered P-FGM plate resting on the Winkler–Pasternak foundation. Fundamental frequencies are obtained for rectangular plates with simply supported edges. Figure 3a illustrated the frequency parameter of the tapered FGM plate resting on the various elastic foundations versus span ratio.

Here, aspect ratio $b/a = 1$ and taper ratio $\eta = 0.25$ with simply supported (SS-SS) boundary condition are kept constant for different span ratios. It can be seen that Winkler foundation has negligible effect while Pasternak foundation has a significant effect on fundamental frequency. An increase in span ratio of plate decreases the fundamental frequency. In Fig. 3b, fundamental frequency increases with increase of taper ratio and a decrease in fundamental frequency with the increase of span ratio for a particular taper ratio. A similar trend is shown in Fig. 4a, b with various aspect

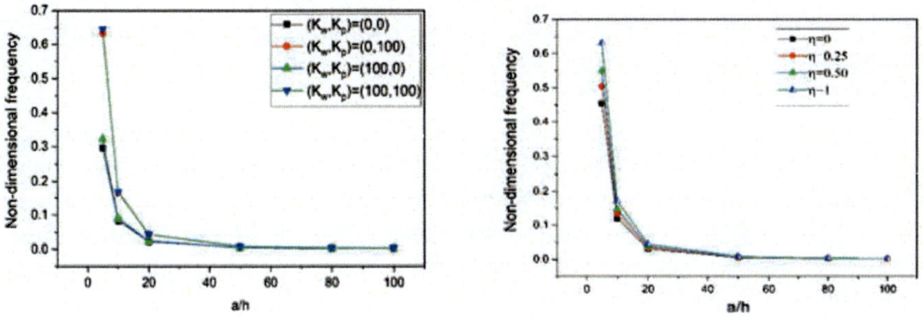

(a) η=0.25 and *p*=4. (b) (K_W,K_P)=(50,50) and *p*=4.

Fig. 3 Non-dimensional frequency parameter of tapered square FGM plates for various span ratios with different **a** elastic foundations and **b** taper ratio

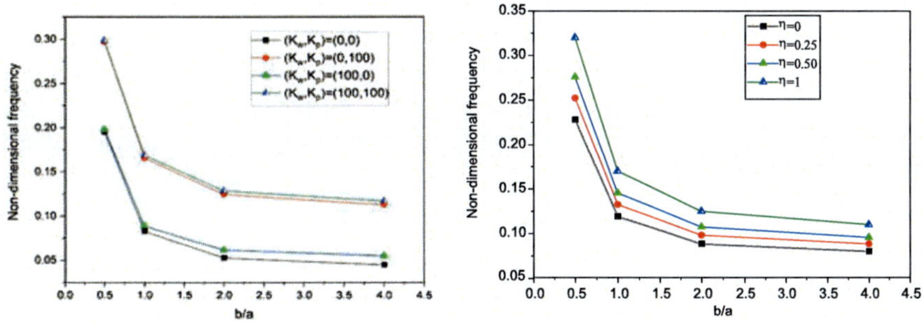

(a) a/h=10, η=0.25 and *p*=4. (b) a/h=10, (K_W,K_P)=(50,50) and *p*=4

Fig. 4 Non-dimensional frequency parameter of tapered FGM plates for various aspect ratios in combination of **a** elastic foundation and **b** taper ratio

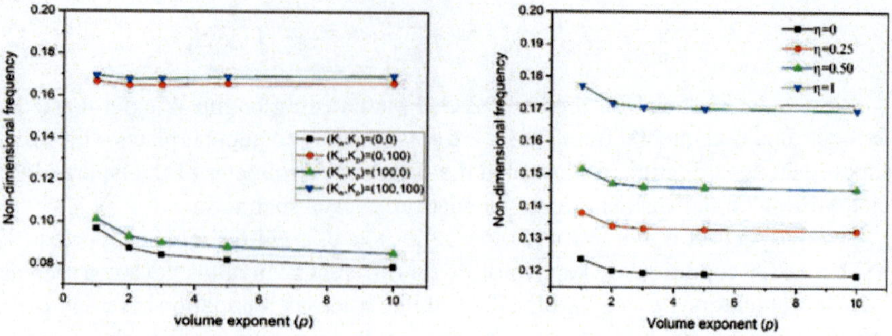

(a) a/h=10, η=0.25

(b) a/h=10, (Kw,KP

Fig. 5 Non-dimensional frequency parameter of tapered FGM square plates for various volume exponent in combination of **a** elastic foundation and **b** taper ratio

ratios. The maximum frequency was obtained when K_{w}, K_{p} have the value 100, 100 in conjunction with aspect ratio.

Figure 5a presents the Pasternak foundation effect on tapered P-FGM plate with volume exponent. The frequency parameter is increased with the incorporation of Pasternak foundation at fixed span ratio. Furthermore, Fig. 5b shows the effect of taper ratio and the volume exponent on the fundamental frequency parameter when taper ratio increases non-dimensional frequency increases and volume exponent shows the reverse effect on fundamental frequency, it is decreasing when volume exponent increases.

7 Conclusions

The Galerkin–Vlasov's solution method is used to investigate the free vibration analysis of tapered P-FGM plate resting on two parameter foundation. Higher-order shear deformation plate theory is implemented with simply supported boundary condition in combination with various span ratio, aspect ratio, one or two elastic foundation effect, and volume exponent. The following outcomes were investigated as mentioned below.

- The fundamental frequency decreases with increase of span ratio.
- It can be seen that the Winkler foundation has negligible effect on fundamental frequency, while Pasternak foundation has shown significant effect on frequency.
- Fundamental frequency was continuously increasing with increase of taper ratio at fixed foundation parameter.
- The fundamental frequency decreases with increase of aspect ratio.
- On the increase of volume fraction index, frequency parameter decreases.

It can be seen from the results that non-dimensional frequency is found maximum when the Pasternak foundation is applied with higher taper ratio. These results may be used as benchmark results for future reference.

References

1. Mortensen A, Suresh S (1995) Functionally graded metals and metal-ceramic composites. 1: Processing. Int Mater Rev 40:239–265. https://doi.org/10.1179/imr.1995.40.6.239
2. Koizumi M (1997) FGM activities in Japan. Compos Part B Eng 28:1–4. https://doi.org/10.1016/S1359-8368(96)00016-9
3. Reddy JN (2000) Analysis of functionally graded plates. Int J Numer Methods Eng 47:663–684 (2000). https:///doi.org/10.1002/(SICI)1097-0207(20000110/30)47:1/3<663::AID-NME787>3.0.CO;2-8
4. Cheng ZQ, Batra RC (2000) Exact correspondence between eigenvalues of membranes and functionally graded simply supported polygonal plates. J Sound Vib. https://doi.org/10.1006/jsvi.1999.2525
5. Hosseini-Hashemi S, Rokni Damavandi Taher H, Akhavan H, Omidi M, Thai HT, Choi DH (2010) Free vibration of functionally graded rectangular plates using first-order shear deformation plate theory. Appl Math Model 101:1276–1291. https://doi.org/10.1016/j.apm.2009.08.008
6. Ferreira AJM, Batra RC, Roque CMC, Qian LF, Jorge RMN (2006) Natural frequencies of functionally graded plates by a meshless method. Compos Struct. https://doi.org/10.1016/j.compstruct.2006.04.018
7. Zenkour AM (2005) A comprehensive analysis of functionally graded sandwich plates. Part 2—Buckling and free vibration. Int J Solids Struct. https://doi.org/10.1016/j.ijsolstr.2005.02.016
8. Xiang Y, Wang CM, Kitipornchai S (1994) Exact vibration solution for initially stressed Mindlin plates on Pasternak foundations. Int J Mech Sci 36:311–316. https://doi.org/10.1016/0020-7403(94)90037-X
9. Zhou D, Cheung YK, Lo SH, Au FTK (2004) Three-dimensional vibration analysis of rectangular thick plates on Pasternak foundation. Int J Numer Methods Eng 59:1313–1334. https://doi.org/10.1002/nme.915
10. Yang J, Shen HS (2001) Dynamic response of initially stressed functionally graded rectangular thin plates. Compos Struct. https://doi.org/10.1016/S0263-8223(01)00122-2
11. Lee WH, Han SC, Park WT (2015) A refined higher order shear and normal deformation theory for E-, P-, and S-FGM plates on Pasternak elastic foundation. Compos Struct 122:330–342. https://doi.org/10.1016/j.compstruct.2014.11.047
12. Amini MH, Soleimani M, Rastgoo A (2009) Three-dimensional free vibration analysis of functionally graded material plates resting on an elastic foundation. Smart Mater Struct 18. https://doi.org/10.1088/0964-1726/18/8/085015
13. Singh SJ, Harsha SP (2019) Exact solution for free vibration and buckling of sandwich S-FGM plates on pasternak elastic foundation with various boundary conditions. Int J Struct Stab Dyn 19. https://doi.org/10.1142/S0219455419500287
14. Lü CF, Lim CW, Chen WQ (2009) Exact solutions for free vibrations of functionally graded thick plates on elastic foundations. Mech Adv Mater Struct 16:576–584. https://doi.org/10.1080/15376490903138888
15. Malekzadeh P, Haghighi MRG, Atashi MM (2011) Free vibration analysis of elastically supported functionally raded annular plates subjected to thermal environment. Meccanica 46:893–913. https://doi.org/10.1007/s11012-010-9345-5
16. Baferani AH, Saidi AR, Ehteshami H (2011) Accurate solution for free vibration analysis of functionally graded thick rectangular plates resting on elastic foundation. Compos Struct. https://doi.org/10.1016/j.compstruct.2011.01.020

17. Singh SJ, Harsha SP (2019a) Nonlinear dynamic analysis of sandwich S-FGM plate resting on Pasternak foundation under thermal environment. Eur J Mech A Solids 76:155–179. https://doi.org/10.1016/j.euromechsol.2019.04.005
18. Thai HT, Choi DH (2012) A refined shear deformation theory for free vibration of functionally graded plates on elastic foundation. Compos Part B Eng 43:2335–2347. https://doi.org/10.1016/j.compositesb.2011.11.062
19. Atmane HA, Tounsi A, Mechab I, Bedia EAA (2010) Free vibration analysis of functionally graded plates resting on Winkler-Pasternak elastic foundations using a new shear deformation theory. Int J Mech Mater Des 6:113–121. https://doi.org/10.1007/s10999-010-9110-x
20. Singh SJ, Harsha SP (2019) Static analysis of functionally graded plate using nonlinear classical plate theory with von Karman strains: a complex solution analysis. In: Lecture notes mechanical in engineering, vol 23, pp 1–20. https://doi.or/10.1007/978-981-13-6469-3_1
21. Thai CH, Zenkour AM, Abdel Wahab M, Nguyen-Xuan H (2016) A simple four-unknown shear and normal deformations theory for functionally graded isotropic and sandwich plates based on isogeometric analysis. Compos Struct 139:77–95. https://doi.org/10.1016/j.compstruct.2015.11.066
22. Gupta A, Talha M, Chaudhari VK (2016) Natural frequency of functionally graded plates resting on elastic foundation using finite element method. Procedia Technol 23:163–170. https://doi.org/10.1016/j.protcy.2016.03.013
23. Singh SJ, Harsha SP (2019b) Buckling analysis of FGM plates under uniform, linear and non-linear in-plane loading. J Mech Sci Technol 33:1761–1767. https://doi.org/10.1007/s12206-019-0328-8
24. Joshan YS, Grover N, Singh BN (2017) A new non-polynomial four variable shear deformation theory in axiomatic formulation for hygro-thermo-mechanical analysis of laminated composite plates. Compos Struct 182:685–693. https://doi.org/10.1016/j.compstruct.2017.09.029

Free Vibration Analysis of Sandwich Plate with Honeycomb Core and FGM Face Sheets

S. J. Singh and S. P. Harsha

1 Introduction

The sandwich materials are introduced in the field of structures to enhance the performance of mechanical components, since the performance of the components is largely affected by its weight and inherent elastic properties. With the advancement in materials and fabrication processes, the shortcoming of the sandwich plate, namely delamination of face sheets, is eliminated. This drawback is overcome by the introduction of new advanced composite material, viz. functionally graded material (FGM) where the materials are continuously graded in the thickness direction. Thus, sandwich FGM plate is honeycomb core, and FGM face sheets are recently introduced in various engineering applications where lightweight is the vital requirement such as aerospace, ship building and automotive industries.

As far as honeycomb structures are concerned, a lot of research on sandwich plates with isotropic face sheets and honeycomb core has been done. In this context, Scarpa and Tomlinson [1] determined theoretically the vibration characteristics of sandwich plate with in-plane negative Poisson's ratio value. They found that the bending stiffness of the sandwich plate increases with the introduction of auxetic honeycomb core. Ruzzene et al. [2] analyzed the wave propagation and vibration characteristics of sandwich plate with periodic auxetic core. Liu et al. [3] analyze the sandwich panels with square honeycomb cores for bending, buckling and free vibration analysis. It was found that the height of a core largely affects the stiffness of the

S. J. Singh (✉)
Department of Mechanical Engineering, Pandit Deendayal Petroleum University, Gandhinagar, Gujarat 382009, India
e-mail: simran.singh@sot.pdpu.ac.in

S. P. Harsha
Mechanical and Industrial Engineering Department, Indian Institute of Technology Roorkee, Roorkee, Uttarakhand 247667, India
e-mail: surajfme@iitr.ac.in

V. H. Saran and R. K. Misra (eds.), *Advances in Systems Engineering*,
Lecture Notes in Mechanical Engineering,
https://doi.org/10.1007/978-981-15-8025-3_85

panels rather than the spacing between core sheets. Jweeg [4] determined the analytical solution for free vibration of honeycomb structure having regular hexagonal cell structure. In the similar way, a lot of research on sandwich honeycomb structure with isotropic face sheets are analyzed for free vibration [5–8] and dynamic analysis [9–14].

Further, with the advancement of composites, sandwich plate with isotropic face sheets is replaced by functionally graded face sheets due to certain advantages such as delamination and working under harsh environment. The research on FGM face sheets and honeycomb core is very scarce, and only one paper has been reported in the literature to the best knowledge of the author. In this framework, Gunes et al. [15] examined the honeycomb structure subjected to ballistic impact. The honeycomb structure is the sandwich between two FGM face sheets consisting of silica and aluminum phases. The two phases are varied across the thickness in accordance with Mori–Tanaka scheme.

Thus, it is observed that the analysis of sandwich FGM honeycomb structure in the available literature is scarce. Therefore, in the present paper, sandwich structure consisting of honeycomb core and FGM face sheets is considered and studied for free vibration analysis. This analysis for the aforementioned structure is done to find out the resonance frequencies which may occur during high excitations. The material properties in the FGM face sheets are assumed to vary in accordance with power law function (P-FGM). The exact solution is obtained using Navier's method. The effect of span-to-thickness ratio, aspect ratio and volume fraction exponent is studied in conjunction with geometric parameters of the honeycomb structure for free vibration of sandwich P-FGM honeycomb structure.

2 Material Properties for FGM Face Sheets and Honeycomb Core

The linear constitutive relation is defined as [16, 17],

$$\begin{bmatrix} \sigma_{xx} \\ \sigma_{yy} \\ \tau_{yz} \\ \tau_{xz} \\ \tau_{xy} \end{bmatrix} = \begin{bmatrix} Q_{11} & Q_{12} & 0 & 0 & 0 \\ Q_{12} & Q_{22} & 0 & 0 & 0 \\ 0 & 0 & Q_{66} & 0 & 0 \\ 0 & 0 & 0 & Q_{66} & 0 \\ 0 & 0 & 0 & 0 & Q_{66} \end{bmatrix} \begin{Bmatrix} \varepsilon_{xx} \\ \varepsilon_{yy} \\ \gamma_{yz} \\ \gamma_{xz} \\ \gamma_{xy} \end{Bmatrix} \tag{1}$$

where Q_{ij}s are reduced stiffness and defined based on different layers of sandwich honeycomb structure.

The sandwich honeycomb plate with honeycomb core and P-FGM face sheet is modeled by modifying the power law function in such a way that the continuity of the material properties at the interface always conserves. The volume fraction (V_t) of ceramic constituent is given as follows:

(a) Bottom P-FGM face sheet:

$$V_t = \left(\frac{z - \hbar_1}{\hbar_0 - \hbar_1}\right)^p \quad \text{for } \hbar_0 \leq z \leq \hbar_1 \tag{2.a}$$

(b) Top P-FGM face sheet:

$$V_t = \left(\frac{z - \hbar_2}{\hbar_3 - \hbar_2}\right)^p \quad \text{for } \hbar_2 \leq z \leq \hbar_3 \tag{2.b}$$

The effective material properties for the ith layer are calculated as follows [16, 17]:

$$\wp^{(i)}(z) = \wp_c V_t^{(i)} + \wp_m\left(1 - V_t^{(i)}\right) \tag{3}$$

where subscripts c and m signify the ceramic and metal constituent of FGM, respectively, and p denotes the volume fraction exponent (≥ 0). The constituent material for FGM face sheets is considered as isotropic. Therefore, reduced stiffness is defined as follows [16–20]:

$$Q_{11} = Q_{22} = \frac{E(z)}{1 - \nu^2}, \; Q_{12} = \frac{\nu E(z)}{1 - \nu^2} = \frac{\nu E(z)}{1 - \nu^2}, \; Q_{66} = \frac{E(z)}{2(1 + \nu)} \tag{4}$$

where $E(z)$ is the elastic modulus given by Eq. (3), and Poisson's ratio ν is constant.

The material properties for honeycomb structure are orthotropic which further depends upon the geometry of the basic unit cell of honeycomb structure as shown in Fig. 1 and is given as follows [8]:

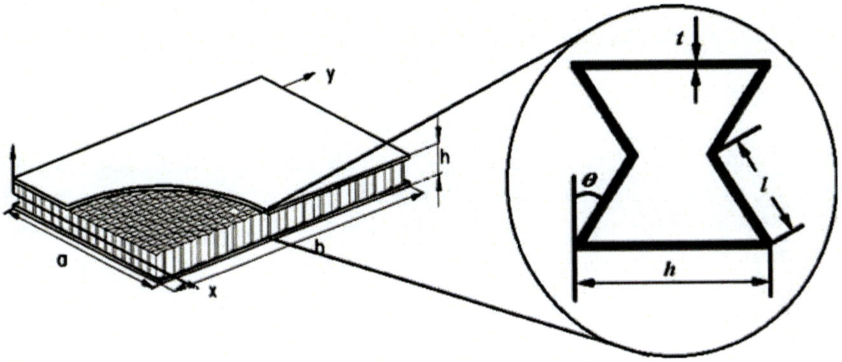

Fig. 1 Sandwich P-FGM plate with honeycomb core and the exaggerated view of unit cell of concave hexagonal honeycomb core [8]

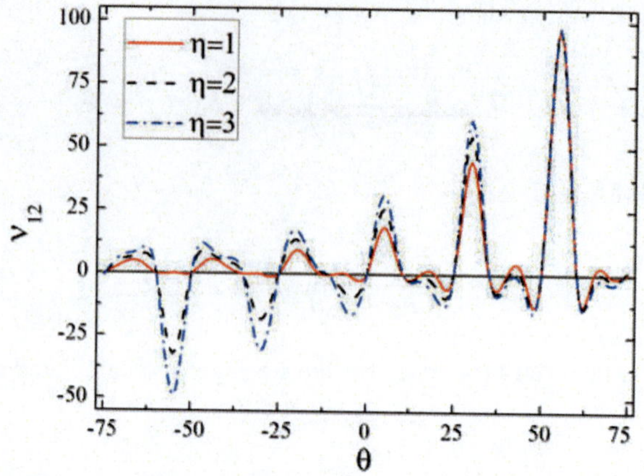

Fig. 2 Effect of angle (θ) on Poisson's ratio (ν_{12}) of the honeycomb core structure for different values of (η).($\xi = 0.1$)

$$\left\{\begin{array}{l} E_1 = E_s \frac{\xi^3(\eta - \sin\theta)}{\cos^3\theta\left[1 + \left(\tan^2\theta + \eta\sec^2\theta\right)\xi^2\right]}, \\ E_2 = E_s \frac{\xi^3}{\cos\theta(\eta - \sin\theta)\left(\tan^2\theta + \xi^2\right)}, \\ G_{12} = E_s \frac{\xi^3}{\eta(1+2\eta)\cos\theta}, \; G_{23} = G_s \frac{\xi\cos\theta}{\eta - \sin\theta}, \\ G_{31} = G_s \frac{\xi}{2\cos\theta}\left[\frac{\eta - \sin\theta}{1+2\eta} + \frac{\left(2\sin^2\theta + \eta\right)}{2(\eta - \sin\theta)}\right], \\ \nu_{12} = -\frac{\sin\theta\left(1-\xi^2\right)(\eta - \sin\theta)}{\cos^2\theta\left[1 + \left(\tan^2\theta + \eta\sec^2\theta\right)\xi^2\right]} \\ \nu_{21} = -\frac{\sin\theta\left(1-\xi^2\right)}{(\eta - \sin\theta)\left(\tan^2\theta + \xi^2\right)}, \; \rho = \rho_s \frac{\xi(\eta+2)}{2\cos\theta(\eta - \sin\theta)} \end{array}\right\} \text{for } \hbar_1 \le z \le \hbar_2$$

The reduced stiffness for orthotropic honeycomb core is defined as follows:

$$Q_{11} = \frac{E_1}{1 - \nu_{12}\nu_{21}}, \; Q_{22} = \frac{E_2}{1 - \nu_{12}\nu_{21}}, \; Q_{12} = Q_{21} = \frac{\nu_{12}E_2}{1 - \nu_{12}\nu_{21}},$$
$$Q_{66} = G_{12}, \; Q_{44} = G_{23}, \; Q_{55} = G_{13} \tag{5}$$

The variation of Poisson's ratio (ν_{12}) is shown in Fig. 2 with angle (θ). It is observed that the variation is periodic, since Poisson's ratio depends on the periodic function. This results in changing the behavior of honeycomb structure from conventional honeycomb structure (positive Poisson's ratio) to auxetic honeycomb structure (negative Poisson's ratio) and vice versa.

3 Theoretical Formulation

A honeycomb sandwich plate of dimension ($a \times b \times h$) is considered to be simply supported along the edges. The linear strain–displacement relation is assumed in conjunction with the following displacement field [21]:

$$u(x, y, z, t) = \overline{u}(x, y, t) - z\left(\frac{\partial \overline{w}}{\partial x}\right) + f(z)\Phi_x(x, y, t)$$
$$v(x, y, z, t) = \overline{v}(x, y, t) - z\left(\frac{\partial \overline{w}}{\partial y}\right) + f(z)\Phi_y(x, y, t)$$
$$w(x, y, z, t) = \overline{w}(x, y, t) \tag{6}$$

where $f(z) = \left(\tanh^{-1}\left(\frac{rz}{\hbar}\right) - z\frac{r/\hbar}{1-r^2/4}\right)$.

The displacement of the mid-surface in the plane of the plate is $\overline{u}$, $\overline{v}$ and out-of-the plane is $\overline{w}$. The rotations of the mid-surface or the coefficients of the shear function are Φ_x and Φ_y. The shape function in a simplified manner can be written as $f(z) = \lambda(z) + z\vartheta$, where $\lambda(z) = \tanh^{-1}(rz/\hbar)$ and $\vartheta = -(r/\hbar)/(1 - r^2/4)$, $r = 0.088$.

The equilibrium equation can be derived using the principle of virtual displacement given by the equation,

$$\int_{t_1}^{t_2} (\delta U - \delta K)\mathrm{d}t = 0 \tag{7}$$

where the variation in strain energy (δU) in terms of stress and moment resultants is given as

$$\delta U = \int_A \begin{bmatrix} \left(N_{xx}\delta\varepsilon_{xx}^{(0)} + N_{yy}\delta\varepsilon_{yy}^{(0)} + N_{xy}\delta\gamma_{xy}^{(0)}\right) \\ +\left(M_{xx}^{\mathrm{b}}\delta\varepsilon_{\mathrm{b}}^{(1)_{xx}} + M_{yy}^{\mathrm{b}}\delta\varepsilon_{\mathrm{b}}^{(1)_{yy}} + M_{xy}^{\mathrm{b}}\delta\gamma_{\mathrm{b}}^{(1)_{xy}}\right) \\ \left(M_{xx}^{\mathrm{s}}\delta\varepsilon_{\mathrm{s}}^{(1)_{xx}} + M_{yy}^{\mathrm{s}}\delta\varepsilon_{\mathrm{s}}^{(1)_{yy}} + M_{xy}^{\mathrm{s}}\delta\gamma_{\mathrm{s}}^{(1)_{xy}}\right) \\ +\left(M_{xx}^{\mathrm{q}}\delta\Phi_x + M_{yy}^{\mathrm{q}}\delta\Phi_y\right) + \vartheta\left(Q_x\delta\Phi_x + Q_y\delta\Phi_y\right) \end{bmatrix} \mathrm{d}x\mathrm{d}y \tag{8}$$

where $\begin{Bmatrix} \mathbf{N} \\ \mathbf{M^b} \\ \mathbf{M^s} \end{Bmatrix} = \begin{bmatrix} \mathbf{A} & \mathbf{B} & \mathbf{F} \\ \mathbf{B} & \mathbf{D} & \mathbf{H} \\ \mathbf{F} & \mathbf{H} & \mathbf{J} \end{bmatrix} \begin{Bmatrix} \boldsymbol{\varepsilon} \\ \boldsymbol{\varepsilon}_\mathbf{b} \\ \boldsymbol{\varepsilon}_\mathbf{s} \end{Bmatrix} \begin{matrix} \mathbf{Q} = (\vartheta\mathbf{A} + \mathbf{K})\boldsymbol{\gamma}_\mathbf{s} \\ \mathbf{M^q} = (\vartheta\mathbf{K} + \mathbf{L})\boldsymbol{\gamma}_\mathbf{s} \end{matrix}.$

$$\left\{A_{ij}, B_{ij}, D_{ij}, F_{ij}, H_{ij}, J_{ij}, K_{ij}, L_{ij}\right\} = \int_{-h/2}^{h/2} Q_{ij}(z)$$
$$\left\{1, z, z^2, \lambda(z), z\lambda(z), \lambda(z)^2, \lambda'(z), \lambda'(z)^2\right\}\mathrm{d}z$$

The kinetic energy due to inertia offered by the plate in variational form can be expressed as follows:

$$\delta K = \int_A \begin{bmatrix} I_0(\dot{\overline{u}}\delta\dot{\overline{u}} + \dot{\overline{v}}\delta\dot{\overline{v}} + \dot{\overline{w}}\delta\dot{\overline{w}}) \\ -I_1\left(\dot{\overline{u}}\frac{\partial\delta\dot{\overline{w}}}{\partial x} + \frac{\partial\dot{\overline{w}}}{\partial x}\delta\dot{\overline{u}} + \dot{\overline{v}}\frac{\partial\delta\dot{\overline{w}}}{\partial y} + \frac{\partial\dot{\overline{w}}}{\partial y}\delta\dot{\overline{v}}\right) \\ +I_2\left(\frac{\partial\dot{\overline{w}}}{\partial x}\frac{\partial\delta\dot{\overline{w}}}{\partial x} + \frac{\partial\dot{\overline{w}}}{\partial y}\frac{\partial\delta\dot{\overline{w}}}{\partial y}\right) \\ +J_1(\dot{\overline{u}}\delta\dot{\Phi}_x + \dot{\Phi}_x\delta\dot{\overline{u}} + \dot{\overline{v}}\delta\dot{\Phi}_y + \dot{\Phi}_y\delta\dot{\overline{v}}) \\ -J_2\left(\frac{\partial\dot{\overline{w}}}{\partial x}\delta\dot{\Phi}_x + \dot{\Phi}_x\frac{\partial\delta\dot{\overline{w}}}{\partial x} + \frac{\partial\dot{\overline{w}}}{\partial y}\delta\dot{\Phi}_y + \dot{\Phi}_y\frac{\partial\delta\dot{\overline{w}}}{\partial y}\right) \\ +J_3(\dot{\Phi}_x\delta\dot{\Phi}_x + \dot{\Phi}_y\delta\dot{\Phi}_y) \end{bmatrix} \mathrm{d}x\mathrm{d}y \tag{9}$$

where $\{I_0, I_1, I_2, J_1, J_2, J_3\} = \int_{-\hbar/2}^{\hbar/2} \rho\{1, z, z^2, f(z), zf(z), f(z)^2\}\mathrm{d}z$.

The governing differential equations using Eqs. 7, 8 and 9 in terms of displacements and rotations can be obtained as follows:

$$\begin{aligned} &A_{11}\frac{\partial^2\overline{u}}{\partial x^2} + A_{12}\frac{\partial^2\overline{v}}{\partial x\partial y} + A_{66}\left(\frac{\partial^2\overline{v}}{\partial x\partial y} + \frac{\partial^2\overline{u}}{\partial y^2}\right) - B_{11}\left(\frac{\partial^3\overline{w}}{\partial x^3} - \vartheta\frac{\partial^2\Phi_x}{\partial x^2}\right) \\ &- B_{12}\left(\frac{\partial^3\overline{w}}{\partial x\partial y^2} - \vartheta\frac{\partial^2\Phi_y}{\partial x\partial y}\right) - B_{66}\left(2\frac{\partial^3\overline{w}}{\partial x\partial y^2} - \vartheta\frac{\partial^2\Phi_x}{\partial y^2} - \vartheta\frac{\partial^2\Phi_y}{\partial x\partial y}\right) \\ &+ F_{11}\frac{\partial^2\Phi_x}{\partial x^2} + F_{12}\frac{\partial^2\Phi_y}{\partial x\partial y} + F_{66}\left(\frac{\partial^2\Phi_x}{\partial y^2} + \frac{\partial^2\Phi_y}{\partial x\partial y}\right) = I_0\ddot{\overline{u}} - I_1\frac{\partial\ddot{\overline{w}}}{\partial x} + J_1\ddot{\Phi}_x \end{aligned} \tag{10.a}$$

$$\begin{aligned} &A_{12}\frac{\partial^2\overline{u}}{\partial x\partial y} + A_{22}\frac{\partial^2\overline{v}}{\partial y^2} + A_{66}\left(\frac{\partial^2\overline{v}}{\partial x^2} + \frac{\partial^2\overline{u}}{\partial x\partial y}\right) - B_{12}\left(\frac{\partial^3\overline{w}}{\partial x^2\partial y} - \vartheta\frac{\partial^2\Phi_x}{\partial x\partial y}\right) \\ &- B_{22}\left(\frac{\partial^3\overline{w}}{\partial y^3} - \vartheta\frac{\partial^2\Phi_y}{\partial y^2}\right) - B_{66}\left(2\frac{\partial^3\overline{w}}{\partial x^2\partial y} - \vartheta\frac{\partial^2\Phi_x}{\partial x\partial y} - \vartheta\frac{\partial^2\Phi_y}{\partial x^2}\right) \\ &+ F_{12}\frac{\partial^2\Phi_x}{\partial x\partial y} + F_{22}\frac{\partial^2\Phi_y}{\partial y^2} + F_{66}\left(\frac{\partial^2\Phi_y}{\partial x^2} + \frac{\partial^2\Phi_x}{\partial x\partial y}\right) = I_0\ddot{\overline{v}} - I_1\frac{\partial\ddot{\overline{w}}}{\partial y} + J_1\ddot{\Phi}_y \end{aligned} \tag{10.b}$$

$$\begin{aligned} &B_{11}\frac{\partial^3\overline{u}}{\partial x^3} + B_{12}\left(\frac{\partial^3\overline{v}}{\partial x^2\partial y} + \frac{\partial^3\overline{u}}{\partial x\partial y^2}\right) + B_{22}\frac{\partial^3\overline{v}}{\partial y^3} + 2B_{66}\left(\frac{\partial^3\overline{u}}{\partial y^2\partial x} + \frac{\partial^3\overline{v}}{\partial y\partial x^2}\right) \\ &- D_{11}\left(\frac{\partial^4\overline{w}}{\partial x^4} - \vartheta\frac{\partial^3\Phi_x}{\partial x^3}\right) - D_{22}\left(\frac{\partial^4 w}{\partial y^4} - \vartheta\frac{\partial^3\Phi_y}{\partial y^3}\right) \\ &- D_{12}\left(2\frac{\partial^4 w}{\partial x^2\partial y^2} - \vartheta\frac{\partial^3\Phi_y}{\partial x^2\partial y} - \vartheta\frac{\partial^3\Phi_x}{\partial x\partial y^2}\right) \\ &- 2D_{66}\left(2\frac{\partial^4\overline{w}}{\partial x^2\partial y^2} - \vartheta\frac{\partial^3\Phi_x}{\partial x\partial y^2} - \vartheta\frac{\partial^3\Phi_y}{\partial x^2\partial y}\right) + H_{11}\frac{\partial^3\Phi_x}{\partial x^3} + H_{22}\frac{\partial^3\Phi_y}{\partial y^3} \\ &+ H_{12}\left(\frac{\partial^3\Phi_y}{\partial x^2\partial y} + \frac{\partial^3\Phi_x}{\partial x\partial y^2}\right) + 2H_{66}\left(\frac{\partial^3\Phi_x}{\partial x\partial y^2} + \frac{\partial^3\Phi_y}{\partial x^2\partial y}\right) \\ &= \left\{\begin{array}{l} I_0\ddot{\overline{w}} + I_1\left(\frac{\partial\ddot{\overline{u}}}{\partial x} + \frac{\partial\ddot{\overline{v}}}{\partial y}\right) \\ -I_2\left(\frac{\partial^2\ddot{\overline{w}}}{\partial x^2} + \frac{\partial^2\ddot{\overline{w}}}{\partial y^2}\right) + J_2\left(\frac{\partial\ddot{\Phi}_x}{\partial x} + \frac{\partial\ddot{\Phi}_y}{\partial y}\right) \end{array}\right\} \end{aligned} \tag{10.c}$$

$$\left\{\begin{array}{l} (F_{11}+\vartheta B_{11})\frac{\partial^2\overline{u}}{\partial x^2}+(F_{12}+\vartheta B_{12})\frac{\partial^2\overline{v}}{\partial x\partial y} \\ -(H_{11}+\vartheta D_{11})\left(\frac{\partial^3\overline{w}}{\partial x^3}-\vartheta\frac{\partial^2\Phi_x}{\partial x^2}\right) \\ -(H_{12}+\vartheta D_{12})\left(\frac{\partial^3 w}{\partial x\partial y^2}-\vartheta\frac{\partial^2\Phi_y}{\partial x\partial y}\right) \\ +(J_{11}+\vartheta H_{11})\frac{\partial^2\Phi_x}{\partial x^2}+(J_{12}+\vartheta H_{12})\frac{\partial^2\Phi_y}{\partial x\partial y} \\ +(F_{66}+\vartheta B_{66})\left(\frac{\partial^2\overline{u}}{\partial y^2}+\frac{\partial^2\overline{v}}{\partial x\partial y}\right) \\ -(H_{66}+\vartheta D_{66})\left(2\frac{\partial^3\overline{w}}{\partial x\partial y^2}-\vartheta\frac{\partial^2\Phi_x}{\partial y^2}-\vartheta\frac{\partial^2\Phi_y}{\partial x\partial y}\right) \\ +(J_{66}+\vartheta H_{66})\left(\frac{\partial^2\Phi_x}{\partial y^2}+\frac{\partial^2\Phi_y}{\partial x\partial y}\right) \\ -\left(L_{55}+\vartheta^2 A_{55}+2\vartheta K_{55}\right)\Phi_x \end{array}\right\} = J_1\ddot{\overline{u}} - J_2\frac{\partial\ddot{\overline{w}}}{\partial x} + J_3\ddot{\Phi}_x \quad (10.d)$$

n

$$\left\{\begin{array}{l} (F_{12}+\vartheta B_{12})\frac{\partial^2\overline{u}}{\partial x\partial y}+(F_{22}+\vartheta B_{22})\frac{\partial^2\overline{v}}{\partial y^2} \\ -(H_{12}+\vartheta D_{12})\left(\frac{\partial^3\overline{w}}{\partial x^2\partial y}-\vartheta\frac{\partial^2\Phi_x}{\partial x\partial y}\right) \\ -(H_{22}+\vartheta D_{22})\left(\frac{\partial^3 w}{\partial y^3}-\vartheta\frac{\partial^2\Phi_y}{\partial y^2}\right) \\ +(J_{12}+\vartheta H_{12})\frac{\partial^2\Phi_x}{\partial x\partial y}+(J_{22}+\vartheta H_{22})\frac{\partial^2\Phi_y}{\partial y^2} \\ +(F_{66}+\vartheta B_{66})\left(\frac{\partial^2\overline{u}}{\partial x\partial y}+\frac{\partial^2\overline{v}}{\partial x^2}\right) \\ -(H_{66}+\vartheta D_{66})\left(2\frac{\partial^3\overline{w}}{\partial x^2\partial y}-\vartheta\frac{\partial^2\Phi_x}{\partial x\partial y}-\vartheta\frac{\partial^2\Phi_y}{\partial x^2}\right) \\ +(J_{66}+\vartheta H_{66})\left(\frac{\partial^2\Phi_x}{\partial x\partial y}+\frac{\partial^2\Phi_y}{\partial x^2}\right) \\ -\left(L_{44}+\vartheta^2 A_{44}+2\vartheta K_{44}\right)\Phi_y \end{array}\right\} = J_1\ddot{\overline{v}} - J_2\frac{\partial\ddot{\overline{w}}}{\partial y} + J_3\ddot{\Phi}_y \quad (10.e)$$

n

4 Analytical Solution

Consider a simply supported sandwich rectangular plate. The boundary conditions in accordance with the aforementioned problem can be given as follows:

Dirichlet conditions:

$$\overline{u}(x,0)=\overline{u}(x,b)=0;\ \overline{v}(0,y)=\overline{v}(a,y)=0$$
$$\overline{w}(x,0)=\overline{w}(x,b)=\overline{w}(0,y)=\overline{w}(a,y)=0$$
$$\phi_x(x,0)=\phi_x(x,b)=0;\ \phi_y(0,y)=\phi_y(a,y)=0$$

Neumann conditions:

$$M_y(x,0)=M_y(x,b)=M_x(0,y)=M_x(a,y)=0$$
$$N_y(x,0)=N_y(x,b)=N_x(0,y)=N_x(a,y)=0$$

The displacement fields satisfying the boundary conditions in terms of undetermined coefficients $\left(U^{mn}, V^{mn}, W^{mn}, \theta_x^{mn}, \theta_y^{mn}\right)$ and trigonometric functions are given as follows:

$$\{\overline{u}(x, y), \phi_x(x, y)\} = \sum_{m=1}^{\infty}\sum_{n=1}^{\infty}\left\{U^{mn}, \theta_x^{mn}\right\}\cos(\alpha x)\sin(\beta y)$$

$$\left\{\overline{v}(x, y), \phi_y(x, y)\right\} = \sum_{m=1}^{\infty}\sum_{n=1}^{\infty}\left\{V^{mn}, \theta_y^{mn}\right\}\sin(\alpha x)\cos(\beta y)$$

$$\overline{w}(x, y) = \sum_{m=1}^{\infty}\sum_{n=1}^{\infty} W^{mn}\sin(\alpha x)\sin(\beta y) \tag{11}$$

where $\alpha = \frac{m\pi}{a}, \beta = \frac{n\pi}{b}$.

The solution for free vibration is obtained by representing the coupled partial differential equations in a matrix form by substituting Eq. (11) into Eq. (10). Thus, the set of coupled linear equations for any values of m and n in matrix form is given as

$$\left(\kappa_{ij} - \omega^2 M_{ij}\right)\partial_j = 0 \quad (i, j = 1, 2, 3, 4, 5) \tag{12}$$

$$\kappa_{ij} = \begin{bmatrix} \kappa_{11} & \kappa_{12} & \kappa_{13} & \kappa_{14} & \kappa_{15} \\ \kappa_{21} & \kappa_{22} & \kappa_{23} & \kappa_{24} & \kappa_{25} \\ \kappa_{31} & \kappa_{32} & \kappa_{33} + \psi & \kappa_{34} & \kappa_{35} \\ \kappa_{41} & \kappa_{42} & \kappa_{43} & \kappa_{44} & \kappa_{45} \\ \kappa_{51} & \kappa_{52} & \kappa_{53} & \kappa_{54} & \kappa_{55} \end{bmatrix}$$

$$M_{ij} = \begin{bmatrix} I_0 & 0 & -I_1\alpha & J_1 & 0 \\ 0 & I_0 & -I_1\beta & 0 & J_1 \\ -I_1\alpha & -I_1\beta & I_0 + I_2\left(\alpha^2 + \beta^2\right) & -J_2\alpha & -J_2\beta \\ J_1 & 0 & -J_2\alpha & J_3 & 0 \\ 0 & J_1 & -J_2\beta & 0 & J_3 \end{bmatrix}$$

where

$$\kappa_{11} = A_{11}\alpha^2 + A_{66}\beta^2$$
$$\kappa_{12} = (A_{12} + A_{66})\alpha\beta;$$
$$\kappa_{13} = -\left[B_{11}\alpha^2 + (B_{12} + 2B_{66})\beta^2\right]\alpha$$
$$\kappa_{14} = (\vartheta B_{11} + E_{11})\alpha^2 + (\vartheta B_{66} + E_{66})\beta^2$$
$$\kappa_{15} = [\vartheta(B_{12} + B_{66}) + (E_{12} + E_{66})]\alpha\beta$$
$$\kappa_{22} = A_{66}\alpha^2 + A_{22}\beta^2$$
$$\kappa_{23} = -\left[B_{22}\beta^2 + (B_{12} + 2B_{66})\alpha^2\right]\beta$$

$$\kappa_{24} = [\vartheta(B_{12} + B_{66}) + (E_{12} + E_{66})]\alpha\beta$$
$$\kappa_{25} = (\vartheta B_{66} + E_{66})\alpha^2 + (\vartheta B_{22} + E_{22})\beta^2$$

$$\kappa_{33} = D_{11}\alpha^4 + (2D_{12} + 4D_{66})\alpha^2\beta^2 + D_{22}\beta^4)$$
$$\kappa_{34} = -\left[\alpha^2(F_{11} + \Delta D_{11}) + \beta^2((F_{12} + 2F_{66}) + \vartheta(D_{12} + 2D_{66}))\right]\alpha$$
$$\kappa_{35} = -\left[\beta^2(F_{22} + \vartheta D_{22}) + \alpha^2((F_{12} + 2F_{66}) + \vartheta(D_{12} + 2D_{66}))\right]\beta$$
$$\kappa_{44} = \left[\left(\vartheta^2 D_{11} + 2\vartheta F_{11} + H_{11}\right)\beta^2 + \left(\vartheta^2 D_{66} + 2\vartheta F_{66} + H_{66}\right)\alpha^2 + \vartheta(\vartheta A_{55} + 2K_{55}) + L_{55}\right]$$
$$\kappa_{45} = [\vartheta(\vartheta D_{12} + F_{12}) + (\vartheta F_{12} + H_{12}) + \vartheta(\vartheta D_{66} + F_{66}) + (\vartheta F_{66} + H_{66})]\alpha\beta$$
$$\kappa_{55} = \left[\left(\vartheta^2 D_{22} + 2\vartheta F_{22} + H_{22}\right)\beta^2 + \left(\vartheta^2 D_{66} + 2\vartheta F_{66} + H_{66}\right)\alpha^2 + \vartheta(\vartheta A_{44} + 2K_{44}) + L_{44}\right]$$
$$\alpha = \frac{m\pi}{a}, \beta = \frac{n\pi}{b}.$$

5 Numerical Results and Discussion

The free vibration analysis is performed for functionally graded sandwich honeycomb plate simply supported along edges. The analysis is executed to determine the effect of geometric parameters of plate in conjunction with the geometric parameters of the honeycomb structure on the fundamental frequency of the sandwich plate. It is unless otherwise specified, the following parameters and material properties are considered for Al/Al_2O_3 FGM face sheets and honeycomb core (Al):

FGM face sheets: Aluminum: $E_m = 69$ GPa, $\rho_m = 2700$ kg/m^3.

Alumina: $E_c = 380$ GPa, $\rho_m = 3800$ kg/m^3.

Honeycomb core: Aluminum: $E_m = 69$ GPa, $\rho_m = 2700$ kg/m^3, $\eta = 1.5$, $\xi = 0.1$, $\theta = 30°$.

Geometrical parameters:

$$a/\hbar = 10, b/a = 1, p = 4, \hbar_1 = \hbar_3 = 5\hbar/12 \text{ (face sheets thickness)},$$
$$\hbar_2 = 5\hbar/6 \text{ (honeycomb core thickness)}$$

Since the material properties of honeycomb core are implemented first time in this study for calculating the fundamental frequency of the sandwich honeycomb plate, therefore, the results for sandwich plate with soft homogenous (aluminum) are calculated and verified with the results of Bennoun et al. [22]. It is depicted from Table 1 that the results are consistent with [22] and average error is less than 0.5%. The dimensionless parameter is defined as follows: $\widehat{\omega} = \omega a^2\sqrt{\rho_0/E_0}/\hbar$, $\rho_0 = 1$ kg/m^3, $E_0 = 1$ GPa. Young's modulus and density of the constituent material

Table 1 Comparison of non-dimensional fundamental frequency ($\widehat{\omega}$) for Al/Al_2O_3 P-FGM square sandwich plate

Configuration	Theory	$a/\hbar$		
		5	10	100
1-0-1	**Present**	**1.5171**	**1.7257**	**1.8203**
	Bennoun et al. [22]	1.50596	1.72227	1.82031
2-1-2	**Present**	**1.455**	**1.6838**	**1.7917**
	Bennoun et al. [22]	1.43325	1.67437	1.79163
1-1-1	**Present**	**1.403**	**1.6397**	**1.7539**
	Bennoun et al. [22]	1.38242	1.63053	1.75379
2-2-1	**Present**	**1.3618**	**1.5789**	**1.6819**
	Bennoun et al. [22]	1.34203	1.57037	1.68184
1-2-1	**Present**	**1.3281**	**1.561**	**1.6749**
	Bennoun et al. [22]	1.32129	1.55788	1.6749
1-8-1	**Present**	**1.1433**	**1.3076**	**1.3833**
	Bennoun et al. [22]	1.14451	1.30825	1.38331

are 70 GPa and 2707 kg/m^3, respectively, for aluminum (Al), and 380 GPa and 3800 kg/m^3, respectively, for alumina (Al_2O_3). Poisson's ratio is assumed as constant (0.3).

The fundamental frequency decreases with increase in volume fraction exponent as shown in Fig. 3, since, increase in volume fraction exponent increases the proportion of metal content in the face sheets. This increases in metal content reduce the overall stiffness of the honeycomb sandwich structure. The increase in η results in an increase in fundamental frequency irrespective of exponent (p). It is observed that the effect of $\eta \geq 1.5$ is negligible when p varies from 1 to 2 and after that the fundamental increases or decreases depend upon different volume fraction exponent.

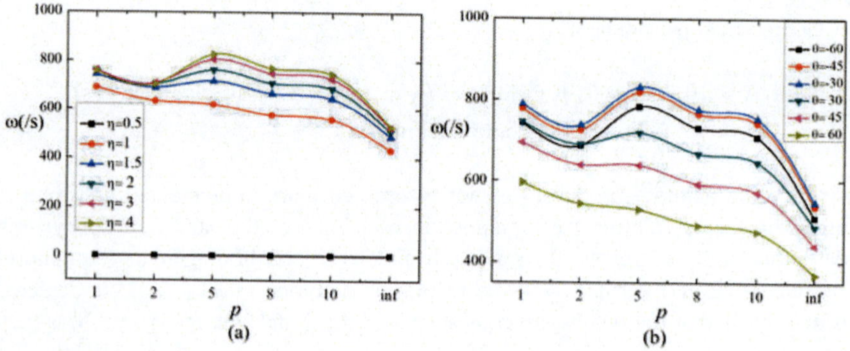

Fig. 3 Effect of volume fraction exponent (p) on fundamental frequency (ω) for different values of (**a**) η and (**b**) θ

As far as effect of angle θ is concerned, the fundamental frequency always increases with decrease in modulus of angle ($|\theta|$) as depicted from Fig. 3b. In addition, the fundamental frequency is the highest for exponent $p = 5$ for $\eta \geq 1.5$ and when θ is negative.

Figure 4 depicts the variation of fundamental frequency with span-to-thickness ratio. As expected, the fundamental frequency decreases with increase in span-to-thickness ratio irrespective of parameters θ and η. Since with increases in span-to-thickness ratio, the thickness of the plate reduces which results in reduction in flexural stiffness of the plate. However, there is an unexpected increase in frequency occurs at $a/\hbar = 20$ when $\eta = 4$ and at $a/\hbar = 10$ when $\theta = -60°, -45°$.

Next, the effect of aspect ratio on fundamental frequency is studied as shown in Fig. 5. The fundamental frequency decreases with increase in aspect ratio regardless of the parameters θ and η. This is due to effect of reduction in overall thickness of plate with increase in in-plane dimensions (a and b) which reduces the stiffness of

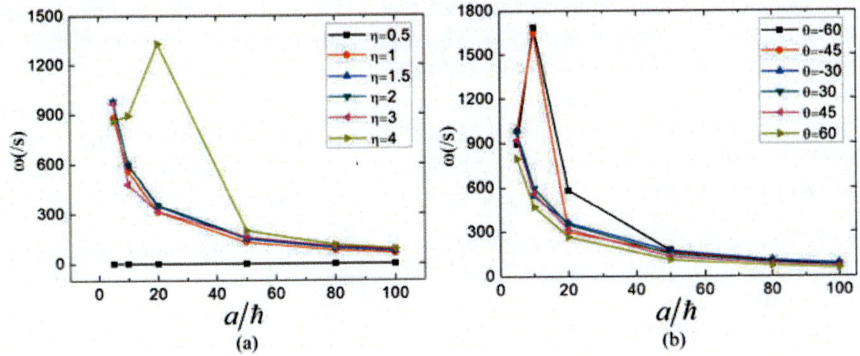

Fig. 4 Effect of span-to-thickness ratio ($a/\hbar$) on fundamental frequency (ω) for different values of (**a**) η and (**b**) θ

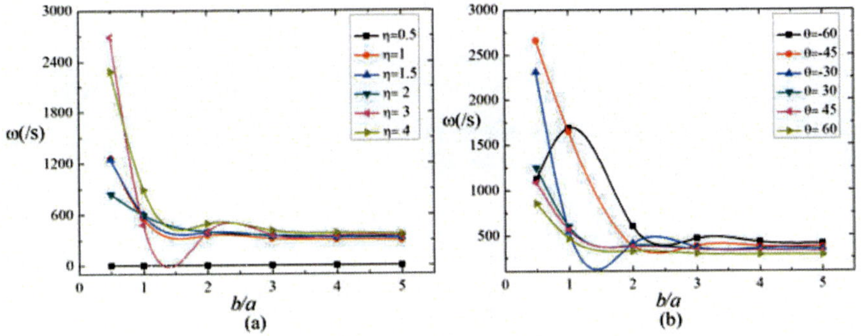

Fig. 5 Effect of aspect ratio (b/a) on fundamental frequency (ω) for different values of (**a**) η and (**b**) θ

the plate, and hence, fundamental frequency is reduced. In addition, there is a sudden increase in frequency is observed when $\theta = -60°$ for square plate.

It is observed form Figs. 3a, 4 and 5a that effect of parameter η is significant when η increases from 0.5 to 1. As for $\eta = 0.5$, there is minor change in fundamental frequency witnessed.

6 Conclusions

The present study is focused on depicting the effect of geometric parameter of honeycomb structures on fundamental frequency of sandwich plate with P-FGM face sheets and honeycomb core. The equation of motion is derived using Hamilton's principle in variational form. The exact solution is then obtained using Navier's method for simply supported sandwich plate. The validation study is also presented to validate the present formulation for sandwich plate, and results are found to be in good agreement with the previous studies. The results are obtained to study the effect of span-to-thickness ratio, aspect ratio and volume fraction exponent on fundamental frequency for different values of parameters θ and η. It is observed that typically fundamental frequency decreases with increases in span-to-thickness ratio, aspect ratio and volume fraction exponent. But due to the presence of honeycomb structure in the core of sandwich plate, an unexpected result can occur:

- Frequency is the highest for exponent $p = 5$ when $\eta \geq 1.5$ and when θ is negative.
- Upsurge in frequency occurs at $a/\hbar = 20$ when $\eta = 4$ and at $a/\hbar = 10$ when $\theta = -60°, -45°$.
- Sudden increase in frequency is observed when $\theta = -60°$ for square plate.

References

1. Scarpa F, Tomlinson G (2000) Theoretical characteristics of the vibration of sandwich plates with in-plane negative Poisson's ratio values. J Sound Vib 230:45–67. https://doi.org/10.1006/jsvi.1999.2600
2. Ruzzene M, Mazzarella L, Tsopelas P, Scarpa F (2002) Wave propagation in sandwich plates with periodic auxetic core. J Intell Mater Syst Struct 13:587–597. https://doi.org/10.1106/104538902031865
3. Liu J, Cheng YS, Li RF, Au FTK (2010) A semi-analytical method for bending, buckling, and free vibration analyses of sandwich panels with square-honeycomb cores. Int J Struct Stab Dyn 10:127–151. https://doi.org/10.1142/S0219455410003361
4. Jweeg MJ (2016) A suggested analytical solution for vibration of honeycombs sandwich combined plate structure. Int J Mech Mechatron Eng 16:9–17
5. Petrović Z, Lazarević I (2016) Design and analysis of the flat honeycomb sandwich structures. Sci Tech Rev 65:50–56. https://doi.org/10.5937/str1501050p
6. Mukhopadhyay T, Adhikari S (2016) Free-vibration analysis of sandwich panels with randomly irregular honeycomb core. J Eng Mech 142. https://doi.org/10.1061/(ASCE)EM.1943-7889.0001153

7. Marythraza M, Anitha D, Dash PK, Ravi Kumar P (2018) Vibration analysis of honeycomb sandwich panel in spacecraft structure. Int J Mech Prod Eng Res Dev 8:849–860. https://doi.org/10.24247/ijmperdjun201890
8. Zhu X, Zhang J, Zhang W, Chen J (2018) Vibration frequencies and energies of an auxetic honeycomb sandwich plate. Mech Adv Mater Struct 6494:1–7. https://doi.org/10.1080/15376494.2018.1455933
9. Strek T, Jopek H, Nienartowicz M (2015) Dynamic response of sandwich panels with auxetic cores. Phys Status Solidi Basic Res 252:1540–1550. https://doi.org/10.1002/pssb.201552024
10. Jin X, Wang Z, Ning J, Xiao G, Liu E, Shu X (2016) Dynamic response of sandwich structures with graded auxetic honeycomb cores under blast loading. Compos Part B Eng 106:206–217. https://doi.org/10.1016/j.compositesb.2016.09.037
11. Nguyen DD, Pham CH (2016) Nonlinear dynamic response and vibration of sandwich composite plates with negative Poisson's ratio in auxetic honeycombs. J Sandwich Struct Mater 109963621667472. https://doi.org/10.1177/1099636216674729
12. Duc ND, Seung-Eock K, Cong PH, Anh NT, Khoa ND (2017) Dynamic response and vibration of composite double curved shallow shells with negative Poisson's ratio in auxetic honeycombs core layer on elastic foundations subjected to blast and damping loads. Int J Mech Sci 133:504–512. https://doi.org/10.1016/j.ijmecsci.2017.09.009
13. Hu LL, Zhou MZ, Deng H (2018) Dynamic crushing response of auxetic honeycombs under large deformation: theoretical analysis and numerical simulation. Thin-Walled Struct 131:373–384. https://doi.org/10.1016/j.tws.2018.04.020
14. Zhang J, Zhu X, Yang X, Zhang W (2019) Transient nonlinear responses of an auxetic honeycomb sandwich plate under impact loads. Int J Impact Eng. https://doi.org/10.1016/j.ijimpeng.2019.103383
15. Gunes R, Arslan K, Apalak MK, Reddy JN (2019) Ballistic performance of honeycomb sandwich structures reinforced by functionally graded face plates. J Sandwich Struct Mater 21:211–229. https://doi.org/10.1177/1099636216689462
16. Singh SJ, Harsha SP (2018) Static analysis of functionally graded plate using nonlinear classical plate theory with Von-Karman strains. Int J Appl Mech Eng 23:707–726. https://doi.org/10.2478/ijame-2018-0039
17. Singh SJ, Harsha SP (2019) Buckling analysis of FGM plates under uniform, linear and non-linear in-plane loading. J Mech Sci Technol 33:1761–1767. https://doi.org/10.1007/s12206-019-0328-8
18. Singh SJ, Harsha SP (2019) Nonlinear dynamic analysis of sandwich S-FGM plate resting on Pasternak foundation under thermal environment. Eur J Mech A/Solids 76:155–179. https://doi.org/10.1016/j.euromechsol.2019.04.005
19. Singh SJ, Harsha SP (2018) Exact solution for free vibration and buckling of sandwich S-FGM plates on Pasternak elastic foundation with various boundary conditions. Int J Struct Stab Dyn 19:S0219455419500287. https://doi.org/10.1142/S0219455419500287
20. Singh SJ, Harsha SP (2018) Nonlinear vibration analysis of sigmoid functionally graded sandwich plate with ceramic-FGM-metal layers. J Vib Eng Technol 18. https://doi.org/10.1007/s42417-018-0058-8
21. Joshan YS, Grover N, Singh BN (2017) A new non-polynomial four variable shear deformation theory in axiomatic formulation for hygro-thermo-mechanical analysis of laminated composite plates. Compos Struct 182:685–693. https://doi.org/10.1016/j.compstruct.2017.09.029
22. Bennoun M, Houari MSA, Tounsi A (2016) A novel five-variable refined plate theory for vibration analysis of functionally graded sandwich plates. Mech Adv Mater Struct 23:423–431. https://doi.org/10.1080/15376494.2014.984088

Design and Static Structural Analysis of Brake Caliper for ATV

Sushant Satputaley, Kshitij Waghdhare, Karan Ksheersagar, Bijay Sankhari, and Yash Jichkar

1 Introduction

An all-terrain vehicle is designed for a rough surface areas or for a hilly regions. There are several competitions held like BAJA SAE India [1] where the performance of this vehicle is checked by performing the events like manoeuverability, rock crawl, acceleration and finally an endurance race. Effective braking is one of the key parameters in determining the performance of a vehicle and hence ensuring the driver's safety. This research deals with the designing and analysis of a brake caliper which is an important aspect of braking system. Braking system converts the kinetic energy of a vehicle to heat energy by applying clamping force on a brake disc. The components of a caliper are

1. Main body
2. Inlet port
3. Bleed port
4. Piston

S. Satputaley (✉) · K. Waghdhare · K. Ksheersagar · B. Sankhari · Y. Jichkar
Department of Mechanical Engineering, St. Vincent Pallotti College of Engineering & Technology, Nagpur, Maharashtra 441108, India
e-mail: satputaley@gmail.com

K. Waghdhare
e-mail: waghdharekshitij@gmail.com

K. Ksheersagar
e-mail: karanksheersagar9@gmail.com

B. Sankhari
e-mail: bjaysankhari87@gmail.com

Y. Jichkar
e-mail: yashjichkar03@gmail.com

V. H. Saran and R. K. Misra (eds.), *Advances in Systems Engineering*, Lecture Notes in Mechanical Engineering,
https://doi.org/10.1007/978-981-15-8025-3_86

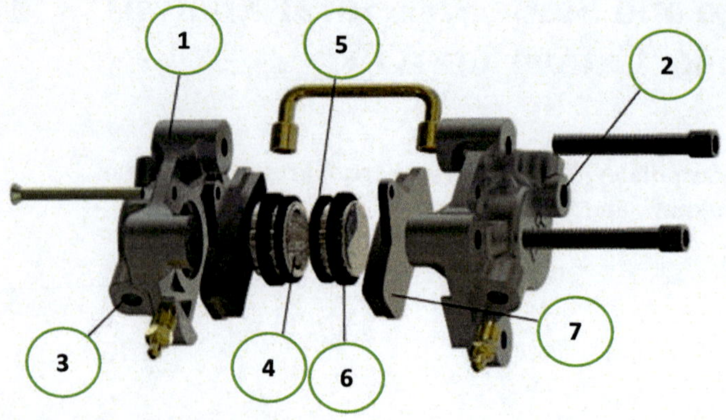

Fig. 1 Exploded view of custom caliper

5. Scrapper seal
6. Retraction rubber seal
7. Friction pads (Fig. 1).

The brake caliper comprises friction pads which comes in contact with the disc when the force is applied on the pedal, and the clamping force is directly proportional to the force exerted by the driver on the pedal. As the caliper weight results in increase of unsprung weight of the vehicle, it is recommended that the design should be optimised in such a way that the weight should be minimised without compromising the strength and stiffness.

2 Design Methodology and Calculations

Brake caliper is a three-dimensional structure. The calculation of a clamping force plays a vital role in designing and material selection for the caliper. Following are the steps for designing the caliper (Fig. 2).

2.1 Calculation of Clamping Force

When the driver applies pedal force, the equivalent pressure is generated in the master cylinder, and this pressure further is converted into the clamping force. The magnitude of clamping force depends upon the piston diameter, number of piston and bore diameter of master cylinder. Also, the clamping force can be increased by increasing the pedal ratio.

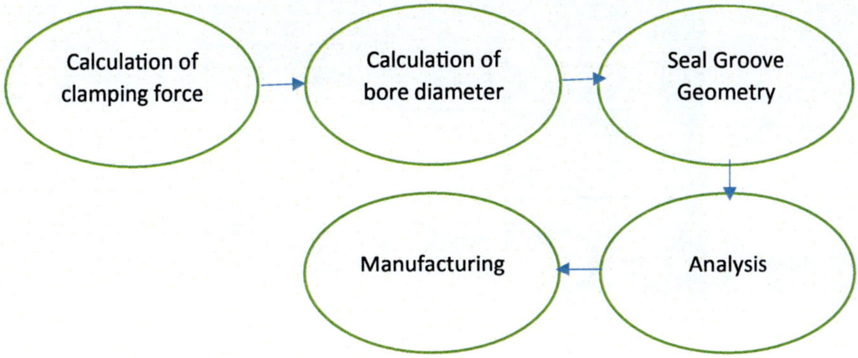

Fig. 2 Steps for designing of caliper [2]

The clamping force can be calculated [3] by the following expression

$$F_c = \frac{T_r}{\mu \times R_e} = \frac{82.27}{0.4 \times 0.095} = 2165\ \text{N}$$

where

F_c Clamping force
T_r Required torque
μ Coefficient of friction between disc and pads
R_e Mean effective radius of disc.

2.2 *Calculation of Bore Diameter*

The piston bore diameter decides the magnitude of the clamping force. The diameter of piston is directly proportional to the magnitude of the clamping force, and more the diameter of the piston, the greater will be the clamping force and vice versa.

The bore diameter can be calculated by [3] the following expression

$$D = \sqrt{\frac{4 \times F_c}{\pi \times n \times P_c}} = \sqrt{\frac{4 \times 2165}{\pi \times 2 \times 2.2 \times 10^6}} = 0.025\ \text{m}$$

where

D Bore diameter
n Number of pistons
P_c Generated pressure by master cylinder.

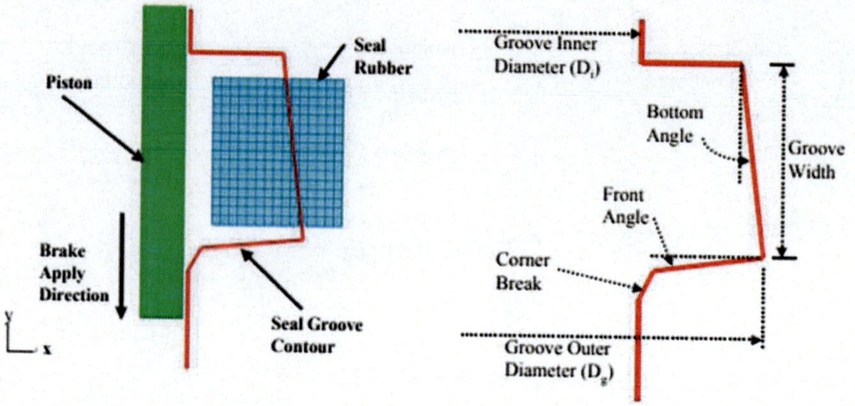

Fig. 3 Seal groove geometry [4]

2.3 *Seal Groove Geometry*

Seal groove geometry plays an important role in designing of caliper. The primary purpose of the seal groove geometry is to fit the seal within it. There are two seal grooves provided inside the cylinder of caliper: one is for the retraction of the piston and other is for to prevent leakage.

When the brakes are applied, the pressure builds inside the cylinder of caliper which forces the piston to move outwards; hence, the retraction of the piston is necessary after the release of brake pedal since it prevents unwanted contact of the pads with the disc.

When the piston slides outwards, the retraction seal extrudes into the corner break, and the extrusion of rubber seal into the corner break depends upon the front angle as well as the friction between rubber seal and the piston. For effective retraction, the front angle is kept between 6 and 10° and corner break is kept in between 0.3 and 0.7 mm [4] (Fig. 3).

3 Material Selection

Material selection is an important aspect in considering the safety of the caliper. Hence, the material selection process is carried out considering the mechanical and chemical properties of various materials. The material is to be chosen for caliper body and piston. Considering the strength, reliability and most importantly weight of the brake caliper, the following materials are recommended [5] (Tables 1 and 2).

Table 1 Materials for caliper

	AI 6061-T6	Al 7075-T6
Density (kg/m^3)	2700	3100
Young's modulus (GPa)	69	70
Poisson's ratio	0.33	0.32
Yield tensile strength (MPa)	370	480
Ultimate tensile strength (MPa)	410	525

Table 2 Materials for piston

	AISI 1018	AISI 1026
Density (kg/m^3)	7900	7900
Young's modulus (GPa)	190	190
Poisson's ratio	0.29	0.29
Yield tensile strength (MPa)	317	470
Ultimate tensile strength (MPa)	480	550

4 Static Structural Analysis of Caliper

Analysis of the caliper is done by using ANSYS Workbench 16.0 to ensure the safety at various load conditions. The static structural type of analysis is performed on the caliper.

a. **Meshing**
Meshing elements like tetrahedral, prism, brick and pyramid are considered, and parameters like skewness, Jacobian and aspect ratio were also considered for an effective meshing. The element size is taken as 5 mm as the size of element directly varies with the error percentage and inversely varies with the solving time. Element type is considered as 3D element (Fig. 4).

b. **Load Conditions**
When the caliper works under dynamic condition, various loads act on the caliper, and for analysis maximum values load is considered to ensure the safety.

In dynamic condition, typically two forces act on the caliper as follows [6]:

1. Force due to the inbuilt pressure.
2. Reaction force due to the clamping action.

Moment force can also be taken into the consideration by calculating the braking torque. Braking torque can be calculated as [7]

$$T_{\mathrm{r}} = \frac{W_{\text{dynamic}}}{2} \times \frac{D}{2} \times \text{max decleration}$$

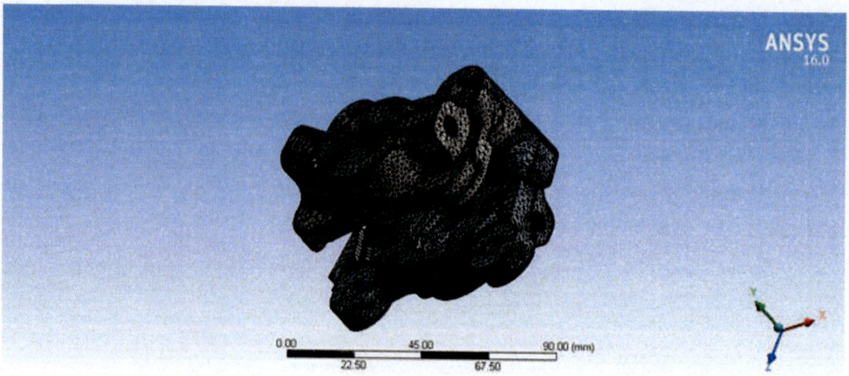

Fig. 4 Meshing of caliper

where

W_{dynamic} Dynamic front axle weight
D Diameter of wheel.

By giving fixed support at the mounting point and applying various load conditions, the following results are obtained (Figs. 5, 6, 7 and 8).

Factor of safety	Deformation	Stress
1.47	0.198 mm	187.5 MPa

From the results of analysis, it is found that the stress is 187.5 MPa which is less than the allowable tensile stress of material; hence, the design is safe having the factor of safety of 1.47 and deformation of 0.198 mm.

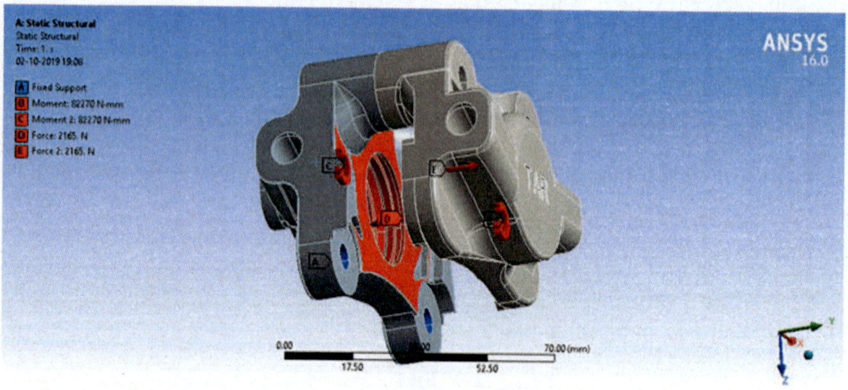

Fig. 5 Loading conditions on caliper

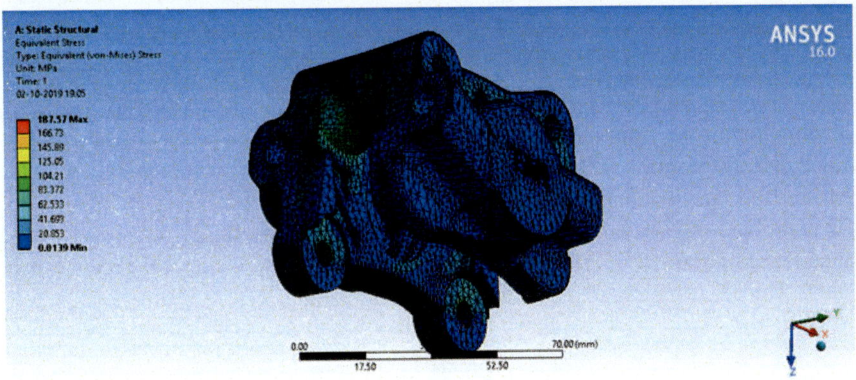

Fig. 6 Stress analysis of caliper

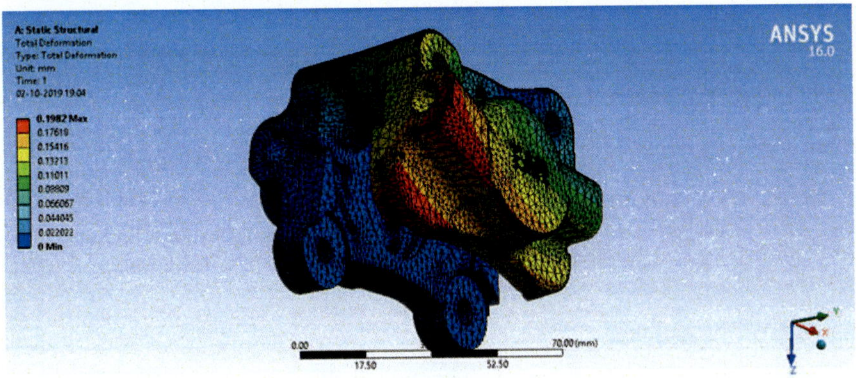

Fig. 7 Deformation analysis of caliper

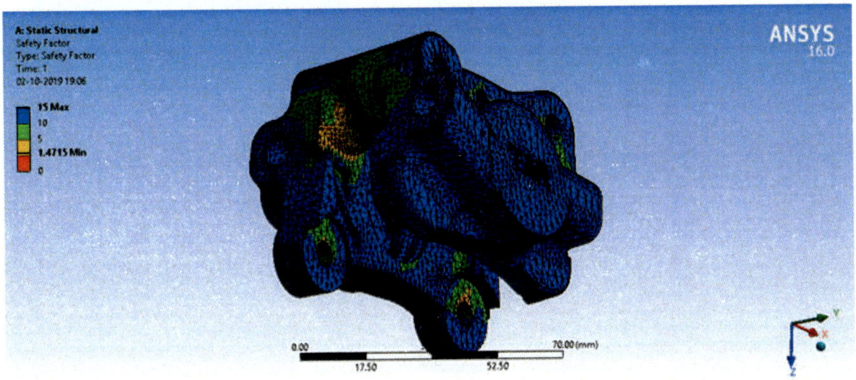

Fig. 8 Factor of safety of caliper

5 Conclusion

The designing and analysis of the brake caliper are performed under maximum loading conditions, and it is coming out to be safe by achieving factor of safety of 1.47. Since structural analysis is done for the maximum load conditions, hence it ensures the safety of caliper under critical circumstances. The design is kept simple for manufacturing as well as for assembly constraints point of view, and the maximum shear stress obtained is 187.5 MPa which is less than that of the allowable tensile stress. Hence, from the above analysis and results, the design of caliper is safe.

References

1. BAJA SAE India Rulebook 2020
2. Ingale S, Kothawade S, Patankar A, Kulkarni R (2016) Design and analysis of a brake caliper. Int J Mech Eng Technol (IJMET) 7(4):227–233, Article ID: IJMET_07_04_024
3. Rudolf L (2011) Brake design and safety. Society of Automotive Engineers, Inc.
4. Anwana OD, Cai H, Chang HT (4–7 Mar 2002) Analysis of brake caliper seal-groove design. In: Delphi Automotive Systems, SAE 2002 World Congress Detroit, Michigan
5. Kashyap PK, Arya D, Gupta K, Kumar K, Khan S (2019) Design and analysis of single piston floating brake caliper. IJERT 8(5). ISSN: 2278-0181
6. Shah R, Shah C, Thigale S (2017) Design and analysis of a hydraulic brake caliper. IMJET
7. Schmidt M (2012) University of Cincinnati SAE Baja Race Team Brake System